MUSHROOMS of Cascadia

A comprehensive guide to fungi of the Pacific Northwest

NOAH SIEGEL and **CHRISTIAN SCHWARZ**

BACKCOUNTRY PRESS
Humboldt County, CA

First Edition
9 8 7 6 5 4 3 2 1

Front cover images: (top) Mount Rainier; (bottom left to right) *Pholiota adiposa, Cortinarius violaceus, Ramaria* sp.
Back cover images: (top to bottom) *Morchella brunnea, Russula americana, Laetiporus conifericola, Cantharellus formosus, Hygrocybe flavescens, Gliophorus psittacinus, Chlorociboria aeruginascens, Cortinarius subfoetidus*

Book layout and cover design by Backcountry Press

Published by Backcountry Press • Kneeland, CA

ISBN 978-1-941624-19-7
Library of Congress Control Number: 2024937744

Order this book online:
www.backcountrypress.com

Book Resources

Created and Printed in the USA

MIX
Paper | Supporting responsible forestry
FSC www.fsc.org
FSC® C005010

Table of Contents

Introduction to Cascadia

The name *Cascadia* harkens to falling waters—and it is this water that sets the Pacific Northwest apart from other bioregions in the United States. Snow-capped mountain peaks, hundreds of rushing rivers, some of the wettest rainforests on the continent, and the temperate influence of the Pacific Ocean create a beautiful mosaic of habitats.

At the biome level, this region is classified as a temperate rainforest. But compared to other temperate rainforests globally, much of what makes Cascadia unique has to do with its ancient geology—particularly the complex interplay of the underlying tectonic plates. Along the northwestern edge of the continent, the turbulent subduction of the Juan de Fuca plate below the North American Plate has resulted in the formation of a chain of steep, rugged volcanic mountains (the Cascade Range) as well as formidable non-volcanic accretionary mountains (e.g., the Olympic Peninsula and Klamath Mountains). Across Cascadia, the close proximity of high-elevation topography to the predominant westerly storm tracks off the Pacific Ocean results in a pronounced orographic rainfall pattern, in which clouds cool as they rise over the mountains and rapidly precipitate large amounts of moisture along the western slopes of these ranges. Consequently, by the time these storms pass east of the crests, they are 'wrung out' and have lost much of their moisture content. This results in a strong contrast between the wetter, more temperate west-facing slopes and the arid eastern slopes which experience more extreme fluctuations between hot and cold temperatures throughout the year.

For the purposes of this book, we use *Cascadia* to refer broadly and imprecisely to the lands extending from southwest British Columbia east to the Pend Oreille River of Washington, southward along a band of the Washington and Oregon coasts, east to (and including) the Cascade Mountains, and finally dissipating messily in the complex 'Klamath Knot' of northern California. The southernmost point of our area of focus approximately coincides with the southern range extent of Sitka spruce (*Picea sitchensis*) in Mendocino County, California. As with all bioregions, our definition of Cascadia is somewhat arbitrary, since the boundaries of species' ranges are usually diffuse (this is even more true for whole habitats).

To the north, the temperate conifer forests of Cascadia meet the colder and more markedly seasonal boreal forests of Alaska and northern Rocky Mountains, while to the south, these same forests transition into the progressively warmer and drier Mediterranean-type climate of California. The Great Basin Desert lies east of the Cascades, and its aridity represents a barrier to fungal dispersal, isolating the wet coastal strip from the rest of the continent.

Thanks to its varied terrain, abundant rainfall, and dense forests, Cascadia is home to some of the richest mushroom assemblages in the country—we hope that this book helps to bring light to some of this wonderful diversity.

What are mushrooms?

The word *mushroom* is an imprecise but useful term for the visible reproductive organs of certain types of fungi. Fungi are their own kingdom of organisms, on the same order of divergence as plants and animals. This group comprises many millions of species of living organisms, ubiquitous in our everyday environments and abundant in nearly every natural habitat. Although fungi are exceptionally important in the function of global ecosystems, we mostly go about our lives unaware of their presence. By paying closer attention to fungi, we develop a heightened awareness that makes it easier to understand the roles that fungi play in human affairs—both as allies and adversaries. But what exactly are fungi?

Broadly, they are *eukaryotes*. This group also includes plants and animals because all have cells with discrete interior structures—the largest and most important of which is the nucleus. The nucleus contains the chemical blueprint for an organism's structure, chemistry, function, and behavior called DNA –. When two organisms have evolved from a common ancestor in the relatively recent past, we say that they are closely related and are likely to share many of the same encoded instructions in their DNA. If two organisms are descendants of a more ancient common ancestor, we say that they are more "distantly" related and generally have less genetic content in common. Surprising to most, the portobello mushrooms on your dinner table are more closely related to the yeast in your bread dough than to the sunflowers in your yard. Perhaps, even more surprisingly, those

Morphology

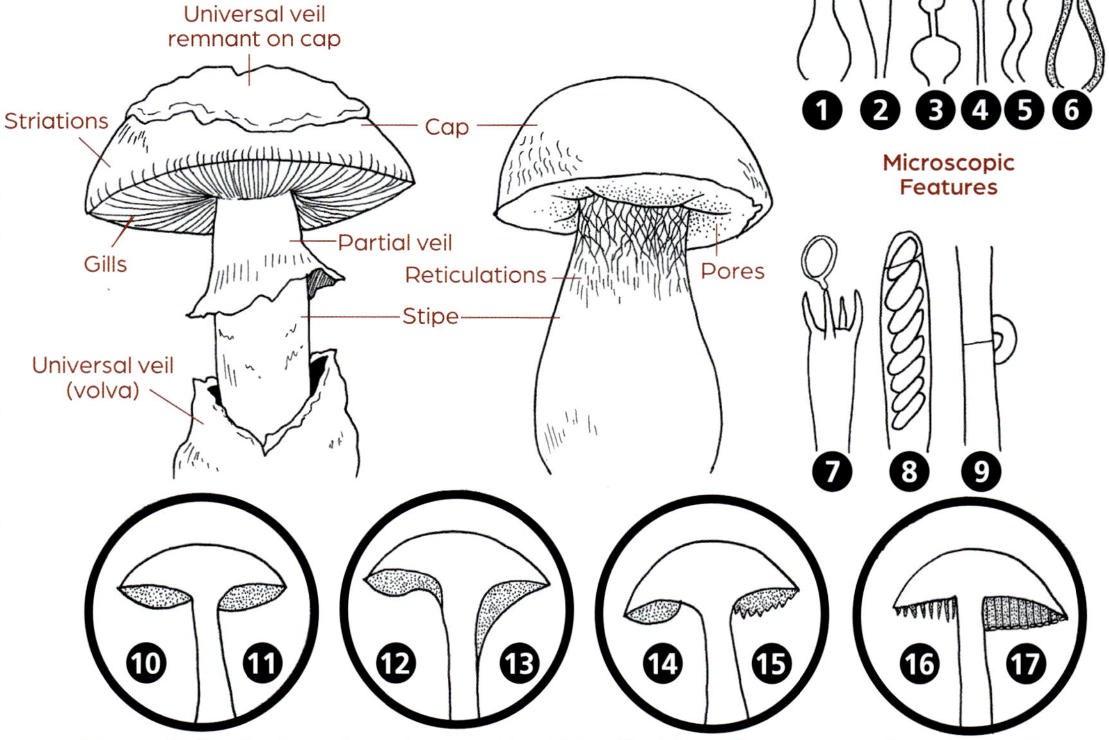

Types of gill attachemnts and some other kids of fertile surfaces as seen in cross-section

1. Urn-shaped **2.** Club-shaped **3.** Bowling pin shaped **4.** Capitate **5.** Sinuous **6.** Crystal-tipped cystidium **7.** Basidium with one spore attached **8.** Ascus with eight internal spores **9.** Two hyphae joined by a clamp connection **10. Narrowly attached 11.** Broadly attached **12.** Sinuate/notched **13.** Decurrent **14.** Free **15.** Serrated edge **16.** Teeth/spines **17.** Tubes and pores

Lifecycle

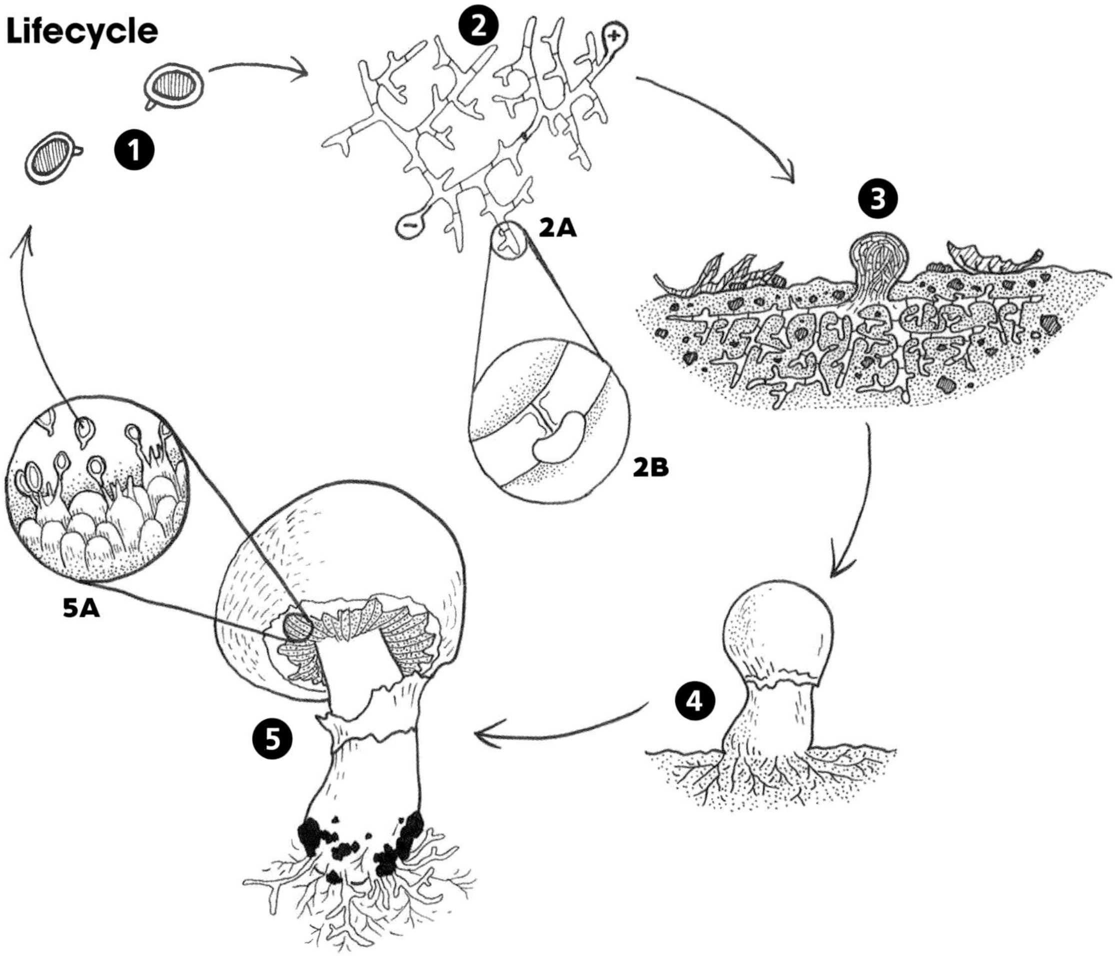

1. Fungal spores floating in the air and are moved around by wind currents.
2. Spores that have landed on a suitable surface germinate into a primary mycelium – a network of branching hyphae. Each hypha (2a) is a microscopic, cylindrical cell with cell walls made of chitin. Clamp connections (2b) are sometimes present, bridging the wall between two adjacent hyphae – these structures help regulate the distribution of nuclei between individuals. The presence or absence of clamps can be useful in identification. If two mycelia encounter each other and are of compatible mating types (+ and – , for example), they can fuse into a mycelium with two different sets of genetic information; one from each parent. This is called karyogamy — exchange of genetic material without fusing – a precursor to sexual recombination.
3. After a period of growth and given appropriate conditions (including plenty of water), a mature mycelium can produce a tiny 'primordium', a knot of hyphae that eventually grows into a mushroom.
4. The primordium grows into a 'button' or 'pin', and then expands into a mature fruitbody.
5. The fertile surface (gills, in this case) is covered in basidia (5a) or asci. The final steps in sexual reproduction (genetic recombination and separation into haploid spores) take place in these cells.

It's important to remember that mushrooms are not representative of the whole fungal organism: they are generally short-lived structures that are produced to achieve sexual reproduction and dispersal, similar to the flowers, fruits, and seeds of a plant. For this reason, mushrooms are referred to as the fruitbodies of the fungi that produce them. A fungus may thrive for many years as a mycelium without ever producing a fruitbody visible to humans!

portobello mushrooms are more closely related to *you* than they are to plants, since fungi and animals share a more recent common ancestor!

While fungi and plants appear morphologically similar in some ways, they are evolutionarily and ecologically quite distinct. Plants are *autotrophs*, meaning that they produce their own food through photosynthesis. In this chemical reaction, plants use energy from sunlight, carbon dioxide, and water to make the sugars that fuel their growth. Fungi, on the other hand, cannot photosynthesize—they are heterotrophs and rely on organic matter produced by other organisms to sustain their metabolism. Fungi are notable in that they perform *extracellular digestion* by exuding powerful enzymes into their environment that break down organic matter and then absorb the byproducts.

Fungi differ from animals in structural composition as well. Animal cells are surrounded by a layer of lipids (fatty acid molecules), while most fungi have cell walls made of a carbohydrate called *chitin*. This same compound is the main component of the exoskeletons (but not the cells) of shrimp, dragonflies, and all other arthropods. The primary cells of the fungi in this book are tube-like structures called hyphae which, in many species, are strung together in long chains and branching networks that form a sort of cellular carpet called the *mycelium*. Although this can be a very large structure, it is usually invisible to humans because it is buried, or embedded, in whatever substrate the fungus is growing in and digesting (whether soil, wood, or animal tissue). The thin, elongate cells and highly branched structure of the mycelium has a high surface-area-to-volume ratio. This anatomy efficiently enables diffusion of digestive enzymes and uptake of water and nutrients.

It's important to remember that mushrooms are not the whole fungal organism but rather a generally short-lived structure produced to achieve sexual reproduction and dispersal (similar to the flowers, fruits, and seeds of a plant). For this reason, mushrooms are referred to as the *fruitbodies* of the fungi that produce them. A fungus may thrive for many years as a mycelium without ever producing a fruitbody!

Fungi are a diverse group of organisms, but only a small fraction of these species produce structures that we call mushrooms. The majority are microscopic, perhaps visible to the human eye during a brief part of their life cycle when they manifest as discolored splotches or powdery coatings on plant tissues. Fungi play many roles in natural ecosystems, ranging from harmless decomposers to virulent agents of disease to mutualistic symbionts of plants.

Ecology of Fungi

Most fungi decompose nonliving organic material to fuel their metabolism. Fungi with this nutritional mode are called saprobes and are found almost everywhere including old straw bales, forest humus, and even the Tupperware at the back of your fridge. Saprobic (or saprotrophic) fungi are extremely important in releasing carbon and nitrogen back into local and global nutrient cycles.

Other mushroom-forming fungi are involved in either parasitic or mutualistic symbioses with other organisms. The symbiotic relationships between plants and fungi, called mycorrhizae, are particularly relevant to those seeking to find, identify, and understand the mushrooms in this book.

The term mycorrhiza is a combination of the Greek words *myco*, "a fungus," and *rhiza*, "roots." It refers to a mutualistic relationship in which the hyphae of a fungus live in close physical contact with the root tips of a plant. In these relationships, the fungal partner exchanges water and nutrients for sugars produced by the plant host. This reciprocity is particularly elegant because it pairs the physiological abilities and limitations of each partner in a complementary way. The cobwebby mycelium of the fungal partner has a high surface area to gather water and powerful enzymes to liberate nutrients from the soil, but cannot produce its own sugars. Conversely, the plant partner can produce sugar but doesn't have the same capacity to efficiently gather resources from the soil.

Scientists have identified and classified many variations of the mycorrhizal relationshipwith two main categories: *endomycorrhizae* and *ectomycorrhizae*. Endomycorrhizal fungi penetrate the root cells of their host plants. Ectomycorrhizal fungi form sheaths around the outside of root tips and between root cells (not inside them). The two groups also have chemical and ecological differences.

Endomycorrhizal fungi are often better phosphorus scavengers, while ectomycorrhizal fungi tend to be more adept at gathering nitrogen. Although there are exceptions, ectomycorrhizal fungi reach their greatest abundance and highest diversity in temperate forests.

Only a fraction of mushroom-producing fungi are mycorrhizal and most of those in this book are of the ectomycorrhizal type (often abbreviated as "ECM"). However, not all ectomycorrhizal fungi produce large fruitbodies. Many form thin crusts under logs or inconspicuous spore-bearing structures in the soil. Although many species are more ecologically flexible, some of these relationships are *obligate*, meaning that the fungi and plants involved cannot complete their life cycles without each other. Mycorrhizal fungi enable many plants to thrive in habitats where they would otherwise not survive due to drought stress or soil nutrient deficits.

These mutualistic relationships are restricted, to some extent, by the number of plant species with which any particular fungus can form a symbiosis (and vice versa). For example, the Death Cap (*Amanita phalloides*) can form mycorrhizae with dozens of different host plants—including coniferous and broad-leaved—while Slippery Jacks (*Suillus* spp.) are found almost exclusively with one or a few conifer species. Most mycorrhizal fungi are at least moderately specific in that they grow primarily with one or two host tree or shrub species, at lower frequency, with a smaller number of other hosts.

The ecological impacts of these symbioses extend well beyond the individual tree and fungus. In some cases, parasitic plants (such as *Corallorhiza* orchids) exploit the mycelium of a mycorrhizal fungus, absorbing sugars that were originally produced by the tree with which the fungus is partnered. In Cascadia's old-growth forests, Spotted Owls (*Strix occidentalis*) prey upon voles (*Clethrionomys* spp.), which in turn feed heavily on the hypogeous fruitbodies of mycorrhizal *Rhizopogon* fungi. The spores of these truffle-like fungi pass through the guts of both animals and emerge on the other end as viable spores in the owl's pellets and droppings. Thus, the owls disperse the fungi that sustain populations of their own prey. Plants also communicate with each other via chemical signaling conducted through fungal mycelium that connects them. This has been observed in both forest and experimental settings and led some researchers to describe mycelial networks as a "Natural Internet" through which individuals communicate and reallocate resources at large scales.

At the other end of the symbiotic spectrum are the parasitic fungi, whose growth causes negative effects on the health of their host organisms. Honey Mushrooms (*Armillaria* species) and conifer root rot (*Heterobasidion* spp.) cause natural disease outbreaks—sometimes killing entire stands of trees.

Fungi also parasitize animals with entire groups of mammals somtimes affected at a landscape scales. The health of bat populations in northeastern North America is at serious risk due to White Nose Syndrome, a disease caused by the fungus *Geomyces destructans*—which has now been detected in the western USA. Some of the most spectacular animal-parasite fungi specialize on invertebrates including *Cordyceps* species which parasitize ants, moths, beetles, and other insects. One of these species, *Ophiocordyceps sinensis*, is a prized medicinal mushroom in Asia. Some fungi even specialize in digesting other fungi! Many of these "mycoparasitic" fungi distort, discolor, or otherwise transform the fruitbodies of their hosts. Most notable in Cascadia are the Lobster Mushroom (*Hypomyces lactifluorum*) and the rare Body-Snatcher Mushrooms (*Squamanita* species).

Finally, we should mention that some fungi exist in a mutualistic relationship with green algae or cyanobacteria—such species are said to be lichenized. Although the majority of lichenized fungi don't produce typical mushroom-shaped fruitbodies, some do, and there is growing evidence that many members of the family Hygrophoraceae are involved in such relationships (Seitzman et al., 2010).

Humans and Fungi

Around the world, humans and fungi have diverse array of relationships. Some are serious pathogens, others produce valuable medicine and drugs, while still others are important commercial crops.

Many fungi are our allies, including the blue cheeses (and many other cheeses) made with the help of ascomycete molds or *Saccharomyces* yeasts that ferment plant sugars into alcohol. Thanks to fungi we have leavened bread and cold IPAs. The famous antibiotic compound penicillin (used to treat many bacterial infections) was first isolated from a fungus called *Penicillium*.

The genus *Amanita* contains the fabled and feared Death Cap (*A. phalloides*) and Destroying Angel (*A. ocreata*, *A. bisporigera*) but many other genera also contain toxic species. With appropriate caution, effort, and the help of knowledgeable friends, you can easily keep yourself from joining the ranks of these unfortunate people. There is a broad range of symptoms caused by mushroom poisoning, dependent on the species and dose involved, as well as the health of the affected individual.

Many fungi impact our daily lives while remaining practically invisible. Diseases like yeast infections, ringworm, valley fever (coccidioidomycosis), and aspergillosis are all caused by fungi. A variety of plant-pathogenic species do serious damage to crops around the world. The fungus *Claviceps purpurea* produces toxic "ergot" on cereal grains, and *Phragmidium* species produce bright orange rusts on the roses in your garden. The global economic impacts of these fungi are considerable, and much research is devoted to mitigating their effects. Likewise, woodr-rotting fungi that parasitize trees can have huge economic impacts on the timber industry.

Dry rot fungi (*Serpula* species) compromise the structural integrity of wood in our buildings. Other fungi are more minor nuisances like an ascomycete called *Rhizopus* that produces the blackish fuzz we find on old bread.

In some cultures, particularly in parts of Eurasia and Mexico, psychotropic fungi play an important role in spiritual and religious practices. Some of the most famous of these fungi are *Amanita muscaria* and various species of magic mushrooms in the genus *Psilocybe*. There has been a recent resurgence of research in the therapeutic uses of psilocybin (the main psychoactive compound in *Psilocybe* fruitbodies) to treat cluster headaches (Sewell, Halpern, and Pope, 2006), mitigate end-of-life anxiety associated with terminal illness (Grob et al., 2011), and to disrupt patterns of alcohol and narcotic abuse. Traditional uses of these mushrooms include cultural and spiritual practices that inform the experiences they catalyze. Without such context, the magic mushroom experience can be upsetting or even mentally destabilizing. These paths should not be pursued carelessly.

Perhaps of even greater concern are the exploitation and cultural damage caused by the capitalist tendency to appropriate, simplify, and commodify indigenous practices. Soon after Gordon Wasson's famous article brought *Psilocybe* to the attention of Americans, a small industry emerged in central Mexico that threatened to reduce the experience to a tourist attraction. Although decriminalization of natural entheogenic substances is well-justified, corporate interests have sought to turn psychotropic compounds into 'silver bullets' for the symptomatic treatment of problems more deeply rooted in our civilization. Although we consider the therapeutic uses of psilocybin promising and worth pursuing, we also believe that great care should be taken to respect traditional relationships that have developed around these mushrooms.

Fungi are all around us. Their spores are abundant in the air we breathe, their fruitbodies are among the most prized delicacies in cuisine, and many other products derived from or processed by fungi are ubiquitous in our everyday environments. Fungi are deeply involved in many aspects of our lives and this is a compelling reason to learn more about them.

The Cascadian Bioregion

Ecological Zones

To make the species accounts easier to interpret, we divide Cascadia into the sub-regions described below. Some of these correspond well with natural biogeographic boundaries, but many species occur across these regions and even more broadly across the continent. Although not covered in this guide, the vast, uninterrupted boreal forests are connected to the northern margins of Cascadia by way of the Rocky Mountains and eastward to Maine and Newfoundland. These cold, dense spruce, larch, birch, and poplar forests are a connective tissue for the dispersal and genetic intermingling of fungal associates across the continent.

Coastal Forests are dominated by salt- and wind-tolerant Sitka spruce *(Picea sitchensis)* and shore pine (*Pinus contorta* subsp. *contorta*). Coastal forests exist in a narrow band along the edge of the Pacific from south-central Alaska into northern California. They often occur on sandy soils with a variety of Ericaceous shrubs including huckleberry *(Vaccinium* spp.), salal (*Gaultheria shalon*), and kinnikinnick (*Arctostaphylos uva-ursi*). The fall produces spectacularly abundant mushroom fruitings.

The Coast Range is a dark, wet place predominated by mixed conifer forests. Beginning near the Columbia River, west of the Willamette Valley, the Coast Range forms a contiguous mix of lower elevation mountains that extend south through Oregon and into California. Western hemlock (*Tsuga heterophylla*), Douglas-fir (*Pseudotsuga menziesii*), grand fir (*Abies grandis*), western redcedar (*Thuja plicata*), red alder (*Alnus rubra*) and bigleaf maple (*Acer macrophyllum*) intermingle to create lush, mossy, and dense vegetation communities.

The Cascade Range are snow-capped volcanic mountains that form the backbone of the entire bioregion; running from southern British Columbia through Washington and Oregon into northern California. Their rugged topography is defined by three distinct habitat types with notably different mushroom communities:

- **The foothills and west slopes** are temperate and wet, and thus similar in composition to the Coast Range.
- **High elevation forests** are found wherever increasing altitude results in frequent freezing temperatures and a winter snow regime. Here, forests of fir and hemlock predominate.
- **East slopes** lie in a rain shadow, over the crest from the high elevation forests. These drier forests are more open and dominated by pines, firs, and larch.

The Olympic Peninsula and its natural habitats are stunningly varied due to the sharp elevational gradient created by the Olympic Mountains. The consequent rain-shadow effect, moving east from the ocean toward the Puget Sound, is noteworthy for its distinct vegetation and mushroom communities. The eastern half of the peninsula is warmer and much drier, with some elements of the plant and fungal communities more reminiscent of the eastern Cascades. This is in stark contrast to the rain-drenched old-growth forests around the Hoh, Sol Duc, and Quinault rivers in the protected lands of Olympic National Park. These ancient trees support an exceptionally diverse community of mushrooms. These old-growth forests are also some of the best remaining examples of Cascadian rainforest. The margins of this large and isolated peninsula are covered in Douglas-fir and hemlock timber plantations.

The Interior Valleys include the flat lowlands of the Willamette Valley and Columbia River Gorge. These places once boasted biodiverse oak woodlands but are today home to fragmented riparian forests of cottonwood and willow abutting remnant oak woodlands fringing farmlands and urban areas. A number of polypores, 'natural' morels, and some fleshy gilled mushrooms are more common in these habitats than anywhere else in Cascadia.

The Klamath Mountains (including the Siskiyou Mountain sub-range) are often described as a "knot," which well encapsulates their topography as well as their significance from a biogeographic perspective. These rugged mountains comprise a kind of tangle of different habitats, in which a staggering diversity of conifers come together. Despite this richness of tree diversity, mushrooms can be difficult to find here due to the steep, rocky terrain and relatively narrow fruiting season between the hot summer and snow-smothered winter.

NOTE: For the purposes of this book, '*Cascadia*' refers to the entire bioregion, while '*The Cascades*' refers specifically to the volcanic mountain range that defines the eastern flank of the region.

Trees of Cascadia

Cascadia is home to some of the most awe-inspiring trees in the world. The copious rainfall, infrequent disturbance cycles, and rich volcanic soils of the lowland rainforests support gargantuan old-growth spruce, cedar, and Douglas-fir. The branches and canopies of these trees are often densely cloaked with dripping garlands of lichens and moss—a primeval setting that inspires feelings of humility and awe. Spending time in quiet contemplation of these ancient ecosystems is rejuvenating and life-affirming.

From a practical perspective, knowing how to recognize and identify these trees is extremely useful when searching for and identifying mushrooms. For instance, some similar species of *Lactarius* are more easily told apart by their host tree than by their morphology! In other cases, you can rule out entire groups of fungi based on habitat and geography. For example, it will be hard to find the conifer-associated Slippery Jacks (*Suillus* spp.) in riparian corridors of maple and alder (unless you find a stray Douglas-fir in their midst).

But trees are only part of the story! Newcomers are often surprised by how difficult it is to find particular mushrooms—even common species can elude casual foragers for years. This speaks to the complexity of the ecological concept of habitat. Most species are sensitive to fine-scale variations in their surroundings and subtle aspects of microclimate and soil chemistry may cause a species to be absent from apparently suitable habitat. A forest's age is an important variable to pay attention to. Mature spruce forests often host a plethora of *Cortinarius* and *Hydnellum* while younger groves are typically richer in *Amanita* and *Laccaria*. Hyper-diverse communities of invertebrates, bacteria, and viruses, which play major roles in giving soils their ecological character, need time to mature and are most complex in older forests.

The tree treatments that follow are arranged first with the ectomychorrhizal conifers. Non-ectomychorrihzal conifers and flowering trees are treated next. Descriptions end with ectomychorrizal flowering trees. Species pictures, when included, precede the treatments.

Ectomycorrhizal Cone-Bearing Trees

Douglas-fir (*Pseudotsuga menziesii*) *is the* most widespread and abundant tree species in Cascadia. It is found from the Coast Ranges all the way to the east slopes of the Cascades and beyond. They are easily recognized by their soft, light green needles that radiate in all directions around the twig. Their medium-sized cones often litter the ground and show distinctive three-tipped "tongues" of papery tissue poking out between the scales. The bark of young trees is slightly rough and grayish while th trunks of mature trees usually become deeply furrowed.

Western hemlock (*Tsuga heterophylla*) is widespread throughout Cascadia. The distinctive, drooping tips of hemlocks are unlike most other conifers. Needles are short, flat, and of differing lengths on the same twig. Each needle has two whitish lines on the paler underside. The bark is grayish with shallow furrows, and the cones are small with relatively large scales. **Mountain hemlock (*Tsuga mertensiana*)** is similar, but often has a bushier appearance and larger cones. It co-occurs with western hemlock across the higher elevation forests in Cascadia.

Sitka spruce (*Picea sitchensis*) are found adjacent to the Pacific Ocean from far northern California northward across Cascadia into British Columbia and Alaska. They are easily recognized by the by sharp-tipped, three-sided needles (roll them between your fingers) with silvery green upper surfaces, green undersides, and thin-scaled hanging cones. Their grayish overlapping bark scales, swooping branches, and crowns with upright tips are also distinctive. This fast-growing species is often found in dense monodominant stands, but larger old growth trees are becoming increasingly rare. **Engelmann spruce (*Picea engelmannii*)** grow at higher elevations on the eastern slopes of the Cascades and host a similar suite of fungi to coastal spruce forests.

Western larch (*Larix occidentalis*) are tall trees of mid- and high-elevation forests on the drier eastern slopes. Their pointed, open crowns and short, spiky-tufted needles make for a distinctive profile. Larches are the only deciduous conifers in Cascadia, and are especially easy to recognize in their brilliant yellow fall colors. A unique assemblage of slippery jacks (*Suillus* spp.) are exclusively associated with larch.

True firs (*Abies* spp.) are classically "Christmas-tree shaped" conifers, with a pointed leader and whorled branches that splay out nearly flat. When young, their bark is distinctively smooth, often covered with sap blisters, becoming furrowed in age (except in subalpine fir, *Abies lasiocarpa*). Needles usually lay flat relative to the stem and are longer (up to 3 cm) than those of the similar hemlock species. Their large cones are borne upright on upper branches, but are rarely seen as they are a favorite food source for a variety of forest creatures. Occasionally, partially eaten cones are found on the ground. The fungal communities of fir habitats overlap, to some degree, with those of hemlock forests and are often rich in *Russula*, *Cortinarius*, boletes, and other mycorrhizal groups.

Grand fir (*Abies grandis*) is common at low- to mid-elevations west of the Cascade Crest. Upper sides of needles are bright green and undersides are paler with two whitish lines. Needles are usually splayed flat in two distinct rows on the lower branches.

Noble fir (*Abies procera*) is common at mid- to high-elevations of the Cascade Range and in the higher elevations of the Coast Range and Olympic Mountains.,The bluish colored needles have a hockey stick shaped twist as they leave the stem. This species is gradually replaced by Shasta fir (*Abies magnifica* var. *shastensis*) in the southern Cascades and Klamath Mountains.

Pacific silver fir (*Abies amabilis*) is a mid- to high-elevation tree of the Cascade Range and Olympic Mountains reaching its southern range limit in the Klamath Mountains. It is similar in appearance to grand fir but with flaring and upright needles—giving it a more half-round look (rather than the flattened splay of grand fir). There is a significant degree of overlap in the mushroom assemblages of high elevation firs and spruce, more so than with the lower elevation grand fir and western hemlock fungi.

Subalpine fir (*Abies lasiocarpa*) occurs at high elevations up to tree-line in the Olympic Mountains and Cascade Range. It is recognized by its curved needles with white lines on the top and bottom surfaces. It usually occurs sympatrically with Engelmann spruce.

Pines (*Pinus* species) have short to long needles in bundles. Pines generally occur in habitats with poor soil and plenty of sun light, as they are shade-intolerant. Habitats such as this usually lack overstory competition from other conifers. Although species diversity in Cascadia is low, a unique assemblage of mushrooms are found where pines do occur.

Shore pine (*Pinus contorta* subsp. *contorta*) is found along the immediate coast throughout much of the southern part of Cascadia; especially along sandy or dune-dominated coastlines. Shore pines are small-statured with short needles in pairs and small, ovate, dark cones. During wet fall seasons, the huckleberry and salal understory of this habitat is often teeming with mushrooms.

Lodgepole pine (*P. contorta* subsp. *latifolia* and *P. contorta* subsp. *murrayana*) grow on the eastern slopes of the Cascade Range in Washington and Oregon respectively. They also share a diverse assemblage of associated mushroom species, with a distinct set of species compared to the shore pines.

Ponderosa pine (*Pinus ponderosa*) is common in the dryer eastern slope forests in the Cascades and scattered through the Willamette Valley and Klamath Mountains. It is easily recognized by the long needles in bundles of three, and the triangular to egg-shaped cones.

Western white pine (*Pinus monticola*) is scattered in the mid- to high-elevations across Cascadia. It has needles in bundles of five and elongate cones four to six inches in length.

Whitebark pine (*Pinus albicaulis*) is restricted to the highest elevations of Cascadia and has shorted needles in clusters of five and smaller cones.

Ectomycorrhizal Flowering Trees

Oaks (*Quercus* spp.) are keystone species in some of the most biodiverse communities in Cascadia. They are exceptional mycorrhizal hosts, with a huge diversity of associated fungi. Groves of deciduous **Oregon white oak (*Quercus garryana*)** are found in warmer, drier microhabitats away from the immediate coast. They support rich communities of *Cortinarius*, as well as a set of boletes, *Amanita*, and *Russula* (and many others) not found in conifer forests. A more diverse assemblage of oaks occurs in the low- and mid-elevations of southwestern Oregon and northern California. These include the Klamath Mountain endemic deer oak (*Q. sadleriana*) and canyon live oak (*Q. chrysolepis*), both of which are evergreen. Black oak (*Q. kellogii*) and valley oak (*Q. lobata*) are leafless in winter. All have some fungal associates in common, with subtle differences often corresponding to deciduous vs. non-deciduous, physiology, and microhabitat preference.

Tanoak (*Notholithocarpus densiflorus*) is restricted to the southwestern part of Cascadia. The bark starts off smooth and gray, then develops dark, vertical furrows as the tree matures. The leaves are large, green, and oblong with a distinctly toothed margin and pronounced diagonal veins and extensive tan fuzz on the underside. The acorn cups are hairy-scaly looking—distinct from most of the true oaks. Mushrooms associated with tanoak include a wide array of coral mushrooms, boletes, Black Trumpets, as well as many *Russula* and *Lactarius*.

Madrone (*Arbutus menziesii*) is found throughout the low elevations mixed-evergreen forests of Cascadia, but is rarely a dominant part of the forest. They are easily recognized by the dark reddish-brown, finely-scaled trunk and smooth, and shiny yellowish or orangish upper branches. There is a limited but distinctive set of fungi associated with madrone, including a number of *Russula*, *Lactarius*, *Tricholoma* and *Leccinum*.

Red alder (*Alnus rubra*) is found along riparian corridors or in disturbed habitat. Their deciduous oval leaves are subtly toothed around the margin with grooved, diagonally-arranged veins. Their pale, smooth bark and unusual cone-like fruits set them apart. A small but distinctive set of ectomycorrhizal mushrooms associate with red alder roots and many polypores and ascomycetes are found on their woody debris.

Black cottonwood (*Populus trichocarpa*) is typically found along riparian corridors, and other wet areas across all but the high elevations of Cascadia. They have deciduous heart-shaped to somewhat triangular leaves with a glossy green top and pallid underside, The bark becomes progressively more furrowed with age. **Quaking aspen (*Populus tremuloides*)** is an occasional inhibitor of east-side forests, with smaller heart-shaped leaves and smooth gray bark. They both associate with a small group of specialist fungi including *Hebeloma*, *Leccinum*, *Inocybe*, *Tricholoma* and *Russula*, while the duff often has abundant Psathyreallaceae, and the decaying trunks will have *Pholiota* and allies, and *Pleurotus*.

Non-Ectomycorrhizal Trees

Western redcedar (*Thuja plicata*) has flat sprays of tiny, neatly-arranged scales and unusual upright cones that distinguish them from other conifers in our area. As with all other members of the Cupressaceae family, these trees do not support ectomycorrhizal fungi. Their tissues are defended with strong volatile chemicals and thus their duff decays slowly and forms an "allelopathic" layer that suppresses the growth of other plants and fungi. Although the diversity of mushrooms under these trees is lower than in other conifer forests, there is a certain set of species that thrive in these low-competition environments, including many coral mushrooms, *Hygrocybe*, *Leptonia*, and earthtongues.

Yellow-cedar (*Callitropsis nootkatensis*) has similar drooping flattened sprays, but has small round cones and flaky to stringy gray bark. It grows in higher elevations in Cascadia.

Incense-cedar (*Calocedrus decurrens*) has "duck-bill" shaped cones, and reddish furrowed bark when mature. More common in the southern portions of Cascadia; it occurs in the Willamette valley, lower to mid-elevations on the Cascade Range in Oregon, and the Siskiyou Mountains. Although the diversity of mushrooms under these trees is lower than in other conifer forests, there is a certain set of species that thrive in these low-competition environments, including many coral mushrooms, *Hygrocybe*, *Leptonia*, and earthtongues.

Bigleaf maple (*Acer macrophyllum*) is easily recognized during most of the year by their enormous, palmate leaves. When completely leafless in winter, their large size, open crown structure, tightly vertically-furrowed gray-brown bark aid in identification. Their debris on the forest floor hosts a number of small decayers like *Mycena* and *Collybiopsis* while dead trunks and branches can host large flushes of Oyster Mushrooms (*Pleurotus* spp.) and polypores. In coastal forests, moss-mats are commonly found on their upper branches. This biomass slowly develops into a layer of soil suspended in the canopy which supports an additional (albeit rarely seen) community of mushrooms.

Finding Mushrooms

The combination of year-to-year fluctuations, elevation range, and longitudinal span across Cascadia makes a simple description of a typical mushroom season nearly impossible. The boreal forests north of Cascadia have a relatively short and intense summer mushroom season, while the coastal redwood forests to the south are nearly the opposite—with a protracted mushroom season spanning the cool, wet winter months. Cascadia exhibits an intermediate pattern, with pronounced fall and spring peaks in mushroom fruiting, but with an additional component of summer fruitings in the montane forests and the wettest coastal regions (e.g., the Olympic Peninsula). Extensive summer-fruiting in the coastal zones occur irregularly and are correlated with the movements of oceanic masses of warm water which bring increased summer rains.

The knowledge and intuition required to find specific mushroom species takes time to develop. Different species have favored fruiting times within the mushroom season coupled with differing habitat preferences. If you intend to search for a particular species of fungus, research into seasonality and habitat preferences will greatly increase your chance of success.

The seasonality descriptions herein are intended to be useful generalizations. If you look hard enough, macrofungi of some kind or another are found fruiting *somewhere* during any month. The onset and persistence of the mushroom seasons vary, depending on many factors—particularly the occurrence of dry spells, heat waves, and cold snaps. The mushroom fruiting season also varies by the shape of the land. Valleys and west- or north-facing slopes are cooler and moister while ridgetops and south- or east-facing slopes are warmer and drier.

A core principle of habitat-level biogeographic theory is that higher latitudes (areas further north and south of the equator) mimic higher altitudes (areas higher above sea level) and vice versa. For example, the northern reaches of Cascadia experience very cold winters, during which mushroom fruiting shuts down completely. The higher elevations of mountains in the southern part of Cascadia experience a winter climate regime more like the northern zones—with prolonged periods of subfreezing weather and snow accumulation. This means that in any given location, the exact pattern of mushroom fruitings will vary and predicting a productive season is a matter of building local, firsthand experience.

In general, the beginning months of the year are a period of low productivity in the Cascadian mushroom calendar but, by March, some groups of ascomycetes begin to fruit abundantly in low-lying zones. Once the warmer spring months arrive, a variety of ascomycetes and some spring-fruiting basidiomycete mushrooms are found, including the snowbank-associates (usually starting sometime in June). Summer fruitings can continue into September across the mountains with persistent snowpack and occasional summer storms. No two mushroom seasons are alike, and the actual pattern of macrofungal fruiting in a given year will deviate significantly.

An afternoon spent dashing about on a forested hill above the wind-whipped Pacific while uncovering King Boletes from beneath deep carpets of pine needles is primally thrilling. Equally so is the slow, steady, and methodical gathering and trimming of a rich flush of chanterelles—filling a basket soothes the mind and the heart in many ways. And then there's the magic and marvel when, after years of seeking and hoping, you finally encounter rare, unusual, and strikingly beautiful creatures like *Dendrocollybia racemosa*, *Hygrocybe virescens*, or *Hydnellum cyanopodium*.

Although the seasonal patterns of mushroom fruitings can seem mysterious, after a few seasons of careful observation and experience you'll develop a sense for when and where different species appear. Joining a local mycological society or amateur mushroom club (see the list on p. 550) is a great first step. Members of these clubs have years of accumulated knowledge and can provide you with support, companionship, guidance, and validation as you go about building and refining your knowledge of mushrooms. Your learning process will also be accelerated by taking field notes! Record-keeping is an especially important realm of natural history, because there a little existing data on fruiting dates and biogeography of macrofungi. Websites like iNaturalist (see Resources, p. 550) are very useful for consolidating and exploring this kind of information.

Risks of Mushrooming

There are many challenges inherent in mushroom hunting, not the least of which have to do with weather. In Cascadia, you are nearly guaranteed to get wet and chilly at some point in your searches for mushrooms. Check weather conditions and dress appropriately before heading out. As with any outdoor adventure, carry plenty of water and snacks and tell someone where you're going and when you expect to be back. Plan ahead by carrying maps and/or a GPS unit if you are unfamiliar with the area. Getting lost in the mountains during a cold and wet winter storm can be fatal. You're unlikely to encounter mountain lions (*Puma concolor*), and even less likely to be threatened by them, but they are certainly worth keeping in mind when you are walking alone in the woods.

Collecting Mushrooms

If you are gathering mushrooms to eat—assuming that you are *absolutely certain* of their identity—the pace of your collecting can become rather frenzied as you dash from one productive area to the next. Once you've gathered enough to make a meal (or two, or five), the next step will be cleaning and preparing your mushrooms for cooking or storing.

If you are searching for rare and unusual fungi, or surveying to find as many species as possible, the pace of your collecting should be much slower. Minuscule and inconspicuous fungi could be missed by a hurried collector. Pause every minute to stop and look closely at the area around your feet. Don't neglect the backsides and undersides of rotting logs, or the trunks and branches of trees above you.

Please keep in mind that mushroom collecting is unlawful in many places, and that there aren't consistent regulations across every park and forest. For example, some state parks have a limit for the harvest of wild mushrooms, while others ban the activity outright. National parks are similarly varied. Likewise, private property boundaries can be difficult to determine in the woods, and you may inadvertently end up speaking with an angry landowner. In all cases, keeping the woods clean, making sure the duff layer is relatively undisturbed as you harvest mushrooms, and respecting property boundaries will go a long way toward ensuring that we retain the access we have to collect mushrooms on public wildlands.

Identifying Mushrooms

Beginners should stick to large, distinctive species when learning to identify mushrooms. Attempting to classify LBMs (little brown mushrooms) is very difficult for novices. With a great deal of practice, however, many little brown mushrooms can be identified to genus or even species without a microscope. Inevitably, the littlest, brownest mushrooms (as well as some larger ones) will require a microscope to positively identify.

There is a general set of procedures you should follow to maximize your chances of identifying the mushrooms you find. This set of procedures begins while you are in the field. The following is a simple protocol geared to help you develop your identification skills.

1. Collect a small number of species to start. Beginners will often come home with a basket brimming with mushrooms and find themselves overwhelmed, discouraged by the seemingly impenetrable variability and diversity. Keep identification manageable by bringing home fewer than five species on your first outings.
2. Even though you might only collect five species, this does not mean you should only collect five mushrooms! Try to collect multiple fruitbodies of the same species. Most often, a single species will produce multiple fruitbodies in the same area, allowing you to collect several in good condition and of multiple developmental stages (see the chart on p. 7).
3. Take brief field notes—date, location, surrounding trees, and substrate. The substrate is the material from which the mushroom was fruiting. Soil, wood, leaf litter, and other mushrooms are examples of different types of substrates. Digging up the very base of the mushroom can sometimes reveal buried wood—investigate carefully. Substrate is just one piece of habitat information. Other important information includes the kinds of surrounding trees, other vegetation, and the elevation.
4. Take photos of your mushrooms in the field. This is probably the single most useful step to aid later identification efforts. See the section on mushroom photography (p. 22) for more tips and techniques.

5. Keep collections of different species separate from each other. Wax paper bags work well in relatively dry weather, or aluminum foil in wet weather; otherwise tackle boxes are ideal, especially for small fruitbodies.

6. When you return home, spread your collections out, keeping different species separate as best you can tell. Dedicate one or two of the more mature fruitbodies in each collection to making a spore print.

7. Write a simple description for each of the collections, perhaps including line drawings for any features of the mushroom you find distinctive. It is especially important that you write down notes on odor and taste at this point since these characteristics often disappear rapidly.

8. Please note that tasting mushrooms is safe if a small, clean piece of the cap is nibbled and held on the tongue, and then spit out completely (distinctive tastes should be detectable within 20 seconds or so).

9. After writing a brief description of your mushroom, use the Key to Groups p. 28. We've decided not to include more extensive keys to make space for more photos. After you've identified the general group to which your mushroom belongs, page through the species descriptions looking for photos that closely match your mushrooms. Then read the text descriptions and pay attention to the similar species mentioned in the comments. Once you feel fairly sure you've identified your mushrooms to species, seek verification from an experienced person (try online mushroom enthusiast groups, but beware "Google Experts" who may seem authoritative, but lack robust knowledge of identification and often make subtle but important mistakes.

Making Spore Prints

1. A spore print is a critical first step in identifying practically any mushroom you may find. The goal is to assess the color of the spore deposit, which is simply a thick layer of spores. Follow these steps to make a spore print of a gilled or pored mushroom.

2. After removing the stipe (if necessary), place the gills, pores, teeth, or other fertile surface face down on a piece of glass, foil, clear plastic, or white paper. Let the mushroom sit undisturbed for six to eight hours. Smaller mushrooms may benefit from being covered by a glass or plastic container to prevent moisture loss and wind disturbance.

3. Carefully lift the mushroom off the surface to reveal the spore deposit. White or pale-colored spore prints can be difficult to see on white paper (tilting the paper can help) but are obvious if glass or clear plastic is placed on a dark background.

4. Observe the color of the spore deposit, preferably on a white or black background. Colored backgrounds can distort color perception dramatically.

5. Be aware that there is some subtlety in interpreting spore print colors—general terms are somewhat useful, but more subtle distinctions can be necessary, especially between different shades of brown.

Tools of the Trade

A number of tools are useful for collecting and identifying macrofungi.

- A wicker basket is the best way to carry many collections at once. Relatively flat baskets are best, since they maximize the number of mushrooms that can be carried in a single layer (minimizing the risk of crushing). Cloth tote bags collapse when full, crushing the mushrooms under their own weight.
- Field notebooks are useful for anyone seriously interested in making collections. Water-resistant notebooks (Rite in the Rain,

for example) work well in wet weather but are relatively expensive.

- Tackle boxes are the best system for keeping collections separate and for keeping small, delicate specimens in good shape. Downsides include the awkward shape of the compartments when attempting to store larger fruitbodies, and poor breathability, which can accelerate bacterial decomposition (so don't keep your collection in plastic longer than necessary).
- Wax paper bags are the best receptacle for larger fruitbodies, since they can better accommodate irregular shapes. Downsides include the difficulty of transporting multiple bags full of mushrooms (unless you've got a basket) as well as their tendency to disintegrate in wet weather. Aluminum foil works well in wet conditions, but like tackle boxes, it can hasten decay.
- Hand lenses are fun in the field, and indispensable back home when examining the smaller features of mushrooms, including details of gill edges and surface textures of many different structures. A 10x lens is usually strong enough for such purposes.
- KOH (potassium hydroxide) is available as pellets from science supply companies. When combined with water (3–10 g KOH per 95 mL water), the solution can be stored in a glass dropper bottle for about a year (after which it starts to become cloudy and weak). The tissues of some mushroom species change color when KOH is applied; these color changes can be very useful in identification. Be sure to take photographs of and make notes about any such color changes.
- The importance of a camera for learning macrofungal identification is hard to overstate. A completely serviceable camera can be bought for less than two hundred dollars, and even many smartphones have good enough built-in cameras to get you started. See the section on Photography for more information.
- Serious collectors will need a food dehydrator in order to desiccate specimens to be deposited in personal or institutional herbaria. See the section on Making Collections.

Making Voucher Specimens

Making preserved collections (also known as voucher specimens) is one of the most valuable ways in which amateur mycologists can contribute to our understanding of the diversity and distribution of mushrooms. Mushroom fruitbodies are preserved by drying them at relatively low heat (30–45°C 85–115°F) until completely crisp by using a commercial food dehydrator. It is often helpful to cut fruitbodies in half (top to bottom) to speed up the process. The dried collection can then be sealed in a ziploc bag with a small bag of silica gel, and a slip of paper listing a basic set of data: 1) Species name (if known), 2) location and date, 3) a collection number, as well as any additional notes about habitat, smell and taste, morpholgy, etc.

The most crucial step is to pair a unique *collection number* to link the physical specimen with photographs and identification notes. We *strongly* recommend using a web-supported approach – making an iNaturalist observation to go with every one of your vouchered specimens, and using the iNaturalist observation number (found at the end of the URL) to link all relevant data. This neatly encodes the date, location, images, identification work, and ongoing discussion, and allows for directly embedding microscopic details and DNA sequence data if it becomes available.

Even the simplest systems for making these vouchers can be difficult to manage when dealing with a large volume of specimens, so be sure to figure out a method that works for you and practice it with a small number of collections first. Once you've got the hang of it, stick to it! Writing a short document explaining your system will help other workers search for, request, and receive your specimen.

Collections management (storing, organizing, and making your specimens available) presents a long-term problem for which there isn't an obvious solution. Historically, local museums, academic institutions, botanical gardens, and research-focused field stations have maintained large collections tended by curators and support staff. However, these kinds of collections tend to be quite exclusive, and don't necessarily accept collections made by non-professional enthusiasts (although some motivated amateurs have had success establishing 'affiliate' status with universities).

To make matters worse, many such institutions are shrinking or becoming severely space-limited. Thus, for many collectors, the best approach will be to devise simple at-home storage solutions using stacking or nesting plastic storage containers and prioritizing efficient use of space and low-cost materials rather than the bulky archival boxes and materials used by established institutions. It's also important to realize that these specimens have a limited lifespan – even well-preserved collections decay or degrade after a few decades, so it's a good idea to front-load microscopic investigation and DNA sequencing efforts soon after vouchers are made.

Photographing Mushrooms

There are innumerable styles of mushroom photography, but photographers looking to document mushrooms for identification purposes should follow a few basic principles:

1. All parts of the mushroom should be visible, either by taking a series of photos of one fruitbody or by placing multiple fruitbodies of the same species at different angles in a single photo.
2. A range of young and old fruitbodies should be shown. A cross-sectioned fruitbody can also be helpful to show the color of the flesh, and before-and-after photos showing KOH or other chemical reactions are also useful.

While automatic modes often do a fine job of determining camera settings, learning to use your camera's manual mode will give you a lot more control in exposing correctly and capturing color accurately. In general, a low ISO setting (less than 640) and high f-stop (greater than 10) are desirable. A tripod is the single most valuable accessory you can pair with your camera to increase the quality of your pictures. Using a tripod allows for greater depth of field, more even natural lighting, and more accurate colors than handheld exposure – this is especially important since many mushrooms fruit in dark habitats beneath dense forest canopies.

Taxonomy and Cladistics

When we identify mushrooms by using the keys in a field guide, consulting websites, or examining microscopic features to compare against technical references, we are engaging in the science of taxonomy. A taxonomic system is a hierarchical arrangement of units called taxa (singular: taxon), which are grouped together according to shared characteristics. Carl Linnaeus (1707–1778) is referred to as the father of biological taxonomy because his system of naming and classifying organisms quickly became the standard in the world of academic biology. The great Swedish mycologist Elias Fries (1794–1878) took this broader system and refined it for use in the classification of macrofungi. Among the most important characters in Fries' system are the shape of the fertile surface and the color of the spore deposit as well as the colors and texture of the fruitbody.

In modern biological taxonomy, each species of living organism is given a two-part name. The first word in a binomial is the name of the genus to which the species belongs and is always capitalized. In the case of the white button mushroom found at your grocery store, the binomial name is *Agaricus bisporus*. A genus is a collection of species that are similar to each other with regard to a defined set of characters. The second word—*bisporus*—is the name of the particular species within the genus. What defines a species is a very complicated topic, particularly for fungi (more on that later in this section).

In the case of *Agaricus bisporus*, what set of characters define the genus? With the naked eye, we can recognize *Agaricus* by their free gills, dark brown spores, and presence of a partial veil (see diagram of fruitbody morphology, p. 6). All members of this genus, despite the specific differences between them, share these characters. If we were trying to identify a mushroom that didn't show these features, we would have very good reason to rule out the genus *Agaricus* as a candidate. Listing the characters that distinguish the species *bisporus* from the rest of the members of the genus *Agaricus* is a larger and more complicated task, and the species descriptions in this book are dedicated to just such endeavors.

Every named species also belongs to a named group at each of the taxonomic levels above it. In the case of *Agaricus bisporus*, we can trace the taxonomic hierarchy upward, with each category being more general (and thus more species rich) than the preceding one. The genus *Agaricus* belongs to the family Agaricaceae (which also includes *Lepiota*, *Chlorophyllum*, and *Leucoagaricus*), which belongs to the order Agaricales (non-Russuloid mushrooms), a subset of the class Agaricomycetes (which includes *Russula* and *Lactarius*), itself a member of the division Basidiomycota (all fungi that produce spores on basidia). All are members of kingdom fungi.

This morphological approach—which classifies mushrooms by shared physical characteristics—is powerful and especially useful for amateurs and field workers. However, it has been superseded in academic biology by the cladistic approach, which groups species according to genetic data. The cladistic system groups organisms based on their shared evolutionary histories rather than by shared morphological characters. Although these two criteria are not independent (because a species' evolutionary history strongly influences its morphology), they are far from equivalent. For example, two organisms that are closely related may appear quite different to our eyes. Dolphins barely resemble bats at all, but do appear quite similar in body shape, behavior, and habitat to sharks. But bats and dolphins are both mammals and are therefore more closely related to each other than either is to a shark. This phenomenon, in which distantly related organisms show similar morphology, is called *convergent evolution*. This occurs when different organisms adapt and evolve in response to similar environmental pressures over very long periods of time.

Convergent evolution is a common phenomenon in the world of fungi and is one of the main reasons that cladistic and morphological approaches to classification don't always align well. For example, although most mushrooms of the genus *Russula* have a cap, a stipe, and spore-producing gills, they are not closely related to the majority of other gilled mushrooms. This similarity in fruitbody shape is due to convergent evolution, not recent shared ancestry.

But how do we know this? What information requires us to reject our intuition about the classification of *Russula*? Most often, it is the use of genetic evidence obtained by sequencing the organism's DNA. The "blueprints" for many aspects of an organism's morphology, biochemistry, and behavior are encoded in the specific order of nucleotides in its DNA molecules. But the order of these nucleotides (and thus the information they contain) differs from species to species and, to a lesser degree, between individuals of the same species. For this reason, DNA can be used to illuminate similarities and differences between organisms that would otherwise have been hidden by the effects of divergent or convergent evolution. Generally speaking, the greater the percentage of shared DNA sequence between two species, the more closely related they are.

Determining the order of nucleotides that make up part of the DNA of a particular organism is called sequencing. After we obtain such a sequence, careful interpretation can provide us with clues about an organism's ancestors—giving us another method of building a family tree. When we say that two organisms are each other's closest relatives, we are implying that they share a more recent common ancestor with each other than with other organisms. This kind of evidence is central to our understanding of biological phenomena and is of especially great influence when we go about revising taxonomic arrangements. The complexities of cladistics are beyond the scope of this book, but it is important to keep in mind that cladistic classifications often conflict with our expectations when using morphology-based methods to classify mushrooms. While cladistics is better for building accurate family trees, morphological approaches are often much more useful when identifying mushrooms. Each system has its place, and deciding how to use them depends largely on your priorities.

The practical outcome of all this is that you'll soon become familiar with mushroom identifiers saying "that's what we call it here," "it needs a better name," "our local version of the species," or "it used to be called so-and-so." These expressions reflect one of the most frustrating facts of taxonomy and systematics: the scientific names of organisms, especially mushrooms, frequently change. Why is this?

Scientists may give mushrooms new names for a variety of reasons including that the original name may have been incorrect in some way. For example, a single mushroom species may have been given two names at different times or the same name may have been mistakenly given to two different species. Much of the time, however, there is no error involved, but rather the new names reflect the reclassification of a group of species, genera, or family based on new data (usually genetic). It is worth noting that the degree of difference that separates two genera in one part of the tree of life might be equivalent to the difference between two families in another group of organisms. Keep in mind that, at their core, these distinctions are arbitrary.

Strictly using only formally published names can be a sticky issue. Many of the species found in this book occur only in the western United States but were formally described by mycologists familiar with European species. What we have been left with are dissimilar mushrooms different morphology, microscopic features, ecology, and genetics from these European namesakes. Frustratingly, until a new name is published, the most effective way of referring to them is with an incorrect epithet.

We recommend taking a practical view of things. "Incorrect" names can still be used to convey information and, if they succeed in doing so without creating too much new confusion, don't worry about it. The need for new names and critical study of Cascadia's mushrooms is widely acknowledged and is slowly being addressed. With luck, the rapidly growing community of enthusiasts at all levels will usher in a renaissance in our understanding of these organisms over the coming years.

How to use this book

While this book does not include a dichotomous key, there is a 6-step workflow we recommend to most effectively navigate this guide. After bringing mushrooms home from the field, set a few fruitbodies up to take spore prints right away – this process can take 6-12 hours. While you wait, you can make an educated guess at the spore color or focus on using other features to follow these steps:

1. **Start with the Pictorial Key to Groups** on p. 28, look for the best match to the fruitbodies you've collected, focusing on the color of the spore print (if you've already obtained one) and structural details (e.g. presence or absence of partial veil, textures of surfaces, gill attachment). Once you've decided on a promising group, move to the next step.
2. **Read the Intro Page** that opens the chapter that your group belongs in. This background material will help you focus on the details that are most useful for that group specifically, as well as provide important ecological context. Spending a little extra time reading these will deepen your knowledge and will help speed up the rest of the identification process.
3. **Picture matching.** Next, carefully page through the species accounts in the chapter you've arrived at, looking closely at your mushrooms and comparing them to the photos for each species account. As with *Mushrooms of the Redwood Coast*, this book is arranged loosely by evolutionary relationships and, more importantly, by visual similarity. This order is intended to put similar species as close to one another as possible, and thus minimize the amount of page-turning needed to find promising candidates. The most important tip at this stage is *not to stop too soon.* Even after you've found a good visual match, flip a few more pages to make sure there's not an even better one!
4. **Read the Comments.** Although we've provided systematic technical descriptions throughout, the comments section provided for each species is the quickest way to decide whether you might be on track or off-base. The most essential identifying features for each species are listed here, as are the most frequently-confused species and tips for separating them. Next, read the **Ecology** section. Details of habitat association and biogeography can often provide quick insight about whether or not your identification is correct.
5. **Compare against other images and resources.** The internet is a vast trove of knowledge for people interested in biodiversity. Although online information is not always of uniform quality or reliability, by comparing against multiple examples you should be able to build a richer picture of a given species. A printed paper resource like this book is space-limited—thus we have to choose one or two images to represent a species and can never represent a species' full range of variation.
6. **Ask someone for confirmation.** Ultimately, there's nothing quite so helpful as getting feedback from a real person. Consulting a knowledgeable friend or an online acquaintance whose expertise you trust will go a very long way towards speeding up and strengthening your learning process. Don't let your hard work come to a dead end! Seek

out a community of folks who will give you positive feedback about your successful identifications—as well as constructive pointers about the mistakes you may have made along the way.

Reading the species descriptions

Our species descriptions are written in a standard format to make it easier for readers to find specific pieces of information about the characters of each species. For example, you can expect to find cap dimensions or notes on ecology to appear in approximately the same place on every page. The descriptive terminology we use is mostly consistent across groups of mushrooms but some have such different morphology that they require their own set of categories and terms.

The dimensions (size measurements) that we list for the various parts of fungal fruitbodies are an approximation intended to give readers a rough idea of the typical size range for a species. In most cases we list an average range, along with any extreme dimensions of specimens we've seen noted in parentheses. Our use of general measurements rather than highly specific ones reflects the fact that fungal fruitbodies are dramatically affected by the conditions under which they are formed—a single mycelium may produce miniature and oversized fruitbodies depending on whether fruiting was preceded by drought or torrential rain. Species that produce larger fruitbodies generally show a larger range of variation in size, while those that produce smaller fruitbodies are usually more consistent. While we've intentionally kept our descriptions of size fairly general, our descriptions of colors tend to be a bit more subtle and specific. The text descriptions of colors are important, since a single photo can't hope to depict the full range of variation in a species. Keep in mind that some species never show much deviation from a specific palette of colors, and others seem to be able to produce a rainbow of pigments.

Mushrooms are sensitive to the environmental conditions in which they develop. They are remarkably tolerant of adversity and are often able to produce and release spores even after being subjected to parching air, soaking rain, freezing temperatures, insect infestation, and bright sun. Under such conditions, you can expect that fruitbodies will look a bit worse for the wear—differing from what might be considered 'typical'. What this means for you, esteemed mushroom identifier, is that you should always try to take into account the effect that such factors may have had on the shape, size, textures, and colors of a given fruitbody. Rain often washes out pigments, making them lighter or leaching them into other parts of the fruitbody; wind often dries out slime layers and causes caps to crack; sun often causes pigments to become paler; drought often results in miniature fruitbodies. As you become familiar with the mushrooms in your area, you'll develop an intuitive sense for these variations.

Species descriptions Format

Name and author: Immediately after the name of each species, we list the authors of that name—the people responsible for the original concept of the species and for its current generic placement. The rules governing how species are described and how these 'authority names' are formatted are complex and mostly not relevant for the average mushroom identification enthusiast. For those interested in a deeper dive into the subject, Nick Turland's free eBook *The Code Decoded* is an extremely helpful guide to International Code of Botanical Nomenclature and its implementation. For some species accounts, we use 'Group' to indicate that to our best understanding, there are a number of different species going by the same name, and for which we have yet to develop reliable distinguish criteria. Sometimes researchers propose names for species but stop short of formally publishing them. These are *provisional names* and are denoted by the abbreviation 'nom. prov.' They are usually given a Latinized epithet, and often come with fairly well-established species concepts. In other cases, we use an even less-formal kind of name called "Tag Names."These are similar to provisional names, but are given in plain English (intended to be easier to remember, like a nickname). They are called 'tag' names because they are often used on paper tags at mushroom fairs and displays to label "known unknowns." These species are recognizable and familiar to the local community but without an appropriate scientific name. Lastly, because of the European naming conundrum mentioned above we emphasize discrepancies with 'sensu PNW' after their names.

This is a Latinized abbreviation meaning 'in the sense used in the Pacific Northwest.' Otherwise, we continue to use the names based on European taxa authors believe that such species will likely be renamed after proper research.

Cap: Most of the species in this book produce fruitbodies which have some sort of clear cap or 'head' that is described in this section. Those mushrooms with a different structures will have other headings to better reflect their shape.

Gills and other fertile surfaces: This section will have many different names depending on the morphology of the species in question: spines, tubes, pores, or simply fertile surface.

Stipe: If the fruitbody in question has a stipe (whether clearly defined or not), it is described in this section. Some species produce fruitbodies that lack a distinct stipe and won't have an entry for this section.

Veil (partial veil & volva): Veil tissues protect the developing fruitbody. Universal veils (or volva) enclose the entire fruitbody as it grows while partial veils cover only the fertile surface until it matures enough to release spores. Only a few genera of mushrooms produce fruitbodies with universal veils however mushrooms with partial veils are common and found in many genera. The texture of these tissues ranges from thick and membranous (most often) to thin and fibrous, cobwebby, cottony, hairy, or fluffy. In many cases, descriptions of the veil tissue will also be included in other sections—as warts or a patch when found on the cap or as a volva if it encloses the base of the stipe in age. Referring to the diagram on p. 6 should clarify things in case of confusion.

Flesh: Color and texture characteristics are noted in this section. Flesh is best observed by cutting a fruitbody in half and breaking the stipe or pulling apart the flesh. This is destructive, so be sure to take note of other features before you mangle your mushroom! Note that by cutting and rubbing the tissues of a fruitbody, you expose the cell contents to oxygen in the air—in some species this will produce a distinctive (and sometimes surprising!) color change. In such cases, pay attention—this is an important identification character. A lot of mushrooms darken to some degree when damaged, but you'll soon learn to distinguish trivial bruising from more pronounced color changes. Some common reactions are bluing, reddening, yellowing, and browning with blackening and purple staining less common.

Odor and Taste: These are very important characteristics for any mushroom. Unfortunately, this character is variably sensed person to person and thus variably described. In general, we have found that descriptions of taste seem to be more consistent between people than descriptions of smell, which often vary dramatically. Even if you don't agree with our descriptions of a smell or taste, it's worth "calibrating" your senses to these descriptions, since much of the terminology we use is in line with commonly established descriptions in prior literature. If you've confirmed beyond doubt that the mushroom you're holding is *Agaricus augustus* (described as smelling like almonds in almost all literature), but you find it to smell more like hair gel, keep this in mind when you encounter "almond" in other species descriptions—you may again interpret the species as smelling like hair gel. Remember that practically all mushrooms will begin to smell unpleasant in age as they begin to decay. Overall, try to tune your senses and make your descriptions consistent and make comparisons with common scents (fruits, vegetables, herbs and spices, cleaning products, familiar chemicals). Don't balance your entire identification on your interpretation of the smell. To safely taste a mushroom, bite off a small, clean piece of the cap flesh and nibble it on the tip of your tongue for 5-10 seconds—then spit it out completely. It is safe to taste mushrooms (even poisonous ones) in this manner, as long as you aren't ingesting the tissue.

KOH and other chemical reactions: The fruitbodies of some mushrooms change color when exposed to various chemicals (often called reagents). We've included potassium hydroxide (KOH) reactions for many species, since this chemical is relatively easy to get a hold of and is often useful in identification. When our photo shows mushrooms that have been treated with KOH (usually visible as a darker or differently-colored spot or streak on the cap and/or flesh of a cross-sectioned fruitbody) we've added an asterisk after the abbreviation (KOH*). Although there's a wide variety of other chemicals that have been used by mushroom taxonomists, we only mention a few in our descriptions—many are hard to find, extremely toxic, and/or less useful for the average identifier. Keep in mind that the intensity of the reaction varies based on a number of factors including age of the fruitbody, degree of rain soaking, and strength of the chemical being used (KOH often becomes weak after a year or two).

Spore deposit: The color of the spore deposit is a very important feature when you are first starting to learn mushroom identification—see p. 20. However, it takes a while to make a spore print, so it won't be something you can determine while out in the field.

Microscopy: Since most people don't have access to microscopes, we've included just a few of the most useful kinds of microscopic details. Spore shape and size are usually the most crucial features to note. For basic definitions and illustrations of microscopic structures and chemical reactions (amyloid, inamyloid, dextrinoid) see the diagram on p. 6 and the Glossary. For a more detailed look at mushroom microscopy, consult Largent's *How to Identify Mushrooms to Genus, Vol. III*; References (p. 551). There's still much to be learned about the range and significance of variation in microscopic features of mushrooms—our data are representative but not definitive. Please seek out measurements from other sources and tell us what you learn!

Ecology: This section includes a summary of what the authors understand about the range, habitat, and seasonality of the species. Our accounts are based on our own experiences but also on data from primary publications, discussions we've had with knowledgeable field mycologists, and, increasingly, on the picture painted by the consortium of observations submitted to community science platforms. These are by no means definitive descriptions, and you are encouraged to report discrepancies between our accounts and your own observations to public forums like inaturalist.org. The same goes for our interpretations of commonality—many species are reported as rare until their habitat preferences become well known. Likewise a species may be under-reported if its identification poses difficulties for most people (or if it lacks a name!).

Comments: This important section summarizes the most useful identification features for a given species and gives a rundown of similar species that you should consider and compare. It also includes occasional additional information unrelated to identification (interesting natural history notes, cultural uses, and the like).

Synonyms and misapplied names: Our process of describing and naming species is far from perfect. Sometimes this leads to a single species of mushroom being given multiple names. Ideally, only one name should be used for each species. In this section, we list some of the commonly used or well-known (but incorrect, imprecise, or otherwise suboptimal) names for the species. Unfortunately, many species have undergone name changes recently and older, "incorrect" names may be much more familiar than the most scientifically accurate name. For some undescribed mushrooms, we've used "tag names" to provide a descriptive but informal way to refer to them. In some descriptions, we use "group" names to refer to complexes of multiple species. In these cases, precise identification is likely to be difficult, complicated, or impossible given our current state of knowledge. We will keep track of nomenclatural updates relevant to the species found in this book on our website: www.redwoodcoastmushrooms.org/cascadia-name-changes. Expect changes to come frequently—things change fast in the world of mushroom taxonomy and identifications!

Pictorial Key to the Major Sections

Chanterelles, Gomphoids, and others—Spores white to ocher-buff. Gills often shallow, ridge-like and blunt, and/or forking repeatedly, veined, or nearly absent and fertile surface smooth. Fruitbodies triangular or trumpet- or vase-shaped, often stout but some thin-fleshed. Colors ranging from bright orange to yellow, beige, creamy-white, violet, bluish or gray to black. Texture from thin and brittle-cartilaginous to tough or firm and rubbery with stringy flesh. See p. 37.

Amanita—Spores white. Small to large mushrooms, most often with a cup-like, scaly rings, or sac-like volva or other veil remnants around the base of the stipe, a patch or warts of veil tissue on the cap (sometimes bare), and a ring or skirt on the stipe (except in one subgroup). Gills free or narrowly attached. Cap colors range from dull to very bright. See p. 48.

Cystoderma* and *Cystodermella—Spores white (brown in *Cystoderma aureum).* Mostly small to medium-sized mushrooms (*C. aureum* is large) with narrowly attached, whitish (or brown) gills, a partial veil on the stipe (sometimes only an obscure veil zone), and a granular, powdery, or scaly cap texture (often present on stipe as well). Cap colors range from ocher to golden-tan, brown, or pinkish red. See p. 63.

***Lepiota* and Allies**—Spores white (grayish green in *Chlorophyllum molybdites*). Small to large mushrooms with distinctly free gills, a partial veil present as a collar, skirt, or obscure annular zone, but lacking universal veil tissue (cap scales can be confused for warts but are more tightly attached). Colors mostly whitish to beige to gray, some with darker gray, black, or brown scales, a few with pinkish to purple cap colors. See p. (63) 65.

Agaricus* and *Melanophyllum—Spores chocolate brown in *Agaricus*, grayish olive to reddish brown in *Melanophyllum. Agaricus* are small to very large mushrooms with distinctly free gills, and have a partial veil which often leaves a skirt- or collar-like ring on the stipe. They often have distinct staining reactions when cut and/or distinctive odors. *Melanophyllum* have rather small fruitbodies with powdery gray brown caps, vivid red to dull reddish gills, and a partial veil (often obscure). See p. 77.

Dark-spored Mushrooms—Spores dark purplish gray, purplish black, or black. Fruitbodies range from very small and fragile to medium sized and fleshy-fibrous. Gills attached to obscurely free, often pale to grayish, becoming dark gray or black. Especially common in urban habitats or disturbed areas, often on wood chips or dung. See p. 89.

Gilled Bolete Relatives—Spores olive brown to dark grayish olive or black. Gills decurrent (and often widely spaced), thick to thin and regular or wavy and branched or forking. Most species with medium-sized fruitbodies; but a few species are smallish and slender, a few others are quite large. On ground or on woody debris. See p. 119.

Brown-spored Decomposers—Spores dull brown to rusty brown. Fruitbodies tiny to huge, dull to colorful, and with or without partial veil tissue. If your mushroom has brown or rusty spores and is growing on wood, it should be found in this section. Otherwise, check the following short descriptions for *Inocybe*, *Hebeloma*, *Phaeocollybia*, and *Cortinarius* in this pictorial key. If none of those seem promising, your mushroom should be in this section. See p. 128.

Mycorrhizal Brown-spored Mushrooms (*Inocybe*)—Spores dull to medium brown. Small to medium sized, with dry, silky, scaly, or shaggy, often conical caps. Stipes cylindrical, often powdery-looking, many with a small bulb at base. Odor frequently spermatic or like green corn; others sweet and fruity, like cinnamon, fishy, or indistinct. Partial veil present and silky to cobwebby, but usually sparse and often visible only on young fruitbodies. Growing on ground. See p. 156.

Mycorrhizal Brown-spored Mushrooms (*Hebeloma*)—Spores dull brown to reddish brown. Small to fairly large mushrooms with mostly smooth, often viscid caps and attached gills (often beaded with droplets when young). Odors often pungent, radish-like, sweet, or chocolatey-musty. Stipes cylindrical to club shaped, often scurfy. Growing on ground. See p. (156) 163.

Mycorrhizal Brown-spored Mushrooms (*Phaeocollybia*)—Spores rusty to ocher-brown. Small to large mushrooms with distinctive long, tapering, rooting, and cartilaginous stipes (often filled with pith). Caps typically conical or bell-shaped, often viscid. Fruitbodies dull to colorful. Veil absent. On ground, found only in forest settings. See p. (156) 169.

Cortinariaceae—Spores rusty (bright orange-brown). A large and varied group of species united by the presence of a cortina—a silky or cobwebby partial veil (membranous in one species). Beware that the cortina can be inconspicuous or disappear in age. Fruitbodies range from very small to very large, with colors ranging from very dull to very colorful. Caps conical, rounded, convex to plane, dry to viscid. Found in forest habitats, growing on ground. See p. 176.

Dermocybe—Fruitbodies small to medium-sized. Cap moist to dry, appressed-fibrillose to smooth. Gills often brightly pigmented. Yellow, orange, red, olive-brown pigments predominate. Often showing strong KOH and/or UV reactions.

Telamonia—Cap and stipe dry. Surfaces usually smooth to silky-fibrillose. Caps usually hygrophanous. Most are brown or drab colors, but some are brighter, usually silvery-blue to purple.

Phlegmacium & Calonarius—Cap viscid, often brightly colored. Stipe dry, often large, clavate or distinctly bulbous. Some have very distinctive KOH reactions (pink, yellow, red, orange).

"Myxacium"—Cap and stipe both viscid, often brightly colored. Stipe usually tall and cylindrical. KOH reactions usually indistinct.

Large Pinkgills (Entolomataceae and others)—Spores pinkish tan to salmon. Fruitbodies medium sized to large, caps dry to viscid, gills attached and notched, or decurrent. Colors white to gray or brown. Veil absent. Found in forest settings, growing on ground. See p. 213.

Small Pinkgills and Entolomataceae—Spores pinkish tan to salmon. Fruitbodies small to medium sized, with attached gills. Caps dry, smooth, finely scaly, or silky. Colors ranging from dull brown to rosy pink, yellowish tan, blue-gray to iridescent purple, or blue to blackish blue. Veil absent. Found in forest settings, on ground, or rarely on wood. See p. 213.

Free-gilled Pink-spored: *Pluteus* and Allies—Spores pinkish to dingy salmon-tan. Gills distinctly free. Caps dry to moist, smooth to distinctly granular, wrinkled, or silky-fibrillose. Mostly on wood, but others grow on rich organic matter and lignin-rich debris, or on decaying mushrooms. Partial veil always lacking, but a few species have universal veil sheathing the base of the stipe (volva). See p. 225

Brittlegills (*Russula*)—Spores white to cream, yellowish, or ocher. Often stout, squat mushrooms with convex to uplifted caps. Texture usually chalky, brittle, or crumbly and lacking any 'milk' (latex) when broken. Some can have hard, firm flesh, but will still break like chalk rather than splitting into fibers like most other gilled mushrooms. Colors dull to bright. Veil absent. Growing on ground. See p. 231.

Milk Caps (*Lactarius*)—Spores white to cream to light yellowish. Often stout and squat with uplifted caps that are centrally depressed. Texture crumbly, brittle, or chalky, usually producing "milk" (latex) when broken or cut. This feature can be hard to see, so check fresh fruitbodies and look closely. The milk can be clear, cloudy, white or brightly colored, sometimes changing color when exposed to air, or staining the surrounding tissue. Veil absent. Growing on ground. See p. (231) 249.

Large Waxy Caps—Spores white. Gills attached, relatively thick-edged and slightly greasy or waxy–feeling, often well-spaced. Caps dry to very slimy, dull to bright. Stipes slender to robust, but always solid. Partial veil present or absent. Found in forest settings. This group can be difficult to distinguish from other white-spored mushrooms (also check Small Waxy Caps and the large species in White-Spored Multitude). See p. 265.

Small Waxy Caps—Spores white. Gills attached, relatively thick, often waxy feeling and well-spaced. Caps dry to very slimy, mostly brightly colored (red, orange, yellow, pink, green), a few duller (gray, brown, white). Stipes usually slender and hollow. Partial veil absent. Found mostly in forest settings, commonly under cedar. See p. (265) 276.

The White-spored Multitude (large species)—Spores white. Fruitbodies medium to large. Gill attachment variable—decurrent to broadly or narrowly attached or notched. Partial veil present or absent, colors variable (most dull, but some bright pinkish, orange, or purplish). Growing on ground or wood. Also check the Pleurotoid section. See description of white-spored groups on p. 283.

The White-spored Multitude (small species)—Spores white. Fruitbodies tiny to small (a few are medium sized but with thin stipes); caps usually less than 2.5 cm across. Caps often translucent-striate; colors range from white to brown to dark gray or brightly colored. Mostly litter decomposers, found in both urban settings and forest habitats. See description of white-spored groups on p. (283) 325.

Pleurotoid Mushrooms—Spores white to pink or brown. Caps fan- or spatula- shaped, and either lacking stipes or with stipes that are lateral, off-center, or short and rudimentary. Most with true gills, but some with poorly defined ridges or wrinkles. Fruitbodies very small to very large. Most grow directly on wood, but some on moss, grass stems, twigs, or woody debris. See p. 369.

Boletes—Spores olive-brown (yellowish brown, reddish brown or vinaceous brown in a few species). Undersides of caps with a spongelike layer of tubes with round or angular pores; stipes usually well developed and central. Texture of flesh ranges from soft to firm, but usually distinctly fleshy (not tough or leathery). Growing on ground or well decayed wood, typically in forest settings. See p. 380.

Stalked and Rosette-forming Polypores and Allies—Spores whitish or brown. Underside of cap with pores, stipe off-center to central, sometimes short or rudimentary. Flesh thin to thick and firm, usually tough or leathery (unlike softer-fleshed boletes). Fruitbodies small and dainty to very large and robust. Colors range from whitish gray or creamy, to brown or reddish brown, orange to bright yellow, or even sky blue or royal blue. On wood or in humus in forest settings. See p. 413.

Shelf-like, Bracket, and Conk-like Polypores—Spores whitish or brown. Underside with distinct pores. Often forming conks, fan-shaped brackets, or lobed clumps, singly or in overlapping shelves; typically lacking a distinct stipe. Flesh tough and leathery to corky or hard and woody. Most fruiting directly from wood (sometimes buried). See p. (413) ~422.

Smooth or Veined Brackets or Rosettes—Spores whitish or brown. Underside smooth, veined or winkled pores. Often forming rosettes or overlapping shelves; some with a central to off-center stipe. Flesh tough and leathery. Fruiting on wood or soil. See p. 442.

Crust Fungi—Spore prints not usually obtainable. A ubiquitous group comprising many unrelated lineages. Fruitbody structures simple but variable; usually flat to slightly shelving and usually irregular in outline. Fertile surfaces smooth, pimply pored, or toothed. Colors dull to brilliant. Often fruiting inconspicuously on the undersides of rotting logs, on standing dead timber, or on other fungi. See p. 445.

Tooth Fungi—Spores white or brown. Fertile surface composed of hanging, icicle-like 'teeth'. Some have distinct or rudimentary stipes, others lack stipes. Some grow on ground (often in rosettes), while others emerge directly from wood and lack any stipe. *Hericium* form large clumps with downward-pointing spines attached to a central mass, whereas *Mucronella* are tiny and highly reduced, composed only of teeth hanging from deadwood. Also check the section on Crust Fungi. See p. 449.

Coral Fungi—Spores pale to ocher-tan or brown, prints often difficult to obtain. Fruitbodies coral-like, with branches and upward-pointing tips (or in one species ruffly, flattened lobes). Degree of branching and shape of tips variable; colors range from white to dull tan to brilliant pinks, purples, yellows, oranges or even greenish. Textures range from thin and fragile to tough or firm and fleshy, or sometimes slightly gelatinous. Mostly found in forest habitats. See p. 465.

Club Fungi—Spores prints usually difficult to obtain. Fruitbodies simple, upright, unbranched or sparsely branched. Some are cylindrical, fruiting singly or in clusters, while others have distinct round, oval, paddle-shaped, or irregular heads. Textures range from thin and fragile to tough, with surfaces that are velvety to pimply, powdery, or wiry. Colors range from tan to purple, black, yellow, orange, green, or white. See p. (465) 477.

Puffballs, Earthballs, and Earthstars—Spores forming a solid or powdery mass inside the fruitbody. Spores yellowish to olive, rusty, brown, purplish, or black. Shape variable. Puffballs and earthballs have round to pear-shaped fruitbodies. Earthstars have a thin-skinned spore sac perched on a star- or saucer-shaped pedestal. *Tulostoma* has it's spore mass elevated on a tall central stipe. See p. 489

Bird's Nests—Spore prints not obtainable. Fruitbodies are tiny to small leathery cups or vase-like goblets with flat, lentil-shaped spore packets ('eggs') in the middle. Textures range from smooth to hairy, granular, or grooved. Colors always dull, mostly beige, gray, brown, and blackish. On dung, rich soil, or woody debris; found in both disturbed urban areas as well as in forest habitats. See p. 497.

Stinkhorns—Spore print not obtainable. Spores are immersed in a thick, dark, usually foul-smelling slime held on strangely-shaped fruitbodies. Some are cylindrical and upright, others are cage-like, tentacle-like, or obelisk-like. Most emerge from whitish, rubbery eggs. Mostly found in gardens, wood chips, trail sides and disturbed settings. See p. 501.

Truffles—Included here are those species that have lost most of their adaptations for fruiting aboveground. Instead of gills, pores, or teeth elevated on a stipe, they form highly reduced "sequestrate" fruitbodies. These are often rounded or knobby and have complex, convoluted inner fertile tissue (sometimes powdery or gelatinous). Many different lineages of mushrooms have trufflelike representatives. See p. 502.

Jelly Fungi—Spore prints usually difficult to obtain. Fruitbodies are tongue- shaped, or amorphous, blobby, brain-like clumps, or gelatinous, bumpy, and sheetlike masses. Most share a distinctive watery- or rubbery-gelatinous texture. Colors range from black to yellow, orange, or brown; many are slightly translucent. (A few gelatinous species with barrel- or bowl-shaped fruitbodies are in the Cup Fungi section.) See p. 516.

Morels, False Morels, and Elfin Saddles—Fruitbodies with clearly-defined, smooth or fluted and grooved stipes and caps with thimble-like, honeycombed, saddle-like, or brain-like structures. Species that produce fruitbodies with stipes and simple cup-shaped "caps" are in Cup Fungi section. See p. 522.

Cup Fungi—Spore print often difficult to obtain; ranging from pale to black. Fruitbodies have upward-facing cup- or bowl-shaped fertile surfaces; some have distinct stipes. Size, color, and texture quite varied. Most are thin fleshed and relatively small; others are robust with thick, rubbery, or gelatinous flesh. Spores produced inside cylindrical cells (asci), usually in groups of eight. See p. 533.

What is a species?

It might come as a surprise to our readers that the authors of this book (which is *about* mushroom species) have such a difficult time answering this. But the question has been a foundational problem in biology for centuries! A full investigation into the philosophy of species is beyond the scope of this book, but we do want to communicate two basic ideas: What a species is, and what a species *isn't*.

We can start roughly defining the concept by describing a species as "*a specific kind of organism.*" A more scientific way of stating this might be to say that a species is "a named unit of living creatures."

But right away there's a problem with this definition. For example, "mammals" are a specific kind of organism, but a mammal is not a species—this word refers to a larger category containing many different kinds species—the formal taxonomic level for mammals is called a *Class*. So we need to more carefully define our level of inclusion. A species is something like 'cats' or 'dogs' —these are named groups of organisms larger (more inclusive) than the level of an individual cat or dog, but below the level of the extended family trees of cats and dogs (e.g. jaguars, mountain lions, coyotes, jackals).

For many familiar organisms, units at this level of inclusion aren't too difficult to recognize intuitively. But for most of life's diversity, this is not the case—we don't have a strong immediate sense about where to draw boundaries between obscure beetles, birds with multiple plumage forms, or highly variable mushrooms.

To address these problems, and provide some formal footing for the concept of a species, scientists have tried to establish conceptual models called *species concepts*. The most common and perhaps most important of these is called the Biological Species Concept (BSC), in which the interfertility of two individuals is the ultimate rubric for determining where one species ends and another begins. Under this model, if two individual organisms can successfully mate and produce viable offspring, this is taken (by definition) to be proof that they are of the same species.

But this seemingly objective approach also runs into problems. It is impractical (and usually downright impossible) to observe the mating of two individual organisms every time we want to determine whether they belong to the same species. Conversely, it's also the case that two organisms of the same species can fail to successfully mate for other reasons. Even more problematically, some organisms don't ever reproduce sexually, and for them, the whole concept falls apart!

The pragmatic reality is that we define species by using *multiple lines of evidence*. This might include things like similarity in physical appearance, genetic similarity, ecological similarity, and chemical similarity. By triangulating between different kinds of evidence, we can make stronger cases for our arguments about what groups of biological organisms should be considered to belong to the same named units. Remember: *A species name is a hypothesis*, and thus, open to testing, refutation, and alteration over time.

It might seem strange to spend time explaining what a species isn't, but understanding this helps to ease some of the tensions which arise in problematic species concepts. To this end, it is important to acknowledge and accept that our ideas about species are *not objective truths about nature*. Lineages of biological organisms are not static and clear cut—the very nature of evolution involves constant change in space and time. Groups of organisms will always display significant variation in any measurable feature, with certain average tendencies which we can arbitrarily draw boundaries between and choose to recognize as species. Species exist within human systems of language, not 'out there in nature.' Although our choices about how to apply names to natural phenomena can draw on measurable evidence, some degree of subjective decision-making is always required in the end.

This should not be taken as a denigration of the concept of species—we need names to allow us to communicate with one another about the living phenomena we witness, marvel at, and depend upon for food, companionship, and a healthy planet. That's what we hope to contribute to with this book.

1 - Chanterelles and Gomphoids

Cantharellus is a genus of well-known and well-loved mushrooms found in forests around the world. These are the 'true' chanterelles, and they share a distinctive fertile surface made of shallow, vein-like gills that run deeply down the stipe. The fruitbodies are usually trumpet shaped or triangular in cross section, and the caps are often ruffled and wavy. Chanterelles are extremely popular as edibles because of their often-fruity aroma, delicate flavors, and their abundance. There are relatively few species of chanterelles in Cascadia, but there has been a recent surge in the description of new species from the United States, and it's possible that undescribed species remain to be found here. All are ectomycorrhizal–including some which form associations with conifers, others with broadleaf trees, and even some with shrubs.

Craterellus species have tubular fruitbodies with a hollow running from the center of the cap through the stipe base and much thinner flesh than the true chanterelles. The fertile surface ranges from completely smooth to branching veins and ridges. *Craterellus* are ectomycorrhizal with broadleaf and conifer trees.

Polyozellus are famous for their astoundingly rich blue and purple colors. Although they are sometimes called blue chanterelles due to their similarly-structured fruitbodies, they are not closely related to chanterelles but instead evolved from a lineage including *Thelephora*. *Polyozellus* are ectomycorrhizal with conifers.

Turbinellus and ***Gomphus*** look like chanterelles, but tend to be funnel-shaped, with deeply decurrent gills but lacking a distinct stipe. The woolly-scaly cap textures of *Turbinellus* and the dingy, lilac-purple fertile surface of *Gomphus* help distinguish these genera from any of the true chanterelles. Both groups form ectomycorrhizae with conifers.

Stereopsis is an unusual genus not closely related to any of the groups mentioned above, but are included here because of their superficial resemblance to a small chanterelle. The fruitbodies are small and pale, with a smooth to finely veined fertile surface and slender stipe. They are members of the Polyporales and are presumed to be decomposers.

Hygrophoropsis species are relatively small and slender, with thin, often flabby flesh, finely velvety cap surfaces, forking gills that are wavy, and colors ranging from beige-whitish to orangey to brick-red or brown. They are 'gilled' members of Boletales.

Important identification characters to note:

1. Shape and stature (stocky or slender?)
2. Coloration of all surfaces (especially when young
3. Surrounding trees and habitat

Cantharellus formosus

Corner

Pacific Golden Chanterelle

Cap: 3–10 (15) cm across, bun shaped with an inrolled margin and often a depressed center at first, becoming broadly convex but margin often strongly downcurved, expanding to flat and then uplifted with a depressed center and strongly wavy margin. Surface moist to dry, smooth but with matted-fibrillose scales that sometimes lift up in dry weather. Bright yellow to yellow-orange or dull brownish beige, often with ocher-brown scales. Much duller when young and dry, brighter when wet. **Gills:** Deeply decurrent, blunt edged, but relatively thin and fairly broad for a chanterelle, with few cross veins when young (more in age). Dull whitish pink to light yellow when young, pinkish buff to dingy yellowish beige in age. **Stipe:** 4–12 cm long, 1–2.5 cm thick, cylindrical or slightly tapered toward base, usually straight and tall, sometimes slightly curvy but still upright. Surface dry, often breaking into chevrons over upper stipe. Whitish cream to pale yellow or orange yellow. **Flesh:** Fairly thick, solid, stringy. White, bruising slowly yellow and then brown when cut. **Odor:** Indistinct or fruity-fragrant. **Taste:** Indistinct. **Spore deposit:** Creamy or pale yellowish. **Microscopy:** Spores 7–9×4.5–6 µm, ellipsoid, smooth. Basidia 4- to 6-spored

Ecology: Solitary, scattered, or gregarious in large arcs and troops in moss and duff of conifer forests. Ectomycorrhizal with conifers, with a preference for younger to middle age forests. Especially common in Douglas-fir and western hemlock forests throughout much of our range and far more common west of the Cascade crest. First fruiting in summer (especially on the Olympic Peninsula), continuing into late fall or early winter on the coast.

Edibility: Edible and very good, usually much cleaner than other chanterelles due to their habit of fruiting above the duff, which also makes them easier to find. Younger, less-waterlogged specimens typically have fruitier flavors.

Comments: The tall, straight, slender stipe, decurrent gills, yellow to brownish ocher cap, and growth under conifers help identify this chanterelle. *C. formosus* is by far the most common chanterelle in the Pacific Northwest; it is often abundant in lower elevation plantation forests. *C. cascadensis* has a brighter yellow cap, whitish gills, and a short, chunky, club-shaped stipe. *C. roseocanus* has brighter orange gills, a pinkish orange cap, and a shorter, chunkier stature. *C. californicus* is much larger and chunkier, and grows under oaks; it's rare in the PNW. *Craterellus tubaeformis* can look similar but has a thin-fleshed, hollow stipe with a cartilaginous texture. *C. formosus* is occasionally confused with *Chroogomphus tomentosus*, which has a velvety-shaggy cap margin and olive blackish spores. *Lactarius olympianus* has an orange cap, whitish gills and bleeds latex when damaged. Genetic data suggest that there are a few different species going by the name *C. formosus*, but little is known about morphological differences between them.

Cantharellus cascadensis

Dunham, O'Dell & R. Molina

Cascade Chanterelle, Hybrid Chanterelle

Cap: (4) 7–15 cm across, convex and lobed with an inrolled margin when young, expanding to broadly convex or flattened and uplifted with a depressed center, usually quite wavy with a thin, ruffled margin. Surface moist, smooth to slightly matted-scaly at the center in age. Bright, clear yellow to egg-yolk yellow, often mottled with watery yellow or creamy patches, occasionally extensively creamy yellow if buried in the duff. **Gills:** Deeply decurrent, blunt, variably shallow to somewhat deeper (but not "true" blade-like gills), with many cross veins and wrinkles. Whitish to creamy, sometimes with a light pinkish hue in age or when dry. **Stipe:** 6–10 cm long, 2–4 cm thick, chunky, squat, often narrower at apex and distinctly club shaped with a rounded base (up to 5 cm thick). Whitish to creamy, fibrous to smooth. **Flesh:** Fibrous, firm, solid, stringy. White (a bit yellowish in cap), bruising slowly yellow and then ocher-tan after being cut. **Odor:** Indistinct to fruity-fragrant. **Taste:** Indistinct. **Spore deposit:** Whitish to creamy. **Microscopy:** Spores 7–13×5–8 µm, ellipsoid to subglobose, smooth. Basidia 4- to 8-spored.

Ecology: Fruiting in fall in small groups or scattered in large arcs in deep humus under a wide variety of conifers. Uncommon but widespread throughout the Pacific Northwest; more likely to be found in Oregon and northern California than Washington. This species does not appear to have a strong preference for a particular forest type like our other chanterelle species. We have seen *C. cascadensis* in young Douglas-fir forests in the Coast Range, in mixed Douglas-fir, fir and pine forests in the Cascades, as well as in old-growth forests in the Siskiyou Mountains.

Edibility: Edible and excellent; often with a fruity flavor more typically found in eastern North American chanterelles.

Comments: This species has been referred to as the "Hybrid Chanterelle," since it appears to have the bright cap of *C. formosus*, but the pale underside of *C. subalbidus*, but is not a true biological "hybrid" of these species. The egg-yolk yellow cap with wavy-ruffly edges, contrasting whitish, decurrent veiny gills, stout, club-shaped stipe, and growth with conifers help identify this species. *C. subalbidus* is similar, but usually has a creamy white to ivory cap (can show yellowish stains). *C. formosus* has a yellow-orange cap and yellowish to pinkish buff gills, and is usually smaller, with a distinctly tall, slender stipe, and a tapered, not club-shaped base. *C. roseocanus* often has pinkish tones on the cap when young and has bright peach-orange to orange gills.

Cantharellus roseocanus

(Redhead, Norvell & Danell) Redhead, Norvell & Moncalvo

Rainbow Chanterelle

Cap: 3-10 (15) cm across, convex with an inrolled and lobed margin when young, expanding to plane or slightly uplifted and wavy, usually ruffled in age. Surface moist, smooth, with a few matted scales in age, center of cap often with a mass of lumpy tissue, frequently with small "rosecombs" of gill tissue as well. Orange with an extensively rosy pink margin when young and wet; pale grayish with a whitish bloom when young and dry, becoming more yellowish orange or dingy ocher, margin frequently fading to grayish and remaining inrolled. **Gills:** Shallow, ridge-like, deeply decurrent, with extensive wavy cross veins. Bright peachy orange to apricot orange when young, becoming bright orange to yellow-orange. Sometimes paler yellow when young. **Stipe:** 4–10 cm long, 1.5–3 cm thick, equal or with a club-shaped base. Surface moist to dry, smooth when young, wrinkled and bumpy in age, often with chevrons near the apex. White to yellowish or dingy ocher. **Flesh:** Fibrous, solid, stringy. White, bruising slowly yellowish and then ocher-brown when cut. **Odor:** Fruity, fragrant. **Taste:** Indistinct. **Spore deposit:** Pale creamy to yellowish. **Microscopy:** Spores (6) 7.5 - 10 (11.5)×4.5–5.5 µm, ellipsoid, smooth.

Ecology: Often in clumps or scattered in troops, rarely solitary. Fruiting in duff and moss under conifers, especially spruce. Locally common in coastal Sitka spruce forests, typically fruiting from early summer into fall, and occasionally with Engelmann spruce in higher elevation and east-side Cascade forests in late summer and early fall.

Edibility: Edible and excellent, especially if clean young specimens are found.

Comments: *Cantharellus roseocanus* is often called the Rainbow Chanterelle due to the array of pink to orange cap colors when young; older fruitbodies are typically concolorous. The rich peachy orange to bright orange of the gills is quite different from other chanterelles—which helps distinguish it. *C. formosus* is taller and often more slender, and has duller, light beige-yellow gills, and yellow to ocher-brown cap. *C. cascadensis* is similar in shape, but has whitish gills, and an egg-yolk yellow cap.

Cantharellus subalbidus

A. H. Sm. & Morse

White Chanterelle

Cap: 6–15 (20) cm across, convex when young with a lobed and inrolled margin, expanding to broadly convex with a downcurved margin, becoming flat to uplifted and irregular in age with a wavy, ruffled margin. Surface matted-tomentose when young, soon moist and smooth or bumpy. Whitish, creamy, becoming dingy ivory with yellowish blotches and ocher streaks in age. **Gills:** Deeply decurrent, blunt edged with extensive cross veins. Whitish to light cream or beige with a light pinkish wash. **Stipe:** 5–12 cm long, 2–4 cm thick, cylindrical or slightly tapered downward, usually rather stout, sometimes curvy or sinuous. Surface smooth to fibrillose. White to ivory-creamy, becoming dingy beige, with yellow to ocher blotches in age. **Flesh:** Fibrous, thick, stringy. White, bruising yellowish, and then orangish brown. **Odor:** Mild to pleasantly fruity-fragrant. **Taste:** Indistinct. **Spore deposit:** Whitish. **Microscopy:** Spores 7–9×5–5.5 µm, ellipsoid, smooth. Basidia 4- to 6-spored.

Ecology: Solitary or in clusters, often in arcs and large troops in duff. Occurring throughout Cascadia but more common in eastern Cascade forests. Although by no means restricted to mature forests, it appears to have a strong preference for mature Douglas-fir stands. Fruiting from late summer through fall.

Edibility: Edible and very good.

Comments: The creamy to pale ivory colors of the entire fruitbody distinguish it from other chanterelles. Occasional specimens become extensively yellow bruised in age and can be confused with *C. cascadensis*, which shares the pallid gills and stout stature. Other stout, white mushrooms that grow in clumps and arcs in forest settings have thinner, blade-like gills, and a few smell distinctly "unfruity" (*Leucopaxillus albissimus* and *Clitocybe nebularis*).

Craterellus tubaeformis

sensu PNW

Yellowfoot

Cap: 2–6 (8) cm across, convex with a small dimple at the center of the cap when young, expanding to broadly convex with a strongly downcurved margin, wavier and wrinkly-ruffled around margin; more deeply depressed at center and uplifted in age. Surface dry, wrinkled, obscurely matted-tomentose to smooth, margin often finely fringed or fuzzy. Brown to ocher brown or yellowish brown, light orange to pale yellowish near the edge (in one form bright orange-yellow when young, becoming pinkish beige). **Gills:** Subdecurrent to decurrent, shallow, blunt, forking with cross veins. Color variable, usually light pinkish when young to dingy orange-beige, sometimes light whitish gray, usually strongly contrasting against upper stipe. **Stipe:** 6–10 (12) cm long, 0.5–1.5 cm thick, tall and upright, straight to slightly curved, usually irregularly shaped and strongly grooved, base slightly enlarged. Surface smooth. Bright yellow-orange, or at times dull beige-yellow, whitish at extreme base. **Flesh:** Very thin and pliable in cap, stipe cartilaginous and entirely hollow. **Odor:** Indistinct. **Taste:** Indistinct. **Spore deposit:** Whitish. **Microscopy:** spores 9–11 x 5.5–7 µm, ellipsoid, smooth.

Ecology: Scattered in clumps, arcs, and large troops in needle duff and moss around stumps and woody debris. Ectomycorrhizal with a range of conifers, especially western hemlock and Sitka spruce. Fruiting from late summer through winter, peaking in late fall through early winter.

Edibility: Edible and underrated. This species is often abundant, easy to pick, and very clean.

Comments: The small, brownish yellow, dimpled caps, shallow, decurrent, vein-like gills, tall, hollow, orange-yellow stipes, and growth in troops under conifers distinguish this species. *Cantharellus formosus* is somewhat similar in color and often found in the same areas but is larger and has solid white flesh. *Chrysomphalina chrysophylla* can look remarkably similar, but typically has a thinner, less cartilaginous stipe and brighter yellow gills. *Xeromphalina* species can look similar and are found in large troops on woody debris or duff, but have thin, solid stipes and deeper, thinner-edged gills; they often show a tuft of ginger-colored hairs at the stipe base. The true *C. tubaeformis* is a European species not known to occur in the Pacific Northwest. The western North American Yellowfoot represents a distinct species. Although the name *C. neotubaeformis* has been proposed, it hasn't been officially published, and there may be more than one species under that name.

Hygrophoropsis aurantiaca group

False Chanterelles

Cap: 2–8 cm across, convex with an inrolled margin when young, becoming plane with a depressed center, to broadly funnel shaped in age. Surface dry, velvety when young, more matted-tomentose to smooth in age. Bright to pale orange to creamy orange, at times with a paler margin. **Gills:** Decurrent, crowded, forking repeatedly, often quite wavy and ruffled. Bright to dull orange or creamy orange. **Stipe:** 2–7 cm long, 0.3–1 cm thick, more or less equal, occasionally tapered toward a slightly rooting base. Surface dry, nearly smooth, base with whitish fuzzy mycelium and white rhizomorphs. Orange to creamy orange, typically colored similar to cap, but usually with darker orange tones. **Flesh:** Thin, soft, stipe more fibrous. Whitish to pale orange. **Odor:** Indistinct to distinctly herbal. **Taste:** Indistinct. **Spore deposit:** White **Microscopy:** Spores 5–8 × 3–5.5 μm, ellipsoid, smooth, dextrinoid.

Ecology: Solitary or scattered on well-decayed logs, stumps or humus in forested settings, or scattered in troops or clustered on wood chips, especially in urban areas. Very common and widespread, fruiting throughout the mushroom season.

Edibility: Best avoided, it has caused mild adverse reactions in some people.

Comments: The most telltale features of this group are the decurrent, often wrinkled gills that fork repeatedly and the growth on woody debris or coarse humus. Despite the common name of "false chanterelle," these mushrooms only vaguely resemble the delectable *Cantharellus*, which are much larger, have thick, stringy white flesh, and blunt gill-ridges (not thin, narrow, forking gills).

Genetic data indicate that at least five species of *Hygrophoropsis* are present in the Pacific states, and four of them are undescribed. At least two of these closely match the above description of *H. aurantiaca*, but we lack criteria to separate *H. aurantiaca* in the strict sense from the "impostor." A darker orange-brown to red-capped species with darker orange gills goes by *H. aff. rufa;* it is common in urban areas on wood chips. A dark brown, olivaceous-brown to orange-brown capped species with whitish to pale orange gills is common in west-side Cascadian forests. Although the name *H. aurantiaca* var. *pallida* has been applied to the creamy-tan capped, whitish-gilled variant, this likely represents a fully distinct species (not just a variety).

Turbinellus floccosus

sensu PNW

Woolly False Chanterelle

Fruitbody: 5–15 (20) cm across, shaped like pointed cylinders when extremely young, but usually found after flattening to a truncate cylinder, then becoming deeply depressed at center, tall and vase shaped at maturity. Margin thin and slightly wavy. Surface slightly viscid when wet, otherwise moist to dry, with large matted-tomentose scales, breaking into concentric rings of curved scales around central hollow, more matted near margin. Dark red to reddish orange when young, soon orange, margin often bright yellow-orange, sometimes duller orange-beige. **Underside:** With deeply decurrent ridges, crinkly looking, branched, and extensively cross veined. Pale creamy to light beige, occasionally bright yellow when very young. **Stipe:** 8–16 cm long, 3.5–5.0 cm thick, vase shaped, tapered toward base. Upper portion covered with veins, smooth to fibrillose lower. Creamy to beige or with yellow-orange tones, with dingy vinaceous brown areas in age. **Flesh:** Thin in cap, stipe thick, fibrous. Off-white. **Odor:** Indistinct. **Taste:** Indistinct. **Spore deposit:** Pale ocher. **Microscopy:** Spores 11–20.5×6–10 µm, elongate-ellipsoid. Ornamented with broad, fused warts up to 0.5 µm tall.

Ecology: Occasionally solitary, but more often in fused clumps or clusters (often with small, pointed primordia in star-shaped clusters at the base), in scattered troops, arcs, and rings in duff in conifer forests. Very common and widespread, occurring throughout Cascadia; fruiting from late summer through fall, and into winter on the coast.

Edibility: Not recommended, causing gastrointestinal distress in some people.

Comments: This bizarre mushroom is instantly recognizable by its bright red to orange, shaggy-scaly cap, tall vase-shaped fruitbodies, and wrinkled, veiny beige fertile surface. Faded specimens can be confused with *T. kauffmanii*, but that species is usually pale beige to tan, with even more dramatic blocks of scales in concentric rings on the cap. There is considerable variation within the *T. floccosus* complex, and species delimitations still need to be defined. While some forms are distinct, others seem to have transition forms. We can easily recognize *Turbinellus (Gomphus) bonarii*; which is typically smaller, and occurs in fused clusters in drier, eastern or southern Cascade forests. It tends to be less funnel-shaped, and has a solid stipe. The Rocky Mountains has at least one species in this complex, with a more peg-like to blocky fruitbody. True chanterelles have smoother, paler orange to yellow caps and are not vase-shaped.

Synonym: *Gomphus floccosus* (Schwein.) Singer

Turbinellus kauffmanii

(A. H. Sm.) Giachini

Scaly Vase False Chanterelle

Fruitbody: 5–20 (30) cm across, cylindrical with a flat or slightly depressed cap when young, soon developing deeply depressed center and becoming broadly convex, then uplifted to vase shaped. Margin thin, often a bit wavy in age. Surface dry, with concentric rings of blocky to flat, pointed scales, those deeper in sunken center of cap much larger and often curled back, more flattened near margin. Older specimens can lose the pronounced upright-blocky scaly look. Dull beige to tan, sometimes whiter when young or darker in age. **Underside:** With deeply decurrent ridges, crinkly looking, branched, and extensively cross veined. Whitish when young, soon creamy beige to tan. **Stipe:** 7–15 cm long, 2–5 cm thick, cylindrical or with a tapered or club-shaped base. Almost entirely covered in gill tissue, pale beige to tan. **Flesh:** Thin in cap, stipe thick, firm, fibrous. Whitish or with orangish stains in age. **Odor:** Indistinct. **Taste:** Mild. **Spore deposit:** Ocher-buff to apricot. **Microscopy:** Spores 12.5–17.5×6–8 µm, elongate-ellipsoid, ornamented with short warts and branched ridges.

Ecology: Solitary or more often in clumps in needle duff of conifers. Fairly common and widespread, occurring across much of the Pacific Northwest, fruiting in late summer and fall.

Edibility: Unknown, likely causing gastric upset like *T. floccosus.*

Comments: The large, vase-shaped fruitbodies, beige-tan to light brownish cap with dramatic blocky, upright scales, and pallid veiny "gills" distinguish this species. It is quite abundant in years with summer or early fall rains in the Cascade Range but is less common than *T. floccosus.* Specimens that have become more matted-scaly could be confused with faded *T. floccosus* but that species usually has traces of orange colors somewhere on the cap. *Turbinellus* is closely related to the Pig's Ear *(Gomphus clavatus)*, coral fungi in the genus *Ramaria*, and the hypogeous genus *Gautieria*; they are not closely related to true chanterelles *(Cantharellus* spp.).

Synonym: *Gomphus kauffmanii* (A. H. Sm.) R. H. Petersen

Gomphus clavatus

sensu PNW

Pig's Ear

Cap: 4–20 cm across, flat when young, soon becoming uplifted, usually with overlapping fan-like lobes forming rosettes, margin thin, wavy, ruffled. Surface finely matted at first, velvety to smooth in age, moist to dry. Brownish at first, often with pale lilac patches near center, soon dingy tan, olive, ocher-olive, beige when dry or in age. **Underside:** Covered in ridges running to base of fruitbody; ridges often poorly defined or crinkly, blunt-edged, and with many cross veins. Purple to lavender at first, fading to light lilac-gray or dingy beige in age and often dusted with ocher-tan spores. **Stipe:** 5–10 (15) cm long, 2–4 cm thick (up to 8 cm thick where fruitbodies are fused), wedge shaped, tapered to base. Mostly covered in gill tissue. Lower part a bit velvety or smooth, deep royal purple to amethyst near base; extreme base white. **Flesh:** Thick, fibrous, solid-stringy, whitish to dingy tan. **Odor:** Indistinct. **Taste:** Indistinct. **Spore deposit:** Ocher-tan. **Microscopy:** Spores 10–15×4–7.5 µm, elongate-ellipsoid, ornamented with discrete warts.

Ecology: In fused clusters, often in troops in duff and moss in conifer forests. Ectomycorrhizal, with an apparent preference for hemlock and fir. Common west of the Cascade Crest, occasional on the east slopes. Fruiting in fall.

Edibility: Edible. Some people consider this species a delicacy, while others find them bland and not worth collecting. Young specimens are more tender.

Comments: This unique species is practically unmistakable–recognized by its purple-veined or ridged undersides, tan to olive-beige caps, and tendency to fruit in clumps and rosettes. A few *Thelephora* are brownish purple, but are more dry and thin-fleshed. *Polyozellus* is only vaguely similar – even faded specimens are much more deeply colored, from dark royal blue to purple-black. Although *G. clavatus* are sometimes marketed under the name Violet Chanterelle, it is more closely related to coral mushrooms in the genus *Ramaria. Another surprising example of convergent evolution in macrofungi.*

Taxonomic note: Our western *Gomphus "clavatus"* is distinct from the European species.

Polyozellus marymargaretae

Polyozellus atrolazulinus

Polyozellus spp.

Blue False Chanterelle, Bluefans

The following description encompasses three different Polyozellus species in the Pacific Northwest.

Fruitbody: Typically in fused clusters 5–30 cm across, of multiple overlapping fan-shaped, shoehorned to spatula-shaped caps 2–8 (10) cm across, margin thick and often downcurved when young, thinner, wavy and often ruffled in age. Surface moist to dry, finely pubescent when young, matted-fibrillose to smooth in age. Light blue, blue-gray, bluish purple, blue-black to dark purple black, typically zonate with a paler margin when young, concolorous in age. **Underside:** With decurrent shallow ridges and veins that are strongly cross veined and wrinkled. Light to dark blue, bluish gray, bluish-purple, purplish gray to nearly black. Bruising blue-black when damaged. **Stipe:** 2.5–10 cm long, 0.5–2.5 cm thick, but often fused into a mass >10 cm thick, central to off-center, irregular, often poorly defined from fertile portion. Concolorous, or darker than underside. **Flesh:** Fleshy fibrous, blue, purple to gray when young, darker with age. **Odor:** Indistinct. **Taste:** Mild to earthy. **Spore deposit:** Whitish. **Microscopy:** Spores subglobose to broadly ellipsoid, with irregular nodules. Spores of this genus are an important identifying feature and allow for specimens to be quickly assigned to either the small-spored group (<7 µm) or the large-spored group (>7 µm). Taken as a group, spores of this genus can be as small as 4.8 µm long and up to 9.5 µm; spore widths range from 4–7.7 µm.

Ecology: Often in scattered clusters or small clumps, fruiting from moss and duff. Ectomycorrhizal with conifers, we have not been able to pinpoint a tree association preference with the three Pacific Northwest species. All show an affinity for old-growth forests and are most often found at higher elevations throughout Cascadia, but this may be because these are the only locations where suitable habitat remains. Fruiting from late summer through fall.

Edibility: Edible, but with mixed reviews. Some find these "blue chanterelles" pleasant, while others find them muddy tasting. It's possible that some species in this complex taste better than the others, but most are not identified to species when they are collected for the table.

Polyozellus purpureoniger

Comments: The three species currently understood to be present in the Pacific Northwest are *Polyozellus atrolazulinus*, *P. marymargaretae*, and *P. purpureoniger*. Until 2017, the entire group was collectively called *Polyozellus multiplex* – a name that is still in use, but which refers specifically to an eastern American species. Distinguishing members of this genus is difficult, especially when dealing with older specimens: *P. atrolazulinus* ranges from sky blue to deep blue, and typically retains these blue colors in age. However, it is reliably distinguished from *P. marymargaretae* by its smaller spores (4.8–7.7 × 3.9–6.7 µm, avg. ~ 6.1 × 5.3 µm). *P. marymargaretae* starts off blue-colored, at times with an olive or buff sheen, but soon darkens to bluish black to purplish black, and has larger spores (5.8–9.6 × 4.8–7.7 µm, avg. 7.9 × 6.7 µm). The other large-spored species in the PNW is *P. purpureoniger*, whose spores measure (6.7–9.6 × 5.8–7.7 µm, avg. 8.0 × 6.9); this species tends to lack blue tones, and is more purple to purple-black colored, but older specimens can be practically indistinguishable from *P. marymargaretae*. All three are highly prized dye fungi, yielding dark green to blue-green colors on wool and silk. We recommend picking them sparingly if at all, as they appear to be rare.

Misapplied name: *Polyozellus multiplex* (Underw.) Murrill

Craterellus calicornucopioides

D. Arora & J. L. Frank

Black Trumpets

Fruitbody: 3–10 (13) cm across, at first tubular, soon expanding to broadly convex to funnel or trumpet-shaped; center of cap deeply depressed, hollow down the stipe. Margin wavy and ruffled. Surface dry, innately fibrillose, with flattened matted-tomentose scales, more ragged in age. Dark blackish gray to dark grayish brown to black, at times warm brown, in drier weather gray to beige-gray. **Underside:** Smooth to slightly wrinkled, not distinct from stipe. Dark bluish gray, brownish gray, or light gray, with a pale sheen. **Stipe:** 4–15 cm long, 1–5 cm thick, (but with a large central hollow), at first cylindrical, soon trumpet shaped, straight to irregular, often grooved or compressed in cross section. **Flesh:** Very thin, somewhat tough, stipe hollow, slightly thicker and tougher at base. **Odor:** Strong and fragrant, slightly fruity or floral and pungent when fresh (especially strong when many are together in a bag or on one's hands after a few minutes of picking). **Taste:** Mild. **Spore deposit:** Off-white to pale creamy. **Microscopy:** Spores 11–15 (20) × (5) 7–11 µm, ellipsoid, smooth.

Ecology: Solitary, in tufts, tight clusters, or rosettes, often scattered in large troops or arcs in duff. Typically with hardwoods, some Pacific Northwest records may be with conifers. They are especially fond of oak and tanoak. Rare in much of Cascadia, found mostly at lower elevations in Oregon white oak stands, or in mixed forests with madrone. More common in southwest Oregon into California with tanoak. Fruiting from late fall into early spring.

Edibility: Delicious! One of the best mushrooms for the table, with a strong, fragrant, earthy to cheesy flavor.

Comments: This well-loved species is one of the best mushrooms for beginning identifiers to learn: It is exceptionally tasty and has no toxic look-alikes. Because it is rare in the PNW people often head to northern California in the winter where it can be abundant in wet years. The hollow, trumpet-shaped fruitbodies and dark colors distinguish it from all but *C. atrocinereus*, which has a veined fertile surface and is not tubular when young; it's an equally good edible, but less common.

Misapplied name: *Craterellus cornucopioides* (L.) Pers.

Stereopsis humphreyi

(Burt) Redhead & D.A. Reid

Ghost Funnel

Cap: 0.5–3 (4) cm across, convex, plane to undulating, often with glove funneling down on stipe side, or kidney-shaped, occasionally uplifted and wavy in age. Surface dry, silky or tomentose when young, to smooth. White, pale gray to whitish tan, obscurely zonate at times. **Underside:** Smooth or wrinkled at first, becoming finely veined. White to off-white, or cream. **Stipe:** 1–3 (4) cm long, 0.1–0.4 cm thick, off-center to lateral, rarely central, equal or enlarged towards base. Surface dry, lower portion with silky hairs. White, developing tan tones from base up in age. **Flesh:** Very thin, soft in cap, fleshy-fibrous in stipe. **Odor:** Indistinct. **Taste:** Indistinct. **Spore deposit:** None obtained, presumed to be pallid. **Microscopy:** Spores 6.5–9×3.5–5.5 µm, broadly ellipsoid to ovate, smooth.

Ecology: Scattered or in gregarious troops from duff or mossy humus; often in depressions or low-lying areas in coastal Sitka spruce and western hemlock forests. Most of the places we have seen it are mature forests, but in areas with disturbance (campgrounds, trails and reclaimed roads). Uncommon and apparently restricted to coastal forests, currently known from south-central Oregon to southeast Alaska; fruiting in fall.

Edibility: Unknown.

Comments: *Stereopsis humphreyi* is recognized by its small size and shape, pallid colors, and smooth to veined underside. *Arrhenia auriscalpium* is a smaller, spatula-shaped species with gray to gray-brown colors; it occurs in mossy poor soil or barren areas; typically in alpine or Arctic regions. *Arrhenia lobata* is gray-brown to buff, lacks a distinct stipe, and has more veiny to gill-like underside. *Cotylidia diaphana* is typically more vase-shaped to funnel-shaped, with a shorter stipe, and thin leathery flesh. At first glance, some *Clitocybe* species, and *Pleurocybella porrigens* look similar, but have distinct gills.

Aphroditeola olida

(Quel.) Redhead & Manfr. Binder

Pink Bubblegum Mushroom

Cap: 1–3 (4.5) cm across, broadly convex to nearly plane with an inrolled margin and a depressed center at first, becoming irregular, more broadly funnel-shaped, with a wavy or lobed margin. Surface dry to slightly tacky, finely tomentose to smooth. Pink to pale coral pink, occasionally with dark pink splotches and stains, fading to pinkish white to pinkish buff in age. **Gills:** Decurrent, quite shallow, forking repeatedly, often wrinkled and with cross veins. Whitish to pale pinkish. **Stipe:** 1–3.5 (5) cm long, 0.2–0.5 cm thick, equal or tapered downward, occasionally rooting. Surface dry, smooth to innately fibrillose. Pale pinkish to whitish, often spotted with darker pink stains. **Flesh:** Thin, fibrous; white. **Odor:** Very pleasant, fruity, like bubblegum and/or root beer. **Taste:** Indistinct. **Spore deposit:** White. **Microscopy:** Spores 3–6 × 2.5–4 µm, ellipsoid to spindle-shaped, smooth, inamyloid.

Ecology: Solitary or scattered in small groups in duff or moss under conifers. Occasional at mid-to high elevations of the Cascades and in humid microclimates of east-side forest. Uncommon in the Coast Ranges. Fruiting in the fall.

Edibility: Unknown.

Comments: This is a delightfully distinctive mushroom with a pinkish cap, gills that are slightly wavy, forking, and decurrent with a fruity, bubblegum, or root-beer-like odor. *Contumyces rosellus* has more widely spaced gills, thin, almost translucent flesh, and lacks an odor. It is common at lower elevations, found in disturbed grassy areas and trail sides. Members of the *Hygrophoropsis aurantiaca* group also have forked gills, but typically have orange to brown colors (some specimens have creamy white gills) and tend to lack any distinctive odor. *Paralepista* have larger, more orange-colored fruitbodies that lack a sweet odor.

Synonyms: *Hygrophoropsis olida* (Quel.) Metrod, *H. morganii* (Peck) H.E. Bigelow

2 - Amanita and Allies

Amanita are some of the most beautiful and most conspicuous mushrooms in the world. This genus is both famous and notorious for its wide range of biochemistry: Some species are deadly toxic, some are excellent edibles, and others still are potently psychotropic. For example, *Amanita muscaria* has played a central role in practices of shamanism across Europe and Asia for millennia.

Although they can seem remarkably different from one another, some unifying characteristics include universal veil remnants present on the cap (as warts or a skull patch) and/or at the base of the stipe (as a volva, cup, or scaly rings); a partial veil (in most species, at least when young); and a white spore deposit. Most *Amanita* are ectomycorrhizal, with the notable exception of the 'Grass Amanitas' (*Saproamanita*).

Important identification characters include overall coloration and stature, presence and distribution of universal (UV) and partial veil (PV) tissues, presence or absence of sulcate grooves around the cap margin, shape of the base of the stipe, and tree species associations. The size, shape, and amyloid reactions of the spores, as well as the presence or absence of clamp connections are the most important microscopic features. The genus *Amanita* has eight major sections, familiarizing yourself with recognizing these subgroups will make the job of species-level identification much easier. As a bonus, they roughly correspond with "natural" groupings of evolutionarily-related taxa. Rod Tulloss (abbreviated RET in these descriptions) is one of the world's foremost experts on the genus and maintains an exhaustively researched website (www.amanitaceae.org) which includes profiles for many provisionally named species, some of which we have referenced here. With a little effort, it is possible to recognize many species of western *Amanita* in the field.

Genus *Amanita*

Section *Amanita* – The iconic Fly Agaric (*Amanita muscaria*) is in this section. Most of these species have strongly warted and richly colored caps. Volva tissue is usually concentrically ringed and scaly or tight and cup- or collar-like around a swollen stipe base. Most species in this group are toxic, containing ibotenic acid and/or muscarine. Spores are not amyloid. (*A. alpinicola, A. aprica, A. farinosa, A. gemmata, A. muscaria, A. pantherinoides, A. 'pseudobreckonii'*)

Section *Amidella* – This group is absent in Cascadia. Most species have a sac-like volva and hanging PV tissue around the margin of the cap, and often bruise reddish or pinkish when cut. Spores are amyloid.

Section *Caesareae* (Caesars) – Large, robust fruitbodies with striate-sulcate cap margins, hollow stipes, and a sac-like volva at the base of the stipe (which is not strongly swollen or bulbous). All are edible, with caution (care must be taken not to confuse them with poisonous *Amanita* species). Spores are amyloid. (*A. calyptroderma, A. vernicoccora*)

Section *Lepidella* – Fruitbodies are most often large, usually entirely white, and without any contrasting pigments, but often discoloring yellowish in age. Universal and partial veil tissues are poorly developed, often powdery, and messy. Spores are amyloid. Note: *A. silvicola* seems to fit here, but genetic data place it in section *Validae*. (*A. smithiana*)

Section *Phalloideae* (Death Caps) – Caps are usually bald and not sulcate, volva usually sac-like but thin, base of stipe is often strongly swollen and bulbed. Spores are amyloid. These are the deadliest mushrooms in the genus due to the presence of molecules called amatoxins. (*A. ocreata, A. phalloides*)

Section *Vaginatae* (Grisettes)– Edges of caps sulcate-striate, usually some shade of gray to brown (although can be many other colors, including ocher-tan, whitish, or pink).

Partial veil is absent or very rudimentary; volva and universal veil tissue stain rusty orange in some species. Spores are not amyloid. Highly diverse in Cascadia with many unnamed species. (*A. 'castellae'*, *A. constricta*, *A. friabilis*, *A. fulva*, *A.* "Gray Warts", *A. pachycolea*, *A. populiphila*, *A. vaginata* group, *A. velosa*)

Section *Validae* – Fruitbodies are pale or colorful, often staining reddish or brownish. Volva is usually weakly developed or fragile, variably warty, membranous, or inconspicuous. *A. silvicola* more strongly resembles a *Lepidella*, but recent genetic evidence places it in this group. Spores amyloid. (*A. augusta*, *A. cruentilemurum*, *A. novinupta*, *A. porphyria*, *A. rubescens*, *A. silvicola*, *A. sponsus*)

***Saproamanita* ('Grass Amanitas')** – This group is sometimes treated as a separate genus. Their small size, dingy colors, and growth in grassy or open habitats (rather than in forest settings) makes them difficult to recognize as relatives of other *Amanita*. There is also a fundamental difference in their ecology—they are non-mycorrhizal! This is the ancestral condition in the genus, and these species provide a window into the early stages of *Amanita* evolution. Spores are amyloid. (*A. prairiicola*, *A. pruittii*)

Genus *Limacella*

(Slime Caps) – Closely related to *Amanita* but fruitbodies lack any membranous UV tissues (warts, patches, and volva are absent) and they are non-mycorrhizal. Recently, some authors split *Limacella* into three separate genera: *Limacella*, *Limacellopsis*, and *Zhuliangomyces*. In Cascadia, the thinly-viscid to moist or dry species pertain to *Limacella*, while those with extremely slimy cap and stipe surfaces would be placed in *Zhuliangomyces*. For the purposes of this book, we retain a broad concept of *Limacella*. The dry-capped species are most likely to be confused with *Leucoagaricus*, but lack collar-like partial veils, among other differences. The slimy species of *Limacella* might be confused with *Hygrophorus* but are usually much taller and more slender.

Amanita aprica

J. Lindgr. & Tulloss

Sunshine Amanita

Cap: 3–12 cm across, rounded to convex when young, becoming broadly convex to flat, or occasionally uplifted in age. Margin not sulcate or obscurely so. Surface slightly viscid to dry, smooth, covered in universal veil tissue that sometimes breaks up into well-defined squares but more often remains aggregated and poorly differentiated in a frosty-looking layer. Bright citrine yellow to apricot orange to nearly peach (pinkish orange); usually darkest at center, fading in sun or with age. **Gills**: Narrowly attached to free, close to crowded, edges often fringed with veil tissue. White, becoming dull creamy white in age. **Stipe**: 5–10 cm long, 1.5–3 cm thick at apex, 1.5–6 cm thick at base, usually relatively short, cylindrical to slightly enlarged or bulbous and swollen toward base. White. **Partial veil**: Hanging, membranous, white skirt, usually quite thin and ragged, often disappearing entirely in age. **Volva**: Present as a small, collared cup held tightly to the base, often with a few concentric rings of tissue above. **Flesh**: Firm to soft, fibrous in stipe, white. **Odor**: Indistinct. **Taste**: Indistinct. **Spore deposit**: White. Microscopy: 9.5–13.0 × 6.5–8.5 µm ellipsoid or elongate, inamyloid. Basidia rarely clamped.

Ecology: Solitary or in small, scattered troops, fruiting in soil and duff under mountain conifers. Common in the eastern Cascades in spring, or summer at higher elevations. Uncommon on the west side and in coastal forests, fruiting in spring or occasionally fall.

Edibility: Poisonous.

Comments: The bright yellow to orange cap, frosty universal veil tissue, and stocky stature help distinguish this species. *A. muscaria* var. *formosa* and faded specimens of *A. muscaria* subsp. *flavivolvata* are usually taller and more gracile, have more distinct warts on the cap, and multiple concentric rings of pointed scales around the stipe base. Members of the *A. gemmata* group usually have duller yellow-tan caps and more discrete warts (not frosty looking). *A. alpinicola* is somewhat similar species with smaller, stout-statured fruitbodies and dull yellow cap. It fruits in spring and summer in mountain-conifer habitats.

Amanita muscaria

(L.) Lam.

Fly Agaric

Cap: 5–20 cm across, round to convex when young, becoming broadly convex, plane, or occasionally uplifted in age. Surface slightly viscid to dry, covered in coarse, discrete off-white to yellowish universal veil warts; these are textured like oatmeal flakes or slightly powdery, sometimes rather solid and angular. Margin distinctly sulcate in age, obscurely so when young, often covered by universal veil warts. Typical caps bright red to cherry red, sometimes duller when young. Pigments in the cap are water soluble and light sensitive, so exposed caps often become more orange, or in more extreme case yellow or even whitish. In general, cap color remains darker and redder near the center. **Gills:** Narrowly attached (less commonly free), close to crowded. Edges of gills covered in dense, pointed fuzz when very young and tightly packed, and a fine ragged layer of cells in age. White at first, becoming creamy white or yellowish in age. **Stipe:** 7–20 (25) cm long, 1.5–4 cm thick at apex, but the swollen base may be substantially wider. Straight and usually cylindrical over most of its length, gradually enlarged toward bulb. Bulb usually distinctly swollen and egg shaped or tapered downward. Upper stipe (above partial veil) usually white and smooth, lower stipe creamy white to yellowish, surface often bumpy or ragged. **Partial veil:** Present as a whitish, membranous hanging skirt, usually with adhering patches of white or yellow universal veil tissue around edge. **Volva:** Whitish to yellow concentric rings of square or pointed scales on lower stipe. **Flesh:** Firm to soft, pale whitish; stipe solid. **Odor:** Indistinct. **Taste:** Mild or slightly sweet. **Spore deposit:** White. **Microscopy:** Spores 9–13 × 6.5–8.5 µm, broadly ellipsoid, occasionally rounder or more elongate, inamyloid. Basidia clamped.

Ecology: Solitary or in arcs, rings, or scattered in troops; often in large numbers. Very common throughout Cascadia and found in a wide range of forested habitats. This species has the remarkable ability to associate with just about any ectomycorrhizal plant, including non-native trees. Within a given habitat and elevation zone, this species usually produces a single, pronounced flush which usually occurs during mid-fall at lower elevations, and mid-summer to early fall at higher elevations. Fruitings in urban areas are more irregular, and small numbers are found throughout much of the year. Experienced mushroom pickers know that the highly conspicuous fruitbodies of this species are often a helpful indicator for King Bolete (*Boletus edulis*) patches. There is also an association between this species and the Peppery Boletes (*Chalciporus* spp.): The latter are most often found fruiting near *A. muscaria* (this holds true even in New Zealand, where neither species is native!). The exact nature of this relationship remains unclear, but is presumed to involve some form of parasitism by the *Chalciporus* (similar to the *Chroogomphus*/*Gomphidius* association with *Suillus*).

Amanita muscaria var. *formosa*

Edibility: If consumed without special preparation, this species is quite toxic. Muscimol is the primary compound that causes sweating, nausea and vomiting, diarrhea, and variable psychotropic alterations ranging from delirium to coma-like sleep (often accompanied by vivid dreams). However, when detoxified properly, this species can be made into a tasty edible. The procedure involves making thin slices of the fruitbodies and boiling these in a large volume of water for two rounds of seven minutes. The toxin is water soluble and thus dissolved into and discarded with the water after each round. (Note: Most poisonous mushrooms DO NOT have water soluble toxins; this method only applies to *A. muscaria).* The mushrooms can then be sautéed and used in normal recipes; most people find them delicious when prepared this way. *Caution is warranted.* Please read Rubel and Arora's 2008 article (see References p. 553).

Comments: The bright red cap with whitish or pale yellowish patches and concentric rings of scaly volva tissue distinguish this species in our area. While the yellowish-veiled *A. muscaria* subsp. *flavivolvata* is widespread throughout the Pacific States, *A. muscaria* var. *muscaria* is common in urban areas and may also account for most of the specimens found in coastal spruce forests. This latter variety tends to have veil tissue that is pure white. Other variants of *A. muscaria* include the yellow-capped form, often called *A. muscaria* var. *formosa*, which is similar in all respects except for the lemon yellow to light yellow cap color and a white variety (*A. muscaria* var. *alba*) with an eggshell-white or pure white cap. *A. aprica* is more common in the mountains and in the spring; it has a small cup-like volva and a yellowish to peachy orange cap (never truly red). Members of the *A. gemmata* complex are sometimes confused with yellow forms of *A. muscaria*, but they have a cup-like volva and are often more beige-tan, with less yellow pigment.

Amanita augusta

Bojantchev & R. M. Davis

Western Yellow-veiled Amanita

Cap: 5–15 cm across, round when young, convex to flat in age. Surface slightly viscid to greasy or dry, covered in warts that are easily removed by rain or handling. Color quite variable, from dark brown to bright lemon yellow, although older fruitbodies can become dull beige-brown or tan. "Typical" mature caps show a warm brown center, become golden brown outward, and are yellow at the margin. **Gills:** Narrowly to broadly attached (rarely free), close to crowded. White or creamy ivory at first, soon with yellow tones. **Stipe:** 8–15 cm long, 1.5–3 cm thick at apex, swollen or enlarged toward base. Base swollen (but not abruptly bulbous), extreme base tapered.. Creamy white, but lower stipe (below partial veil) covered with many lines of yellow, pointed, curved scales. Often developing reddish stains at base. **Partial veil:** Creamy white membranous skirt, usually with a ring of yellow universal veil scales adhering to edge. **Volva:** Sometimes indistinct, otherwise as concentric rings of flat, yellow scales, often as a fused "sheet" that detaches from base of stipe and remains in soil. **Flesh:** Firm to soft, moderately thin, whitish. Bruising slowly dull reddish or the color of oxidized apple flesh, especially in lower stipe. **Odor:** Indistinct. **Taste:** Indistinct. **Spore deposit:** White. **Microscopy:** 8.5–9.5 × 6–7 µm, broadly ellipsoid to ellipsoid, amyloid. Basidia not clamped at base.

Ecology: Solitary or in small groups in duff and soil under conifers, fruiting from late summer through late fall. Quite common throughout Cascadia, especially in coastal spruce forests, but also under mixed montane conifers and occasionally in hardwood habitats.

Edibility: Edible but not recommended. Local experience with this species is growing, but still quite recent; this species should always be thoroughly cooked.

Comments: *A. augusta* is recognized by the variable dark brown to bright yellow cap, creamy gills, skirt-like partial veil, yellowish universal veil remnants, and reddish staining at the stipe base. *A. gemmata* and her sisters might cause confusion, but they are rarely as brightly colored and have white (not yellow) universal veil patches.

Amanita 'pseudobreckonii'

Amanita pantherinoides

(Murrill) Murrill

Western Panther

Cap: 4–10 (15) cm across, rounded to convex, expanding to plane in age. Surface slightly viscid or dry to nearly metallic, smooth; margin weakly to distinctly sulcate. Universal veil warts white, often evenly spaced across entire cap, sometimes washing off. Color variable: pale brownish tan to creamy-beige or yellowish, usually darkest and with more brown tones at center; at times extensively brown. **Gills:** Narrowly attached or variably breaking away from stipe and appearing free, broad, crowded. Edges finely scurfy, especially noticeable when young. White to pale cream. **Stipe:** 5–12 cm tall, 1–2 cm thick, straight, narrowest at apex, enlarged below, with a bulbous base. Surface dry, smooth or with small uplifted pointed scales over lower part. White to creamy. **Partial veil:** A membranous tissue forming a weakly defined skirt or ragged band on mid- to upper stipe. **Volva:** Forming a narrow appressed rim-like collar around the bulb. **Flesh:** Thin, fleshy-fibrous, stipe solid or weakly hollowing in age. White. **Odor:** Indistinct. **Taste:** Indistinct. **Spore deposit:** White. **Microscopy:** Spores 8.5–11.2 × 6.3–7.7 µm, mostly ellipsoid, but somewhat variable in length-width ratio.

Ecology: Solitary or in small groups on the ground, rarely in larger numbers. Typically found in forested habitats, mostly associated with conifers, especially fond of Douglas-fir. Occasional in urban areas with conifers and hardwoods. Quite common in spring, but also found in fall.

Edibility: Highly toxic. Like most relatives of *A. muscaria* it contains muscimol, but apparently in higher concentrations.

Comments: Long called *A. pantherina* in the Pacific Northwest, it appears that this species is best referred to as *A. pantherinoides.* Although quite variable in color (within a range from cream to beige to tan or brown), the thick white warts and rimmed bulb help distinguish it. Yellower forms are difficult to distinguish from species in the *A. gemmata* complex, but *A. pantherinoides* tends to have 'cooler' tones, at least over the disc.

Amanita gemmata group

Cap: 5–12 (18) cm across, rounded when young, becoming broadly convex to plane. Margin finely sulcate. Surface slightly viscid to dry, covered with variable amounts of small whitish warts or patches of flat, felty, or frosted universal veil tissue. These slide off easily, leaving the cap bare at times. Color variable, warm yellow to yellowish orange or creamy yellow. **Gills:** Narrowly to broadly attached, rarely truly free, close to crowded. Edges fringed with veil tissue when young. White to creamy. **Stipe:** 5–11 cm long, 1–2 cm thick at apex, gradually enlarged downward to a swollen or bulbous base. Surface dry, smooth above the partial veil, smooth or with fine tufts or scales below. White to creamy. **Partial veil:** A thick, membranous, skirt-like ring on the upper stipe, often falling off. White. **Volva:** A rimmed or short-collared membranous cup around the bulbous stipe base. **Flesh:** Firm to soft, fleshy-fibrous, solid in stipe. White. **Odor:** Indistinct. **Taste:** Indistinct. **Spore deposit:** White. **Microscopy:** Spores typically 9–13 × 7–8 (10.5) µm, ellipsoid to elongate, inamyloid, but there is much variation within the group.

Ecology: Solitary or scattered in small troops. Different members of the group grow with a wide range of both hardwoods and conifers. Common throughout Cascadia. Fruiting in fall. There is a spring-fruiting species in the eastern Cascades as well.

Edibility: Toxic.

Comments: Members of the 'Gemmatoid' group of *Amanita* are recognized by their creamy yellow caps with whitish warts, skirt-like partial veils, and rimmed volva tissue around a bulbous stipe base. There appear to be at least four unnamed species in the *A. gemmata* complex, there is much overlap in color with *A. pantherinoides*. Two common species in this group that grow in west-side forests include a large, pale yellow species with thick white veil tissue, and a small to medium-sized species with yellow-orange cap tones and a tendency for the veil tissue to rub off the cap, leaving it smooth. *A. 'pseudobreckonii'* has a pale straw-yellow cap with pinkish- tan tones and an indistinct or absent partial veil; it is most common with spruce and pine along the coast. There is also a common spring-fruiting species with fir in the Cascades.

Amanita cruentilemurum

Tulloss et al. nom. prov.

Bloody Ghost Amanita

Cap: 5–12 cm across, rounded or tightly convex when young, becoming broadly convex to flat in age. Surface dry to slightly viscid, white or off-white with pinkish blotches, darkening to pinkish or reddish overall with age. Covered with scattered white warts, sometimes extensively so. **Gills:** Finely to broadly attached (rarely truly free), close to crowded. White at first, becoming dull beige in age; at all stages may have blotches of pinkish to reddish coloration. **Stipe:** 5–10 cm long, 1.5–2 cm thick at apex, cylindrical to a bulbous base. Surface dry, smooth or with small scales over lower portion. White to cream, at times with pinkish or reddish stains. **Partial veil:** A membranous skirt-like ring on upper stipe, or breaking as the cap expands. White. **Volva:** Unusually indistinct, often present only as an obscure pigmented line at the top of the basal bulb of the stipe. **Flesh:** Firm, pale whitish, bruising pink or dull reddish very slowly, then becoming dark, dingy reddish brown (especially visible in larva tunnels). **Odor:** Indistinct. **Taste:** Mild. **Spore deposit:** White. **Microscopy:** Spores 8–11 x 5–6.5 µm, ellipsoid or slightly elongate, amyloid.

Ecology: Solitary or in scattered in small groups under conifers, especially Sitka spruce. Locally common in southern portions of our range. Occasionally with Douglas-fir in western Cascadia and with pine in eastern Cascadia. Fruiting in fall.

Edibility: Edible, but not recommended.

Comments: The whitish to pinkish cap, membranous partial veil, and tendency to stain pinkish to reddish in age help distinguish the blushing *Amanita*. The growth under conifers and smooth stipe help distinguish A. *cruentilemurum* from the spring-fruiting oak-associated *A. novinupta.* There is also a darker capped, fall-fruiting species in the *A. rubescens* group which grows with introduced oaks. *A. sponsus* nom. prov. is another similar oak associated species with a white cap, but stains and discolors brownish. Members of the *A. gemmata* group and *A. pantherinoides* lack the pinkish staining.

Amanita smithiana

Bas

Smith's Amanita

Cap: 5–15 cm across, rounded to tightly convex when young, becoming broadly convex to flat in age. Surface dry to moist, never viscid, with poorly defined radially-arranged fibrillose scales or small, fluffy warts. Specimens growing in drier conditions often have a smooth cap with a metallic silvery gray color. Margin with hanging triangular or ragged, fluffy, disintegrating veil tissue when young, often with a broad sterile band in age. Pure white when young, often developing creamy or pinkish tones in age. **Gills:** Narrowly to broadly attached, often with a fine line of tissue descending the upper stipe, close to crowded. Short gills present and often truncate. Pure white when young, becoming dull beige-yellow in age. **Stipe:** 6–20 (30) cm long, 2–3.5 cm thick at apex, enlarging toward a swollen, bulbous base. Bulb with a swollen upper portion and a long, tapered "root" that is easily broken off when collected. Surface scruffy, covered with flocculose scales. White to creamy white. **Partial veil:** Usually remaining attached to cap margin, with some ragged remnants usually visible on upper stipe, very powdery (often covering your fingers when handled). **Volva:** Indistinct, usually appearing as bands of irregular, ragged, white to creamy scales on lower stipe. **Flesh:** Thick, fleshy, fragile, whitish. **Odor:** Usually indistinct to fishy or foul at maturity, reportedly to occasionally smell sweet-spicy (like cinnamon or Matsutake) when young. **Taste:** Not sampled. **Spore deposit:** White. **Microscopy:** Spores 8.5–12 × 5.5–8 µm, broadly ellipsoid or elongate, amyloid. Basidia clamped.

Ecology: Solitary or in small groups under conifers, fruiting primarily in fall and early winter. Generally scarce but can be locally common (although usually not in large numbers). Found primarily in the mid- to higher elevations of the Cascades, with occasional fruitings near the coast.

Edibility: Seriously toxic, causing severe gastrointestinal distress and in some cases kidney failure.

Comments: This *Amanita* can be recognized by its tall stature, bulbous and elongate tapering base, white color, and ragged veil remnants ornamenting most of the fruitbody. The definition between the smooth cap cuticle and the powdery universal veil warts is more distinct than in many other *Amanita* in section *Lepidella*. It is most similar to *A. silvicola*, which is typically stouter-statured, with an abruptly bulbed or truncate (not strongly rooting) stipe base. *A. sponsus* nom. prov has a white cap with pyramidal warts, and a cylindrical stipe with a rounded bulb which develops brown stains in age. It grows with oaks.

Amanita farinosa

sensu PNW

Floury Amanita

Cap: 3–8 cm across, rounded to convex when young, becoming broadly convex to flat, or uplifted in age. Margin finely sulcate when young, distinctly so in age. Surface dry, completely covered in powdery, gray to grayish tan universal veil tissue that binds to the surface, but can be sparse in old or weathered specimens. **Gills:** Narrowly attached or notched, close to subdistant, partial gills numerous and abruptly truncate, edges often ragged or finely frosted. White to creamy white. **Stipe:** 4–9 cm long, 0.8–2 cm thick at apex, club shaped, gradually enlarged downward to swollen or bulbous base. Bulb is round to elongate, sometimes with a distinct rim at top. Surface dry, covered with whitish scales or punctations. White to pale creamy white with gray to grayish tan veil remnants at base. **Partial veil:** Present only when very young, consisting of a powdery tissue, disappearing as soon as the cap expands, leaving no sign of a ring on stipe. **Volva:** A powdery rim of tissue or fragile patches around the top of the bulbous stipe base. **Flesh:** Thin, soft, fibrous, and hollow in stipe, white. **Odor:** Indistinct. **Taste:** Indistinct. **Spore deposit:** White. **Microscopy:** Spores 7.5–9.5 × 5.5–7.5 µm, ellipsoid, inamyloid.

Ecology: Solitary or scattered, rarely abundant, on ground or in duff in mossy conifer woods. Mostly found along the coast and lower elevations of the west slope of the Cascades; fruiting in fall, often early in the season.

Edibility: Unknown. Possibly poisonous.

Comments: This uncommon *Amanita* is easily identified by its small size, powdery gray cap, and stout stipe with a club-shaped or slightly bulbous base. Although the western taxon is distinct, a name for it has yet to be published. Another small species, *A. pruittii*, is vaguely similar but has more wart-like veil tissue and paler white to creamy or yellowish colors and grows in grassy or open-ground habitats (including urban areas); it is part of the '*Saproamanita*' group. Another member of this group, *A. prairiicola* has a taller, straight-cylindrical stipe and is known from high-desert sagebrush steppe or grassland east of the Cascades.

Amanita silvicola

Kauffman

Woodland Amanita

Cap: 3.5–10 cm across, rounded when young, becoming broadly convex to nearly plane. Surface dry, cottony when young, soon matted, rarely truly smooth. Sometimes with discernible low warts of matted fibrils. Margin with an often-ragged overhanging band of sterile tissue. Evenly white colored, slightly grayish or creamy in age. **Gills:** Narrowly to more broadly attached, close, edge finely fringed. White, becoming creamy in age. **Stipe:** 4–12 cm long, 1.5–3.5 cm thick at apex, up to 5 cm at swollen base, cylindrical with a distinctly enlarged rounded bulbous or truncate base, rarely with a noticeable rim. Surface dry, floccose, with cottony chevrons as it elongates. White to creamy overall. **Partial veil:** Fragile, powdery-cottony tissue usually clinging to cap margin or as a ragged ring zone on stipe. **Volva:** Usually indistinct, powdery. **Flesh:** Moderately thick, soft. Whitish to cream. **Odor:** Indistinct at first, unpleasant in age. **Taste:** Indistinct. **Spore deposit** White. **Microscopy:** Spores 7–10 (12.5) × 4–6 (8.5) µm, elongate-ellipsoid, amyloid. Basidia lacking clamps.

Ecology: Solitary or in small groups, occasionally clustered in conifer forest, especially with western hemlock, but also with Sitka spruce and Douglas-fir. Less often with pine or tanoak. Locally common across Cascadia, fruiting in fall.

Edibility: Likely toxic.

Comments: Recognized by the short, stout stature with bulbous or truncate (not rooting) stipe bases, and fluffy-snowy to matted texture of the cap surface (even in age). Despite this species looking like an iconic example of the "Lepidella" group, genetic data have indicated its placement within section *Validae*. However, its most similar look-alike, *A. smithiana*, is in fact a Lepidella. That species is recognized by its strongly tapered, often rooting stipe base (difficult to see if not carefully removed from the soil), smoother cap with more obviously-separate warts with age, and often creamy or pinkish cap tones.

Amanita phalloides

(Vaill. ex Fr.) Link.

Death Cap

Cap: 3–12 (20) cm across, egg shaped to tightly convex when young, becoming broadly convex to plane. Margin usually not sulcate, but can occasionally appear so on old, rain-beaten or dried-out fruitbodies due to collapse of cap tissue onto gills. Surface slightly viscid when wet, quickly becoming dry, smooth, appearing metallic in dry weather. Occasional fruitings may show large (but thin) universal veil patches on the caps. Color quite variable, ranging from pale greenish yellow to yellowish white to dingy grayish olive or grayish green when young, often with darker fibrils or streaks. Older caps often duller yellowish green, grayish green, brownish yellow to medium brown. Caps often radially mottled or with obscure dark streaks in any combination of the above colors. Margin usually remaining dingy white or at least paler. **Gills:** Narrowly attached to free, close to crowded. White at first, becoming pale creamy white, often with pale yellowish blushes around margin or with an obscure tinge of pink in age. **Stipe:** 5.5–13 cm long, 1.5–2 cm thick at apex, tapered upward from a distinctly bulbous, round, swollen base. Upper stipe (above ring) smooth, usually white, pale creamy white to pale yellowish. Lower stipe variably white, creamy yellow to yellowish green, often with darker chevrons. **Partial veil:** Thin, membranous hanging skirt, sometimes only present as a ragged collar when veil membrane remains attached to cap margin. Underside yellowish green, upper surface white to pale creamy white. **Volva:** Spacious, sac-like, as an extension of the bulbous stipe base, relatively thin. Exterior white, interior usually yellowish. **Flesh:** Firm, relatively thick to thin, soft, whitish or yellow just under cap skin. Stipe solid or rarely partially hollow in age; whitish. **Odor:** Mild when fresh, distinctive but difficult to describe. Some people call it sickening sweet; others musty metallic or like rotting potatoes, becoming stronger and unpleasant in age. **Taste:** Not sampled. **KOH:** No reaction. **Spore deposit:** White. **Microscopy:** Spores 8–10 × 6–8 µm, nearly round to ellipsoid, amyloid. Basidia not clamped.

Ecology: Fruiting in scattered troops or large flushes in fall and early winter, but can fruit almost any time of year, especially in irrigated areas with oak and non-native hardwoods. This species was accidentally introduced to central California from Europe and has been spreading rapidly ever since. In most of its western range it is most abundant with oaks but has been found associated with an astonishing range of other trees and even shrubs. In much of Cascadia, Death Caps are most common with planted European hardwoods (e.g. linden, chestnut, hornbeam, oak and beech).

Edibility: Deadly poisonous due to the presence of molecules called *amatoxins*. If you suspect you have eaten this species, don't wait for symptoms to develop further—contact a poison control center and your local hospital right away. Prompt treatment and accurate identification of the cause of poisoning are essential. There are occasional human deaths due to consumption of this species, but dogs are frequent victims as well.

Comments: This stately *Amanita* is easily identified by the combination of the variable but usually bare, yellowish green cap, whitish gills, and bulbous stipe base encased in a thin white sack of universal veil tissue. Perhaps the most similar species is *Volvopluteus gloiocephalus*, which sometimes has a greenish gray, metallic cap but does not have a partial veil at any stage, and the gills are distinctly free and soon become pinkish tan—it is fairly common in wood chips and disturbed soils. *A. phalloides* var. *alba* is identical in all respects except that it is pure white on all parts. *A. ocreata* is a similar white to pale pinkish or yellowish species that fruits with oaks in spring. It is common in California, but rare in Cascadia. Some pickers worry about confusing Death Caps with the Coccora (*Amanita calyptroderma*, p. 58). See the comments section under that species for tips on separating these two species.

Amatoxins

Amatoxins are a group of molecules responsible for the toxicity of *Amanita phalloides* and *A. ocreata*, the species involved in the most serious cases of mushroom poisoning in our area. Amatoxins are also found in *Galerina marginata* (which has caused some poisonings due to confusion with *Psilocybe* spp.), some *Lepiota* species, and species in a few other genera).

On the molecular level, amatoxins inhibit the ability of mammal cells to create new RNA, which is necessary for protein synthesis. When this happens, metabolism slows to a halt, and the cell breaks down; death is usually attributed to resultant massive liver and kidney damage.

Victims of amatoxin poisoning often don't realize that anything is amiss until 12–24 hours after eating toxic mushrooms, when gastrointestinal symptoms set in. Unfortunately, these initial symptoms often disappear for an "intermission" lasting 12–24 hours, and thus many victims don't seek treatment. When this "false recovery" period ends, severe damage to the liver and kidneys has already occurred, and more severe symptoms set in. This can lead to a slow death over the course of the next week or longer.

It's extremely important to seek treatment IMMEDIATELY in cases where ingestion of amatoxin-containing mushrooms is suspected. Treatment usually involves massive intravenous injection of fluids to help hydrate and rid the body of amatoxins. More recently, IV administration of silibinin (an extract of milk thistle) has proven effective in mitigating organ damage and preventing fatalities.

Amanita porphyria

Alb. & Schwein.

Gray-veiled Amanita

Cap: 2.5–10 cm across, rounded to round-conical when young, becoming convex to nearly flat with a low umbo or slightly uplifted in age. Surface slightly greasy to dry, with thin, pallid, grayish universal veil tissue. Color variable, but distinctive; ranging from dull grayish lilac with a pale sheen to dusky purple., Washed-out older caps can show more tan or beige, while a small percentage of fruitbodies show redder or even yellow-brown colors. **Gills:** Narrowly attached to obscurely free, close to subdistant. White or slightly creamy in age. **Stipe:** 5–15 cm long, 1–2 cm thick at apex, more or less equal, or often slightly enlarged at apex and base (thinnest at middle). Base of stipe with abrupt rimmed or cleft bulb. White with pale, grayish or lilac bands or chevrons (especially lower stipe). **Partial veil:** A thin, membranous skirt, with a grayish purple undersurface. **Volva:** Forming a short collar (often indistinct) around the upper edge of the abruptly-bulbous stipe base. **Flesh:** Solid, fibrous, whitish. **Odor:** Like raw potatoes or radish. **Taste:** Mild to strong, radish-like or earthy. **KOH:** Sometimes turning cap surface bright lilac. **Spore deposit:** White. **Microscopy:** Spores 8–9.5 × 7.5–9 µm, globose to subglobose, amyloid.

Ecology: Solitary or scattered in troops (rarely in large numbers) in duff under conifers, especially Sitka spruce. Most common along coast, where it is often found in summer or early fall, with scattered fruitings through fall or again in spring. Occasional in the foothills and mid-elevations of the west slope of the Cascades in fall.

Edibility: Unknown, not likely to be seriously toxic.

Comments: The slender stature, grayish lilac to brownish cap, gray partial and universal veil, and blocky, abruptly bulbous stipe base distinguish this species from other *Amanita*. Browner or faded specimens with more wart-like, discrete universal veil patches on the cap could be confused with *A. gemmata* and *A. pantherinoides*. Fruitbodies of the species in those groups tend to be stouter and should have white (not gray) veil tissue and distinctly sulcate margins at maturity.

Amanita calyptroderma

Amanita vernicoccora

Amanita calyptroderma

G. F. Atk. & V. G. Ballen

Coccora

Cap: 6–20 (30) cm across, round when young, becoming broadly convex to flat or uplifted in age. Surface smooth, slightly greasy to dry, center typically covered with a strikingly large and thick patch of white universal veil tissue (a "skullcap"), this patch occasionally breaking into smaller warts. Margin with distinct radial grooves (sulcate). Color somewhat variable but usually warm, radiant coppery brown with golden brown tones outward and yellow around the margin. Fruitbodies with olive or darker colors, or even pigmentless (white) forms are sometimes found. **Gills:** Narrowly to broadly attached, close to crowded. Creamy white to pale yellowish cream at first, developing yellower tones in age. **Stipe:** 8–15 (20) cm long, 2.5–4.5 cm thick at apex, up to 6 cm thick at base, cylindrical or weakly tapered upward from base, which is enclosed in the volva. Uniformly creamy white to yellowish cream. **Partial veil:** Thin but persistent membranous skirt, often collapsing against the stipe in age, not usually wearing off. **Volva:** Persisting as a large, thick, sock-like sock on stipe base. **Flesh:** Fairly firm, whitish. Stipe predominantly hollow or filled with cottony fibers ("pith") or a clear gel. **Odor:** Indistinct when young, becoming fishy in age. **Taste:** Mild at first, becoming slightly metallic or fishy with age. **Spore deposit:** White to creamy white. **Microscopy:** Spores 9–12 × 5.5–7 µm, ellipsoid, inamyloid. Basidia clamped.

Ecology: Solitary, scattered, or in large groups, arcs or troops in forest duff; common in mixed hardwood forests in southwestern Cascadia, uncommon to locally common with Douglas-fir in the Coast Range and on the west slopes of the Oregon Cascades; generally uncommon or absent in Washington outside of the Columbia River Gorge. Fruiting in short-lived flushes in early fall.

Edibility: Edible; the firm texture appeals to most everyone, but the slightly fishy flavor is more polarizing. Some collections also have a metallic aftertaste. Don't collect this species for food until you are very familiar with the range of variation of *A. phalloides*—the deadly poisonous Death Cap.

Comments: Coccora are some of the loveliest and most impressive mushrooms in the west. They are easily recognized by their thick, enclosing shell of veil tissue, warm, coppery caps with sulcate margins, and robust, hollow stipes. The Spring Coccora (*A. vernicoccora*) differs by its consistently pale yellow cap and tendency to fruit during spring. Cases in which Death Caps were confused for Coccora are rare, and primarily due to a lack of familiarity with both species. The two are reliably told apart in all but the most exceptional cases by using a *combination* of the following characters: The Coccora has a warmer-colored cap that is darker at the center and brighter at the margin, overall dominated by yellow-brown tones, whereas that of the Death Cap is a cooler, pale greenish yellow to whitish at the margin (rarely olive, but can be brownish with age). The Coccora usually shows a single large, thick, cleanly-peeling universal veil patch on the cap while the Death Cap is usually bald (when a universal veil patch is present in Death Caps, it tends to be small, flat, thin, and tightly pressed-on). The stipe is almost always hollow in the Coccora, but solid in the Death Cap (rarely partially hollow). The margin of the cap in the Coccora is distinctly sulcate even when young, whereas the Death Cap has a plain cap margin (although it can develop "pseudostriations" in age as the cap tissue collapses against the gills).

Amanita section *Vaginatae*

Grisettes, Ringless Amanitas

The following account describes a general suite of traits shared by a group of similar species; any given representative of this group will differ from some of the specific details listed here.

Cap: 5–15 cm across, egg shaped to rounded-conical when young, soon convex to nearly flat, sometimes with a weak umbo. Surface smooth, dry, usually with a single large patch of white universal veil tissue. Margin with distinct radial grooves. Dull, medium gray to pale warm gray, sometimes more grayish tan. **Gills:** Narrowly attached to free, close to crowded. White at first becoming pale cream in age. **Stipe:** 7–15 cm long, 1.5–2.5 (3) cm thick at apex, cylindrical and rather slender in most cases, although some variants are quite stout. Surface smooth or evenly covered in obscure white or pale gray fibrils or chevrons over a whitish to creamy ground color. **Partial veil:** Absent. **Volva:** Relatively thin, forming a sac-like cup on stipe base. In some members of this group, the UV tissue stains orange, while in others the inside of the volva is grayish. **Flesh:** Soft, thin, stipe stuffed with pith when young, often becoming hollow. Whitish. **Odor:** Usually indistinct, some are fishy in age. **Taste:** Mild. **Spore deposit:** White. **Microscopy:** Most species with spores 9–12 µm long, globose to teardrop shaped, inamyloid.

Ecology: Solitary or scattered in groups in duff under both conifers and hardwoods. Various representatives are commonly encountered throughout Cascadia. Fruiting from late summer through fall, although some species in this group are more common in spring.

Edibility: Edible. Although the flavor is generally highly regarded, the texture is rather soft and the mushrooms are perishable. Although the grisettes are the most easily recognized group of edible *Amanita*, they are not recommended for beginning mushroom pickers.

Comments: Species of *Amanita* that lack partial veil tissue and have markedly sulcate cap margins and cylindrical, non-bulbous stipes are placed in Section Vaginatae and collectively called "Grisettes." Although they are easily recognized as a group, they are very diverse, difficult to identify, and many are nameless. The core group of grisettes in this account are recognized by their grayish or dull gray-brown sulcate caps, and unconstricted sac-like volva tissue at the stipe bases. Some of the other species in this section (i.e., *A. pachycolea*, *A. constricta*, *A. fulva*, and *A.* 'Gray Warts') are treated separately because they have one or more additional features that help distinguish them. *A. pachycolea* is usually taller, with a darker and more deeply sulcate cap, and has orange-staining universal veil tissue. *A. constricta* often has extensive gray ornamentation on the stipe and a volva that flares upward above a constricted midpoint that "chokes" the stipe base.

Amanita constricta

Thiers & Ammirati

Constricted Grisette

Cap: 5–15 (25) cm across, buttons egg shaped to convex, becoming broadly convex and soon flat, or even slightly depressed at the center with an uplifted margin. Surface dry to slightly greasy, smooth, sometimes with a flat, membranous universal veil patch over the center. Margin with distinct grooves (sulcate), opaque. Dark grayish or brownish gray, sometimes appearing slightly dusted with whitish powder. **Gills:** Narrowly attached and soon seceding, less commonly truly free or broadly attached, close to crowded. Off-white, becoming pale grayish. **Stipe:** 8–15 (25) cm long, 1–3 cm thick, equal or narrowest at apex. Whitish to off-white, often covered in variably dense parallel lines of gray scales, sometimes small and inconspicuous or restricted to base of stipe, or pronounced over entire length of stipe and darker gray. **Partial veil:** Absent. **Flesh:** Thin, rather soft, whitish. Stipe usually at least partially hollow or filled with cottony pith. **Volva:** A membranous but somewhat fragile sheath tightly adhering to the stipe base but with a tight 'collar' and prominently flaring upper edge. Inner surface grayish, outer surface white, darkening slightly with age. **Odor:** Mild when young, rather fishy and unpleasant at maturity. **Taste:** Mild or slightly sweet. **Spore deposit:** White. **Microscopy:** Spores 9.5–13 × 8–11.5 µm, globose to broadly ellipsoid, inamyloid.

Ecology: Solitary or in scattered troops under both hardwoods and conifers. Common throughout most of our area, but status and distribution unreliable due to scant and often confusing taxonomy of this section. Primarily fruiting in late fall and winter.

Edibility: Edible and tasty, but rather perishable.

Comments: The diagnostic feature separating this species (or species complex) from other grisettes is the constricted volva with a flaring upper portion. If this has disappeared (and it often does, being quite fragile), separation from species in the *A. vaginata group* is very difficult. Furthermore, there are multiple species in Cascadia going by the name "*constricta*" with differences in size, coloration, and possible host associations.

Amanita sp. 'Gray Warts' group

Cap: 5–10 cm across, round when young, becoming broadly convex to plane, margin typically sulcate. Surface dry, often covered in small felty warts, occasionally smooth. Light to dark gray, with whitish to dark gray warts. **Gills:** Narrowly attached or seceding from the stipe, close, partial gills numerous. White but often with adhering bands of grayish floccose edges (remnants of the same tissue left as lines or chevrons on the stipe). **Stipe:** 5–15 cm tall, 0.6–1.5 cm thick, cylindrical, narrowest at apex. Whitish to grayish and sometimes decorated with vertical lines of fine scales or small chevrons. **Partial veil:** Absent. **Volva:** Thin to fairly thick, felty or more membranous, disintegrating around stipe base, and leaving warts on cap. Variably whitish to rather dark gray, sometimes also with ocher staining. **Flesh:** Soft, fibrous, thin in cap, stipe hollow or stuffed with pith. Whitish to pale gray. **Odor:** Indistinct or slightly fishy in age. **Taste:** Mild to slightly sweet. **KOH:** Not recorded. **Spore deposit:** White. **Microscopy:** Spores 10.5–14.5 µm, globose, smooth, inamyloid.

Ecology: Like most *Amanita*, it is often restricted to forest settings and mostly found with conifers, but also occurring with hardwoods (and in one case even alder—see comments). Since the members of this group are poorly known and often occur in mixed habitats, more specific host associations may eventually be discovered.

Edibility: Edible.

Comments: This description encompasses a subgroup of the 'grisettes' that have gray-tinted universal veil tissue on their caps. Although mostly unnamed and quite variable, some are more easily recognized. The one pictured here is a common and often fairly large species which occurs with Douglas-fir. One beautiful species close to *A. friabilis* has pronounced scaly gray rings and chevrons over the entire length of stipe, as well as many gray, felty warts across the cap which often coalesce into a single patch over the disc. It is quite rare, fruiting in small troops in the fall in association with alder. *A. protecta* is similar but more frequently associated with willows and has less neatly-separating veil tissue on the cap (messier-looking overall).

Amanita pachycolea group

Cap: 5–10 cm across, round-conical to bell-shaped, soon flattening or slightly uplifted in age, usually with a low, round umbo. Surface smooth, slightly greasy when young, becoming dry. Colors variable, with browns, grays, beige and tan in a complex pattern: Usually darkest at disc surrounded by a pale ring, followed by another dark band ('halo') at the inner edge of the long grooves of the cap margin, then paler again around the extreme edge. **Gills:** Narrowly attached, narrow to broad, close to crowded. Whitish at first, but soon taking on a gray cast and becoming cool light gray in age. **Stipe:** 9–20 cm tall, 1–3 cm thick, straight, cylindrical, narrowest at apex. Surface smooth, whitish, often with fine grayish scales over part of the surface. **Partial veil:** Absent. **Volva:** Thick, membranous. Forming a spacious sheathing sack around the stipe base and often leaving a single patch on the cap as well. White, staining orange to ocher. **Flesh:** Soft, fibrous, thin in cap, hollow in stipe at maturity. Whitish to faintly gray. **Odor:** Indistinct to slightly fishy in age. **Taste:** Mild. **KOH:** Not recorded. **Spore deposit:** White. **Microscopy:** Spores 11.5–14 ×10–12 µm; globose to ellipsoid, smooth, inamyloid.

Ecology: Solitary or in small scattered groups in duff of a variety of conifers. Quite common across Cascadia, often appearing early in fall and continuing into early winter.

Edibility: Edible and good, with a rich flavor and a somewhat more substantial texture than other grisettes.

Comments: The large fruitbodies, zonate-looking cap pattern, and lack of a strongly constricted collar on the volva help characterize the conifer-associated *A. pachycolea.* Members of this group have thick, membranous volva which develops orange to ocher stains. A very similar species, *A. 'castellae'* (RET nom. prov.), has warmer yellow-tan to brown tones in the cap, and perhaps smaller fruitbodies on average.

Amanita fulva

sensu PNW

Cap: 4–12 cm across, distinctly round-conical when young, expanding to broadly bell-shaped or flat with a low, round umbo in age. Surface smooth, thinly greasy when moist, otherwise dry and slightly metallic; sometimes with a single patch of thick, white UV tissue, margin distinctly sulcate. Apricot-orange, tan-orange, or ocher-brown. **Gills:** Narrowly attached, deep, closely spaced; edges minutely fringed when young. White to pale cream or with a slight pinkish tint. **Stipe:** 7–15 cm tall, 0.6–2 cm thick, straight, cylindrical, narrowest at apex. Surface dry, smooth or with fine scales. White or slightly creamy, at times with orangish chevrons. **Partial veil:** Absent. **Volva:** Thick, membranous, forming a spacious sheathing volva at the base of the stipe and sometimes present as a single patch on the cap. White, unchanging or with orangish stains. **Flesh:** Very thin in cap, soft, fibrous. Stipe stuffed with a pith or becoming hollow. White. **Odor:** Indistinct or slightly fishy in age. **Taste:** Indistinct or slightly sweet. **KOH:** Not recorded. **Spore deposit:** White. **Microscopy:** Spores 11–13 × 10.5–12.5 µm, globose to subglobose, smooth, inamyloid.

Ecology: Solitary or in small groups in mossy duff under a range of conifers. Uncommon to locally frequent in fall, mostly west of the Cascade crest.

Edibility: Edible, see comments under *Amanita* section *Vaginatae*.

Comments: The lovely light orangey-brown cap of this grisette renders it distinct from all others in Cascadia. Although the epithet *'fulva'* has been used for decades, that name is based on specimens from Europe; the western North American species is morphologically and ecologically distinct and needs its own name. The somewhat similar *A. velosa* has a pinkish cap with patches of white UV tissue; it grows in spring under oaks around meadow edges. Although locally common in south-central Oregon, it is otherwise rare in Cascadia, mostly found at low elevations around inland valleys. *A. populiphila* has a warm tan cap and grows with aspen.

Limacella sp. 'Pink'

Cap: 3–8 cm across, broadly round-conical or rounded, becoming broadly convex to nearly plane in age. Surface with broad, appressed patches, slightly viscid when young, then greasy to dry in age. Dull pinkish to tan, often darker on disc. **Gills:** Narrowly attached, beige to dull pinkish. **Stipe:** 3.5–10 cm tall, 0.6–1.5 cm thick, equal to narrowly club-shaped, often with a tapered or rooting base. Whitish above veil zone, often covered in pinkish or orange-tan chevrons below. **Partial veil:** Present only as an indistinct ring zone. **Volva:** Absent. **Flesh:** Thin, soft and fragile in cap, stipe fibrous. Dingy pale cream to pinkish beige. **Odor:** Strongly farinaceous. **Taste:** Farinaceous. **KOH:** Not recorded. **Spore deposit:** White. **Microscopy:** Spores 3–5 µm, globose, smooth, inamyloid.

Ecology: Solitary or in small groups (never in large numbers) from duff in conifer forests. Apparently quite rare, found mostly on the west slopes and in coastal areas in fall and early winter.

Edibility: Unknown.

Comments: The combination of medium-large size, pale gills, faint partial veil zone, chevrons of tissue on the stipe, and dry, dull pinkish caps is somewhat distinctive. However, the overall impression tends to leave even fairly experienced collectors confused. The deadly toxic *Lepiota subincarnata* has a distinctly scaly cap and stipe when young but weathered specimens could be easily confused. *Cystodermella* can be vaguely similar in stature and color but have granular surfaces and are usually smaller. Other *Limacella* are slimier or differently colored and *Amanita* have universal veil tissue either on the cap or around the stipe base (or both).

Limacella sp. 'Tan Slime'

Cap: 3–9 cm across, hemispherical or egg-shaped when young, becoming broadly convex to nearly flat in age. Surface smooth and typically glutinous when fresh, sometimes dry in age and then usually with glued-on leaf litter and duff. Light tan from slime layer when young, soon white to creamy, sometimes retaining tan tones at disc. **Gills:** Narrowly attached or nearly free, close, broad, with many partial gills present. White or off-white. **Stipe:** 6–15 cm long, 0.5–2 cm thick, cylindrical, clavate. Surface glutinous when fresh, to dry, smooth. Tan to beige when young, soon whitish overall. **Partial Veil:** None; although the thick layer of slime on the stipe can be interpreted as a proto-veil. **Volva:** Absent. **Flesh:** Thin, soft, fragile. White. **Odor:** Strongly farinaceous. **Taste:** Farinaceous. **KOH:** No reaction. **Spore deposit:** White. **Microscopy:** Spores 4.5–6.5 × 4–6 µm, subglobose, smooth, inamyloid.

Ecology: Solitary or in small groups scattered in duff of a variety of conifers, often in drier east-side forests of the Cascade Range. Rare in Cascadia, fruiting in fall.

Edibility: Unknown.

Comments: Easily recognized by beige to white color scheme and heavy layer of translucent tan slime on both cap and stipe. *Limacella* (*Zhuliangomyces*) *illinita* (in the PNW sense) is quite similar but is pure white and clear slime. It is more likely to be found in coastal forests but is also quite rare. These unusual species are more likely to be mistaken for *Hygrophorus eburneus* than for any closer relatives. That species is shorter and stouter, with thicker, rubbery-fibrous flesh. Other *Limacella* in Cascadia are differently colored or less slimy (or even dry overall). *L. glischra* (at least something going by that name) is also occasionally found in Cascadia; it is also very slimy, with warm ocher-brown colors on both the cap and stipe. *L. roseicremea* is a poorly known and apparently extremely rare species described from Washington: It has a viscid, pale whitish-pink cap, and a large, membranous, skirt-like partial veil. It might be mistaken for an *Amanita* but lacks any universal veil tissue.

3 - Lepiota and Allies

Free gills, presence of a partial veil (usually membranous but occasionally obscure or cortina-like), and a white spore print are characteristic features of this group. While especially diverse in cedar, alder, and maple-dominated forests, *Lepiota* and Allies are found in a variety of forest habitats, amongst the urban interface, and even in cultivated areas.

All members of this group feed on dead or decaying organic matter (saprobic). Although genera can be told apart with practice, the group's nomenclature is in flux. Extensive taxonomic rearrangements (and name changes) are coming soon – most notably the merger of *Leucocoprinus* and *Leucoagaricus*, and a narrowed definition of *Lepiota*.

The most important macroscopic characters to note when identifying Lepiotoid fungi include the texture of the cap surface, the structure of the partial veil, overall stature and size, and any discoloration or staining reactions. Microscopically, the size and shape of the spores, their reactions in Melzer's reagent (dextrinoid or not?), the presence, absence, and shape of cheilocystidia, and in particular, the structure of the cap cuticle are important characters.

Cystoderma and *Cystodermella* belong to Squamanitaceae and are not closely related to the other mushrooms in this section. They are included here because their finely scaled caps, partial veils, narrowly attached white gills, (mostly) white spores, and growth in humus in forest settings make them likely to be confused with *Lepiota*. *Cystoderma aureum* is an outlier, with large fruitbodies and brown spores.

Else Vellinga has studied this group extensively, and contributed a great deal to our knowledge and understanding of Lepiotoid fungi in the western USA. Her work has shown that many of these species remain undescribed, including a number of rare, endemic, and threatened taxa. We encourage readers to consult her technical papers, many of which are freely available online.

Cystoderma aureum

(Schaeff. ex Matt.) Kühner & Romagn.

Alaska Gold

Cap: 7–20+ cm across, obtusely conical at first, becoming convex to plane, often with a rounded umbo. Surface dry, covered with fused granular powder, occasionally nearly smooth in age. Ocher-orange to golden-ocher, at times more pallid, orangish buff. **Gills:** Broadly attached or notched, close, broad; partial gills numerous. Light brown when young, rich brown to orange-brown when mature. **Stipe:** 10–20 (30) cm long, 1.5–3 (5) cm thick, equal or enlarged lower. Surface dry, granular-powdery below ring, smooth above. Orangish buff to ocher-orange below ring, pallid creamy above (becoming orange-brown with age). **Partial veil:** A continuation of the granular-powdery membranous sheath of the lower stipe, forming an upward-flaring ring, adhering to cap margin when young, collapsing in age. Upper surface whitish, soon covered in brown spores. **Flesh:** Thick, firm, fibrous. Whitish, developing orangish streaks with age. **Odor:** Indistinct. **Taste:** Indistinct to unpleasant. **Spore deposit:** Yellow-brown to brown. **Microscopy**: Spores 11–13 × 5–6 µm, ellipsoid, smooth or slightly roughened, thin-walled. Pleurocystidia rare, club-shaped with pointed tips; cheilocystidia absent.

Ecology: Often clustered or in scattered troops from soil in disturbed areas (old roadsides, trails, or meadow edges). Widespread in northern Cascadia but rather uncommon. Fruiting in fall.

Edibility: Not recommended. Although some people eat it with no ill effect, others experience gastrointestinal problems.

Comments: The large size, ocher-orange colors, granular coating of cap and stipe, flaring partial veil, and brown spores make this mushroom quite unmistakable. Although uncommon in Cascadia, it can fruit in large patches. Despite the spore color, it is closely related to the white-spored *Cystoderma. Gymnopilus ventricosus* grows on wood, has a fibrillose-membranous partial veil, a bitter taste, and rusty brown spores. *Cortinarius caperatus* has a whitish bloom to the cap when young, a whitish stipe, and a white membranous partial veil.

Synonym: *Phaeolepiota aurea* (Matt.) Maire

Cystodermella granulosa

(Batsch) Harmaja

Brickbrown Powdercap

Cap: 2–5 cm across, hemispheric when young, becoming broadly convex to plane, sometimes weakly umbonate. Margin slightly inrolled when young, often with small, hanging 'teeth' or patches of silky-fibrillose white partial veil tissue, lobed or slightly wavy in age. Surface dry, densely granular-scaly, smoother when weathered or in age. Color variable, orangey brown to brick-brown, dull brown or tan, often more reddish near disc, paler overall in age. **Gills:** Finely attached, often notched, broad, moderately spaced; edges often slightly eroded or ragged. White at first, soon dingier to grayish or light beige. **Stipe:** 3–7.5 cm long, 0.3–0.6 cm thick, straight, equal. Surface dry, lower portion covered in fluffy scales and chevrons often forming a weak annular zone, with silky pale fibrils above. Brick-red to tan-brown, often darkest at apex. **Partial veil:** A continuation of stipe fibrils, adhering to cap margin, evanescent, leaving remnants on cap margin when young. **Flesh:** Thin, fragile, fibrous in stipe; white. **KOH:** Indistinct or reddish brown on cap. **Odor:** Indistinct. **Taste:** Not sampled. **Spore deposit:** Dull whitish or off-white. **Microscopy:** Spores 3.5–5 × 2–3 µm, ellipsoid, inamyloid. Cheilocystidia absent.

Ecology: Usually in small groups and troops in moss beds and needle duff under west-side conifer forests. Occasional, fruiting from late fall through midwinter at lower elevations and rarely in spring at higher elevations.

Edibility: Unknown.

Comments: The relatively slender stature, granular brownish cap, fluffy-scaly stipe covering, and white gills help distinguish this species. *C. cinnabarina* has a stouter stature, a brighter pinkish orange to brick-red cap, and has cystidia on the gills. The rare *C. adnatifolia* is very similar in appearance to *C. cinnabarina* but is a bit brighter orange in color and has a sparser partial veil. *Cystoderma* species in our area are differently colored (ocher, mustard-beige to tan), have amyloid spores. All lack cheilocystidia.

Cystoderma carcharias var. *fallax*

(A.H. Sm. & Singer) I. Saar

Sheathed Powdercap

Cap: 1.5–5 cm across, domed or bell-shaped, becoming broadly convex, sometimes weakly umbonate. Surface dry, densely powdery-granular. Ocher-tan to mustardy olive-tan or slightly darker dingy brown (a less common form is pale beige to nearly cream-white). **Gills:** Broadly attached or narrowly notched, close, broad. White to creamy or light tan in age. **Stipe:** 3–7 cm long, 0.4–0.7 cm thick, equal or slightly enlarged and/or curved near base. Surface silky-smooth above ring, lower part covered in a layer of dense, fine granular chevrons. White to creamy above ring, becoming tan to dingy brown. Ocher-tan to light leather-brown below ring. **Partial veil:** An upward-flaring, sheathing membranous ring formed by a continuation of the granular sheath of the lower stipe. Upper surface white, smooth to slightly striate. **Flesh:** Fairly thin, white to dingy beige-brown. **KOH:** Dark rusty to reddish-brown or nearly black on cap. **Odor:** Indistinct. **Taste:** Not sampled. **Spore deposit:** White. **Microscopy:** Spores 3.5–5 × 3–4 µm, ovoid to broadly ellipsoid, amyloid. Cap cells round, light brown in KOH.

Ecology: In small groups or scattered in troops, in leaf litter and duff, often in mossy areas—particularly in pine and Douglas-fir forests. Common throughout Cascadia, mostly in west-side forests away from the coast. Fruiting in fall into early winter.

Edibility: Unknown.

Comments: The granular cap and stipe, brownish to tan colors, and large, sheathing ring help distinguish this species. *C. amianthinum* is paler, usually more slender, has a weaker partial veil (not forming a flaring ring), and often with more wrinkled caps. *Cystodermella granulosa* is darker brick-brown, has coarser, flakier scales on the stipe, lacks a flaring partial veil, and has inamyloid spores. A rare, somewhat similar-looking *Leucopholiota* occurs in Cascadia and can be told apart by its more shaggy-scaly surfaces, brighter yellow-brown to orangey colors, and growth on decaying wood. *Cystoderma aureum* looks like a giant version of this species but is more golden-brown to ocher-brown in color and has dull brown mature gills and brown spores.

Cystoderma amianthinum

(Sccp.) Fayod

Common Powdercap

Cap: 1–4 cm across, hemispherical when young, expanding to broadly convex, often with a small but distinct umbo. Surface dry, distinctly finely granular or powdery, often extensively radially wrinkled; margin often with hanging 'teeth' of white veil tissue. Dingy ocher-tan to bright ocher at center, paler buff to light yellowish outwards. **Gills:** Narrowly attached to obscurely free, moderately spaced, broad. White, becoming beige or light tan in age. **Stipe:** 3–7 cm long, 0.3–0.5 cm thick, cylindrical. Surface dry, smooth above the annular zone, but with many pale, buff, or tan chevrons breaking up over the lower portion. Light beige to tan, often extensively covered with silvery white fibrils, dingier pinkish-brown in age. **Partial veil:** A small, indistinct zone formed by the sheathing layer of scales on the stipe, occasionally almost forming a ring. **Flesh:** Thin, soft, white. **Odor:** Musty earth or green corn. **Taste:** Mild. **Spore deposit:** White. **Microscopy:** Spores 5–6 × 3–3.5 µm, ellipsoid, amyloid.

Ecology: Solitary or scattered, sometimes in large troops, almost always in lush beds of moss under conifers. Common throughout Cascadia, fruiting in fall and winter. This species is the host for *Squamanita paradoxa* which is an extremely rare "body-snatcher" parasite that replaces the host's cap with its own purplish-gray, scaly-textured fruitbody.

Edibility: Unknown.

Comments: The small size, grainy ocher-tan cap, white gills, scales or chevrons over the lower stipe, and growth in moss help distinguish this species. *C. jasonis* is uncommon in Cascadia, mostly restricted to high elevations and has a brighter orange-brown cap and dingier brownish stipe.it. *C. carcharias* var. *fallax* has a flaring, persistent ring, and is often larger and slightly darker in color. A few *Inocybe* are superficially similar and some share the green corn odor, but they develop dingy brownish gills and have surfaces that are shaggy-fibrillose-scaly rather than powdery-granular.

Leucoagaricus roseilividus

(Murrill) E. Ludw.

Rosy Parasol

Cap: 1.5–4 cm across, hemispheric or broadly ovoid when young, becoming broadly conical or conical with a low, poorly-defined umbo, finally broadly convex or nearly plane with an umbo. Surface dry, densely velvety at disc, fibrillose overall, with appressed hairs (often in chevrons towards margin). Purplish at center, wine-colored to rosy pink outwards; fibrils pinkish lilac over a whitish or creamy surface. **Gills:** Free, close to crowded, narrow. Whitish. **Stipe:** 4–9 cm long, 0.3–0.9 cm thick, cylindrical, more or less equal, sometimes slightly enlarged at base or club-shaped. Surface dry, smooth. White. **Partial veil:** A thin, membranous ring or band, usually flared upwards, sometimes disappearing in age. White with a distinct, pinkish purple edge. **Flesh:** Thin, fragile in cap; stipe fibrous; white. **Odor:** Indistinct or slightly rubbery. **Taste:** Not sampled. **KOH:** Indistinct or obscurely brownish or green on cap. **Spore deposit:** White. **Microscopy:** Spores 6–9.5 × 4–5.5 µm, almond-shaped or elongate-ellipsoid, dextrinoid. Cheilocystidia abundant, cylindrical to club-shaped or like elongated bowling pins.

Ecology: Solitary or in small groups in rich humus of forest settings—particularly in alder and cedar duff. Uncommon in Cascadia, fruiting whenever there is sufficient warm, moist weather from late summer through spring, but most likely in fall.

Edibility: Unknown.

Comments: Although there are several pinkish to purplish rose *Lepiota* and *Leucoagaricus*, this species stands out by its slender fruitbodies, lack of bruising reactions, and finely-hairy cap with deeper purplish tones at the disc. *Lepiota (Leucoagaricus) decorata* is larger, stouter, and has an evenly rose-pink to pale pink cap with a velvety rather than hairy surface. It occurs in similar habitats across lower elevations in western Cascadia.

Synonym: *Lepiota roseilivida* Murrill

Cystolepiota petasiformis

(Murrill) Vellinga

Duncecap Powderhead

Cap: 1–5 cm across, bell-shaped to convex but covered with a remarkable accumulation of powdery fluff that forms a pointed peak at the center of the cap (sometimes so pronounced as to be slightly ridiculous-looking). In age becoming more modestly granular to nearly smooth when weathered. White to cream-beige, ornamentation slightly tan over disc. **Gills:** Narrowly attached to free, moderately spaced, broad. Whitish to cream, occasionally brownish-tan in age. **Stipe:** 2–7 cm long, 0.2–0.5 cm thick, cylindrical. Surface dry, fluffy-dandruffy. Whitish to cream, with whitish to tan ornamentation. **Flesh:** Thin, fragile, whitish to beige, or light brown in stipe. **Odor:** Indistinct to farinaceous. **Taste:** Farinaceous. **Spore deposit:** White. **Microscopy:** Spores 4–5.5 × 2.5–3 µm, ellipsoid to oblong, slightly spiny. Pileipellis cells irregularly inflated and elongated, clamp connections absent.

Ecology: Solitary or in small groups in thick, damp duff. Often under western redcedar, or in wet, dark alder leaf duff. Widely distributed but rare in Cascadia. Fruiting in fall.

Edibility: Unknown.

Comments: This species is easily recognized by its pallid colors and dunce-cap of powdery fluff. *C. seminuda* is smaller, with a whitish, slightly granular to nearly smooth cap and stipe, and a pinkish stipe base. It is common in mixed alder forests, but often overlooked. *C. fumosifolia* has fawn-brown colors on the cap and stipe and its cheilocystidia are filled with yellow contents. A species with pinkish, matted or fluffy-felty cap and stipe surfaces has been tentatively identified as *C. moelleri*; it appears to be rare in Cascadia. *C. bucknallii* is a tall, slender species with a beautiful lilac-bluish to amethyst color, a finely granular cap, a strong coal-tar or paperwhite narcissus odor, and elongated, dextrinoid spores. It is a rare species found in low elevation forests with maple and alder. *Leucocoprinus cretaceus* is white overall and has dramatic, powdery-fluffy, nearly pyramidal warts on the cap and stipe surfaces (these can wear down in age or rainy weather). It typically fruits on dung or in nutrient-rich humus in warm weather.

Leucocoprinus brebissonii

(Godey) Locq.

Gray-eyed Dapperling

Cap: 1–4 cm across, ovoid at first, expanding to broadly convex, usually with an indistinct, low umbo. Edge of cap usually distinctly striate or pleated. Surface dry, dark gray or mousy tan-gray and finely felty near center, whitish and radially-fibrillose to silky outwards, covered with many flat, light grayish or tan scales and a large, unbroken, dark, circular patch over the umbo. **Gills:** Free, close, broad; edges even. Pale creamy white. **Stipe:** 3–7 cm long, 0.2–0.4 cm thick, cylindrical, slightly enlarged or club-shaped at base. Surface dry, smooth or slightly fibrillose. White to light creamy. **Partial veil:** A thin, free, whitish ring with a collar-like lower part, sometimes remaining attached to cap edge or disappearing. **Flesh:** Very thin, fragile, hollow in stipe; white. **Odor:** Indistinct. **Taste:** Not sampled. **Spore deposit:** White. **Microscopy:** Spores 9–12 × 5.5–7 µm, ellipsoid to almond-shaped, smooth, with a large germ pore. Cheilocystidia irregularly mummy-shaped.

Ecology: Scattered in small troops in humus and wet litter of a variety of forests—especially riparian alder woods where it can be abundant at times. Also found in rich soil of gardens and in urban areas, including indoor plant pots. Common at lower elevations in Washington and coastal habitats across Cascadia with a range expanding southwards. Fruiting in late summer and early fall as one of the first Lepiotoids to appear at the onset of the wet season.

Edibility: Unknown.

Comments: This species is recognized by its pleated whitish cap with a distinct, dark gray patch at the center and tendency to fruit early in the wet season. It could be confused with members of the *Lepiota atrodisca* group but those have stockier fruitbodies (in some species beaded with golden droplets), finely hairy-scaly cap surfaces, and substantially smaller spores without a germ pore. *L. ianthinus* is smaller, lilac-grey colored, and grows in the soil of potted plants indoors. *Leucoagaricus meleagris* has larger, densely clustered fruitbodies that stain yellow to pinkish-red, and fine, dark grayish-brown scales on the cap. It grows in gardens and urban areas, usually in summer. See comments under *Lepiota atrodisca* and *Lepiota cristata*.

Leucocoprinus cepistipes

(Sowerby) Pat.

Cap: 2–6 cm across, tightly bell-shaped or cylindrical to nearly rectangular (with a flattened disc) when young, expanding to broadly bell-shaped or plane with a broad, round umbo or squared-off disc in age. Margin thin, striate to sulcate in age. Surface dry, finely powdery to dandruffy, with flattened patches near the center. Whitish to pale grayish at first; disc grayish to yellowish-tan or beige. **Gills:** Free, close to fairly widely spaced, broad; edges smooth. White to creamy. **Stipe:** 2–9 cm long, 0.3–0.5 cm thick, widened in lower half. Surface dry, smooth overall, or with a finely dandruffy lower stipe. Whitish, developing dingy beige or grayish tones. **Partial veil:** A thin, white membrane forming a loose, free ring on the stipe. **Flesh:** Very thin and fragile in cap; whitish. Stipe fibrous, whitish to grayish cream. **Odor:** Indistinct or faintly rubbery. **Taste:** Not sampled. **Spore deposit:** White. **Microscopy:** Spores 7.5–13 × 6–8 µm, ellipsoid, smooth, with a germ pore, dextrinoid. Cheilocystidia large, club-shaped. Cap cells upright, awl-shaped, with a few scattered, irregularly shaped upright cells.

Ecology: Fruiting in dense clusters and troops in wood chips, rich soil, grassy areas, and less often in humus in disturbed forest areas. Fairly common in urban areas in the inland valleys across Cascadia. Fruiting in summer and fall.

Edibility: Unknown.

Comments: The white colors, dandruffy texture, and growth in clusters help define this species. Other light-colored species are *Lepiota sequoiarum* and allies but they tend to be pure white overall, slender, and with a fibrillose cap surface, growing deep in shaded forests. *Leucocoprinus cretaceus* is white overall with the same stature as *L. cepistipes* but has a dramatically fluffy texture with nearly pyramidal warts and scales on the cap and stipe surfaces. It typically fruits on dung or nutrient-rich humus in warm weather.

Lepiota atrodisca group

Dusky Parasols

Cap: 1.5–5 cm across, ovoid to hemispherical when young, soon expanding to broadly convex to plane, occasionally weakly umbonate. Margin striate to sulcate in age, sometimes with a narrow band of hanging white veil tufts when young. Surface dry, at first densely covered with fine, pointed clumps of blackish-gray to olive-gray fibrils over a white, radially-fibrillose background. In age dark grayish to nearly black at center then light mousy gray outwards and then white at margin. **Gills:** Free, close, narrow; edges smooth or very finely beaded with a white band of cystidia. White to light creamy. **Stipe:** 3–7.5 cm long, 0.2–0.4 cm thick at apex, cylindrical with a slightly enlarged base, straight to curved. Surface smooth or with a few flattened whitish chevrons of tissue just below the ring. White to creamy, becoming dingy beige, occasionally covered with many droplets of golden liquid. **Partial veil:** A thin, white, skirt-like membrane forming an upwards-flaring ring with a grayish to black, finely fuzzy edge. **Flesh:** Thin, whitish, hollow in stipe. **Odor:** Indistinct or slightly rubbery. **Taste:** Not sampled. **Spore deposit:** White. **Microscopy:** Spores 6–7 × 4 µm, ellipsoid, smooth, slightly dextrinoid. Cap surface cells arranged in a flat layer (a cutis).

Ecology: Solitary or scattered in small troops from duff or rich humus. Common in forest settings across lower elevations in western Cascadia. Primarily fruiting in fall, occasional in summer-winter.

Edibility: Unknown; do not experiment.

Comments: The small size, dark blackish-gray fibrillose-scaly cap, and grayish-edged ring on the stipe help distinguish this species group. Genetic data suggests that there are at least five lineages going under this name in the western North America. Currently, they are only reliably distinguished by ITS sequencing but some of them do appear to have corresponding morphology differences including texture of the cap surface and subtleties of color. Two nameless species with golden droplets and dark brown-grey-green colors also occur in our area. *L. cristata* with brownish rather than black or gray tones. The slightly larger, more upright-statured *L. felina* has dark gray to nearly blackish velvety scales on the cap and around the edge of the partial veil.

Lepiota rubrotinctoides

Murrill

Early Ruddy Parasol

Cap: 3–6.5 cm across, ovoid to hemispherical when young, soon broadly convex to plane, occasionally with a wavy slightly uplifted margin in age. Usually with a distinct round umbo as soon as it starts to expand. Surface dry, smooth and shiny or with radially appressed fibrils. Blackish red, wine-red, reddish brown, or pinkish red at center, lighter pinkish red to coral or orangish outwards with a distinctive tendency to split radially extensively white to creamy beige towards margin. The color is quite variable, often appearing much paler in dry weather even when young. **Gills:** Free, close, broad. White. **Stipe:** 4–9 cm long, 0.6–1 cm thick at apex, cylindrical with an enlarged, club-shaped base.Surface dry, smooth or slightly fibrillose. White to creamy; base occasionally staining weakly orangish. **Partial veil:** A prominent, upward-flaring ring; the lower part forming a wide collar-like band. Edge finely fuzzy, sometimes with orange or pinkish colors. **Flesh:** Thin, fragile in cap; stipe fibrous with a silky-stuffed central hollow. White. **Odor:** Indistinct. **Taste:** Mild. **Spore deposit:** White. **Microscopy:** Spores 7–8.5 × 4–5 (5.5) µm, ellipsoid to almond-shaped, often with a rounded nipple at the end, smooth, dextrinoid. Cheilocystidia club-shaped but elongate.

Ecology: Solitary or scattered in small troops in rich leaf humus and needle litter of forested areas. Collectively, species in this complex are common in a wide range of habitats across western Cascadia. Fruiting soon after the first rains in late summer, continuing through fall.

Edibility: Non-toxic.

Comments: The smooth, often slightly umbonate reddish cap, collar-like veil with an orangey to reddish edge, and tall, slender white stipe help identify this attractive mushroom. It is often one of the first species to fruit in the warmer part of the season after the first rains. Genetic results indicate that there are ten or more still-nameless species in this complex in the western USA. Other pink- or reddish-capped *Leucoagaricus* have subtly different colors (more rosy, pinkish, lilac, or purple–hued), and spores and cheilocystidia that are differently shaped and sized.

Lepiota flammeotincta

Kauffman

Flaming Parasol

Cap: 1.5–7.5 cm across, hemispheric when young, soon expanding to broadly convex to plane, splitting in age. Surface dry, innately fibrillose, blackish-brown at center, with dark brown flattened, pointed tufts of fibrils radially arranged outwards, often showing paler whitish beige in between when young, but soon more extensively brown (with little pale color remaining in age). Staining brilliant cherry-red to blood-red when touched, quickly fading to brown. **Gills:** Free, close to crowded, broad. White, not staining when damaged. **Stipe:** 4–9 cm long, 0.4–0.7 cm thick at apex, often tall, straight to curvy, enlarged and club-shaped near base. Surface dry, appressed-fibrillose to finely scale below. Whitish to cream at first, but almost immediately developing tan and brown areas mottled with fibrils. Staining brilliant scarlet to deep orange-red immediately when touched, quickly fading to dark brown. **Partial veil:** A thin, flaring ring on the stipe, whitish above, tan to brown below, staining red. **Flesh:** Fairly thin, fibrous, soft, stipe soon hollow. Whitish cream, staining red when cut. **KOH:** Red on all parts, often obscured by inherent staining. **Odor:** Indistinct or slightly chemical. **Taste:** Not sampled. **Spore deposit:** White. **Microscopy:** Spores 6–9 × 3.5–5.5 µm, ellipsoid or elongate-ellipsoid or slightly tapered at one end, smooth, dextrinoid. Cheilocystidia abundant, cylindrical, walls often distinctly wavy, some narrowly club-shaped.

Ecology: Solitary or in small groups in needle duff or humus under conifers, especially Sitka spruce. Locally common in coastal and Coast Range forests but largely absent from higher elevations, drier habitats, and northern Cascadia. Fruiting in fall.

Edibility: Unknown.

Comments: The dark, fibrillose cap, white gills, and brilliant flaming red reaction of the cap and stipe (but not the gills) help identify this species, which is a distinctive members of a difficult-to-identify species complex. *Leucoagaricus flammeotinctoides* is similar but has gills that stain orangey-red and cylindrical, or narrowly club-shaped, cheilocystidia without wavy walls. *La. pyrrhulus* has a dense, even coating of fine, dark fibrils on the cap, gills that do not stain when damaged (but that turn coppery when dried), and almond-shaped spores. *La. pyrrhophaeus* has gills that stain orange when bruised but remain pale when dried and ellipsoid spores. Both of these tend to be smaller, more slender, and with variably-shaped cheilocystidia (including bottle- and urn-shaped cells as well as club-shaped ones). *La. erythrophaeus* immediately stains reddish on the cap and stipe and orangey on the gills, but is larger and stouter, with gills that are fused into a sort of 'collar' near the stipe. It has long (up to 90 µm) club-shaped cheilocystidia and a cap cuticle with many upright cells (a trichoderm). *Lepiota castanescens* stains less intensely orange-red and has spores with a distinct nipple at the end. Expect to see many of these names reshuffled into the genera *Leucoagaricus* and/or *Leucocoprinus*.

Lepiota cristata

(Bolton) Kumm.

Stinking Parasol

Cap: 2–6.5 cm across, ovoid or rounded-conical when young, expanding to broadly convex, occasionally slightly wavy. Often with a distinct low, round umbo. Surface dry, deep reddish-brown or dark brown when young, but soon cracking into small flat scales (often concentrically arranged), revealing a whitish to creamy, radially fibrillose background, often extensively white and without scales near margin. **Gills:** Free, close, broad. Whitish to creamy. **Stipe:** 3–8 cm long, 0.3–0.5 cm thick cylindrical, base slightly enlarged. Surface dry, finely silky to smooth. White to light beige, developing tan or pale pinkish tan tones, often overlaid with silvery-white fibrils. **Partial veil:** A membranous ring, usually fairly persistent, sometimes remaining attached to cap edge or disappearing entirely. Upper surface whitish, underside often light tan or with a few brownish scales. **Flesh:** Very thin, fragile in cap; stipe fibrous, white. **Odor:** Usually strongly like burnt rubber or burnt hair with a sweet component. **Taste:** Not sampled. **Spore deposit:** White. **Microscopy:** Spores 5.5–8.5 × 2.5–3.5 µm, with a rounded triangular-outline, bullet-shaped with a knob below the apiculus, weakly dextrinoid. Cheilocystidia club-shaped with a narrowed base. Cap cuticle with tightly packed upright pear-shaped cells.

Ecology: In clusters or scattered in troops in humus or wood chips in disturbed areas, especially along trails, roadsides, and in urban habitats. Generally uncommon throughout Cascadia but can be locally frequent. Fruiting in fall and early winter.

Edibility: Unknown. Very similar to some extremely toxic species so do not experiment.

Comments: The somewhat large size (for a *Lepiota*), strongly contrasting cap scales, burnt-rubber odor, ring on the stipe, growth in disturbed areas, and bullet-shaped spores help distinguish this species. *L. castaneidisca* has paler cap scales and is more slender overall, but shares the truncate or bullet-shaped spores. The elegant *L. felina* has brownish to blackish scales on the cap surface, around the margin of the partial veil, and over the lower stipe. Several other, more fragile species, share the brown-centered cap but lack the bullet-shaped spores. *Leucoagaricus ophthalmus* has the cells of its cap surface arranged in an irregular-upright layer (a trichoderm) with scattered clumps of elongated, bottle-shaped cells with brownish walls and short, and club-shaped cheilocystidia. *L. oculata* and *La. paraplesius* also have upright cells in the cap surface but both have more elongate-cylindrical cheilocystidia; those of the latter species are slightly more enlarged at the tip and narrowed towards the base. *La. infuscatus* has sausage-shaped cap cells in chains arranged horizontally (a cutis) and ellipsoid-rounded spores (the latter three species have more almond-shaped spores that are distinctly tapered at both ends).

Lepiota subincarnata

J.E. Lange

Deadly Parasol

Cap: 1.5–6 cm across, convex when young, becoming broadly convex and then expanding to plane or slightly concave, sometimes with a wavy margin. Surface moist to dry, at first entirely felty or matted-tomentose, soon breaking up into concentric rings of fibrillose scales, more matted to nearly smooth in age. Pinkish, or light pinkish brown, often with a narrow, bright white band around cap margin when young, pinkish beige with a creamy background showing in cracks between scales, fading to dull ocher or beige-tan in age or dry weather. **Gills:** Free, close, broad; edges often eroded. White to creamy, becoming light beige. **Stipe:** 2.5–6 cm long, 0.4–0.6 cm thick, equal. Surface dry, silky-smooth above annular zone, with chevrons of brown or pinkish tan tissue near base. Whitish when young, soon light beige-tan to pinkish but covered with silvery-whitish fibrils. **Partial veil:** Silky-cobwebby at first, soon breaking and forming an indistinct annular zone around the middle of the stipe. **Flesh:** Thin, fibrous; stipe hollow; whitish cream. **Odor:** Indistinct to slightly fragrant. **Taste:** Not sampled. **Spore deposit:** White. **Microscopy:** Spores 6–7.5 × 3–4 µm, ellipsoid or elongated-ovoid, smooth, thick-walled, dextrinoid. Cheilocystidia club-shaped.

Ecology: Scattered or in clusters and troops in wood chips, lawns, on nutrient-rich soil in disturbed areas, or in mossy duff under conifers in forested settings. Locally common in urban settings—mostly fruiting in late summer and fall but can appear whenever there is sufficient moisture.

Edibility: Deadly toxic, contains amanitin (same compound in *A.phalloides* and *Galerina marginata*).

Comments: This species is recognizable by its finely felty, pinkish cap with zones of flat scales, often somewhat stout stature (for a *Lepiota*), and lack of a distinct ring or skirt on the stipe. *L. cristata* is similar, but lacks the pinkish colors. Some *Limacella* species look strikingly similar to older specimens of *L. subincarnata* but with strongly farinaceous odors and very fragile flesh. *Echinoderma* species often grow in disturbed, urban habitats but are typically much larger, with ample, white, skirt-like partial veils, and caps with pointed, upright scales.

Lepiota magnispora

Murrill

Fluffstem Parasol

Cap: 2.5–10 cm across, rounded or ovoid when young, becoming convex with a distinct round umbo, to nearly plane with a low umbo, or occasionally with uplifted margin in age. Surface dry, nearly smooth when very young but almost immediately cracking into fairly coarse scales concentrated at center; margin often with cottony whitish to yellow veil remnants. Dull yellowish-beige to pale tan but covered in orangey brown to reddish brown scales; umbo much darker ocher-brown to reddish brown or mahogany-red. **Gills:** Free, close, soft and sometimes eroded. Whitish when young, soon creamy to pale yellowish. **Stipe:** 5–12 cm long, 0.6–1.5 cm thick, cylindrical, base slightly swollen. Surface dry, cottony-fluffy overall or with shaggy bands and chevrons, often with adhering ocher-brown scales bound to soft fibrils, especially around base of stipe. Pale creamy to yellowish-beige when young, ocher-tan in age. **Partial veil:** A copious amount of cottony creamy to whitish tissue between gills and stipe (enclosed in membranous layer of brown scales when very young), soon collapsing and leaving an annular zone on stipe. **Flesh:** Thin, soft, fibrous; stipe with an indistinct central hollow. White. **Odor:** Mild or often like burnt rubber. **Taste:** Not sampled. **Spore deposit:** White. **Microscopy:** Spores 15.5–20.5 × 4.5–6 µm, elongate, spindle-shaped but with a curved pot-bellied side and a straight 'back,' resembling a penguin. Cheilocystidia narrowly club-shaped to cylindrical.

Ecology: Solitary, in pairs, small clusters, or occasionally scattered in larger troops. Common across varied habitats but usually in deep, moist conifer duff (particularly under western red cedar). Fruiting in fall and winter but occasionally in spring.

Edibility: Not recommended because of possible confusion with deadly species.

Comments: The shaggy-cottony stipe, ocher- to red-brown umbo, often ovoid young caps, and yellowish veil and stipe make this a striking denizen of conifer forest. *L. spheniscispora* ihas a more evenly colored golden-tan cap with a less-defined umbo and has a smoother, silky stipe. It fruits earlier in fall and later in the spring.

Lepiota sp. 'Scaly'

Cap: 1–3 (5) cm across, ovoid when young, soon broadly convex, expanding to plane; margin often weakly inrolled when young. Surface dry, densely covered in pointed, hairy tufts when young, becoming more matted-felty; scales flattening towards margin, smoother at maturity. Whitish to light beige, often paler towards margin, but appearing darker due to dense covering of dark reddish brown tufts and scales. **Gills:** Free, close to crowded, narrow. White. **Stipe:** 2.5–5 cm long, 0.6–1.2 cm thick, equal or slightly tapered at base. White to creamy, silky-smooth near apex, lower half with many brick-brown to dark brown, pointed scales, often arranged in weak rings. **Partial veil:** Weak, forming a faint annular zone at the end of scaly band on the stipe. **Flesh:** Thin, fibrous; white. **Odor:** Indistinct or like plastic or rubber. **Taste:** Mild. **Spore deposit:** White. **Microscopy:** Spores 4.5 × 2.5–3 µm, ellipsoid, dextrinoid. Cheilocystidia present. Cap surface with many inflated, thin-walled brownish cells in chains.

Ecology: Scattered or in small groups in rich needle humus, especially western redcedar. Rare, most likely to be found in urban and low-elevation areas of western Cascadia. Fruiting in fall.

Edibility: Unknown.

Comments: The small but stocky fruitbodies, whitish cream colors with many brown scales, rings on the cap and stipe, and lack of a distinct, membranous ring on the stipe distinguish this attractive mushroom. This is one of a confusing group of species which, visually, fall somewhere between *Cystolepiota*, *Echinoderma*, and *Lepiota*. The small spores align it with true *Lepiota*. The most frequently encountered *Echinoderma* in Cascadia are large (caps to 10 cm across), and have distinctive white, ample, cobwebby-membranous partial veils that soon collapse into large, hanging skirts.

Echinoderma asperum

(Pers.) Bon

Freckled Dapperling

Cap: 4–10 cm across; round to drumstick-shaped when young, soon broadly bell-shaped to convex becoming nearly plane or slightly wavy in age. Surface dry, densely covered with fluffy pyramidal scales that become matted-felty or flattened outwards towards margin or in age. Orangish tan to warm brownish, soon exposing whitish to cream colors between scales. **Gills:** Free, markedly crowded, often forked near the stipe, broad. White to pale cream, becoming mottled with grayish tones in age. **Stipe:** 4–12 cm long, 0.5–1.5 cm thick, tall, cylindrical, usually slightly enlarged at base. Surface dry, smooth to finely silky, whitish to pale cream above veil, distinctly silky, whitish, overlaid with fawn to pinkish-beige layer below the veil; lower portion sometimes with rings of flaky veil tissue. **Partial veil:** A striking white, silky-cobwebby membrane when young, collapsing into a large, drooping skirt in age. Underside often beige-tan and covered in patchy scales similar to those on the cap; cap scales often spilling over and remaining attached to the margin of the veil as well. **Flesh:** Moderately thick, fibrous, stipe solid; whitish to beige. **Odor:** Mild to rubbery-plasticky. **Taste:** Not sampled. **Spore deposit:** White. **Microscopy:** Spores 6–10 × 2–3 µm, oblong, dextrinoid. Cheilocystidia broadly club-shaped to round and stalked. Cap with chains of inflated, brownish cells.

Ecology: Solitary, in loose clumps and troops in rich humus, and duff in both urban and slightly disturbed wild habitats. Uncommon to rare, more often found in northern Washington in urban settings. Fruiting in fall and early winter.

Edibility: Unknown, to be avoided.

Comments: This attractive species stands out from other "lepiotoids" by virtue of its large size and unusual surface textures – silky on the veil and stipe, with felty or oatmeal-like flakes and pointed scales on the cap. Likewise, there are a few other very similar looking species, which may differ in cap texture and color (more matted-felty and grayish brown in one form), and/or details of microscopy (shorter, smaller, rounder spores).

Synonym: *Lepiota aspera* (Pers.) Quél.

Leucoagaricus cupresseus

(Burl.) Boisselet & Guinb.

Cap: 4–10 cm across, rounded at first, soon broadly convex to plane with a low round umbo, or slightly square (broad and flat at center). Surface dry, plush-velvety at first, cracking into fine plaques and fine fibrillose scales outwards, nearly smooth in age or when weathered. Often with a narrow white band of veil tissue around the margin when young. Dark brown at center when young, dark reddish-brown outwards, with a mottled appearance as it expands, becoming extensively tan to creamy beige near margin or overall in age. **Gills:** Distinctly free, close, broad. White, becoming creamy, often bruising light yellow and then light orangey red. **Stipe:** 4–12 cm long, 1–2 cm thick, cylindrical, distinctly enlarged into a bulb at base. Surface, dry, smooth. Pale creamy at first but soon extensively dingy tan, developing brown chevrons and streaks; weakly staining orange-red when scraped **Partial veil:** A flaring ring attached to a broad band or collar over the middle of the stipe. White; often contrastingly strongly against the dingy stipe; edge often dark brown. **Flesh:** Thick, fibrous; stipe hollow. White, slowly and irregularly staining orange to reddish. **KOH:** Reddish with a greenish-gray halo on stipe, greenish on gills, slowly green on cap. **Odor:** Indistinct or pleasantly nutty, musty in age. **Taste:** Mild to sweet or slowly peppery. **Spore deposit:** White. **Microscopy:** Spores 7.5–9.5 × 4.5–5 µm, ellipsoid or ovoid, sometimes a bit tapered at one end with an obscure nipple, dextrinoid. Cheilocystidia variable: club-shaped, bowling-pin shaped, or swollen but with a narrow, wavy, cylindrical tip.

Ecology: Solitary or scattered in small groups in rich duff and needle humus. Uncommon, mostly in urban areas in western Cascadia. Fruiting in fall.

Edibility: Unknown.

Comments: Although quite variable in size and stature, typical fruitbodies are relatively large and stout, have plush to suede-like or finely patchy dark brown caps (when young), free white gills, orange-reddish staining flesh, and a band- or collar-like ring. *L. adelphicus* is virtually identical, but perhaps taller and more slender on average and has spores that measure 6–7.5 × 3.5–4.5 µm. There are quite a few similar species known from California, some of which may occur in Cascadia.

Leucoagaricus leucothites

(Vittad.) Wasser

Cap: 3.5–10 cm across, marshmallow-shaped to hemispheric when young, expanding to broadly convex or plane, occasionally concave in age. Surface dry, very finely velvety when young, in some places breaking into tiny scales, but more often appearing perfectly smooth and metallic. Edge of cap sometimes with a narrow, ragged fringe of white veil tissue. Extensively white to light creamy beige, at times cool gray to light grayish or light creamy beige in others. **Gills:** Free, close to crowded, broad, soft. White to creamy or dingy beige, developing pinkish tones, sometimes extensively stained pinkish gray or salmon in age. **Stipe:** 4–10 (12) cm long, 0.7–1.2 cm thick, cylindrical, with an enlarged, club-shaped base. Surface dry, smooth; base often with felty white mycelium and white rhizomorphs. Whitish to gray, developing dingy beige or tan tones, base occasionally with yellow to faintly salmon tones. **Partial veil:** A membranous flaring ring, collapsing in age. **Flesh:** Fairly thick, firm to soft, fibrous; stipe with a central hollow. White, sometimes slowly staining pinkish or yellow in stipe when cut. **Odor:** Indistinct. **Taste:** Mild. **Spore deposit:** White. **Microscopy:** Spores 6.5–8 × 4.5–5.5 µm, ellipsoid to oval, smooth, with a germ pore, dextrinoid. Cheilocystidia club-shaped.

Ecology: Solitary or more often scattered in troops, arcs, and rings. Most frequently in grass lawns but also in short-grass meadows, rich garden humus, and occasionally even wood chips in disturbed areas. Rarely in forested settings and then along edges of trails. Common throughout Cascadia, especially in inland valleys. Fruiting in summer and early fall.

Edibility: Edible and good but can cause allergic reactions, sample cautiously the first time and cook it well. Also keep in mind the possibility of chemical and other contaminants when found in urban areas.

Comments: The whitish to light gray smooth caps, flaring ring on a club-shaped stipe, and growth in grassy areas help distinguish this species. The similar *L. barssii* has a shaggier, often darker grayish cap, is less common, and fruits early in late summer/fall. Some *Agaricus* can appear superficially similar and grow in the same habitat but have pinkish to grayish gills when young, maturing to dark brown, and producing dark brown spore deposits.

Chlorophyllum olivieri

(Barla) Vellinga

Olive Shaggy Parasol

Cap: 6–18 cm across, ovoid, hemispheric or nearly globose when young, soon broadly convex to plane, sometimes with a low, broad umbo. Surface dry, smooth only when very young, then extensively shaggy-fibrillose; margin often with an overhanging band of shaggy whitish tissue. At first with a light brick-brown to tan skin, soon breaking up into clumps of fibrils to show light whitish beige background, rapidly developing olive grayish to dingy brownish tan tones, usually retaining a smooth, pinkish tan to light brown area at center of cap. **Gills:** Free, close, broad, soft; edges even. White to cream, but easily bruised and turning light orange to pinkish. **Stipe:** 8–20 (25) cm long, 0.8–1.5 cm thick, tall and slender, equal to a distinctly swollen to abruptly bulbous base. Surface smooth, dry. Base often with copious cottony mycelium and adhering substrate. White to light beige, developing tan tones. **Partial veil:** A whitish membrane splitting to form a flaring ring with a shaggy, double edge, often grayish tan to brown on the underside and edge. **Flesh:** Thin to fairly thick, soft. Stipe brittle to somewhat fibrous, with a narrow, cylindrical hollow. Whitish, staining orange to pinkish or reddish in all parts when cut or scraped. **Odor:** Mild or pleasantly earthy. **Taste:** Mild. **Spore deposit:** White. **Microscopy:** Spores 8–11 × 5.5–7 µm, ellipsoid, smooth, thick-walled, dextrinoid, with a small germ pore. Cheilocystidia rounded, balloon-shaped with a very narrow base.

Ecology: In troops, arcs, and rings (occasionally solitary), in rich leaf litter and forest humus in disturbed areas, usually producing its largest flushes in fall and early winter. Introduced from Europe, it is now common in Cascadia's inland valleys, particularly near urban areas. It has spread southward to at least central California.

Edibility: Edible and very good. However, *Chlorophyllum* can cause stomach upsets due to allergic reactions, so be sure to eat only fresh specimens, cook them thoroughly, and eat only a small amount the first time.

Comments: The large, tall fruitbodies, strikingly shaggy cap surface, and dingy olive-gray colors distinguish this species. Likely to be confused with *C. rhacodes* and *C. brunneum*, which are thicker-stemmed, slightly shorter, and have more contrastingly dark brown scales on a light whitish-beige cap. The toxic *C. molybdites* has a whitish cap with tan to light brown scales and greenish gray gills at maturity. *Macrolepiota* can closely resemble this species (especially in stature) but have a 'snakeskin' pattern of brownish scales on the stipe. They have been introduced to Cascadia from Europe but so far remain rare.

Lepiota

Chlorophyllum rhacodes

(Vittad.) Vellinga

Shaggy Parasol

Cap: 5–20 cm across, rounded or blocky when young, becoming to plane. Surface dry, medium brown to dull brick-brown when young, cracking into patches to reveal whitish-beige fibrils between the scales, overall dingy brown in age. **Gills:** Free, close, broad, margin even. Whitish when young, becoming dull beige, brownish in age, staining orange when damaged. **Stipe:** 8–20 (25) cm long, 2–3 cm thick at apex, 4–6 cm thick at base, cylindrical, widening downwards to a club-shaped base. Surface dry, smooth; extreme base often covered in felty white mycelium. Creamy to tan, dingy brownish in age, staining orangish when scraped. **Partial veil:** A prominent, flaring, double-edged ring. Whitish to grayish cream or beige; upper surface often paler. **Flesh:** Moderately thick, firm when young, soft in age; stipe fibrous with a narrow, cylindrical hollow. Whitish to creamy beige, staining orangish when cut, becoming dingy red and then brown. **Odor:** Pleasant at first, musty in age. **Taste:** Mild to pleasantly earthy. **Spore deposit:** Whitish. **Microscopy:** Spores 8–12 × 5–8 µm, ellipsoid to almond-shaped with truncate end, thick-walled, dextrinoid. Cheilocystidia round with a short, tapered base.

Ecology: Scattered or in pairs, sometimes in large troops in thick duff of cedar and alder but also found in almost any rich humus. Can be locally abundant in/near urban centers across Cascadia. Fruiting in fall, occasional in spring, rarely in summer in irrigated areas.

Edibility: Edible and excellent but can cause serious GI upsets even in people who have eaten it without problems in the past. Use only fresh caps, cook thoroughly, and don't eat large quantities. Some people are allergic to all *Chlorophullum* species.

Comments: The large fruitbodies, free white gills, brown cap scales, prominent double-edged ring on the stipe, and stipe which gradually enlarges to a bulbous base distinguish this species. It is often confused with the extremely similar *C. brunneum* which is common in Cascadia and fruits in similar habitats has an abruptly widened stipe base and club-shaped cheilocystidia. *C. olivieri* has a more slender stature, a dingier and shaggier cap, and round cheilocystidia. *C. molybdites* is a toxic, grass-dwelling species with greenish-gray gills at maturity; it is rare in Cascadia but expanding northwards in inland urban areas. *Macrolepiota procera* looks somewhat similar but is markedly tall, slender, and with an intricately chevroned "snakeskin" pattern of brown scales on the stipe. It was introduced from Europe to Cascadia where it fruits occasionally in grassy settings in summer.

Chlorophyllum brunneum

(Farl. & Burt) Vellinga

Western Shaggy Parasol

Cap: 5–20 cm across, rounded-convex or blocky when young, soon expanding to broadly convex to plane. Surface slightly greasy when very young, soon dry. Brown to dull brick when young, cracking into plaques of flat brownish scales on top of shaggy whitish beige radial fibrils; becoming dingy brown overall in age. **Gills:** Free, close, broad, edges smooth. Whitish, becoming dull beige, sometimes light greenish near stipe. **Stipe:** 8–20 (25) cm long, 2–3 cm thick at apex, 4–6 cm thick at base, cylindrical; base with an abrupt, often "gutter-edged" bulb. Surface dry, smooth. Creamy to tan, dingy brownish in age; staining orangish when scraped. **Partial veil:** A large, whitish thin ring with a spiky-shaggy edge. **Flesh:** Thin, firm when young, soft in age. Stipe fibrous with a narrow hollow center. Pale creamy beige, turning orangish, dingy red and then brown. **Odor:** Pleasant to musty in age. **Taste:** Mild to pleasantly earthy. **Spore deposit:** Whitish. **Microscopy:** Spores 10–13 × 7–8.5 µm, ellipsoid to almond-shaped, with truncate end, dextrinoid. Cheilocystidia club-shaped, fairly long.

Ecology: Solitary or in pairs, clumps, arcs, or troops. Favors thick duff of many kinds of trees, often in urban settings. Can be locally abundant but distribution is patchy. The largest flushes appear in fall and spring, with scattered fruitings throughout the year.

Edibility: Edible with caution—see comments under *C. rhacodes.*

Comments: The large size, free, whitish gills, brown cap scales, prominent single-edged ring on the stipe, orange-staining flesh, and abruptly bulbous base help distinguish this species from most other mushrooms. Often confused with *C. rhacodes* but that species has a gradually swollen stipe base (not rimmed or abrupt) and a shorter, rounder cheilocystidia and slightly smaller spores. *C. olivieri* has a taller, slender stature and an evenly dingy olive-gray cap. It also has round, balloon-like cheilocystidia. *C. molybdites* has evenly greenish gray gills at maturity and greenish spores; it grows in grass lawns.

Chlorophyllum molybdites

(G. Mey.) Massee

Green-Gill Parasol

Cap: 5–20 cm across, tightly convex or egg-shaped when young, soon broadly convex to plane, often with a distinct, rounded umbo. Surface dry, entirely light brown to tan when young soon breaking up into scales on a white to cream background. Scales of some specimens uplifted around edges and "floating" on small flaky granules, or peeling off almost entirely. Darkening overall, to dingy whitish in age. **Gills:** Free, close, broad. Whitish when young, slowly becoming light bluish green and then dingy olive-gray, browner in old age. **Stipe:** 7–18 cm long, 1–1.5 cm thick at apex, cylindrical, usually quite even and straight; base often distinctly swollen (up to 3 cm thick). White, becoming dingy tan to pale brownish where handled, smooth. **Partial veil:** A neat skirt when young, soon forming a flaring ring with a tattered edge, whitish above with a brown underside. **Flesh:** Thin, soft; stipe firm, fibrous, with a central hollow. Whitish, staining orange and then dark reddish brown. **Odor:** Indistinct or musty, especially in age. **Taste:** Mild. **Spore deposit:** Dull olive-green to gray-green. **Microscopy:** Spores 8–13 × 6.5–9 µm, broadly ellipsoid to ovoid, smooth, dextrinoid, with a small germ pore. Cheilocystidia club- or spindle-shaped.

Ecology: Usually in small, scattered troops or sometimes in impressive arcs or rings with 20+ fruitbodies, rarely solitary, on grass lawns. Currently rare in Cascadia but expanding northwards and likely to become more common as the climate warms. Fruiting in summer and early fall.

Edibility: Poisonous to most, causing severe vomiting and diarrhea. Reportedly the most common cause of mushroom poisoning in the United States.

Comments: This species needs warm nighttime temperatures to fruit and, in recent years, has expanded northward to Cascadia facilitated by climate change. It can be confused with *C. brunneum* and *C. rhacodes* but those species have darker cap scales, white mature gills, and white spore prints; they are much more common in Cascadia and rarely grow in open grassy lawns. The greenish gills can be slow to develop and may not show any hint of color until maturity, leading to confusion with one of the Shaggy Parasols.

4 - Agaricus and Melanophyllum

Although the range of cultivated mushrooms seems to widen every year, portobello, crimini, and white button mushrooms (all variants of *Agaricus bisporus*) still account for a majority of the world's commercially-produced food mushrooms. As such, *Agaricus* are the mushrooms with which Americans are (albeit unknowingly) most familiar. This genus was once the all-encompassing name for any gilled mushroom—a fact still echoed today in our use of the general term *'agaric'* for any mushroom with a cap, gills, and stipe. Although now reduced to a tiny fraction of its former sense, *Agaricus* remains a species-rich genus, broadly distributed around the world, and occurring in all kinds of climates. They are especially diverse in Mediterranean and subtropical regions, and there's even a group of *Agaricus* species adapted to sand dune and desert environments!

Members of this genus are easily recognized by their free gills, dark brown spores, and presence of a partial veil (although this veil can be obscure or disappear in age). *Agaricus* in the Pacific Northwest are found in a variety of settings, including deep forest, lawns and parks, and on heavy clay or silty soil in floodplains. Some species of otherwise-rare *Agaricus* have expanded into urban areas and other disturbed or human-modified settings, suggesting that the rarity of these species may be driven primarily by ecological competition with other groups of fungi that are absent in such habitats.

Melanophyllum species are uncommon to rare in Cascadia and are told apart from *Agaricus* by their small, slender fruitbodies with powdery cap and stipe surfaces, deep reddish gills when young, and greenish olive spore deposits when fresh. They are not particularly closely related to *Agaricus*, but are included in this section because of their similarity in structure and dark spore deposits.

For *Agaricus* species, distinctive odors and staining reactions when damaged are the most important features to examine, followed by the overall coloration of the fruitbody, the structure of the partial veil, and the texture of the cap and stipe surfaces. Habitat details also provide strong identification clues as does KOH reactions.

The genus *Agaricus* is less diverse in Cascadia than in the drier Mediterranean-type climate of California and the deserts of the Southwest. Nevertheless, they are conspicuous, abundant, and more than a little frustrating for beginning identifiers. Keep in mind that there are more species present in the Pacific Northwest than we profile here, many of which are rare, poorly understood, or yet unnamed. Rick Kerrigan's monograph on North American *Agaricus* is an important reference for anyone seriously motivated to make further headway in this difficult group.

Although the following groups are 'artificial' (they don't exactly correspond with our understanding of evolutionary relationships), learning them can make the job of identifying *Agaricus* a little easier:

Phenol-scented yellow stainers (section *Xanthodermatei* & section *Hondenses*) usually smell unpleasant or chemical-like, often described as "phenolic," like Band-Aids, pen ink, or Sharpie markers (although not everyone can detect this smell). The staining reaction of these species is usually clear yellow and often fades away or leaves dull, grayish brown discolorations. Be aware that in older fruitbodies (and even in fresh fruitbodies of a few species) the yellow staining reaction can be slow or nearly absent; in these cases, the reaction is best observed by scraping the surface of the stipe base. KOH reactions are yellow.

Almond-scented golden stainers have a sweet or anise-like aroma, although some can develop horsey, musty, or urine-like overtones in age. The staining reactions of fruitbodies in this group tend to be more gold to orange-yellow and persist or turn amber. Members of section *Arvenses* produce large, robust fruitbodies, while those of section *Minores* are quite small. Many of these smaller species have pinkish or dingy purplish colors on the cap and their aroma can be difficult to detect unless a piece of the cap is crushed. KOH reactions are yellow.

Red stainers discolor blood-red to dark red quickly and intensely when cut, bruised, or scratched. Members of this group are uncommon in our area, often found under cedars and some non-native trees. A few species that turn weakly orangey, dingy reddish, or light pinkish when cut and rubbed (but which do not show immediate striking "bleeding") are treated with the non-staining species.

Non-staining species don't strongly change color when cut and rubbed or scratched, although some may blush pinkish or orangey. Most lack any KOH reaction but some turn gray-green or brownish. They may have salty-briny or archetypally 'mushroomy' smells but are not sweet or phenolic.

Agaricus diminutivus group

Diminutive Agaricus

Cap: 1–3 cm across, egg-shaped to rounded as first, soon broadly convex to plane. Surface dry, smooth to radially fibrillose, interwoven fibrils splitting in age. Margin fringed and hanging past edge of gills. Pale white to pale tan, with orange-tan tones at center, but also with pale pinkish and yellowish to yellow-orange tones. Staining yellowish when bruised. **Gills:** Free, close. Pale pinkish tan at first, becoming pink and then turning brown when spores mature. **Stipe:** 2.5–5 cm long, 0.3–0.7 cm thick, mostly cylindrical but base sometimes enlarged. Surface dry, hairy or cottony texture below veil. Whitish to pale creamy buff with patchy yellow-orange tones. **Partial veil:** A short, thin, hanging skirt; margin often fringed; with yellow-orange tones. **Flesh:** Very thin, fragile; white, bruising faintly yellow. **Odor:** Usually faintly sweet, like anise. **Taste:** Mild. **KOH**: Yellow on cap. **Spore deposit:** Dark brown. **Microscopy:** Spores (4) 5–5.5 (6.5) × 3.5–4.5 µm, broadly ellipsoid, smooth. Partial veil tissue with yellow to golden orange hyphae in KOH.

Ecology: Solitary or in small, scattered groups in duff of wooded areas, usually under conifers. Widespread, but more common in western Cascadia. Fruiting in fall.

Edibility: Edible. It has a pleasant smell but is tiny and rarely abundant.

Comments: The pale pinkish tones of the cap, yellow discolorations, slender, upright stature, and faint anise odor (difficult to detect unless crushed) are important identification features. At first glance, one might mistake it for a *Lepiota*, but the brown mature gills and spores will distinguish it from that group. *A. micromegethus* prefers more open habitat and is not as tall and slender. *A. kerriganii* (likely the appropriate name for most of what had been called *A. semotus* in our area) is paler, more robust and has distinct white rhizomorphs at the stipe base—it most often fruits in spring under hardwoods. Other small, slender *Agaricus* stain reddish or differ in color and/or odor. There are several species going under the name *A. diminutivus*, but reliable distinguishing criteria have not yet been established.

Melanophyllum haematospermum

(Bull.) Kreisel

Red-spored Dapperling

Cap: 1–4 cm across, domed at first, soon becoming broadly convex to plane, margin occasionally uplifted in age. Surface dry, finely matted-felty at first, soon cracked into small, powdery, granular scales. Sometimes developing wrinkles in age. Margin often with triangular 'teeth' of pale, powdery tissue hanging over edge of gills, soon wearing off into a narrow band. Dingy grayish brown to tan or beige, often with olive tones; margin paler. **Gills:** Free, close to well-spaced, sometimes fairly thick and with eroded edges in age. Blood red to dingy coral red, brightest when young, becoming dingy, mottled pinkish brown and then dark purplish brown. **Stipe:** 2–5.5 cm long, 0.2–0.4 cm thick, cylindrical, thin, and usually straight. Metallic coral pink to magenta-red or blood-red at apex, lower part extensively covered with pale grayish tan to dingy brown powdery granules (similar in color and texture to cap surface). **Partial veil:** Whitish membrane covered in pale powder, soon rupturing and adhering to edge of cap, not leaving a ring on stipe, or at most a weak annular zone. **Flesh:** Very thin, fibrous, light beige in cap, reddish on either side of whitish pith filling hollow stipe. **Odor:** Indistinct. **Taste:** Not sampled. **Spore deposit:** Olive-green, supposedly more bluish green when taken from very fresh mushrooms, older prints turning reddish brown to black. **Microscopy**: Spores 4.5–7 × 2.5–3 µm, ellipsoid or slightly elongate, very slightly roughened. Light brown, not changing color in Melzer's reagent. Cap surface a layer of thin-walled, globose, brownish cells.

Ecology: Solitary or more often scattered in small troops in rich soil, compost, or wood chips. Mostly found in gardens, landscaped areas, or disturbed areas along roads and trails. Fairly rare in Cascadia, fruiting in late summer and fall. Perhaps introduced.

Edibility: Unknown.

Comments: The small size, powdery grayish tan cap, and vivid pink to red gills when young help identify this species; very few other mushrooms in our area produce olive-green spore prints. A couple undescribed *Melanophyllum* species occur in our area. One is larger and duller colored, tends to be less powdery, and often shows blue-gray discolorations near the stipe base. Another species has tiny fruitbodies with bright pink gills. Some *Cystolepiota* (which belong in *Melanophyllum)* in the PNW share the powdery cap, small stature, and grow in similar habitats, but all have pale, whitish to cream or tan gills. A species of *Cystoagaricus* was recently discovered among ferns in mixed conifer-redwood forest in California, and probably occurs in Cascadia as well. It has paler, duller pinkish brown gills, a weakly translucent-striate cap, and paler spores.

Agaricus augustus

Fr.

The Prince

Cap: 7–30 cm across, tightly convex or squarish when young, becoming broadly convex, often flatter around center of cap to plane in age. Surface dry, covered in flattened fibrous scales. Dark, warm brown to golden yellow or golden brown with orange blushes, covered with tan to brownish fibrous scales, bruising yellow to golden-yellow to nearly orange and remaining so. Orange to reddish brown when rain beaten. **Gills:** Free, close to crowded. At first pale, dull tan to pinkish gray, maturing to dark brown to almost black. **Stipe:** 7–25 (35) cm long, 2–5 cm thick, cylindrical or tapered toward the top; base sometimes swollen to club-shaped. Smooth above annulus, usually distinctly scaly below. White to pale creamy yellow, developing distinct yellow or even yellow-orange tones. Scales fibrous and dull yellowish. **Partial veil:** A thin membranous ring, often with small, squarish cottony patches on underside. **Flesh:** Fairly thick, firm when young, softer in age. Whitish, bruising yellow to golden-yellow. **Odor:** Usually strong and sweet or pleasant like almond extract or marzipan. **Taste:** Mild to slightly sweet. **KOH:** Yellow. **Spore deposit:** Dark brown. **Microscopy:** Spores 7–9.5 (10.5) × 4.5–6 (7) µm, ellipsoid, smooth.

Ecology: Often clustered or scattered in arcs, rarely solitary; usually along roads and trails or other disturbed areas. In Cascadia, this species has a preference for western redcedar duff, but are found in a wide range of habitats with deep, rich organic litter. Common throughout the Pacific Northwest, fruiting from spring into fall.

Edibility: Considered one of the finest edible mushrooms by almost everyone lucky enough to try it, but some people dislike the strong almondy flavor. Eat small amounts the first time you try this species, as some people experience allergic reactions including swollen lips, itching, and vomiting.

Comments: A handsome species as well as a delicious edible, this mushroom is recognized by its golden yellow to tawny brown colors, scaly lower stipe, and almondy odor. *A. smithianus* is similar but has a smoother cap and stipe and grows primarily under spruce. *A. subrutilescens* has a wooly-scaly lower stipe but is usually brown to lilac-gray or light wine reddish (lacking golden tones), and does not smell distinctly almondy. The toxic *A. deardorffensis* and *A. buckmacadooi* have gray to gray-brown caps with small scales, smooth stipes, and phenol odor.

Agaricus smithianus

Kerrigan & L.A. Parra

Golden Spruce Agaricus

Cap: 5–20 cm across, button stage rather egg-shaped (like tapered ovals), becoming convex or to plane with a broad umbo. Surface dry, often covered with flattened, dull tan-brown fibrils. Usually an attractive orange-brown, dull yellow, gold, or warm tan-brown over a paler base with a darker disc. Base color darkens orangish in age. Bruising weakly yellowish. **Gills:** Free, close to crowded, broad. Pallid when young, becoming pinkish tan and finally brown to dark brown. **Stipe:** 8–15 (20) cm long, 1–2.5 cm thick, quite tall and slender, but with an enlarged, bulbous base, up to 5 cm across. Surface finely scaly or scruffy when young to smooth. Whitish or creamy with golden tan tones. **Partial veil:** Fairly thick, white, with yellow-gold patches, membranous, forming a flaring or hanging skirt set rather high on stipe. **Flesh:** Moderately thick, fragile. Stipe firm, stuffed, fleshy-fibrous, occasionally hollow in age. Whitish. **Odor:** Almondy. **Taste:** Sweet-nutty to mild. **KOH:** Yellow, especially on cap. **Spore deposit:** Dark chocolate brown. **Microscopy:** Spores (6.5) 7.5–8.5 (10) × (4.5) 5–5.5 (6.5) mm, ellipsoid, smooth.

Ecology: Scattered in troops, rings, or arcs in duff in conifer woodlands, almost always under Sitka spruce. Occasional throughout coastal areas of Cascadia. Often produces two or three flushes from late summer into winter.

Edibility: A great edible. The almond flavor that is sometimes overpowering in *A. augustus* isn't as strong in this species, making it more appealing to some people.

Comments: This interesting *Agaricus* is easily recognized by its egg-shaped cap when young, tawny to gold or orange cap colors, tall stipe (often with yellowish rhizomorphs at the base), almondy odor, and growth under Sitka spruce, although it does turn up under other conifers, at least occasionally. It is unusual for an *Agaricus* to have a strong preference for an ectomycorrhizal tree habitat. *A. augustus* has a squared-off or 'marshmallow-shaped' cap when young, more pronounced cap scales, a much scalier stipe, and is more common in urban or disturbed areas.

Agaricus benesii

(Pilát) Pilát

Cap: 4–10 cm across, round or more often square with an inrolled margin, expanding to round-convex and then plane in age. Surface dry, whitish to creamy, covered with flattened, dull brown fibrils. Bruising strikingly red when scraped. **Gills:** Free, close to crowded. Pallid pinkish-tan when young, becoming dark brown when mature, developing bloody tones where cut or damaged. **Stipe:** 5–10 cm long, 1–2.5 cm thick, tall and upright, with an enlarged, sometimes knobby or bulbous base. Surface scaly or scruffy when young, becoming matted-down to nearly smooth in age or when handled. Whitish to creamy. Surface staining dark red where scraped. **Partial veil:** Thick, fluffy-membranous, forming a flaring or hanging skirt set fairly high on the stipe. **Flesh:** Thick, fairly firm, fibrous. Stipe firm, solid, fleshy-fibrous, with a cylindrical hollow in age. Whitish, staining strikingly red when cut. **Odor:** Faintly to intensely mushroomy (like portobello), or musty in age. **Taste:** Indistinct. **KOH:** No reaction. **Spore deposit:** Dark chocolate brown. **Microscopy:** Spores 4.5–5.5 (6.5) x 3.5–4.5 µm, ellipsoid, smooth.

Ecology: Fruiting singly or in small groups in duff of conifers (often spruce) in late summer and fall. Rare in Cascadia, but probably at least somewhat under-reported due to difficulty of identification.

Edibility: Not commonly-eaten, but unlikely to be toxic. Many red-staining *Agaricus* have caused minor allergic reactions, so caution is warranted.

Comments: The red-staining *Agaricus* are poorly known in Cascadia. The overall more whitish specimens illustrated as *A. benesii* in *Mushrooms of the Redwood Coast* look quite different from those here (which are closer to the true species' concept). This whiter "form" of *A. benesii* also occurs in Cascadia, and almost certainly represents a distinct species. *Agaricus thujae* is smaller, more slender, and also has a scaly stipe surface and red-staining, but has a more evenly dark brown cap with fine, scaly fibrils, a weaker partial veil, and spores measuring 4.5–5 × 3–3.5 µm. It is found in conifer needle duff in coastal Cascadia. *Agaricus sylvaticus* subsp. *occidentalis* is another smallish red-staining *Agaricus*, but has less pronounced red-staining than the species mentioned above.

Agaricus subrufescentoides

Murrill

Cap: 3–15 cm across, rounded with an inrolled margin at first, expanding to broadly convex or plane. Dark red-brown over disc, outwards pinkish tan to beige-buff with an even, fairly dense coating of medium reddish brown, tightly flattened fibrils. **Gills:** Free, close to crowded. Whitish or pale grayish tan at first, becoming pinkish tan and then dark brown as spores mature. **Stipe:** 5–15 cm long, 1.5–4 cm thick, equal, often crooked or j-shaped, with a club-shaped or bulbous base. Surface dry, smooth. Whitish or gray-white, sometimes stained pinkish or dark brown in age. **Partial veil:** A large, white, cottony skirt with a fairly thick edge. **Flesh:** Thin to moderately thick, firm when young, becoming soft, pale grayish to white. **Odor:** Slightly phenolic/chemical and unpleasant. **Taste** Mild. **KOH:** Yellow. **Spore deposit:** Dark chocolate brown. **Microscopy:** Spores 5–7 × 3–5 µm, ellipsoid, rather variable in size and shape, smooth.

Ecology: Solitary or scattered, sometimes in small troops in duff or humus of conifer forest. Uncommon, or at least under-reported, fruiting in fall.

Edibility: Presumably toxic for most people, as with most members of section *Xanthodermatei.*

Comments: This *Agaricus* is recognized by its large size, unpleasant smell (chemical-phenolic), and fairly extensive, rich pinkish tan to dark red-brown colored fibrils, and a pronounced partial veil. *A. hondensis* is very similar but much paler and with more sparsely distributed fibrillose scales. *A. subrutilescens* has a shaggy lower stipe, much thinner skirt-like ring, a mild odor, and does not stain yellow. The scarcity of records of this species is likely due to under-identification. Until recently, most collectors (including the authors!) did not have a clear concept for distinguishing this species from *A. hondensis*.

Agaricus subrutilescens

(Kauffman) Hotson & D.E. Stuntz

Wine-Colored Agaricus

Cap: 4–15 cm across, ovoid when young, soon broadly convex. Surface dry to moist, covered with distinct, appressed-fibrillose scales. Dull wine red to pinkish or reddish brown or purplish brown or entirely dark to medium brown, background color off-white to creamy white. **Gills:** Free, close to crowded. Pale pinkish when young, becoming pinkish tan and finally dark brown. Occasionally staining red where damaged. **Stipe:** 6.5–17 cm long, 1–3 cm thick, cylindrical or with a swollen base. Surface dry, shaggy below partial veil. Whitish or with creamy tones. **Partial veil:** A thick, membranous, white skirt, but often collapsing in age. **Flesh:** Thin, fleshy, often soft in age. Whitish. Stipe cottony-stuffed or hollow, whitish, staining slightly pinkish when cut. **Odor:** Indistinct to faintly fruity or sweet. **Taste:** Mild. **KOH:** Greenish gray on cap. **Spore deposit:** Dark chocolate brown. **Microscopy:** Spores 4–6.5 × 3–4 µm, broadly ellipsoid, smooth.

Ecology: Scattered, or in loose troops or occasionally clustered. Common and widespread in a variety of woodland habitats, especially at lower elevations. Fruiting sporadically between late summer and late fall;can be abundant for a short period before cold weather sets in.

Edibility: Edible and very good, with a delicate flavor. Sample a small amount the first time you try it—some people are allergic, while others detect an unpleasant metallic aftertaste.

Comments: The tall stature, shaggy stipe, brownish to vinaceous cap fibrils, and absence of strong odors or bruising reactions set it apart from most other *Agaricus*. The greenish KOH reaction and woodland habitat help clinch the identification. *A. thiersii* is very similar, but has a browner cap with broader patches of flattened scaly fibrils, and possibly a preference for pine and oak duff. *A. hondensis* has a lighter pinkish brown cap, a thick felty ring, a yellow KOH reaction, and a faint phenol odor. *A. augustus* stains golden to amber and has an almondy aroma.

Agaricus hondensis

Murrill

Felt-ringed Agaricus

Cap: 4–15 cm across, round to convex, becoming broadly convex to plane. Surface dry to moist, covered with tightly flattened fibrils and scales. Golden brown when very young, soon a dull dingy whitish background overlaid by pale tawny or pale tan to pinkish brown scales. **Gills:** Free, close to crowded. Very pale pinkish tan when young, becoming pinkish brown and finally dark brown. **Stipe:** 4.5–15 (20) cm long, 1.5–3 cm thick, almost always with an enlarged, club-shaped base. Extreme base often bulbous, irregularly contorted, and knobbed. Surface smooth or with silky appressed fibrils at base. Whitish to grayish white, sometimes shiny, discoloring to dingy brownish at apex in age. **Partial veil:** A thick, whitish, felty-spongy, flaring to hanging skirt; edge usually thick and flat. **Flesh:** Firm, thick, whitish, bruising irregularly pale yellowish in all parts, most readily seen when stipe base is scratched. **Odor:** Faintly to strongly phenolic. **Taste:** Mild. **KOH:** Bright to pale yellow on all parts. **Spore deposit:** Dark chocolate brown. **Microscopy:** Spores (3.5) 4.5–6 × 3–4 (5.5) µm, broadly ellipsoid, smooth.

Ecology: Solitary, in small clusters, or sometimes gregarious in rings or arcs. Found in conifer duff or debris in low elevation forests, fruiting in fall.

Edibility: Like most phenol-smelling *Agaricus*, this species is poisonous for most people.

Comments: Oftentimes the most common *Agaricus* in forest settings, this lovely species is recognized by its pinkish tan caps with appressed fibrils, pronounced felty ring, faint phenolic odor, and weak yellow staining. *A. deardorffensis* and *A. buckmacadooi* have grayish to blackish scales and somewhat more obvious yellow-staining reactions, especially at their stipe bases. *A. xanthodermus* has a smooth whitish cap, stains yellow much more strongly, has a thinner partial veil, often crooked stipe base, and usually grows in urban habitats.

Agaricus buckmacadooi

Kerrigan

Buck's Agaricus

Cap: 7–20 cm across, rounded to convex at first, becoming broadly convex to plane. Surface dry, tan-brown, covered evenly with dull, brownish gray to dark brownish scales and fibrils tightly pressed against the surface. **Gills:** Free, close to crowded. Pale pinkish when young, becoming pinkish tan and finally brown to dark brown at maturity. **Stipe:** 7–15 cm long, 1–4 cm thick, equal or with a club-shaped or irregularly lobed and contorted, bulbous base. Surface dry, smooth. Whitish or off-white with dingy, pale tan tones. **Partial veil:** A thick, somewhat spongy or felty, flat-edged, white collar-like ring. **Flesh:** Thin to moderately thick, off-whitish, marbled with dingy pale tan areas, weakly yellow in all parts, sometimes more strongly so, especially in base of stipe; stained areas slowly oxidizing to reddish brown. **Odor:** Phenolic or chemical. **Taste:** Mild. **KOH:** Bright yellow. **Spore deposit** Dark chocolate brown. **Microscopy:** Spores 3.5–7 × 3–5 µm, ellipsoid, smooth.

Ecology: Solitary, scattered, or in troops, clusters, and arcs. Widespread and common in disturbed forest settings, especially along roads and trails in duff of conifers, with a preference for western redcedar. Fruiting from early fall into early winter, mostly absent in cold weather, occasionally fruiting again in spring.

Edibility: Poisonous for most people, as are most yellow-staining, phenol-smelling *Agaricus*.

Comments: This imposing mushroom is recognized by its large size, dark brown to grayish-brown cap scales, phenolic odor, thick partial veil, and weak yellow bruising. The fruitbodies of *A. deardorffensis* are slightly smaller on average (although still quite large!) and the cap scales are more of a cool inky-gray color. They tend to fruit in larger numbers and more gregarious clusters. The names *A. praeclaresquamosus* and *A. moelleri* had been widely and variably applied to members of this group in recent decades, work by Rick Kerrigan established firm footing for *A. deardorffensis*, *A. buckmacadooi*, and *A. berryessae* as western North American taxa. The latter species is similar but generally has a paler, white to grayish cap and slightly larger scales; it also stains quickly and more brightly yellow. *A. hondensis* is paler, with flattened, pinkish tan to light brown fibrils, more rounded cap when young, and a thick, felty partial veil. *A. augustus* smells sweet (like almonds), is differently colored (paler golden tan to tawny brownish) and has a scaly-shaggy lower stipe.

Agaricus sylvicola group

Woodland Almond Agaricus

Cap: 4–10 (14) cm across, egg-shaped to round at first, expanding to broadly convex to plane. Surface dry, appearing smooth, but with tiny, fibrillose scales. Whitish to creamy beige or washed with light golden yellow. Bruising warm yellow when rubbed or scraped. **Gills:** Free, close to crowded. Pale tan when young, becoming pinkish tan and finally brown to blackish brown. **Stipe:** 4–10 cm long, 1–3 cm thick, cylindrical, enlarged towards base. Surface dry, smooth. Whitish to creamy. **Partial veil:** A thin, white tissue-like skirt often with cogwheel pattern on the underside, then forming a large, hanging skirt, occasionally disappearing in age. **Flesh:** Thick, firm when young, becoming soft, stipe with a small central hollow or solid. Whitish to mottled tan-white, bruising slowly yellow. **Odor:** Sweet and slightly almondy when young, becoming mustier in age. **Taste:** Mild. **KOH:** Yellow. **Spore deposit:** Dark chocolate brown. **Microscopy:** Spores 5.5–6 × 4 µm, ellipsoid, smooth.

Ecology: Solitary or scattered in small numbers in duff or debris, often in areas with an open understory. Occasional to uncommon in most years, but can be locally rather abundant. Fruiting in late summer and fall.

Edibility: Edible and good when young, but the flesh can become soft and the odor musty with age.

Comments: The relatively large fruitbodies, whitish to creamy cap colors, yellow to amber staining, sweet odor, ample but thin veil, and forest habitat help distinguish this species. *A. albolutescens* is very similar, but typically stains darker golden amber. *A. summensis* is often larger, with a slightly darker tawny to golden beige cap and slightly smaller spores (6.5–7 × 4–5 µm); it is rare and poorly understood. *A. crocodilinus* has much larger, stockier fruitbodies without pronounced creamy or white tones, and tends to grow in coastal pastures. *A. moronii* is a stocky, spring-fruiting species in montane conifer forests. The forest-dwelling members of section *Arvenses* (large, gold-to-amber staining, almond-smelling species) are a difficult group. A number of older names may pertain to species in this group in the western USA, for which original descriptions of type collections are ambiguous. And there are some rare and difficult to recognize species in this group.

Agaricus crocodilinus

Murrill

Crocodile Agaricus

Cap: 5–12 cm across in button stage, 20–40(!) cm across when expanded. Buttons rounded or somewhat marshmallow shaped at first, becoming broadly convex to nearly plane in age. Surface dry, nearly smooth or with small, thin scales near margin when young. In dry conditions (or occasionally even in wet weather) the cap surface breaks up into large, angular warts, giving the cap a "crocodile skin" appearance. Whitish at first, soon dull creamy, developing blushes of light yellowish. Bruising yellowish ocher when scraped or rubbed or in age. **Gills:** Free, close to crowded. Pale tan to dingy light pinkish gray when young, reddish brown to blackish brown in age. **Stipe:** 6–20 cm long, 2–4 cm thick, up to 6 cm thick at base. Shape variable, sometimes spindle shaped (with a tapered base and apex) or enlarged and club shaped at base. Upper parts smooth, with rings of small scales below partial veil (becoming obscure in age). Whitish to off-white, with some yellowish gold to dingy ocher tones in age. **Partial veil:** A thin, whitish, membranous skirt with floccose patches on underside, after rupturing persisting as a hanging skirt. **Flesh:** Thick, firm at first, soft in age. White. Bruising weakly yellowish gold, then aging to amber. **Odor:** Sweet, like almonds, sometimes faint. **Taste:** Sweet or mild. **KOH:** Yellow on cap. **Spore deposit:** Dark chocolate brown. **Microscopy:** Spores 9–11 (14) × 6–7.5 µm, ellipsoid to more elongate, smooth.

Ecology: Scattered or in large arcs and rings in coastal prairies with short, sparse vegetation as well as in more lush, grassy livestock pastures. Uncommon and local in Cascadia, primarily in rich meadow settings on the northern part of the Olympic Peninsula and on the Oregon coast. Fruiting primarily in early fall and especially late spring or occasionally in summer, absent in winter.

Edibility: Edible and by all accounts delicious. Sweetish when young, richer in age, with a firm, meaty texture.

Comments: The large to enormous caps, sweet odor, growth in open grassy areas, and large spores set this *Agaricus* apart. The Giant Horse Mushroom (which had been going by *A. osecanus* in our area, but is perhaps more appropriately called *A. arvensis*) is similar, but has smaller spores (6–6.5 × 4.5–5 µm), and less of a tendency to show a scaly cap. *A. fissuratus* is typically smaller, but otherwise similar; spores size falls between the two larger species (7.5–8.0 × 5–5.5 µm). Arora's *Mushrooms Demystified* states these large meadow *Agaricus* were abundant in pasturelands along the Pacific Coast and collected for food every year by a small group of enthusiasts. Since the 1980s, however they appear to have vanished almost entirely from these pasturelands. It is not known whether climate change, changes in cattle husbandry practices, invasion of non-native soil organisms (or some combination of these factors) is responsible for this species' demise.

Agaricus xanthodermus

Genev.

Yellow Stainer

Cap: 3–15 cm across, rounded and with strongly inrolled, and often lobed margin at first, becoming broadly convex to plane. Surface dry, smooth, sometimes becoming distinctly cracked. Pure white to dull whitish, when exposed to sun or in dry conditions becoming light, dull grayish brown to brown, often shiny and metallic in such conditions. Quickly and strongly staining yellow when rubbed or scraped. **Gills:** Free, close to crowded. Pale whitish gray when young, becoming light pink and then brown, dark brown to nearly black. **Stipe:** 6–17 cm long, 1–4 cm thick, usually equal and cylindrical but with a distinctly crooked, enlarged base, often lobed and with rhizomorphs. Surface dry, smooth or wavy. Whitish to grayish white or with light brownish tones; base readily staining yellow when damaged. **Partial veil:** A thick, white, membranous skirt, often with patches around margin, thus appearing to have a 'cogwheel' pattern. **Flesh:** Thin to moderately thick, whitish with dingy areas, staining yellow, especially in stipe base. **Odor:** Strongly chemical or phenolic (some people cannot detect any scent), variously described as Sharpie marker, antiseptic, cleaning products, or Band-aids. **Taste:** Mild. **KOH:** Bright yellow. **Spore deposit:** Dark chocolate brown. **Microscopy:** Spores (4) 5–5.5 (6.5) × 3–4.5 µm, broadly ellipsoid to ellipsoid, smooth.

Ecology: Scattered, clustered, sometime solitary, more often in large troops, generally in urban habitats, lawns, wood chips, and gardens but also beginning to appear in a number of native habitats including riparian corridors and trailsides in forests. Very common, fruiting throughout the year except for the cold midwinter months.

Edibility: Poisonous! Typical of the phenol-smelling *Agaricus*, this species causes brief, but unpleasant, gastrointestinal distress for most people who consume it.

Comments: The intense yellowing reactions, large membranous veil, smooth white to grayish cap, crooked stipe base, and phenol odor distinguish it. *A. californicus* is best distinguished by its weak yellow staining, faint phenol odor, and longer, wider spores. Other phenolic-smelling *Agaricus* typically have more pigmented or finely scaly caps, and do not stain yellow as strongly. Some other strongly staining *Agaricus* smell sweet and almondy and their yellow bruising tends to be more amber-orange and persistent, rather than fading to dingy gray-brown. Data on the presence and abundance of this species are sparse prior to the 1960s, but the human-commensal habitat preferences of the species strongly suggest that it is introduced. This represents one of few invasive macrofungi in the western US. Although the date of its introduction to the Americas is unknown, *A. xanthodermus* was apparently common in central California by the 1960s, and first appeared in the Willamette Valley in the early 1990's; it is now abundant throughout this range. Having recently appeared in Seattle, it seems likely to continue its expansion.

Agaricus braendlei subsp. *annettae*

(Kerrigan) Kerrigan

Cap: 3–7 cm across, broadly ovoid or rounded with an in-rolled margin at first, becoming broadly convex to plane. Margin with a narrow band of thin cottony tissue hanging past edge of gills (disappearing in age). Surface dry to moist, smooth to weakly fibrillose-scaly. Bright silvery whitish to duller white to tan or brownish with scattered dark scales. **Gills:** Free, close to crowded. Bright pink when young (even when cap is still closed), becoming pinkish brown and purple-brown to deep chocolate brown when mature. **Stipe:** 2.5–8.5 cm long, 1–3 cm thick, usually short and slender, straight to slightly curved, almost always with a pointed or tapered stipe base. Surface smooth or slightly scaly, whitish. **Partial veil:** Thin, cottony, forming a small, weak ring, soon disappearing or becoming an inconspicuous spiky or ragged zone. **Flesh:** Fragile, soft, whitish. **Odor:** Like grocery store button mushrooms, but fainter. **Taste:** Mild or earthy like button mushrooms. **KOH:** No reaction. **Spore deposit:** Dark chocolate brown. **Microscopy:** Spores 6.5–8 × 4.5–5 µm, ellipsoid, smooth.

Ecology: Solitary or scattered in troops, arcs, or rings in lawns and short-grass meadows, or in areas with bare soil and sparse weedy herbaceous plants such as *Erodium*. Fairly common, fruiting in fall and spring.

Edibility: Edible and good, although firm fruitbodies are not easy to find in enough quantity to make a meal (the flesh is watery and cooks down dramatically). Care should be taken not to eat specimens from potentially polluted urban lawns, and to avoid confusion with the toxic Yellow Stainer (*A. xanthodermus*).

Comments: This species (and a few similar-looking relatives in our area) have long been referred to collectively as *A. campestris*—a European species not known to occur in Cascadia. They are very difficult to distinguish from one another due to overlap in morphology and a low degree of genetic divergence. As a group, they are distinguished from other *Agaricus* by their small size, pale colors, often pointed stipe bases, ragged or inconspicuous partial veils, and lack of strong odors or staining reactions. *A. porphyrocephalus* has a scaly, brown-toned cap. Although rare in Cascadia, a paler, slightly genetically-distinct variety, *A. p.* var. *pallidus*, is at least somewhat more common and looks very similar to *A. braendlei* subsp. *annettae*. *A. andrewii* looks quite similar and might eventually be found in Cascadia; currently it's known from the Eastern United States. *A. bisporus* is stockier, has a club-shaped stipe base, a more developed partial veil, and is not usually found in grassland habitats. *A. bernardii* is much larger and firmer, stains reddish, and has a briny scent. *A. incultorum* is similar to the browner variants of species in this group, but has a club-shaped stipe base and a scruffy-felty cap surface.

5 - Dark Spored Mushrooms

The mushrooms in this section share dark purplish gray to black spore prints but otherwise exhibit quite a range in size, stature, and coloration; they are not all closely related. The inky caps, whose gills dissolve into an inky, spore-laden liquid in a process called "deliquescence," comprise two distantly related groups. The first includes the few species of "true" *Coprinus* (based around the type species, (*C. comatus*) are in the Agaricaceae (related to *Agaricus*) while all other inky caps have been assigned to other genera (*Coprinellus*, *Coprinopsis*, *Narcissea*, *Parasola*, and *Psathyrella*), all belonging to Psathyrellaceae. However, all of the generic concepts in the latter groups are still being develped, and *Psathyrella* continues to be split into smaller genera. Expect reshuffling to continue for some time to come.

Although identification of species in *Coprinus*, *Stropharia*, and *Hypholoma* is relatively straightforward, other groups of dark-spored mushrooms are difficult or impossible to identify given current knowledge. A microscope is often required to examine details of spores, cystidia, cap cuticle, and veil. *Psathyrella* are notoriously hard to pin down, as are many small inky caps.

Recognition of general groups is facilitated by learning a few of the most distinctive individual species in each genus. An early fall or winter visit to the dung heap of a horse corral or landscaping area with wood chips will often turn up a "sampler" of many genera of dark-spored mushrooms.

Of the deliquescent species, *Coprinus* contains only one common species in our area—the large, distinctive Shaggy Mane (*C. comatus*). With exceptions, the other inky caps have small, fragile, and often flimsy fruitbodies with caps ranging from scruffy-shaggy to smooth or finely flaky (*Coprinellus*, *Coprinopsis*) or pleated-sulcate (*Parasola*). Keep in mind that these are rule-of-thumb generalizations. *Lacrymaria* is a bit of an oddball genus, with dry, spiky-shaggy to scaly-looking, stout fruitbodies with strongly mottled gills that "weep" clear droplets of liquid when young. *Psathyrella* tend to have thin, white, tall, brittle stipes (breaking when bent) and small, often conical caps. They tend to grow in troops on wood chips or in disturbed areas and are often abundant under alder, cottonwood, and maple in forests. *Panaeolus* are similar to species in the latter genus but tend to be slightly less brittle, have mottled gills, and are more likely to be found on dung or in grass.

The following genera tend to have more substantial fruitbodies and frequently have viscid or slightly gelatinized cap surfaces. *Protostropharia* have pale fruitbodies, grow on dung, and are extremely slimy on both cap and stipe. *Deconica*. have small fruitbodies that could be confused with many of the other black-spored genera. *Psilocybe* tend to stain blue when bruised, handled, or in age. *Hypholoma* sp. mostly grow in clusters out of wood or on woody debris. *Leratiomyces* are commonly found on wood chips. *Stropharia* have the largest, fleshiest fruitbodies of any of the genera listed here and usually have a distinct ring and viscid caps.

Coprinus comatus

(O. F. Müll.) Pers.

Shaggy Mane

Cap: 2–6 cm wide, ovoid or cylindrical when young, to 12 cm across and becoming flatter and curling upward as cap and gills dissolve, sometimes with a low umbo; edge often extensively curled up or ragged or hanging down in black "drips." Surface dry, smooth when young and in 'egg' stage, soon fibrillose, with shaggy fibrils clumped into scales. Ocher-tan to light brown when young, soon cracking and becoming extensively white with a brown patch at center with many scattered flat, tan brownish patches outward. Turning pinkish red upward from the margin as gills dissolve. **Gills:** Free or narrowly attached, extremely crowded (often not individually distinguishable). White at first, becoming pink, then grayish, and finally black (often exuding red droplets). In age dissolving into a slightly viscous black liquid. Color change progressing upward from edge of cap. **Stipe** 4–25+ cm long, 1.5–3 cm thick, tall, cylindrical with slightly enlarged base. Surface dry, finely scaly to smooth. White. **Partial veil:** A whitish band with brown edge around cap margin when young, soon detached and sliding as a free ring, often settling at base of stipe. **Flesh:** Stringy, soft, white. Stipe hollow with a thin string running down center. **Odor:** Indistinct or unpleasant in age. **Taste:** Mild. **Spore deposit:** Black, inky. **Microscopy:** Spores 10.5–14 (17.5) × 5.5–7.5 (10) µm, long ellipsoid to nearly cylindrical, smooth, dark grayish black in KOH, with an apical germ pore.

Ecology: Solitary, clumped, or in dense troops in wood chips, lawns, and disturbed soil along trails and roads. Common and widespread across Cascadia. It can fruit any time the substrate is wet enough but is by far most abundant in fall.

Edibility: Edible and popular. However, this species often grows in polluted areas (roadsides and lawns). Such specimens should be avoided since they may bioaccumulate toxic chemicals and heavy metals. Since the fruitbodies dissolve so quickly, young specimens should be picked and cooked the same day. Immersing them in salt water and refrigerating them will slow the self-digestion process. Some chefs have used the "ink" of older specimens like squid ink to color rice or pasta.

Comments: The cylindrical, whitish caps with brownish scales, liquefying gills, and sliding ring on the stipe distinguish this species. It is found in a wide variety of habitats and produces relatively small to large fruitbodies. The Alcohol Inky Caps (*Coprinopsis atramentaria, C. striata,* and *C. romagnesiana)* have smooth to slightly scaly, gray-toned caps. The Shaggy Parasols (*Chlorophyllum* spp.) have rounded caps when young, which expand to convex or plane with age, with white to creamy gills and white spores. The desert-dwelling *Podaxis pistillaris* looks similar to *C. comatus* but its cap never expands and its gills dry out and split irregularly rather than turning to ink.

Coprinopsis atramentaria

(Bull.) Redhead, Vilgalys & Moncalvo

Alcohol Inky Cap

Cap: 2–8 cm across, ovoid to cylindrical at first, soon bell-shaped, convex to plane, often deliquescing while expanding. Surface moist to dry, mostly smooth, with a few scattered brownish scales over disc, and whitish hairs on margin when young, occasionally cracking when dry. Light to dark gray, sometimes with grayish brown disc, darkening in age. **Gills:** Free, very crowded. Whitish at first, soon gray, then black and deliquescing. **Stipe:** 3–15 cm long, 0.3–1.5 cm thick, equal or enlarged toward, or tapering at base. Surface dry, smooth, occasionally with veil remnants on lower stipe; white with brownish fibrillose scales at base. **Flesh:** Relatively thin and fragile; stipe hollow; white when young. **Odor:** Indistinct. **Taste:** Indistinct. **Spore deposit:** Black. **Microscopy:** Spores 6.5–10.5 × 4–6.5 µm, ellipsoid to ovoid, with a broad germ pore, smooth, dark reddish black in KOH. Pleurocystidia and cheilocystidia narrowly cylindrical or constricted with a rounded head.

Ecology: Clustered or in troops, occasionally in small tufts, rarely solitary, usually on buried woody debris in grassy areas, on dead tree roots, around stumps, or dead trees. Widespread and common, especially in urban areas, rarer in natural habitats. Fruiting throughout the year but more likely found in fall or in summer around irrigated areas.

Edibility: Edible with caution! Consumption of alcohol within a few days of eating this mushroom can result in sweating, light-headedness, severe headaches, swelling and tingling in the extremities, as well as nausea and vomiting. In more severe cases, extremely low blood pressure and even fainting can occur. Species in this group contain "coprine," a disulfram-like molecule (the active compound in Antabuse), which inhibits the body's ability to metabolize alcohol. These inky caps have a rich, mushroomy flavor well suited for use in sauces but avoid those found in urban areas or near roadsides.

Comments: The grayish, ovoid to cylindrical cap that expands and deliquesces as it ages, clustered growth habit, and the medium to large size are helpful identification features. This is a species complex with several similar species in Cascadia. *Coprinus atramentaria* appears to be the most common species in urban areas. *Coprinus striata* (=*C. acuminata* sensu PNW) is often found in forested settings, typically smaller, has a more conical, smooth or striate cap with a small, rounded umbo, and smaller spores (5.5–8.5 × 4–5 µm). *Coprinus romagnesiana* has scalier browner caps and often shows bands of small dark scales on the lower stipe. Both species should be treated like *Coprinus atramentaria* in terms of edibility. *Coprinus comatus* has a tall, cylindrical, whitish cap that is usually scaly or shaggy.

Synonym: *Coprinus atramentarius* (Bull.) Fr.

Coprinopsis lagopus group
Rabbit's foot Inky Caps

Cap: 1–5 cm across, ovoid at first, soon expanding to plane. Margin often becoming upturned or incurved, often ragged, splitting, or dissolving when wet. Surface finely pleated or grooved, covered with short whitish to pale gray spiky to matted hairs of veil material, breaking up into irregular patches as cap expands, occasionally scuffing off completely. Grayish at first, sometimes gray with a buff disc, darker gray overall in age; somewhat translucent as all stages. **Gills:** Narrowly attached to free, close, narrow, often deliquescing at least partly in age. Pallid at first, soon gray, then black. **Stipe** 3–9 (12) cm long, 0.2–0.6 cm thick, equal or enlarged downward, occasionally with a bulbous base. Surface dry, floccose at first to smooth with age. White. **Flesh:** Thin and fragile; stipe hollow. **Odor:** Indistinct. **Taste:** Indistinct. **Spore deposit:** Black. **Microscopy:** Spores 10–14 × 6–8.5 µm, ellipsoid to ovoid, with a broad germ pore, smooth, dark reddish black in KOH. Pleurocystidia and cheilocystidia oblong. Veil consists of elongated, sausage-like cells.

Ecology: Often in troops, tufts, or clusters, rarely solitary, usually in wood chips or on woody debris, also on straw or compost piles. Very common in urban areas across Cascadia, rarer in natural settings. Most often fruiting in fall but can be found in small numbers throughout the year.

Edibility: Unknown, likely non-toxic.

Comments: The fuzzy, grayish, "rabbit-foot" young caps, deliquescing gills, and growth in wood chips are good identifying features. Recent genetic work has found a number of lineages in this complex but which occur in Cascadia is unclear. *C. cinerea* is a similar common species in dung and compost piles. *C. jonesii* is typically found in forested areas, especially in the first few years after forest fires; it is generally stockier and slightly browner. *C. friesii* is a very small grass-inhabiting species with scattered, flocculose-granular scales on the young caps.

Synonym: *Coprinus lagopus* (Fr.) Fr.

Coprinopsis nivea
(Pers.) Redhead, Vilgalys & Moncalvo

Snowy Inky Cap

Cap: 1–4 cm across, ovoid at first, becoming bell-shaped, occasionally plane. Margin often slowly dissolving after it opens, occasionally up-curled in age. Surface dry, completely covered in flocculose-granular powder, persisting in age or sometimes washing off with rain. Bright snowy white to pale grayish white in age. **Gills:** Narrowly attached, often becoming free, close, narrow, edges often fringed when young, soon deliquescing. Whitish at first, soon gray, then black. **Stipe:** 3–7 cm long, 0.2–0.4 cm thick, equal or enlarged slightly toward base, fragile. Surface dry, finely powdery to smooth. White. **Flesh:** Very thin and fragile, stipe hollow. **Odor:** Indistinct. **Taste:** Indistinct. **Spore deposit:** Black. **Microscopy:** Spores 12–19 × 11–15.5 µm, lens-shaped to lemon-shaped in frontal view, ellipsoid and narrower in side view, smooth, thick-walled with a germ pore, dark reddish black in KOH. Cheilocystidia sac-like, ellipsoid to nearly cylindrical. Pleurocystidia similar in shape, but larger. Veil made up of round cells.

Ecology: Solitary, scattered, or in small clusters on dung, especially of horse and cow. Widespread and common. Fruiting in spring and fall, or occasionally in summer when substrate is wet. Fruitbodies rarely last more than a day or two.

Edibility: Harmless.

Comments: The white colors, powdery cap, dissolving gills, and growth on dung are distinctive. Of the myriad inky caps on dung, none have the striking white fruitbodies characteristic of this species. *C. cinerea* has a cylindrical to bell-shaped gray cap that is covered with pale, shaggy hairs when young (similar to *C. lagopus*). Most other dung-dwelling species are smaller and listed in the comments under the *C. stercorea* group. The dung-dwelling *Panaeolus antillarum* has a smooth, whitish cap and non-deliquescing gills.

Coprinopsis uliginicola

(McKnight & A.H. Sm.) Örstadius & E. Larss.

Cap: 3–9 cm across, hemispherical when young, becoming convex to nearly plane, sometimes with a broad, round umbo; margin occasionally slightly uplifted and wavy in age. Surface dry, appressed-fibrillose to matted overall, with silky veil tissue present around margin. Light gray to grayish tan, occasionally darker gray. **Gills:** Narrowly attached, close to crowded, narrow. Light gray when young, becoming darker dingy gray as spores mature. **Stipe:** 5–9 cm long, 0.8–2 cm thick, equal or tapered to middle, often swollen at base. Surface dry, finely scurfy or silky to nearly smooth, usually with distinct vertical lines at apex. White to grayish, darkening from spores in age. **Flesh:** Very fragile, whitish to pale grayish brown in cap. Stipe solid at first, soon hollow, white. **Odor:** Indistinct. **Taste:** Indistinct. **Spore deposit:** Dark blackish brown. **Microscopy:** Spores 10–12 (15) × 5–6 µm, ellipsoid to oblong, smooth; germ pore inconspicuous; light brown in KOH. Cheilocystidia bladder-shaped to cylindrical, smaller cells club-shaped or with swollen middles. Pleurocystidia absent.

Ecology: Solitary or scattered from duff and well-decayed cottonwood and aspen. Common in appropriate habitats. Fruiting in spring (especially in eastern Cascade Range) and fall across Cascadia.

Edibility: Unknown, likely non-toxic.

Comments: The gray fibrillose cap, grayish gills, white stipe, and fragile flesh are helpful identification features. Additionally, the growth on aspen and cottonwood duff and debris, and dark spores clinch the ID. This odd species doesn't fit neatly into rule-of-thumb generic concepts. It is larger and more robust than most other *Psathyrella* (more closely resembling *Lacrymaria*) and doesn't deliquesce like most *Coprinopsis*. There is a very similar, unnamed species which grows on oak debris in California; it may also occur in southern Cascadia. Young specimens with pale gills could even be confused with gray *Tricholoma* species but the texture is more fragile and the gills soon turn grayish as the spores mature. Another dark-spored species commonly encountered in cottonwood forests is *Homophron spadiceum*, which has a smooth, distinctly hygrophanous cap and typically grows in clusters on wood.

Synonym: *Psathyrella uliginicola* McKnight & A.H. Sm.

Coprinellus flocculosus

(DC.) Vilgalys, Hopple & Jacq. Johnson

Cap: 1.5–5 (7) cm across, cylindrical to oval at first, becoming bell-shaped and ragged with a deliquescing margin, expanding to plane while dissolving. Surface grooved, irregularly covered with flocculose-felty patches and specks of veil tissue, becoming smooth in age. Buff to honey-brown with whitish to pale buff veil patches at first, soon beige to gray, dissolving to blackish. **Gills:** Narrowly attached to free, crowded, broad. Pallid to gray with whitish to pale gray fringed edges, soon deliquescing to black. **Stipe:** 3–9 cm long, 0.2–0.6 cm thick, equal or slightly enlarging to a bulbous base, with a collar-like volva when young. Surface dry, finely pruinose to smooth; whitish. **Flesh:** Thin and fragile; stipe hollow. **Odor:** Indistinct. **Taste:** Indistinct. **Spore deposit:** Black. **Microscopy:** Spores 9.5–16.5 × 6.5–10 µm, ellipsoid to ovoid, smooth, thick-walled with an eccentric germ pore, blackish in KOH. Pleurocystidia ellipsoid, ovoid, with a swollen middle or cylindrical. Cheilocystidia smaller, more globose to ovoid.

Ecology: Scattered in troops, small clusters, or occasionally solitary on wood chips. Common in urban habitats throughout Cascadia. Fruiting throughout the year in irrigated areas, more widespread in spring and fall.

Edibility: Non-toxic.

Comments: This wood chip-loving species is recognized by its honey-colored to brownish gray, cylindrical to bell-shaped, deliquescing caps with small, flocculent-felty scales. Members of the *C. micaceus* group are often clustered and have pale granular specks (not felty scales) on the young caps. Members of the *C. domesticus* group grow out of an orange, coarsely orange-brown spiky-hairy mat (called an "ozonium") and have persistent, flocculose-granular specks on the cap. *Coprinopsis atramentaria* is a bit larger and stouter and has a gray cap that is smooth or with scattered brownish scales.

Coprinellus micaceus

(Bull.) Vilgalys, Hopple & Jacq. Johnson

Mica Cap

Cap: 1.5–4 (5) cm across, ovoid at first, becoming bell-shaped to convex; margin sulcate, becoming ragged, often splitting, and deliquescing when wet. Surface with scattered glistening whitish specks, which often disappear in age. Yellow-brown, honey-brown with a paler margin, occasionally more ocher-brown to brown on disc, becoming gray, at least at margin in age. **Gills:** Narrowly attached to free, crowded, narrow, partially to completely deliquescing. Creamy buff at first, soon grayish brown to black. **Stipe:** 3–7 (11) cm long, 0.2–0.5 cm thick, equal or slightly enlarged toward base. Surface dry, pruinose to smooth. White to pale buff. **Flesh:** Thin and fragile; stipe hollow. **Odor:** Indistinct. **Taste:** Mild. **Spore deposit:** Black. **Microscopy:** Spores 7–10 × 5–7 µm, ellipsoid or ovoid, with a conical base and truncate apex (mitre-shaped), smooth, thick-walled with a broad germ pore, dark reddish black in KOH. Pleurocystidia oblong to ovoid. Cheilocystidia similar or more rounded. Stipe covered in caulocystidia with swollen base and tapered neck.

Ecology: Generally clustered, scattered in small to large troops, in tufts, or occasionally solitary around stumps, dead trees, rotted logs, or buried woody debris. Very common, occurring throughout Cascadia. Fruiting throughout the year, with peaks in spring and early fall.

Edibility: Edible and flavorful, but practically fleshless. Probably best used for gravy and sauces.

Comments: The clustered growth, glistening whitish specks on honey-yellow caps, and deliquescing gills are helpful identifying features. However, similar species in this group require microscopic examination to distinguish them. Keep in mind that no serious taxonomic work has been done on Cascadia's inky caps and much is unknown about their diversity. The following species lack caulocystidia. *C. saccharinus* has spores measuring 6–9.5 × 5–6.5 µm, with somewhat conical bases and rounded or truncate apices. *C. truncorum* has slightly smaller and more rounded-ellipsoid spores (6.5–9 × 4.5–6.5 µm); it has been found in California but has not been reported from Cascadia. Members of the *C. domesticus* group have granular specks on the cap and emerge from an orangish, hairy mat.

Parasola species

Umbrella Inky Caps

Cap: 0.5–2.5 (3.5) cm across, ovoid at first, soon broadly convex to plane, often becoming depressed at disc in age. Surface dry, compressed into pleats and narrow grooves at first, expanding with grooves extending to disc in age. Buff, light to dark brown, many with a grayish margin and becoming gray overall, others remaining brown. **Gills:** Free to attached, some with a distinct collar around stipe, crowded to close, narrow. Buff to light brown with a paler edge at first, soon darkening; brown to black. **Stipe:** 2–6 cm long, 0.1–0.2 cm thick, equal or enlarged slightly toward base. Surface dry, smooth. Whitish to pale buff. **Flesh:** Very thin and fragile; stipe hollow. **Odor:** Indistinct. **Taste:** Indistinct. **Spore deposit:** Black. **Microscopy:** Spores ~6–15 × 5–8 µm, in most species ellipsoid to oblong, in some heart-shaped to angular, with small germ pore, smooth, dark reddish black in KOH. Pleurocystidia and cheilocystidia present.

Ecology: Solitary, scattered, or in small tufts in grassy areas, on bare soil, wood chips. Many species are common and widespread, others are likely more localized. Fruiting throughout the year in irrigated areas. Most species occurring in natural settings are spring and fall fruiters.

Edibility: Harmless.

Comments: This group of dainty inky caps is recognized by their small size, pleated caps that become deeply grooved in age, and buff, gray or brown caps that lack veil remnants. These species tend not to deliquesce into liquid like other inky caps but instead crumble in age. There are a number of species in Cascadia, many of which remain nameless. *P. lactea* is typically in grass or on rich soil, has a buff to light gray cap, distinctly free gills, and heart-shaped to angular spores measuring 8–12 × 5.5–7 µm. *P. plicatilis* is very similar but is generally slightly smaller, appears to be restricted to grassy lawns, and has longer spores (10–14.5 µm). *P. auricoma* and *P.* aff. *neoplicatilis* have darker ocher-brown to orangish brown colors when young. *P. schroeteri* (with a pleated gray cap) and *P. miser* (a tiny, honey-brown to gray-capped species) occur on dung. Members of the *Coprinopsis stercorea* group occur on dung and have fibrils on the cap when young. *Coprinellus disseminatus* often grows in large troops around woody debris. It has a pallid cap with minute glistening hairs when young, darkening to grayish, and losing the hairs in age.

Coprinopsis stercorea group

Dung-loving Inky Caps

Cap: 0.3–1.2 cm across, ovoid at first, becoming broadly convex to plane, margin pleated, soon deliquescing, occasionally becoming up-curled in age. Covered in whitish to pale gray hairy-flocculose or granular veil scales, becoming gray in age. **Gills:** Narrowly attached, close to subdistant, narrow, soon deliquescing. Grayish at first, soon black. **Stipe:** 1–3.5 cm long, 0.05–0.1 cm thick, equal, very fragile. Surface dry, with scattered fine whitish specks or hairs. Translucent to watery white. **Flesh:** Extremely thin and fragile, stipe hollow. **Odor:** Pungent. **Taste:** Not sampled. **Spore deposit:** Black. **Microscopy:** Spores 5.5–7.5 × 3–4.5 µm, broadly cylindrical to ellipsoid, smooth, thick-walled with a broad germ pore, dark reddish black in KOH. Pleurocystidia oblong or with a slightly constricted neck. Cheilocystidia similar or more rounded. Veil made up of warty, round cells.

Ecology: Solitary, scattered, or in small troops on relatively fresh (not steaming) dung. Usually occurring on dung of grazing animals but also found on bear and dog scat. Widespread and common but the fleeting fruitbodies can come and go in less than a day. Fruiting throughout the year but more common in spring and fall.

Edibility: Unknown.

Comments: This tiny inky cap is one of a handful of very similar species that occur on dung; almost all require microscopic examination to identify. As a group, they are recognized by their fragile, flimsy fruitbodies, grayish colors, and deliquescing gills. *C. stercorea* is the most encountered species in this group. It generally has a cap less than 1 cm wide, is covered with hairy-flocculose or granular scales when young, and has small, ellipsoid spores. *Coprinellus ephemeroides* is a tiny species with a partial veil that leaves a flaring ring on the stipe. *C. heptemerus* has a minutely spiky cap when young. Still smaller (caps 0.2–0.7 cm across) are *C. pellucidus* and *C. heterosetulosus*, which have finely granular to smooth, brownish caps that soon turn grayish. *C. heterosetulosus* has thick-walled cystidia on the cap cuticle, though *C. pellucidus* lacks these. Common, slightly larger, dung-inhabiting species include the following. *Coprinopsis radiata* with an egg-shaped cap at first covered with whitish to pale gray hairs (similar to *C. lagopus*) and a rooting stipe and spores measuring 7.5–8.5 µm. *C. macrocephala* is nearly identical but with wider spores (8–10 µm). *C. cinerea* is similar but with a slightly bulbous stipe base; it often grows in compost piles or in straw mixed with manure (rarely on pure dung).

Psathyrella species

Brittlestems

Psathyrella are quite diverse in Cascadia and most are difficult (or downright impossible) to identify given current knowledge. A.H. Smith's 1972 monograph on *Psathyrella* in North America listed over 150 species from the Pacific Northwest. Recent sequencing work by Jonathan Frank on collections from southern Cascadia, combined with data compiled by Danny Miller, found close to 100 species. Although we highlight a few of the more distinctive species here, many others are macroscopically extremely similar and will require an ITS sequence and/or meticulous microscopic examination for identification. The following is a general Brittlestem description. A few distinctive species are treated separately.

Cap: 0.5–3 (5+) cm across, hemispheric, conical to convex. Surface typically smooth, moist to dry, strongly hygrophanous. Most are light to dark brown or gray-brown. **Gills**: Narrowly attached, close. Typically pallid when young, darkening as spores mature. **Stipe**: 2–10 cm long, 0.1–0.5 (0.8) cm thick, often equal. Usually pallid. **Flesh**: Very thin, fragile; stipe breaking easily. **Odor**: Indistinct to fruity or earthy. **Taste**: Indistinct. **Spore deposit**: Dark purple-brown to nearly black, others dark reddish brown to purplish black, a few species have reddish pink or light brown spores. **Microscopy**: Spores mostly ellipsoid to ovoid, some with distinct germ pore, smooth, moderately thick-walled. Cheilocystidia and pleurocystidia typically abundant.

Ecology: Solitary, scattered, or in large troops. On soil, duff, litter debris, and decaying wood in a wide variety of habitats across Cascadia. Most abundant in alder forests or riparian zones with cottonwood but also abundant in urban areas on wood chips. Primarily fruiting in spring and fall.

Edibility: Unknown, likely non-toxic.

Psathyrella pennata
(Fr.) A. Pearson & Dennis

Carbon Brittlestem

Cap: 1.5–4 (6) cm across, ovoid-conical first, becoming broadly convex to plane. Surface dry, smooth to wrinkled, covered with silky whitish fibrils when young, restricted to margin or disappearing in age. Dark to light brown, gray-brown to cinnamon-brown, hygrophanous, faded when dry. **Gills:** Narrowly attached, close, narrow. Light brown to grayish brown when young, becoming dark brown; edge often paler. **Stipe:** 3–7 cm long, 0.2–0.5 cm thick, equal, fragile. Surface dry, covered with silky white fibrils or tufts with a pruinose apex, nearly smooth in age. White to pale buff, darkening when handled. **Partial veil:** Leaving silky fibrils on cap and a faint annular zone on stipe when young. **Flesh:** Very thin and fragile; stipe hollow. **Odor:** Indistinct. **Taste:** Indistinct. **Spore deposit:** Dark brown to dark vinaceous brown. **Microscopy:** Spores 6.5–7.5 × 3–4.5 µm, ellipsoid to ovoid, with an indistinct germ pore, smooth, dark brown in KOH. Pleurocystidia and cheilocystidia abundant, narrowly to broadly spindle-shaped, tapered towards pointed tip.

Ecology: Scattered, in small clusters or tufts, occasionally solitary in burnt duff and soil, also occurring in campfire pits or on burn piles. Widespread, common in eastern Cascadia for two to five years following forest fires. Fruiting mostly in spring, occasionally in fall and winter.

Edibility: Non-toxic.

Comments: The brown colors, whitish fibrils adorning young fruitbodies, fragile texture, and growth on burnt soil are distinctive. *P. frustulenta* has a paler honey-brown to grayish cap covered with silky-flocculose hairs and tufts. It grows on unburnt duff, mostly at lower elevation in west-side forests. *P. hirta* has a darker brown cap and grows on dung.

Synonym: *Psathyrella carbonicola* A.H. Sm.

Psathyrella frustulenta
(Fr.) A.H. Sm.

Fluffy Brittlestem

Cap: 0.8–2 (3) cm across, ovoid-conical to bell-shaped at first, becoming broadly bell-shaped to convex. Honey-brown, brown to grayish brown, covered with white, silky-flocculose hairs and tufts, becoming more gray with age, retaining whitish hairs around margin, or occasionally smooth in age. **Gills**: Narrowly attached or seceding, close, narrow. Pale cinnamon-buff at first, soon gray, then blackish when spores mature. Occasionally with paler, sterile patches. **Stipe**: 3–7 cm long, 0.2–0.4 cm thick, equal, very fragile. Surface dry, covered with silky white fibrils or tufts, upper portion pruinose, to nearly smooth overall in age. White to pale buff in age. **Partial veil**: Silky fibrous, leaving remnants on cap margin. **Flesh**: Very thin and fragile; stipe hollow. **Odor**: Indistinct. Taste: Indistinct. **Spore deposit**: Deep reddish black. **Microscopy**: Spores 7–8.5 × 4.5–5.5 µm, ellipsoid to ovoid, germ pore indistinct, smooth, bright ochraceous to pale reddish cinnamon in KOH. Cheilocystidia and pleurocystidia abundant, narrowly to broadly spindle-shaped, to nearly cylindrical, with pointed or slightly rounded tip.

Ecology: Scattered in small troops or solitary in duff or moss under conifers. Fairly common in western Cascadia, especially in coastal and Coast Range forests, but often overlooked. Fruiting in fall.

Edibility: Non-toxic.

Comments: The honey-brown to grayish cap covered with striking silky-flocculose hairs and tufts, small size, and fragile fruitbody are helpful clues to the identity of this *Psathyrella*. A nearly identical species, *P. hirta*, has a darker brown cap and grows on dung. *P. subcaespitosa* is also similar but often grows in clusters and has whitish to brown gills. *P. fibrillosa* has more appressed hairs and slightly narrower spores (7–9 × 4–5 µm). *P. cortinarioides* has tan fibrils on the cap when young and larger spores (7–10 × 5–6 µm). *P. suavissima* is slightly larger, has a tan to light brown cap with silky white fluff, and a sweet, fruity odor; it grows in garden soil and on wood chips. *P. pennata* has a gray to dark brown cap, with scattered, matted white fibrils when young, and grows on burnt soil.

Psathyrella microrhiza group

Rooting Brittlestems

Cap: 1.5–4 cm across, hemispherical to bell-shaped at first, becoming broadly convex to nearly plane. Surface smooth, often with veil fibrils scattered over the surface when young, restricted to margin or disappearing in age. Hygrophanous, translucent-striate when wet, dull, opaque when dry. Dull deep brown to dingy ocher-brown when wet, fading to grayish brown to beige when dry. Occasionally more pinkish brown. **Gills:** Narrowly attached with a broad notch and decurrent tooth, close, broad. Beige to light brown at first, soon brown to dark brown as spores mature. **Stipe:** 5–15 (20) cm long, 0.2–4 cm thick, tall, slender, equal; base often narrowed and elongated into tapered "root" covered in spiky white fibrils. Surface dry, silky-fibrillose to smooth. White to tan. **Flesh:** Very thin, fragile; stipe hollow. **Odor:** Indistinct. **Taste:** Indistinct. **Spore deposit:** Dark purplish brown to nearly black. **Microscopy:** Spores 10–13.5 × 5.5–7 µm, elongate-ellipsoid, with a germ pore, smooth. Cheilocystidia abundant and crowded, spindle-shaped or swollen below with an elongate neck and an obtuse to pointed neck. Pleurocystidia scattered, similar.

Ecology: Scattered or in large troops on wood chips and in other disturbed areas, occasionally in forests, especially under alder. Abundant in suitable habitats across Cascadia. Fruiting in fall and early winter.

Edibility: Non-toxic.

Comments: This species complex includes many similar small-capped *Psathyrella* with thin, tall, straight stipes which grow in wood chips or on woody debris. Members of the *P. microrhiza* group have brown to ocher-brown caps when fresh, pallid gills which become dark brown as the spores mature, and long, often rooting stipes. These fruit in large troops on wood chips. A very similar species, *P. corrugis*, tends to be slightly more gracile in stature, lacks veil fibrils on the cap, and has more scattered cheilocystidia. *P. longicauda* is another common woodchip species with a consistently dark brown to dark gray-brown cap when wet. *P. prona* has a pinkish to brown cap, pallid stipe that darkens in age, and grows from rich soil. *P. bipellis* is larger (caps 0.5–3.5 cm across) and stockier species with a wine-red cap which fades to pinkish and then buff, and a fruity odor.

Psathyrella longipes

(Peck) A.H. Sm

Fringecap Brittlestem

Cap: 1.7–5 cm across, rounded-conical to nearly cylindrical when young, becoming bell-shaped. Surface moist to dry, smooth; margin with hanging "teeth" of thin white veil tissue soon stained blackish by spores. Strongly hygrophanous; dull grayish tan or brown when wet, fading outward to beige or cream as it loses moisture. **Gills:** Obscurely attached or seceding, close to crowded, broad. Off-white to pallid gray, becoming dark brownish gray to nearly black. **Stipe:** 5–15 cm long, 0.2–0.7 cm thick, cylindrical with a slightly enlarged base. Surface smooth, sometimes with chevrons of tissue around lower part, sometimes with extensive mycelium at base. White to creamy, often shiny. **Partial veil:** A thin sheet of white tissue leaving hanging "teeth" on cap margin. **Flesh:** Thin, fragile, beige in cap; stipe hollow, white. **Odor:** Indistinct. **Taste:** Indistinct. **Spore deposit:** Dark purple-brown. **Microscopy:** Spores 10–13 × 7–10 µm, ellipsoid to ovoid, germ pore distinct, smooth, brown in KOH. Cheilocystidia spindle-shaped to narrowly club-shaped. Pleurocystidia absent.

Ecology: Solitary or in small groups in humus or well-decayed leaf litter, especially of bigleaf maple. Occasional across western Cascadia. Fruiting in spring and fall.

Edibility: Edible.

Comments: The slender stature, tall grayish brown cap with hanging patches of veil tissue, comparatively large size, and fragile white stipe help identify this species. *Coprinopsis marcescibilis* (=*Psathyrella elwhaensis*) is very similar but may tend to be slightly smaller, is less restricted in habitat preferences, and has longer, narrower spores (12–15 × 6–8.5 µm). A number of other *Psathyrella* are vaguely similar but usually lack the hanging veil tissue on the cap margin, are smaller, or more richly colored.

Taxonomic note: This species belongs in the genus *Coprinopsis*.

Psathyrella longistriata

(Murrill) A.H. Sm.

Ringed Brittlestem

Cap: 2.5–8 (10) cm across, conical to bell-shaped at first, becoming broadly convex to nearly plane, often with a low, broad umbo, and becoming wavy in age. Surface moist to dry, with scattered, fine whitish hairs when young, otherwise smooth. Hygrophanous, translucent-striate when wet, fading when dry. Color variable, often reddish brown, brown to dark grayish brown, but can be pale creamy beige to yellow-brown. **Gills:** Narrowly attached to free, close, narrow. Grayish buff, becoming grayish brown to dark purplish brown as the spores mature. **Stipe:** 4–12 cm long, 0.3–1 cm thick, often tall and rather straight, equal, with a swollen base. Surface dry, covered with fine, appressed, floccose scales when young, nearly smooth in age. White to pale gray. **Partial veil:** Membranous, often forming a conspicuous skirt-like ring on stipe that collapses in age, occasionally breaking up and leaving remnants on cap margin. **Flesh:** Thin, very fragile, buff to light brown in cap. Stipe fibrous but fragile, hollow, whitish. **Odor:** Mushroomy. **Taste:** Indistinct. **Spore deposit:** Dark purplish brown to purplish black. **Microscopy:** Spores 6.5–9 × 4–5 µm, ellipsoid to ovoid, with a small, indistinct germ pore, smooth, moderately thick-walled, dark brown in KOH.

Ecology: Solitary or scattered, rarely in large numbers, on well-decayed wood or duff, especially under alder and bigleaf maple. More rarely on conifer duff or debris. Common across Cascadia, especially in lower elevation west-side forests. Primary flush in fall but fruiting continues through winter into spring.

Edibility: Harmless.

Comments: The hygrophanous brownish to beige cap, dark spores, and fragile flesh and stipe help identify this species to genus level. The membranous ring sets it apart from the horde of other brown *Psathyrella*. That being said, there are at least two species in Cascadia going under this name. One may be more likely to fruit in spring on conifer debris but morphological features to distinguish the two remain undefined.

Psathyrella piluliformis

(Bull.) P.D. Orton

Clustered Brittlestem

Cap: 1.5–6 cm across, ovoid or dome-shaped, becoming convex or occasionally plane in age. Surface often with a conspicuously contrasting band of silky white fibrils at margin, moist to dry, smooth or radially wrinkled. Reddish brown to brown, strongly hygrophanous, becoming brownish tan on drying out, often with a beige "halo" as it loses moisture. **Gills:** Narrowly attached, occasionally notched, close to crowded, narrow. Pallid beige to pale pinkish tan, becoming darker as spores mature. **Stipe:** 3–10 cm long, 0.2–0.7 cm thick, equal, usually curved. Surface smooth to silky, with chevrons or patches of veil tissue. White to beige or pale yellowish tan over lower part. **Partial veil:** A thin white cortina leaving remnants around cap margin, becoming covered with dark spores. **Flesh:** Thin, fragile, pale. Stipe becoming hollow. **Odor:** Indistinct. **Taste:** Indistinct. **Spore deposit:** Dark purplish brown. **Microscopy:** Spores (4) 5–7 × 3–4.5 µm, ellipsoid to ovoid, smooth; germ pore sometimes obscure; brown in KOH. Cheilocystidia and pleurocystidia with an elongated, narrow apex.

Ecology: In small to large clusters of densely packed fruitbodies, rarely in small tufts or solitary. Usually found near well-decayed logs, stumps, or decaying roots of hardwoods, especially bigleaf maple. Common in western Cascadia, occasional to rare on the east side of the Cascade Range. Fruiting at any time of year, often producing multiple flushes in the same spot; most likely to be found in fall and spring.

Edibility: Edible but not recommended. Similar to other toxic species, such as *Galerina marginata*.

Comments: The clustered growth habit, rich brown caps with contrasting white bands around the margins when young, and occurrence near rotting wood are characteristic. *P. carinthiaca* may tend to grow in sparser clusters and have slightly smaller spores; these distinguishing features are tentative.. *Homophron spadiceum* also occurs in clusters near rotting wood but is usually paler, with fleshier stipes, and lacks the pronounced veil. *Kuehneromyces lignicola* is more fleshy-fibrous and has cinnamon-brown spores. *Galerina marginata* has warmer ocher-brown colors and rusty brown spores.

Candolleomyces candolleanus

(Fr.) D. Wächt. & A. Melzer

Common Brittlestem

Cap: 1.5–7 cm across, hemispheric at first, becoming convex to nearly flat or concave, margin wavy and often splitting in age. Surface smooth, dry, often extensively covered in patches of whitish veil tissue when young; margin frequently with hanging "teeth" of veil remnants even in age. Caramel-brown to ocher-tan at first (but often with white patches of veil), strongly hygrophanous and soon fading to yellowish beige to nearly white. **Gills:** Narrowly to broadly attached, close to crowded, narrow. Whitish to pale grayish beige when young, becoming lilac-gray to dark purplish brown in age. Edge often paler. **Stipe:** 2–9 cm long, 0.3–0.7 cm thick, equal, fragile. Surface silky-fibrillose to scurfy when young, appressed-scurfy to nearly smooth in age. White. **Partial veil:** Thin, silky-membranous, typically remaining on cap margin, sometimes leaving an annular zone on stipe. **Flesh:** Thin, fragile, often splitting and breaking; stipe hollow. **Odor:** Indistinct. **Taste:** Indistinct. **Spore deposit:** Dark purplish brown. **Microscopy:** Spores 7–9 × 4–5 µm, ellipsoid, with germ pore, smooth, dark brown in KOH. Cheilocystidia sac-like, club-shaped or cylindric with rounded apex. Pleurocystidia absent.

Ecology: In clusters, clumps, and troops, rarely solitary. Widespread and common in urban areas, usually disturbed, nutrient-rich soils and roots of dying trees. Most abundant in spring and fall but can fruit throughout the year.

Edibility: Edible.

Comments: The fragile fruitbodies, hanging veil remnants from the cap margin, caramel to beige colors, purplish gray gills, and growth in clusters in urban areas help identify this species. *P. rogueiana* is described as a similar species from southern Oregon, differing by having pinkish buff veil remnants and spores 8–11 µm long (it may end up being synonymized with *C. candolleana*). *Homophron spadiceum* has a similar stature, grows in clusters and troops on woody debris (especially that of cottonwood and aspen), but has thicker, slightly less brittle flesh, lacks partial veil tissue, and has grayer cap colors.

Synonym: *Psathyrella candolleana* (Fr.) Maire

Lacrymaria velutina

(Pers.) Konrad & Maubl.

Weeping Widow

Cap: 3–9 cm across, ovoid or hemispheric at first, sometimes with a rounded umbo, becoming broadly convex to nearly plane. Margin incurved at first, occasionally uplifted and wavy in age. Surface dry to moist, silky-fibrillose to fine-scaly when young, to nearly smooth or slightly wrinkled in age; margin often with hanging patches of veil tissue. Color variable, ocher-beige to warm orangey brown, becoming tan to creamy beige when dry. **Gills:** Notched to broadly attached, close, usually adorned with clear to milky droplets when young. Beige to pinkish tan, dark grayish brown mottled with black, often with a reddish brown hue in age. Edges whitish fringed. **Stipe:** 4–10 cm long, 0.7–2 cm thick, equal or slightly enlarged at base. Surface dry, silky-fibrillose to scurfy near apex, in lower portions fairly smooth or fibrillose, sometimes with shaggy, pale chevrons. Whitish beige to grayish tan, ocher, or with yellowish or orangey brown tones in age. **Partial veil:** Forming a cottony, cobwebby, or spiky ring zone near stipe apex, soon collapsed and obscure. **Flesh:** Thin, fragile; stipe hollow hollow in age; beige to brownish or with reddish tones in age. **Odor:** Indistinct or musty. **Taste:** Indistinct. **Spore deposit** Dark purplish brown. **Microscopy:** Spores 10–12 × 6.5–8 µm, lemon-shaped with a protruding, chimney-like germ pore, roughened with dark warts, blackish brown in KOH. Pleurocystidia and cheilocystidia clustered, often capitate.

Ecology: In small clusters and troops on wood chips, nutrient-rich soil, or in disturbed areas with woody debris, such as logging roads. Occasional; more frequent in urban areas in western Cascadia. Fruiting primarily in fall or spring but also in summer in irrigated areas.

Edibility: Non-toxic.

Comments: The silky-fibrillose, ocher-brown cap, mottled gills often adorned with droplets, dark spores, and fragile texture help define this species. Distinguishing the various species of *Lacrymaria* is difficult. *L. echiniceps* has tufts of radially aligned fibrils on the cap and often pointed clumps of fibrils on the stipe. Microscopically, its spores average less than 9.5 µm long and are weakly ornamented to nearly smooth. This species is rare in Cascadia. We have seen it on hardwood debris at lower elevations. *L. rigidipes* is smaller and more slender, has spores that are greater than 9.5 µm in length with less pronounced warts, and without a protruding germ pore. It appears to be more common at mid-elevations of the Cascade Range in conifer forests. *Stropharia scabella* can look quite similar but is usually more squat with a firmer texture and without droplets on the gills. *Armillaria* species are structurally somewhat similar but have white spores.

Synonym: *Psathyrella velutina* (Pers.) Singer. Some authors treat *L. lacrymabunda* as a synonym but it appears to be distinct based on genetic data and is not yet known from Cascadia.

Panaeolus foenisecii

(Fr.) J. Schroet.

Lawnmower's Mushroom

Cap: 1–3 (4) cm across, hemispheric at first, becoming broadly convex. Surface moist to dry, smooth, or with fine cracks in dry weather. Dark brown to reddish brown when wet, hygrophanous, fading from disc often with a darker band at margin, becoming tan to grayish beige when dry. **Gills:** Broadly to narrowly attached, close, broad. Light brown at first, becoming mottled and darker to dark brown in age; edges paler. **Stipe:** 3–9 cm long, 0.2–0.4 cm thick, more or less equal. Surface dry, pruinose to smooth. Whitish buff to light brown, darkening from base up in age. **Flesh:** Moderately thin, fragile in cap; stipe fibrous. **Odor:** Indistinct to musty. **Taste:** Indistinct. **Spore deposit:** Dark brown. **Microscopy:** Spores 14–17 (22) × 7.5–11 µm, lemon-shaped, finely wrinkled-roughened, with a projecting germ pore, dark golden brown in KOH. Cheilocystidia spindle-shaped or more swollen, with an obtuse apex. Pleurocystidia typically absent.

Ecology: Solitary or scattered in troops in lawns. Very common and widespread, occurring throughout our range. Fruiting during the warmer months, especially in watered lawns.

Edibility: Non-toxic. Reports of it being mildly hallucinogenic are unsubstantiated and may have been the result of confusion with *P. cinctulus.*

Comments: The hygrophanous, brown to tan cap, often with a different-colored band around the margin, mottled gills, dark brown spores, and growth in manicured lawns are helpful identifying features. The name *P. castaneifolius* has been applied to collections with narrower spores (rarely more than 9.5 µm wide), but it may be conspecific. *P. cinctulus*, a mildly hallucinogenic species, is very similar. The differences between the two are subtle including a slightly "cooler" colors, a stockier stature, and jet-black spores of *P. cinctulus*. Although the latter species can grow in grassy areas, it is more often found in compost or garden beds. *P. fimicola* has a dark grayish brown to black, convex to nearly flat cap and a dark stipe with a pale pruinose coating. *P. acuminatus* has a conical to bell-shaped cap, is generally dark reddish brown to dark brown at first, and grows on dung or in grassy areas. Sand dune habitats with dune grass host two similar dark-spored species: *Psathyrella ammophila* and an apparently unnamed *Panaeolus.* These are difficult to distinguish in the field but the *Panaeolus* typically has grooves at the stipe apex, shows mottled gills as the spores mature, and tends to be slightly stockier.

Synonym: *Panaeolina foenisecii* (Pers.) Maire

Panaeolus fimicola group

Grass Mottlegills

Cap: 1–3 (4.5) cm across, broadly conical, bell-shaped to convex. Surface smooth, moist to dry. Grayish to gray-brown at first, darkening to gray or nearly black and often coated with black spores when wet, paler if dry. **Gills**: Narrowly attached, often notched, close to crowded, broad. Light gray-tan when very young, soon gray with darker spots and blotches, nearly black mottled with paler spots in age. Edges finely fringed, pale. **Stipe**: 3–8 cm long, 0.1–0.4 cm thick. Equal, tall, slender. Surface dry, distinctly pruinose, often with faint longitudinal lines at apex. Tan when young, covered with whitish powdery specks, soon grayish to dingy gray-brown. The pruinose bloom easily wiped off when handled. **Flesh**: thin, fragile; pallid when young, dark gray in age. **Odor**: Indistinct. **Taste**: Earthy. **Spore deposit**: Black. Microscopy: Spores 11–15 × 6.5–10 µm, smooth, lemon-shaped, with a germ pore.

Ecology: Solitary or scattered in grassy areas, typically around dung-enriched areas (but rarely directly on dung). Widespread but not commonly reported. Fruiting in fall and spring.

Edibility: Unknown. Closely related species are hallucinogenic.

Comments: The grayish to nearly black cap, mottled gills, tall, slender stature, black spores and growth in grassy areas help with identification. It belongs to a confusing group of species, for which appropriate names remain highly uncertain; *P. ater* may be a synonym. *P. acuminatus* tends to have more reddish-brown tones to the cap, and can grow in grass or from dung. Species in the *P. olivaceus* group have greenish or olive-gray tones to the cap. In Cascadia, one member of this group grows in grass, while another seems to prefer wood chips. *P. cinctulus* has a convex, brown to cinnamon-brown cap, often with a darker band around the margin.

Panaeolus papilionaceus

(Fr.) Quél.

Bell-capped Mottlegill

Cap: 1–4 cm across, conical at first, becoming bell-shaped, occasionally expanding to convex in age. Margin with a band of sterile tissue, breaking up into tooth-like remnants hanging from the margin. Surface moist to dry, smooth, or with fine cracks in dry weather. Color variable, generally gray to grayish brown, at times dark gray, olivaceous gray, pinkish gray, or even pale grayish beige when dry. **Gills:** Broadly to narrowly attached, close to subdistant, broad. Grayish at first, becoming mottled with dark grayish spots, to blackish in age; edge paler. **Stipe:** 4–12 cm long, 0.2–0.6 cm thick, more or less equal, often quite tall and slender. Surface pruinose at first, becoming smooth, occasionally with clear droplets when young. Buff to light reddish brown, with a whitish bloom, darkening when handled and in age. **Flesh:** Thin, fragile in cap, fibrous in stipe. **Odor:** Earthy. **Taste:** Mild to earthy. **Spore deposit:** Black. **Microscopy:** Spores 11–18 × (7.5) 8.5–12 µm, lemon-shaped to ellipsoid, with a germ pore, smooth, thick-walled, dark brown to blackish brown in KOH. Cheilocystidia club-shaped or cylindrical, often with an "S-curved" neck. Pleurocystidia absent.

Ecology: Scattered in troops or solitary on dung, most often that of cows and horses. Common, occurring throughout our range. Fruiting throughout wet periods in fall and spring.

Edibility: Non-toxic; similar species are mildly hallucinogenic.

Comments: This species can be readily identified by the dingy colors, conical to bell-shaped cap with small "teeth" around the margin, mottled gills, and black spores. A similar dung dweller is *P. semiovatus*, which is generally stockier and has a partial veil that forms a ring on the stipe. *P. acuminatus* also grows on dung or in grassy areas; it lacks the toothed cap margin. *P. cinctulus* occurs in similar habitat but generally has a dark band around the cap margin and a more convex cap. It is mildly hallucinogenic.

Deconica coprophila

(Bull.) P. Kumm.

Dung-loving Deconica

Cap: 0.5–4 cm across, hemispheric when young, soon bell-shaped to convex, flattening only in age. Surface smooth, thinly viscid at first, then greasy to moist. Margin translucent-striate, often decorated with fine white patches, a narrow band, or triangular "teeth" of veil tissue. Bright reddish orange when young, but soon dingier brick-red to dull, dark brown, then fading as it loses moisture, becoming beige to tan. **Gills:** Broadly attached or slightly decurrent, widely spaced. Beige when young, then ocher-brown, soon mottled dark brown with purplish gray tones as spores mature; edge contrastingly fringed whitish. **Stipe:** 1–5 cm long, 0.2–0.5 cm thick, usually curved. Surface covered in pale silvery veil tissue, breaking up into chevrons in age; apex slightly scurfy. White to beige or tan, darkening grayish to blackish upward from base. **Flesh:** Thin and fragile, watery, dark. **Odor:** Indistinct. **Taste:** Indistinct. **Spore deposit:** Dark purplish brown to nearly black. **Microscopy:** Spores 10–14 × 5.5–8.5 µm, slightly angular to nearly ellipsoid, with a broad germ pore, thick-walled, smooth, dark brown in KOH. Cheilocystidia with a slightly swollen base and elongated neck. Pleurocystidia absent.

Ecology: Solitary or in clumps and troops on dung of grazing animals—primarily cows and horses. Widespread in appropriate habitat throughout our area. More common in the eastern portion of Cascadia because of the higher cattle rangelands density there. Generally fruiting in spring but fruiting whenever substrate is moistened thoroughly.

Edibility: Non-toxic, not hallucinogenic.

Comments: The small size, reddish brown cap, slightly decurrent gills that are dark in age, and dung habitat are helpful clues in identification. The very similar *D. subcoprophila* has larger (15–20 µm long) spores that are ellipsoid or almond-shaped. Two other species fruiting on dung produce slightly larger fruitbodies with paler, less reddish caps and often more pronounced ring zones on the stipe. *D. merdicola* has longer (> 12 µm) ellipsoid spores. *D. merdaria* has more angled (hexagonal) shorter spores (< 13 mm long).

Synonym: *Psilocybe coprophila* (Bull.) P. Kumm.

Deconica montana

(Pers.) P.D. Orton

Moss-loving Deconica

Cap: 0.5–2 cm across, hemispheric to dome-shaped at first, becoming broadly convex. Surface smooth, thinly viscid when wet, typically drying out. Margin translucent-striate, sometimes with scattered silvery fibrils of veil tissue. Cap cuticle gelatinous, easily separated. Dark brown to dark reddish brown when young, lighter brownish tan in age; fading to beige or paler outward from center as it loses moisture. **Gills:** Broadly attached with a decurrent tooth or slightly decurrent, moderately spaced. Pale beige-tan becoming warm brown, then dark purplish brown as spores mature. Edge whitish fringed. **Stipe:** 1–5 cm long, 0.2–0.4 cm thick, equal. Surface dry, fibrous, vertically striate, surface covered in pale silvery streaks and chevrons of veil tissue. Dull beige or tan, soon browner overall and darkening from reddish brown or grayish to nearly black upward from base. **Flesh:** Thin, fibrous, dark brown. **Odor:** Indistinct. **Taste:** Indistinct. **Spore deposit:** Dark purplish brown. **Microscopy**: Spores 7–8.5 (10) × 4.5–5.5 µm, slightly angular to nearly ellipsoid, with a broad germ pore, thick-walled, smooth, dark brown in KOH. Cheilocystidia with a swollen base and elongated neck. Pleurocystidia absent.

Ecology: Solitary or in scattered troops in moss beds, sometimes weakly clustered (but not fused). Widespread and common in Cascadia but more abundant at higher elevations. It is found throughout the year but more commonly winter at lower elevations and spring in the mountains.

Edibility: Probably non-toxic, not hallucinogenic.

Comments: The rather dark reddish brown young caps, upward-darkening stipe covered in silvery veil remnants, and growth in moss beds are distinct. There are a large number of *Deconica* in Cascadia, many of which remain unnamed. Members of the *D. inquilina* group are reddish brown to ocher-brown when young, soon becoming butterscotch or caramel-colored, often in grass or on small woody debris. Members of the *D. phyllogena* group have an angular, four-sided ("rhomboid") look to the spores. *D. coprophila* and allies grow on dung.

Synonym: *Psilocybe montana* (Pers.) P. Kumm.

Protostropharia semiglobata

(Batsch) Redhead, Moncalvo & Vilgalys

Dung Roundhead

Cap: 2–6 cm across, hemispheric to broadly convex, occasionally becoming plane, at times with a low umbo. Surface smooth, viscid to tacky when wet, shiny when dry. Pale golden-yellow, yellowish beige to creamy yellow. In shaded areas becoming darker golden-brown, and bleached to pale creamy white in sunny areas, occasionally with purplish brown spores sticking to the cap. **Gills:** Broadly attached, often with a shallow notch at the stipe, close, broad. Pallid at first, soon gray to lilac-gray, often with a slight olive tone, to dark gray to purplish black in age. **Stipe:** 4–15 cm long, 0.2–0.6 cm thick, tall, slender, more or less equal with an enlarged or bulbous base. Surface smooth, viscid to tacky when wet, shiny when dry. Pale creamy yellow to yellowish gold, often darker toward base, discoloring golden when handled or in age. **Partial veil:** Thin, membranous, often covered in slime when young, forming a flaring ring at first, becoming purplish brown with spores, constricting to a faint ring on upper stipe in age. **Flesh:** Thin, somewhat firm, creamy yellow. Stipe tough, fibrous. **Odor:** Faint, musty-earthy. **Taste:** Mild to slightly farinaceous. **Spore deposit:** Purplish black. **Microscopy:** Spores 15–19 × 8–10 µm, ovoid, ellipsoid to cylindrical, with a broad central germ pore, smooth, thick-walled, dark brown in KOH. Cheilocystidia abundant, club-shaped, or with a swollen middle and long neck, with yellowish content in KOH. Pleurocystidia scattered, more club-shaped, often pointed at tips, with yellowish content in KOH.

Ecology: Solitary, scattered, or in small clusters or troops on dung, especially cow patties. Widespread but more common on the east side of the Cascade Range. Fruiting soon after the first rains in early fall and in spring.

Edibility: Non-toxic.

Comments: The hemispheric, viscid, creamy-yellow cap, long, viscid stipe with a faint ring, purplish black spores, and growth on dung are distinguishing features for this group of species. *P. dorsipora* is similar and cannot be distinguished macroscopically—if anything, it is slightly stockier overall. The only way to reliably distinguish them is to look at the spores. *P. dorsipora* has a small, off-center germ pore, whereas *P. semiglobata* has a wide germ pore centrally positioned at the top of the spore. *P. alcis* is also quite similar. It grows on moose dung, often has a strikingly tall stipe, and has smaller spores (12–17 × 7–8.5 µm). *Stropharia semigloboides* grows on duff and litter in forested habitats, has a rooting stipe, and small spores (8–10 × 4–5 µm). Other dung-dwelling species include *Deconica merdaria*, which has a dry to thinly viscid cap that is often yellow-brown to gray-brown, and a dry stipe. *Panaeolus* have dry caps with more gray tones and black spores. The dung-loving *Agrocybe semiorbicularis* has a golden-brown, thinly viscid to dry cap, and brown spores.

Synonym: *Stropharia semiglobata* (Batsch) Quél.

Psilocybe cyanescens

Wakef.

Wavy Caps

Cap: 1–6 (8) cm across, rounded-conical or bell-shaped with an incurved margin when young, expanding to irregular convex to plane, sometimes with a low, round umbo. Margin usually strongly and neatly wavy when mature, sometimes lobed even when young. Surface smooth, viscid to moist or dry; margin translucent-striate in age or when wet; margin sometimes with wispy white veil remnants when young. Orange-brown, caramel to chestnut-brown or reddish brown and yellower towards margin. Strongly hygrophanous, usually fading outward from center to light brown to beige, soon developing blue-green blotches. **Gills:** Broadly attached to notched or with a decurrent tooth. Pallid pinkish beige to yellowish brown, often with greenish stains, mottled to darker brown as spores mature. **Stipe:** 5–10 cm long, 0.4–0.7 cm thick, equal or enlarged towards base. Surface dry, silky-fibrillose to smooth. Usually bright whitish when young, becoming dingy beige-tan, mottled with bluish bruising at any age. **Partial veil:** A relatively thin, whitish cortina, leaving wispy hairs on stipe when young, soon disappearing. **Flesh:** Thin, somewhat rubbery in cap, tough, fibrous in stipe. Whitish to dingy tan. Bruising slowly blue-green to dark dingy blue. **Odor:** Mild, farinaceous or earthy. **Taste:** Bitter to farinaceous. **Spore deposit:** Dark purple-brown. **Microscopy:** Spores (9) 11–13 (15.5) × 6–8 µm, elongate-ellipsoid, smooth, moderately thick-walled, with a broad germ pore, brown in KOH. Cheilocystidia swollen with a narrowed, cylindrical tip, occasionally forked or branched. Pleurocystidia present, but sparse.

Ecology: In troops, small clusters or occasionally solitary, most commonly on wood chips or on woody debris, less commonly in disturbed, grassy areas. Also among dune grass (*Ammophila* spp.) and in dune edges along the coast. Common and often abundant in urban areas; widespread across western Cascadia. Fruiting in fall, typically peaking after first frost, with stragglers into early winter, and rarely in spring.

Edibility: Strongly hallucinogenic.

Comments: This attractive species is distinguished by the tough white stipe, blue-green staining, dark purplish brown spores, and a caramel-colored, wavy-edged cap. The deadly poisonous *Galerina marginata* sometimes grows right alongside this species but is told apart by its hemispheric to broadly convex cap, orangish stipe that darkens to gray from the base up in age, and rusty brown spores. A few other *Psilocybe* are quite similar. *P. allenii* is rarely wavy-capped, but otherwise is difficult to distinguish. *P. azurescens* is very similar but with a broadly conical or peaked-umbonate cap. *P. stuntzii* and *P. ovoideocystidiata* have more pronounced partial veils and browner (less orange) colors.

Psilocybe azurescens

Stamets & Gartz

Azzies

Cap: 1–8 (10) cm across, conical at first, becoming convex to nearly plane, often with a pronounced obtuse to pointed umbo. Margin down-curved to plane, rarely wavy. Surface smooth, viscid to dry; translucent-striate when wet. Orange-brown, caramel- to chestnut-brown when wet; strongly hygrophanous, usually fading outward from center to light brown to beige, readily developing blue-green blotches and stains. **Gills:** Attached or notched, close to subdistant, narrow. Pallid pinkish beige to brownish, mottled to darker brown as spores mature, staining bluish green when damaged. **Stipe:** 5–20 cm long, 0.3–0.6 cm thick, equal or enlarged towards base, tall and often slightly twisting. Surface dry, silky-fibrillose, at base with dense white or bluish hairs. Whitish when young, becoming dingy beige-tan to brown, can be mottled with bluish bruising at any age. **Partial veil:** A thin, whitish cortina, leaving wispy hairs on the stipe, often forming a faint annular zone. **Flesh:** Thin, somewhat rubbery in cap; tough and fibrous in stipe. Whitish to dingy brown. Bruising slowly blue-green to dark dingy blue. **Odor:** Slightly farinaceous. **Taste:** Bitter. **Spore deposit:** Dark purple-brown. **Microscopy:** Spores 12–13.5 × 6.5–8 µm, ellipsoid, smooth, moderately thick-walled, with a broad germ pore, brown in KOH. Cheilocystidia and pleurocystidia swollen, with a short, narrow neck.

Ecology: Scattered, in pairs or small clusters around clumps of dune grass, also sandy soil and grass debris. Locally common but restricted in its natural habitat of coastal dunes in southern WA and northern OR. Occasionally naturalized in woodchip beds across western Cascadia—often appearing clustered in these settings. Fruiting in fall, typically peaking late in the season, with stragglers into winter.

Edibility: Strongly hallucinogenic.

Comments: The orange-brown to caramel-colored hygrophanous cap, white stipe, blue staining, and purple-black spores help distinguish this as a *Psilocybe*. Tall stature, umbonate caps, and growth among dune grass clinch the species. *P. cyanescens* differs by having a rounded-conical to irregular convex or plane cap with a distinctly wavy margin, and lacks (or at most) has a low, broad umbo.

Psilocybe allenii

Borov., Rockefeller & P.G. Werner

Turpid Psilly

Cap: 1.5–9 cm across, hemispheric to convex at first, soon broadly convex to plane; margin plane. Surface smooth, viscid to dry, strongly hygrophanous, usually fading outward on drying, translucent-striate at margin when wet. Dull orange-brown to caramel, often with a slightly paler margin when wet, yellowish brown, light brown to beige when dry, soon developing blue-green blotches and becoming more dingy colored in age. **Gills:** Broadly attached to notched, close, narrow. Creamy beige at first, becoming light brown and then mottled purple-brown overall as spores mature. **Stipe:** 3–9 cm long, 0.2–0.7 cm thick, equal or with a slightly enlarged base. Surface dry, silky-fibrillose, pruinose to smooth at apex; base with white, binding rhizomorphs. Whitish at first, soon discoloring dingy brownish, can be mottled with bluish bruising at any age. **Partial veil:** Thin, whitish cortina, leaving wispy hairs on the stipe when young, soon disappearing. **Flesh:** Fibrous, rather firm, somewhat rubbery in cap, tough in stipe. Whitish to dingy tan; bruising slowly dark dingy blue. **Odor:** Mild, farinaceous or earthy. **Taste:** Farinaceous. **Spore deposit:** Dark purple-brown. **Microscopy:** Spores 11–14 × 6.5–8 µm, elongate-ellipsoid, smooth, thick-walled, with a broad germ pore, brown in KOH. Cheilocystidia swollen with narrowed, cylindrical tip. Pleurocystidia similar.

Ecology: In small clusters or troops in wood chips, dune grass, or disturbed areas. Occasional to locally common in western Cascadia, mostly in urban areas and coastal habitats. Fruiting in late fall.

Edibility: Strongly hallucinogenic.

Comments: The convex to plane, orange-brown to caramel cap, whitish to dingy-colored stipe, and blue staining help define this species. It is very similar in appearance to *P. cyanescens* and can be difficult to distinguish. The primary morphological difference is the convex to plane cap that does not become strongly wavy as in *P. cyanescens*. *P. allenii* also tends to have a stockier, stouter stature. The deadly poisonous *Galerina marginata* has a similarly-shaped cap but has an orangish stipe that darkens to blackish (not blue) from the base up in age, and rusty brown spores.

Psilocybe stuntzii

Guzmán & J. Ott

Blue Ringer

Cap: 1–5 cm across, broadly conical to bell-shaped when young, becoming convex to plane. Margin often wavy and occasionally uplifted in age. Surface smooth, slightly viscid or lubricious, hygrophanous, margin translucent-striate when wet. Color variable: olivaceous brown, dark brown to grayish brown, generally fading from center outward, becoming ochraceous brown to yellowish brown when dry. **Gills:** Broadly attached, moderately spaced, pallid grayish at first, soon medium brown with a purplish cast and often covered in dark brown speckles as spores mature to dark purplish brown in age. **Stipe:** 2–7 cm long, 0.2–0.5 cm thick, equal or enlarged slightly toward base, often squatter when growing in wood chips, taller when growing in grass. Surface finely fibrillose to smooth. Whitish, buff to grayish brown, darkening in age; occasionally with bluish stains. **Partial veil:** Membranous to cortina-like, leaving a flaring and often bluish ring when young, collapsing in age. **Flesh:** Thin; stipe fibrous; whitish to pallid brown, weakly bluing when young. **Odor:** Farinaceous. **Taste:** Farinaceous. **Spore deposit:** Dark purple-brown. **Microscopy**: Spores 11–13 × 6–8.5 µm, elongate-ellipsoid, thick-walled, smooth, with a distinct broad germ pore, dark yellowish brown in KOH. Cheilocystidia abundant, spindle- to flask-shaped, with long, thin necks. Pleurocystidia similar, with pointed necks.

Ecology: Usually clustered, sometimes solitary or scattered on wood chips, grassy areas, and on garden soil. Occasional to rare, primarily coastal and urban in western Cascadia. Fruiting in fall or early winter.

Edibility: Mildly hallucinogenic.

Comments: The olivaceous brown to grayish brown cap, flaring ring with bluish tones, and dark purple-brown spore deposit help identify this species. *P. ovoideocystidiata* is similar but often has a paler, light brown to buff cap, a whitish, evanescent veil, readily develops greenish blue stains overall, and fruits in clusters on wood chips in spring. It is common in western Cascadia and is often planted in woodchip beds. The deadly poisonous *Galerina marginata* also has a ring on the stipe but has rusty brown spores, a stipe that darkens from the base up in age, and doesn't stain blue.

Psilocybe baeocystis

Singer & A.H. Sm.

Blue Bells

Cap: 1.5–3.5 (5.5) cm across, broadly conical to bell-shaped and often irregular and somewhat knobby when young, becoming convex to plane. Margin sometimes wavy but rarely uplifted. Surface smooth, slightly viscid to dry. Color variable, caramel-brown to dark brown, hygrophanous and yellowish beige or whitish when dry; most often appearing caramel-brown heavily mottled with bluish green stains, although sometimes appearing entirely dark blue-green even when young. **Gills:** Broadly attached, moderately spaced, grayish at first, soon medium brown with a purplish cast and often covered in dark brown speckles as spores mature. **Stipe:** 3–7 cm long, 0.2–0.4 cm thick, equal or enlarged slightly toward base. Surface smooth or twisted-fibrous. Whitish to pale grayish or tan-brown, very soon becoming dingy, dark blue black to greenish blue in blotches. **Partial veil:** Indistinct, a thin cortina when young. **Flesh:** Pallid brown. Bruising dark greenish blue to nearly black. **Odor:** Mild to farinaceous. **Taste:** Farinaceous. **Spore deposit:** Dark purple-brown. **Microscopy**: 8.5–14 (17) × 5–7 µm, ellipsoid, thick-walled, smooth, dark brown in KOH. Cheilocystidia abundant, spindle- to flask-shaped, with long, thin necks. Pleurocystidia generally absent.

Ecology: Solitary, scattered in groups, or in small clusters on wood chips or in grassy areas. Often intentionally introduced into woodchip beds but not naturalizing as readily as *P. cyanescens*. Rare; low elevations of western Cascadia. Fruiting in fall.

Edibility: Hallucinogenic, similar to other *Psilocybe*. Only species of *Psilocybe* that has been implicated in a fatal poisoning, which involved a young child who may have also consumed other grass-dwelling mushrooms (perhaps including *Pholiotina*, a more likely source of such serious toxicity).

Comments: The fruitbodies of this species are, for lack of a gentler word, ugly. Rarely does this mushroom appear fresh, even when it is! The tendency for the fruitbodies of this species to dry to whitish overall is unusual for *Psilocybe* in our area. *P. ovoideocystidiata* has a distinct cortina-like ring on the stipe, while *P. stuntzii* usually has a blue-stained, flaring, membranous veil.

Psilocybe semilanceata

(Fr.) P. Kumm.

Liberty Cap

Cap: 0.5–2.5 (3.5) cm across, conical to bell-shaped, often with a peaked umbo. Surface smooth, thinly viscid to dry; margin translucent-striate when wet, opaque when dry. Cap skin gelatinous, peelable. Dark brown to ochraceous, generally with a paler, somewhat translucent umbo when wet, occasionally with olivaceous tones, strongly hygrophanous, fading to pale yellowish brown to tan when dry. More dingy-colored in age, occasionally with bluish stains. **Gills:** Broadly to narrowly attached, often notched, occasionally becoming free in age, close, narrow. Pale grayish buff when young, darkening to grayish brown and then purplish brown as spores mature; edges whitish. **Stipe:** 3–10 (14) cm long, 0.1–0.3 cm thick, often quite tall, slender and wavy, more or less equal. Surface dry, pruinose at apex, with sparse, silky hairs over lower portion when young, smooth in age. Whitish to pale buff at first, darkening to brownish in age. Staining blue when damaged. **Partial veil:** Indistinct, consisting of sparse, silky hairs, not forming a ring. **Flesh:** Very thin, fragile in cap, fibrous in stipe, concolorous, staining blue when damaged. **Odor:** Slightly farinaceous. **Taste:** Mildly farinaceous. **Spore deposit:** Purple-brown. **Microscopy:** Spores 11–14 (18) × 7–9 µm, ellipsoid, occasionally slightly lemon-shaped, smooth, moderately thick-walled, with a broad germ pore, brown in KOH. Cheilocystidia abundant, swollen in lower portion, or bottle-shaped with elongated, occasionally forked or branched necks.

Ecology: Solitary or scattered in troops in rich grassy pastures, usually in clumps of taller grass, generally around, but not on, dung. Occasional in appropriate habitat at lower elevations in western Cascadia, especially near the coast. Fruiting in fall or early winter with scattered flushes in spring.

Edibility: Hallucinogenic.

Comments: The peaked, conical to bell-shaped cap, purplish brown mature gills, grassland habitat, slender stipe that stains blue when damaged, and purple-brown spores are distinctive. *P. pelliculosa* is fairly similar but grows in forest settings. Some *Panaeolus* are frequently found fruiting alongside this species but they have bell-shaped to convex caps and black spores and don't bruise bluish.

Psilocybe pelliculosa

(A.H. Sm.) Singer & A.H. Sm.

Woodland Psilly

Cap: 0.5–3 cm across, conical to bell-shaped, often with a rounded umbo, occasionally more convex. Surface smooth, thinly viscid to dry. Hygrophanous, margin translucent-striate when wet, fading and opaque when dry. Yellowish brown to brownish orange, occasionally reddish brown, fading to pale yellowish brown to tan when dry, becoming more dingy in age, often with grayish or greenish blue tints. **Gills:** Broadly attached, often notched; close, narrow. Pale grayish buff when young, darkening to grayish brown and then purplish brown as spores mature. **Stipe:** 3–10 cm long, 0.1–0.3 cm thick, equal, or with an enlarged or slightly bulbous base. Surface dry, pruinose at apex, with sparse, silky hairs over lower portion. Whitish to pale buff with a brownish base, darkening in age. Base occasionally with bluish stains. **Partial veil:** Indistinct, consisting of sparse, silky hairs; not forming a ring. **Flesh:** Thin, fragile in cap, stipe fibrous; concolorous. **Odor:** Slightly farinaceous. **Taste:** Mild to slightly farinaceous. **Spore deposit:** Purple-brown. **Microscopy:** Spores 9.5–13 × 5–7 µm, ellipsoid to ovoid, smooth, moderately thick-walled, with a broad germ pore, brown in KOH. Cheilocystidia abundant, spindle-shaped or more elongate with pointed tips. Pleurocystidia absent.

Ecology: Solitary, scattered, or troops in duff, soil, or small woody debris in disturbed areas. Occasional in western Cascadia, more common coastal and in the Coast Range. Fruiting from late fall into winter.

Edibility: Mildly hallucinogenic.

Comments: Recognized by its markedly translucent-striate, yellowish brown caps, purple-brown spores, and thin stipe that often has bluish stains at the base. Similar but poorly known species reported from the Pacific Northwest include *P. sylvatica* with smaller spores (6.5–9.5 × 4–5.5 µm), *P. fimetaria* which supposedly grows on dung, and *P. cyanofibrillosa* with a conical to plane cap. Further study is needed. *P. semilanceata* has a peaked-conical cap and it grows in grasslands. *Galerina* have paler gills and rusty brown spores. *Hypholoma dispersum* is generally larger, has silvery whitish chevrons covering the stipe and a domed cap that isn't as strongly translucent-striate.

Hypholoma dispersum

(Fr.) Quél.

Cap: 1–5 cm across, domed to bell-shaped, becoming convex. Surface moist to dry; margin often with silky patches or a narrow ring of whitish veil fibers when young, soon smooth overall. Ocher-brown to pinkish tan or with yellowish brown colors, paler to beige or white near the margin. Fading as it dries out, becoming extensively pale beige in age. **Gills:** Narrowly attached with a small notch, close to crowded. Pale cream when young, then light yellowish, soon mottled with gray and becoming brownish gray in age. **Stipe:** 3–10 cm long, 0.2–0.5 cm thick, cylindrical, base slightly enlarged; usually rather straight and upright. Surface dry, whitish overall when young, soon darkening, tan to brown, covered in shiny chevrons of silky veil tissue, often forming a striking snakeskin pattern that contrasts strongly with darker base. **Flesh:** Thin, stipe tough and fibrous; whitish to dingy brown. **Odor:** Indistinct. **Taste:** Mild to bitter. **Spore deposit:** Dark grayish with purple-brown tones. **Microscopy:** Spores 7–10 × 4–5 µm, ellipsoid, with a central germ pore, smooth, dark brown in KOH. Cheilocystidia and pleurocystidia present, some yellow in KOH.

Ecology: Solitary, in troops, small clusters, or arcs in deep needle litter, moss beds, or among woody debris or woodchip beds. Widespread and common across western Cascadia, rarer on the eastside. Fruiting in late fall and winter.

Edibility: Unknown.

Comments: The domed ocher-brown cap, upright stature, and snakeskin pattern of silvery chevrons on the stipe readily distinguish this species. *Phaeonematoloma myosotis* has a viscid, olive-brown to grayish olive cap, a long stipe, dark brown spores, and is found in moss in wet areas. *Bogbodia uda* has a smooth, moist to dry, orange-brown cap, more purplish brown gills and larger, almond-shaped, finely roughened spores. *H. elongatum* grows on moss or wet soil and has a yellow-brown to straw-brown cap. *Psilocybe pelliculosa* has a distinctly translucent-striate cap. Some *Galerina* have silvery bands on the stipe but ocher-brown spore deposits.

Synonym: *Hypholoma marginatum* J. Schröt.

Mythicomyces corneipes

(Fr.) Redhead & A.H. Sm.

Cap: 1–3 cm across, obtusely conical to bell-shaped at first, becoming broadly convex, occasionally with an umbo. Surface smooth, moist to dry; translucent-striate when wet. Orange to orange-brown, fading to ocher-brown, strongly hygrophanous, fading to tan when dry. **Gills:** Narrowly attached with a slight notch to seceding, close, broad. Pallid to beige, becoming tan, often with olivaceous tones in age. **Stipe:** 3–6 cm long, 0.1–0.3 cm thick, equal, often curved or twisted. Surface moist to dry; apex finely scaly when young, soon smooth; lower portion with sparse coarse hairs; base with ocher mycelium. Yellow-brown to orange-brown at apex, transitioning to dark reddish brown, blackening from the base up in age. **Flesh:** Thin, brownish in cap. Stipe tough, cartilaginous-fibrous. **Odor:** Indistinct to faintly geranium-like. **Taste:** Mild to slightly bitter. **Spore deposit:** Pale purple-brown. **Microscopy:** Spores 6–8.5 × 4–5.5 µm, ovoid to ellipsoid, ornamented with scattered short ridges and projections, dextrinoid. Cheilocystidia and pleurocystidia abundant, spindle-shaped or more swollen in middle, with rounded tips often encrusted with amyloid crystals.

Ecology: Solitary or scattered in soil, moss, duff, or grass thatch along stream banks or near wet seeps with an affinity for old growth conifer forest. Locally common in low or mid-elevation old growth forests, rare elsewhere. Fruiting in late summer and fall.

Edibility: Unknown.

Comments: The hygrophanous orange to ocher-brown cap, beige to tan gills which develop an olive cast, cartilaginous stipe that darkens from the base up, and pale purple-brown spores help set this species apart from a plethora of other little brown mushrooms. *Stagnicola perplexa* is very similar macroscopically and can be difficult to distinguish but it is usually slightly paler and has a noticeably bitter taste, it lacks encrusted cystidia, and has smooth spores. *Galerina* spp. look similar but all have rusty brown spores. *Phaeocollybia attenuata* has a rooting, wire-like stipe base, a reddish brown spore deposit, roughened, lemon-shaped spores, and it lacks pleurocystidia.

Hypholoma fasciculare

(Huds.) P. Kumm.

Sulfur Tuft

Cap: 1–10 cm across, rounded-conical to hemispheric when young, occasionally with a conical umbo, soon broadly convex, often wavy and uplifted in age. Surface smooth, dry, often with patches of silky veil remnants clinging to margin. Pale yellowish green to lemon-yellow, orange-yellow, or ocher, sometimes nearly reddish brown, especially over disc or when young. **Gills:** Attached, fairly close, narrow. Greenish yellow to bright yellow or pale, dull yellowish beige when young, maturing to brownish or purplish brown to grayish black, but often retaining an underlying greenish yellow hue. **Stipe:** 3–10 cm long, 0.4–0.7 cm thick, mostly cylindrical, often tapered to base and fused into moderate sized clusters. Surface dry, appressed-fibrillose to smooth. Pale yellow, greenish yellow to ocher-brown, especially browner near base. **Partial veil:** Cobwebby, sometimes forming a nearly membranous-fibrillose sheet over young gills, often leaving a dark annular zone near stipe apex that collects purplish black spores. **Flesh:** Thin, fibrillose, yellowish in cap, often with brownish stains in stipe. **Odor:** Indistinct. **Taste:** Bitter. **KOH:** No reaction. **Spore deposit:** Variable, dark purple-brown to dark purplish black. **Microscopy:** Spores 6–8 × 3.5–5 µm, ellipsoid, smooth, with a germ pore. Pleurocystidia with swollen bases and elongated necks, with golden-yellow content in KOH. Cheilocystidia similar, often broader.

Ecology: In small groups, arcs, and clumps or in tremendous clusters and rows on logs, stumps, or others woody debris of both hardwoods and conifers. Extremely common and cosmopolitan. Fruiting throughout the year but by far most abundant in fall.

Edibility: Toxic, causing severe gastrointestinal distress.

Comments: The yellow to orangish cap, yellowish green gills that become grayish purple as spores mature, clustered growth on wood, and bitter taste set this species apart. *H. capnoides* is quite similar but lacks the yellow-green tones in the gills (and is usually less bright overall), tastes mild, and grows only on conifer wood. A few smooth-capped yellow *Pholiota* and *Flammula* are quite similar but lack greenish tones, taste mild, and have brown spores.

Hypholoma capnoides
(Fr.) P. Kumm.

Conifer Tuft

Cap: 1.5–10 cm across, convex to broadly convex with a low umbo, to plane in age, sometimes wavy. Surface smooth, viscid when wet, otherwise dry, hygrophanous, not striate. Margin often with thin patches of adhering cobwebby veil. Dull or bright yellow-orange to honey-tan, ocher-brown or nearly reddish brown. **Gills:** Broadly attached to slightly decurrent, close, narrow. Creamy to grayish beige, soon smoky gray to dark gray as spores mature. **Stipe:** 4–12 cm long, 0.3–1.2 cm thick, cylindrical, somewhat tapered to base, often loosely fused or clustered. Surface dry, covered in inconspicuous pale chevrons or fibrils. Whitish to creamy beige or dingy brownish tan (especially over lower parts). **Partial veil:** A thin cortina, cobwebby and nearly membranous when young, soon splitting, leaving remnants on cap margin, sometimes collapsing on upper stipe and collecting dark spores, but often disappearing entirely. **Flesh:** Thin, fibrous, tough in stipe. Concolorous with exterior. **Odor:** Indistinct. **Taste:** Mild. **KOH:** Light orange-brown on cap. **Spore deposit:** Dark grayish purple to purplish black. **Microscopy:** Spores 6–7.5 × 3.5–5 µm, ellipsoid, smooth, with a small germ pore. Pleurocystidia club-shaped or swollen with a pointed tip, with a golden-yellow mass in KOH. Cheilocystidia spindle-shaped.

Ecology: In clusters and troops on decaying conifer logs and stumps. Common and widespread, occurring across Cascadia in most forested habitats. Fruiting in fall and winter, with stragglers in spring.

Edibility: Edible but not frequently collected for the table.

Comments: The Conifer Tuft is recognized by its yellow-brown to ocher-brown caps, smoky gray gills, growth on conifer wood, and dark grayish purple spore deposit. *H. fasciculare* is usually brighter yellow with distinctly greenish yellow gills when young and a bitter taste. *Pyrrhulomyces astragalinus* is somewhat similar but has a beautiful pinkish peachy cap, tastes bitter, has brown spores, and sometimes slowly bruises ash-gray.

Hypholoma tuberosum
Redhead & Kroeger

Cap: 2–6 cm across, conical to bell-shaped at first, becoming broadly bell-shaped or convex with a broad umbo. Surface thinly viscid to dry, smooth, except for veil remnants near margin when young. Dull orange, orange-brown to yellow-brown. **Gills:** Broadly attached, subdistant, broad. Pale grayish when young, becoming yellowish gray, then purplish gray to purplish black as spores mature. **Stipe:** 4–8 cm long above ground 0.2–0.5 cm thick, equal or enlarged lower, then tapered and rooted to buried sclerotium. Surface dry, appressed-fibrillose, with scattered fibrillose scales below, smoother in age. Pallid at apex, yellowish to orangish below. Sclerotium irregularly lobed, hard, woody or corky; exterior smooth to roughened, dark brown; interior densely marbled beige, light brown to orangish brown. **Partial veil:** A thin, fibrillose membrane, leaving remnants on the cap margin and a silky annular zone on the stipe. Whitish to pale yellowish, soon covered in purplish black spores. **Flesh:** Thin to thick, firm, fibrous in stipe; light orangish brown. **Odor:** Slightly farinaceous; sclerotium earthy. **Taste:** Flesh bitter; sclerotium radish-like. **Spore deposit:** Dark purple-brown. **Microscopy:** Spores 9.5–11.5 × 5–7.5 µm, ellipsoid, with small germ pore, smooth, grayish brown in KOH. Cheilocystidia cylindrical with wavy necks. Pleurocystidia spindle-shaped, with yellow contents in KOH.

Ecology: Solitary or scattered, often in pairs or tufts, arising from a sclerotium in wood chips or other nutrient-rich soil. Uncommon; urban or developed areas across western Cascadia. Fruiting in fall.

Edibility: Unknown.

Comments: The orange-brown, umbonate cap, dark purple-brown spores, and growth in rich soil or wood chips are distinct. If the sclerotium is excavated, identification is easy, however unwary collectors usually break the stipe when collecting, leaving the sclerotium unnoticed. *Agrocybe arvalis* has a straw-yellow to ocher cap, brown gills and spores, and a hard, round to ovoid, black sclerotium. *Leratiomyces ceres* is a common woodchip species with a bright red to orange-red cap. *L. squamosus* occurs in duff under conifers, has an ocher-orange cap, membranous partial veil, and a tufted-scaly stipe – both *Leratiomyces* species lack a sclerotium.

Leratiomyces ceres

(Cooke & Massee) Spooner & Bridge

Chip Cherries

Cap: 2–7.5 cm across, bell-shaped when young, becoming convex to broadly convex, often with a round umbo; margin often uplifted or wavy in age. Surface slightly viscid when young, soon dry, margin often with small patches of whitish veil fibrils, soon smooth overall. Dark red to orange-red, dingier in age. **Gills:** Broadly attached or with a slight decurrent tooth, close, narrow. White when young, becoming dingy gray-tan, then cool gray mottled with dark brown; can be spotted or extensively blushed with orange to red at any age. **Stipe:** 3–8 (11) cm long, 0.5–0.8 cm thick, equal or slightly enlarged towards base. Surface dry, slightly scaly or covered in chevrons over lower portion, smooth in age. Base with white rhizomorphs. White at first, becoming dingy cream, reddish at base and becoming orange-red streaked upward in age. **Partial veil:** Cottony-fibrillose, white and pronounced when young, but soon collapsing to an indistinct fibrillose zone that becomes darkened with spores in age. **Flesh:** Thin, fibrous, fairly tough in stipe; pallid. **Odor:** Indistinct. **Taste:** Indistinct. **KOH:** Orangish on cap. **Spore deposit:** Dark purplish brown. **Microscopy:** Spores 10–13 × 6–7.5 µm, ellipsoid, smooth, dark brown in KOH. Cheilocystidia with yellow contents in KOH.

Ecology: Scattered, in troops, or small clusters on wood chips and nutrient-rich soil in landscaped areas throughout Cascadia. Fruiting at any time of year in irrigated areas with the primary flush in late fall.

Edibility: Unknown, possibly toxic.

Comments: The red cap, clustered or trooping growth, frequently red-streaked stipe, and dark purplish brown spores identify this species. It sometimes fruits by the thousands in wood chip beds, often in the company of *L. percevalii*, as well as various species of *Agrocybe*, *Psilocybe*, and *Psathyrella.* Thought to be native to Australia, it is expanding throughout North America. *L. squamosus* has a slender, upright stature and distinctly tufted-scaly cap and stipe. It grows in duff and among fine woody debris at edges of meadow and forest settings, often at high elevations. *Hypholoma tuberosum* has an orange-brown cap and arises from a hard, irregularly lobed sclerotium.

Leratiomyces squamosus

sensu PNW

Cap: 2–6 cm across, obtusely conical at first, becoming convex to plane, often with a broad umbo. Surface thinly viscid to dry, with upright tufts of veil remnants when young, usually smooth in age; hygrophanous. Ocher-orange to orange-brown. **Gills**: Broadly attached, close to subdistant, broad. Pale grayish when young, becoming purplish gray to purplish black as spores mature; edges paler and often fringed. **Stipe**: 4–10 cm long, 0.3–0.7 cm thick, more or less equal. Surface dry, extensively tufted-scaly below veil, pruinose to smooth above veil. Whitish to yellowish, developing brown tones from base up in age. **Partial veil**: A flaring, membranous skirt, collapsing in age. Upper side grooved, whitish, soon purplish black with spores, underside scaly. **Flesh**: Moderately thin, firm, stipe fibrous; light orangish brown. **Odor**: Indistinct. **Taste**: Indistinct. **Spore deposit**: Dark purple-black. **Microscopy**: Spores 8–12 × 4.5–7 µm, ellipsoid (with a small germ pore), smooth, grayish brown in KOH. Cheilocystidia narrowly cylindrical. Pleurocystidia absent.

Ecology: Solitary or scattered in pairs or tufts in duff and other debris under conifers. Rare; occurring across western Cascadia from low to high elevations. Fruiting in fall.

Edibility: Unknown.

Comments: The ocher-orange cap (often adorned with scales when young), scaly stipe that is tall and slender, membranous partial veil, and dark purple-brown spores make this beautiful mushroom quite distinctive. There are also reports of darker red-capped collections from urban areas, which may pertain to the European *L. squamosus* sensu stricto. *Hypholoma tuberosum* has an orange-brown, umbonate cap and scattered fibrillose scales on the stipe. It grows in rich soil or wood chips. *L. ceres* has a bright red to orange-red cap and a finely scaly to nearly smooth whitish stipe that is often blushed with orange-red tones.

Leratiomyces percevalii

(Berk. & Broome) Bridge & Spooner

Mulch Maids

Cap: 2–12 cm across, bell-shaped to convex with a round or slightly conical umbo (sometimes rather pointed), plane or wavy in age. Surface moist to dry, sometimes finely scaly. Margin often with patches or hanging "teeth" of beige-yellow or whitish veil tissue. Light beige to pale golden-yellow, sometimes ocher or darker tan at center. **Gills:** Broadly attached, close to subdistant. Pale cream at first, soon dingy grayish beige, mottled with lilac-gray, to dark gray in age; edges whitish-fringed. **Stipe:** 5–12 cm long, 0.3–1 cm thick, equal or enlarged towards base. Surface dry, smooth to finely scaly over lower half. Whitish cream to yellowish beige, dingier and darker in age, developing orangey brown tones. **Partial veil:** Fibrillose to slightly membranous. Usually only present as a spiky or narrow, flattened ring zone, whitish, soon darkened with grayish spores. **Flesh:** Rather tough, fibrous in stipe, beige to dingy ocher-brown. **Odor:** Indistinct. **Taste:** Indistinct. **Spore deposit:** Dark purplish gray. **Microscopy:** Spores 12–14 × 7–9 µm, ellipsoid to almond-shaped, smooth, brown in KOH. Pleurocystidia club-shaped to nearly cylindrical, wavy.

Ecology: Solitary, scattered, or in small clusters and troops on wood chips. Very common in urban areas or along roads and trails throughout Cascadia. Also in riparian areas under cottonwood and alder in forested settings. Fruiting throughout the year but most abundant in fall and spring.

Edibility: Non-toxic. Some have reported a good texture but poor taste. Often in contaminated areas.

Comments: Fairly distinctive by virtue of the beige to golden cap, lilac-gray gills, tough, pale stipe with indistinct annular zone, and growth in wood chips. Two species of *Leratiomyces* described from western North American may represent *L. percevalii*; *L. riparius* (Olympic Peninsula) and *L. magnivelaris* (southeast Alaska) with a supposed membranous partial veil. *Agrocybe* species often grow alongside but have brown spores and gills and either skirt-like rings on the stipe or lack the shaggy annular zone. *Stropharia ambigua* is larger, with a viscid cap when young, a shaggier white stipe, and fluffy veil tissue often fringing the cap margin.

Stropharia ambigua

(Peck) Zeller

Ambiguous Stropharia

Cap: 3–10 (15) cm across, rounded-conical to convex at first, expanding to plane or wavy in age. Margin incurved at first, expanding and covered with large hanging white veil tissue around the margin, becoming bald and occasionally slightly uplifted in age. Surface thinly viscid to lubricious, smooth, except for white veil remnants near margin when young. Dull yellow to yellow-brown, often fading to creamy beige in age. Young caps can be more ocher-yellow to butterscotch colored; older caps are often dusted in purplish brown spores, especially when growing in clusters. **Gills:** Broadly attached, often with a shallow notch and a decurrent tooth, close to crowded; partial gills numerous. Gray at first, soon darkening to purplish gray and then purplish black as spores mature. **Stipe:** 5–20 cm long, 0.5–2.5 cm thick, club-shaped at first, more or less equal, tall and slender in age. Surface dry; when young covered in fine white tufts near apex and larger white scales below, nearly smooth in age. Base with white rhizomorphs binding bits of duff and litter. White, occasionally with buff to orangish brown stains over lower portion. **Partial veil:** Flocculose-membranous, adhering to cap margin, typically in hanging triangular pieces when young, often disappearing in age, rarely leaving a small, ragged ring zone on the stipe. **Flesh:** Moderately thin to thick, soft; whitish. **Odor:** Mild to earthy. **Taste:** Earthy. **Spore deposit:** Dark purple-brown to purplish black. **Microscopy:** Spores 11–14 × 6–7.5 µm, ellipsoid, truncate with an apical germ pore, smooth, thick-walled, grayish brown in KOH. Cheilocystidia abundant, bottle-shaped with a rounded head, cylindrical to club-shaped. Pleurocystidia embedded and inconspicuous, broadly spindle-shaped with pointed tips, with yellow contents in KOH.

Ecology: Solitary or scattered in small troops when fruiting in leaf litter and duff in natural habitats. Patches growing in urban areas often fruit profusely in large troops and clusters, especially when on wood chips. Very common throughout Cascadia. Fruiting in large flushes in late fall, with scattered fruitings throughout winter and spring.

Edibility: Edible but (very) poor from a culinary perspective.

Comments: The thinly viscid to slimy yellowish caps, fluffy white veil tissue (often disappearing in age), tall, scaly to nearly smooth white stipes, and dark purple-brown spores make this ubiquitous species easy to identify. Older specimens with spores covering the caps might look similar to *S. hornemannii* but that species has a skirt-like ring and a more distinctly scaly stipe. *S. coronilla* is a similarly colored, stocky, grassland species with a membranous veil. *Leratiomyces percevalii* is a somewhat similar species commonly found on wood chips but has smaller fruitbodies with fibrillose partial veils and dry caps.

Stropharia rugosoannulata

Farl. ex Murrill

Wine Cap, Garden Giant

Cap: 4–12 (20) cm across, rounded to convex at first, becoming broadly convex to plane, at times with a broad umbo. Surface dry, with veil remnants at margin when young, soon smooth, often cracking in age or when dry. Dark wine red, burgundy or red-brown, fading with age, pale reddish brown to straw-brown, or dingy cream. Some forms are cream to pale yellow even when young. **Gills**: Broadly attached or notched, close, broad; partial gills numerous. Pale grayish at first, soon becoming deep vinaceous gray to blackish when spores mature, edges paler. **Stipe**: 5–15 cm long, 1–4 cm thick, equal or club-shaped with an enlarged base. Surface dry, smooth, fibrillose to finely scaly. Base with white rhizomorphs. Whitish to light tan, darkening slightly with age. **Partial veil**: Thick, membranous, often with a radially split "cogwheel" of tissue on the underside when young, forming a flaring skirt-like ring with a grooved upper surface. Whitish, soon dusted with purplish black spores. **Flesh**: Thick, firm, meaty. White. **Odor**: Pleasant. **Taste**: Mild. **Spore deposit**: Dark purplish black. **Microscopy**: Spores 9.5–12.5 × 7–9 µm, ellipsoid to ovoid, with a broad germ pore, smooth, thick-walled, purplish brown in KOH. Cheilocystidia club-shaped, or more swollen lower with narrow necks, with yellow content in KOH. Pleurocystidia swollen at middle, some with yellow content in KOH.

Ecology: Solitary, scattered, or in small clusters on wood chips or garden soil. Introduced to Cascadia, restricted to lower elevation urban areas. Fruiting in spring and fall, more rarely in summer.

Edibility: Edible, with a meaty texture and great flavor. Use caution with specimens from potentially polluted (e.g. urban) areas.

Comments: This species is easily recognized by its large size, wine-red cap which fades to straw-yellow, thick partial veil with a cog-wheel pattern on the underside, and dark purple-black spores. It is a commonly cultivated species which has 'escaped' to woodchip beds in urban areas across western Cascadia. *S. hornemannii* has a lilac-brown to reddish brown cap which fades to brownish buff, a shaggy white stipe below the ring, and it grows in forests.

Stropharia hornemannii

(Fr.) S. Lundell & Nannf.

Cap: 4–12 (15) cm across, rounded-conical at first, becoming convex to plane with a broad umbo. Margin incurved and with whitish veil specks when young, often uplifted and wavy in age. Surface viscid to lubricious, soon smooth. Color variable, usually lilac-brown to reddish brown at first, fading to grayish brown or smoky buff in age. **Gills:** Broadly attached, often with a decurrent tooth, close, broad; partial gills numerous. Pallid at first, soon gray to vinaceous gray, becoming deep vinaceous gray to blackish as spores mature. **Stipe:** 6–11 (15) cm long, 0.5–2 cm thick, equal or club-shaped with an enlarged base. Surface dry, covered with white floccose scales below veil, pruinose above. **Partial veil:** Thick, membranous, forming a flaring skirt-like ring high on the stipe, collapsing in age. **Flesh:** Thick, soft; whitish to buff, discoloring ocher-brown around larva tunnels and in age. **Odor:** Mild to unpleasant. **Taste:** Disagreeable. **Spore deposit:** Dark purple-brown to purplish black. **Microscopy:** Spores 11–13 × 6–6.5 µm, ellipsoid to almond-shaped, with a broad germ pore, smooth, thick-walled, grayish brown in KOH. Pleurocystidia with a swollen base and slender necks. Cheilocystidia club-shaped.

Ecology: Solitary, scattered, or in small clusters on or near well-rotted wood or in thick duff. Widespread in Cascadia but uncommon. Mostly found in high elevation and east-side forests, rare at lower elevations. Fruiting in late summer and fall.

Edibility: Unknown.

Comments: This large *Stropharia* is recognized by its usually lilac-brown to reddish brown cap (soon fading to brownish buff), cool gray gills, shaggy white stipe with a skirt-like ring, and dark purple-brown spores. *S. ambigua* is typically taller, with a yellowish cap and a fluffy partial veil that leaves ragged remnants on the cap margin. *S. rugosoannulata* has a wine-red, slightly viscid to dry cap that fades to brownish, and thick, cogwheel-patterned partial veil. It is a commonly cultivated species that has become naturalized in urban areas on wood chips or rich soil.

Stropharia cyanea group

Blue Stropharia

Cap: 1–6 cm across, rounded-conical at first, becoming convex to plane, occasionally umbonate. Surface thinly viscid, with whitish specks around margin when young, soon smooth. Turquoise blue to bluish green at first, soon fading from the center outward to pale blue to creamy yellow with a bluish margin, occasionally yellowish beige or milky overall in age. **Gills:** Broadly attached, close to subdistant, broad. Pallid at first, developing vinaceous brown tones as spores mature, becoming brown in age. **Stipe:** 3–10 cm long, 0.3–1 cm thick, equal or tapering downward, often tall and slender. Surface moist to dry, with whitish fibrillose scales below veil, pruinose above, nearly smooth in age. Bluish at first, fading to whitish, then buff in age. **Partial veil:** Fleeting, forming a thin ring on stipe and adhering to cap margin when young, disappearing in age. **Flesh:** Moderately thin, soft, bluish when young, buff in age. **Odor:** Indistinct. **Taste:** Mild. **Spore deposit:** Dark purple-brown to purple-black. **Microscopy:** Spores 7–9.5 × 4.5–6 µm, ellipsoid, smooth, thick-walled, grayish brown in KOH. Cheilocystidia and pleurocystidia abundant, with swollen midsections and short, slender necks or bottle-shaped, with yellowish contents in KOH.

Ecology: Solitary, scattered, or in small clusters in grassy areas, leaf litter, or occasionally in rich soil at garden edges or in wood chips throughout Cascadia. Most frequent at lower elevations. Fruiting in fall.

Edibility: Unknown.

Comments: Although members of this group are easily recognized by their beautiful turquoise colors and purplish brown to black spores, differentiating these species from one another is difficult. ITS sequence data has shown that there are two species in the *S. cyanea* group in Cascadia with small to medium-sized fruitbodies, blue colors that slowly fade to creamy yellow, and an evanescent partial veil. *S. aeruginosa* is generally larger, often grows in wood chips, retains the blue color longer, has a more membranous partial veil, and whitish fringed gill edges (due to cylindrical, capitate cheilocystidia). *S. pseudocyanea* is smaller and has a conical umbonate cap and its colors fade more quickly than those of the other two groups. *Psilocybe* species are more slender and are never as brightly colored blue.

Stropharia scabella

(Zeller) E.J. Tian & M. Gordon

Cap: 4–10 (15) cm across, convex to broadly convex, occasionally expanding to plane; margin incurved at first, often with whitish remnants. Surface dry, covered with recurved scales becoming more matted-fibrillose in age. Brown, buff to honey-brown, covered with dark brown or yellow-brown scales. **Gills:** Narrowly attached, often seceding, close to crowded; partial gills numerous. Gray to lilac-gray, becoming deep brown as spores mature. **Stipe:** 5–10 cm long, 1–3 cm thick, more or less equal. Surface dry, covered with recurved scales below veil. Base often with white rhizomorphs. White to pale buff, lower portion often with pale orange scales, and discoloring orangish in age. **Partial veil:** Membranous, adhering to cap margin and forming a short skirt- to collar-like ring on stipe, collapsing or disappearing in age. **Flesh:** Thick, soft, whitish. **Odor:** Mild to unpleasant, rancid. **Taste:** Slightly disagreeable. **Spore deposit:** Dark purple-brown. **Microscopy:** Spores 6–8 × 4–4.5 µm, ellipsoid, smooth, thick-walled, grayish brown in KOH. Cheilocystidia nearly cylindrical to spindle-shaped. Pleurocystidia club-shaped with pointed tips to spindle-shaped, with golden content in KOH.

Ecology: Solitary, scattered in leaf litter and duff of alder and bigleaf maple, or more rarely in conifer duff or around well-rotted woody debris. Rare but widespread in Cascadia. Fruiting in fall.

Edibility: Unknown.

Comments: The dry, scaly cap, lilac-gray to brown gills, dark purple-brown spores, and often disagreeable odor make this rare mushroom distinctive. At first glance one might mistake it for an *Agaricus* but the gills are attached, not free.

Taxonomic note: *Stropharia kauffmanii* is likely a synonym.

Dark Spored

6 - Gilled Bolete Relatives

Given their gilled morphology, it is surprising to learn that the mushrooms in this section are closely related to boletes—a group defined by their pores! This paradox is explained by of convergent evolution—a common pattern in which similar adaptations arise in distantly related lineages of organisms. Although these macroscopic structures might be misleading, a quick look under the microscope shows that the spores of many 'gilled boletes' have the same elongate shape found in the more typical members of the Boletales. This is another reminder that when contemplating the deep-time scales of evolution, there is always more to the story than meets the eye.

Gomphidius have decurrent, somewhat thick-edged gills that start out pale but become smoky gray or blackish as the spores mature. They often have a bright yellow stipe base with pale whitish to grayish flesh. Most distinctively, these 'Slime Spikes' have sticky-viscid caps and partial veils made of slime. They are found in association with Slippery Jack boletes (*Suillus* spp.) in various non-pine conifer forests.

Chroogomphus are similar in stature to *Gomphidius* but have orange flesh and are mostly found in pine forests (one species is found with hemlock). The partial veils of *Chroogomphus* are made of fine silky fibrils (cortina-like) but are often obscure and only noticed when young or when forming a wispy ring zone. Both genera appear to be parasitic on the mycelia of various boletes (not mycorrhizal, as was once thought). Each species appears to prefer a particular bolete host, which is a helpful identification feature (see the table below).

Parasite	**Mycorrhizal Partner**	**Tree**
Chroogomphus tomentosus	*'Boletus' mirabilis*	Hemlocks
Chroogomphus ochraceus	*Suillus tomentosus*	2-3 needle pines
Chroogomphus 'Small Dark'	*Suillus flavidus*	Shore/lodgepole pine
Chroogomphus albipes	*Rhizopogon* spp.*	2-3 needle pines
Gomphidius maculatus	*Suillus clintonianus**	Larch
Gomphidius subroseus	*Suillus lakei*	Douglas-fir
Gomphidius smithii	*Suillus lakei*	Douglas-fir
Gomphidius glutinosus	Generalist (or cryptic ecotypes?)	Spruce, Douglas-fir
Gomphidius oregonensis	*Suillus caerulescens, S. ponderosus*	Douglas-fir

*presumed

The ectomycorrhizal *Paxillus*, and the wood decaying *Tapinella* have medium-sized to large fruitbodies with brown colors, close to crowded decurrent gills and brown spores. *Tapinella* also have a distinctly tomentose stipe.

Phylloporus have velvety-plush, olive-green to brownish cap surfaces, brilliant yellow gills that often stain slowly blue, and dull olive-brown spores. They are ectomycorrhizal, and closely related to Suede Boletes *(Xerocomus)*.

Chroogomphus tomentosus

(Murrill) O. K. Mill.

Woolly Pine Spike

Cap: 2–10 cm across, peg-like when young with a rounded cap, becoming convex to plane, often with a broad umbo. Margin inrolled at first, often ruffled in age. Surface moist to dry, appressed-tomentose to matted fibrous or finely scaly in dry conditions. Beige-orange to buff-orange, slightly darker toward center, sometimes developing vinaceous or purple stains in age. **Gills:** Subdecurrent to decurrent, thick, widely spaced. Buff-orange, orangish tan to brownish in age. **Stipe:** 5–15 cm long, 0.5–1.5 cm thick, more or less equal, but tapers at base. Surface moist to dry, finely tomentose to appressed-fibrillose. Beige-orange to buff-orange. **Partial veil:** Silky fibrillose when young, disappearing in age. **Flesh:** Thin, fibrous, pale beige-orange. **Odor:** Indistinct, although may get an odor of rubbing alcohol in age. **Taste:** Mild. **KOH:** Purple on all parts. **Spore deposit** Olive-black. **Microscopy:** Spores 15–25 × 6–9 μm, long-ellipsoid, smooth.

Ecology: Scattered or in small clusters on ground, often next to rotting logs. Found in western hemlock forests in association with *"Boletus" mirabilis*. Very common throughout Cascadia, fruiting from summer into early winter, even in dry periods.

Edibility: Edible. Creative chefs will undoubtedly find ways to intentionally exploit both the sliminess and the brilliant purple color of the cooked fruitbodies.

Comments: The beige-orange colors, widely spaced decurrent gills, fibrillose veil when young, and black spores are important features for identifying this mushroom. Contorted specimens with poorly formed or loculate gills have been called *C. loculatus*, but genetic data suggest that these are simply developmental variants rather than a separate species. Although *Cantharellus formosus* can be similar in color, the deeply decurrent, shallow, thick-edged gills and whitish fibrous flesh help differentiate it. Other *Chroogomphus* species are darker colored and grow under pines with *Suillus*. *Lactarius* have brittle flesh and exude "milk" when broken.

Chroogomphus ochraceus

(Kauffman) O. K. Mill.

Olive Pine Spike

Cap: 2.5–7 cm across, hemispherical to broadly convex with an incurved margin, becoming nearly plane, often with an acute to rounded umbo at all ages. Surface smooth, dry to moist or viscid in wet weather. Appearing metallic in dry conditions. Dingy ocher-olive to grayish tan (often with a lilac cast), usually mottled with a combination of all these colors. Edge of young caps often with a band of ocher fibrillose partial veil, especially in dry weather. **Gills:** Decurrent, fairly widely spaced, sometimes forked near stipe, edges smooth. Dull ocher when young, becoming dull grayish ocher and then dark olive-gray. **Stipe:** 5–11 cm long, 1–2 cm thick, cylindrical or tapered to base, straight or slightly curved, sometimes slightly irregular or rooting into duff. Surface dry, upper portion sometimes with a slight ring zone of darker fibrils (soon disappearing). Yellowish ocher to orange, sometimes marbled or streaked looking. **Flesh:** Fibrous, distinctly ocher. **Odor:** Indistinct. **Taste:** Mild. **KOH**: Deep magenta. **Spore deposit:** Dark olive-gray. **Microscopy:** Spores 15–20 × 4–7 µm, long-ellipsoid, smooth. Cystidia long, abundant, thin-walled.

Ecology: Solitary or scattered in large groups near patches of *Suillus tomentosus* under two- and three-needle pines. Locally common in Cascadia, fruiting in fall into early winter on the coast.

Edibility: Edible.

Comments: *C. ochraceus* is a small to medium species with a grayish olive to pale ocher cap, often with a pointed umbo. The association with *S. tomentosus* is helpful, although the fruitbodies of both species are not always present at the same time. DNA sequence data suggests multiple species under this name, but distinguishing features of the cryptic species have not yet been established. *C. vinicolor* var. *californicus* and *C. pseudovinicolor* have rich wine-red or mahogany tones. *C. tomentosus* is entirely dull orange-yellowish and has a floccose-scaly cap surface; it is associated with "*Boletus mirabilis*" under western hemlock. *Chroogomphus* 'Small Dark' has dark fruitbodies and grows with *S. flavidus.*

Chroogomphus sp. 'Small Dark'

Slim Pine Spike

Cap: 1.5–6 cm wide, hemispherical when young, becoming bell-shaped with a round or pointed umbo, broadly convex in age. Surface smooth or with broad, appressed-fibrillose scales, becoming metallic-shiny age. Dark mahogany to nearly blackish, sometimes paler with more red-brown or wine-colored tones. **Gills:** Broadly attached to subdecurrent, thick-edged, widely spaced, with few short gills. Pale beige-orangey when young, maturing to dark brownish-black. **Stipe:** 4–10 cm tall, 0.6–1 cm wide, cylindrical with a tapered stipe base. Surface smooth to finely fibrous, dry. Pale orangey-beige to yellowish-beige. **Flesh:** Fairly firm, fibrous, orange. **Odor:** Indistinct. **Taste:** Indistinct. **KOH**: Deep magenta. **Spore deposit:** Dark olive-gray. **Microscopy:** Spores 20.5–24 x 6–7 µm, long-ellipsoid, smooth. Cystidia long, abundant, thin walled.

Ecology: Solitary or more often in scattered troops under shore pine and lodgepole pine with its presumed host, *Suillus flavidus.* Common in appropriate habitat along the coast, occasional in montane areas, fruiting in fall and early winter.

Edibility: Edible.

Comments: The genus *Chroogomphus* is quite easy to recognize even for beginners, and within the genus this species stands out by the combination of its small, slender stature, dark cap colors, and tripartite association with shore pine and the Slim Jack (*Suillus flavidus*). It is most similar to *C. vinicolor* var. *californicus*, which primarily grows with *S. pungens* under 3-needle pines in California.

Gomphidius maculatus

(Scop.) Fr.

Larch Slime Spike

Cap: 2–6 cm across, tight-hemispheric when young, expanding to round or broadly convex in age; margin inrolled at first. Surface smooth, viscid to glutinous. Creamy to pale grayish-beige, to tan-brown; darker over disc, becoming dingier grayish or with black spots in age. **Gills:** Broadly attached to decurrent, thick-edged, moderately widely spaced, clean and neat. Dingy whitish when young, soon pale gray to smoky gray or sooty olive at maturity. **Stipe:** 3–10 cm tall, 0.7–1.5 cm wide; straight, cylindrical or with a tapered stipe base. Surface smooth or with small tufts of fibrils, sticky-slimy, soon drying out. Dingy whitish to pale grey, becoming blotched with tan or brownish stains in age. Bruising red and then blackish when cut. **Partial veil:** Obscure; only present as a thin continuation of cap slime when young, lacking a fibrillose zone. **Flesh:** White to pale grey, yellow tones largely absent in the stipe base, slowly blackening when cut. **Odor:** Indistinct. **Taste:** Indistinct. **Spore deposit:** Sooty olive-gray to nearly black. **Microscopy:** Spores 14–22 × 6–9 µm, elongate, smooth.

Ecology: Scattered or in small groups under larch, restricted to high elevations and eastern Cascades. Uncommon to rare, fruiting late summer into fall. Which of the larch-specialist *Suillus* it associates with remains unclear—maybe restricted to one, or maybe a generalist occurring with multiple species.

Edibility: Non-toxic, but gooey, bland, and possibly a concentrator of heavy metals.

Comments: This unusual *Gomphidius* is recognized by its viscid to gooey cap surface, pale overall coloration that develops brown to grayish or black stains, lack of a yellow stipe base, smoky spore deposit, and lack of even a faint fibrillose partial veil. Combined with the larch habitat, this species is readily recognizable. The gills remain pale into maturity and might lead to confusion with white-spored *Hygrophorus*. *G. smithii* has pale pinkish colors when young but grows with *Suillus lakei* under Douglas-fir. Pale *Gomphidius* lacking yellow stipe bases occur with larches in the circumboreal zone, on multiple continents, and it has not been resolved whether the minor genetic differences between them warrant species-level recognition.

Gomphidius smithii

Singer

Smith's Slime Spike

Cap: 1.5–7 cm across, rounded with an inrolled margin when young, becoming domed to broadly convex, expanding to flat and then uplifted and often slightly wavy in age. Surface smooth, very viscid and gooey when wet, sticky-slimy when dry. Pinkish, pinkish buff, pinkish lilac with a paler margin at first, becoming more dingy, to ochraceous in age. **Gills:** Decurrent, slightly thick edged, widely spaced. White or creamy when young, soon becoming light gray or dingy beige-gray, and then smoky gray in age. **Stipe:** 5–8 cm long, 0.5–2 cm thick, cylindrical or tapered downward. Smooth to fibrous, but covered in an extensive layer of sticky slime. White or off-white, base of stipe lacking, or just faintly washed with yellow, soon developing dark stains from base up in age. **Partial veil:** Translucent layer of slime covering gills when young, soon rupturing into fibrils forming a ring around top of stipe that becomes blackish as it collects dark spores. **Flesh:** Thin, fibrous, firm to soft. Whitish throughout, sometimes with a hint of yellow in stipe base, staining blackish slowly when cut. **Odor:** Indistinct. **Taste:** Indistinct. **Spore deposit:** Sooty gray to black. **Microscopy:** Spores 14–18.5 × 4.5–6 µm, elongate, smooth.

Ecology: Solitary or in scattered groups in duff or soil under Douglas-fir with *Suillus lakei* mycelium nearby. Widespread throughout Cascadia but generally less abundant than other *Gomphidius*. Fruiting in fall and early winter.

Edibility: Non-toxic, but gooey, bland, and known to concentrate heavy metals.

Comments: The pinkish to buff cap (occasionally with lilac tones) and stipe that either lacks a yellow base or just has a blush of yellow help distinguish it from the brighter pink-capped *G. subroseus* and the smoky-purple *G. glutinosus*. The *Suillus lakei*/Douglas-fir association distinguish it from *G. maculatus*, which grows with larch. Older specimens with a stained stipe could be mistaken for *G. oregonensis*, but that species is typically larger, often grows in clusters, develops conspicuous black splotches on the cap, and has an extensively yellow stipe base.

Gomphidius subroseus

Kauffman

Rosy Slime Spike

Cap: 1.5–7 cm across, rounded with an inrolled margin when young, becoming domed to broadly convex, expanding to flat and then uplifted and often slightly wavy in age. Surface smooth, very viscid and gooey when whet, sticky-slimy when dry. Dark rosy pink, reddish pink or pastel pink, often quite evenly colored overall, sometimes more extensively creamy or grayish pink, paler toward margin and fading in age. **Gills:** Decurrent, slightly thick edged, widely spaced. White or creamy when young, soon becoming light gray or dingy beige-gray, and then smoky gray in age. **Stipe:** 5–8 cm long, 0.7–1.5 (2.5) cm thick, cylindrical or tapered downward. Smooth to fibrous, but covered in an extensive layer of sticky slime. White or off-white, base of stipe washed with yellow, usually fairly muted but sometimes brighter, becoming more yellowish upward in age. **Partial veil:** Translucent layer of slime covering gills when young, soon rupturing into fibrils forming a ring around top of stipe that becomes blackish as it collects dark spores. **Flesh:** Fibrous, firm to soft. White throughout except for yellow in stipe base, slowly darkening, grayish when cut. **Odor:** Indistinct. **Taste:** Indistinct. **Spore deposit:** Sooty gray to black. **Microscopy:** Spores 15–20 × 4.5–7 µm, elongate, smooth.

Ecology: Solitary or in scattered groups in duff of conifer forest under Douglas-fir with *Suillus lakei* mycelium nearby. *Gomphidius* are thought to be host-specific parasites on *Suillus*. Widely distributed and common throughout Cascadia, often fruiting in a large flush in early fall.

Edibility: Non-toxic, but gooey, bland, and known to concentrate heavy metals.

Comments: This beautiful species is identified by the rosy cap, pallid to grayish, decurrent gills, slimy-gooey cap and stipe, and yellow flesh near the base of the fruitbody. *G. smithii* is similar and also occurs with *S. lakei*. It has a dingier or paler pink to beige-pink cap and mostly lacks yellow colors on the stipe base. Members of the *G. glutinosus* group exhibit more smoky purple cap colors. The poorly known *G. pseudoflavipes* was described from under pine and true fir in the mountains of California but may occur in southern Cascadia. It has an ocher-brown cap and bright yellow lower stipe, but its enormous spores (18–35 × 6–8 µm) are its most distinctive feature.

Gomphidius glutinosus group

Purple Slime Spike

Cap: 4–7 cm across, rounded with an inrolled margin when young, expanding to bell shaped or broadly convex, then uplifted and wavy in age. Surface smooth, extremely gooey-viscid when wet, and with a thick moist, sticky layer even when dry. Purple-gray to smoky gray when very young, becoming purplish brown to lilac-brown, or occasionally more bluish gray, to grayish buff in age. At times with blackish spots or stains. **Gills:** Decurrent, slightly thick edged, widely spaced, quite clean and neat looking. Waxy white, soon light gray, slowly becoming smoky gray. **Stipe:** 4–9 cm long, 1–3 cm thick, cylindrical or tapered toward the base. Surface smooth to scaly-fibrous but extensively covered in a thick layer of slime except near apex. White with a chrome yellow stipe base, yellow extending farther upward in age, then developing a few dingy ocher-brown splotches. **Partial veil:** A thin layer of slime concealing gills when young, usually clear or milky translucent, soon rupturing and forming a ring of fine cobwebby fibers covered in dark spores around top of stipe. **Flesh:** Fibrous, firm to soft, white throughout except for brilliant yellow in base of stipe. **Odor:** Indistinct. **Taste:** Indistinct. **Spore deposit:** Dark gray to black. **Microscopy:** Spores 15–20 × 4–6 µm, elongate, smooth.

Ecology: Solitary or scattered in troops in duff and humus. One species in this group fruits in coastal spruce forests with no apparent *Suillus* or *Rhizopogon* present, while another is more common in higher elevation mixed conifer forests with *Suillus lakei.* Occasional in Cascadia, fruiting in fall.

Edibility: Not recommended. Although non-toxic, it is extremely gooey and known to concentrate heavy metals.

Comments: The name *G. glutinosus* appears to be applied to at least two species in Cascadia, neither of which is likely to be the same as the European species that the named is based on. *G. largus* was described from spruce habitat in New Mexico and may be the same as one of the Cascadian taxa. As a group, these are recognized by their slimy purplish caps, decurrent, widely-spaced, pale to smoky gills, and viscid white stipes with yellow bases. Separating paler specimens from *G. oregonensis* can be tough, but that species often develops blackish blotches in age and is more likely to grow in clusters. Pinkish-lilac variants of *G. glutinosus* can resemble *G. smithii*, which lacks the yellow stipe base, and *G. subroseus*, which is more deeply and evenly light pinkish to rose colored and is often less brilliantly yellow at the stipe base.

Gomphidius oregonensis

Peck

Blackening Slime Spike

Cap: (2) 4–15 cm across, rounded with an inrolled margin when young, expanding to broadly convex with a strongly incurved margin, then becoming flat or uplifted and often slightly wavy in age. Surface smooth, gooey-slimy when wet, often drying out and becoming shiny and metallic looking. Color extremely variable, ranging from dingy beige-white to tan with olive streaks when young, then becoming pinkish beige to tan, reddish tan, ocher-brown, usually developing sooty olive or blackish blotches and streaks, sometimes almost entirely black in age. **Gills:** Decurrent, somewhat thick edged and widely spaced. Whitish to cream when young, becoming dingy gray and often blackening entirely in age. **Stipe:** 6–15 cm long, 1–4 cm thick, cylindrical or variably tapered or slightly enlarged at base, often curved or sinuous. Smooth to fibrillose, covered in thin viscid layer or a sticky sheath of slime. White to dingy grayish beige at first and bright yellow over much of stipe base; developing sooty olive to black splotches. **Partial veil:** Thin membrane of translucent slime covering gills when young, breaking and then forming a ring of fibrils embedding in slime around apex of stipe that is soon blackish as it becomes covered in spores. **Flesh:** Thick, firm to soft, stipe fibrous, solid. Whitish throughout except for extensive yellow area near base of stipe. **Odor:** Indistinct. **Taste:** Indistinct. **Spore deposit:** Sooty gray to black. **Microscopy:** Spores 10.5–14 × 4.5–8 µm, elongate, smooth.

Ecology: Often in clumps with fused bases, or occasionally solitary in small troops in duff under Douglas-fir where *Suillus caerulescens* mycelium is present (the two sometimes fruit out of the same spot in the ground at the same time). Common throughout Cascadia, often fruiting in a large flush in early fall, with scattered fruitings into late fall.

Edibility: Non-toxic, but rather gooey and often downright unattractive. Also known to concentrate heavy metals.

Comments: This rather variable species is recognized by the medium to large fruitbodies, decurrent white to blackish gills, bright yellow stipe base, and tendency to develop extensive black splotches in age. The color of the cap can overlap with other *Gomphidius* species, but none of the others are as likely to grow in fused clusters or develop black stains in age.

Paxillus cuprinus

P. Jargeat, H. Gryta, J. P. Chaumeton & Vizzini

Coppery Pax

Cap: 3–10 cm across, convex with a dramatically inrolled margin when young, expanding to plane, often with a depressed center, margin remaining distinctly inrolled. Surface densely tomentose when very young, soon matted-felty and often breaking up in a ribbed pattern near margin; usually becoming smooth and shiny in age. Dull brown, olive-brown, or yellow brown often mottled, becoming coppery with ocher stains in age. **Gills:** Broadly attached to decurrent, forking multiple times, occasionally forming irregular "pores" around top of stipe; scraping off easily. Whitish cream when young, becoming dingy beige or mustard colored, staining reddish brown when handled. **Stipe:** 3–7 cm long, 1.5–2.5 cm thick, evenly cylindrical, stout. Smooth to fibrillose, covered with a pale bloom. Dingy cream to beige or with olive-yellow tones, often light pinkish near apex, soon stained brown. **Flesh:** Thick, firm, solid, marbled whitish and tan, quickly darkening to dark reddish brown when bruised or cut. **Odor:** Indistinct. **Taste:** Indistinct. **Spore deposit:** Dingy ocher to reddish brown. **Microscopy:** Spores 7–9 × 5–6 µm, almond-shaped, smooth.

Ecology: In small groups, clumps, or large arcs and rings on ground and in lawns surrounding planted silver birch (*Betula pendula*) and possibly with linden (*Tilia* spp). Fruiting from late summer through fall, largely restricted to urban areas.

Edibility: Toxic, but sometimes slow to manifest (some people eat it for years before experiencing negative effects; others react badly right away). Contains a compound that hypersensitizes the immune system, leading it to attack and destroy the body's red blood cells.

Comments: The distinctly inrolled margin, squat stature, coppery colors, and pale, forked gills that stain dramatically reddish brown separate it from most other species. There are a few other *Paxillus* species in Cascadia, distinguished by subtle aspects of color, fruitbody size, host association, and ITS sequence. Most often referred to as *P. involutus*, which is a smaller, uncommon to rare species in Cascadia that occurs with both urban birch and conifers in native habitats (at least in the northern portion of our range). There are at least two other species with large (sometimes downright huge!), pale whitish-beige fruitbodies growing with hardwoods (primarily cottonwoods and oaks). The names *P. ammoniavirescens* and *P. obscurisporus* have been applied, but both names are originally European and the Cascadian species are genetically slightly different.

Misapplied name: *Paxillus involutus* (Batsch) Fr.

Tapinella atrotomentosa

(Batsch) Šutara

Velvet-footed Tap

Cap: 5–15 (20) cm across, convex with an inrolled margin at first, becoming uplifted, wavy and often lobed in age. Surface dry, velvety to matted-tomentose, often cracking in age. Light brown, brown to rusty brown. **Gills:** Decurrent, close, often forking, narrow and scraping off easily. Creamy buff to light yellowish brown. **Stipe:** 2–10 cm long, 1.5–3.5 cm thick, central to off-center, rarely lateral, stout. Surface dry, distinctly velvety to matted-tomentose. Creamy beige to light brown at first, soon dark brown. **Flesh:** Thick, firm, somewhat fibrous. Pale yellowish to light brown, occasionally with vinaceous stains. **Odor:** Indistinct. **Taste:** Mild to bitter. **KOH:** Olive-yellow on all parts. **Spore deposit:** pale brown to yellowish brown. **Microscopy:** Spores 4.5–6 × 3–4.5 µm, broadly ellipsoid, smooth, yellowish.

Ecology: Solitary, scattered, or in clusters on rotting conifer stumps, logs, or buried roots. Quite common throughout Cascadia, typically fruiting in summer and early fall, with scattered fruitings later in fall.

Edibility: Not recommended, often bitter tasting. Commonly used by fiber artist to produce wool dyes.

Comments: The large size, velvety brown cap, fuzzy dark brown stipe, and growth on wood help identify this species. *Paxillus* species generally grow on the ground, stain brown, and lack the distinctive stipe texture. *Panus conchatus* has violet colors when young, but fades to beige. It has tough, leathery flesh.

Synonym: *Paxillus atrotomentosus* (Batsch) Fr.

Phylloporus arenicola group

Western Gilled Boletes

Cap: 1–3.5 cm across, convex to nearly plane, slightly uplifted and wavy in age. Surface dry, finely tomentose and suede-like, becoming appressed and smooth in age. Dark olive green to lighter forest green with tan areas, older specimens more reddish brown, becoming tan-brown in drier weather. **Gills:** Variably attached, notched to slightly decurrent, more rarely nearly free. Thick, widely spaced, often forked near attachment to stipe, inter-veined or anastomose, occasionally forming areas with pores. Bright yellow, duller in age, occasionally staining blue when damaged. **Stipe:** 3–9 cm long, 1–2.5 cm thick, cylindrical with a tapered base, often curved overall, base often with a tuft of fine yellow rhizomorphs. Surface dry, smooth or often with fine dark clumps of cells. Bright to dull yellow, often streaked with tan or brown, and brownish-punctate at apex. **Flesh:** Thin, firm to soft, yellow, occasionally bluing. **Odor:** Indistinct. **Taste:** Mild. **Spore deposit:** Dull olive-brown. **Microscopy:** Spores 9–12 × 4–5 µm, elongate-ellipsoid to somewhat spindle shaped, smooth.

Ecology: Often solitary or scattered in small groups, rarely prolific. Occurring throughout Cascadia in mixed woods, perhaps favoring habitats with Douglas-fir. Fruiting in fall and early winter on the west slopes and along the coast as well as in spring and summer in mountain areas.

Edibility: Edible.

Comments: The greenish to olive-brown caps and strongly contrasting bright yellow gills that often bruise slowly blue distinguish this species. The gill morphology of the fertile surface obscures the fact that this species is a member of the bolete family; however, the colors, velvety cap texture, and yellowish rhizomorphs are reminiscent of *Xerocomus subtomentosus*, a close relative. Preliminary ITS sequence data have suggested surprising species-level diversity in this group, with at least four species in western North America. However, stable macro-characters to distinguish them have not been established.

7 - Brown-spored Decomposers

The various genera included here are not all closely related, as evidenced by their wide range of fruitbody morphologies and ecological strategies. *Pholiota*, *Hemipholiota*, and *Flammula* fruitbodies are usually medium-sized to large, most often grow in troops or clusters, usually on wood or woody debris, and have dull brown spore deposits. Many have viscid, scaly caps and/or stipe surfaces, but a few are smooth. The stipes of some species have a partial veil that is usually slimy, weakly membranous, or evanescent and cortina-like. *Kuehneromyces* have distinctly hygrophanous caps and grow in clusters. Our single species of *Pyrrhulomyces* has a pinkish orange cap, a wispy partial veil, and dull brown spores. The single species of *Crassisporium* is readily recognized by its small, reddish brown to ocher fruitbodies, reddish brown spores, and growth in burnt areas. The single brown-spored *Cystoderma* (formerly *Phaeolepiota*) included here produces large fruitbodies with granular surfaces and a remarkable sheathing veil on the stipe.

Gymnopilus are distinguished by their rusty orange-brown spore deposits, typically bitter taste, and growth on lignin-rich humus or more often directly on dead wood.

Galerina have orangey brown spore deposits and produce tiny to small fruitbodies with bell-shaped or conical, translucent-striate caps. They are particularly common in moss beds. The deadly toxic *G. marginata* produces larger fruitbodies with a small, thin membranous ring on the stipe and grows on decaying wood.

Tubaria have small, dull pinkish tan to brick-brown or wine-colored fruitbodies, and caps that are usually translucent-striate. *Agrocybe* have opaque, yellowish to tan or dull ocher caps. Only a few *Agrocybe* produce fruitbodies with membranous partial veils but many species in these two genera show some form of partial veil tissue. Both genera produce dull brown spore deposits. *Cyclocybe* have dark brown (often slightly viscid) caps, membranous skirt-like partial veils, and dark brown spores.

Conocybe, *Bolbitius*, and *Pholiotina* produce small, slender, fragile fruitbodies and have orangey brown spore deposits. They are most often encountered on lawns, in manure piles, and in urban habitats with wood chips. Some species in each genus occur in forest settings. *Bolbitius* are easily distinguished from most *Conocybe* by their distinctly slimy caps. Most *Pholiotina* are hard to tell apart from *Conocybe*, with the exception of *P. rugosa* and its relatives, which have membranous partial veils.

Three common genera of beige- to brown-spored wood decayers with small fruitbodies are told apart by their textures and colors: *Simocybe* are typically olive-colored and have a finely velvety cap, *Flammulaster* are yellowy to beige or tan and distinctly granular or fluffy, and *Phaeomarasmius* are dark red-brown with finely hairy to hairy-scaly caps.

Identification of species within most of these genera is difficult and often requires examination of microscopic features. *Gymnopilus*, *Pholiota*, *Agrocybe*, and *Bolbitius* are common, conspicuous, and somewhat easier to put names to. *Galerina*, *Pholiotina*, and *Conocybe* are among the hardest to identify.

Crassisporium funariophilum

(M.M. Moser) Matheny, P.-A. Moreau & Vizzini

Cap: 1–3 cm across, broadly convex to plane, at times with an obtuse to broad umbo. Surface moist to dry, often with fibrillose veil remnants on margin when young, soon smooth, strongly hygrophanous. Reddish brown, orange-brown to ocher, occasionally evenly brown, often with paler veil flecks when young. **Gills:** Narrowly attached, notched, or seceding, close to subdistant, moderately broad; edges finely fringed. Yellow-brown when young, ocher-brown to more orange when mature. **Stipe:** 2–5 cm long, 0.2–0.4 cm thick, equal. Surface dry, silky to appressed-fibrillose below veil zone, smooth above. Ocher-brown to orange-brown, dingier with age. **Partial veil:** A fibrillose cortina which leaves a silky annular zone on the stipe and fibrils on the cap margin when young. **Flesh:** Thin, fibrous in stipe, orangish to brownish. **Odor:** Indistinct. **Taste:** Indistinct. **Spore deposit:** Reddish brown. **Microscopy:** Spores 8–11.5 × 5.5–7 µm, vaguely diamond-shaped to ovoid in front view, ellipsoid in side view, smooth, thick-walled, truncate with a germ pore, dark reddish cinnamon in KOH. Pleurocystidia absent, cheilocystidia abundant, spindle-shaped with a rounded to pointed apex, hyaline.

Ecology: Scattered, in troops, or in loose clusters on burnt ground, duff, and soil. Scattered fruitings may flush during the first spring following a forest fire but huge numbers appear two to three years post-fire. Especially common in the eastern Cascade Range but will appear anywhere in Cascadia within a 1–5 years after a forest fire. First fruiting as the snow melts (sometimes fruiting through the edge of the snowbank!) and continuing into mid-spring.

Edibility: Unknown.

Comments: The small size, reddish brown to ocher colors, reddish brown spores, and tendency to fruit in spring in a post-fire landscapes identify this otherwise nondescript 'LBM.' Fruitings carpet the ground as far as the eye can see in burned landscapes. It often grows alongside the pyrophilous *Pholiota* species which are distinguished by their viscid caps (when wet), duller brown spores, and non-hygrophanous caps. *Deconica* species are often smaller, have a translucent-striate cap and purplish black spores.

Synonym: *Pachylepyrium carbonicola* (A.H. Sm.) Singer

Pholiota brunnescens

A.H. Sm. & Hesler

Cap: 1.5–7 cm across, broadly convex to plane, at times with a low umbo. Surface glutinous when wet, to tacky, shiny and often with adhered debris when dry, with fine scales around margin when young, soon smooth. Dark brown to orange-brown, at times paler to yellow-brown, typically more red-brown to vinaceous brown when dry. **Gills:** Broadly attached or notched, close to crowded; partial gills numerous. Pallid at first, becoming grayish brown; cinnamon to yellow-brown when mature. **Stipe:** 2–7 (9) mm long, 0.5–1 mm thick, more or less equal. Surface dry, apex pruinose, fibrillose-scaly below veil. Pale yellow when young, darkening slightly orangish with age, staining brownish when handled. **Partial veil:** Thin, fibrillose, leaving remnants on cap margin, occasionally forming a ring on stipe. **Flesh:** Thin, fleshy-fibrous, stipe becoming hollow. Off-white to dingy yellowish. **Odor:** Indistinct. **Taste:** Indistinct. **KOH:** Unknown. **Spore deposit:** Dull brown. **Microscopy:** Spores 6–7.5 × 4–5 µm, ellipsoid to ovate, smooth, moderately thick-walled, germ pore small, yellowish brown to brown in KOH. Pleurocystidia swollen in middle, with long cylindrical necks or tapered outward, thin-walled, hyaline to pale yellowish. Cheilocystidia similar but smaller.

Ecology: Clustered or in troops on burnt soil or wood. Common for two to five years following forest fires, in campfire pits, or burn piles. Fruiting in fall - spring.

Edibility: Unknown.

Comments: The glutinous to tacky brown cap, pale gills, fibrillose-scaly stipe, and growth from burnt areas are helpful identification features. It occurs in large patches, often mingled with other burn *Pholiota* following forest fires. *P. molesta* often has a paler pinkish tan to orange-brown cap, a stouter stature, and a dingy yellow stipe which doesn't discolor when handled. *P. carbonaria* is a slender species with a light brown to orange-brown cap, often showing reddish veil remnants. Similar *Pholiota* from unburnt forests include: *P. velaglutinosa*, with a distinctly glutinous veil and growing from pine debris; *P. avellaneifolia*, with a yellowish, somewhat membranous veil and grows from conifer debris at higher elevations in summer and early fall.

Pholiota molesta

A.H. Sm. & Hesler

Cap: 1–5 cm across, convex, broadly convex to plane. Surface viscid to dry, appressed-fibrillose to smooth, often with fibrillose veil remnants on margin when young. Pinkish tan to orange-brown when young, developing brown tones with age. **Gills:** Broadly attached or notched, close to crowded; partial gills numerous. Pallid at first, becoming grayish brown, then cinnamon to dull brown when mature. **Stipe:** 1.5–7 cm long, (0.3) 0.5–1 cm thick, equal. Surface dry, pruinose at apex, fibrillose-scaly below veil zone. Whitish to pale yellowish ground color and at apex, lower portion with orangish appressed scales in age. **Partial veil:** Fibrillose cortina, leaving an annular zone on stipe and fibrils on cap margin when young. **Flesh:** Thin, firm, fibrous in stipe. Whitish, watery yellow to yellow-brown. **Odor:** Indistinct to somewhat pungent. **Taste:** Disagreeable. **KOH:** Unknown. **Spore deposit:** Brown. **Microscopy:** Spores 6–8 × 4–4.5 µm, ellipsoid to ovoid, with a small germ pore, smooth, thick-walled, cinnamon in KOH. Pleurocystidia abundant, broadly spindly-shaped with obtuse or pointed apex, thin-walled, hyaline. Cheilocystidia more clavate.

Ecology: Scattered in troops or loose clusters on burnt ground or in charcoal. Often seen in small numbers in campfire pits or burnt brush piles but can occur in huge numbers after forest fires. Especially common in the eastern Cascade Range but can appear anywhere the first five years after a forest fire. Fruiting in large flushes in spring and fall with stragglers whenever there is enough moisture in summer, or warm enough in winter. A closely related species, *P. carbonaria*, has been found growing vegetatively as an endophyte within lichen and tree tissue in unburned forests, and fruiting in profusion after fires. *P. molesta* probably has a similar ecology.

Edibility: Unknown.

Comments: This *Pholiota* is one of a handful of species that thrive in burnt areas. After forest fires, thousands of fruitbodies carpet the ground in otherwise barren areas. Among the species found in burnt habitats, *P. molesta* typically has a stout stature and paler colors when young, however, there is much overlap with *P. carbonaria (=P. highlandensis).* Typical specimens of *P. carbonaria* have a yellowish buff cap with yellowish to reddish ocher veil remnants on the margin, pale gills that darken in age, and a slender stipe covered with reddish ocher pointed scales. However, not all specimens exhibit the reddish ocher scales and *P. molesta* is occasionally slender. To make matters more confusing, these two species often grow intermingled with *P. brunnescens*, a larger, stockier species with a glutinous brown, orange-brown to reddish brown cap, and yellowish veil bands on the stipe. *Crassisporium funariophilum* is a smaller species with a smooth cap, ocher to reddish brown or brown colors, and with distinctive, vaguely diamond-shaped spores. It fruits in spring in burnt forests.

Hemipholiota populnea

(Pers.) Bon

Cap: 5–15 (20) cm across, rounded-convex when an inrolled margin when young, becoming broadly convex. Surface thinly viscid to dry, covered with floccose scales and fibrils, more appressed with age. Color variable, from white, creamy beige, light brown to cinnamon-brown, typically darker over disc. Scales white when young, remaining so, or darkening slightly with age. **Gills:** Broadly attached or notched, close, broad; partial gills numerous. Whitish at first, soon developing gray-brown tones as spores mature, brown in age. **Stipe:** 4–15 cm long, 1–3 cm thick at apex, typically club-shaped with a swollen base, occasionally more equal, often pinched at attachment point. Surface floccose, covered with erect to appressed scales. Whitish to light brown, with whitish scales; base developing brown tones with age. **Partial veil:** Thick, cottony-floccose, forming an annular zone, and leaving scales on stipe and cap. **Flesh:** Thick, firm, stipe often quite tough. Whitish or with brown tones in age. **Odor:** Indistinct. **Taste:** Unpleasant. **KOH:** Unknown. **Spore deposit:** Brown. **Microscopy:** Spores 7–9.5 × 4–5.5 µm, ovoid to ellipsoid, with a germ pore, thick-walled, smooth, cinnamon-brown in KOH. Cheilocystidia occasional to abundant, cylindrical to narrowly club-shaped, thin-walled, colorless; pleurocystidia absent.

Ecology: Solitary or clustered on cut ends of logs, or from wounds from living and dead cottonwood trees, logs, and other woody debris. Widespread in Cascadia, typically in riparian zones. Fruiting from late summer into winter, with the primary flush in fall.

Edibility: Unknown.

Comments: The creamy to light brown cap and stipe (which are covered with white scales), brown spores, and growth from cottonwood are distinctive. *Hemistropharia albocrenulata* has an orange-brown to dark reddish brown cap and yellowish to brownish floccose scales on the cap and stipe. The darker colors and smaller size help distinguish it from *H. populnea*. *Hypsizygus* grows in a similar manner but has a smooth cap and stipe with white spores.

Synonym: *Pholiota populnea* (Pers.) Kuyper & Tjall.-Beuk., *Pholiota destruens* (Brond.) Gillet

Pholiota scamba

(Fr.) M.M. Moser

Cap: 1–3 cm across, convex, broadly convex to plane, occasionally with an umbo. Surface viscid to tacky when wet, with appressed fibrils or tufts towards margin when young, becoming nearly smooth in age. Pallid, whitish to beige at first, darkening slightly to pinkish buff in age. **Gills:** Broadly attached, or with a slight notch and decurrent tooth, close to subdistant, narrow; partial gills numerous. Whitish to pale yellow at first, grayish to pale olive-brown as spores mature. **Stipe:** 1.5–3 cm long, 0.1–0.3 cm thick, equal, often curved. Surface dry, floccose with scales below veil, pruinose to nearly smooth above. Whitish to yellow, darkening slightly from base up in age. **Partial veil:** A relatively thin, cottony cortina leaving remnants on the cap margin, and an annular zone on stipe (occasionally an evanescent ring). **Flesh:** Thin, whitish to pale yellowish. Stipe fibrous, hollow in age, whitish, developing brownish tones in age. **Odor:** Indistinct to pleasant. **Taste:** Indistinct. **KOH:** Unknown. **Spore deposit:** Dark brown. **Microscopy:** Spores 7–9 × 4.5–5.5 µm, ellipsoid to nearly ovoid, with a small germ pore, smooth, moderately thick-walled, brown in KOH. Pleurocystidia spindle-shaped or swollen in the middle with a short neck and an obtuse apex, cheilocystidia similar, both thin-walled, hyaline in KOH.

Ecology: Solitary or scattered on woody debris or decaying bark of conifers. Uncommon, easily overlooked, widespread (more frequent in west-side and high elevation regional forests). Fruiting in fall.

Edibility: Unknown.

Comments: The small size, pallid colors, floccose-scaly stipe, and dark brown spores help distinguish this species. *P. lenta* is like a larger version of *P. scamba*, with caps 4–10 cm across; it occurs in thick duff and debris under conifers and hardwoods. The dull beige to creamy pinkish or orangish-capped *P. olivaceophylla* grows on woody debris at the edges of snowbanks in montane areas in spring and early summer. Although generally rare in the Pacific Northwest, it is known primarily from the southern Oregon Cascades and is rather common in the Mount Shasta region of California. *P. gummosa* typically grows in clusters from soil and has a distinctly scaly cap when young.

Pholiota decorata

(Murrill) A.H. Sm. & Hesler

Cap: 2–8 cm across, convex at first, becoming convex to plane, occasionally with a low, broad umbo. Surface viscid to dry, appressed-fibrillose, often with scales over the disc at first, soon nearly smooth. Creamy beige with a brown center or extensively pinkish brown, radially streaked with darker fibrils. **Gills:** Broadly attached to slightly notched, close to subdistant, broad; partial gills numerous. Creamy beige to yellowish tan at first, darkening to yellowish gray or pallid clay-colored. **Stipe:** 3–9 cm long, 0.3–1 cm thick, equal. Surface dry, covered with scruffy white tufts and hairs, flocculent when young, appressed in age. Base often with fuzzy mycelium and rhizomorphs. Whitish as first, soon with brownish streaks and stains, darkening from base up. **Partial veil:** A thin, flocculent membrane, leaving tissue on the cap margin and stipe. **Flesh:** Thin, fibrous in stipe, whitish to brown. **Odor:** Indistinct. **Taste:** Indistinct. **KOH:** Orange-brown on all parts. **Spore deposit:** Dark brown. **Microscopy:** Spores 5.5–8.5 × 3.5–4.5 µm, ovoid to ellipsoid in face view, flattened on one side in side view, with a tiny germ pore, smooth, thick-walled, rusty brown in KOH. Pleurocystidia with a slender base and swollen middle, apex rounded, thin- to thick-walled, hyaline to yellow in KOH. Cheilocystidia spindle-shaped to clavate.

Ecology: Solitary or scattered on small logs, branches, and surrounding duff of both conifers and hardwoods. Common across Cascadia. Fruiting in fall.

Edibility: Unknown.

Comments: This is one of the more easily identified *Pholiota* in our area thanks to the combination of a creamy beige cap with a brown, scaly center, white stipe with pronounced cottony scales, and growth on dead branches. *P. gummosa* is often clustered, fruiting from soil or buried debris (often in disturbed areas), and with creamy to pale yellow-brown cap with whitish scales. *P. terrestris* has a scalier cap, darker scales on the stipe, and growth in clumps and clusters on the ground or in wood chips.

Pholiota gummosa

(Lasch) Singer

Cap: 1–6 (8) cm across, broadly conical to convex with an incurved margin at first, becoming broadly convex to plane, occasionally with an umbo. Surface viscid when wet, with scattered appressed to tufted scales at first, becoming nearly smooth in age. Creamy to grayish, soon developing yellow and brown tones, sometimes extensively pale citrine yellow when young. **Gills:** Broadly attached, or with a slight notch and a decurrent tooth, close to subdistant, narrow; partial gills numerous. Creamy to pale yellow, darkening with age. **Stipe:** 3–7 cm long, 0.3–0.8 cm thick, equal or tapering toward base. Surface slightly viscid to dry, with pointed tufts and scales below veil at first, becoming more sparse and matted. Whitish to pale yellow, darkening slightly at maturity. **Partial veil:** Thin, membranous, leaving ragged remnants on cap margin, occasionally an evanescent ring on stipe. **Flesh:** Thin, firm, watery whitish, yellowish to light brown, darkening in age. Stipe fibrous, hollow in age. **Odor:** Indistinct or slightly sweet. **Taste:** Indistinct. **KOH:** Unknown. **Spore deposit:** Brown. **Microscopy:** Spores 6–7.5 × 3.5–4 µm, ellipsoid to nearly ovoid; with a small germ pore, smooth, moderately thick-walled, brown in KOH. Pleurocystidia clavate with ochraceous content in KOH. Cheilocystidia narrow with capitate head.

Ecology: Scattered in small clusters, rarely solitary, often in disturbed ground, lawns, and roadsides; occasionally in wood chips or on buried woody debris. Occasional in Cascadia, most records are from urban areas. Fruiting in fall.

Edibility: Unknown.

Comments: The pale yellowish colors, whitish scales on cap and stipe when young, brown spores, and growth in clusters from soil are important identification features. The very similar *P. terrestris* differs in having a tan to dark brown, distinctly scaly cap and stipe (when young). It is common in western Cascadia, also in similar disturbed habitats. See *P. scamba* for other pallid species. *Lacrymaria velutina* has an orange-brown, dry fibrillose-scaly caps and often grow in clusters, it is quite fragile and have black warted spores.

Pholiota avellaneifolia

A.H. Sm. & Hesler

Cap: 2–8 cm across, convex with an inrolled margin at first, becoming broadly convex to plane. Surface viscid to dry, with veil fibrils around margin and occasional tufts over disc when young, to smooth in age. Dark brown to reddish brown with pale yellowish fibrils around the margin when young, fading on margin, developing rusty tones with age. **Gills:** Broadly attached, or with a decurrent tooth, close to subdistant, broad; partial gills numerous. Dingy yellowish at first, developing smoky tones, grayish brown in age. **Stipe:** 3–8 cm long, 0.5–1.5 cm thick, equal or slightly enlarged toward base. Surface dry, covered with fibrillose, often chevron-like scales below the veil, smooth or longitudinally lined above. Pale yellowish to pale citrine yellow when young, soon developing orange-brown discoloration from base up in age. **Partial veil:** A fairly thick fibrillose membrane, leaving fibrils on cap margin and an annular zone on stipe. **Flesh:** Moderately thin, firm, whitish to yellowish, stipe fibrous, yellowish buff to brownish, darkening in age. **Odor:** Indistinct. **Taste:** Indistinct. **KOH:** Unknown. **Spore deposit:** Brown. **Microscopy:** Spores 6–7 × 3–4 µm, ovoid to ellipsoid, with a small germ pore, smooth, moderately thick-walled, cinnamon in KOH. pleurocystidia broadly spindle-shaped, or with a narrow base and swollen middle, thin-walled. Cheilocystidia swollen in middle, cinnamon-brown in KOH.

Ecology: Scattered in troops or clusters, rarely solitary, in duff or on well decayed woody debris under spruce, fir, and hemlock. Common at higher elevations. Fruiting in summer and fall.

Edibility: Unknown.

Comments: The brown cap, pale yellowish veil (which is unusually membranous for a *Pholiota)*, yellowish to grayish brown gills, and growth in conifer duff at higher elevations are helpful in identifying this otherwise non-descript mushroom. *P. velaglutinosa* is quite similar but has a glutinous veil when young and is more common in southern Cascadia, often on pine debris.

Pholiota flammans

(Batsch) P. Kumm.

Flaming Pholiota

Cap: 3–7 (10) cm across, convex with an incurved margin at first, becoming broadly convex to nearly plane, occasionally with an umbo. Surface dry, covered with tufts or curved fibrillose scales when young, occasionally washing off and becoming nearly smooth in age. Brilliant golden yellow when young, developing orange tones at it ages. **Gills:** Broadly attached, at times with a shallow notch and a decurrent tooth, close, narrow; partial gills numerous. Bright yellow at first, becoming rusty yellow as spores mature. **Stipe:** 3–8 (10) cm long, 0.4–1 cm thick, equal, or enlarged downward, often with a pointed base. Surface covered with bands of curved and sometimes pointed scales, becoming more matted in age; smooth above the veil. Bright yellow, discoloring only slightly in age. **Partial veil:** A ragged ring zone on the upper stipe, leaving remnants on the cap margin when young, almost completely disappearing in age. **Flesh:** Moderately thick, firm, fibrous in stipe. Yellow to greenish yellow. **Odor:** Indistinct. **Taste:** Indistinct. **KOH:** Orange-brown. **Spore deposit:** Brown. **Microscopy:** Spores 4–5 × 2.5–3 µm, oblong to ellipsoid, smooth, thick-walled, ocher in KOH. Pleurocystidia abundant, club-shaped or pointed, thin-walled, with yellowish content in KOH. Cheilocystidia similar or small and cylindrical.

Ecology: Solitary, scattered, or in small clusters on decaying logs and stumps. Often on conifer wood but occasionally on alder. Common but rarely abundant in western Cascadia, rarer on the east side. Fruiting from late summer into early winter with the primary flush in fall.

Edibility: Reportedly non-toxic, but not recommended.

Comments: The striking golden yellow colors, scaly cap and stipe, and growth on wood make this beautiful *Pholiota* easy to recognize. *P. adiposa* has a viscid, yellow to golden orange cap covered in tufted to appressed scales. *P. squarrosoides* has a viscid, beige to orange cap covered with upright or curved orangish scales. *P. squarrosa* has a dry, beige to brown, scaly cap and similarly scaly stipe.

Pholiota adiposa group

Golden Scaly Pholiota

Cap: 4–10 (15) cm across, convex with an inrolled margin at first, becoming broadly convex to plane with a broad umbo. Surface thinly viscid when wet, covered with tufted to appressed scales when young, more appressed and sparse with age. Golder yellow when young, soon developing more orange to ocher tones with age. **Gills:** Broadly attached or with a decurrent tooth, close to subdistant, broad; partial gills numerous. Yellowish to straw yellow, developing brownish tones as spores mature. **Stipe:** 5–12 cm long, 0.5–1.5 cm thick, equal or enlarged slightly toward base. Surface slightly viscid to dry, finely floccose to smooth above annular zone, covered with fibrillose upcurved to flattered scales below the veil. Yellowish to yellow-brown, developing orange-brown discoloration from base up in age. **Partial veil:** A fibrillose membrane, leaving fibrils on the cap margin and as a ring zone when young. **Flesh:** Thin to moderately thick, firm, yellowish, stipe fibrous, yellowish buff to brownish, darkening in age. **Odor:** Indistinct to slightly sweet-pungent. **Taste:** Indistinct. **KOH:** Unknown. **Spore deposit:** Brown. **Microscopy:** Spores 5–9 (11) × 3–6 µm, ovoid to ellipsoid, with a small germ pore, smooth, moderately thick-walled, cinnamon in KOH. Pleurocystidia of two types; spindle-shaped, thin-walled, dark brown in KOH, or spindle-shaped or more broad in middle, with yellowish content in KOH.

Ecology: Scattered in troops or clusters, rarely solitary, on logs, stumps, and dead trees—also from wounds on living trees. Typically on conifers but occasionally on hardwoods. Common across Cascadia but more abundant in the eastern Cascades. Fruiting from early summer through fall.

Edibility: Unknown.

Comments: The golden cap with appressed scales, yellowish gills, brown spores, and growth in clusters on wood help outline this species complex. The names *P. adiposa*, *P. limonella*, *P. aurivella*, *P. abietis*, *P. connata*, *P. filamentosa*, and *P. hiemalis* have been applied to Pacific Northwest collections matching the above description. These have been split mostly based on spore size, the extent of scaliness, and details of color (all of which are widely variable features). It remains to be seen how many species in this group occur in Cascadia, but many of these names will likely end up being treated as synonyms of *P. adiposa*. *P. squarrosoides* is similar in many regards, but has erect, often curved scales on the cap and stipe. *P. squarrosa* has a dry, scaly, beige to brown cap and stipe.

Pholiota spumosa group

Slender Pholiota

Cap: 2–6 cm across, conical to convex with an incurved margin at first, becoming broadly convex to plane, often with an umbo. Surface viscid to tacky, fibrillose-streaked to smooth. Color variable, generally yellow-brown with an ochraceous disc and an overall olive cast when young, becoming more ochraceous brown and dingy in age. Occasionally lacking ochraceous tones and more olive-yellow in color. **Gills:** Broadly attached, or with a slight notch and a decurrent tooth, close to subdistant, narrow; partial gills numerous. Yellowish green at first, soon beige, darkening to brown as spores mature. **Stipe:** 3–6 (10) cm long, 0.3–0.5 (0.7) cm thick, equal or enlarged slightly toward base. Surface dry, finely fibrous to smooth below veil, pruinose above. Pale yellow to pale greenish yellow as first, developing stains and streaks and darkening to dingy orangish brown from base up in age. **Partial veil:** Thin, cobwebby, adhering to cap margin when young, leaving sparse silky fibrils on stipe. **Flesh:** Thin, firm, stipe fibrous, watery greenish in cap. Stipe becoming hollow, yellowish green at first, darkening in age. **Odor:** Indistinct. **Taste:** Indistinct. **KOH:** Pale orange-brown. **Spore deposit:** Brown. **Microscopy:** Spores 5.5–8 × 4–4.5 (5.5) µm, ovoid, with a small germ pore, smooth, moderately thick-walled, golden brown in KOH. Pleurocystidia spindle-shaped with obtuse apex, thin-walled, with a yellowish content in KOH, cheilocystidia similar, with some more club-shaped.

Ecology: Scattered in troops or small clusters, often on wood chips or on and around well-decayed logs and stumps of both conifers and hardwoods. Common and widespread across Cascadia. Fruiting in fall and early winter.

Edibility: Unknown.

Comments: The viscid, yellow-brown cap with an ochraceous disc, light yellowish gills with a greenish cast when young, silky partial veil, and growth on rotting wood (most often wood chips) are helpful identifying features of this *Pholiota.* There are a number of species in this complex described from North America, but modern taxonomic work (and sequencing type collections) is needed to help sort them out. They may all end up being treated as synonyms of the European species *P. spumosa. Flammula alnicola* often grows in clusters on dead alder stumps and logs; it has a smooth yellow cap when young that darkens slightly in age, pallid to straw-yellow gills, and a fragrant odor. *Hypholoma fasciculare* has a yellowish cap (occasionally olive or ocher toned), greenish yellow gills that turn gray, and purplish black spores. *H. capnoides* lacks the greenish yellow colors of the latter species and has pallid gills that turn gray, and a smoky gray spore deposit. Similar *Gymnopilus* species have rusty orange spores and brighter orange colors, and most have a bitter taste.

Flammula alnicola

(Fr.) P. Kumm.

Cap: 3–8 (12) cm across, conical to convex at first, becoming broadly convex to plane with a broad umbo. Surface thinly viscid to dry, with scattered fibrils towards margin when young, soon smooth. Yellow to golden at first, developing ocher to orange tones, dingier in age. **Gills:** Broadly attached or with a decurrent tooth, close to subdistant, broad; partial gills numerous. Bright yellow to straw-yellow when young, soon developing ocher to brown tones as spores mature. **Stipe:** 4–12 (16) cm long, 0.4–1.2 cm thick, often disproportionately tall, equal or enlarged slightly toward base. Surface finely floccose, appressed-fibrillose to smooth. Yellowish at first, soon developing brown tones from the base up, all brown in age. **Partial veil:** A thin fibrillose membrane, typically disappearing with age. **Flesh:** Moderately thin, firm, yellowish; stipe hollow, fibrous, yellowish to brownish when young, dingy brown in age. **Odor:** Indistinct to sweet. **Taste:** Mild. **KOH:** Unknown. **Spore deposit:** Brown. **Microscopy:** Spores 8–10 × 4–5 µm, ovoid to ellipsoid, with a small germ pore, smooth, yellow-brown in KOH. Cheilocystidia bladder-, spindle- to club-shaped, thin-walled, colorless; pleurocystidia absent.

Ecology: Often in clusters and troops, rarely solitary, on stumps, roots, logs and other woody debris. Abundant on alder, less frequently on other hardwoods, occasionally on conifers. Widespread in western Cascadia with a preference for lower elevation forests, largely absent from the east side of the Cascades. Fruiting from late summer through winter, with the primary flush in fall.

Edibility: Unknown.

Comments: The yellow to golden colors (when young), tall stature, growth on wood (which is sometimes buried), and brown spores help define this species. It can be quite abundant in fall in stands of alder. It has also been identified as *Flammula* (or *Pholiota) malicola* and *F. flavida*; both are likely synonyms. Most similar is *Hypholoma fasciculare*, which has brighter citrine yellow gills when young and dark purple-gray spores; it tends to be a bit smaller and fruits in dense tufts.

Synonym: *Pholiota alnicola* (Fr.) Singer

Pyrrhulomyces astragalinus

(Fr.) E.J. Tian & Matheny

Pink Pholiota

Cap: 1–5 cm across, rounded-conical to convex, soon flattening or with wavy margin in age. Surface dry to moist, appressed-fibrillose to smooth, often with veil remnants around margin when young. Pinkish orange to peachy orange with a paler margin at first, fading to dull peachy with a whitish to buff margin, occasionally developing dingy stains in age. **Gills:** Broadly attached to slightly notched, close to subdistant, broad. Pinkish to pale yellow at first, darkening to yellow-brown or light brown. **Stipe:** 4–12 cm long, 0.3–0.7 cm thick, equal, base often tapered or rooted into substrate. Surface dry, with fine scruffy white fibrils, becoming matted in age. Whitish as first, with a pinkish buff base color, becoming more orange in age. **Partial veil:** Thin, whitish wispy fibrils, leaving patches on stipe and cap margin when young. **Flesh:** Thin, relatively fragile, orangish to brownish in cap; stipe fibrous, hollow in age, whitish to pale orangish brown. All parts slowly turning grayish to blackish when damaged. **Odor:** Indistinct. **Taste:** Bitter. **KOH:** Orange-brown on all parts. **Spore deposit:** Brown. **Microscopy:** Spores 6–8 × 4–5 µm, ovoid to ellipsoid in face view, with a small germ pore, smooth, thick-walled, rusty brown in KOH. Pleurocystidia spindle-shaped to club-shaped with a pointed tip, thin-walled, content bright yellow in KOH. Cheilocystidia with a swollen base and a long cylindrical neck, yellowish in KOH.

Ecology: Solitary, scattered, or in small clusters on rotting conifer logs and stumps. It prefers, but not restricted to, decaying western hemlock. Common in western Washington into central Oregon, rare elsewhere. Fruiting in fall, rare winter and spring.

Edibility: Unknown.

Comments: The peachy pink cap, pale pinkish or yellowish gills that become brown, pale stipe, and bitter taste identify this species. It could be confused for a *Hypholoma*. *H. fasciculare* has a yellower cap (occasionally orangey) and bright yellowish green to greenish gray gills. *H. capnoides* has a paler, dull orangey beige cap and light grayish gills. Both produce dark grayish purple spore deposits.

Synonym: *Pholiota astragalina* (Fr.) Singer

Kuehneromyces lignicola

(Peck) Redhead

Cap: 1–3 (4.5) cm across, typically conical to bell-shaped at first, becoming convex to plane, often with an umbo. Surface moist to dry, with scattered silky fibrils towards margin when young, soon smooth. Strongly hygrophanous, translucent-striate near margin when wet, fading and becoming opaque when dry. Dull dark yellow-brown when wet, straw-brown to pale yellow-brown when dry. **Gills:** Narrowly attached, crowded, narrow, edges at times eroding and ragged in age. Dingy tan, yellowish brown to dingy brown. **Stipe:** 3–6 cm long, 0.2–0.4 cm thick, equal. Surface dry, silky to appressed-fibrillose below annular zone, smooth above to more or less smooth in age. Yellowish brown to dingy brown, soon darkening dark brown tones from base up. **Partial veil:** Thin, cortina-like, leaving fibrils on the cap margin, and occasionally a silky annular zone on stipe. **Flesh:** Thin, firm, fragile, stipe hollow. Brownish in cap, dark dingy brown in stipe, especially with age. **Odor:** Indistinct. **Taste:** Mild. **Spore deposit:** Cinnamon-brown. **Microscopy:** Spores 5.5–7.5 × 3–4.5 µm, ovoid to ellipsoid, with a small germ pore, smooth, cinnamon-brown in KOH. Cheilocystidia spindle-shaped, or with a swollen middle and elongate neck, thin-walled; pleurocystidia rare.

Ecology: In clusters or large troops from conifer logs, stumps, and other woody debris. Possibly hardwoods. Abundant across Cascadia in a wide variety of habitats. Fruiting soon after snow melt into summer; fall and winter fruitings are rare.

Edibility: Unknown, do not experiment.

Comments: The hygrophanous cap with light to dark yellow-brown colors, narrow, crowded gills, stipe darkening from the base up and clustered growth on logs and other woody debris are helpful. *K. obscurus* is similar, but supposedly broader, more widely spaced gills and larger spores (7–10 × 3.5–4.5 µm). *K. mutabilis* is often larger, with membranous partial veil leaving a ring, and scurfy scales on the stipe below ring. *Galerina marginata* has a hemispherical to convex cap, a membranous partial veil leaving a small ring, and rusty brown spores. *Psathyrella piluliformis* and *Hypholoma* have purplish-black spores.

Synonym: *Pholiota lignicola* (Peck) Jacobsson

Kuehneromyces mutabilis

(Schaeff.) Singer & A.H. Sm.

Cap: 3–6 cm across, obtusely conical to bell-shaped at first, soon convex to nearly plane, typically with a broad umbo. Surface thinly viscid to moist, smooth with veil flecks when young. Strongly hygrophanous, translucent-striate near margin when wet; colors often two-toned and highly variable. Ocher-brown, orange-brown yellow-brown, stray-yellow to tan. **Gills:** Narrowly attached, close, broad. Pallid to tan at first, cinnamon-brown as spores mature. **Stipe:** 4–10 cm long, 0.3–1.2 cm thick, equal or tapered downward. Surface dry, distinctly scaly or scrufy when young below veil, remaining so or becoming smooth overall in age. Light brown to yellowish brown with pallid scales when young, soon darkening from base up, to dark brown. **Partial veil:** Membranous-fibrillose, collapsing to a ring or annular zone on stipe. **Flesh:** Thin, firm, stipe fibrous, brownish. **Odor:** Pleasant. **Taste:** Mild. **Spore deposit:** Cinnamon-brown. **Microscopy:** Spores 5.5–7.5 × 3.5–4.5 µm, ovoid to ellipsoid, with a small germ pore, smooth, cinnamon-brown in KOH. Cheilocystidia abundant, nearly cylindrical to spindle-shaped, some with rounded heads, thin-walled, colorless; pleurocystidia absent.

Ecology: In clusters or large troops from logs, stumps, and other woody debris of both hardwoods and conifers. Abundant across Cascadia in a wide variety of habitats. Fruiting throughout the year, with peaks in fall and spring.

Edibility: Not recommended. Too easily confused with the Deadly Galerina.

Comments: This common species is identified by the hygrophanous and often two-toned ocher to tan cap, scaly stipe below the partial veil, and clustered growth on logs and other woody debris. *K. lignicola* shares the clustered growth habit but is typically smaller and lacks the stipe scales. The deadly toxic *Galerina marginata* has a hemispherical to convex cap, a fibrillose to smooth stipe, and rusty brown spores.

Synonym: *Pholiota mutabilis* (Schaeff.) P. Kumm

Gymnopilus luteofolius

(Peck) Singer

Purple Gym

Cap: 3–12 cm across, convex with an inrolled margin at first, to nearly plane in age. Surface dry to moist, densely covered with upright scales and tufts when young, becoming radially appressed-fibrillose and often metallic looking in age. Margin often with whitish patches of veil, soon covered with rusty spores. Purple or maroon-red at first, soon duller red to reddish pink or purplish pink (often with a contrasting yellow or whitish margin), then showing yellow-orange between scales, often predominantly orange overall in age. Variably marked with blue-green dots or blotches. **Gills:** Broadly attached with a fine decurrent tooth, close. Greenish yellow when young, becoming bright orange to brownish orange as spores mature. **Stipe:** 5–10 cm long, 1–2.5 cm thick, cylindrical with an enlarged base, often curved. Surface dry, smooth or vertically striate. Pinkish purple to yellowish orange, with a silky white sheen when young, often appearing vertically marbled or variegated. Erratically bruising blue-green. **Partial veil:** Pronounced white cortina (sometimes nearly membranous) set high on stipe, soon covered in rusty spores, sometimes present only as a sparse ring zone in age. **Flesh:** Firm, fibrous, yellowish orange, mottled with purplish red in cap. **Odor:** Indistinct. **Taste:** Strongly bitter. **KOH:** Orangish. **Spore deposit:** Bright orange-brown. **Microscopy:** Spores 5.5–8.5 × 3.5–4.5 µm, ellipsoid to nearly ovoid, flattened on one side in side view, finely warted, dextrinoid. Cheilocystidia bottle-shaped to flask-shaped, with a slender neck, with or without a round head.

Ecology: In large clumps and troops on wood chips or scattered on conifer logs or debris. Fairly common in urban areas, occasional elsewhere. Fruiting in fall.

Edibility: Mildly hallucinogenic, very bitter.

Comments: The caps phase from purple to maroon-red, scaly young to soon duller red with yellow-orange showing in the cracks, to all orange in age. Other helpful characters are greenish yellow young gills, whitish partial veil, orange-brown spores, and growth on wood. *G. aeruginosus* may be a more appropriate name for the taxon profiled above, but until this assertion is backed by more conclusive evidence, we have chosen to continue using the more familiar name, *G. luteofolius*. *Tricholomopsis rutilans* can look very similar and sometimes grows in wood chips, but its fruitbodies taste mild, lack a cobwebby partial veil, and have white spores.

Gymnopilus ventricosus

(Earle) Hesler

Western Jumbo Gym

Cap: 5–30+ cm across, rounded with an inrolled margin at first, becoming convex to plane, sometimes wavy in age. Surface dry to moist, smooth, innately fibrillose, sometimes with scattered, pale, appressed-fibrillose patches. Margin often with adhering pale patches of partial veil. Yellowish orange to orange, often with pale streaks, in age extensively streaked with reddish brown. **Gills:** Notched with a fine decurrent tooth, sometimes more broadly attached, close; edges irregular. Pale yellow when young, becoming bright ocher, sometimes with distinct pale edges; darker brown in age, bruising brownish when damaged. **Stipe:** 10–30 cm long, 2.5–10 cm thick, curved, with a strongly swollen lower portions, tapering at base, especially when fruiting in dense clusters. Surface fibrillose, often ridged, and sometimes scaly near apex. Whitish to yellowish or with ocher-brown vertical streaks, bruising brownish when handled. **Partial veil:** Conspicuous, a thick cortina or membranous. Whitish yellow when young, eventually collapsing and becoming covered in rusty spores. **Flesh:** Thick, rubbery-firm, fibrous. Marbled yellowish to light orange throughout. **Odor:** Indistinct. **Taste:** Very bitter. **KOH:** Dark red-brown on cap, orangey red on gills. **Spore deposit:** Rusty orange-brown. **Microscopy**: Spores 7.5–9 × 4–5.5 µm, ellipsoid, roughened, slowly dextrinoid, ocher-brown in KOH. Pleurocystidia swollen or bottle-shaped. Cheilocystidia bottle-shaped with a rounded, often swollen head.

Ecology: Scattered or clustered, rarely solitary at the base or higher up trunk of dead conifers, and on stumps or large downed logs. Common across Cascadia. Fruiting in late summer and fall.

Edibility: Inedible, very bitter. Similar species in eastern North America are hallucinogenic, but ours are not.

Comments: The large, orange fruitbodies in clusters on wood, bitter taste, and the rusty brown spores help distinguish this species. Quite similar is *G. voitkii,* which is typically slightly smaller and may have a preference for fir and hemlock. This contrasts *G. ventricosus* which seems to prefer (but is not restricted to) pine. Until better macroscopic features are defined to help distinguish *G. voitkii*, it is best identified by the larger spores; 8–10 × 5.5–7 µm. Both species were known as *G. junonius (=G. spectabilis)* in the past. Jumbo Gyms are great dye mushrooms, yielding butter-yellow colors on wool and silk.

Gymnopilus aurantiophyllus

Hesler

Cap: 3–11 cm across, convex with an incurved margin, expanding to nearly plane. Surface dry to moist, edge opaque, often with adhering remnants of whitish partial veil, becoming covered in spores. Orange to ocher at first, usually marbled and streaked with brick red, brownish red and yellowish, and whitish yellow toward margin. **Gills:** Narrowly attached with a distinct notch, close, narrow, edge even or irregularly saw-toothed in age. Dull pale beige to yellowish beige, becoming more orange to ocher brown in age; edge contrastingly pale, bruising brownish when damaged. **Stipe:** 5–13 cm long, 1–2.5 cm thick, cylindrical with an enlarged base. Surface dry, longitudinally striate, fibrous or ridged. Base with whitish mycelium attached to woody substrate. Pale yellowish white to beige, vertically streaked, darkening to ocher-brown or reddish brown upward in age or when handled. **Partial veil:** Thick white cortina, sometimes nearly membranous when young, set high on stipe, soon covered in rusty spores, often collapsing and sparse in age. **Flesh:** Fibrous; marbled whitish yellow. **Odor:** Indistinct to slightly farinaceous. **Taste:** Bitter. **Spore deposit:** Bright orange-brown. **Microscopy:** Spores 7–9 × 4–5.5 µm, almond-shaped, slightly roughened, slightly thick-walled, dextrinoid. Cheilocystidia densely packed, vase-shaped with a rounded head.

Ecology: Solitary, scattered, or in clumps on wood chips, branches, small logs, and other woody debris of conifers, especially pine. Common across Cascadia, especially in pine forest. Fruiting in fall.

Edibility: Inedible, very bitter.

Comments: Species in this group of *Gymnopilus* are recognized by their yellowish orange to ocher, medium-sized fruitbodies with conspicuous partial veils and growth on well decayed logs, wood chips, and other lignin-rich debris. However, identification to species is a real challenge, primarily due to vague and sometimes competing original concepts but also due to the genuine similarity to one another. Most prior authors used the name of a European taxon, *G. sapineus*, as a catchall for this group. *G. penetrans* has an orange-brown cap and often gets rusty spots on the gills. It prefers small hardwood debris but will also grow on conifers.

Gymnopilus oregonensis

Murrill

Cap: 1.5–5 cm across, rounded to convex, sometimes with a broad umbo. Surface moist to dry, smooth to finely scurfy or appressed scaly. Margin fringed with a narrow overhanging band of translucent tissue. Reddish brown, brick-red to orange-brown. **Gills:** Finely attached, close, and rather narrow. Pale creamy yellow to yellow when young, soon dull orange-tan, rusty to brown as spores mature. **Stipe:** 3–8 cm long, 0.2–0.7 cm thick, more or less equal, often curved. Surface dry, finely powdery. Dull orange-tan to brown with a whitish bloom, base darkening in age. **Flesh:** Thin, fibrous, orange-brown; stipe hollow. **Odor:** Indistinct. **Taste:** Slightly bitter. **KOH:** Dark brown to almost black on cap. **Spore deposit:** Rusty orange-brown. **Microscopy:** Spores 7.5–9 × 4.5–5.5 µm, ellipsoid, roughened to spiny-warted, ocher-brown in KOH. Cheilocystidia flask-shaped with a round head, pleurocystidia similar.

Ecology: Scattered or in troops, rarely solitary on decaying conifer wood, especially large, moss-covered logs. Common across western Cascadia. Fruiting from late summer through fall.

Edibility: Unknown.

Comments: The small size, nearly smooth cap, reddish brown to brick colored cap, growth on wood and lack of a partial veil help distinguish this species. Reports of *G. picreus* in Cascadia likely pertain to this species. *G. rufescens* is typically slightly smaller, paler, and has considerably smaller spores (4–5.5 × 2.5–3.5 µm). *G. penetrans* has more orange colors, and typically grows on hardwood debris.

Gymnopilus punctifolius

(Peck) Singer

Blue-green Gym

Cap: 3–10 cm across, rounded to broadly convex with an inrolled edge, often very irregular and wavy or contorted. Surface dry, silky-fibrillose, sometimes cracking into small plaques in age; covered in a fine bloom that rubs off when handled. Color variable within a distinctive range. Dull sky blue when young, becoming mottled and streaked with green, yellow, olive, and ocher-tan, often all on a single cap. Generally duller and more brownish in age, brighter greenish yellow to blue-gray when young. **Gills:** Narrowly attached or notched, close to subdistant, narrow. Bright yellow-green when young, olive-tan to ocher or brown in age; often with a pale and uneven edge, bruising brown when damaged. **Stipe:** 5–16 cm long, 0.5–1 cm thick, cylindrical, curved, often strongly sinuous with a tapered, rooting portion buried in wood. Surface with a pale bloom (sometimes vertically striate); apex scurfy with short yellow tufts. Bright greenish yellow to powdery blue-gray or dingy olive overall; base with patches of lilac mycelium. **Flesh:** Thin, cap fragile, stipe fibrous. Yellowish olive, bruising brown. **Odor:** Indistinct. **Taste:** Bitter. **KOH:** No reaction. **Spore deposit:** Rusty brown to darker ocher-brown. **Microscopy:** Spores 4–6.5 × 3.5–5 µm, ovoid to broadly ellipsoid, finely warted, dextrinoid, brownish in KOH. Pleurocystidia slender or swollen in the middle, with a swollen head. Cheilocystidia similar.

Ecology: Solitary or in clumps or dispersed small groups inside hollowed cut ends and undersides of old-growth conifer logs. Locally common in suitable habitat across western Cascadia, rarely occurring outside old-growth forest. Fruiting in fall.

Edibility: Unknown. Reports of psilocybin content are unsubstantiated.

Comments: The bitter taste, growth on wood, and rusty spores place this species in *Gymnopilus*. The variable sky-blue, green, yellow, and olive tones coupled with the lilac basal mycelium are unique. Smaller, yellower forms could be confused with *Hypholoma fasciculare* but that species produces a purplish gray spore deposit.

Simocybe serrulata

(Murrill) Singer

Cap: 1–3 cm across, rounded when young, soon convex to nearly plane in age; sometimes umbonate. Surface dry, at first distinctly finely velvety, but can be nearly smooth in age. Dark olive-green or dark olive-brown, soon paler to olive-drab or brownish tan, almost always with a pale sheen. Margin narrowly whitish when very young, extensively beige or ocher in age. **Gills:** Narrowly attached, close to subdistant, narrow. Dull olive to pale brownish tan, sometimes dark brown in age; edges fringed or ragged, sometimes also with clear, milky, or brown droplets when young. **Stipe:** 1–6 cm long, 0.1–0.5 cm thick, cylindrical, often curved, slightly enlarged toward base. Surface entirely covered with pale tiny scurfy scales, smoother and often vertically striate in age. Olive to grayish tan or yellowish olive; base often with a conspicuous tuft of white mycelium. **Flesh:** Very thin, fibrous, brownish. **KOH:** Reddish or orange on cap. **Odor:** Indistinct. **Taste:** Indistinct or slightly bitter. **Spore deposit:** Dull brown. **Microscopy:** Spores 7–10 × 4–5.5 µm, bean-shaped or ellipsoid, lacking a germ pore. Cheilocystidia with broad, round heads and tapered bases

Ecology: Solitary, scattered in small troops or clusters on decaying hardwood branches and logs, especially oak and bigleaf maple. Locally common at lower and mid-elevations across western Cascadia, rare elsewhere. Fruiting throughout the year but most abundant in fall and spring.

Edibility: Unknown.

Comments: *Simocybe* are recognized by their small size, finely velvety caps, olive colors, fringed gill edges, and growth on wood. There is uncertainty about the number of species in Cascadia and which names to apply. *S serrulata* appears to be the most common and largest of the known species; its cheilocystidia have broad, round heads. *S. centunculus* (sensu PNW) is slightly smaller, has brighter forest greenish colors when young, and cylindrical cheilocystidia up to 60 µm long. There is also a species that produces tiny fruitbodies with distinctly translucent-striate caps; it is close to the European concept of *S. haustellaris. Callistosporium* have smooth, yellow-olive caps, have white spores and are usually larger.

Phaeonematoloma myosotis

(Fr.) Bon

Cap: 1.5–4 cm across, conical to bell-shaped at first, becoming broadly convex. Surface glutinous to viscid when wet, shiny when dry, smooth, with silky veil remnants near margin when young. Olive-brown, grayish olive to gray-brown, margin opaque or obscurely striate. **Gills:** Broadly or narrowly attached with a small notch; subdistant, broad. Whitish to grayish beige when young, developing an olivaceous tone, then maturing to light gray-brown. **Stipe:** 5–20 cm long, 0.2–0.5 (0.7) cm thick, tall, slender, equal or slightly enlarged below, base rooting deeply into moss and often decaying. Surface moist, powdery at apex, covered in chevrons of darker fibrils below, often creating a striking snakeskin pattern. Whitish to pale yellowish, covered with olivaceous brown to grayish brown fibrils, soon darkening to dingy brown from base up. Bruising brownish when handled. **Flesh:** Thin, fibrous in stipe; whitish to dingy brown. **Odor:** Indistinct. **Taste:** Indistinct. **Spore deposit:** Dull dark brown. **Microscopy:** Spores 14–17 x 7–9 µm, ovoid, somewhat truncate, with a germ pore, smooth to wrinkled, thick-walled, brown in KOH. Pleurocystidia spindle- to club-shaped with pointed tips; cheilocystidia similar, some with yellow content in KOH.

Ecology: Solitary or scattered in troops in wet areas in sphagnum. Widespread in Cascadia but restricted to a limited habitat. Fruiting in late summer and fall.

Edibility: Unknown.

Comments: Tall stature, viscid olivaceous to gray-brown cap, long stipe adorned with brown fibrils, dark brown spores, and growth in sphagnum are distinctive. At first glance it resembles *Hygrophorus fuscoalboides* which is stockier with broad white gills and white spores. A similar bog dweller is *Bogbodia uda*, with orange-brown, smooth, moist to dry cap, more purplish brown gills, and almond-shaped, roughened spores. *Hypholoma elongatum* is also found on moss or wet soil, but has a yellow-brown to straw-brown cap. *H. dispersum* has a domed, ocher-brown cap, and strong snakeskin pattern formed by silvery chevrons on a dark stipe. Latter three species all have purple-black spores.

Synonym: *Hypholoma myosotis (Fr.) M. Lange*

Stropharia albovelata

(Murrill) Norvell & Redhead

Cap: (2) 4–8 cm across, obtusely convex as first, soon convex to plane, often with a low, broad umbo. Surface slightly viscid to greasy, smooth. Vinaceous brown, pinkish brown to warm brown over disc, paler towards margin and in age. **Gills:** Attached, or with a shallow notch and decurrent tooth, close to subdistant, narrow. Whitish at first, soon grayish brown to brown. **Stipe:** 4–10 cm long, 0.3–1 cm thick, equal. Surface dry, smooth to finely appressed-fibrillose above veil, scurfy below; base with white rhizomorphs. White with scattered orangish stains. **Partial veil:** Persistent, membranous, flaring when young, collapsing in age; upper surface grooved; underside floccose. White, becoming brown with spores. **Flesh:** Thin, soft, fibrous in stipe. Whitish. **Odor:** Fungoid. **Taste:** Mild. **Spore deposit:** cinnamon-brown to dark yellow-brown. **Microscopy:** Spores 7–9 × 4–5.5 µm, ellipsoid to ovoid, with tiny germ pore, smooth, dull cinnamon in KOH. Pleurocystidia abundant, club-shaped to pointed, thin-walled, content yellow in KOH. Cheilocystidia thread-like with slightly rounded heads and wavy bases.

Ecology: Solitary or scattered in mossy duff and other debris under conifers. Uncommon, mostly in undisturbed and old-growth forests in western Cascadia. Fruiting in fall.

Edibility: Unknown

Comments: The smooth, viscid, pale to dark vinaceous brown cap, pallid gills that become grayish brown in age, slender stipe with a persistent, membranous, white partial veil, and yellowish brown spores readily identify this uncommon mushroom. The cinnamon-brown spore color is unusual for a *Stropharia;* most have dark purple-brown spores. *S. hornemannii* is similar but larger, with a distinctly scurfy-scaly stipe below the veil, and darker spores.

Synonym: *Pholiota albovelata* Murrill

Cyclocybe erebia group

Cap: 2–6 cm across, convex to nearly plane, with a broad umbo. Surface viscid to dry, margin with many veil flakes when young, soon smooth. Margin translucent-striate in age, shiny when wet, dull when dry. Brown to dark reddish brown when young, hygrophanous and fading to dull brown to clay-colored when dry. **Gills:** Broadly attached, often notched, well-spaced; edges finely pale-fringed. Dull brown to clay-brown when young, darker in age. **Stipe:** 3–8 cm long, 0.3–0.8 (1) cm thick, more or less equal. Surface dry, fibrous, scurfy or with chevron-like tufts below veil, longitudinally-lined or nearly smooth above. Dingy whitish to brownish, darkening from base up in age. **Partial veil:** Relatively thick, membranous; at first flaring upward, then skirt-like, in age collapsing to form a band on stipe, occasionally falling off completely. **Flesh:** Thin to thick and firm, fibrous in stipe, light brown, darkening in stipe, especially in age. **Odor:** Indistinct. **Taste**: Mild. **KOH:** Black on cap. **Spore deposit:** Dark brown. **Microscopy:** Spores 9–13 (15) × 6–7 µm, elongate-ellipsoid, many with a snout-like end, without a germ pore, smooth, dark brown. Basidia 2-spored. Pleurocystidia scattered to abundant, gourd- to bottle-shaped with an obtuse head, hyaline. Cheilocystidia variable, ovoid, spindle-shaped with a swollen middle or club-shaped.

Ecology: Solitary or scattered, sometimes in small clusters, from duff, rich soil, or other herbaceous debris. Widespread but more common in lowland forest across western Cascadia. Primarily fruiting in fall but can be found in spring and summer.

Edibility: Unknown.

Comments: The dark cap, brownish gills, skirt-like ring, and dark brown spores help define this species complex. There appear to be three species in this complex in Cascadia: The 'real' *C. erebia* appears to be stockier, is more likely to have a scurfy stipe below the veil, with lines above, grows in rich soils, and has spores with a distinct 'snout'. The other two are difficult to distinguish but are typically smaller, have a fibrillose stipe surface below the veil, and their spores lack the distinct snout. There are two species described from Seattle in the early 1900's: *Agrocybe washingtonensis* which seems to match the current concept of *C. erebia*, and *Pholiota subnigra* which represent one of the smaller species mentioned here. *Meottomyces dissimulans* has tan to light brown caps, widely spaced gills, and reddish brown spores. It grows on duff and woody debris, typically under cottonwood and willow. *Agrocybe* have paler colors, lighter brown spores, and often a musty or farinaceous odor.

Synonym: *Agrocybe erebia* (Fr.) Kühner ex Singer

Agrocybe praecox

(Pers.) Fayod

Spring Fieldcap

Cap: 2–8 (15) cm across, convex with an incurved margin at first, becoming broadly convex to plane, often with a low umbo or depressed at center, margin occasionally uplifted and/or splitting in age. Surface dry, smooth, often splitting in dry weather. Light brown, creamy brown to yellowish beige, fading or browning in age. **Gills:** Broadly attached, close to crowded, broad. Whitish to pale clay-colored when young, darkening to brown as spores mature. **Stipe:** 4–12 cm long, 0.5–2 cm thick, cylindrical or club-shaped with an enlarged base. Surface dry, fibrous to smooth, base often with white rhizomorphs. Whitish to tan, darkening slightly in age. **Partial veil:** Relatively thick, membranous, skirt-like. In age collapsing to form a band on stipe, or breaking up and adhering to cap margin, or falling off completely. **Flesh:** Thin to thick and firm, fibrous in stipe; pallid. **Odor:** Farinaceous, earthy or like stale chocolate. **Taste:** Earthy-farinaceous. **Spore deposit:** Brown. **Microscopy:** Spores 8–11 × 5–7 µm, ellipsoid, with a germ pore, smooth, thick-walled, yellow-brown. Pleurocystidia scattered, gourd- to spindle-shaped. Cheilocystidia bladder-shaped to broadly spindle-shaped.

Ecology: Solitary, scattered, or in small to large clusters. Generally on wood chips or in grassy urban areas; occasionally also found in disturbed lowland forest. Common in urban areas across Cascadia. Fruiting in spring, rarely in fall

Edibility: Not recommended.

Comments: This *Agrocybe* is highly variable in size and stature; even single patches can produce small, slender specimens next to large, stocky ones. This species is easily recognized by the smooth, light brown to yellowish beige cap, brown mature gills, and membranous partial veil. When exposed to sun or dry weather, the cap often develops deep cracks. Although the name *A. dura* has sometimes been applied to larger specimens, cracked-capped specimens, or collections in which the ring remains attached to the cap margin, we believe all such forms represent the same species in Cascadia. *Agrocybe sp.* 'Olive Spring' has a dingy greenish brown cap when young and grows scattered from duff in conifer forests. *Cyclocybe erebia* has a dark reddish brown, viscid cap, dark brown spores, and a persistent, rather thick, partial veil.

Agrocybe arvalis

(Fr.) Singer

Tuberous Fieldcap

Cap: 1–3 cm across, convex to plane, sometimes with a broad umbo. Surface smooth, slightly greasy to dry, shiny, smooth or wrinkled. Straw-yellow to ocher when young, darkening slightly when wet, fading to yellowish or creamy beige when dry. **Gills:** Attached, close to subdistant, broad. Yellowish buff when young, darkening to brown as spores mature. **Stipe:** 5–10 cm long, 0.2–0.4 cm thick, more or less equal, rooting into a hard, round or ovoid, black sclerotium 1–3 cm across. Surface dry, finely powdery when young to smooth, often twisting and fibrous. Whitish to yellowish buff, darkening slightly with age. **Flesh:** Thin, fibrous in stipe, whitish to pale buff. **Odor:** Indistinct. **Taste:** Bitter. **Spore deposit:** Dull brown. **Microscopy:** Spores 9–10.5 × 5–6 µm, ellipsoid in face view, flattened on one side in side view, smooth, thick-walled with germ pore, brown. Pleurocystidia club-shaped with finger-like projections.

Ecology: Often in scattered troops, occasionally solitary, from wood chips and garden beds, or other landscaped areas. Rarely reported in Cascadia but perhaps misidentified or overlooked. Fruiting in fall.

Edibility: Non-toxic.

Comments: The small size, smooth, yellowish to ocher cap, growth in landscaped areas (often with wood chips), and brown spores are helpful identification features; but the black tuber is the clincher. *A. pediades* is quite similar but typically grows in grass or compost and doesn't have the sclerotium. *Hypholoma tuberosum* is larger, has an orange-brown cap, gray-brown spores, and a larger, multi-lobed sclerotium.

Agrocybe putaminum

(Maire) Singer

Cap: 2–8 cm across, irregular-conical to convex with an inrolled and often ribbed margin at first, becoming broadly convex with a broad umbo, or wavy and irregular in age. Surface dry, velvety to finely pubescent at first, becoming smooth. Pale ocher-brown to yellow-brown, often fading to light brown to tan. Sometimes darker to orange-brown in age. **Gills:** Broadly attached or with a narrow notch, close to crowded, broad. Whitish to pale clay-colored when young, darkening to clay-brown to brown as spores mature. **Stipe:** 2–8 cm long, 0.4–1 cm thick, club-shaped with an enlarged base to nearly equal. Surface dry, velvety to finely pubescent, longitudinally grooved, at least over upper portion; base often with white rhizomorphs. Pale ocher-buff to yellowish brown, darkening slightly in age. **Flesh:** Thin to fairly thick, firm, fibrous in stipe; whitish to buff. **Odor:** Musty. **Taste:** Bitter-farinaceous. **Spore deposit:** Brown. **Microscopy:** Spores 9.5–15 × 6–8 µm, ellipsoid to oblong, with small germ pore, smooth, thick-walled, yellow-brown. Pleurocystidia club- to bladder-shaped, thin-walled. Cheilocystidia gourd- to bladder-like, caulocystidia similar.

Ecology: Clustered or in scattered troops in wood chips, or on woody debris in gardens or other rich soils. Abundant in Cascadian urban, fruiting primarily in spring, but occasionally at other times of the year—especially during summer in irrigated areas.

Edibility: Non-toxic.

Comments: Recognized by its velvety, ocher-brown to yellow-brown cap, velvety stipe with pronounced longitudinal lines, and growth in wood chips. It was introduced to western North America and now is a common member of our wood chip funga. *A. arvalis* is much smaller and more slender, and grows out of a large, dark sclerotium. Microscopically, it has swollen pleurocystidia that are divided into 'fingers' at the tip. *Hypholoma tuberosum* has a reddish brown cap, smooth stipe, dark gray-brown spores emerge from a knobby, brown, hard sclerotium.

Taxonomic note: *Agrocybe putaminum* and *Agrocybe smithii* Watling & H.E. Bigelow are likely synonyms of *Agrocybe flexuosipes* (Peck) Beker & U. Eberh (a forgotten name, originally described as a *Hebeloma*). Misidentified as *A. sororia.*

Agrocybe pediades

(Fr.) Fayod

Hemispheric Fieldcap

Cap: 1–4 cm across, rounded to hemispherical at first, becoming convex, margin incurved when young, often remaining so in age. Surface smooth, slightly viscid to dry, shiny, occasionally cracking in age or when dry. Golden brown to pale ocher when young, fading to yellowish or creamy beige. **Gills:** Broadly attached, close to subdistant, broad. Yellowish buff when young, darkening to brown as spores mature, to dark brown in age. **Stipe:** 2–5 cm long, 0.1–0.5 cm thick, equal or enlarged slightly toward base. Surface dry, finely powdery when young, often smooth, fibers appearing twisted in age. Whitish as first, darkening to yellowish buff. **Flesh:** Thin, fibrous in stipe; whitish to pale buff. **Odor:** Earthy-farinaceous. **Taste:** Mild to farinaceous. **Spore deposit:** Dull brown. **Microscopy:** Spores 10.5–14 × 6.5–8 µm, broadly ellipsoid in face view, asymmetrically flattened in side view; smooth, thick-walled with a germ pore, brown. Basidia 4-spored. Cheilocystidia scattered to abundant, variable, wider lower, or swollen in middle, at times with long necks and rounded heads or with irregular, finger-like projections, thin-walled, colorless; pleurocystidia similar but rare.

Ecology: Solitary to scattered in grassy habitats, often in lawns or fields, occasionally in areas of nutrient-poor soil. Common throughout Cascadia, especially in urban areas. Typically fruiting in spring but scattered fruitings can occur throughout year.

Edibility: Non-toxic, but small and can be confused with other little brown, toxic mushrooms.

Comments: The small, slender fruitbodies with hemispherical, often shiny, golden brown to beige caps, brown gills, and growth in grassy areas help identify this common lawn mushroom. The name *A. semiorbicularis* has been applied to a slightly larger and darker ocher-brown capped species with 2-spored basidia; it usually grows in nutrient-rich areas, often near herbivore dung, and it may end up being treated as a synonym of *A. pediades*. A similar (apparently undescribed) *Agrocybe* is common on the roots of dune grass (*Ammophila* spp.) in spring on the coastal dunes; it is similarly colored but is larger and stockier. Another similar species occurs in montane meadows and other 'natural' grassy and herbaceous habitats; it is smaller and often olive-tinged. *A. arvalis* typically grows in gardens or from rich soil; it has a convex to plane cap (not as strongly hemispheric) and its stipe arises from a black sclerotium. *Protostropharia* are similar in stature but have viscid caps and stipes, often-flaring partial veils when young, purplish black spores, and they grow on dung. Small specimens of *Leratiomyces percevalii* could also be mistaken for this *Agrocybe* but they have grayish mature gills and purplish black spores.

Tubaria furfuracea

(Pers.) Gillet

Fringed Tubaria, Scurfy Twiglet

Cap: 1–4 (5) cm across, convex when young, soon broadly convex to plane, often with a depressed center and wavy margin in age. Surface with fine silky fibers when young, often with a ring of white veil flecks around the margin, bare in age. Moist to dry, smooth, translucent-striate when wet. Cap strongly hygrophanous, thus color quite variable, depending on moisture content and age. Brown, reddish brown to cinnamon-brown when wet, to light brown, tan or whitish beige when dry. **Gills:** Broadly attached, close to subdistant, moderately broad. Light tan to cinnamon-brown. **Stipe:** 1–4 (6) cm long, 0.1–0.4 cm thick, equal or slightly enlarged toward base. Surface dry, often with silky longitudinal fibers. Whitish cream to tan with a silky white sheen and white fluffy basal mycelium. **Partial veil:** Faint, silky, generally leaving flecks of tissue on the cap margin, rarely a slight ring on stipe. **Flesh:** Very thin and fragile. Light brown to buff. **Odor**: Indistinct. **Taste:** Indistinct. **Spore deposit:** Pale brown to cinnamon-brown. **Microscopy:** Spores 6.5–9.5 × 4–5.5 µm, ellipsoid to nearly cylindrical, smooth, light yellow, without a germ pore. Cheilocystidia cylindrical to bottle-shaped, or with wavy necks, heads occasionally slightly rounded and swollen; pleurocystidia absent.

Ecology: Solitary, in clusters, often scattered in large troops, generally in urban or disturbed areas, especially on wood chips, disturbed soil, or overgrown grassy areas. Very common throughout our range. Fruiting throughout wet season, primarily in fall, with scattered fruitings through winter and again flushes in wet springs.

Edibility: Non-toxic.

Comments: This rather nondescript mushroom is recognized by the translucent-striate cap that fades dramatically as it dries, whitish veil flecks around the margin, and cinnamon-brown gills. *T. confragosa* (sensu PNW) is similar but the caps are darker reddish brown when young and has a more distinct ring on the stipe. Another species found on hardwood branches has a small, light brown to tan caps lacking veil remnants, and often has strigose hairs at the stipe base; it might be *T. abramsii*, although this has not yet been confirmed with a type sequence. Another small species, *T. conspersa*, often grows in gregarious troops on woody debris. It has a brownish to pinkish tan cap with extensive dandruffy clumps of pale fibrils. *Laccaria* share similar colors but have tough fibrous stipes and white spores. *Galerina marginata* has a hemispherical to convex cap, a membranous partial veil when young, and rusty orange spores. Some *Psathyrella* are similar but have dark brown to blackish spores.

Tubaria vinicolor

(Peck) Ammirati, Matheny & Vellinga

Wine-colored Twiglet

Cap: 1–5 cm across, convex at first, becoming broadly convex to nearly plane, often with a slight umbo. Surface moist to dry, finely velvety or appressed-fibrillose when young, becoming smooth, moist to dry, hygrophanous, translucent-striate around margin when wet. Deep red to burgundy or maroon when wet, duller when dry, fading to reddish brown to orange-brown in age. **Gills:** Broadly attached to slightly decurrent, close to subdistant, broad. Pinkish to wine red when young, darker red to brownish maroon in age, with a pallid edge. **Stipe:** 3–7 cm long 0.2–0.5 cm thick, equal or slightly enlarged at base. Surface dry, often with silky longitudinal fibers. Pinkish red with a whitish, silky fibrous sheen when young, darkening when handled and in age. Basal mycelium white. **Partial veil:** Scant, silky, forming a sparse annular zone when young, disappearing completely in age. **Flesh:** Thin, fibrous, brittle. Watery red to reddish brown. **KOH:** No reaction. **Odor:** Indistinct. **Taste:** Indistinct. **Spore deposit:** Pale brown to cinnamon-brown. **Microscopy:** Spores 7–9 × 4–5.5 µm, bean-shaped to ellipsoid, smooth, thick-walled without a germ pore, cinnamon in KOH. Cheilocystidia clustered, irregularly cylindrical to narrowly club-shaped, pleurocystidia absent.

Ecology: Solitary or in scattered groups or clusters from small woody debris, often in urban areas. Rare; known from lower elevation in western Cascadia. Fruiting in fall.

Edibility: Non-toxic.

Comments: The burgundy colors when young, brown spores and growth from woody debris identify this striking *Tubaria*. *T. punicea* is indistinguishable except by habitat; it grows at the base of rotted-out madrone trees, tends to have deeper burgundy colors, especially in age. *T. confragosa* (sensu PNW) is more reddish brown when young with a distinct ring on the stipe. Also compare with the burgundy-red *Cortinarius* species which grow on the ground under conifers, have colored basal mycelium, and red to dark purple KOH reactions. Red-colored *Hygrocybe* grow in duff or on the ground, have waxy gills, and white spore deposits.

Synonym: *Naucoria vinicolor* Peck

Tubaria conspersa

sensu PNW

Cap: 0.7–2.5 cm across, convex, broadly convex to plane; disc often slightly depressed in age. Surface dry, covered with scurfy fibrils and tufts, more appressed and occasionally smooth in age. Typically pinkish tan to light brown or cinnamon-brown, but can be dark brown, hygrophanous, translucent-striate in age when wet. **Gills:** Broadly attached, close to subdistant, broad. Tan to cinnamon-brown. **Stipe:** 2–4 cm long, 0.1–0.3 cm thick, equal to club-shaped. Surface dry, covered with silky-floccose hairs, occasionally smooth in age. Tan to light cinnamon-brown. **Partial veil:** Thin, scurfy-floccose tissue, often adhering to cap margin when young. **Flesh:** Very thin and fragile, pallid to tan. **Odor:** Indistinct. **Taste:** Indistinct. **Spore deposit:** Pale brown. **Microscopy:** Spores 7–9 × 5–6 µm, ellipsoid, smooth, thin-walled, without a germ pore, nearly colorless. Cheilocystidia scattered to abundant, highly variable, cylindrical to bottle-shaped, heads occasionally rounded and swollen; pleurocystidia absent.

Ecology: Scattered, often in gregarious troops on woody debris or duff, typically around hardwoods. Fairly common but often overlooked, occurring across western Cascadia. Fruiting in wet periods throughout the year, most abundant in fall and spring.

Edibility: Unknown.

Comments: The small size, whitish fibrils and tufts covering the cap and stipe when young, and brown colors help identify this *Tubaria*. The PNW species is distinct from the European species and there are probably multiple species in Cascadia. Some collections have lighter fruitbody colors and longer, narrower spores (9–12 × 4.5–5 µm). Most similar is *Flammulaster rhombosporus* (sensu PNW), which is slightly pale, but of similar size and stature. It has a granular-powdery cap interspersed with silky white hairs. *F. muricatus* is also similar but has darker ocher-brown to ocher-buff colors and granular scales on the cap and stipe. Also compare with the slightly larger *T. furfuracea* which mostly occurs on wood chips or disturbed urban areas.

Flammulaster muricatus group

Cap: 0.5–2 cm across, rounded with an incurved margin at first, soon convex to broadly convex. Surface dry, covered with granular to finely floccose powder, becoming less apparent in age, to nearly smooth. Ocher-orange, with slightly darker granular powder at first, slowly fading to cinnamon, then pale ocher-buff in age. **Gills:** Broadly to narrowly attached, close, broad; partial gills numerous; edges often fringed. Creamy buff when young, becoming pale ocher-buff to pale cinnamon-brown. **Stipe:** 1–3 cm long, 0.1–0.3 cm thick, equal. Surface dry, covered with scruffy-floccose powder below annular zone. Ocher-orange to ocher-buff. **Partial veil:** Thin, scruffy-floccose tissue, more like a continuation of cap surface, leaving jagged remnants on cap margin. **Flesh:** Very thin and fragile, fibrous in stipe. **Odor:** Indistinct. **Taste:** Indistinct. **Spore deposit:** Rusty brown to dark brown. **Microscopy:** Spores (6) 7–9 × 4–5 µm, ellipsoid to bean-shaped, smooth, thin to moderately thick-walled, with a germ pore, rusty orange. Cheilocystidia cylindrical with a broadened apex.

Ecology: Solitary or scattered on or even inside decaying hardwood logs and stumps; uncommon in western Cascadia but often overlooked. Fruiting in fall.

Edibility: Unknown.

Comments: The ocher-orange to buff-tan colors, granular to finely floccose powdery covering on cap and stipe, and the habitat on rotting wood make this mushroom fairly easy to recognize. *F.* aff. *limulatus* is slightly larger and has a bright orange cap. *Phaeomarasmius erinaceus* is dark red-brown and hairy-scaly (not powdery-granular scaly), stout-statured, and often growing on small dead branches of oaks. *F. rhombosporus* is much paler with more drab colors and a whitish, granular to finely floccose covering. All these names were based on specimens from Europe and the species going under these names in Cascadia appear to be distinct.

Phaeomarasmius rimulincola

(Rabenh.) Scherff.

Cap: 0.2-0.7 (1) cm across, convex to broadly convex, depressed at disc, with pleated margin. Surface dry, finely hairy to nearly smooth. Dark to light brown, at times more orange-brown. **Gills:** Broadly attached, widely spaced; edges fringed. Light brown to cinnamon; edges pale cream. **Stipe:** 0.2–1 cm long, 0.1 cm thick, equal, often curved. Surface dry, finely scaly, with sparse hairs at base. Light brown at apex, darker at base. **Flesh:** Very thin; stipe tough. **Odor:** Indistinct. **Taste:** Indistinct. **Spore deposit:** Scant, brown. **Microscopy:** Spores 11–14 × 7–8 µm, ellipsoid, lacking a germ pore, thin-walled, smooth, yellow in KOH. Cheilocystidia cylindrical to narrowly club-shaped, thin-walled.

Ecology: Scattered in troops on the bark of living oak trees. Uncommon but likely under-reported in Cascadia. Fruiting from late fall into spring.

Edibility: Unknown.

Comments: The fibrillose-scaly pleated brown cap, widely spaced gills with conspicuous pale edges, brown spores, and growth on bark of living oaks is distinctive. Because of its small size and unusual niche, you are unlikely to find this mushroom unless you are specifically looking for it. That said, it can be quite abundant in appropriate habitats. *P. erinaceus* is slightly larger, with a convex to plane, fibrillose-scaly reddish brown cap; it is fairly common in Cascadia, typically on elevated, small-diameter dead branches of hardwoods. *P. proximans* is slightly larger, with a yellow-brown cap and appressed scales. Members of the *Flammulaster muricatus* group have orange to orange-brown, granular-scaly caps and stipes. Both latter species grow on decaying hardwood logs. The vinaceous gray to bluish gray *Mycena pseudocorticola* and the pinkish brown to nearly white *M. meliigena* are similar in size and shape and also grow on bark of standing oaks but are more likely to be found on bigleaf maple; both lacks hairs/scales on the cap and have white spores.

Taxonomic note: More work is needed to see if the *P. rimulincola* in Cascadia is conspecific with the European species. If our turns out to be distinct, the epithet '*liratus*' (based on *Agaricus liratus* Berk. & M.A. Curtis), described in 1860 from California could be used.

Bolbitius titubans

(Bull.) Fr.

Sunny Side Up

Cap: 0.7–2.5 cm wide when young, 2–7 (10) cm when expanded, at first egg-shaped or cylindrical, soon expanding to round-conical, broadly convex (sometimes with an umbo), finally plane; margin soon translucent-striate, often pleated in age, sometimes curling inward. Surface viscid at first, sometimes becoming dry, faintly translucent-striate when wet, often sulcate in age. Color variable, bright lemon-yellow to deep golden yellow when young, fading outward to pallid dingy grayish yellow or ocher, to dingy cream or whitish. Often olive or with orange to red tones, and strongly reticulate in *B. titubans* var. *olivaceus*. **Gills:** Narrowly attached, often secding close, edges finely fringed. Pallid creamy yellow when young, becoming ocher-brown as spores mature. **Stipe:** 3–15 cm long, 0.2–0.5 (0.8) cm thick, enlarged towards base. Creamy whitish to yellow, covered with fine whitish floccose tufts, coarsest toward base. **Flesh:** Very thin and fragile; stipe stuffed to hollow. **Odor:** Indistinct. **Taste:** Indistinct. **Spore deposit:** Ocher-brown. **Microscopy:** Spores 12–13 (15) × 6–7 µm, ellipsoid, smooth, thick-walled with a germ pore, ocher-brown.

Ecology: Solitary, scattered, or clustered in grassy areas, on wet straw, in compost piles, or on wood chips, occasionally dung. Common in western Cascadia but can be found throughout the region. Fruiting after wet periods from fall until spring.

Edibility: Non-toxic.

Comments: Recognized by the fragile, bright yellow, and viscid cap; pale stipe and preference for nutrient-rich, disturbed, or urban. Fruitbodies in grass can be very slender, thin fleshed, and flimsy, while those in dung-rich habitat or in wood chips are often more robust and stocky. *B. aleuriatus* grows on well-decayed hardwood logs and has a grayish cap. The rarer *B. coprophilus* has a pinkish cap that becomes grayish; it is found primarily on dung. *B. lacteus* shares a similar stature, but has a whitish cap, and grows in irrigated lawns in the summer. *Conocybe apala* is similar to the latter species in appearance and habitat, but has a dry, conical to bell-shaped cap.

Synonym: *Bolbitius vitellinus* (Pers.) Fr.

Conocybe apala

(Fr.) Arnolds

White Dunce Cap

Cap: 1–3.5 (5.5) cm across, conical to bell-shaped; margin often flaring in age. Surface smooth, dry. Whitish to creamy white, often developing beige to pale buff colors in age. **Gills:** Narrowly attached to free, close to crowded, narrow. Whitish to beige at first, soon darkening to light rusty brown. **Stipe:** 4–11 cm long, 0.1–0.2 cm thick, cylindrical with a bulbous base. Surface dry, finely scruffy to smooth. Whitish to cream. **Flesh:** Very thin and fragile, stipe hollow. **Odor:** Mild to slightly garlicky. **Taste:** Indistinct. **Spore deposit:** Light rusty brown. **Microscopy:** Spores 10–14 × 7.5–8.5 µm, broadly ellipsoid to ovoid, smooth, thick-walled, with large germ pore, ocher-brown. Cheilocystidia and caulocystidia bowling pin-shaped.

Ecology: Solitary, scattered, or in troops in grass. Common in lawns across Cascadia. Fruiting from spring to fall, most abundant in summer in watered lawns.

Edibility: Potentially dangerous. Some collections have been found to contain phallotoxins. Although these are not usually directly implicated in *amanitin* poisonings because they are not absorbed through the human gut, phallotoxins are highly toxic to liver cells *in vitro*.

Comments: The white to pale buff, conical to bell-shaped cap, ocher-brown spores, and growth in grass (usually in the summer months) are distinctive features. These mushrooms often come up during the night and wilt as soon as the morning dew dries up and the sun starts to shine. Another summer fruiting species found in the same habitats is *Bolbitius lacteus*, which has a whitish to pale grayish, conical to bell-shaped cap that becomes plane or uplifted in age. *B. titubans* has a viscid yellow cap when young but can fade to whitish buff in age; it fruits from fall through spring.

Synonyms: *Conocybe albipes* Hauskn., *Conocybe lactea* (J.E. Lange) Métrod

Conocybe species

Brown Dunce Caps

Cap: 0.5–4 cm across, conical to bell-shaped, margin often flaring in age. Surface smooth, dry. Cinnamon-brown, orange-brown to brown, occasionally more gray-toned, especially when dry. Hygrophanous, faintly to distinctly translucent-striate when wet, opaque when dry. **Gills:** Narrowly attached to free, close to crowded, narrow. Cinnamon-brown to rusty brown. **Stipe:** 4–15 cm long, 0.1–0.2 cm thick, cylindrical with a bulbous base. Surface dry, finely scruffy to smooth. Cinnamon, orange to brown. **Flesh:** Very thin and fragile, stipe hollow. **Odor:** Indistinct. **Taste:** Indistinct. **Spore deposit:** Ocher-brown. **Microscopy:** Spores (6) 10–13 (16) × (5) 6–8 µm, broadly ellipsoid to ovoid, smooth, thick-walled, with large germ pore, ocher-brown. Cheilocystidia and caulocystidia bowling pin-shaped.

Ecology: Solitary, scattered, or in troops in duff, grassy areas, or on small debris. Especially common under cottonwood and other broadleaf trees. Widespread across Cascadia, especially abundant in spring, but members of this complex can be found throughout the year.

Edibility: Potentially dangerous; similar species are highly toxic.

Comments: The orange to brown, conical to bell-shaped cap, tall, fragile stipe, and ocher-brown spores help define these species. Although long referred to as *C. tenera* in field guides, there are at least ten species in Cascadia which would fit the description given here. Based on preliminary data and sampling by Danny Miller and Yi-Min Wang, *C. semiglobata* (=*C. subpubescens*?) appears to be the most common species in Cascadia but a lot more work is needed before we get have a handle on what species occur here. A similar confusing mess of *Pholiotina* species look similar but typically have more convex caps and are often more slender.

Conocybe aurea

(Jul. Schäff.) Hongo

Golden Dunce Cap

Cap: 1–4 (6) cm across, conical to bell-shaped, occasionally pointed, or broadly convex with an umbo; margin often flaring, sometimes uplifted in age. Surface smooth, thinly viscid when young, soon dry. Dark ocher-orange to ocher-yellow at first, sometimes dark reddish ocher when wet. Hygrophanous, becoming light golden beige when dry. Slightly translucent-striate when wet, opaque when dry. **Gills:** Narrowly attached, close to crowded, broad. Creamy tan when young, very soon light cinnamon-brown to rusty brown as spores mature. **Stipe:** 2–7 cm long, 0.1–0.5 cm thick, equal or widening toward a small basal bulb. Surface dry, finely pruinose. Buff to light ocher-brown, darkening in age. **Flesh:** Very thin and fragile, stipe hollow in age. **Odor:** Indistinct. **Taste:** Indistinct. **Spore deposit:** Dark ocher-brown to chestnut brown. **Microscopy:** Spores 10–13 × 6–7 µm, broadly ellipsoid to ovoid, smooth, thick-walled, with a large germ pore, dark ocher-brown. Cheilocystidia and caulocystidia bowling pin-shaped.

Ecology: Solitary, scattered, or in troops in compost-rich soil or occasionally wood chips. Common in urban areas. Fruiting in fall or spring, or occasionally in summer in watered gardens.

Edibility: Unknown; do not experiment. Closely related species are known to be toxic.

Comments: The large size and relatively short stipe (for a *Conocybe*), ocher-orange to golden brown cap, rusty brown gills, and growth in compost-rich soil are helpful. This species often shows up in houseplant pots or in gardens. *C. rickenii* shares the same habitat but is generally smaller and light brown, tan, or beige. A whole suite of similar looking species are found on dung, and while many are cosmopolitan, there have been no in-depth studies on this group in Cascadia. Species which have been documented from California include: *C. magnispora*, with relatively small (<1 cm wide), orange-brown to yellow-brown caps when young, and large spores (13.5–20.5 × 7.5–11 µm); *C. macrospora*, with larger fruitbodies (caps >1 cm wide), 2-spored basidia, and large spores as well (17–20 × 10–11 µm). Another dung-dwelling species, *C. antipus*, has a rooting stipe base up to 4 cm long.

Pholiotina rugosa group
Deadly Pholiotina

Cap: 0.5–3 (5) cm across, conical to bell-shaped at first, expanding to broadly convex or plane. Surface wrinkled to smooth, moist to dry. Reddish brown, ocher-brown to orange, often with a paler margin. Hygrophanous, fading to ocher-buff or creamy when dry. Translucent-striate when wet, opaque when dry. **Gills:** Narrowly attached, close, narrow. Buff when young, becoming cinnamon-brown to brown as spores mature. **Stipe:** 1.5–5 (8) cm long, 0.1–0.4 cm thick, equal or slightly swollen toward base, rather tall and straight. Surface dry, finely scurfy. Pallid or yellowish, with an orange-brown base when young, darkening upward in age. **Partial veil:** A rather thick, felt-like, flaring, circular disc with a lined or grooved upper surface, occasionally detaching and sliding down stipe, or falling off altogether. Whitish to cream-colored. **Flesh:** Very thin and fragile, pallid. **Odor:** Indistinct. **Taste:** Not sampled. **Spore deposit:** Dark ocher-brown. **Microscopy:** Spores 8–12 × 4–6 µm, elongate-ellipsoid to almond-shaped, smooth, thin to moderately thick-walled, with a large germ pore, ocher-yellow in KOH. Cheilocystidia abundant, typically with a swollen base and tapering outward to gourd-shaped, apex pointed.

Ecology: Solitary, scattered, or in small clusters or troops in a wide variety of habitats. Often in wood chips or compost-rich soil but also in duff or disturbed soil in woodland habitat and commonly in soil of potted houseplants. Locally common in western Cascadia. Fruiting throughout the year, with the primary flushes in spring and fall.

Edibility: Extremely poisonous due to the presence of amatoxins (the same toxins found in the Death Cap, *Amanita phalloides*, and *Galerina marginata*).

Comments: This small mushroom has a hygrophanous reddish brown cap that fades to ocher-buff and a conspicuous, white, disc-like ring on the stipe. It is part of a species complex, with at least five lineages in Cascadia. There are a number of similar *Pholiotina* that have thinner veils that leave remnants only at the cap margin, including a fairly common, pallid, spring-fruiting species on alder debris.

Synonyms: *Conocybe rugosa* (Peck) Watling, *C. filaris* (Fr.) Kühner, *Pholiotina filaris* (Fr.) Singer.

Galerina semilanceata
(Peck) A.H. Sm. & Singer

Cap: 0.4–2 (2.5) cm across, conical to bell-shaped when young, becoming broadly convex with a rounded umbo and a flared margin in age. Surface moist to dry, smooth. Strongly hygrophanous, translucent-striate when wet, opaque when dry. Yellowish to ocher when young, soon fading around margin to pale cinnamon-buff with a darker center, or buff to tan when dry. **Gills:** Broadly to narrowly attached, close to subdistant. Pallid at first, then cinnamon-brown, often with darker rusty blotches as spores mature. **Stipe:** 3–7 (10) cm long, 0.01–0.3 (0.5) cm thick, enlarging downward with a club-shaped base. Surface dry, covered with fine, appressed silky hairs over lower portion, pruinose over upper portion. Whitish to cinnamon-brown; base covered with fine white hairs. **Partial veil:** Very faint, easily overlooked; consisting of a few silky white hairs leaving sparse remnants on stipe. **Flesh:** Very thin, pale. **Odor:** Indistinct. **Taste:** Indistinct. **Spore deposit:** Light rusty brown. **Microscopy:** Spores 8–10 × 5–6 µm, ovoid, wrinkled-roughened, lacking a germ pore, pale ocher-brown. Basidia 4-spored. Cheilocystidia abundant, ventricose, with slender necks and rounded heads.

Ecology: Solitary, scattered, or in troops in moss, or mossy lawns. Common and widespread across Cascadia. Fruiting from fall into spring, but most prolific in winter.

Edibility: Unknown; do not experiment.

Comments: The bell-shaped, yellowish buff to ocher-orange, translucent-striate cap, pallid to cinnamon-brown gills, tall stipe covered in whitish fuzz, rusty brown spore deposit, and growth in moss are helpful in identification. Microscopically, the capitate cheilocystidia are distinctive. It has also gone by *G. clavata*, which is actually a distinct, less common species most easily distinguished by microscopy: its spores are larger, averaging 12 µm long. *G. paludosa* has orange-brown colors, a conical to broadly convex cap, and grows in sphagnum moss. A number of *Conocybe* are similar but are generally more fragile and have smooth rusty brown spores. A few *Mycena* are similar in shape and stature but have white spores.

Galerina stylifera group

Cap: 0.5–3.5 cm across, bell-shaped to convex at first, expanding to broadly convex or nearly plane, at times with a low, broad umbo. Surface smooth, moist to dry; margin translucent-striate when wet. Strongly hygrophanous (thus color variable); cinnamon-brown to ocher when young and wet, soon fading around margin to pale cinnamon-brown. Pale tan to beige if dry. Gills: Broadly attached, occasionally with a decurrent tooth, close to subdistant, edges often fringed. Cinnamon-brown to light ocher-brown. **Stipe**: 2–6 cm long, 0.2–0.4 cm thick, more or less equal. Surface dry, covered with fine, longitudinally-appressed silky hairs below annular zone; pruinose above when young to nearly smooth in age. Pale cinnamon-brown over much of length, with a darker brown base, covered with whitish fibrils over lower portion when young, darkening from base up in age. **Partial veil**: Faint, often overlooked, consisting of a few silky white hairs and silky remnants on stipe. **Flesh**: Very thin. **Odor**: Indistinct. **Taste**: Indistinct. **Spore deposit**: Light rusty brown. **Microscopy**: Spores 6.5–8.5 × 4–5 µm, ellipsoid to ovoid, smooth, lacking a germ pore, colorless in KOH. Basidia 4-spored. Cheilocystidia abundant, bowling pin-shaped. Pleurocystidia absent.

Ecology: Solitary, scattered, or in troops in duff, soil, moss, or on small woody debris. It seems to prefer slightly disturbed areas, such as road- and trail-sides, or old logging roads with a thin layer of moss or duff. Common and widespread throughout Cascadia. Fruiting in fall.

Edibility: Unknown. Do not experiment!

Comments: The strongly hygrophanous, cinnamon-brown to tan, convex cap, silky-fibrillose stipe, and smooth spores help outline the *G. stylifera* group. This complex of difficult-to-distinguish species has also been identified as *G. sideroides.* The most distinctive member of the group is *G. mammillata* (pictured above), which often has an acute to obtuse umbo on the cap. *G. badipes* looks quite similar macroscopically and occurs in the same habitat but has a slightly darker, ocher-brown to tan cap, and dark brown stipe. Microscopically, it has larger spores (10–13 × 5–7 µm) with a wrinkled surface, abundant cheilocystidia and pleurocystidia, and 2-spored basidia. Several *Deconica* are similar but produce purplish black spore deposits.

Galerina vittiformis group

Cap: 0.3–1 (1.5) cm across, conical to bell-shaped when young, becoming convex with a rounded to pointed umbo. Surface smooth, moist to dry. Ocher-buff to pale cinnamon-brown, with a paler, almost white margin. Strongly translucent-striate, with dark lines most of the way to a translucent umbo. **Gills:** Broadly attached, close to subdistant. Whitish buff to pale cinnamon-brown. **Stipe:** 2.5–6 (9) cm long, 0.07–0.2 cm thick, tall, thin, equal or with a slightly enlarged base. Surface dry, covered with short, fine, hair-like projecting caulocystidia. Reddish brown over lower portion, cinnamon-brown in midportion and whitish to pale buff at apex when young; darkening from base up, to dark reddish brown to blackish brown, with a paler apex. **Flesh:** Very thin and fragile, concolorous. **Odor:** Indistinct. **Taste:** Indistinct. **Spore deposit:** Light rusty brown. **Microscopy:** Spores (8) 10–12.5 × 5–6.5 µm, ovoid to almond-shaped, wrinkled-roughened with a distinct smooth area, light brown in KOH. Pleurocystidia and cheilocystidia often abundant, both with a narrow base, swollen midsection, and variably pointed or rounded tip. Cystidia absent from cap.

Ecology: Scattered or in troops in moss. Very common, especially in mossy forests across western Cascadia. Fruiting in fall, with stragglers in winter and early spring.

Edibility: Unknown, do not experiment.

Comments: This species is one of many tiny, moss-dwelling species that have cinnamon-brown, translucent-striate caps and thin stipes that darken from the base upward. This is a complex of species, best identified by their ITS sequences. *G. atkinsoniana* is similar in many regards but has cystidia on the cap which give it a pruinose look. *G. badipes* has a reddish brown to grayish black lower stipe but is larger, has a silky partial veil, and grows in duff (not moss). *G. sahleri* has more of a bell-shaped cap with a rounded umbo and is almost always on moss-covered trunks or logs of Douglas-fir. There are plenty of other *Galerina* reliably separated only by microscopic examination. A number of white-spored *Mycena* species have the same general stature and habitat preference but most are not ocher-colored. They usually have taller stipes that don't darken upward in age.

Galerina marginata group

Deadly Galerina

Cap: 1.0-1.5 (8) cm across, hemispherical to convex at first, expanding to broadly convex to nearly plane, occasionally with a low, broad umbo. Surface smooth, moist to dry, margin translucent-striate when wet. Color variable, strongly hygrophanous; ocher-orange to ocher-brown, or paler; cinnamon-brown, but can be dark reddish brown. Fading around margin to pale cinnamon-brown to pale tan. **Gills:** Broadly attached, occasionally with a decurrent tooth, close to subdistant; edges often fringed. Beige to buff at first, darkening to cinnamon to ocher-brown as spores mature, but often retaining a paler edge. **Stipe:** 2–8 cm long, 0.3–1 cm thick, equal or enlarged toward base. Surface dry, covered with fine, appressed-silky hairs below annular zone, pruinose above. Beige to cinnamon-brown, covered with whitish fibrils at first, soon darkening, dark brown to reddish brown from the base up. **Partial veil:** Membranous, forming a flaring ring at first, soon collapsing and leaving a small ring of tissue on the stipe that is often covered in rusty brown spores. **Flesh:** Thin, brownish in cap, cinnamon in stipe when young, dark brown to black in age. **Odor:** Indistinct. **Taste:** Indistinct. **Spore deposit:** Rusty brown. **Microscopy:** Spores 8–11 × 5–6.5 µm, ovoid, wrinkled-roughened except for a bare spot ('plage'), lacking a germ pore, ocher-brown. Basidia 4-spored. Pleurocystidia sparse, spindle-shaped, with wavy necks and pointed tips. Cheilocystidia abundant, similar in shape but smaller.

Ecology: Solitary, scattered, or in troops on rotten logs, stumps, or woody debris of both hardwoods and conifers. It is often clustered when on wood chips. Occasionally in duff or moss. Common and widespread throughout Cascadia, quite abundant on the west side. Fruiting in late summer and fall, occasionally in winter and spring.

Edibility: Deadly poisonous, containing amatoxins! Avoid consuming little brown mushrooms ('LBMs').

Comments: The hygrophanous, ocher-brown to tan cap, membranous partial veil (sometimes collapsed into a faint ring), growth on wood, and rusty spores are important features. Make an effort to learn to recognize this mushroom since it is fairly common and one of our few deadly poisonous species. This is especially relevant for those seeking *Psilocybe* species—the two often grow alongside one another in wood chips. *Psilocybe* differ by their dark purplish black spores and blue staining when damaged. The dark grayish to black lower stipe of older fruitbodies of *G. marginata* is sometimes interpreted as bluing by over-eager pickers. *G. marginata* belongs to a group of difficult-to-distinguish, closely related species, including one apparently undescribed species which grows on Douglas-fir (and other conifers) and has much larger fruitbodies with flatter caps. A closely related species, *G. castaneipes* is occasional in Cascadia. It tends to have thinner, more evanescent partial veils, and although there is some overlap, typically has shorter spores (6–9.5 µm long; usually <9 µm). It likely continues the same toxins.

Synonym: *Galerina autumnalis* (Peck) A. H. Sm. & Singer

Stagnicola perplexa

(P.D. Orton) Redhead & A.H. Sm.

Cap: 0.4–2.5 cm across, conical at first, becoming bell-shaped to convex with a prominent, blunt umbo, margin often flaring in age. Surface smooth, moist to lubricous; margin translucent-striate when wet, dull when dry. Brown over disc, paler; orange-brown to yellow-brown towards margin; becoming more yellowish orange to yellowish overall in age. **Gills:** Narrowly attached, at times seceding, close, narrow, edges even or eroded. Pale olivaceous gray when young, developing golden tones, to cinnamon-gray to cinnamon when mature. **Stipe:** 1.5–5 cm long, 0.1–0.2 cm thick, equal or tapering towards a curved base. Surface smooth, moist to dry; base with ocher mycelium. Yellow-brown to ocher at apex, transitioning to a reddish brown to blackish base. **Flesh:** Very thin, fragile in cap, in stipe fibrous, concolorous with surface. **Odor:** Indistinct. **Taste:** Mildly to strongly bitter. **Spore deposit:** Tan-brown to hazel. **Microscopy:** Spores 5–6.5 × 3–4 µm, ellipsoid to nearly kidney-shaped, smooth, lacking a germ pore, nearly colorless to pale yellow in KOH. Cheilocystidia abundant, cylindric to narrowly spindle-shaped, occasionally forked, thin-walled.

Ecology: Solitary or scattered on rotten, often saturated wood, duff, or on decayed plant debris. Typically in moist depressions, dried pools, or other seasonally flooded areas, or near springs and seeps. Mostly found in old-growth forests. Rare in Cascadia. Fruiting in fall.

Edibility: Unknown.

Comments: *Stagnicola* along with *Mythicomyces* belong to the small, recently established Mythicomycetaceae. Although rather nondescript, S. perplexa can be recognized by its umbonate, brownish to yellow-brown cap, olivaceous gray to cinnamon gills, and fibrous stipe with a reddish brown to black base and a pale apex. Additionally, the bitter taste, smooth spores (tan-brown in deposit), and cylindric cystidia help distinguish this species from the otherwise similar *Mythicomyces corneipes*. Both species share the same habitat, but *M. corneipes* typically lacks the distinct umbonate cap, tends to be slightly more orange in color, has a pale purple-brown spore deposit, and has a mild to slightly bitter taste; microscopically the spores are ornamented with scattered short ridges and projections, and the gills have scattered large cystidia with amyloid encrustations. *Rhodophana nitellina* has a convex to plane cap, typically brighter orange colors, a strong cucumber odor, and teardrop-shaped roughened spores that are pinkish in deposit. Many *Galerina* species are somewhat similar but all have rusty brown spores and are typically more translucent, paler in color, and often more delicate. *Phaeocollybia attenuata* can look similar but has a more evenly dark, wiry stipe with a tapered, rooting base, reddish brown spore deposit, and roughened, lemon-shaped spores.

ECM Brown-spored

8 - Mycorrhizal Brown-spored Mushrooms

This group contains some of the most anonymous and difficult-to-identify of all mushrooms (*Inocybe* and relatives) as well as some of the most attractive and interesting (*Phaeocollybia)*. We treat *Cortinarius* and its allies in a separate section because of their distinctly rusty (bright orange-brown) spore deposits and exceptional diversity. The following four groups constitute the backbone of this chapter:

Alnicola – Fruitbodies are small and drab brown with relatively smooth caps, stipes often thin and twisted with silky veil remnants. Most grow with alder.

***Inocybaceae* (Fiberheads)** – Members of this group have long been contained within a single expansive concept of the genus *Inocybe*, but were recently divided into a number of smaller genera based primarily on phylogenetic criteria. In addition to *Inocybe*, representatives of these new segregate genera found in Cascadia include *Inosperma, Pseudosperma, and Mallocybe*. This is emblematic of the taxonomic problems facing many groups of mushrooms in the age of phylogenetic primacy and generic 'oversplitting' (see the note on p. 547).

Taxonomic strife aside, the Fiberheads are inherently difficult to identify because they are poorly known, quite diverse, and have a habit of growing intermingled with one another (care is required to avoid mixed collections). Although it can be frustrating that microscopy is usually needed to assess definitive features, these features are usually quite interesting, often including beautiful crystal-tipped, thick-walled cystidia called metuloids, and spores which range from ellipsoid to strikingly knobby or angular.

***Hebeloma* (Poison Pies)** – Of the ~40 species in Cascadia, most have medium to large fruitbodies but a few are quite small. The caps are often smooth and viscid when fresh but appear shiny when dry. Many have strong odors: most often radish-like, while others are fruity or smell like stale chocolate. When young, the gills of many species are adorned with clear, milky, or brownish droplets — a feature strongly indicative of this genus. Most have dull brown to medium brown spores with some being more reddish brown. Identification to genus is straightforward but species determination is difficult due to a high degree of variation in size and color. Microscopic examination is often necessary. An online identification key to species is available at www.hebeloma.org.

***Phaeocollybia* (Phaeos)** – These mushrooms have cartilaginous stipes that usually taper strongly and root deeply into the soil—a distinctive character among brown-spored mushrooms. The caps are often viscid and the fruitbodies of many species are rather colorful. Most are restricted to old-growth conifer forests and many rare or endangered due to habitat loss.

Alnicola escharioides group

Brown Alder Mushrooms

Cap: 1–4 cm across, convex at first, becoming broadly convex to plane, margin occasionally uplifted, wavy and splitting in age. Surface moist to dry, silky looking when young, appressed-fibrillose to smooth in age. Dark to light brown, beige with an olive cast, or orangish brown. Hygrophanous, fading when dry. **Gills:** Broadly attached or with a decurrent tooth, close or subdistant. Beige to brown. **Stipe:** 1.5–4 cm long, 0.1–0.4 cm thick, equal or slightly enlarged and often twisted or curved toward base. Light brown to ocher-brown, darkening from base up in age. **Partial veil:** A silky cortina, leaving irregular chevrons or patches on stipe in age (often disappearing). **Flesh:** Concolorous, thin, fragile, fibrous in stipe. **Odor:** Indistinct. **Taste:** Mild to slightly bitter. **Spore deposit:** Brown. **Microscopy:** Spores 9.5–13 x 5–7 µm, almond shaped to broadly ellipsoid, finely roughened, inamyloid, yellow-brown.

Ecology: Solitary, scattered, or in small clusters on ground. Mycorrhizal with alders, which are usually found in swampy areas and riparian zones. Common in appropriate habitat throughout our range. Mostly fruiting in fall, occasional in spring, or during summer at higher elevations.

Edibility: Unknown; do not experiment.

Comments: This group of drab little brown mushrooms is recognized by the small size, relatively smooth caps, thin and often twisted stipes with silky veil remnants, and growth under alder. This group is in serious need of modern taxonomic work—there are a number of species in western North America currently passing under the European name *A. escharioides*. While most are alder associates, others in Cascadia include *A. salicis* with willow, and a dark, stocky species close to *A. bohemica* with oak and planted birch. A number of *Inocybe* species are similar but have a different look—generally with more radially-fibrillose or scaly caps and stipes, and/or a stockier stature (thicker stipes). Some small *Cortinarius* can be similar but generally have strongly hygrophanous caps, more prominent cobwebby veils when young, and rusty brown spores.

Synonym: *Naucoria escharioides* (Fr.) P. Kumm.

Inocybe calamistrata group

Cap: 1–3 (5) cm across, conical to bell shaped with an incurved margin at first, becoming broadly convex to nearly plane, often with a low, broad umbo. Margin becoming wavy, uplifted, and often splitting in age. Surface dry, covered with fibrillose tufts and scales. Brown to reddish brown, covered with reddish brown scales, darkening in age. **Gills:** Narrowly to broadly attached, close, narrow. Grayish at first, becoming grayish buff to brown. **Stipe:** 3–9 cm long, 0.3–1 cm thick, more or less equal or with a swollen base. Surface dry, scaly, fibrous with a pruinose apex. Pallid at first, soon light brown to grayish brown, often with reddish brown scales, and a bluish lower portion. Staining weakly red when damaged. **Flesh:** Moderately thin to thick, fibrous, somewhat fragile in cap, tough in stipe. Whitish to light brown in cap, whitish in stipe, with some bluish gray around base; weakly staining red when cut. **Odor:** Mild to fishy when young, strong, disagreeable, like rotting fish in age. **Taste:** Indistinct. **Spore deposit:** Brown. **Microscopy:** Spores 8–12 x 5–6 µm, ellipsoid to ovoid, smooth, thick walled, brownish. Cheilocystidia subcylindric to club shaped, frequently capitate. Pleurocystidia absent.

Ecology: Solitary, scattered, or in troops in duff or moss under alder. Other members of this group prefer conifers (especially Sitka spruce). Fairly common in appropriate habitat throughout Cascadia. Fruiting in fall, spring, and even early summer.

Edibility: Unknown.

Comments: At least three species in the *I. calamistrata* group are in Cascadia, all of which appear to be going under this misapplied name. As a group they are recognizable due to their fairly large size, scaly surface textures, dark colors, pallid gills when young, bluish-gray stipe base, somewhat fishy odor, and red staining where damaged. *Inocybe hirsuta* var. *maxima* (*Inosperma maximum*) is larger and overall paler, shows somewhat stronger reddish staining when cut, and grows with spruce and other conifers. Outside this group, *I. lanuginosa* is also brown and scaly, but is smaller, lacks the bluish stipe base, and doesn't have a distinctive odor. It grows on well decayed wood under conifers.

Synonym: *Inosperma calamistratum* (Fr.) Matheny & Esteve-Rav.

Inocybaceae (Fiberheads)

Although these mushrooms are abundant in the forests of the Pacific Northwest during fall and winter, their identities remain obscure even to committed mycologists!

Most come in drab colors ranging from brown through tan, beige, and gray, but some of the fully-profiled species in this section highlight the more distinctive whitish, purple, pink, and yellow examples. Although most are small mushrooms and none are truly large, a few species are fairly substantial, particular the shorter, stout-statured species.

Fiberheads are most abundant in young forests, but some species are restricted to more mature or old-growth forests. Likewise, although they are most diverse in conifer forests, there is a distinct assemblage that associates with hardwoods, and there is little overlap between the two groups.

Trying to identify Fiberheads to species-level is truly an endeavor for the most motivated amateurs, requiring engagement with primary literature, microscopy, genetic data, and a willingness to accept uncertainty. If you dive deeply into this realm, you will likely end up being a source of knowledge for the rest of us!

Key features to note:

- Texture of the cap: Scaly? Radially-silky? Ragged margin? Smooth?
- Associated trees and habitats
- Presence or absence of a bulb at the base of the stipe
- Presence or absence of pale lines (of cystidia) at the stipe apex
- Aromas are often strong and distinctive.
- Microscopic features (spores shape, cystidia, clamps).

The recent split of the classic concept of *Inocybe* into smaller subgenera (Matheny et al., 2020) is not uniformly accepted, but if one chooses to follow this 'split' model, it is worth trying to learn to recognize the following genera:

Inocybe — Even if you choose to recognize the 'split' Fiberheads, this is the second most diverse genus of mushrooms in Cascadia (after *Cortinarius*), with over 100 species known (this number includes many unnamed species). They tend to be small, slender, and show a range of textures, from smooth to silky to slightly scaly or ragged.

Inosperma — Comparatively large fruitbodies, often shaggy surfaces and dark colors, often with stipe bases contrastingly bluish green colored. <10 species in Cascadia.

Mallocybe — Often stout fruitbodies, withshaggy-scaly surfaces. Basidia have brownish pigment under the microscope. ~15 species in Cascadia.

Pseudosperma – Relatively large fruitbodies, with round-conical, radially-patterned caps and frequently with green corn/spermatic odors. ~15 species in Cascadia.

Inocybe lanuginosa

(Bull.) P. Kumm.

Cap: 1–5 cm across, convex with a slightly inrolled margin at first, expanding to plane in age, often with a low umbo. Surface dry, densely scaly; scales composed of upright, tufted fibers aggregated into a point. Rather evenly dark to medium brown when young; paler in age with a darker center. **Gills:** Narrowly attached, moderately spaced. Whitish to pale grayish beige, becoming dull brown at maturity, edges finely fringed with white. **Stipe:** 2–7 cm tall; 0.3–0.7 cm thick; straight to slightly curved, equal. Surface dry, fibrillose-scaly, pale beige to tan or light brown in age; apex usually remaining whitish. **Partial veil:** Sparse, fleeting cortina-like. **Flesh:** Thin, fibrous, solid, whitish to pale beige. **Odor:** Indistinct. **Taste:** Indistinct. **Spore deposit:** Dull brown. **Microscopy:** 8–10.5 × 5.5–7.5 µm; ellipsoid and distinctly nodulose. Pleurocystidia pear-shaped to club-shaped with short narrowed bases, sparsely tipped with crystals.

Ecology: Solitary or scattered in small groups in duff or soil under conifers; sometimes found on rotting wood. Occasional throughout Cascadia, fruiting in late fall, winter, and occasionally again in spring.

Edibility: Toxic.

Comments: This distinctive species is recognized by its convex to plane brownish cap and dry, extensively scaly cap and stipe surfaces. Nearly identical is *I. stellatospora*, which is almost always found growing on rotting wood and has longer cystidia (> 50µm). *I. leptophylla* lacks pleurocystidia and has larger, more nodulose spores than the latter two species. *I. lacera* is a stout, scaly species with smooth, elongate spores (12–14 × 4.5–5 µm), and club-shaped to capitate cystidia. *I. cinnamomea* has rather warm-toned fruitbodies (particularly the orange gills when young), associates primarily with Douglas-fir, and has smooth spores. *I. chelanensis* has a medium-brown cap with white veil tissue over the disc and lacks a bulbous stipe base. It is most distinctive due to the combination of spring fruiting with lodgepole pine and fir and unusually elongated bullet-shaped spores (13-19 × 5.5-8 µm). *I. rainierensis* is microscopically similar but has a darker and more uniformly colored cap and a pronounced bulb at the stipe base. Species in the *Mallocybe* group (treated by some as a separate genus, partly based on their lack of pleurocystidia) are uncommon in Cascadia. *I. leucoblema* has thick whitish, fibrillose tissue covering the cap and stipe, a pronounced cortina when young, spores averaging 9.7 × 5 µm, and club or pear-shaped cheilocystidia. *I. leucoloma* is evenly dull brown, with a weakly shaggy-fibrillose cap surface; its spores measure 9-10 × 5.2 µm. It is from sea level to alpine areas, usually growing with willows.

Inocybe geophylla group

White Fiberhead

Cap: 1–3.5 (5) cm across, conical at first, soon convex to nearly plane, with a rounded to pointed umbo. Surface dry, silky-fibrillose to smooth. White to creamy, becoming dingy white to pale grayish in age. **Gills:** Broadly to narrowly attached, close, narrow. Pallid when young, becoming grayish to grayish brown, edges often paler. **Stipe:** 2–7 cm long, 0.2–0.5 cm thick, equal or enlarged downward, occasionally with a slightly bulbous base. Surface dry, silky-fibrillose to smooth. White to off-white. **Partial veil:** Cortina-like, relatively sparse, leaving silky hairs on stipe and cap margin, disappearing in age. **Flesh:** Thin, firm to fragile, fibrous, white. **Odor:** Mildly to strongly spermatic. **Taste:** Mild to unpleasant. **Spore deposit:** Brown in mass, more grayish brown from prints. **Microscopy:** Spores 7.5–9.5 × 4.5–5.5 µm, ellipsoid to ovoid, smooth, thick walled, brownish. Cheilocystidia and pleurocystidia abundant, narrowly spindle shaped to nearly cylindrical, with crystals at apex.

Ecology: Solitary, scattered, or in small clusters and troops on ground, in moss or duff under conifers and hardwoods. Like many *Inocybe*, it has a preference for younger forest. Very common throughout Cascadia, fruiting from fall well into winter.

Edibility: Toxic.

Comments: The dry, white, silky-fibrillose cap, grayish to brown gills, scant, cortina-like partial veil, and spermatic odor are distinctive. The *Inocybe geophylla* complex consists of a couple of undescribed species in the western United States, as well as *I. insinuata*—a chalk-white species with a stocky stature found under pine and oak (occasionally in clusters). *I. ionocephala* and *I. pallidicremea* are similar but show lilac colors (at least when young) that can fade to pale whitish or gray and are thus often confused with members of the *I. geophylla* group. *I. pudica* starts off white but develops pinkish to peachy stains. Some *Hebeloma* are superficially similar but usually have a smooth-skinned, slightly viscid cap and/or radish-like odor.

Inocybe pallidicremea

Grund & D.E. Stuntz

Lilac Fiberhead

Cap: 1–4 cm across, conical at first, soon convex to nearly plane, with a rounded to pointed umbo. Surface moist to dry, silky-fibrillose to smooth, often splitting radially at maturity. Color variable, pale lavender when young and moist; becoming more beige-tan with a lilac wash, usually with ocher tones at disc; splitting to reveal paler ground color. **Gills:** Broadly to narrowly attached, close, narrow. Pale lilac-gray when young, becoming grayish to grayish brown, edges often paler. **Stipe:** 2–6 cm long, 0.3–0.6 cm thick, equal or enlarged downward, occasionally with a slightly bulbous base. Surface dry, silky-fibrillose to smooth. Pale lilac-gray with a yellowish-creamy base. **Partial veil:** Cortina-like, relatively sparse, leaving silky hairs on stipe and cap margin, disappearing in age. **Flesh:** Thin to moderately thick, fibrous, often hollow in age whitish to pale beige. **Odor:** Spermatic. **Taste:** Unpleasant. **Spore deposit:** Brown in mass, more grayish brown from prints. **Microscopy:** Spores 7.5–9.0 × 4.5–5.5 µm, ellipsoid, smooth. Pleurocystidia >50 µm, Pleurocystidia 45–75 × 11–20 µm, swollen at middle and with a narrow base, often lacking crystals at apex or with sparse crystals.

Ecology: Solitary, scattered, or in small to large troops on ground or in moss or duff under a variety of conifers. Fairly common throughout Cascadia, fruiting primarily from late fall well into winter.

Edibility: Toxic.

Comments: The combination of pale lilac colors which fade to grayish white, pallid to brown gills, and faint cortina is somewhat distinctive. However, distinguishing the very similar *I. ionocephala* is difficult—*I. pallidicremea* may have a slightly more slender stature, more ocher-brown umbo on average, and more pronounced yellowish-cream colors at the stipe base (where as *I. ionocephala* has a stockier stipe with a white base). Both are recent described species, and have been referred to as *I. lilacina* in the past. *I. griseolilacina* has a more scaly-brown cap, lilac-gray stipe, and favors pines. Faded specimens of the *I. pallidicremea/ionocephala* group can lose their purple tones almost entirely and thus resemble members of the *I. geophylla* group.

Inocybe pudica

Kühner

Blushing Fiberhead

Cap: 2–8 cm across, conical to bell shaped at first, becoming broadly conical to nearly plane, often with a rounded umbo. Surface dry to moist, smooth at first, except for silky veil remnants on margin, soon with appressed fibrils radiating from the center, margin occasionally splitting in age. Whitish to pale pinkish, becoming blushed with pinkish salmon to peachy orange streaks, spots, and stains, and often becoming dingy pinkish buff in age. **Gills:** Narrowly attached to free, close, narrow. Whitish to pale pinkish salmon when young, becoming grayish to grayish brown; staining pinkish salmon when damaged. **Stipe:** 2.5–8 cm long, 0.3–1 cm thick, equal or enlarged downward. Surface dry, silky-flocculent to smooth. Whitish to pale pinkish, staining or discoloring pinkish salmon to peachy orange. **Partial veil:** Cortina-like, consisting of silky white fibrils, leaving remnants on cap margin and a few silky threads on stipe when young, disappearing altogether in age. **Flesh:** Moderately thin to thick, fibrous but somewhat fragile in cap; fibrous, tough in stipe. Whitish, slowly staining pinkish salmon to peachy orange. **Odor:** Spermatic. **Taste:** Unpleasant. **Spore deposit:** Brown. **Microscopy:** Spores 8.5–10 x 5–5.6 µm, ellipsoid, thick walled, inamyloid, brownish. Pleurocystidia abundant to sparse, rather narrow, spindle-shaped with crystals at apex, walls thick, yellow in color. Cheilocystidia abundant, similar to pleurocystidia, walls bright yellow.

Ecology: Solitary, scattered, or in troops on ground or in duff under conifers. Common and widespread. Fruiting in fall and winter but also in spring in wet years.

Edibility: Toxic.

Comments: The pallid colors with pinkish orange staining and growth with conifers make this *Inocybe* easy to recognize. Members of the *I. geophylla* group also have white to off-white colors but don't develop the pinkish blush.

Inocybe albodiscoides

Matheny

White-disc Fiberhead

Cap: 2–8 cm across, broadly convex to plane, often with a low umbo; sometimes splitting weakly around the margin in age. Surface moist to dry, innately radially fibrillose with small scales, nearly smooth in age. Pale beige, often with a faint lilac-gray tone, often with a whitish disc, fading to nearly white in age. **Gills:** Finely attached, moderately spaced, edges minutely covered in white scurfiness. Pale off-white when young, becoming light gray and eventually dull gray-brown in age. **Stipe:** 3–7 cm long, 0.6–1.2 cm thick, stout when young; tall, straight, and narrowing in age, base with a weakly marginate bulb.. Surface dry, smooth but with very fine vertical pruinose lines. Whitish to beige or with light pinkish tones. **Partial veil:** Sparse, cortina-like, often fringing the cap margin when young. **Flesh:** Thin, stipe solid, fibrillose. Whitish to light beige. **Odor:** Spermatic. **Taste:** Indistinct. **Spore deposit:** Dull brown. **Microscopy:** Spores 6.5–8 × 4.5–6 um, angular-nodulose; pleurocystidia with a short, narrowed base.

Ecology: Scattered in small groups on the ground under a range of conifers and hardwoods. Fairly common in Cascadia, fruiting in late fall and early winter.

Edibility: Toxic.

Comments: The lilac-gray to beige cap with a whitish center, and a pruinose-lined and bulbous-based stipe make this stout Fiberhead easily recognizable. *I. lanatodisca* has thicker, more cottony-floccose tissue over the disc. *I. ceskae* has a dull, pale, beige-tan cap and has a straight-cylindrical (not bulbous) stipe base. Its spores measure 7–11 × 5–7.5 µm and are covered in projecting, rounded warts, while the abundant pleurocystidia are strongly crystal-tipped, urn-shaped, and usually not strongly narrowed at the base. It occurs under Douglas-fir, madrone, and other trees. *I. occulta* is similar but has spores with less-pronounced ornamentation (covered in low, rounded knobs); its abundant crystal-tipped pleurocystidia are variably shaped but usually do not show a distinctly narrowed base.

Misapplied names: *Inocybe albodisca* Peck, *Inocybe grammata* Quél.

Inocybe sororia group

Corn-silk Fiberhead

Cap: 3.5–11 cm across, tall-conical to broadly conical, usually sharply peaked but sometimes rounded, expanding in age to nearly flat but retaining pointed central umbo. Margin incurved at first, but soon radially splitting, especially in age. Surface dry, innately fibrillose and soon splitting radially, appearing like thatched straw. Light yellowish beige to pallid gold, sometimes with a pale greenish wash and metallic looking in dry weather, older caps with brownish ocher blotches. **Gills:** Narrowly attached or weakly sinuate, close, margin slightly irregular. Pale beige to yellowish, becoming grayish tan to light brownish. Edges beaded, contrastingly pale. **Stipe:** 6–12 (15) cm long, 1–2 cm thick, cylindrical with a slightly enlarged base, rather tall and straight. Whitish to pale tan, with silvery white chevrons over most of surface, finely powdery near apex. **Partial veil:** Thick whitish cortina adhering to cap edge, not forming a ring zone on stipe. **Flesh:** Thin, stringy, whitish. **Odor:** Strongly like green corn or corn silk, especially when crushed. **Taste:** Indistinct. **Spore deposit:** Dull brownish. **Microscopy:** Spores 9–12 x 5.5–6.5 (7) µm, smooth, ellipsoid to bean shaped. Cheilocystidia large, swollen with blunt or slightly squared-off ends, without conspicuous crystal incrustation. Pleurocystidia absent.

Ecology: Solitary or scattered in troops on ground under conifers and hardwoods. Widely distributed throughout western North America and common throughout Cascadia. This is usually one of the first *Inocybe* to appear in fall and continues fruiting through early winter.

Edibility: Toxic, containing muscarine.

Comments: The large size (for an *Inocybe*), peaked cap, straight stipe, pallid yellowish colors, and strong green-corn odor render this species practically unmistakable in our area. Although not yet known to occur in Cascadia, *I. aestiva* is a summer-fruiting interior-montane species that grows with conifers. These two species are currently thought to be indistinguishable without sequence data. *I. lanatodisca* is also fairly robust, with an often umbonate or round-conical cap, but is usually warmer golden-yellow with pronounced whitish cottony patches over the disc when young. Other members of this radially-silky conical-capped group (Section Rimosae) in Cascadia include: *I. spuria*, with a light coppery or yellow-brown cap and spores measuring 8.0–11.5 x 4.5-6.0 µm; and *I. niveivelata*, with a yellowish-white cap and larger spores (11.5–18.5 x 6.0–7.5 µm). All of these have smooth, bean-shaped spores, lack pleurocystidia, and grow with a variety of conifers and cottonwood. *I. cercocarpi* is also a member of this group but has rather large fruitbodies and grows with mountain-mahogany (*Cercocarpus* spp.) in dry habitats.

Synonym: *Pseudosperma sororium* (Kauffman) Matheny & Esteve-Rav.

Hebeloma velutipes

Bruchet

Common Poison Pie

Cap: 2–6 (9) cm across, convex to broadly convex, often with a broad umbo. Surface smooth, thinly viscid when wet, often with adhering debris when dry. Color variable, typically creamy to ivory when young, occasionally whitish, light brown to light olivaceous brown, darker and more dingy with age. **Gills:** Attached, often notched, close, narrow, edges with fine whitish fringes, often adorned with tiny droplets when fresh. Whitish to beige when young, becoming light brown to dull cinnamon-brown when mature. **Stipe:** 3–10 cm long, 0.5–1 (1.5) cm thick, typically clavate to bulbous, occasionally equal. Surface dry, scrufy or pruinose, to nearly smooth. White to off-white. **Flesh:** Thin, watery, concolorous in cap, fibrous, white to light brown in stipe. **Odor:** Radish-like to earthy. **Taste:** Indistinct, radish-like to bitter and disagreeable. **KOH:** No reaction. **Spore Deposit:** Medium brown. **Microscopy:** Spores 8–15 x 5–8 µm, almond- to lemon-shaped, weakly ornamented, yellow-brown in KOH, dextrinoid. Cheilocystidia clavate or with a swollen base and narrow neck.

Ecology: Solitary, scattered, or in small to large troops from duff, moss, and soil in a wide range of habitat types from lowlands and urban areas to subalpine forests. Abundant; by far the most commonly-encountered *Hebeloma* in Cascadia. Fruiting in fall.

Edibility: Toxic.

Comments: *H. velutipes* is one of many species which have been lumped under the name *H. crustuliniforme* in North America. Members of this group are recognized by their pallid cream or light beige caps which darken with age, whitish to pale cinnamon-brown gills, clavate to bulbous stipe, medium brown spore deposits, and radish-like odor and taste. Curiously, the number of gills attached to the stipe (as opposed to partial gills or 'lamellulae' per cap) has been used as an important identification feature for *Hebeloma*. *H. velutipes* typically has ~50–70 complete gills. Distinguishing *Hebeloma* species requires careful microscopic examination, as macro-morphology within the genus is highly variable. *H. incarnatulum* is typically more slender, with a distinctly umbonate cap, and bulbous stipe. It prefers wet mossy areas, especially with Sphagnum moss. *H. leucosarx* typically has a slightly darker ocher to warm brown, umbonate cap. *H. crustuliniforme* (sensu stricto) is usually slightly larger, has more crowded gills (60–100 complete), and has non-dextrinoid spores. In Cascadia, collections of *H. crustuliniforme* have been found with cottonwood but it is a widespread species found in a variety of habitats. Also compare *H. cavipes* and *H. hiemale*.

Synonym: *Hebeloma bakeri* Earle.

Hebeloma hiemale

Bres.

Cap: 1.5–5 (8) cm across, convex to broadly convex, occasionally with an umbo. Surface smooth, dry to thinly viscid when wet. Pinkish buff to pale yellowish brown, margin often paler, becoming dingier with age. **Gills:** Notched at attachment, close, narrow, edges with fine whitish fringes. Pallid, with pinkish or yellowish brown tones when young, becoming light brown to brown when mature. **Stipe:** 2–10 cm long, 0.3–0.8 (1.2) cm thick, typically equal, occasionally clavate or bulbous. Surface dry, scurfy with whitish flecks or pruinose, to nearly smooth. Whitish when young, developing brown tones from the base up as it ages. **Flesh:** Thin, firm, whitish in cap, fibrous, white to light brown in stipe. **Odor:** Radish-like. **Taste:** Radish-like, bitter to indistinct. **KOH:** No reaction. **Spore Deposit:** Medium brown. **Microscopy:** Spores 9–14 x 4.5–8 µm, almond-shaped, ornamented, yellow-brown in KOH, not, or weakly dextrinoid. Cheilocystidia variable, clavate with a swollen base and narrow neck, or with a swollen middle.

Ecology: Solitary or scattered, often in wet soil or duff under cottonwood and willow, but like many *Hebeloma*, it can occur in a variety of habitats. Occasional across Cascadia, more common in lowlands. Fruiting in fall.

Edibility: Toxic.

Comments: The smaller and more slender stature, pinkish tones, and growth with cottonwood and willow help distinguish this species. *H. limbatum* is strikingly similar but in Cascadia we have only seen it with Oregon white oak. *H. cavipes* is often (but not always) stockier in stature and tends to exhibit dingier yellowish brown to brownish tones. It is more commonly encountered in urban areas with introduced hardwoods, especially oak and birch.

Hebeloma cavipes

Huijsman

Cap: 2–5 (7.5) cm across, convex to broadly convex, occasionally with an umbo. Surface smooth, dry or thinly viscid when wet. Yellow-brown, pinkish buff to grayish brown, margin often paler, becoming dingier overall with age. **Gills:** Attached, often with a small notch, close, narrow, edges finely fringed, rarely with tiny droplets. Pallid with pinkish or yellowish brown tones when young, becoming light brown to brown when mature. **Stipe:** 2–6 cm long, 0.3–1 (1.5) cm thick, equal to clavate. Surface dry, smooth, or pruinose over upper portion. Whitish when young, developing brown stains and discolorations with age. **Flesh:** Thin, firm, whitish in cap, fibrous, white to light brown in stipe. **Odor:** Indistinct to radish-like. **Taste:** Radish-like, bitter to indistinct. **KOH:** No reaction. **Spore Deposit:** Medium brown. **Microscopy:** Spores 9.5–15 x 5.5–6.5 µm, almond- to lemon-shaped, strongly ornamented, yellow-brown in KOH, dextrinoid. Cheilocystidia variable, clavate with a swollen base and narrow neck, or with a swollen middle.

Ecology: Scattered or in troops, often in lawns and other urban settings under hardwoods (especially oak, birch, and linden) but also occurring in natural areas with cottonwood or conifers. Locally common, fruiting in fall.

Edibility: Toxic.

Comments: *H. cavipes* is yet another pallid, radish-scented "Poison Pie" in Cascadia. Habitat may be one of the better distinguishing features for this species, as it is most often found under introduced hardwoods. *H. limbatum* is another species usually found with hardwoods (white oak); it tends to be more slender-statured and more brightly colored. *H. hiemale* is also comparatively slender, but typically exhibits more warm pinkish buff colors and prefers associating with cottonwood and willow . Also compare *H. velutipes* and kin, which usually have a bulbous stipe base. *H. crustuliniforme* is usually slightly larger, has close to crowded gills, and has non-dextrinoid spores.

Hebeloma sacchariolens

Quél.

Sweet-scented Poison Pie

Cap: 1.5–6 cm across, convex, occasionally with an umbo, at times wavy in age. Surface smooth, tacky to dry. Pinkish buff to pale cinnamon brown, often two-toned with a paler margin when young, at times more pallid, creamy or whitish, becoming dingier overall with age. **Gills:** Notched at attachment, close to fairly well spaced, moderately broad. Pinkish buff to yellowish brown at first, becoming cinnamon brown to light brown. **Stipe:** 2–7.5 cm long, 0.3–0.8 (1.1) cm thick, equal or tapered downward. Surface dry, smooth. Whitish, pinkish buff to light brown when young, darkening as it ages. **Flesh:** Thin, firm, pallid in cap, fibrous, whitish to light brown in stipe. **Odor:** Sweet and chemical, often like grape soda or burnt sugar, or with floral tones. **Taste:** Mild to bitter. **KOH:** No reaction. **Spore Deposit:** Medium brown. **Microscopy:** Spores 8.5–14.5 x 6–8 µm, almond- to lemon-shaped, roughly ornamented, perispore loosening, brown in KOH, strongly dextrinoid.

Ecology: Scattered or in gregarious troops from soil, duff, or mossy lawns. Fruiting in fall under a wide variety of trees including both native and introduced hardwoods and conifers.

Edibility: Toxic.

Comments: This would be just another pinkish buff *Hebeloma* if not for the distinctive odor—reminiscent of a grape soda-chemical cocktail. Although *H. sacchariolens* can be a sparse fruiter some years, other years it can be quite prolific. We have seen thousands of fruitbodies carpeting the forest floor under Oregon white oak. *H. ischnostylum* has the same odor and is difficult to distinguish in the field—it tends to have a more evenly colored whitish to ivory cap when young and appears to be restricted to introduced hardwoods in urban areas. Microscopically, it has smaller, spindle-shaped spores measuring 9.5–13 x 5–7 µm. *H. praeolidum* can have a sweet, fruity to radishy odor but has a scant cortina-like partial veil when young and weakly ornamented almond-shaped spores.

Hebeloma praeolidum

A.H. Sm., V.S. Evenson & Mitchel

Cap: 1.5–5 (6.5) cm across, convex, occasionally with an umbo, margin incurved at first, soon downturned to even. Surface smooth or with appressed-fibrillose veil remnants around margin, tacky to dry. Ocher, orange-brown to brown, often two-toned with a paler margin when young; at times more grayish brown or dingy, darkening with age. **Gills:** Notched, close, narrow, edges finely fringed to even. Beige to dull brown at first, light brown to cinnamon brown as spores mature. **Partial veil:** A scant cobwebby cortina, typically only visible in younger specimens, leaving a faint annular zone on stipe and cap margin. **Stipe:** 2–9 (12) cm long, 0.3–0.8 (1.2) cm thick, equal to clavate. Surface dry, fibrillose to pruinose, and with scattered zone of veil remnants when young. Whitish, buff to light brown at first, darkening from base up as it ages. **Flesh:** Thin, firm, pallid in cap, fibrous, whitish to brown in stipe. **Odor:** Often sweet, like grape juice or generally fruity, or radishy. **Taste:** Mild to slightly bitter. **KOH:** No reaction. **Spore Deposit:** Medium brown. **Microscopy:** Spores 10–14 x 6–8 µm, almond-shaped, weakly ornamented, perispore not loosening, yellow-brown, dextrinoid.

Ecology: Solitary or scattered in moss, duff or soil, from coastal to subalpine conifer forests. Fairly common throughout Cascadia but more frequently encountered in wetter west-side forests. Fruiting in fall.

Edibility: Toxic.

Comments: The smaller stature, ocher to brown caps, scant cobwebby partial veil, and often sweet odor help distinguish this species. Collections with a sweet odor can be mistaken for *H. sacchariolens*, but that species lacks a partial veil and has distinctly ornamented, strongly dextrinoid spores. *H. alpinicola* can look very similar but has slightly less 'warm brown' tones, a mild to radishy odor, and smooth, non-dextrinoid, ellipsoid to ovoid spores. See *H. mesophaeum* for other similar veiled species.

Hebeloma angelesiense

A.H. Sm., V.S. Evenson & Mitchel

Cap: 2–12 (16) cm across, domed to convex, often distinctly umbonate, margin inrolled to downcurved, even to undulating in age. Surface thinly viscid to tacky when wet, otherwise dry, smooth, or appressed-fibrillose around margin. Orangish with a pale bloom when young, soon cinnamon-brown to ocher-brown, typically darker, orange to orange-brown with age. **Gills:** Attached with a notch, close, narrow, edges even. Dull brown at first, brown to dark brown as spores mature. **Partial veil:** A very scant cobwebby cortina, typically only visible in younger specimens. **Stipe:** 5–15 (23) cm long, 0.5–1.5 (3) cm thick, often tapering downward to fused base. Surface dry, pruinose at apex, fibrillose overall. Whitish to tan, darkening when handled or with age. **Flesh:** Thick, firm, pallid to beige when young, tan in age, stipe fibrous, hollow. **Odor:** Sweet, musty or earthy, occasionally like stale chocolate. **Taste:** Mild to slightly bitter or radishy. **KOH:** No reaction. **Spore Deposit:** Dull yellowish brown. **Microscopy:** Spores 7.5–10.5 x 4.5–6.5 µm, ellipsoid, nearly smooth, grayish yellow, not dextrinoid.

Ecology: Clustered or more rarely scattered in deep conifer duff, especially of Douglas-fir. Appears to be rare, most collections come from montane forests. Fruiting in fall.

Edibility: Unknown, likely toxic.

Comments: This *Hebeloma* is easily recognized by the large size, orange to brown cap, and clustered growth. Solitary or scattered fruitbodies could be mistaken for *H. avellaneum*, which lacks a partial veil and has dextrinoid spores.

Hebeloma avellaneum

Kauffman

Cap: 3-7 (12) cm across, convex to broadly convex, with a low broad umbo, margin incurved when young, often undulating in age. Surface tacky when wet, often with a powdery sheen when dry. Brown, reddish brown to orange-brown, often with a paler margin or sheen when young, developing dark blotches or spots around the margin in age. **Gills:** Broadly attached, close, narrow. Pallid when young, tan, light gray brown to light brown. **Stipe:** 3-8 (10) cm long, 7-15 mm thick, equal to clavate, occasionally bulbous or rooting at base. Surface dry, scurfy to pruinose. White, dingy white, or with slight brownish discolorations below. **Flesh:** Firm, thin, stipe stuffed to hollow, brownish. **Odor:** Indistinct, slightly raphanoid to musty. **Taste:** Mild to slightly raphanoid or slightly bitter. **KOH:** Unknown. **Spore Deposit:** Dull brown. **Microscopy:** Spores 7.5–11.5 x 4.5–6 µm, almond-shaped, smooth to weakly ornamented, yellow-brown, dextrinoid. Cheilocystidia cylindrical, narrowly clavate, or clavate with a swollen middle.

Ecology: Scattered to gregarious, occasionally clustered in duff, moss, and soil in mixed conifer forests. Widespread in western Cascadia, but rarely reported, fruiting in fall with an apparent preference for younger fir-hemlock forests of mid- to high-elevations

Edibility: Unknown, likely toxic.

Comments: The brown to brick-colored cap (often with a paler sheen when young) that develops dark blotches in age, pallid to light brown gills, and scurfy whitish stipe help define this species. *H. kelloggense* looks similar but has a more cinnamon to orange-brown cap, often a stale chocolate to musty radish-like odor, and slightly broader spores (8–11.5 x 5–6.5 µm). *H. olympianum* is typically slightly smaller, has an evenly colored brown cap (viscid when wet), and an indistinct to earthy odor. *H. harperi* is a small, chocolate-brown species associated with hardwoods.

Hebeloma olympianum

A.H. Sm., V.S. Evenson & Mitchel

Cap: 1.5–5 (9) cm across, convex to broadly convex, occasionally with an umbo. Surface smooth to appressed-fibrillose around margin, shiny and viscid when wet, dull when dry. Orange-brown, at times with a paler margin, becoming more dingy brown in age. **Gills:** Attached with a notch, close, narrow, edges even. Beige to tan at first, light brown to cinnamon brown when mature. **Stipe:** 2–10 cm long, 0.3–1 cm thick, equal to clavate, rarely tapering. Surface dry, scrufy-fibrillose to pruinose. Off-white, buff to light brown at first, darkening from base up with age. **Flesh:** Thin, firm, light brown to watery brown in cap, fibrous, whitish to brown in stipe. **Odor:** Indistinct to earthy. **Taste:** Mild. **KOH:** No reaction. **Spore Deposit:** Medium brown. **Microscopy:** Spores 7–11 x 4.7–6.7 µm, almond-shaped, weakly ornamented, perispore not loosening, yellow-brown, strongly dextrinoid. Cheilocystidia cylindrical.

Ecology: Solitary, scattered, or in troops in moss, duff or soil under conifers. This species appears to have a fondness for 'nutrient-enriched' spots near campsites. Widespread across western Cascadia, locally common in the Klamath Mountains, and known from the Olympic Peninsula and the Cascade foothills. Fruiting in fall.

Edibility: Unknown, likely toxic.

Comments: *Hebeloma olympianum* is recognized by its viscid to dry orange-brown cap, tan to light brown gills, stipe which darkens with age, and indistinct odor and taste. Microscopically, it has weakly ornamented, strongly dextrinoid spores, and cylindrical cheilocystidia. *H. viscidissimus* has paler pinkish-buff to yellow-brown colors and a bitter taste. *H. immutabile* is a tall, slender species with a yellow-brown to ocrhe-brown cap with a paler margin, a scant cortina and small (8–10.5 x 5–6.5 µm), almond-shaped, distinctly ornamented dextrinoid spores. *H. parcivelum* has a dingy, grayish buff cap with darker spots and blotches.

Hebeloma albomarginatum

Hesler

Cap: 5–15 cm across, domed, margin strongly inrolled and often lobed and distinctly ribbed when young, becoming broadly convex to nearly flat but remaining inrolled well into maturity, occasionally broadly umbonate. Surface thinly viscid to dry, smooth, often with stuck-on leaves and debris when dry. Dingy reddish brown to pinkish brown at center with radiating brown streaks, extensively beige near margin, inrolled part often with a whitish bloom or pale fibrils. **Gills:** Broadly attached, notched, close to crowded, edges finely fringed or serrate and eroded, conspicuously dotted with clear to brown droplets when young and fresh. Light, dull grayish beige at first, becoming tan and then brown in age. **Stipe:** (4) 6–15 cm long, 1–3 cm thick, equal, clavate, often distinctly bulbous at base (up to 4 cm thick). Surface dry, covered in large shaggy or pointed scales, especially dense near apex, lower parts smooth or vertically fibrillose. Base often with obvious white rhizomorphs. Whitish to cream, becoming beige to tan. **Flesh:** Thin to thick, firm, fibrous. White marbled with tan. **Odor:** Radishy. **Taste:** Indistinct. **Spore deposit:** Dull brown. **Microscopy:** Spores 10–14 x 6.5–8.5 µm, elongate almond shaped to lemon-shaped, thick walled, coarsely roughened.

Ecology: Solitary, scattered or in troops from soil and duff. Widespread but occasional to locally common in Cascadia. Growing on ground in many forest types including young to old-growth conifers, aspen and cottonwood, or even with oak or madrone. Fruiting in fall.

Edibility: Unknown, likely toxic.

Comments: This species is recognized by its hefty fruitbodies, strongly scaly stipe with a bulbous base, rosy brown to beige cap, and droplets on the gills. *H. albomarginatum* is a widespread species which also occurs in a broad range of habitats; it is especially common in the Rocky Mountains. *Cortinarius* are also similar but lack the droplets on the gills and have brighter rusty brown spores.

Taxonomic note: We called this species *Hebeloma* 'Big Scaly' in *Mushrooms of the Redwood Coast*. *H. insigne* is a synonym. It has also been misidentified as *H. sinapizans*, which is a European species.

Hebeloma mesophaeum

(Pers.) Quél.

Cap: 1.5–4 cm across, domed when young, expanding to convex or umbonate. Surface dry or thinly viscid when wet, smooth over disc, often with appressed partial fibrils around the margin. Color variable, evenly colored or two-toned: Typically, red-brown, orange-brown or yellow-brown over the disc with a paler margin. Some collections can have light to dark brown, gray-brown or beige caps. **Gills:** Attached with a small notch, close, broad. Whitish to yellowish at first, becoming dingy tan, grayish brown to light brown. **Stipe:** 2.5–10 cm long, 0.3–0.7 cm thick, equal. Surface dry, fibrillose to pruinose. White to beige, darkening from base up in age. **Partial veil:** Fibrillose or cobwebby, leaving a sparse silky annular zone on stipe, and cap margin, occasionally thicker and persistent. **Flesh:** Thin, often hollow in stipe; white to beige. **Odor:** Radish-like. **Taste:** Radishy to slightly bitter. **Spore deposit:** Dull brown. **Microscopy:** Spores 6.5–12 x 4–7 µm, ellipsoid to ovoid, nearly smooth, non-dextrinoid. Cheilocystidia flask-shaped.

Ecology: Scattered, in small groups or loose clusters, from soil, moss and duff. Often in disturbed areas in conifer forests, but also found with alder, willow, and cottonwood. Occasional in Cascadia, fruiting in fall.

Edibility: Unknown, likely toxic.

Comments: *H. mesophaeum* belongs to a group of closely related species which are very difficult to distinguish from one another. As a group, they are recognized by their small to medium size, often two-toned caps with paler margins, cobwebby partial veils, and radish-like odor. Microscopically, the nearly smooth, ellipsoid to ovoid non-dextrinoid spores help distinguish these from other veiled *Hebeloma.* Species in the *H. mesophaeum* group include *H. excedens,* which may be more evenly colored and slightly darker overall (this may be wishful thinking). *H. alpinicola* is also similar and at times indistinguishable. It tends to have a more unicolored cap, a stockier stature, and typically grows in higher elevation forests. *H. velatum* has drab brown colors, with a scant partial veil. We have found it in the spring under cottonwoods in the eastern Cascades. *H. paludicola* is a small species with a two-toned brown, dark brown to red-brown cap with a paler margin, and a scant cortina. It grows in wet areas with willow and has almond-to lemon-shaped, weakly ornamented, dextrinoid spores. *H. palustre* is like *H. paludicola,* but has a slightly larger, dark brown, olivaceous-brown to red-brown two-toned cap, and a more fibrillose stipe. Some *Inocybe* are similar but have more radially fibrillose caps and/or bulbous stipe bases and are microscopically different. *Pholiota* species tend to grow directly on wood and have smooth spores with a (sometimes inconspicuous) germ pore.

Phaeocollybia kauffmanii

(A.H. Sm.) Singer

Cap: 5–15 (20) cm across, conical with an inrolled margin at first, becoming broadly conical to convex with a broad umbo; margin becoming wavy and sometimes slightly uplifted in age. Surface smooth, shiny, viscid, covered with thick slime when wet, dull and moist in dry weather. Orange-brown, ochraceous brown to warm brown, often with a slightly darker center, developing darker stains or blotches when damaged or in age. **Gills:** Narrowly attached or appearing free, close to crowded. Buff to tan when young, becoming dingy golden brown in age. **Stipe:** 4–12 cm long above ground, 20–30 cm (up to 40 cm) long overall, 1–3.5 cm thick, equal, swollen in middle or enlarged toward lower stipe with a tapering "root" underground. Surface moist to dry, appressed-fibrillose to smooth, often cracking longitudinally in age or dry weather. Pinkish buff to cinnamon buff on upper stipe, gradually becoming orange-brown lower when young. Entire stipe darkening in age; cinnamon brown to reddish brown at apex, orange-brown to reddish brown upwards from base. **Flesh:** Thin to moderately thick, watery brown when wet, pinkish buff in cap when dry. Stipe stuffed with pinkish ivory pith that slowly stains orangish brown and a fibrous to cartilaginous, pale pinkish brown to orangish brown rind-like skin. **Odor:** Slightly farinaceous. **Taste:** Strongly farinaceous to slightly bitter. **KOH:** Brownish. **Spore deposit:** Cinnamon brown. **Microscopy:** Spores 7.5–10 x 4–6 µm (averaging 8.8 x 5.2 µm), almond shaped or lemon shaped in side view, roughened, but with a smooth beak. Cheilocystidia narrowly club-shaped, thin- walled.

Ecology: Solitary, scattered, or in small clusters, often in rows and arcs under conifers, especially in mature and old-growth forests. Widespread but uncommon across western Cascadia, fruiting in fall.

Edibility: Unknown, likely toxic.

Comments: The large, viscid, orange-brown to warm brown cap, solid, rooting stipe, and farinaceous taste help identify this *Phaeocollybia*. *P. chefensis* also has an orange to ocher-brown cap but has tibiiform (cylindrical base, long thin neck and rounded head) cheilocystidia. *P. ammiratii* is generally smaller and more slender and has a brownish yellow to butterscotch-colored cap, orangish cream gills when young, and a distinctive burgundy blush on the lower stipe. Microscopically, it differs from others in this group by having easily observed clamp connections. *P. ochraceocana* differs in its tawny ocher-orange, dry to moist (not viscid) cap, creamy to buff gills, and a pale creamy buff to pale ochraceous salmon stipe; its spores measure 5.8–8.9 x 3.8–5.6 µm. *P. luteosquamulosa* is another ocher-yellow species that looks almost identical to *P. ochraceocana* but can be easily told apart by its larger spores (8.7–12 x 5–7.2 µm).

Phaeocollybia benzokauffmanii

Norvell

Smoky Phaeo

Cap: 4–15 cm across, obtusely conical with an inrolled margin when young, expanding to convex or nearly plane with broad umbo, margin becoming wavy and uplifted. Surface smooth, viscid to glutinous when wet, tacky when dry. Color variable, often variegated with various shades of smoky violet, pinkish, lilac to purplish brown, typically with a darker disc, and dark spots or splotches. As it ages, brownish and oranges tones become more prominent, turning vinaceous brown to dark brown. **Gills**: Narrowly attached to free, crowded, broad. Whitish to pale pink or grayish when very young, soon developing smoky tones, dark yellowish gray in age. **Stipe**: 3.5–10 cm long above ground, up to 25 cm long overall, 0.7–3 cm thick at apex, equal or swollen towards ground level, tapering to an often deeply rooted pseudorhiza underground. Surface dry, smooth to appressed-fibrillose, longitudinally lined (occasionally the cartilaginous rind will crack longitudinally). Pale pinkish gray to purple-gray at apex, transitioning to a deep purple-brown or red-brown below when young, darkening from the base upward in age. Pseudorhiza deep red brown overall. **Flesh**: Firm, thin except at disc; whitish, tinged with pink or purplish tones in cap, slowly staining cinnamon. Stipe solid with a dark cartilaginous rind and firm, pallid to whitish pith, slowly staining cinnamon brown when cut. **Odor**: Slightly floral, or sweet-farinaceous. **Taste**: Farinaceous to slightly bitter. **Spore deposit**: Dull yellowish brown. **Microscopy**: Spores 8.2–10 x 5–6 µm, averaging 9 x 5.5 µm, fusoid-ellipsoid in face view, somewhat lemon-shaped in side view, coarsely roughened, Cheilocystidia a dense gelatinous layer on gill edge; variable in length and shape; a mix of narrow and broadly clavate elements.

Ecology: Solitary, scattered, or in troops in areas with deep duff under conifers, typically in mature and old-growth forests. Uncommon and likely declining due to habitat loss. Fruiting in fall.

Edibility: Unknown, likely toxic.

Comments: *P. benzokauffmanii* is a large mushroom with a viscid, smoky violet to dark vinaceous brown cap. It has pallid whitish to yellowish gray gills and a deeply rooted stipe with a pinkish gray to purple-gray apex. *P. redheadii* has reddish brown to dark brown caps (even when young), long rooting stipes stuffed with cream-colored pith, and larger spores; (9–13 x 5–7.3 µm, avg. 10.5 x 6 µm). *P. spadicea* has a dark brown to gray-brown cap that develops vinaceous tones in age; it is typically smaller (caps 2.5–12 cm across), has darker gills, slightly smaller spores (8 x 5 µm), and has 'tibiiform' cheilocystidia. *P. oregonensis* has a grayish to grayish brown cap, smoky-white to dark yellowish brown gills, a stipe with a grayish pink apex stuffed with a whitish pith, and dull yellowish cinnamon spores. Microscopically, it has small, nearly smooth spores (6.8 x 4 µm), and thin-walled, narrowly clavate cheilocystidia.

Phaeocollybia attenuata

(A. H. Sm.) Singer

Little Phaeo

Cap: 1–5 cm across, conical to broadly conical, often with a pointed umbo when young, becoming bell shaped to convex. Surface smooth, moist to slightly viscid. Dull orange-brown to clay brown or yellowish brown, occasionally with a slightly darker or lighter margin as it dries, sometimes developing reddish brown spots or stains in age. **Gills:** Narrowly attached or appearing free at times, close to moderately crowded. Light warm buff to pinkish buff when young, soon darkening to buff-brown, becoming dingy orange-brown in age. **Stipe:** 2–6 (8) cm long aboveground, 20+ cm long overall (but tough to extract from ground without breaking), 0.2–0.5 cm thick, more or less equal aboveground, tapering to a long, brittle, wire-like pseudorhiza (root) underground. Clay brown to yellowish brown at first, darkening to orangish brown and gradually becoming dark brown to blackish brown from the base upward in age. **Flesh:** Very thin, creamy ivory when young, darkening in age. Stipe loosely stuffed with whitish fibrous pith at first, soon becoming hollow. **Odor:** Indistinct to strong; often radish-like with sweet floral undertones and a disagreeable burnt or fishy odor. **Taste:** Mildly to strongly unpleasant and often bitter. **KOH:** Dark brown on all parts. **Spore deposit:** Reddish brown. **Microscopy:** Spores 7–10 x 4.5–7 μm, average 9 x 7.8 μm, lemon shaped or almond shaped with a long beak. Surface strongly roughened, warted, apiculus and beak smooth. Cheilocystidia numerous, thread-like or narrowly club shaped.

Ecology: Scattered in small to large troops, from moss, duff, and soil in mixed conifer forests. Locally common in lower elevation, west-side forests. Fruiting in fall.

Edibility: Unknown, likely toxic.

Comments: The clay brown to orange-brown cap, slender stipe that darkens from the base up, and very small size (for a *Phaeocollybia*) help identify this mushroom. Other similarly colored small *Phaeocollybia* include: *P. radicata*, with a brighter yellow-orange cap and pale orange to cinnamon buff gills when young, and lacking the dark wiry root; the spring-fruiting *P. pleurocystidiata*, which has an ocher-yellow to tawny yellow cap, yellowish buff gills, and a salmon-colored, deeply rooting stipe; and *P. phaeogaleroides* which is typically smaller and more slender in stature, with a translucent-striate cap and translucent stipe. ITS sequences suggest that the latter name may refer to a cryptic species complex.

Phaeocollybia scatesiae

A.H. Sm. & Trappe

Kit's Phaeo

Cap: 2–6 cm across, obtusely conic when young, becoming broadly bell-shaped, with a broad to pointed umbo. Margin inrolled when young, becoming down-curved to even, or wavy in age. Surface smooth, glutinous to viscid, hygrophanous. Yellowish brown when young, typically darker on disc, becoming orange-brown, dark reddish brown or blackish brown. **Gills:** Narrowly attached, crowded, narrow. Pale yellowish to pale pinkish white at first, developing cinnamon color, becoming dingy yellow-brown in age. **Stipe:** 4–7 (10) cm long above ground, up to 25 cm overall, 0.4-1.2 cm thick at apex, equal, pseudorhiza tapering downwards to a fused point. Surface smooth, dry to lubricous. Pale pinkish tan to orange buff at apex, cinnamon to orang lower, becoming dark orange-brown from base up in age. Pseudorhiza salmon-colored. **Flesh:** Thin, buff. Stipe hollow with a cartilaginous rind. **Odor:** Indistinct, slightly floral or slightly farinaceous in age. **Taste:** Mild to slightly bitter. **Spore deposit:** Yellowish brown. **Microscopy:** Spores 7.2–9.5 x 4.5–5.6 µm, averaging 8.5 x 5 µm, ovate in face view, lemon-shaped with a beak in side view, roughened-warty, pale rusty brown in KOH, non-dextrinoid in Melzer's reagent. Cheilocystidia flask- to bowling pin-shaped, base ~20 µm long, ~5 µm wide, necks ~3 µm wide, thick-walled.

Ecology: Fruiting in small to large clusters in soil, typically in areas with thick duff, with a preference for mature and old growth forests. Uncommon, mostly in the Coast Range of Oregon, with scattered sites in the western Cascades and Olympics. Fruiting in fall.

Edibility: Unknown, likely toxic.

Comments: This *Phaeocollybia* is recognized by its growth in large clusters, conic to bell-shaped, yellowish brown to dark reddish brown or blackish brown cap, yellowish, pinkish to cinnamon-colored gills, a slender, hollow, rooting stipe (tapering to a fused point), and large, roughened, lemon-shaped spores. *P. californica* has an ocher, orange, to red-brown cap, and grows in clusters without the fused stipes. *P. gregaria* has a golden-tan to brown, conical to bell-shaped cap, smoky white to pale cinnamon gills, and a tall, slender, rooting stipe with an ivory to tan apex and orange base. Microscopically, it has variably clavate, thin-walled cheilocystidia, and large spores (9.6 x 5.7 µm). *P. olivacea* has an olive-green cap, yellowish green gills and stipe apex, and coarsely roughened spores. *P. spadicea* has a darker brown to gray-brown cap that develops vinaceous tones in age; it also has darker gills and slightly smaller spores (8 x 5 µm).

Phaeocollybia sipei

A.H. Sm.

Cap: 2.5–12 cm across, conical at first, becoming broadly conical to bell-shaped, often with a broad umbo; margin slightly inrolled to downcurved, occasionally undulating in age. Surface smooth, shiny, slightly viscid to greasy when wet. Orange-brown to orange. **Gills:** Narrowly attached or appearing free, close to crowded. Yellowish buff to tan when young, becoming yellow-brown in age. **Stipe:** 4–8 cm long aboveground, up to 40 cm long overall, 0.5–1.5 cm thick, typically equal, then swollen at ground level and quickly tapering into a "root". Surface moist to dry, fibrillose to smooth. Yellowish-buff to cinnamon buff on upper stipe, gradually becoming orange lower when young, darkens in age; cinnamon-brown, orange to orange-brown. **Flesh:** Thin, tan in cap. Stipe with a cartilaginous orangish exterior; stuffed with a whitish pith when very young, soon hollow, fragile. **Odor:** Indistinct when young, disagreeable in age. **Taste:** Indistinct to slightly farinaceous. **KOH:** Orange-brown on cap. **Spore deposit:** Cinnamon brown. **Microscopy:** Spores 6–7.5 x 3.3–5 µm (averaging 6.5 x 3.8 µm), broadly ellipsoid to nearly ovoid, finely roughened. Cheilocystidia abundant, thread-like to narrowly club-shaped, at times with rounded heads, thin walled. Lacking clamp connections.

Ecology: Scattered troops or arcs, more rarely solitary or in small clusters, under conifers, especially in mature and old-growth forests. Widespread in western Cascadia but uncommon. Fruiting in fall.

Edibility: Unknown, likely toxic.

Comments: The medium size, orange colors, fragile stipe that soon becomes hollow, and small spores are characteristic. *P. dissiliens* is generally slightly smaller and more brightly colored; microscopically, it has slightly wider spores (6.7 x 4.3 µm on average) and clamp connections. Another small orange species is *P. piceae*, which has a pith-filled stipe (often gets hollowed near the base by insect larvae), and a gradually tapered "root." It often has a bitter taste, and has larger almond-shaped spores (9.4 x 6 µm on average). *P. radicata* has a yellow-orange cap, a long-rooting stipe which is stuffed with pith when young, and smaller spores (5.8 x 3.2 µm on average). *P. californica* typically has darker orange to red-brown colors and grows in clusters.

Phaeocollybia lilacifolia

A. H. Sm.

Lilac-gilled Phaeo

Cap: 2–6 (10) cm across, conical to broadly conical when young, becoming bell shaped to convex with a low, obtuse umbo; margin inrolled to plane, often undulating in age. Surface smooth, viscid when wet, shiny when dry. Ocher-brown to dark brown. **Gills:** Narrowly attached or appearing free, close to moderately crowded. Lilac-gray to violet when young, developing rusty brown tones in age. **Stipe:** 3–6 cm long aboveground, up to 18 cm long overall (but difficult to extract without breaking), 0.5–1.5 cm thick, equal or swollen aboveground, tapering underground. Surface moist to dry, fibrillose to nearly smooth. Cinnamon-brown with pinkish tones at apex when young, darkening slightly in age. **Flesh:** Thin, dull brown in cap, stipe stuffed with whitish to cinnamon pith and a darker cartilaginous outer skin. **Odor:** Floral-earthy when young, often unpleasant in age. **Taste:** Indistinct. **Spore deposit:** Cinnamon-brown. **Microscopy:** Spores 6.5–9 x 4–5.5 µm, lemon to almond shaped with a beak in side view, spindle-shaped in face view, surface marbled, apiculus smooth. Cheilocystidia abundant, club-shaped to cylindrical.

Ecology: Solitary, scattered or in troops from moss and duff, typically in old growth conifer forests. Rare, known from scattered locations in the Coast Range and Cascade foothills. Fruiting in fall.

Edibility: Unknown, likely toxic.

Comments: The ocher-brown to dark brown cap, lilac to violet young gills, rooting, solid stipe, and medium size help distinguish this species. *P. fallax* also has violet gills when young but has a greenish cap and a hollow stipe. *P. benzokauffmanii* is typically larger, has pallid gills when young, and has a smoky-violet cap.

Synonym: *Phaeocollybia riffipes* Norvell. ITS sequence data has shown that the slender, small-spored, violet-gilled specimens are conspecific with *P. lilacifolia*.

Phaeocollybia olivacea

A. H. Sm.

Olive Phaeocollybia

Cap: 3.5–10 (12) cm across, conical with an incurved margin at first, becoming broadly conical to convex, often with a low umbo; margin becoming wavy and slightly uplifted in age. Surface smooth to slightly wrinkled, viscid and shiny when wet, occasionally becoming dull and dry. Color variable and hygrophanous, dark olive to olive-green or forest green when young, becoming olive-green, orangish green to brownish olive or even reddish olive in age. **Gills**: Narrowly attached or appearing free at times, close to crowded. Dingy yellow to pale greenish yellow when young, often with reddish spots and stains, becoming yellowish brown to dirty reddish brown in age. **Stipe**: 7–14 cm long aboveground, up to 22 cm long overall, 0.8–2 cm thick, equal or slightly enlarged toward lower stipe and tapering to a long root underground. Drab yellowish to greenish buff on upper portion, gradually becoming darker, lower stipe orange-red when young, in age turning reddish to orange-brown from base up, often remaining paler buff-brown at apex. **Flesh**: Thin, greenish in cap, stipe stuffed with whitish, silky-fibrillose pith that stains reddish brown, and a fibrous to cartilaginous reddish to watery gray rind-like skin. **Odor**: Farinaceous to cucumber-like. **Taste**: Indistinct. **KOH**: Orange on cap, reddish brown on flesh. **Spore deposit**: Rusty brown. **Microscopy**: Spores 5–7 x 8–11 µm, average 6 x 10 µm, broadly ovoid with an abruptly projecting snout in face view, roughened or warted with a smooth apical beak. Cheilocystidia thin walled, variable, club shaped, often with thread-like extensions in age.

Ecology: Often in clusters or in large arcs, occasionally solitary or scattered in small groups on ground under conifers. Frequent in coastal forests with Sitka spruce and western hemlock but also found with a wide variety of conifers across lower and mid-elevations in western Cascadia. Fruiting in fall.

Edibility: Unknown, likely toxic.

Comments: The conical greenish cap, dingy yellowish gills, and solid, rooting stipe help identify this mushroom. *P. fallax* is another common green-capped species, distinguished by the beautiful lilac gills when young, smaller fruitbodies, and hollow stipe in age. However, the gill color fades rapidly, and mature fruitbodies of these two species can be very difficult to tell apart, especially since they often grow alongside one another! *P. pseudofestiva* also has a greenish cap and drab yellowish gills but is usually smaller and has a green stipe apex when young; microscopically, it has relatively thick-walled cheilocystidia that are bottle shaped or cylindrical with abruptly rounded heads. The rare *P. gregaria* is very similar in many respects to *P. olivacea* but has a golden tan to warm brown cap.

Taxonomic note: ITS sequence data suggest that *P. olivacea* represents a species complex in the PNW.

Phaeocollybia fallax

A. H. Sm.

Beautiful Phaeo

Cap: 1.5–6 (9) cm across, conical to broadly conical with an incurved margin when young, becoming broadly convex to nearly plane, often with a pronounced pointed umbo; margin becoming wavy and occasionally slightly uplifted in age. Surface smooth, shiny, viscid, margin translucent-striate when wet; occasionally become dull and dry. Color variable, hygrophanous; fading when dry. Dark olive, olive-green to forest green, often with paler blotches when young, becoming dingy olive to olive-brown in age. **Gills:** Narrowly attached or appearing free, close to crowded. Bluish lilac to violet when young, soon fading to grayish lilac and developing orangish spots and stains, becoming dingy orange-brown, but often retaining some grayish color in age. **Stipe:** 4–8 cm long aboveground, up to 25 cm long overall (but tough to extract from ground without breaking), 0.3–1 cm thick, equal, or occasionally slightly tapered toward the base, with a short and abrupt to long tapering root underground. Drab gray to pale lilac-gray on upper portion, gradually becoming darker, to reddish to orange-red on lower stipe when young, then reddish brown to dark brown from base up in age, but often staying paler dingy gray to olive-gray at apex. **Flesh:** Very thin, fragile, watery green to grayish in cap. Stipe often hollow, but occasionally stuffed with whitish fibrous pith when young; outer rind fibrous to cartilaginous. **Odor:** Generally mild; a mix of radish and sweet floral scent with an unpleasant undertone. **Taste:** Mild, slightly radish-like or occasionally bitter. **KOH:** Dingy orangish brown on cap. **Spore deposit:** Pinkish cinnamon to pale rusty brown. Microscopy: Spores 7–10 x 4–6 µm, average 9 x 5.3 µm, lemon to almond shaped in side view, ovoid with an elongated beak and a pointed apiculus, marbled or roughened. Cheilocystidia numerous, thin walled, with small, swollen heads.

Ecology: Solitary, scattered, or in small clusters on ground under conifers. Occasional in coastal and Coast Range forests especially with, but not restricted to, Sitka spruce and western hemlock. Rarer in the Cascade Range with mixed conifers. Fruiting in fall.

Edibility: Unknown, likely toxic.

Comments: The small size, conical green cap, beautiful lilac gills, and rooting stipe make this *Phaeocollybia* unmistakable when young. However, be sure to check each fruitbody carefully, as two other green-capped *Phaeocollybia* often grow intermingled with *P. fallax*: The rare *P. pseudofestiva* shares the same stature, size, and cap color, but has drab yellowish gills and a green stipe apex when young. *P. olivacea* is larger, has yellowish to drab gills, and solid stipe.

Taxonomic note: ITS sequence data indicate that *P. fallax* represents a species complex in the PNW, however at this time, no clear morphological differences have been defined.

9 - Cortinariaceae

Cortinarius is a spectacularly species-rich genus found in temperate forests around the world, and the Pacific Northwest in particular is a diversity hotspot — there are nearly 500 species known from Cascadia! As expected for a group of this size, there is great diversity in the appearance of mushrooms in this genus: some are tiny, others are enormous, a great many are dull, while others are brilliantly colored. They are often difficult to identify, but many are particularly beautiful, and the group can be very satisfying to work with once you develop some familiarity with it. All are ectomycorrhizal, and are primarily found in forested areas (although some species do grow in urban settings).

Most of the *Cortinarius* in our area can be recognized by their *cortinas* (cobwebby partial veils made of fine fibrils), ocher-brown, rusty brown, or bright orange-brown spore deposits, and growth on ground in soil or humus (none appear to be truly or consistently lignicolous). However, the cortina can be inconspicuous, short-lived, or covered in a layer of slime. Lastly, keep in mind that there is a single species in our area which has a membranous veil (*C. caperatus*), and that *C. renidens* lacks a veil entirely!

Important characters to make note of when identifying *Cortinarius* include: coloration of cap and gills (especially when young), texture of the surfaces (dry or viscid? smooth or scaly?), shape of the stipe base, odor, KOH reactions on all parts, bruising reactions, and nearby tree species. Important microscopic features include: size and shape of spores, coarseness of warts, bumps, and wrinkles on the spore surface, and the presence or absence of clamp connections in the hyphae of the cap.

Some authors have proposed splitting the genus, and currently recognize ten genera, seven of which occur in Cascadia (see On the Future of Mushroom Identification, p. 547). Under this classification, a majority of species remain in *Cortinarius*, but Cascadia has many species now in *Phlegmacium*, *Thaxterogaster* and *Calonarius*, as well as a few species each in *Aureonarius*, *Cystinarius*, and *Hygronarius*. The other three genera, *Mystinarius*, *Austrocortinarius* and *Volvanarius* are less diverse, and the latter two are represented only by southern hemisphere taxa.

The split of *Cortinarius* is now infamous for its sweeping scope and contentious reception. Some authors prefer a more conservative approach and would rather see these genera as subgenera. It should be noted that in a few cases in this book, species which were described since (and adhered to) these proposed splits are referred to using the new genus names. Readers with a particular interest in this genus should consult *Mushrooms of the Redwood Coast* and *Mushrooms Demystified* for an understanding of traditional subgroups.

- ***Cortinarius*** – A large assemblage of species with varying appearances. A few major sections of *Cortinarius* are exceptionally diverse in Cascadia, with over 150 species of (mostly brown) *Telamonia*, and ~50 species of *Dermocybe*, with bright gill colors. Other distinctive sections like *Leprocybe* often have scaly caps, and bright UV reactions. *Anomali* are small to medium-sized webcaps with silvery violet to gray-brown colors. ~330 species in Cascadia.
- ***Aureonarius*** – Caps thinly viscid to dry, stipes dry. Brightly colored (orange, ocher, to red-brown). Caps not hygrophanous or weakly so. Similar to *Telamonia* and some *Leprocybe*. ~5 species in Cascadia.
- ***Calonarius*** – Caps viscid, stipes dry, often brightly colored. Many have distinct, brightly colored KOH reactions. Stipes often stout and bulbous. Generally uncommon, and many are habitat specialists. Stature and appearance overlaps considerably with some sections of *Phlegmacium*. ~50 species in Cascadia.
- ***Cystinarius*** – Caps tacky to dry, stipes dry. Fruitbodies medium to very large. Gills have cystidia. 2 species in Cascadia.
- ***Hygronarius*** – Caps moist to dry, strongly hygrophanous, fruitbodies generally small to medium-sized. Partial veil absent. 2 species in Cascadia.
- ***Phlegmacium*** – Caps viscid, stipes dry, often brightly colored. Some have distinctive KOH reactions. Stipes clavate to bulbous. Many species are medium-sized to large. ~60 species in Cascadia
- ***Thaxterogaster*** – A mismatched group with some Phlegmacium-like species, while others would have traditionally been placed in the Myxacium 'morpho-group' (with a slimy cap and stipe).

Cortinarius clandestinus

Cortinarius clandestinus (smooth form)

Cortinarius aff. *veneto-occidentalis*

Cortinarius atrosquamosus

Cortinarius clandestinus

Kauffman

Cap: 4–10 cm across, rounded to convex with an inrolled margin when young, becoming broadly convex to plane; margin becoming wavy and often slightly uplifted. Surface dry, densely to finely covered with dark olive-brown to blackish fibrillose scales over a yellowish to orange-brown base color. **Gills:** Broadly to narrowly attached, often notched, close to subdistant. Straw-yellow to pale olivaceous brown at first, soon orange-brown as spores mature. **Stipe:** 4–10 cm long, 0.5–1.8 cm thick at apex, equal, with an enlarged base, or occasionally tapered below. Surface dry, covered with longitudinal fibrils and silky cortina remnants, at times with darker bands of veil sheaths. Pale stray yellow to buff-yellow, more rusty brown in age. **Partial veil:** A sparse cortina, leaving an annular zone of yellowish to dark olive fibrils on stipe, soon covered in rusty brown spores. **Flesh:** Thin to moderately thick, firm, pale olivaceous yellow to dingy yellowish brown. **Odor:** Radish-like to earthy. **Taste:** Radish-like. **KOH:** Dark brown on cap, red-brown on stipe and flesh. **Spore deposit:** Rusty brown. **Microscopy:** Spores 6.5–8.5 (9.5) × 5–6.5 µm, broadly ellipsoid to subglobose, moderately to coarsely roughened.

Ecology: Solitary, scattered or in small clusters from moss and duff under conifers, especially true fir and hemlock. Widespread in Cascadia, especially common in the Cascade Range. Fruiting in fall and spring (see comments).

Edibility: Unknown, possibly toxic.

Comments: As a group, *Leprocybe* can be recognized by their dry, often scaly caps, yellowish, olive, brown, or rusty orange colors, frequently with dark veil bands on the lower stipe. Most fluoresce brilliant yellow or golden under UV light. Described above is the fall-fruiting form of *C. clandestinus*. The spring-fruiting form tends to lack, or have only sparse scales on the cap, and to show more yellowish ocher colors. Other conifer-associated species include: *C. atrosquamosus*, with a densely scaly or velvety blackish to dark-olive cap, and spores averaging 7.2 × 5.4 µm; *C. fuscoflavidus* also has a dark cap, but tends to be smaller, with brighter yellow gills and slightly shorter spores (6.8 × 5.4 µm). *C.* aff. *veneto-occidentalis* has dark olive colors with spores averaging 7.5 × 6.1 µm. *C. brunneofibrillosus* is a stocky species with a brown cap and widely-spaced gills; it grows with oak. *C. fuscotomentosus* looks quite similar to *C. clandestinus* but grows with tanoak. *C. parkeri* is a spring-fruiting species found under conifers in the eastern Cascades. It has a smooth, orange-brown cap and a slightly upward-flaring veil near the stipe base.

Cortinarius renidens

Fr.

Webless Webcap

Cap: 2–6.5 cm across, broadly conical, bell-shaped to convex when young, becoming broadly convex to plane, often with a low umbo; margin occasionally wavy in age. Surface moist to dry, smooth, strongly hygrophanous. Orange, orange-brown to ocher, fading markedly when dry. **Gills:** Narrowly attached, often notched, close, narrow. Pale yellowish ocher at first, becoming ocher-orange to rusty orangish brown. **Stipe:** 4–7 (11) cm long, 0.4–0.8 cm thick, equal to club-shaped, often twisted or irregular. Surface moist to dry, with faint, appressed longitudinal fibrils. Pale yellowish ocher with whitish to pale yellow fibrils, pale ocher-brown in age. **Partial veil:** Absent. **Flesh:** Thin, fragile in cap, fibrous in stipe. Pale ocher orange to ocher-brown when wet, pallid when dry. **Odor:** Indistinct. **Taste:** Indistinct. **KOH:** Dark brown on cap. **Spore deposit:** Rusty brown. **Microscopy:** Spores 6–7 × 5–5.5 µm, ovoid to broadly ellipsoid, slightly roughened.

Ecology: Solitary or scattered from duff and moss under a wide range of conifers. Widespread in most forested habitats in Cascadia, but rarely abundant. Can be locally common in high elevation fir-hemlock forests in the Cascade Range. Fruiting in late summer and fall.

Edibility: Unknown; do not experiment.

Comments: The small- to medium-sized fruitbodies, smooth, hygrophanous ocher-orange cap, and rusty spores facilitate identification. The recently described *Hygronarius californicus* may tend to show more red-brown colors, and the spores are ever-so-slightly longer (6.5–7.5 µm). Unlike other *Cortinarius*, these species lack a partial veil! *Gymnopilus* can look similar, but typically have a fibrillose to finely scaly cap and grow on wood. *C. infucatus* (and other members of the *C. callisteus* group) have yellow-orange, orange to ocher colors, and cobwebby cortinas. *C. illuminus* has a strongly hygrophanous red-brown cap, a white stipe, and white cortina fibrils.

Synonym: *Hygronarius renidens* (Fr.) Niskanen & Liimat

Cortinarius rubellus

Cooke

Deadly Webcap

Cap: 3–6 cm across, conical to bell-shaped when young, becoming broadly convex to plane with a broad umbo. Surface dry, finely scaly, not hygrophanous. Orange-brown to reddish brown, duller brown when dry. **Gills:** Narrowly attached, with a broad notch, well-spaced. Yellowish brown when young, rusty brown when spores mature. **Stipe:** 4–10 cm long, 1–1.8 cm thick, equal or enlarged slightly below. Surface dry, longitudinally fibrillose, at times with veil bands. Pale ocher brown when young, soon reddish to orange-brown, darkening when handled. **Partial veil:** A thin, evanescent whitish to yellowish cortina. **Flesh:** Moderately thin, fragile in cap, fibrous in stipe. Yellowish tan, developing rusty stains in age. **Odor:** Radish-like. **Taste:** Indistinct to radish-like. **KOH:** Not tested. **Spore deposit:** Rusty brown. **Microscopy:** Spores 9–10 × 6.5–8 µm, subglobose to broadly ellipsoid, finely roughened.

Ecology: Solitary or scattered in small troops from moss, duff or moss-covered decaying logs and stumps in mixed conifer forests. Occasional, and widespread; occurring from coast to the high Cascades. Typically fruiting in the summer on the coast, with occasional spring and fall fruitings.

Edibility: Deadly toxic, due to the orellanine content. Symptoms are slow to develop, occurring 2–14 days after ingestion; these may include stomach pains and vomiting, headaches and muscle aches, followed by kidney failure.

Comments: This species is recognized by its bell-shaped to convex, finely scaly brown cap, well-spaced gills, scant cortina and growth with conifers. It is as unassuming as it is toxic – a prime example of why you shouldn't eat random webcaps! *C. kroegeri* is difficult to distinguish at times, but tends to have a brighter orange, smooth to watery-streaked cap, and a taller stature. *C. infucatus* and its kin typically have rounded to convex, slightly hygrophanous caps with brighter colors. *C. gentilis* typically has tall, slender fruitbodies and yellow veil remnants.

Cortinarius infucatus

Fr.

Cap: 3–6 (8) cm across, rounded to convex, often wavy in age. Surface moist to dry, smooth or finely scaly over disc in age, hygrophanous. Yellow-orange to ocher orange or pinkish-orange, fading when dry. **Gills:** Broadly attached, at times notched, close. Tan to straw-yellow at first, more yellow-brown as spores mature. **Stipe:** 4–10 cm long, 0.8–1.3 cm thick, club-shaped, to more or less equal. Surface dry, with fibrillose veil bands, and patches. Off-white to yellowish beige, covered with ocher to yellow-orange fibrils. **Partial veil:** A rather thick, orangish to yellowish cortina leaving silky remnants on stipe. **Flesh:** Moderately thin, fragile in cap, fibrous in stipe, becoming hollow. Whitish to watery orangish tan, darkening overall in age; slowly staining ocher-brown when cut. **Odor:** Indistinct to radish-like. **Taste:** slightly radish-like. **KOH:** Dark brown on all parts. **Spore deposit:** Rusty brown. **Microscopy:** Spores 7–8.2 × 5.7–6.6 µm, subglobose to broadly ellipsoid, moderately roughened.

Ecology: Scattered or in troops and small clusters from duff, moss and soil in mixed conifer forests. Especially common at higher elevations in Cascadia. Fruiting in late summer and fall.

Edibility: Unknown; do not experiment! Close look-alikes are deadly poisonous.

Comments: This webcap is recognized by its ocher-orange to yellow-orange colors, hygrophanous cap, and dark brown KOH reaction. *C. callisteus* is difficult to distinguish, although it tends to be more orange-yellow to orange-brown in color, and has a supposed odor of "scorched linen" or a "recently extinguished candle". It also has slightly smaller spores (6.7–7.8 × 5.5–6.3 µm). The poorly known *C. neocallisteus* is another difficult to distinguish species in Cascadia, which differs by a lack of an odor, and slightly larger spores (7.4–8.9 × 6–6.7 µm). *C. kroegeri* is taller, has a more bell-shaped cap, and an evanescent yellowish cortina.

Taxonomic note: All the species mentioned here would be in the genus *Aureonarius*, under a "split" model of *Cortinarius*.

Cortinarius uliginosus

Berk.

Orange Willow Webcap

Cap: 1.5–4.5 cm across, broadly conical to bell-shaped at first, soon convex to plane, often with a rounded umbo, at times wavy in age. Surface smooth to finely fibrillose, moist to dry. Deep reddish orange to orange-ocher; margin slightly paler, and duller overall with age. **Gills:** Attached, or with a slight notch, close, somewhat narrow. Yellowish orange to ocher-orange, developing brighter orange stains, and darkening as spores mature. **Stipe:** 2.5–7.5 cm long, 0.2–0.8 cm thick, more or less equal, or with a slightly enlarged base. Surface dry, covered with silky longitudinal fibers. Reddish orange fibrils over a paler ground color, dingier in age. Basal mycelium salmon-orange. **Partial veil:** A sparse cortina of reddish orange fibrils leaving a silky annular zone on stipe, or disappearing altogether. **Flesh:** Thin, olive-orange in cap. Stipe fibrous, hollow, orange to reddish orange. **Odor:** Indistinct. **Taste:** Mild. **KOH:** Dark purple to almost black on cap, purple-brown on stipe, red-brown on stipe flesh. **Spore deposit:** Rusty brown. **Microscopy:** Spores 8.5–11 (14) × 5.5–6.5 µm, ellipsoid to narrowly ellipsoid, roughened.

Ecology: Scattered or in troops on saturated soil in wet areas under willow. Locally common in appropriate habitats across Cascadia. Fruiting in fall.

Edibility: Unknown.

Comments: The reddish orange cap and stipe fibrils, yellowish orange to ocher-orange gills, dark purple KOH reaction and growth under willow are quite distinctive. Since mushroom hunters rarely bash through willow thickets, it is probably more common than records suggest. It often grows alongside *C. cinnamomeoluteus*, a *Dermocybe* with a cinnamon-brown to orange-brown cap, mustard-yellow gills, and a bright red KOH reaction. Two similar *Telamonia* with conifers are the larger *C. californicus*, with a burgundy-red to orange hygrophanous cap, bright reddish to orange gills, and bright purple KOH reaction. And the smaller *C. miniatopus* with an orange-brown cap, pinkish orange gills and a purplish KOH reaction only on the orangish veil fibrils on the stipe.

Cortinarius croceus

Cortinarius thiersii

Cortinarius viridiflavus

Cortinarius zakii

Corts

Cortinarius subgenus Dermocybe
Dye Webcaps

This is a diverse and devilishly difficult group in Cascadia, with over 35 named species and additional undescribed ones. As a group, they are recognizable by their modest size (most caps are between 2–6 cm across), smooth to innately fibrillose honey-brown to dark brown caps, (a few are more brilliantly colored), and in particular their vivid, often contrastingly bright gill coloration. For the purpose of identification, this gill color is best observed at the button stage (before the cap opens), since the gill color of most of these species converges on rusty orange as their spores mature. Although KOH reactions and UV fluorescence can be helpful tools in distinguishing some species of Dye Webcaps, keep in mind that many others are difficult, if not impossible, to reliably identify without an ITS sequence. There is a great deal of overlap in the macro-morphology features of these species, and many remain poorly known. Some of the commoner species include: *C. croceus* which has variable colors, with mustard to orange gills; *C. thiersii* is very similar but mostly fruits in spring in east-side forests; *C. aurantiobasis* with mustard-orange to ocher-orange gills and a reddish lower stipe, found in coastal forests; and the widespread *C. zakii*, with brilliant orange gills. The orange-gilled *C. viridiflavus* is more habitat restricted, but can be locally abundant in coastal dunes. *C. humboldtensis* also occurs in coastal dunes, but is quite rare. Species like *C. chrysolitus* and *C. davemallochii* are birch associates, and in Cascadia, they are mostly restricted to bog habitats. The table provided here has the primarily gill color, however, some with species, this is a highly variable trait.

Cortinarius subgenus *Dermocybe*

Species	Young gill color	Key features
C. birkebakii	Deep red	Dark red, streaked cap
C. neosanguineus	Deep burgundy-red	Slender stature, low elevations with hemlock
C. phoeniceus	Deep burgundy-red	Reddish veil bands on stipe
C. rufosanguineus	Burgundy-red	Stout stature, high elevation hemlock-fir
C. semisanguineus	Blood red	Bright golden UV rxn on sstipe
C. smithii	Blood red	No UV rxn
C. trappei	Maroon to brownish red	Duller red colors, rare
C. cinnamomeus	Orange	Over-applied name, orange-brown cap, red KOH rxn
C. malicorius	Bright orange	Olive flesh
C. pitkinensis	Orange to yellow-orange	Slender stature, orange cap
C. subrufulus	Yellow-orange to reddish orange	*C. croceus*-like, reddish-brown girdles on stipe
C. uliginosus	Yellowish orange to ocher-orange	Reddish orange cap, with willow
C. viridiflavus	Deep orange mixed with brownish	Olive-brown cap, with coastal pines
C. zakii	Bright orange	Common, widespread, brilliant orange gills
C. aurantiobasis	Mustard-orange to ocher-orange	Reddish stipe base, common coastally
C. rufulus	Mustard-yellow to orange	Very close to *C. croceus*
C. croceus	Mustard-yellow to orange	Common, variably-colored
C. thiersii	Mustard-yellow to orange	Common in spring in montane habitats, occasional in fall
C. ceskarum	Cinnamon-brown (mature only)	Only known from mature specimen, southern BC
C. ammiratii	Yellow, then orange-yellow	With hemlock, brown veil girdles on stipe
C. aureifolius var. hesperius	Lemon yellow	*Inocybe*-like, with boletoid spores
C. chrysolitus	Bright yellow to olive	Rare, birch associate, boreal, restricted to bogs in PNW
C. holoxanthus	Bright yellow	Rare, more common northwards
C. tillamookensis	Dingy yellow to olive-yellow	Olive to olive-brown cap, slender stature
C. hadrocroceus	Olive-yellow	More common in boreal forests
C. humboldtensis	Olive-yellow to mustard-orange	Deep purple KOH rxn, rare, coastal only?
C. olivaceoluteus	Olive-yellow	Dull, drab colors, thick cortina, orange-brown veil band
C. olivaceopictus	Dark yellowish olive	Reddish brown cap, reddish KOH rxn
C. tubarius	Yellowish olive	Drab colors, more northerly, olive-brown KOH rxn.
C. aff. cinnamomeoluteus	Light mustard-yellow to olive-yellow	Common with willow
C. davemallochii	Grayish yellow-olive	Boreal species, rare in PNW, with birch in bogs
C. transatlanticus	Brown-yellow to dull orange-yellow	Brownish veil fibrils on stipe
C. ferruginosus	Olive-yellow to olive-brown	Small, *Inocybe*-like, reddish brown cap
C. cascadensis	Dull olive-yellow to brownish olive	Poorly known
C. idahoensis	Yellowish brown to ocher-brown	Dull colors overall

Cortinarius rufrosanguineus

Cortinarius neosanguineus

Cortinarius birkebakii

Cortinarius californicus

Corts

Cortinarius rufosanguineus

S.D. Adams, Ammirati & Liimat.

Bloodred Webcap

Cap: 2–6.5 cm across, rounded when young, soon convex to plane; margin occasionally wavy or slightly uplifted in age. Surface dry, appressed-fibrillose to nearly smooth. Dark red to red-brown, somewhat brighter red on margin when young, light red to red-ocher when dry. **Gills**: Attached, often notched and with a decurrent tooth, close, narrow. Deep blood-red when young, developing rusty brown tones as spores mature. **Stipe**: 2–6.5 cm long, 0.5–1.5 cm thick, equal or enlarged below; base occasionally slightly bulbous. Surface dry, silky-fibrillose to nearly smooth. Dark red to red-brown or coral-red when dry, often with a paler sheen at apex; base developing brown tones in age. Basal mycelium orange. **Partial veil:** A sparse, salmon-pink to orange cortina, leaving silky annular zone on stipe, or disappearing altogether. **Flesh**: Thin to moderately thick, firm; stipe fibrous, solid. Dark red to blood-red when young, often deeper blood-red in lower stipe and developing brown tones overall in age. **Odor**: Indistinct. **Taste**: Indistinct. **KOH**: Dark magenta to vinaceous red on cap and stipe. **Microscopy**: Spores 7–9.5 × 4.5–6 µm, ellipsoid, moderately roughened.

Ecology: Solitary or in scattered in small troops in moss or duff under true fir and hemlock. Rare, occurring in subalpine meadow edges and stunted forests. Fruiting from late summer into early fall.

Edibility: Unknown.

Comments: This beautiful high elevation Webcap is easy to recognize by its dark red to red-ocher cap and stipe, deep blood-red gills, orange basal mycelium and stout stature. *C. neosanguineus* has similar colors, but has a slender stature, ocher-yellow basal mycelium and tends to grow at lower elevations with hemlock. *C. birkebakii* is another slender species with a dark red to orange-red cap (often with blackish streaks), bright red to orange-red gills, and orange basal mycelium. It occurs at lower elevations in mixed hemlock-Douglas-fir forests. *Tubaria punicea* and *T. vinicolor* are similarly colored but grow on woody debris, have strongly hygrophanous caps, and do not have distinct KOH reactions. *C. californicus* is larger, has deep reddish orange colors when young, and has a bright purple KOH reaction.

Cortinarius semisanguineus (honey-capped form)

Cortinarius semisanguineus (red-capped form)

Cortinarius smithii

Cortinarius trappei

Cortinarius semisanguineus

(Fr.) Gillet

Red-gilled Webcap

Cap: (1.5) 3–11 cm across, broadly conical to convex, becoming broadly convex to nearly plane with a low, broad umbo. Surface smooth to finely fibrillose, moist to dry. Typically honey-brown to mustard-yellow, occasionally burgundy-red to maroon. **Gills:** Broadly attached, often with a decurrent tooth, close to subdistant. Deep blood-red to burgundy-red, becoming deep orangish red to rusty orange as spores mature. **Stipe:** 4–12 cm long, 0.3–1.5 cm thick, equal, or with a slightly enlarged base. Surface dry, often with silky longitudinal fibers. Yellow, yellowish orange to more dingy yellowish brown in age. Basal mycelium pale pinkish orange. **Partial veil:** Faint, cortina-like, leaving some silky fibrils or disappearing altogether. Yellow, soon fading. **Flesh:** Thin to moderately thick, fibrous. Pale yellow to olive-yellow, dingy yellow-brown in age. **Odor:** Indistinct. **Taste:** Indistinct. **KOH:** Purple-black on cap, reddish on stipe, pinkish on flesh. **Spore deposit:** Rusty brown. **Microscopy:** Spores 6–8 × 4–5 µm, ellipsoid, roughened.

Ecology: Solitary, scattered, or in small clusters, often in troops in moss and duff under a wide range of conifers; especially in coastal pine stands and hemlock-fir forests. Common and widespread across Cascadia. Fruiting from fall into early winter.

Edibility: Unknown.

Comments: Flipping this drab mushroom over reveals shockingly vibrant blood-red gills, often eliciting an involuntary "*Wow!*" There is still a great deal of taxonomic uncertainty regarding which names to use for western red-gilled, yellow-stiped *Dermocybe.* This species has been going by *C. ominosus* in recent years, and to confuse matters further, it has many features overlapping with *C. smithii.* Although *C. semisanguineus* typically has a brown-toned cap, red forms are common, leading to confusion with *C. smithii,* which is almost always red-capped. The easiest way to distinguish them is with a UV light: the stipe on *C. semisanugineus* glows a brilliant gold, whereas *C. smithii* shows no fluorescent reaction. Additionally, there is an unnamed species close to *C. phoeniceus* in Cascadia, with a burgundy-red cap and gills and a dingy yellowish stipe which is often adorned with reddish veil fibrils. *C. trappei* is quite similar to *C. semisanguineus* in color, but has an ochraceous buff to brown stipe. It occurs at higher elevations, fruiting in late summer or early fall. All of these red *Dermocybe* are wonderful dye mushrooms, yielding vibrant and colorfast, red, orange, and purple colors on wool and silk.

Cortinarius subgenus *Telamonia*

This is by far the largest subgenus of *Cortinarius* in Cascadia, with more than 150 species, of which a one-third are still unnamed! Although some are colorful with vibrant purple, orange or even reddish colors, a vast majority are overall gray to brown. Further confounding identification, they are strongly hygrophanous, and fade dramatically as they dry, and generally 'brown out' in age. You may be lucky enough to collect *Telamonia* specimens that look distinct in the field, but will likely have a pile of similar-colored specimens by the time you get home. Most are fall-fruiting, but a small number are common springtime members of the eastern Cascade mycota. We highlight some of the most colorful and distinctive species here, but keep in mind that for every distinctive species, there are ten cryptic ones.

Key features include:

- Moist to dry caps and stipes (rarely viscid).
- Hygrophanous caps (typically strongly so).
- Often thin or scant cortina, and easy to miss on some species without buttons.
- Most species lack colorful KOH reactions, or turn dingy brown, gray to blackish.
- Most lack an odor, but a few are distinctly cedar-like, fruity, floral or geranium-like. There are also a few with green corn or pungent odors.
- UV light fluorescence can be a helpful identification tool, especially for the spring-fruiting species.

Cortinarius gentilis

(Fr.) Fr.

Goldband Webcap

Cap: 1–4 (6.5) cm across, conical with an obtuse to acute umbo when young, becoming broadly convex to plane, typically retaining the umbo throughout; margin flaring to uplifted in age. Surface moist to dry, silky-fibrillose when young, strongly hygrophanous. Brown to ocher-brown with yellow veil fibrils at first, more red-brown to dark ocher-brown in age, light yellow-brown when dry. **Gills:** Broadly attached, well-spaced, broad. Cinnamon-brown, yellowish brown to reddish brown. **Stipe:** 4.5–12 cm long, 0.3–0.9 cm thick, tall, slender, equal, or tapered at base. Surface moist to dry, with silky bands of yellow veil fibrils over a yellow-brown, ocher to reddish brown ground color, smooth and darker with age. **Partial veil:** A thin yellowish cortina, leaving a silky annular zone near the stipe apex. **Flesh:** Thin, brittle in cap, fibrous in stipe, yellow-brown to pale ocher. **Odor**: Radishy. **Taste:** Earthy. **KOH:** Dingy brown. **Spore deposit:** Rusty brown. **Microscopy:** Spores 7–9.5 × 6–7.5 µm, ovoid to broadly ellipsoid, nearly smooth to slightly roughened.

Ecology: Scattered or in loose clusters and gregarious troops from moss and duff under conifers, with a preference for mature and old growth forests. Widespread and fairly common, most abundant at lower elevations. Fruiting in fall.

Edibility: Unknown. It was once thought to be closely related to the deadly *C. rubellus*, but now understood to belong in a different section of the genus, and likely does not contain orellanine. In any case, do not experiment – it is not worth the risk!

Comments: The small to medium size, tall stature, yellow veil bands on the stipe, strongly hygrophanous, ocher-brown to reddish brown cap, and growth with conifers are helpful identification clues. *C. paludosaniosus* has a similar stature, with a distinctly umbonate, orange- to yellow-brown cap which fades to tan or beige when dry, grows with cottonwood and willow, and has narrower spores (8–9 × 5–5.5 µm). *C. saniosus* is smaller (< 2cm) with a reddish brown cap, and grows in mixed forest. Both of these species have yellow veil remnants. *C. incisior* has a brown to tan umbonate cap, and grows with oaks in urban areas of Cascadia. *C. obtusus* has an orange-brown, translucent-striate, bell-shaped cap with a rounded umbo, and a white stipe and veil; it is common in pine forests. *C. acutus* is a tiny species with a pointed umbo found in spruce forests.

Cortinarius boulderensis

A.H. Sm.

Cinnamon-ringed Webcap

Cap: 1.5–4 cm across, conical when young, becoming broadly conical to bell-shaped to nearly plane, with a well-defined umbo at all stages. Surface finely fibrillose to smooth, faintly translucent-striate when moist. Reddish brown to vinaceous brown at first, medium brown to gray-brown in age. Hygrophanous, fading to light brown or buff-brown as it dries. **Gills:** Narrowly attached, with a distinct notch, close. Lilac-gray when young, grayish cinnamon to brown when mature. **Stipe:** 3–8 cm long, 0.3–0.7 cm thick, tall, slender, equal or swollen towards base. Surface dry to moist; with pale violet-gray to whitish appressed longitudinal fibrils, and reddish to orangish veil chevrons, irregular patches and a band under the cortina. **Partial veil:** A sparse whitish cortina, leaving a few silky fibrils on the stipe. **Flesh:** Thin, brittle in cap; stipe fibrous, becoming hollow. Watery vinaceous gray to grayish tan, developing brownish tones in age. **Odor:** Indistinct. **Taste:** Indistinct. **KOH:** Brown on all parts. **Spore deposit:** Rusty brown. **Microscopy:** Spores 7–9 × 4–5.5 µm, ellipsoid, finely roughened.

Ecology: Solitary, scattered or in small troops on ground in mossy conifer forests. Widespread but uncommon, with a preference for old growth forests in the Cascade Range and on the Olympic Peninsula. Fruiting in fall.

Edibility: Unknown.

Comments: This would be just another unrecognizable brown Telamonia, were it not for the reddish to orangish veil remnants on the stipe. Other features are the faintly translucent-striate vinaceous to reddish brown cap and pale lilac gills when young. Members of the *C. spilomeus* group have lilac-gray to gray-brown caps, and brighter red chevrons or rings on the stipe. Also compare members of section *Anomali*, which tend to have more yellowish veil bands and silvery gray to light brown caps.

Cortinarius bridgei

Ammirati, Niskanen, et al.

Cap: 2–6 (8) cm across, conical to rounded when young, becoming broadly convex to plane; margin often wavy in age. Surface moist to dry, finely fibrillose to smooth. Watery dark red-brown to brown when young, fading slightly to reddish brown, then developing ocher colors in age. Hygrophanous, duller when dry. **Gills:** Broadly attached, close to subdistant, narrow. Light brown to cinnamon-brown when young, ocher-brown to orange-brown in age. **Stipe:** 3.5–8.5 cm long, 0.5–1.2 cm thick, equal or swollen towards base. Surface dry, silky fibrillose. Whitish to pale watery brown, darkening with age. Lower stipe and base with yellowish universal veil remnants. **Partial veil:** A sparse whitish cortina, leaving a few silky fibrils on stipe. **Flesh:** Thin, brittle in cap; stipe fibrous, becoming hollow. Whitish to watery brown, in stipe often paler when young. **Odor:** Earthy. **Taste:** Radishy-earthy. **KOH:** Brown on cap, bright red on stipe base. **Spore deposit:** Rusty brown. **Microscopy:** Spores 7–9 × 4–4.5 µm, 8 × 4.4 µm on average, almond-shaped, moderately to coarsely roughened.

Ecology: Scattered in troops or small clusters (rarely solitary) in duff under conifers, especially fir. Often at the edges of melting snowbanks at higher elevations and in east-side forests of the Cascade Range. Fruiting in spring and early summer.

Edibility: Unknown.

Comments: This is one of our more common spring Webcaps, fruiting as soon as the snow cover melts. The reddish brown to brown cap, light brown gills which develop ocher-orange colors, and yellowish veil remnants at the stipe base get you in the ballpark, but to confirm the species, microscopic details may be necessary. It's in the group where UV fluorescence is helpful: In *C. bridgei*, the basal mycelium fluoresces bright orange, while the veil remnants of the cap and stipe fluoresce pale yellowish. The widespread *C. flavobasilis* is very similar, but often with a brighter yellow stipe base; it has the same UV fluorescence, but larger spores (9.4 × 5 µm on average). *C. vernalishastensis* tends to be larger, with a brown to grayish brown cap that often has whitish veil remnants, and pale yellow veil at the stipe base which fluoresces light orange in UV.

Cortinarius ahsii

McKnight

Cap: 2–8 (10) cm across, rounded-conical to bell-shaped when young, expanding to broadly convex to plane; margin irregularly undulating in age. Surface moist to dry, finely fibrillose to smooth. Brown to reddish brown to ocher-brown, typically with yellowish veil remnants when young, becoming more bronze and shinier in age. **Gills:** Narrowly attached, close to subdistant, narrow. Brown to cinnamon-brown, to rusty brown as spores mature. **Stipe:** 2.5–8 cm long, 0.5–2 cm thick, equal or tapered towards base, rarely club-shaped. Surface dry, fibrillose to smooth. Light brown, with a whitish fibrillose covering, and with yellow veil bands on middle- and lower portions. **Partial veil:** A thin, whitish cortina, leaving a silky annular zone on the upper stipe. **Flesh:** Thin to thick, firm, fleshy-fibrous. Watery brown when wet, light brown to whitish when dry. **Odor:** Indistinct. **Taste:** Indistinct. **KOH:** Red on the yellow portions of the veil. **Spore deposit:** Rusty brown. **Microscopy:** Spores 6.5–9.5 × 4.5–5.5 µm, ovoid to ellipsoid, moderately roughened.

Ecology: Solitary, scattered, or in troops and small clusters from moss, duff, and soil under conifers. Common and often abundant in regional east-side forests, occasionally western in higher elevations. Fruiting after snowmelt.

Edibility: Unknown.

Comments: The reddish brown to ocher-brown cap which often shows yellow veil remnants when young, brown gills, and the yellow-banded and/or blushed stipe are diagnostic for this common spring species. The bright UV fluorescence on the universal veil is especially noticeable when young; on old, weathered specimens it is usually restricted to the lower stipe. The similar *C. colymbadinus* shows a lot of overlap in color, but is typically slightly darker overall when young, and all parts of the mushroom fluoresce yellow in UV light. *C. flavobasilis* can also be easily mistaken for the latter two species, but has orange UV fluorescence on the lower stipe. The spring form of *C. clandestinus* has a smooth, yellowish to ocher cap, pallid to tan gills which can become more orange as they mature, and a pallid stipe with a thin membranous-flaring ring. It shows brilliant yellow UV fluorescence on the stipe and gills.

Cortinarius brunneovernus

Niskanen, Liimat. & Ammirati

Cap: 3.5–10 cm across, rounded-conical to bell-shaped when young, expanding to broadly convex, plane to wavy, with a broad, rounded umbo. Surface dry, smooth or finely fibrillose. Light brown to brown, often with paler streaks, developing ocher tones with age, more gray-brown when dry. **Gills:** Broadly attached, often notched, close to subdistant, broad. Tan to brown when young, cinnamon-brown to orange-brown in age. **Stipe:** 5–12 cm long, 1–2.2 cm thick, equal to club-shaped or slightly bulbous. Surface dry, light brown covered with whitish fibrils, becoming streaked, overall brown in age. **Partial veil:** A thin, whitish cortina, leaving a silky annular zone on upper stipe. **Flesh:** Thick, firm, fleshy-fibrous. Brown marbled with tan, slowly darkening when cut. **Odor:** Earthy. **Taste:** Indistinct. **KOH:** No reaction. **Spore deposit:** Rusty brown. **Microscopy:** Spores 7.5–9 × 5.5–7 µm, subglobose to ellipsoid, moderately to coarsely roughened.

Ecology: Scattered in troops and clusters from duff and soil under conifers. Common in east-side fir-pine forests of the Cascade Range. Fruiting in spring, soon after snowmelt.

Edibility: Unknown.

Comments: Despite the overall lack of distinctive features, the medium to large size, brown colors, and spring fruiting make this Webcap fairly easy to recognize. Other large spring fruiting *Telamonia* include: *C. politus* with a similar size and brown colors, but with a violet-blushed stipe; *C. truckeensis* with a whitish to grayish cap, tan to brown gills and a whitish stipe; and a large, unnamed species with a gray-brown to brown cap, and brown stipe. All of these species occur in the drier forests in the eastern Cascades. They also all lack UV fluorescence, which helps distinguish them from *C. ahsii*, *C. bridgei* and other species mentioned in those entries.

Cortinarius rubiginosus

Ammirati, Bojantchev, Niskanen & Liimat.

Cap: 3–9 cm across, rounded to broadly bell-shaped when young, becoming broadly convex to plane or irregular. Surface dry to tacky, smooth or with appressed fibrils. Creamy beige to yellowish beige at first, developing reddish streaks and stains, becoming ocher to orange-brown overall in age. **Gills:** Broadly attached, close to subdistant. Beige at first, darkening to tan or light brown, developing orangish brown stains and becoming orangish to rusty brown as spores mature. **Stipe:** 4–8 (12) cm long, 0.5–2 cm thick, often club-shaped, enlarging downward to a swollen lower portion, and often with a tapered, rooting base below that. Surface dry, slightly fibrillose. Off-white to creamy beige at first, developing ocher-orange streaks and discoloration from the base up, becoming ocher-orange to orange-brown overall in age. **Partial veil:** Scant, cortina-like, leaving an annular zone of silky whitish to pale ocher fibrils on the stipe that are soon covered with rusty brown spores. **Flesh:** Thin to moderately thick, fleshy in cap, fibrous in stipe. Whitish to buff. Bruising bright yellow-orange, slowly becoming ocher-red to reddish orange on all parts. **Odor:** Indistinct. **Taste:** Indistinct. **KOH:** No reaction. **Spore deposit:** Rusty brown. **Microscopy:** Spores 6–7 (7.5) × 4–5 µm, almond- to lemon-shaped, finely roughened. Pleurocystidia and cheilocystidia spindle-shaped.

Ecology: Solitary, scattered, or in small clusters on ground. Common and abundant in coastal spruce forests, occasional elsewhere. Fruiting in fall

Edibility: Unknown.

Comments: The best identifying features are its pale creamy beige cap and bright yellow-orange to ocher-red or reddish staining. Although it superficially resembles a number of other *Cortinarius* in Cascadia, none show this color-staining reaction. The presence of cystidia is unusual in *Cortinarius*, one of the reasons some have split off the genus *Cystinarius*. Its sister species, *C. crassus* has a dry, light brown, fibrillose to tomentose cap, and pallid to light brown gills and stipe; it does not stain.

Misapplied name: *Cortinarius rubicundulus* (Rea) A. Pearson is a European species. *Cystinarius rubiginosus* (Ammirati, et al.) Liimat. & Niskanen is a synonym.

Cortinarius laniger group

Cap: 3–10 cm across, rounded to conical when young, expanding to broadly convex or nearly plane, becoming wavy in age, often with a low, broad umbo. Surface dry, often with small, matted fibrils when young, becoming smooth or nearly so in age. Cinnamon-brown to ocher-brown or light reddish brown when wet, often with a whitish band around margin when young, developing darker streaks, then "browning out" in age. Hygrophanous, fading when dry. **Gills:** Narrowly attached with a distinct notch, close to subdistant. Cinnamon at first, becoming bright orange-tan to ocher, then browner as spores mature. **Stipe:** 4–12 cm long, 0.7–2 cm thick at apex, equal or with a swollen to distinctly bulbous base. Surface dry, streaky, covered in silky fibrils below cortina. Usually extensively whitish, becoming pale cinnamon, bruising cinnamon-tan when handled and developing grayish tan discolorations in age. **Partial veil:** Scant, cortina-like, leaving a faint, evanescent zone of silky whitish fibrils on the stipe and on the cap margin. **Flesh:** Thick to moderately thin, firm, fleshy in cap, fibrous in stipe. Cinnamon-brown, light brown to pale tan. **Odor**: Faint; like peas, sweet, earthy, or radish-like. **Taste:** Indistinct to earthy. **KOH:** Indistinct to faintly greenish gray on cap. **Spore deposit:** Rusty brown. **Microscopy:** Spores 8.5–10.5 (11.5) × 5–6.5 µm, ellipsoid, moderately roughened.

Ecology: Solitary or scattered from duff and moss under conifers, rarely with tanoak. Common and widespread, occurring in most forested habitats of Cascadia. Fruiting in late summer and fall.

Edibility: Unknown.

Comments: The large size, cinnamon-brown to ocher-brown cap, rather bright ocher-cinnamon gills, and whitish stipe help identify this species group. We are lumping *C. laniger*, *C. distortus*, *C. subbalaustinus*, and *C. solis-occasus* into a single group here: although they supposedly show slight differences in morphology, they have practically identical ITS sequences. More collections and detailed morphological notes are needed to see which features, if any, distinguish them. *C. bivelus* is generally smaller, with similar colors, and a thick white cortina. It grows with birch, and is far more common in boreal forests. *C./Th. vespertinus* has slightly paler colors and subglobose spores.

Cortinarius nauseosouraceus

Niskanen, Liimat. & Ammirati

Cap: 3–8.5 cm across, rounded-conical when young, becoming broadly convex to plane, at times with a low umbo. Surface moist to dry, faintly fibrillose around margin to smooth; hygrophanous. Dark brown to red-brown, with a pale yellowish band around the margin when young, more evenly colored in age, paler when dry. **Gills:** Attached, often notched, close to subdistant. Yellowish brown to light brown when young, darkening as spores mature, but retaining a paler edge. **Stipe:** 5–13 cm long, 0.5–1.5 (2) cm thick, more or less equal, often slightly rooting. Surface dry, fibrillose-silky. Grayish brown to brown with yellowish to olivaceous fibrils and sheen. **Partial veil:** Scant, cortina-like, leaving an annular zone of silky yellowish fibrils on stipe or disappearing altogether in age. **Flesh:** Thin, fleshy-fibrous; brown. **Odor:** Earthy or nauseatingly strong green corn. **Taste:** Earthy. **KOH:** No reaction. **Spore deposit:** Rusty brown. **Microscopy:** Spores 7.5–9 × 4.5–5.5 µm, narrowly ellipsoid to ellipsoid, moderately roughened.

Ecology: Scattered or in clusters and troops in duff and moss under conifers. Common, and in some years abundant in the Cascade Range and Olympic mountains. Fruiting in fall.

Edibility: Unknown.

Comments: One of many non-descript *Telamonia* in Cascadia, this species is recognized by its brown cap with a pale yellowish band around the margin, and olive-yellow sheen to the stipe. About one in fifteen people detect a raunchy, nauseating, green corn odor from this mushroom, while others perceive a modest earthy aroma. Older specimens are practically impossible to distinguish from the plethora of similar species. *C. gentilis* has more yellow-brown colors, and a bright yellow cortina which often leaves veil bands on the stipe.

Cortinarius lucorum

(Fr.) E. Berger

Cap: 3–7 (10) cm across, rounded when young, becoming broadly convex to plane, at times with a low, rounded umbo. Surface moist to dry, often with a silky sheen of appressed fibrils and a band of partial veil remnants around the margin when young. Hygrophanous. Color variable, typically gray-brown with a violet cast, and often streaked or marbled when young; soon losing the violet tones and often showing an evenly brown spot on the disc and a paler margin; in age brownish or sometimes whitish to pale grayish tan overall. **Gills:** Broadly attached, with a small notch, close to subdistant. Purplish to reddish purple when young, fading to grayish purple, and developing brown tones as spores mature. **Stipe:** 5–10 (14) long, 0.7–2 cm thick, club-shaped or swollen and bulbous at base. Surface dry, with a felty-fibrillose sheath below the annular zone. Grayish violet, whitish to gray. **Partial veil:** Thick, cortina-like, leaving an annular zone of silky whitish fibrils on stipe, becoming matted and covered in rusty brown spores in age. **Flesh:** Moderately thick, fleshy in cap, fibrous in stipe. Violet, marbled or streaked, losing the violet tones and more gray with age, developing brown tones in stipe. **Odor:** Indistinct to radishy. **Taste:** Indistinct to slightly earthy. **KOH:** Indistinct or darkening slightly. **Spore deposit:** Rusty brown. **Microcopy:** Spores 9–11 × 5–7 µm, ellipsoid, distinctly roughened.

Ecology: Scattered or in clusters (rarely solitary) on ground, typically in open areas under aspen and cottonwood. Common and often abundant in appropriate habitats across Cascadia. Fruiting in fall.

Edibility: Unknown.

Comments: The marbled or streaked grayish brown cap with a violaceous cast when young, and a brown disc as it matures, purplish young gills and growth with aspen and cottonwood make this a fairly easy species to recognize. The colors are quite variable, but the overall gestalt is distinctive. *C. fructuodorus* also shows the streaky gray-brown cap and violet tones throughout, but has a distinctive, ripe pear odor and grows with conifers. Also with cottonwood is, *C. saturninus* with a similar gestalt, but it lacks the violet tones (it is tan to light brown when young).

Cortinarius section *Anomali*

Cascadia has at least twelve species in this group of small to medium-sized Webcaps. They are recognized by their violet, bluish, silvery gray to brown caps, often bluish to violet young gills, slender stipes which are often adorned with bands of veil fibrils, and nearly round to ellipsoid spores. The bluish and violet colors soon fade to gray or brown tones, making identification of older specimens difficult. A few species are distinctive and can be easily identified in the field, while others have overlapping features, and require microscopic examination to pin down to species. Odor and taste are indistinct or slightly earthy-radishy. Unless otherwise mentioned, they have subglobose to broadly ellipsoid, moderately to coarsely roughened spores. With one exception (*C. rarus*), they fruit in fall. Edibility of all species is unknown.

Cortinarius albidipes

Peck

Cap: 4–10 cm across, smooth or fringed with silvery white fibrils around margin. Pale silvery gray to pale buff, more brownish, cinnamon in age. **Gills:** Attached, with a decurrent tooth, close. At first pale violet to pale bluish gray, pale brown to cinnamon in age. **Stipe:** 4–10 cm long, 0.8–1.5 cm thick, equal to club-shaped. Surface dry, with a whitish, silky-fibrillose covering over a lilac-violet to pale grayish lilac background color. Lower stipe with pale yellow-buff to pale ochraceous veil zones and patches. **Flesh:** Thin, firm, mottled violet and gray, whitish in age. In stipe fibrous, pale bluish to pale violet at first, whitish to dingy in age. **Microscopy:** Spores 8–9.5 (11) × 6.7–7.8 µm, averaging 9 × 7.5 µm.

Ecology: Solitary or in scattered troops, often in wet areas under hardwoods, especially willow. Widespread, occasional.

Comments: The fairly squat stature, pallid cap and stipe, gills that retain their violet colors well into maturity, growth with willow, and large spores are the identifying features for this species. *C. clintonianus* has fleeting violaceous gray colors when young, soon becoming whitish to light brown, with pallid to buff veil remnants, and small spores (6.7–8.1 × 5.2–6.3 µm). It is widespread, with a preference for mature conifer forests.

Cortinarius anomalodelicatus

Ammirati, Liimat., Niskanen & Dima

Cap: 2–2.5 cm across, smooth, moist to dry. Grayish brown to pale vinaceous brown over disc, grayish lilac towards margin when young, extensively brown in age. **Gills:** Broadly attached, close, grayish purple when young, soon grayish brown to brown as spores mature. **Stipe:** 5–7 cm long, 0.4–0.6 cm thick, tall, slender, swollen or bulbous at base. Surface dry, finely fibrillose to smooth, at times with buff veil patches on lower stipe when young. Pale lilac to lilac-gray when young, dingy brownish in age. **Flesh:** Thin, somewhat fragile, stipe fibrous. Whitish to violet-gray under cap surface, purplish to lilac in upper stipe when young, whitish to dingy grayish in age. **Microscopy:** Spores 7–9 (10.5) × 6–7.5 (8) µm, averaging 8.5 × 6.7 µm.

Ecology: Solitary or scattered in duff, moss or soil in mixed forests of true fir, hemlock, Douglas-fir and alder. Also known from northern spruce forests. Occasional.

Comments: This species is recognized by its small, slender stature, fairly drab colors and large spores; young specimens often show yellowish buff veil remnants on the stipe.

Cortinarius anomalovelatus

Ammirati, Berbee, Harrower, Liimat. & Niskanen

Cap: 0.5–4.5 cm across, rounded conical when young, becoming broadly convex. Surface dry, covered with matted silky fibrils. Silvery blue to silvery lilac, covered with silvery white fibrils; often developing an orangish center with age. **Gills:** Broadly attached, close to crowded. Bright bluish to lilac when young, developing gray tones and then becoming light rusty orange as spores mature. **Stipe:** 3–10 cm long, 0.3–0.7 cm thick, long, slender, enlarged slightly downward to a swollen base. Surface dry, with tufts or bands of whitish to pale buff, silky veil fibrils or with patches of felty veil fibrils. Silvery whitish, often more lilac near apex. **Flesh:** Thin, brittle in cap, fibrous in stipe. Bluish gray to grayish lilac in cap, bright bluish purple to purplish gray in upper stipe, transitioning to whitish or pale buff toward base. **Microscopy:** Spores 7.5–9.5 × 6.3–7.4 µm

Ecology: Scattered or in small groups and clusters of two to three fruitbodies on ground or on well decayed mossy logs and stumps under hemlock. Fairly common in coastal forests and lower elevations across western Cascadia. Fruiting in fall, into winter or early spring on the south coast.

Comments: This is probably the most recognizable species of section *Anomali*, due to the small size, slender stature, slivery blue cap, blue to lilac-gray gills, and silky-felty veil remnants forming bands on the stipe. Compare with *C. barlowensis*, which has a darker cap.

Cortinarius barlowensis

Ammirati, Berbee, Harrower, Liimat. & Niskanen

Cap: 1.5–3 cm across, smooth or with veil remnants near margin. Grayish purple to vinaceous brown. **Gills:** Attached with a shallow notch, close to crowded. Bluish lavender to dull gray-brown, or light brown as spores mature. **Stipe:** 5–10 cm long, 0.3–0.5 cm thick, enlarged towards a club-shaped to slightly bulbous base. Surface dry, with tufts or bands of whitish to pale buff, silky or felted veil fibrils. Upper portion bluish to lavender, lower part bluish gray, darkening overall in age. **Flesh:** Thin, watery gray-violet in cap. Stipe bluish violet in upper stipe, paler below, whitish to grayish brown in age. **Microscopy:** Spores 8.5–12.5 (13) × 5.8–7 µm, averaging 10.8 × 6.5 µm, ellipsoid to broadly ellipsoid.

Ecology: Solitary, scattered or in small clusters from duff, moss or on well decayed logs, typically in old growth forests, with a wide range of conifers. Uncommon.

Comments: The grayish purple to vinaceous brown cap, long, slender stipe with distinct veil bands, and large, ellipsoid spores help distinguish this species. *C. latiodistributus*, is similar and occurs in the same habitats, but has deeper purple colors (especially in the flesh), and considerably smaller spores (7–10 × 5–6.5 µm, averaging 8 × 6 µm). *C. clackamasensis* has a vinaceous brown to purplish brown, slightly hygrophanous cap, gray to light brown gills, buff to brownish veil remnants on the stipe, and fairly large spores (9–11× 6.5–7.5 µm). It appears to prefer high elevation forests in the Cascade Range.

Cortinarius nettieae

Ammirati, C.L. Cripps, Liimat., Niskanen & Dima

Cap: 4–12 cm across, thinly viscid to dry, smooth. Typically gray to gray-brown when young, occasionally more silvery lavender to pale lilac, soon brown to reddish brown over disc, to extensively brown in age. **Gills:** Narrowly attached, with a distinct notch, close to crowded. Bluish lilac to pale violet when young, soon developing grayish tones, then light brown as spores mature. **Stipe:** 4.5–17 cm long, 0.6–2 cm thick, tall, slender, equal or enlarged slightly towards base. Surface dry, distinctly fibrillose when young, often with faint veil bands or chevrons over lower portions. Whitish to pale violet when young, more dingy grayish in age. **Flesh:** Watery violet-gray to grayish brown. Stipe solid, fibrous, violet in upper portion when young, pale grayish brown below, and overall in age. **Microscopy:** Spores 7–9 (9.5) × 6–6.5 µm, averaging 8 × 6.3 µm.

Ecology: Scattered or in troops from duff, especially common under shore pine in coastal dunes, rarer elsewhere with a range of conifers.

Comments: The large size, fleetingly silvery lavender or gray-brown cap, bluish lilac gills, tall stipe and growth under pine distinguish this species. *C. rarus* is another large member of section *Anomali*; it has a grayish blue, tan to brown cap, bluish gray young gills, and most distinctively, fruits in spring in eastside forests of the Cascade Range. *C. caesiifolius* tends to have buff to brown cap colors, and distinct yellowish to ochraceous veil bands on the stipe. *C. caninus* has a whitish to violaceous gray cap when young (soon becoming extensively brown), pale violaceous gray to dull brown gills, a nearly smooth stipe, and spores averaging 8.5 × 7 µm. It occurs in coastal spruce forests.

Cortinarius substriatus

Kauffman

Cap: 2–8 cm across, conical when young, becoming broadly conical to convex, often with a low, broad umbo; margin wavy and occasionally uplifted in age. Surface smooth, shiny, moist to slightly viscid when wet, dull, with scattered appressed fibrils when dry. Hygrophanous, color extremely variable: Typically violet to violet-gray when young, soon losing the violet tones, and becoming gray-brown to brown. **Gills:** Broadly attached or with a small notch, close, broad. Grayish violet at first, soon cinnamon tan to brown as spores mature. **Stipe:** 5–12 cm long, 0.7–1.5 cm thick, equal or slightly enlarged downward. Surface dry to slightly moist, covered with silky fibrils when young, smooth in age. Pale grayish violet at apex, whitish below, then losing the violet color, and developing dingy tones in age. **Partial veil:** Thin, sparse, cortina-like, binding to cap margin and leaving a silky annular zone of whitish fibrils on stipe. **Flesh:** Thin, brittle, violet to violet-brown in cap when young, fading and browning with age. Stipe fibrous, solid to stuffed, violet at apex, whitish to tan below. **Odor**: Indistinct to radishy. **Taste:** Indistinct to earthy. **KOH:** Brownish on cap. **Spore deposit:** Rusty brown. **Microscopy**: Spores 9–10.5 × 5–6 µm, narrowly ellipsoid to almond-shaped, roughened.

Ecology: Solitary, scattered, or in small clusters from duff and moss under conifers (especially hemlock). Common and widespread in western Cascadia in most forested settings. Fruiting from late summer at elevation until late fall at lower elevations.

Edibility: Unknown.

Comments: This is a variably colored species: the cap ranges from deep purple to brown, but is typically grayish violet; this and the relatively thick white cortina, and growth with conifers help set it apart. There are a couple similar unnamed species, which can apparently only be distinguished genetically. The late fall and winter fruiting *C. olsoniae* is markedly purple when wet to silvery blue when dry; it tends to retain the violaceous colors and appears to be restricted to coastal spruce forests. *C. bibulus* is a small, slender species with a deep purple to purplish brown umbonate cap and grows in wet areas with alder.

Cortinarius oregonensis

A.H. Sm

Egg-yolk Webcap

Cap: 1.5–4 (7) cm across, rounded to conical at first, becoming broadly convex to nearly plane, at times with a low, broad umbo. Surface viscid to tacky, smooth. Silvery lilac to lavender with a pale tan to orangish disc when young, lilac tones fading, and the disc becoming more egg-yolk yellow in age. **Gills:** Broadly attached, close to crowded, narrow. Lavender to lilac when young, fading to grayish, then light brown as spores mature. **Stipe:** 4–9 cm long, 0.4–1.3 cm thick, equal to club-shaped. Surface viscid below cortina, dry above; lower portion with a fibrillose sheath. Pale lilac young, soon whitish, developing orangish tones from base up in age. **Partial veil:** A thin, cortina, leaving a faint annular zone on stipe. **Flesh:** Thin, firm, stipe fibrous. Whitish to watery tan, with lilac tones in upper stipe when young, developing orangish stains in age. **Odor:** Indistinct. **Taste:** Indistinct. **KOH:** No reaction. **Spore deposit:** Rusty brown. **Microscopy:** Spores 7–8 × 4–5 µm, ellipsoid, finely roughened to nearly smooth.

Ecology: Solitary, scattered or in small clusters from duff and moss under conifers, especially hemlock and fir. Widespread but rarely fruiting in large numbers. Fruiting in fall.

Edibility: Unknown.

Comments: This small Webcap is recognized by the pale lilac to lavender cap with an orangish disc and lilac young gills. Older specimens will lose most of the lilac colors, and the central disc spot will become brighter yellow to golden. *C. leucophanes* is often smaller, with a viscid white to creamy cap and clay-colored gills. *C. occidentalis* has dark purple to grayish purple colors when young (fading with age), and purple-staining flesh and gills.

Synonym: *Thaxterogaster oregonensis* (A.H. Sm.) Niskanen & Liimat.

Cortinarius alboviolaceus

(Pers.) Fr.

Silvery-violet Webcap

Cap: 3–7 (10) cm across, rounded to convex when young, becoming broadly convex to nearly plane or occasionally wavy in age. Surface thinly viscid to dry, silky and often with small tufts of fibrils when young, more appressed-fibrillose to smooth in age. Silvery violet to silvery lilac, becoming silvery gray to grayish lilac, developing some blotchy orangish coloration in age. **Gills:** Broadly to narrowly attached, often with a distinct notch, close to subdistant. Lilac at first (but can be light brown), becoming lilac-gray and developing rusty tones as spores mature. **Stipe:** 5–10 cm long, 1–2 cm thick at the apex, enlarging downward to a swollen base. Surface thinly viscid to dry, silvery violet to pale lilac, covered with silvery white, silky to somewhat felty fibrils on lower portion. **Partial veil:** Cortina-like, leaving a zone of whitish to pale grayish fibrils on stipe that become covered in rusty brown spores; often leaving a band of silky fibrils around cap margin. **Flesh:** Thick, firm to soft, whitish gray marbled with lilac, especially in stipe apex, occasionally becoming orangish beige in lower stipe. **Odor:** Indistinct. **Taste:** Indistinct. **KOH:** No reaction. **Spore deposit:** Rusty brown. **Spores** 7–10 × 4.5–6.5 µm, ellipsoid to almond-shaped, slightly roughened.

Ecology: Solitary or scattered from duff and moss under conifers and occasionally hardwoods. Common and widespread, occurring in most forested portions of Cascadia. Fruiting in late summer and fall.

Edibility: Unknown. Likely non-toxic.

Comments: The silvery violet cap, lilac-gray to brownish gills, rusty brown spores, whitish gray to lilac-marbled flesh and lack of a distinctive odor help identify this beautiful species. *C. camphoratus* is quite similar, but can easily be told apart by its awful odor of rotting potatoes. *C. traganus* has a richer violet-purple cap, marbled tan to orange flesh, and a sweet, fruity odor. *C. alboglobosus* has a whitish, grayish to brown cap, and tan to light brown gills and flesh. *C. occidentalis* generally has darker purple colors, and stains purple. *Clitocybe nuda* has a smooth, greasy cap, pale pinkish buff to lilac-buff (not rusty brown) spores, and no cortina.

Cortinarius alboglobosus

Kytov., Liimat., Niskanen & Ammirati

Snowglobe Webcap

Cap: 4–12 (17) cm across, convex to broadly convex, at times with a broad umbo; margin often wavy in age. Surface dry, innately fibrillose, often cracking. Whitish when young, soon grayish buff to brown. **Gills:** Broadly attached, close to moderately spaced, broad. Tan to light brown when young, darkening and developing rusty tones as spores mature. **Stipe:** 4–12 cm long, 1–4.5 cm thick, equal, club-shaped or swollen below and tapered at base. Surface dry, fibrillose. White to light brown, darkening when handled or with age. **Partial veil:** A thick, cobwebby cortina, leaving an annular zone on the stipe and fibrils on the cap margin. White to off-white, soon covered with rusty spores. **Flesh:** Thick, firm, fleshy; tan to light brown. **Odor:** Pleasant, somewhat fruity. **Taste:** Earthy. **Spore deposit:** Rusty brown. **Microscopy:** Spores 7.5–9 × 5–6 µm, ovoid to lemon-shaped, finely to moderately roughened.

Ecology: Scattered or in troops on ground under conifers, especially hemlock and fir. Widespread in Cascadia, but typically in montane forests. Common and often abundant in the Cascade Range, occasional elsewhere. Fruiting in summer at high elevations, continuing into late fall.

Edibility: Unknown, likely non-toxic.

Comments: The large size, whitish colors which soon become grayish to brown, brown gills and light brown flesh make this a fairly distinctive species. Collections from the southern part of Cascadia and California tend to be smaller and brighter white when young. *C. fructuodorus* is quite similar, but often smaller, with a more upright and slender stature. It may exhibit violet colors (typically grayish to brown), and can have a mottled cap in age. It has a strong pear-like odor, similar to *C. traganus*. Old, faded specimens of *C. traganus* which have lost their violet colors might be mistaken for the above two species, but the stronger, overripe pear odor helps distinguish it from *C. alboglobosus*, and darker orange-brown flesh distinguishes it from both. *C. alboviolaceus* has more silvery violet colors, whitish to pale lilac flesh and lacks an odor.

Cortinarius traganus

(Fr.) Fr.

Lilac Webcap

Cap: 4–10 (12) cm across, rounded to convex when young, becoming broadly convex to nearly plane, occasionally wavy in age. Surface dry, with silky, matted fibrils when young, often cracking into irregular patches or upturned scales over center in age or when dry. Lilac-purple to bluish lilac when young, fading to lilac-gray, occasionally grayish overall, or at times extensively ocher. **Gills:** Broadly to narrowly attached, often with a distinct notch, close to crowded. Light brown to cinnamon-brown at first, becoming rusty brown. **Stipe:** 4–10 cm long, 0.7–2 cm thick at apex, 1–3.5 cm thick at base, club-shaped, with a swollen or distinctly bulbous base. Surface dry, covered with silky to flocculent fibrils that become matted in age, below a silky cortina. Lilac-purple to grayish lilac, covered with irregular belts or patches of white to lilac fibrils on lower portion, developing orangish discolorations from base up in age. **Partial veil:** A thick cortina, leaving an annular zone of whitish to lilac fibrils on stipe which soon become covered in rusty brown spores; often also leaving a band of silky fibrils on cap margin. **Flesh:** Thick, firm, fleshy to fibrous. Rusty brown to orange-brown, with creamy or tan marbling. **Odor:** Usually strong, a pleasant, sweet, fruity odor often described as "pear-like" when young, becoming stronger and disagreeable in age. **Taste:** Earthy. **KOH:** No reaction. **Spore deposit:** Rusty brown. **Microscopy:** Spores 7–10 × 5–6 µm, ellipsoid to almond-shaped, roughened.

Ecology: Solitary or scattered on ground in a wide variety of forest types. Widespread, especially common in conifer forests in the Cascade Range and in coastal spruce forests. It also occurs with tanoak in southern Cascadia. Fruiting from late summer at higher elevations into late fall on the coast.

Edibility: Unknown, likely non-toxic

Comments: The dry, lilac-purple cap, marbled orange-brown flesh, and distinctive pear-like fruity odor are distinctive. *C. calopus* has the same odor, but is more slender, and has violet to pallid flesh. Older *C. traganus* can fade to silvery and develop an unpleasant odor similar to *C. camphoratus*, but the odor is usually still somewhat sweet and slightly less disgusting. *C. camphoratus* also differs by having violet-purple gills and violet-gray marbled flesh in the cap (sometimes marbled rusty brown in the lower stipe). *C. alboviolaceus* has a silvery violet cap, violet to lilac-gray gills, violet-marbled flesh, and lacks a distinctive odor. *C. alboglobosus* is quite similar in stature and appearance, but is gray-brown colored. These violet Webcaps are sometimes confused with *Clitocybe nuda*, a popular edible which has a smooth, greasy cap, pale pinkish buff to lilac-buff (not rusty brown) spores, and no cortina.

Cortinarius camphoratus

(Fr.) Fr.

Stinking Webcap

Cap: 5–10 (12) cm across, rounded to convex when young, becoming broadly convex to nearly plane; margin often wavy and irregular in age. Surface viscid to dry, smooth, occasionally with appressed fibrils or tiny scales. Grayish lilac, silvery violet to silvery gray, slowly losing lilac colors as it ages, becoming silver to gray and developing orangish tones at center. **Gills:** Broadly to narrowly attached, often with a distinct notch, close to crowded. Violet-purple at first, becoming violet-gray to lilac-gray and developing rusty brown tones as spores mature. **Stipe:** 5–12 cm long, 1–2 cm thick, variable in shape, often club-shaped with a swollen base, but can be distinctly bulbous or even tapered to a pointed base. Surface dry, covered with silky fibrils that become matted in age. Silvery violet to violet, covered with longitudinal whitish fibrils and bruising violet when handled, becoming dingy grayish violet and developing orangish brown stains in age. **Partial veil:** Cortina-like, leaving an annular zone of whitish to lilac fibrils on the stipe (covered in rusty brown spores in age), and often a band or patches of silky fibrils on cap margin. **Flesh:** Thick, firm. Violet to violet-gray marbled with whitish, lower parts of stipe developing tan to orangish brown colors in age. **Odor:** Strong, putrid and unpleasant, especially in age. Often compared to rotting potatoes. **Taste:** Unpleasant. **KOH:** No reaction. **Spore deposit:** Rusty brown. **Microscopy:** Spores 9.5–10.5 × 6–6.5 µm, ellipsoid to almond-shaped, slightly roughened.

Ecology: Solitary, scattered, or in small clusters on ground, moss or duff under conifers, especially true fir and hemlock. Widespread and common, fruiting in late summer and fall.

Edibility: Unknown.

Comments: The grayish lilac cap, violet-purple gills that fade to grayish lilac, marbled violet-gray flesh, and putrid odor get you to the ballpark. The similar smelling *C. putorius* is nearly identical, but tends to have brighter violet colors, a taller stature and slightly smaller spores (8.8–9.5 × 5–5.7 µm). The memorable odor has been described in many different ways, from "burnt flesh or burnt horn" (Brandrud) to "cold mashed potatoes and goat cheese" (Courtecuisse), to simply "smelling like goats." It can become so nauseating that we have thrown collections out of the car in disgust. *C. traganus* has a pear-like, fruity odor when young that becomes more unpleasant in age; it is readily told apart by its light brown to cinnamon-brown gills and marbled rusty brown flesh. *C. alboviolaceus* has a silvery violet cap, lilac-gray to brownish gills, pale to lilac-gray marbled flesh, and lacks a distinctive odor.

Cortinarius violaceus

(L.) Gray

Violet Webcap

Cap: 4–10 (13) cm across, rounded to broadly conical when young, becoming broadly convex to nearly plane. Surface dry, somewhat velvety, covered in upturned pointed tufts of hairs that become matted in age. Occasionally cracking when dry. Deep blackish violet to deep purple, becoming brighter royal purple or graying slightly in age. **Gills:** Broadly attached with a distinct notch, moderately close at first, becoming well-spaced. Deep blackish violet to deep purple when young, becoming grayish purple and finally dusted with rusty brown spores when mature. **Stipe:** 6–15 (18) cm long, 1–2.5 cm thick at apex, often club shaped with a swollen or bulbous base (up to 4 cm across). Surface dry, covered with longitudinal fibrils and a silky annular zone. Deep blackish violet to deep purple with a sheen of paler grayish violet fibrils, darker purple where handled. Basal mycelium bright violet to whitish. **Partial veil:** A cobwebby cortina, leaving a prominent zone of fibrils on stipe that becomes covered in rusty brown spores. **Flesh:** Thick, fibrous to somewhat brittle in stipe. Dark violet to violet-gray, becoming grayer in age, staining deep purple when cut, especially in stipe when young. **Odor:** Faint, somewhat cedar-like. **Taste:** Indistinct. **KOH:** Deep blood-red on all parts. **Spore deposit:** Reddish brown to orange-brown. **Microscopy:** Spores 13–17 × 7–10 µm, ellipsoid to almond-shaped, distinctly roughened. Pleurocystidia and cheilocystidia abundant, bottle-shaped with elongated necks.

Ecology: Solitary, scattered, or in pairs and small clusters from duff, moss or well decayed woody debris. Widespread and locally common, especially in middle to higher elevations in the Cascade Range and in coastal forests. Fruiting in late summer and fall.

Edibility: Edible but reportedly "earthy" or with a slight bitter, metallic flavor.

Comments: This beautiful Webcap can easily be recognized by its dry, velvety to coarsely hairy cap, blackish violet to deep purple coloration, and rusty brown spores. Some *Leptonia* can have similar colors and cap textures, but are smaller and have pinkish buff spores and paler gills. Recent findings suggest that *C. violaceus* is a widespread and variable species that encompasses the larger conifer-associated forms often identified as *C. hercynicus* (= *C. violaceus* var. *hercynicus*).

Cortinarius glaucocephalus

M.M. Moser, Ammirati & Halling

Green Streaked Webcap

Cap: 4–10 cm across, convex with an inrolled margin when young, becoming broadly convex to plane. Surface viscid, innately fibrillose to smooth. Color variable, typically a combination of dark turquoise green, olive-green, grayish olive, or with violet tones and dark streaks when young, developing ocher-brown tones over the disc, and fading overall in age. **Gills:** Attached, close to crowded. Typically, dark blue to bluish lilac when young, at times more lavender-gray, then fading to grayish and developing brown tones as spores mature. **Stipe:** 3.5–10 cm long, 0.9–2.5 cm thick at apex, equal or enlarged towards an abruptly bulbous base. Surface dry, longitudinally fibrillose. Bluish gray at apex, more grayish green below, becoming grayer and dingier in age. **Partial veil:** A cobwebby, bluish gray cortina, attached over most of the lower stipe, soon collapsing and leaving a silky annular zone. **Flesh:** Thick, firm, fleshy-fibrous. Blue to blue-gray, often marbled when young, more grayish and developing brown tones in the stipe in age. **Odor:** Indistinct. **Taste:** Indistinct. **KOH:** Reddish brown on cap surface. **Spore deposit:** Rusty brown. **Microscopy:** Spores 8–9.5 × 4–5 µm, lemon-shaped, coarsely roughened.

Ecology: Solitary or scattered from soil, moss and duff under both conifers and oaks. Occasional at lower elevations in west-side forests. Also in urban areas with planted oaks. Fruiting in fall.

Edibility: Likely non-toxic.

Comments: This species is recognized by its variably colored cap, which often shows greenish tones and has dark streaks when young, blue gills, a bulbous stipe, and lemon-shaped spores. Most similar are members of the *C. glaucopus* group, of which there are at least seven members in Cascadia, (most of which remain unnamed). They typically don't have dark streaks on the cap, and have almond-shaped to broadly ellipsoid spores. Most similar is a common, but unnamed oak-associated species with bluish to violet gills. *C. subfoetens* has lilac gills and a violet-blushed to ocher-brown cap; it grows with conifers in drier portions the Cascade Range. *C. alticaudis* has the same shape, but light brown to ocher colors. *C. glaucopus* has a yellowish brown cap and lavender gills. *C. subolivascens* is a similarly stout-statured species found in drier portions of Cascadia, fruiting in spring. All species mentioned here are in *Phlegmacium* if you split *Cortinarius.*

Cortinarius subolivascens

A.H. Sm.

Cap: 3–8.5 cm across, convex with an inrolled margin when young, becoming broadly convex to plane or slightly wavy and lobed in age. Color extremely variable: typically dull violet to lilac-gray (more creamy buff to pale violet if buried under duff), and often streaked or speckled when young. At times more olivaceous to gray-brown, or two-toned with a darker margin. Developing ocher-brown to brownish tones in age. **Gills:** Attached, close, fairly narrow to crowded. Variably colored: bluish to dull violet when young, soon becoming grayish to tan, to cinnamon-brown as spores mature. **Stipe:** (2) 3–8 cm long, 1–2 cm thick at apex, equal or enlarged slightly to a basal bulb. Bluish white to lilac when young, becoming creamy to yellowish tan. **Partial veil:** A thin cobwebby cortina, leaving a silky annular zone on stipe or nearly disappearing. **Flesh:** Thick, firm, fibrous-fleshy. Whitish to bluish gray with bluish or pale violet in upper stipe, paler overall in age. **Odor:** Indistinct. **Taste:** Indistinct. **KOH:** Orange-brown to reddish brown on cap and stipe. **Spore deposit:** Rusty brown. **Microscopy:** Spores 9–12 × 5–6 µm, almond-shaped, roughened.

Ecology: Solitary, scattered or in small clusters from soil, duff and moss under conifers, especially fir. Common in east-side forests and drier parts of the Olympic Mountains. Fruiting in spring and early summer, a few weeks after snowmelt.

Edibility: Non-toxic.

Comments: This variably colored species is best recognized by its stout stature, bluish violet tones when young, bulbous stipe base, and most importantly, fruiting in spring and early summer. *Phlegmacium vernalidistinctum* is a very similar spring-fruiting species which appears to be more common in the southern portions of the Cascade Range. It tends to be slightly paler (less likely to show violet tones), has an equal to clavate (not distinctly bulbous) stipe, and has slightly narrower spores. Members of the *C. glaucopus* group fruit later in summer and fall, and have smaller spores.

Synonym: *Phlegmacium subolivascens* (A.H. Sm.) Niskanen & Liimat.

Cortinarius calyptratus

A.H. Sm.

Cap: 4–8 cm across, rounded when young, becoming broadly convex to plane. Surface thinly viscid to dry, appressed-fibrillose to nearly smooth, covered with irregular patches of pale violet to whitish veil tissue. Dark purple to grayish purple with paler streaks when young, more grayish in age. **Gills:** Broadly attached, or with a slight notch, close to crowded. Lilac when young, becoming grayish lilac to grayish brown as spores mature. **Stipe:** 4–7.5 cm long, 1–1.8 cm thick, equal or enlarged downward toward a bulbous base. Surface dry, silky-fibrillose to nearly smooth. Violet to lilac when young, grayish lilac to gray in age. **Partial veil:** A thin to moderately thick, lilac cortina which leaves a silky annular zone on the stipe. **Flesh:** Thick, firm, fibrous. Creamy yellow with violet tones in stipe apex, orangish in lower stipe, more orangish brown in age. **Odor:** Indistinct. **Taste:** Indistinct. **KOH:** Reddish brown to rusty on the cap. **Spore deposit:** Rusty brown. **Microscopy:** Spores 8–10 × 4.5–5.5 µm, almond-shaped, roughened.

Ecology: Scattered or in small troops from moss and duff under conifers. Rare, we have seen it in mature and old growth forests in the Coast Ranges and in the Cascade Range. Fruiting in fall.

Edibility: Unknown.

Comments: This beautiful Webcap is recognized by its purple cap with whitish to pale violet veil patches, lilac gills and a bulbous stipe. *Phlegmacium albocalyptratum* is very similar, but tends to have paler, bluish gray gills, and grows in mixed forests in the southern part of Cascadia, it may be associated with tanoak. *C. glaucocephalus* has a darker, bluish to greenish, streaked, cap without veil patches. *Phlegmacium pastorinoi* has a violet cap, pale violet gills and a whitish, clavate stipe. It is known from coastal forests in California, but likely occurs in Cascadia.

Synonym: *Phlegmacium calyptratum* (A.H. Sm.) Niskanen & Liimat.

Cortinarius subfoetidus A.H. Sm.

Aromatic Lavender Webcap

Cap: 3–12 cm across, rounded-convex with an inrolled margin when young, expanding to broadly convex to nearly plane, at times with broad, rounded umbo. Surface viscid to tacky, smooth or with scattered veil remnants. Lavender-blue to lilac-blue, often with paler streaks, fading overall in age. **Gills:** Broadly attached, close to crowded, broad. Pallid, often with a bluish wash (especially near margin and edges when young) to grayish buff, cinnamon-brown when mature. **Stipe:** 4–9 cm long, 2.5–5 cm thick, often rather tall, cylindrical, with a swollen or tapered base. Surface moist to dry, covered with a silky-membranous sheath from base to veil, which occasionally breaks into a chevron-like pattern; smooth above veil. Bluish to whitish over lower portion, whitish above veil. **Partial veil:** A sparse, lilac to whitish cortina leaving an evanescent annular zone. **Flesh:** Thick, firm; white. **Odor:** Indistinct to sweet but unpleasant. **Taste:** Indistinct. **KOH:** No reaction. **Spore deposit:** Rusty brown. **Microscopy:** Spores 7.5–10.5 × 4.5–6 µm, ellipsoid to almond-shaped, moderately roughened.

Ecology: Solitary, in pairs, or scattered (rarely in large numbers), from moss, duff, and soil under conifers. Uncommon, prefers old growth forests at mid to high elevations in the Cascade Range and Olympic Mountains, rare elsewhere. Fruiting in fall.

Edibility: Unknown, unlikely to be poisonous.

Comments: This species is recognized by its lavender-blue cap, pallid gills, and bluish to white stipe. A few vaguely similar, closely related species occur in Cascadia. They are all larger, have stockier cylindrical to club-shaped stipes and a yellow KOH reaction on the flesh. These include: *C. balteatus*, with a violet to violet-gray cap when young, associated with conifers; and *C. lilacinocolossus*, with a bluish violet to violet-brown cap. *C. balteatocumatilis* has a pale violet to grayish brown cap and fruits under introduced hardwoods. *C. citriolens* is another large species with a brighter violet cap when young which fades to grayish brown. It fruits in spring under conifers in drier habitats in the Cascade Range. All species mentioned here would belong in *Phlegmacium* if you choose to split *Cortinarius.*

Cortinarius olympianus A.H. Sm.

Olympic Webcap

Cap: 3–7 (10) cm across, rounded to convex when young, becoming plane; margin inrolled at first, remaining downcurved, or plane. Surface smooth, viscid when wet. Pale, cool violet, lilac to grayish lavender at first, fading to whitish lilac, occasionally developing yellowish tones in age. **Gills:** Narrowly attached, often notched, close to crowded, narrow. Pallid in button stage, soon violet to lilac, retaining color well into maturity, with rusty dusting of spores in age. **Stipe:** 3.5–6 cm long, 0.8–1.5 cm thick at apex, cylindrical, with abrupt basal bulb. Surface dry, covered with silky veil fibrils. Pale violet to lilac when young, fading and developing brownish stains in age. **Partial veil:** A scant to moderately thick cortina, pale lilac to whitish, leaving a silky annular zone on stipe. **Flesh:** Moderately thick, firm, fleshy. White to pale grayish tan. **Odor:** Indistinct. **Taste:** Indistinct. **KOH:** Bright rosy pink on cap. **Spore deposit:** Rusty brown. **Microscopy:** Spores 8–10 × 5–6 µm, almond- to lemon-shaped, roughened.

Ecology: Solitary or scattered from soil or duff, often in small patches in old growth forests under conifers. Widespread but rare. Most collections come from mid- to high-elevation forests in the Cascade and Coast Ranges, and on the Olympic Peninsula. Fruiting in fall.

Edibility: Unknown.

Comments: This rare species is easily recognized by the beautiful pale violet to lilac colors, viscid cap, stout stipe with an abruptly bulbous base, and bright rosy pink KOH reaction. *C. alboviolaceus* has a dry, silvery violet cap, lacks the bulbous stipe base and doesn't have a distinctive KOH reaction. *Phlegmacium amethysteum* has a viscid, pale amethyst cap which develops pale yellowish tan coloration over the disc, amethyst gills and a dry, pallid stipe; it lacks a KOH reaction. *C. perplexus* has a pale creamy ocher cap that turns bright rosy pink in KOH, and has pale grayish lilac gills that develop purple colors in age; it grows with oaks. Another oak associate, *C. albidolilacinus* has a variably whitish to lilac or lavender cap, and an ocher-brown KOH reaction.

Synonym: *Calonarius olympianus* (A.H. Sm.) Niskanen & Liimat.

Cortinarius occidentalis

A.H. Sm.

Western Purple-staining Webcap

Cap: 3–8 cm across, rounded to convex when young, becoming broadly convex to nearly plane. Surface viscid to tacky, smooth. Color variable: dark purple, bluish purple to purplish gray, often with darker streaks and blotches, fading overall in age or when dry. **Gills:** Broadly attached with a shallow notch, close to crowded, narrow. Purple to bluish purple when young, soon grayish purple, developing rusty tones as spores mature. Bruising dark purple when damaged. **Stipe:** 4–8 (11) cm long, 0.7–1.4 cm thick, club-shaped when young, more or less equal in age. Surface dry, covered with longitudinal fibrils and a silky annular zone. Grayish purple to purple, covered with paler fibrils, darkening when handled, and staining dark purple when damaged. **Partial veil:** A thin cortina, leaving an annular zone of fibrils on stipe that becomes covered in rusty brown spores. **Flesh:** Thin to moderately thick, firm, stipe fibrous. Purple when young, soon grayish purple to pale grayish in age. Slowly staining dark purple when cut. **Odor:** Indistinct. **Taste:** Indistinct. **KOH:** No reaction. **Spore deposit:** Rusty brown. **Microscopy:** Spores 7–9 × 4.5–5.5 µm, broadly ellipsoid to ovoid, finely roughened.

Ecology: Solitary, scattered or in small clusters from duff and moss under conifers, especially hemlock and fir. Common at middle to higher elevations in the Cascade Range and on the Olympic Peninsula, occasional southward in coastal forests. Fruiting in fall.

Edibility: Unknown.

Comments: The slimy, often streaked or blotchy cap, purple to grayish colors and dark purple staining are quite distinctive. Other purple-staining Corts include *Thaxterogaster americanoporphyropus* with paler colors (often a grayish to pale lilac cap), and *C. purpurascens*, with dark dingy purple to brown cap. The non-staining *Th. obscurovibratilis* has a glutinous to moist, bluish violet cap that fades to brown, grayish beige to tan gills and a bitter taste.

Synonym: *Thaxterogaster occidentalis* (A.H. Sm.) Niskanen & Liimat., *Cortinarius mutabilis* A.H. Sm.

Thaxterogaster americanoporphyropus

Niskanen, Liimat. & Ammirati

Cap: 3.5–10 (13) cm across, rounded to convex when young, becoming broadly convex to plane; margin often wavy in age. Surface viscid to tacky, smooth. Grayish tan to pale ocher, often blotched or mottled when wet, more ochraceous in age. **Gills:** Broadly attached, close to subdistant, narrow. Lilac to purplish gray when young, grayish brown to light brown as spores mature. Bruising purple when bruised. **Stipe:** 4–8 cm long, 0.6–2 cm thick, equal to club-shaped. Surface dry. Appressed-fibrillose to smooth. Off-white to dingy gray, at times with a violet blush. Slowly staining bright purple when handled. **Partial veil:** A thin cortina leaving a faint annular zone on stipe. **Flesh:** Moderately thick, firm, stipe fibrous. Mottled purple when young, more whitish to watery gray in age, slowly staining purple when cut. **Odor:** Indistinct or honey-like. **Taste:** Indistinct. **KOH:** No reaction. **Spore deposit:** Rusty brown. **Microscopy:** Spores 8.5–9.5 × 5–6 µm, almond-shaped to ovoid, distinctly roughened.

Ecology: Often clustered or scattered in duff and moss under conifers and willows. Common at higher elevations across northern Cascadia. Fruiting in summer and early fall.

Edibility: Unknown.

Comments: This common high elevation species is recognized by its grayish, tan to ocher cap, lilac young gills which become grayish brown, and most importantly, the distinct purple staining of the gills and stipe when bruised. *C. purpurascens* has a dark dingy purple to ocher brown cap, and bluish violet to grayish gills which stain dark purple. *C. occidentalis* has a purple to grayish purple cap, purple gills and stains dark purple. Members of the *C. cyanites* group are larger, with bluish to violet-gray caps, bluish young gills and purple to vinaceous red-staining flesh.

Misapplied name: *Cortinarius porphyropus* (Alb. & Schwein.) Fr.

Cortinarius talus

Fr.

Honey-scented Webcap

Cap: 3.5–11 (15) cm across, rounded to convex when young, becoming broadly convex to nearly plane. Surface viscid when wet, smooth. Creamy to pale yellowish buff, often with darker streaks, occasionally with spots or blotches. **Gills:** Broadly attached or notched, close to moderately crowded. Off-white, cream to grayish, becoming tan to light brown as spores mature. **Stipe:** 3.5–9 cm long, 0.8–2 cm thick at apex, enlarging downward to a fairly large, rounded bulb. Surface dry, with longitudinal fibrils and silky cortina remnants. White to cream when young, soon discoloring brownish. **Partial veil:** A sparse cortina which leaves an zone of white fibrils on the stipe when young. **Flesh:** Thick, firm. Whitish, slowly staining orange-brown, especially around larva tunnels. **Odor:** Sweet, honey-like. **Taste:** Indistinct. **KOH:** No reaction. **Spore deposit:** Rusty brown. **Microscopy:** Spores 7–8.5 × 4.5–5.5 µm, ellipsoid to almond-shaped, nearly smooth to finely roughened.

Ecology: Solitary, scattered or in troops from duff and moss under a wide range of trees. Especially common under shore pine on the coast and true fir at higher elevations, but also with aspen on the east side of the Cascade Range. Fruiting in late summer and fall.

Edibility: Unknown.

Comments: The creamy to yellowish buff cap with darker streaks, pallid gills, whitish stipe which browns as it ages, sweet odor, and lack of a KOH reaction on the cap help to identify this common species. *C. albofragrans* has a white to cream cap, yellowish veil bands on a white stipe, and a sweet, spicy odor. It is common under oaks. *C. multiformis* and its kin have darker caps, with orange-brown to carmine-red KOH reactions. *C. patrickensis* also has pallid colors, but has a more slender, upright stature and a green corn odor.

Synonym: *Thaxterogaster talus* (Fr.) Niskanen & Liimat.

Cortinarius riederi

(Weinm.) Fr.

Cap: 4–11 cm across, rounded when young, becoming broadly convex to plane. Surface thinly viscid when wet, shiny when dry, smooth to innately fibrillose. Slightly hygrophanous. Ocher-brown to brown, occasionally with a slight violaceous cast typically with darker streaks or blotches when young. **Gills:** Attached, often with a small notch, close to crowded. Deep violet to violet-gray when young, fading to bluish gray to grayish brown as spores mature. **Stipe:** 4–11 cm long, 0.7–2 cm thick at apex, enlarged below; basal bulb up to 5 cm thick. Surface dry, often silky-fibrillose when young, smoother in age. Base with bluish white mycelium. Pale violet at apex, grayish violet below when young, becoming dingier, and developing a shiny, bronzed surface in age. **Partial veil:** A cobwebby cortina, leaving silky remnants on the stipe, soon covered in rusty brown spores, sometimes disappearing in age. **Flesh:** Thick, firm, solid. Whitish to pale violet in cap and lower stipe, violet in stipe apex, developing brown discolorations in age. **Odor:** Indistinct. **Taste:** Indistinct. **KOH:** Slight brownish on cap. **Spore deposit:** Rusty brown. **Microscopy:** Spores 11.5–14 × 7.5–9 µm, broadly ellipsoid to almond-shaped, distinctly roughened.

Ecology: Solitary or scattered from soil, duff and moss under conifers, especially true fir and hemlock. Common in western Cascadia, especially in montane forests. Fruiting in late summer and fall.

Edibility: Unknown.

Comments: Although common, this Webcap can be quite perplexing at first, with an unassuming and variable set of features. However, with some practice, it can be recognized in the field by its ocher-brown cap, violet young gills, cylindrical stipe with a basal bulb and typically upright stature. *C. burlinghamiae* has a yellowish to ocher cap, pale bluish gills when young, and smaller spores (11–13 × 7–8 µm). It appears to be widespread, but rare in drier forests with Douglas-fir. Members of the *C. purpurescens* group stain distinctly violet when damaged.

Synonym: *Thaxterogaster riederi* (Weinm.) Niskanen & Liimat. All species mentioned in the comments would be *Thaxterogaster* if you recognize a split *Cortinarius*.

Cortinarius multiformis

Cortinarius talimultiformis

Cortinarius callimorphus

Cortinarius rufoallutus

Corts

Cortinarius multiformis

Fr.

Cap: 3.5–8 cm across, rounded to convex when young, becoming broadly convex; margin incurved to down-curved well into maturity. Surface viscid when wet, smooth or wrinkled. Yellowish ocher to ocher when young, more ocher-brown in age, often with whitish streaks if dry. **Gills:** Broadly attached or with a small notch, close to moderately crowded. Off-white to pale grayish when young, becoming clay-tan to light brown as spores mature; sometimes with rusty stains in age. **Stipe:** 3.5–8 (10.5) cm long, 1–2 cm thick at apex, enlarging downward to an often abrupt, rounded to slightly elongate bulb. Surface dry to slightly moist, with longitudinal fibrils and silky cortina remnants. White when young, developing rusty ocher stains. **Partial veil:** A relatively sparse cortina, often with a viscid covering when young, leaving an annular zone of white fibrils on stipe and occasionally fibrils on cap margin when young. **Flesh:** Thick, firm. Whitish, slowly staining orange-brown, especially around larva tunnels. **Odor:** Indistinct. **Taste:** Indistinct. **KOH:** Indistinct to light orange-brown on cap. **Spore deposit:** Rusty brown. **Microscopy:** Spores 8–9.5 × 5–5.5 µm, ellipsoid to almond-shaped, slightly roughened.

Ecology: Scattered in small to large troops in moss and duff under conifers, especially spruce in coastal forests, and true fir at elevation. Common and widespread in Cascadia. Fruiting in fall.

Edibility: Unknown.

Comments: This common species is recognized by the yellowish ocher to ocher cap, pallid gills, bulbous, white stipe and pale orange-brown KOH reaction. *C. talimultiformis* may average a slightly taller stature, but is practically indistinguishable, and best identified genetically. *C. callimorphus* is quite similar, but has a darker, ocher-orange cap with a carmine-red KOH reaction, and lemon-shaped spores. It is rare, restricted to coastal spruce forests. The common *C. talus* has a creamy to pale yellowish cap with darker streaks, pallid gills, an orange-brown KOH reaction, and small (7–8.5 × 4.5–5.5 µm), relatively smooth, almond-shaped spores. It grows with a wide range of trees, both conifers and hardwoods. Additionally, there are two rarer species with darker caps: *C. rufoallutus*, with a streaked orange to red-brown cap and a short bulbous stipe, and *C. caesiolamellatus* with a red-brown to ochraceous brown cap, and a bluish tint to the gills and stipe flesh when young. All of these species would be placed in *Thaxterogaster* if one chooses to recognize a split *Cortinarius*.

Cortinarius violaceonitens

(Rob. Henry) Moënne-Locc

Cap: 3–8 (10) cm across, rounded to convex when young, becoming broadly convex to plane or wavy in age, often with a low, rounded umbo. Surface smooth, viscid to dry. Often distinctly two toned, and slightly hygrophanous. Yellowish brown, ocher-brown to reddish brown over disc with a darker margin, occasionally more uniformly colored in age. **Gills**: Broadly attached, or with a slight notch, close to crowded. Pale greenish to pale olive-gray when young, soon developing buff tones before becoming rusty orange to brownish. **Stipe**: 4–12 cm long, 1–2 cm thick, equal or enlarged downward toward a bulbous base. Surface dry, covered with silky longitudinal fibrils. Pale violet at apex when young, whitish to yellowish below, developing orangish tones in age. Basal mycelium bright greenish yellow when first collected, soon fading. **Partial veil**: Sparse, cortina-like, leaving evanescent silky fibrils on stipe. Flesh: Thick, firm, fibrous. Pale yellowish, at times marbled with tan in the cap, often with violet tinges in stipe, developing ocher stains around larva tunnels and in age. **Odor**: Indistinct. **Taste**: Indistinct. **KOH**: Brownish on cap. **Spore deposit**: Rusty brown. **Microscopy**: Spores 9.5–10.5 × 5.5–6.5 µm, broadly ellipsoid, minutely roughened.

Ecology: Scattered or in small troops in moss and duff under conifers. Especially common in the Cascade Range under fir and hemlock, but distribution poorly known due to confusion with similar species. Fruiting in late summer and fall.

Edibility: Unknown.

Comments: The combination of two-toned caps, greenish yellow to violet-green gills, rusty spores, and bulbous stipes with yellow mycelium around the bulb help identify webcaps belonging to section Scauri, but identifying them to species is considerably harder! There are at least five species in Cascadia, and a few remain unnamed. These include: *C. violaceonitens* as described above (it is usually more reddish to ocher than others); *C. scaurus*, with a duller brown cap and brighter yellow-olive gills; the poorly-known *C. montanus*, which appears to have a dark olive-brown cap and slightly larger spores (9.5–12 µm); and the *C. sphagnophilus* group, which tends to have more olive tones to the cap and distinctly violet flesh. The larger *C. fuligineofolius* may also occur in Cascadia; it's common in coastal forests in California. All species mentioned here would be placed in *Thaxterogaster*, if one recognized a split *Cortinarius*.

Cortinarius oliveopetasatus

M.M. Moser

Cap: 5–10 cm across, rounded to convex when young, becoming broadly convex, plane to slightly concave. Surface often with veil scales over disc when young, smoother in age, viscid when wet. Olive-green to forest-green when young, at times with yellowish towards margin, more olive-brown in age. **Gills:** Attached, close to crowded, broad. Grayish green to dull yellowish green when young, developing brown tones as spores mature. **Stipe:** 5–10 cm long, 1–2.5 cm thick at apex, up to 4.5 cm thick at bulbous base. Nearly cylindrical or more often club shaped toward abruptly bulbous base. Surface dry, fibrillose to nearly smooth. Dull yellowish, developing orange-brown stains in age. **Partial veil:** A cobwebby cortina of yellowish to greenish fibers, leaving an annular zone on upper stipe, soon covered in rusty brown spores. **Flesh:** Firm, fibrous, solid. Creamy to pale yellowish, developing brown stains around larva tunnels and from base up in age. **Odor:** Fragrant, like fresh cut grass. **Taste:** Radishy. **KOH:** Brown to orange-brown on cap. **Spore deposit:** Rusty brown. **Microscopy:** Spores 11–14 × 5.5–7.5 µm, almond- to lemon-shaped, roughened.

Ecology: Solitary or scattered in small patches, rarely more than a few fruitbodies under conifers, especially Douglas-fir. Rare, and possibly restricted to old growth forests in the Cascade Range and Olympic Peninsula. Fruiting in fall.

Comments: This beautiful species is easily recognized by its olive-green cap, grayish green to dull yellowish green gills, and brown KOH reaction. *C. subsulfurinus* is very similar and occurs in similar habitats, but tends to show more olive-yellow to dingy yellow colors; older fruitbodies of these two species are practically indistinguishable. *C. luteicolor* also has an olive-yellow to dull yellow cap; it differs by having a thick white cortina and an orange-red KOH reaction on the cap. There are a number of similar species in oak woodlands; most striking (and similar) is the grass-green *C. viridirubescens*, which stains reddish in the flesh and on the lower stipe, and has a red-brown KOH reaction. *C. anetholes* has a yellowish to olive-yellow cap, pale lilac to brownish lilac gills and a vinaceous KOH reaction. *C. xanthodryophilus* has dull yellow colors, and a light brown KOH reaction. *C. adonis* has a greenish yellow to reddish brown cap, yellow to vinaceous-toned gills, violet stipe flesh (mottled yellowish in the cap and stipe base) and a dark raspberry-purple KOH reaction. All of these species are rare in Cascadia, being restricted to relatively undisturbed oak woodlands. They may be threatened by increased nitrogen deposition due to human emissions into the atmosphere. All the species mentioned above have been placed in the genus *Calonarius.*

Cortinarius saxamontanus

Cortinarius wiebeae

Cortinarius verrucisporus

Cortinarius velatus

Cortinarius saxamontanus

Fogel

Cap: 3–7 (10) cm across, rounded to convex when young, becoming broadly convex to plane. Surface dry to slightly tacky, appressed-fibrillose to smooth. Yellow to greenish yellow, occasionally more golden yellow, developing ocher to orangish brown tones over the disc. **Gills:** Typically obscured by the veil, at times visible in age or where veil is torn. Decurrent to broadly attached, close to crowded, somewhat ragged. Yellowish to straw-yellow, soon rusty brown as spores mature. **Stipe:** 2–7 cm long, 1–2.5 cm thick, enlarged towards a bulbous base, encased in veil tissue. Surface dry, fibrillose to smooth. Light yellow, developing golden to orangish stains, becoming covered with spores in age. **Partial veil:** A persistent, silky-membranous, radially fibrillose curtain (cortina) extending from the cap margin to the top of the stipe bulb. Stipe remaining completely encased well into maturity; veil developing a few radial tears in age or upon drying. Yellow to greenish yellow, soon covered with rusty spores on the upper surface. **Flesh:** Thick, firm, fleshy-fibrous. Whitish to pale yellowish when young, soon developing violet to purple tones. **Odor:** Indistinct. **Taste:** Indistinct. **KOH:** Dark red to vinaceous on cap. **Spore deposit:** Copious, rusty brown. **Microscopy:** Spores 10.5–14.5 × 8–9.5 µm, almond-shaped, distinctly roughened with coarse warts.

Ecology: Solitary or scattered under duff or in soil; rarely exposed, typically only breaking the surface in areas with hard-packed ground or a thin duff layer. Rarely reported, known from drier conifer forests in eastern Cascadia. Fruiting in spring and early summer.

Edibility: Unknown.

Comments: This species is recognized by its yellowish colors, thick, persistent veil, bulbous stipe base and whitish to yellowish flesh that develops violet to purple tones. *C. verrucisporus* has a short, stout stature, slightly paler colors, whitish to pale yellowish flesh which lacks violet tones and smaller spores (9–11.5 × 6–7.5 µm). It is more common in the southern Cascades and into California. There is another unnamed species with a taller stature, and a thick, but more cobwebby (less sheet-like) cortina that soon tears and becomes ragged. *C. wiebeae* (= *C. magnivelatus)* has a white cap that can discolor yellowish in age, and a thick white veil. *C. bigelowii* has a leather-brown cap with a thin, dense layer of whitish veil fibrils. *C. velatus* has a pale violet cap, with a white to violet veil; it is rare, known from the southern Cascades. These 'velate' Webcaps seem to be in evolutionary progression to a secotioid form, and are rarely found above ground. However, unlike fully secotioid mushrooms, they still forcibly discharge their spores, (often a great amount) but most spores are caught behind the persistent veil.

Corts

Calonarius flavipavonius

S.D. Adams, Bojantchev et al.

Yellow Peacock Webcap

Cap: (3) 5–15 (20) cm across, convex, becoming broadly convex to plane; margin inrolled when young, plane to uplifted and often undulating in age. Surface viscid when wet, smooth. Colors variable, rosy brown to rusty red-brown over disc, with lilac tones towards edge; margin often with glutinous yellow veil, becoming ocher to vinaceous-brown in age. **Gills:** Narrowly attached, close to crowded, narrow. Pale citrine yellow to yellow-olive when young, soon clay-yellow, then light olivaceous brown as spores mature. **Stipe:** 3–7.5 cm long, 1–2 (4) cm thick at apex, cylindrical, enlarged below with a rimmed basal bulb 1.5–4 cm across. Surface viscid to dry, silky-fibrillose. Whitish, often with lilac to yellow tones at apex, discoloring in age. **Partial veil:** A thick, pale yellow cortina, typically covered in yellow gluten when young, leaving an annular zone on stipe. **Flesh:** Thick, firm, fleshy; whitish to pale citrine yellow, often with violet tones in cap and bulb, more yellow in stipe with age. **Odor:** Indistinct. **Taste:** Indistinct. **KOH:** Deep maroon-red on cap and stipe base, slight purple on cap flesh in lower stipe. **Microscopy:** Spores 9–11 × 5–6.5 µm, lemon-shaped, moderately roughened.

Ecology: Solitary, scattered or in small troops from moss and duff under conifers. Locally common in mixed fir-hemlock forests at mid- to high elevations in the Cascade Range, rare elsewhere. Fruiting in late summer and fall.

Edibility: Unknown.

Comments: The beautiful variation of cap colors, in combination with the yellow gluten and maroon-red KOH reaction make this an easy species to recognize. The spring-fruiting *C. saxamontanus* has a thick veil, brighter yellow colors and develops distinctly violet flesh. *C. cupreorufus* has a yellowish grey to olive-grey cap when young (often with lilac tones), soon developing warm brown to red-brown colors. Its young gills are brighter yellow, and it has a more reddish brown KOH reaction. *C. cacodes* has a violet-brown cap when young, develops ocher-brown colors in age, lilac-buff to grayish gills, and a brownish KOH reaction on the cap.

Synonym: All of the species mentioned here are in *Calonarius*, if one splits *Cortinarius.*

Cortinarius aureofulvus

M.M. Moser

Cap: 4–8 (10) cm across, rounded to convex with an inrolled margin, becoming broadly convex to nearly plane. Surface thinly viscid to dry, smooth or with veil patches over the disc. Bright yellow when very young, soon golden orange to ocher, more orange-brown in age. **Gills:** Narrowly attached with a notch, close to crowded, broad. Pale citrine yellow to grayish yellow, more rusty as spores mature. **Stipe:** 4–9 cm long, 0.8–2 cm thick at apex, cylindrical with a rounded basal bulb. Surface dry, silky-fibrillose. Whitish to cream or light yellowish. Bulb often with yellow gluten on upper portions. **Partial veil:** A prominent cobwebby cortina broadly covering the middle of stipe. Whitish to pale yellowish, becoming rusty brown with spores. **Flesh:** Thick, firm; whitish to pale grayish. **Odor:** Indistinct. **Taste:** Mild. **KOH:** Vinaceous red to red-brown on cap and stipe base. **Spore deposit:** Rusty brown. **Microscopy** Spores 10–13 × 6–7 µm, almond- to lemon-shaped, distinctly roughened.

Ecology: Solitary or scattered in duff, moss, and soil. Uncommon, occurring in old growth conifer forests in the Cascade Range. Fruiting in fall.

Edibility: Unknown.

Comments: The yellow to ocher cap, citrine gills, thick cortina, bulbous stipe and growth with conifers help identify this vibrant species. Other species we see in similar old-growth conifer habitats include: *C. elegantio-occidentalis*, which tends to be larger, with similar but duller colors, and a bright red KOH reaction on the cap and stipe base; *C. luteicolor*, with an olive-yellow to dull yellow cap which turns orange-red in KOH; and *C. metarius*, with a golden yellow to ochraceous cap, pale lilac gills and an orange-red KOH reaction on the cap (brownish on the stipe base). There are a number of similar oak-associated species, which are rare in Cascadia, but common in California; these include *C. vellingae*, with a yellow to ocher cap, yellow gills, a stout stature, a large, bulbous stipe base and a blood-red KOH reaction. *C. amabilis* is similar to *C. vellingae* but has lilac-beige gills when young. All of these species mentioned here are in *Calonarius*, if one splits *Cortinarius.*

Synonym: *Calonarius aureofulvus* (M.M. Moser) Niskanen & Liimat.

Cortinarius superbus

A.H. Sm.

Superb Webcap

Cap: 4–15 cm across, rounded to convex when young, becoming broadly convex to plane; margin often wavy and occasionally uplifted in age. Surface glutinous when wet, appressed-fibrillose under the layer of gluten, finely scaly when dry. Dull straw-yellow to pale ocher-yellow when young, soon with a smoky, vinaceous brown cast over the disc, to reddish-brown in age. **Gills:** Narrowly attached, often with a distinct notch, close to crowded. Light yellow, straw-yellow, yellowish brown to dull ocher-brown when mature. **Stipe:** 6–15 cm long, 1–3 cm thick at apex, enlarging downward to a swollen base. Surface dry to viscid near base, covered with a flocculent-fibrillose sheath of veil, and copious cortina remnants. Light yellow, covered with reddish to vinaceous brown bands of veil tissue, which darken to vinaceous gray in age. **Partial veil:** A thick yellow cortina, leaving an annular zone of fibrils on the stipe that become covered in rusty brown spores. **Flesh:** Thick, firm, fleshy-fibrous. Yellow, stipe staining over reddish brown. **Odor:** Typically pungent, like musty green corn, occasionally more fruity-fragrant. **Taste:** Indistinct. **KOH:** Not tested. **Spore deposit:** Rusty brown. **Microscopy:** Spores 11–13 × 5.5–7 µm, almond-shaped, roughened.

Ecology: Solitary or scattered, rarely in large numbers. Growing in duff, moss, and soil under conifers, especially Douglas-fir. Widespread but rare—strong preference for old growth forests. Fruiting in fall.

Edibility: Unknown.

Comments: The straw-yellow cap which develops darker colors over the disc, yellow gills, copious yellow cortina, and brown to vinaceous veil bands on the stipe readily identify this species. The pungent green corn odor allows one to smell it before they even see it! *C. citrinifolius* has a greenish yellow, ocher-orange cap, greenish yellow gills, bright yellow flesh, and a sweet, spicy-fruity odor. *C. albofragrans* is an oak-associated species with a fragrant odor. It has a creamy white to creamy buff cap, pale gills, and whitish flesh. *C. patrickensis* has a dry, finely scaly, pale ochraceous buff cap, grayish buff gills and a tall, slender stipe. It has a fruity odor when young, but soon develops hints of green corn, and becomes quite rancid in age.

Cortinarius variosimilis

Cortinarius pallidifolius

Phlegmacium variocracens

Cortinarius pini

Corts

Cortinarius variosimilis

M.M. Moser & Ammirati

Cap: 5–12 cm across, rounded-convex with an inrolled margin when young, expanding to broadly convex to nearly plane, at times with broad umbo. Surface viscid when wet, often with scattered fibrillose veil remnants. Ocher-brown to yellow-brown, typically paler towards margin. At times with a whitish band of veil tissue around margin. **Gills:** Broadly attached to notched, close to crowded, broad. Beige, light brown to clay-brown, at times lilac-blushed, becoming more rusty brown as spores mature. **Stipe:** 4–15 cm long, 1.5–4 cm thick, often rather tall, equal or club-shaped. Surface moist to dry, covered with a silky-membranous sheath from base to veil which occasionally breaks into a chevron-like pattern; smooth above veil. White, developing orangish tan discolorations in age. **Partial veil:** A thick white cortina, leaving silky fibrils on stipe and cap margin. **Flesh:** Thick, firm; white. **Odor:** Indistinct. **Taste:** Indistinct. **KOH:** Yellow on flesh. **Spore deposit:** Rusty brown. **Microscopy:** Spores 9–11 × 6–7 µm, ellipsoid to almond-shaped, roughened.

Ecology: Solitary, scattered, or in pairs or small clusters from duff and moss under conifers, especially fir. Common in east-side forests of the Cascade Range, less often on the west side. Fruiting in late summer and fall.

Edibility: Unknown, unlikely to be poisonous.

Comments: The large size, ocher-brown cap, thick white cortina, and silky-membranous sheath on the stipe and growth with conifers help identify this species. It belongs to a group of very closely related and difficult-to-distinguish species. *Phlegmacium variocracens* tends to be more slender, and has slightly larger, more almond-shaped spores. It is widespread, but is more likely to be found in west-side forests than *C. variosimilis*. *C. pini* is a small species with similar colors, but lacks the veil sheath on the stipe, and has a yellow-ocher KOH reaction on the flesh. *C. pallidifolius* has a dull ocher-brown cap and stipe, off-white to buff gills and an olive-brown KOH reaction. *C. ponderosus* is a giant species (caps frequently 30 cm across) with a mustard-yellow to straw-yellow cap, which often has small appressed scales, and a hefty cylindrical to club-shaped stipe. It is more common in southern Cascadia and northern California.

Synonym: *Phlegmacium variosimile* (M.M. Moser & Ammirati) Niskanen & Liimat.

Cortinarius caperatus

(Pers.) Fr.

Wrinkled Cort

Cap: 4–12 (15) cm across, rounded to egg-shaped when young, becoming broadly convex with a low, broad umbo, often wavy and irregular in age. Surface dry to moist, dull, often radially wrinkled. Tan to light brown with a whitish sheen or bloom when young, remaining so or becoming honey-brown to ocher-brown at center. **Gills:** Broadly attached or with a small notch, close to crowded. Light beige to tan, becoming more rusty brown. **Stipe:** 5–12 (15) cm long, 1–2.5 cm thick, equal or enlarged below, with a club-shaped or weakly bulbous base. Surface dry, with longitudinal striations or faint chevrons, slightly silky to smooth. White to dingy whitish, developing pale orangish tan stains at base. **Partial veil:** Thick and membranous, leaving a rimmed, often grooved, collar-like ring near the midpoint of the stipe. **Flesh:** Moderately thick to thin, firm but brittle in cap, fibrous and sometimes tough in stipe. Whitish to tan, orangish tan around larva tunnels. **Odor**: Indistinct. **Taste:** Mild to slightly unpleasant. **KOH:** No reaction. **Spore deposit:** Ocher-brown. **Microscopy:** Spores 10–14 × 7–9 µm, ellipsoid to almond-shaped, moderately roughened.

Ecology: Scattered in small to large troops under conifers, especially Sitka spruce in coastal forests and true firs at high elevations. Common and often abundant in coastal and montane forests, rare elsewhere. Fruiting in late summer and fall.

Edibility: Edible and good, one of the few popularly eaten *Cortinarius*, although not recommended for beginners.

Comments: This atypical *Cortinarius* can be identified by its tan to light brown cap with a whitish bloom, pale to rusty gills, and thick membranous partial veil that leaves a membranous collar-like ring or skirt on the stipe. No other *Cortinarius* in the western United States has such a thick, membranous partial veil. Members of the *C. multiformis* group superficially resemble it, but have a silky to cobwebby partial veil.

Synonym: *Rozites caperata* (Pers.) P. Karst. Segregation of *Rozites* was based on the membranous partial veil (differing from the typical cortina of most *Cortinarius*). Genetic studies have supported treating these species as *Cortinarius*, as the membranous veil appears to have evolved multiple times.

Corts

Cortinarius causticus

Cortinarius aff. *pluviorum*

Cortinarius vibratilis

Thaxterogaster mitipes

Corts

Cortinarius causticus

Fr.

Caustic Webcap

Cap: 1.5–4.5 cm across, rounded-conical when young, becoming broadly convex to plane, often with a rounded umbo. Surface viscid when wet, usually shiny when dry, smooth. Bright ocher-orange to peachy orange, fading in age. **Gills:** Narrowly attached, often with a distinct notch, close. Whitish to grayish when young, becoming brown as spores mature. **Stipe:** 3.5–8 cm long, 0.3–0.6 cm thick, often tapered below a swollen midpoint; base often soft or decayed. Surface viscid below an annular zone, dry to moist above. White at first, often becoming dingy white and developing a rusty band of spores on veil zone and orangish brown discolorations in age. **Partial veil:** Cortina-like, leaving a scant annular zone that becomes covered in rusty brown spores. **Flesh:** Thin, fibrous to fleshy; watery orangish in cap, whitish to beige in stipe. **Odor:** Indistinct. **Taste:** Very bitter, especially the slime on the cap. **KOH:** No reaction. **Spore deposit:** Rusty brown. **Microscopy:** Spores 6–7 × 4–5 µm, ellipsoid, moderately roughened.

Ecology: Solitary or scattered in moss or duff under conifers, especially true fir and hemlock. Widespread and especially common in western Cascadia. Fruiting in fall.

Edibility: Unknown.

Comments: This group of small to medium-sized Webcaps belongs to section *Vibratiles*. They are easily identified by their slimy ocher, orange, yellowish or creamy caps, pallid gills and stipe, and a bitter-tasting slime. They are very diverse in Cascadia, and difficult to identify beyond " Section Vibratiles". *C. causticus* stands out as it often has a bright ocher-orange cap. *C. vibratilis* has an orange-brown to honey-brown cap which soon turns light brown to tan, and has broadly ellipsoid, finely roughened spores measuring 6.5–8.5 × 4–4.5 µm. *Thaxterogaster mitipes* often is shorter and stockier, has a honey-brown cap, a thinly viscid to dry stipe, and moderately roughened, almond-shaped spores measuring 6.3–7.5 × 4–4.8 µm. *Th. tremens* is a larger, peachy orange to orange-brown species with relatively large, almond-shaped spores (7–8 × 4.5–5 µm). *C. microspermus* is another large, pale yellowish orange species, with a slightly viscid cap and stipe, a slightly bitter taste, and small spores (4.8–5.7 × 3.5–4 µm). An unnamed member of the *C. pluviorum* group which grows under shore pine has a paler, creamy yellow cap and ellipsoid to slightly ovoid, finely roughened spores measuring 6–7 × 4–4.5 µm. All of these species would be in *Thaxterogaster* if you split *Cortinarius.*

Cortinarius brunneoalbus

Cortinarius collinitus

Cortinarius trivialis

Cortinarius mucosus

Corts

Cortinarius brunneoalbus

Ammirati, Liimat. & Niskanen

Cap: 3.5–8 (11) cm across, rounded-convex when young, becoming convex to plane, often undulating in age. Surface smooth, glutinous, often dripping slime when wet and usually remaining viscid even in drier weather. Rich dark brown to warm brown, slightly paler in age and around margin. **Gills:** Narrowly attached, close to crowded, edges pale and slightly serrate or heavily eroded in age. Light brown to warm brown as spores mature. **Stipe:** 3–7 (9) cm long, 0.8–2.5 cm thick, cylindrical, equal or enlarged slightly towards base. Surface dry and somewhat scurfy or with fine chevrons above veil, with a viscid sheath below veil. Whitish to pale buff, at times with a grayish wash on slimy portion. **Partial veil:** Inconspicuous, embedded in a layer of slime spanning from cap margin to stipe when young, soon more or less obliterated, leaving only a faint annular zone around the middle of the stipe where the viscid layer ends or becomes markedly thinner. **Flesh:** Thick, firm, fleshy-fibrous in stipe, whitish to pale tan, darkening slightly in age. **Odor:** Indistinct. **Taste:** Indistinct. **KOH:** No reaction. **Spore deposit:** Rusty brown. **Microscopy:** Spores 13–15 × 7.5–8.5 µm, lemon-shaped, finely to moderately roughened.

Ecology: Scattered or in troops from moss and duff, typically at higher elevations under true fir and hemlock. Known from the Olympic Mountains and the Cascade Range south into central Oregon. Common and at times abundant. Fruiting in late summer and fall.

Edibility: Unknown, likely non-toxic.

Comments: The glutinous brown cap, light brown gills, whitish stipe and short stature help identify this species. There are several similarly colored Webcaps in Cascadia with viscid caps and stipes; habitat and cap color are helpful with macro identification of these species. The common *C. collinitus* (= *C. muscigenus*) has an orange to orange-brown cap and a white to violet-blushed stipe. It prefers montane fir-hemlock habitat. *C. mucosus* has an orange to reddish brown cap and a white stipe; it grows with lodgepole and shore pine. *C. septentrionalis* has similar colors to *C. mucosus*, but has a more northerly range, growing with birch and alder. *C. trivialis* has a yellowish ocher cap, distinct glutinous veil girdles on the stipe and grows with aspen and cottonwood. *C. vernicosus* is a rare spring- and early summer-fruiting species with an ochraceous to orange-brown cap; it is often found in drier eastside forests. *C. absarokensis* (= *C. alpinus*?) has a yellow-brown to gray cap, and grows with willow. Also see *C. seidliae* and kin for slimy-capped species with a violet stipe.

Cortinarius corpulentus

Bojantchev, S.D. Adams, Liimat. & Niskanen

Cap: 4–13 cm across, rounded to convex when young, becoming broadly convex to plane. Surface smooth, glutinous when wet, usually remaining viscid, or shiny if dry. Smoky pale violet to more purple near margin when young, soon developing orange coloration over disc, occasionally extensively dingy gray-tan to dull orange-brown in age. **Gills:** Narrowly attached, often with a distinct notch, close. Pale clay-tan to gray or bluish gray at first, soon grayish tan, to rusty brown as spores mature. **Stipe:** 5–13 cm long, 1–3.5 (5) cm thick, typically swollen in middle with a tapered base, occasionally club-shaped to more or less equal. Surface glutinous when wet, appressed-fibrillose to nearly smooth. White when young, the gluten layer soon develops orange tones. **Partial veil:** Cortina-like, forming a slight membranous collar, white, soon covered in rusty brown spores. **Flesh:** Thick, firm, stipe fibrous. White to pale gray, developing orange-brown stains in age. **Odor**: Indistinct to earthy. **Taste:** Indistinct. **KOH:** No reaction. **Spore deposit:** Rusty brown. **Microscopy:** Spores 7–9.5 × 5–6.5 µm, subglobose, moderately roughened.

Ecology: Solitary or scattered in moss or duff with a wide range of conifers. Occasional, widespread from coast to Cascades, but rarely fruiting in large numbers. Fruiting in fall.

Edibility: Unknown.

Comments: Although this glutinous mushroom is quite variable in stature, it is easily recognized by the light smoky-violet cap which soon develops orange tones, and a white stipe with a clear to orange gluten layer. This species was recently described, and is sister to the European *C. delibutus*. A closely related, but unnamed species close to *C. emunctus* (it also has been referred to as *C. salor* in Cascadia) is smaller, with lilac to violet colors when young, fading to gray and developing tan colors in age. It is fairly common in Cascadia, in both coastal and montane conifer forests. *Thaxterogaster obscurovibratilis* has a glutinous, royal purple cap, which soon fades to gray-brown, and a distinctly bitter taste. It occurs in coastal spruce forests.

Cortinarius seidliae

Ammirati, Niskanen & Liimat.

Cap: 3–10 cm across, rounded to conical and often with a crimped margin at first, soon bell-shaped, conical-convex or nearly plane, usually with an umbo. Surface strongly wrinkled, but covered in a smooth layer of glutinous slime when wet, otherwise viscid or tacky. Dark brown at first, paler toward margin, becoming cinnamon to beige in age. Paler forms, light brown to beige, are often encountered. **Gills:** Narrowly attached, often with a broad notch, close to crowded. Pale grayish beige to pale tan at first, becoming cinnamon-brown to light rusty brown. Edges often paler. **Stipe:** 4–15 cm long, 1–2 cm thick, equal, swollen at middle, or tapering toward base. Surface covered in a glutinous slime layer below annular zone, dry above. Variably smooth or broken into chevrons. Light to dark violet, apex and base often whitish. Occasionally losing the violet tones and becoming whitish overall in age. **Partial veil:** A clear slime layer, disappearing in age. **Flesh:** Thick, firm, stipe tough and fibrous. Whitish to pale buff, often developing ocher stains in lower stipe. **Odor**: Mild to earthy. **Taste:** Indistinct. **KOH:** No reaction. **Spore deposit:** Rusty brown. **Microscopy:** Spores 12.5–15 × 8–9 µm, lemon-shaped, distinctly roughened.

Ecology: Solitary, scattered, or in troops, in moss or duff under conifers, and especially common with hemlock and spruce. Widespread, occurring across western Cascadia. Fruiting in fall, with stragglers into early winter on the coast.

Edibility: Unknown.

Comments: The brown cap with wrinkles under the viscid layer, slimy violet stipe, and pallid gills get you close, but final determination may be more difficult. In the past, we identified species matching this description as *C. vanduzerensis*, but it turns out that the common and widespread species is *C. seidliae*. *C. vanduzerensis* appears to be a rare species with a darker (blackish to dark brown) cap, and slightly narrower spores. It is known from coastal spruce forests in Oregon. There is another unnamed species in this group with larger fruitbodies and a consistently dark purple stipe. It is more common in the Coast Range and Cascade foothill forests with fir. *C. collinitus* (=*C. muscigenus*) has an orange to orange-brown cap and a white to violet-blushed stipe.

10 - Entolomataceae

The species in this chapter are all members of an well-supported evolutionary lineage—the family Entolomataceae. Although their fruitbodies range from tiny to large, drab to colorful, and show a wide array of statures, they are unified by their attached gills, pinkish spore deposits, and ridged or angular spores under the microscope. Partly due to its great diversity, this Family remains quite poorly known in the United States. Further complicating matters, taxonomists disagree about how to arrange the species in this family into genera. Globally, there are two basic approaches: One which recognizes fewer (but more species-rich) genera and the other which divides the Family into a greater number of genera, with fewer species in each one.

Most European authors (e.g. Reschke et al., 2023, Morgado et al., 2013, Co-David et al., 2009), have taken the "lumping" approach, with a majority of species placed in a broad-concept *Entoloma*, and splitting out *Clitopilus* (with elongate-ridged spores,) and *Rhodocybe* (with bumpy-warty spores). However, other authors (Lan-He et al., 2019, Karstedt et al., 2019, Baroni et al., 2011, Largent 1994) recognize many additional separate genera, and continue to add new ones. These smaller genera include (so far): *Alboleptonia*, *Clitocella*, *Clitopilopsis*, *"Cyanula"*, *Entocybe*, *Inocephalus*, *Leptonia*, *Lulesia*, *Nolanea*, *Pouzarella*, *Rhodophana*, and *Trichopilus*. This "splitting" approach comes with the benefit of providing a more fine-grained vocabulary for field workers but inevitably creates cases of confusion regarding the criteria for recognition of (at least) 16 genera!

The Entolomataceae is not unique in these frustrations — see "On the Future of Mushroom Identification" (p. 547) for more discussion. Although we acknowledge that the generic assignments we use here are problematic from the point of view of some systematists but we think they are useful for recognizing species in Cascadia and other parts of western North America—our main goal for this book. Of the hundreds of species of Entolomataceae found in the western USA, we include here only a small sampling of distinctive, well-known species and a few "groups" that should help with general recognition where specific identification is difficult or impossible. Keep in mind that there are many species in this group that are impossible to identify with confidence given current resources, as well as others which are still undescribed!

Alboleptonia have small fruitbodies that are white overall (except for the pinkish mature gills), while *Rhodophana* have small, slender, orange-red fruitbodies and usually have a strong cucumber odor.

Claudopus are small to medium-sized, gray-brown mushrooms that lack a stipe (or have small, off-center stipes), and often have distinctive white rhizomorphs and sheeting mycelium. Due to their "pleurotoid" stature, they are included in that section (p. 369).

Clitopilus are variable in stature. The most commonly-encountered species in this genus have distinct central stipes, strongly decurrent gills and pale, medium-sized fruitbodies. Others are tiny, thin-fleshed, and lack stipes (or have reduced, irregular stipes). These latter species can resemble *Claudopus* (but are paler in color and have differently shaped spores). These are included in the Pleurotoid chapter.

Entoloma – We use this genus name for species which produce robust, fleshy, tricholomatoid, generally dull-colored fruitbodies, some of which have mealy or bleachy odors. If and when the many-genus model becomes the norm, some of the species we include in this group would be rearranged into the segregate genera "*Prunuloides*" and "*Rhodopolium*".

Entocybe are often metallic-looking, with bluish, purplish, or gray to brownish colors and smooth textures overall. Fruitbodies range from small and slender to large and robust.

Leptonia and ***"Cyanula"*** – Although "*Cyanula*" has never been recognized as a full-fledged genus, recognizing *Leptonia* at that rank necessitates this additional group. Generally speaking, these are small, slender mushrooms (with a few robust, stout species in *Leptonia*), which have colors ranging from yellowish to brownish to blue, blue-black, grayish blue, purple, or even pink. These are species-rich groups in the western United States, and although they can be found in any habitat, they are most abundant and diverse in redcedar and maple forests across Cascadia, as well as in redwood, cypress, and bay-laurel forests further south. Although *Leptonia* and "*Cyanula*" are not particularly closely related within the Entolomataceae, they are often visually similar and precise criteria for distinguishing them are quite technical. However, a few rules of thumb can help: *Leptonia* often have evenly scruffy, scaly, or hairy caps and stipes, while "*Cyanula*" typically have smooth or finely-scaly cap surfaces and smooth stipes. Microscopically, *Leptonia* have clamp connections on the cells of the cap cuticle (with the exception of a couple species), while "*Cyanula*" lack such clamp connections.

Nolanea – Most species in this group have tall, slender fruitbodies with conical to bell-shaped caps that are translucent-striate and hygrophanous, and stipe surfaces that are often silky-silvery, with twisting longitudinal striations. Brown, tan, and beige colors predominate in this group.

Rhodocybe have a subtle gestalt impression, with opaque caps, and broadly attached, slightly decurrent gills that are often widely spaced. *Inocephalus* tend to have quite straight, slender, slightly cartilaginous stipes and boxy, cube-shaped spores. One local species of *Clitopilopsis* has small, slender, pale grayish-tan fruitbodies with umbilicate caps and widely-spaced gills.

Pouzarella species are generally quite rare; they often have somber-colored, bell-shaped to conical caps and a slender stature, as well as their "trademark" finely-hairy to scaly-scurfy or silky textures.

For all these groups, important identification features include the texture of cap and stipe, translucence of the cap margin, subtleties of color, and details of structure and stature. Definitive identification usually requires microscopic examination—the size and shape of the spores (especially the number of angles as seen in side view), presence or absence of clamp connections in various tissues, type and distribution of pigments, and size and shape of cystidia (if present) are critical features to look for.

Alboleptonia adnatifolia

(Murrill) Largent & R.G. Benedict

Pallid Pinkgill

Cap: 1–6 cm across, domed to convex with an incurved and often scalloped margin, expanding to broadly convex; occasionally slightly umbilicate, in others broadly conical with a small umbo. Surface dry, fibrillose-silky, sometimes with small scales. Whitish to dull beige or with yellowish tones near center. **Gills:** Broadly attached, often with a short decurrent tooth, moderately to fairly widely spaced; edge usually serrate or eroded. Whitish at first, soon creamy and then pinkish to salmon as spores mature. **Stipe:** 2–9 cm long, 0.2–0.5 cm thick, cylindrical, base sometimes slightly enlarged. Surface smooth to silky or with small, obscure chevrons of silvery fibrils near apex. White to creamy or light dingy tan near base. **Flesh:** Thin, fragile; stipe hollow; white. **Spore deposit:** Pinkish. **Microscopy:** Spores 9–10 × 7–8 µm, with 5 or 6 sides in side view, distinctly angled. Cheilocystidia scattered, club-shaped. **Ecology:** Solitary or scattered, or in large troops in humus and needle litter, moss, or on gravelly soil under western redcedar and in other forest habitats. Widespread but uncommon. Fruiting from late fall through early spring.

Edibility: Unknown.

Comments: The silky white cap and stipe, and pinkish gills and spore deposit are characteristic. It could be confused with members of the *Inocybe geophylla* group, which have a silky white cortina when young and grayish tan to dull brown mature gills. Specimens with yellowish tones may pertain to *A. sericella* var. *lutescens*, which is distinguished by its longer spores (averaging > 10 µm). This variety may warrant elevation to full species status. *A. ochracea* develops somewhat darker, dingy orange colors, has longer spores (up to 13 µm long), and a densely scurfy upper stipe (with abundant club-shaped caulocystidia). A small and pale species similar to the European *Entoloma cremeoalbum* (which genetically groups into *Claudopus*) grows in saturated moss. It has a tan-gray to pale brownish cap and widely-spaced, slightly thick-edged gills.

Synonym: *Entoloma adnatifolium* (Murrill) Blanco-Dios. **Misapplied Name:** *Alboleptonia sericella* (Fr.) Largent & R.G. Benedict.

Clitopilus prunulus

(Scop.) P. Kumm.

The Spy, The Miller

Cap: 2–7 (10) cm across, convex with an inrolled margin when young, becoming broadly convex to irregularly uplifted and wavy in age, often retaining an inrolled margin. Surface dry, often somewhat lumpy, slightly velvety to smooth. Dingy whitish gray to grayish tan, covered in a dusty white bloom when young, sometimes disappearing entirely in age. **Gills:** Slightly to deeply decurrent, close, narrow; partial gills present. Pallid whitish to dingy tan, becoming pinkish tan as spores mature. **Stipe:** 2–5 (10) cm long, 0.8–1.8 cm thick, often swollen near midpoint or enlarged lower, then tapered at extreme base. Surface finely velvety to smooth; base often with a large white "boot" of mycelium. Dingy whitish gray with tan areas. **Flesh:** Thin, fibrillose-fleshy, whitish. **Odor:** Strongly farinaceous. **Taste:** Mealy. **Spore deposit:** Dingy pinkish tan. **Microscopy:** Spores 9–12 × 5–7 µm, long-ellipsoid to nearly cylindrical (but tapered at one end), longitudinally ridged (but this can be hard to see).

Ecology: Solitary, in clumps, or scattered in arcs and troops in duff on ground in forested habitats throughout Cascadia, most common along the coast and in the higher elevation fir-hemlock forests in the Cascade Range. Fruiting in late summer and fall. The common name refers to this species' habit of fruiting in same area and about a week earlier than the prized "royalty" boletes (*Boletus edulis*, *B. mottiae*, and *B. fibrillosus*) – thus, savvy foragers use this as an indicator species. *C. prunulus* is probably involved in a parasitic relationship with these boletes; the fruitbodies are often present rather close to the boletes, sometimes sharing practically the same spot in the ground!

Edibility: Edible but extreme caution is warranted; *Clitocybe rivulosa* is a toxic look-alike (see below).

Comments: This species is fairly easy to recognize by its dry, velvety, grayish white cap with a whitish bloom, decurrent gills that eventually develop a pinkish tone, strong farinaceous odor, and pinkish tan spores. Some *Rhodocybe* in our area are similarly colored but usually have thinner stipes and most lack the strong farinaceous odor. A few poorly known *Clitocella* species tend to have striking concentric wrinkles in the cap surface, grayer gills, and dingier gills overall. These seem to be rare in Cascadia but may be overlooked in aspen, cottonwood, and other montane habitats. Perhaps most similar is the toxic *Clitocybe rivulosa* which is often smaller, grows in arcs and troops in grassy areas, and has a paler spore deposit. *Clitocybe nebularis* is much larger, with a smoother cap that is darker (often gray), white spores, and a skunk-like odor. *Lepista subconnexa* grows in clusters and has a whitish cap and paler, buff-colored spores.

Entoloma species

Pinkgills

This description applies to a group of species which are difficult to distinguish. It should be interpreted broadly; see comments for more detailed potential species concepts.

Cap: 3–8 (12) cm across, convex to plane, often undulating in age. Surface moist to dry, smooth. Colors variable between species, but generally dull. A few are very pale beige to nearly white, and a few others are quite dark olive-gray; however most fall into the range of dull medium gray to tan or brown or even bluish. **Gills:** Narrowly attached, often notched, close to well-spaced; edges even when young, becoming irregularly ragged. Whitish to cream, becoming pinkish as spores mature. **Stipe:** 4–7 (12) cm long, 0.5–2 cm thick, cylindrical or slightly enlarged lower, then tapered or pointed at extreme base. Surface dry, obscurely striate but nearly smooth. Whitish to creamy. **Flesh:** Thin to moderately thick in cap, fairly firm, stipe fibrous, becoming hollow in age. **Odor:** Indistinct to bleach-like, farinaceous or like celery. **Taste:** Often farinaceous. **Spore deposit:** Dull pinkish. **Microscopy:** Spores variable across the group; averaging around 7–9 µm, isodiametric (nearly as long as wide) to heterodiametric (longer than wide), with 5–6 sides in side-view, distinctly angled.

Ecology: Solitary, in small clusters, arcs or troops, fruiting on the ground in humus or on mossy ground in forested settings, with a variety of trees and shrubs, widely distributed across Cascadia. Most members of the group fruit in fall and winter.

Edibility: Most are unknown, some are toxic. Do not experiment.

Comments: These are fairly nondescript mushrooms, best recognized by their relatively robust, fleshy-fibrous fruitbodies, lack of any veil tissue, caps ranging from dull-colored, pale to dark gray or brown, and attached gills that become dingy salmon-pink at maturity. Some members of the group have strong odors (bleach, celery, farinaceous) while others are odorless. The following species are all poorly known but most were originally described from Cascadia and thus are likely valid names awaiting more widespread interpretation and usage. The name *E. rhodopolium* has often been used (likely misapplied) to any medium-sized gray to brown-capped species with a farinaceous odor and isodiametric spores. Likewise the name *E. lividoalbum* has been a catchall for stout, gray-capped, farinaceous species. *E. majaloides* is a large, robust species with yellowish to honey-tan caps; it occurs in conifer forests. *E. heracleodora* has a greenish gray cap and a strong celery-like odor; it was described from grassy habitat with birch. *E. subsaundersii* has large, fleshy fruitbodies with wrinkled gray-brown caps and a tendency to fruit in clusters; its spores measure 7–11 × 6–10 µm. *E. pseudolividum* has honey-tan to beige caps, pale yellowish gills when young, and large, slightly elongated spores (7–12 × 6–10 µm); it was described from mixed conifer forest in WA. *E. lyophylloidium* has dark blackish brown caps, a farinaceous odor, fruits in clusters, and has spores about as long as they are wide; it was described from a late summer fruiting in true fir forest on Mt. Rainier. *E. pseudocostatum* has a dry, opaque, fairly dark brown cap, a whitish stipe, and farinaceous odor; microscopically, it has clamp connections in the cap cuticle and spores about as long as they are wide; it was described from a lawn and is also known from grassy areas near conifers. *E. griseo-avellaneum* has a grayish, finely hairy-scaly cap, a fragrant aroma, and spores measuring 6.5–9 × 5–7.5 µm; it was described from red alder forest, fruiting in large troops. Many resemble gray *Tricholoma* species, which have white spores.

"Leptonia" tjallingiorum var. *laricina*

Largent

Blue Wood-Leptonia

Cap: 2–7 cm across, broadly bell-shaped with a low, round umbo, becoming convex to plane; margin slightly in-rolled when young. Surface dry, finely hairy to radially silky-scaly with larger, darker scales near center, becoming matted-fibrillose in age. Metallic navy-blue to lavender-blue or lavender-gray with brown tones, sometimes much paler beige-tan overall, or dull gray to gray-brown in age. **Gills:** Narrowly attached, notched, close to moderately spaced. White to pale cream, becoming washed with salmon-pink as spores mature. **Stipe:** 3–10 cm long, 0.5–1 cm thick, equal or slightly enlarged towards base. Surface dry, finely fuzzy or with tufts of darker scales, Deep blue to purplish blue, often metallic, often with a whitish band near apex. **Flesh:** Thin, fragile; stipe fibrous; whitish. **Odor:** Indistinct. **Taste:** Indistinct. **Spore deposit:** Dull salmon pinkish. **Microscopy:** Spores 9–13.5 × 6.5–8 µm; 5–6 angled in side view.

Ecology: Solitary or in small clusters on rotten wood, typically of old growth conifers. Widespread but uncommon throughout Cascadia, with a preference for foothills and higher elevations. Fruiting in fall.

Edibility: Unknown.

Comments: This beautiful mushroom is recognized by its combination of metallic blue fuzzy-scaly stipe surface, scaly-hairy cap ranging from pale beige-gray to bluish to metallic gray-brown, whitish to pinkish gills, and growth on rotting conifer wood. When *Leptonia* is eventually recognized as a separate genus (again) and this variety is raised to species rank, it would most likely leave us the name *Leptonia laricina* but there are many ways in which this name might end up being formulated otherwise — only time will tell. *Leptonia lampropus* var. *insueta* (= *L. insueta*) is similar but has a pale to brownish gray cap, a deep navy-blue stipe with dark blue-black scales, and pale gills. *L. occidentalis* is similar in texture but has stunning evenly blue and purple tones throughout the cap and stipe, and gills ranging from whitish to dusky bluish or dull purple-gray; it grows from duff, typically under western redcedar. *L. dichroa* is another stout, fibrillose-hairy, entirely blue species. *Leptonia carnea* (=*Entoloma subcarneum*) is similar in stature and texture but has a consistently deep purple-blue cap and stipe and more bluish gray gills. It grows on humus and duff of coast redwoods in northern California but so far has not been recorded in Oregon or Washington. Additional species with blue-toned, finely-hairy caps which grow on decomposing conifer wood include *L. subeuchroa* and *L. subgracilis*, but both of these are significantly more slender and often show gray to brownish caps in age. All of these are examples of the "core" or strict-sense *Leptonia* group.

Synonyms: *Leptonia cyanea* var. *occidentalis* Largent, *Entoloma tjallingiorum* var. *laricinum* O.V. Morozova, Noordel., Vila & E.S. Popov

Entolomataceae

Entocybe nitida

(Quél.) T.J. Baroni, Largent & V. Hofst.

Shining Pinkgill

Cap: 2–6 cm across, domed when young, expanding to rounded-conical or broadly convex with a rounded umbo, sometimes slightly uplifted to wavy in age. Surface thinly viscid to moist or dry, slightly wrinkled to smooth and metallic-looking. Deep navy-blue to royal blue or pale cobalt, sometimes paler sky-blue towards margin or in age. **Gills:** Narrowly attached or deeply notched with a decurrent tooth, fairly widely spaced; margins relatively thick, even to eroded. Whitish, sometimes becoming pale tan with a bluish to gray wash, finally becoming extensively salmon-colored as spores mature. **Stipe:** 4–10 cm long, 0.5–1.2 cm thick, cylindrical with a slightly enlarged base, usually somewhat twisted and irregular, wavy and bumpy. Surface smooth but distinctly vertically striate, often with a silvery, twisted appearance. Pale bluish, sky blue to metallic royal blue, with extensive whitish area over base. **Flesh:** Thin, fragile in cap; stipe fleshy-fibrous, becoming hollow in age. Whitish to pale gray or bluish. **Odor:** Indistinct. **Taste:** Not sampled. **Spore deposit:** Dull pinkish tan to salmon. **Microscopy:** Spores 7–8.5 × 6.5–8 µm, isodiametric, obscurely bumpy with 6–9 angles in side view; clamp connections present in cap cuticle.

Ecology: Solitary or scattered in small troops, in needle litter and moss under conifers. Uncommon along the coast and in the mountains of northern Cascadia; rare or absent south of central Oregon. Fruiting from fall through early winter.

Edibility: Unknown.

Comments: Although vaguely similar in color to some of the bluish *Cyanula* and *Leptonia*, the cap surface of *E. nitida* is smoother and the stipe is usually thicker and more distinctly silvery striate. Microscopically, the rather round spores with many obtuse angles are a helpful distinguishing character. The common *E. trachyospora* is typically smaller and more slender, with purple to gray or brownish tones in the cap which often shows a translucent-striate margin. *Entoloma medianox* is similar in color but has a more robust, stocky stature, and often has more deeply wrinkled cap. It is more common further south, where *Entocybe nitida* becomes scarce or absent. An apparently undescribed species of *Entocybe* with a deeply striate, lavender-blue cap, widely spaced pale gills, and a short, pale stipe has been found growing on decaying logs of white oak. A number of striking metallic-bluish *Leptonia* occur in Cascadia (e.g., *L. occidentalis*) but most have distinctly scaly or scurfy cap and stipe surfaces.

"Cyanula" species

Cyanixes

Cap: 1–6.5 cm across, domed to bell-shaped, becoming broadly convex, occasionally plane or wavy in age, often umbilicate (in some species deeply depressed and caving in to the hollow upper stipe during rainy weather), rarely if ever umbonate. Surface dry, often finely tomentose to granular, mostly remaining so at the center, with outer areas becoming smooth, flattened-scaly, or splitting radially. Color variable across the group, ranging from nearly black to deep royal blue, navy blue, or bluish gray, some species more beige to gray-tan or even slightly golden-tan. Occasionally strikingly ruby-red or pink forms are encountered. Margin ranging from opaque to obscurely translucent to deeply striate. **Gills:** Mostly narrowly attached or appearing free, in a few species with a small decurrent tooth, close to well-spaced. Color variable, ranging from bright white to creamy, or washed with bluish to light gray (often blotchy), in some species the gills are "marginate", with distinct dark edges; mature gills becoming dingy salmon-pinkish. **Stipe:** 3–8 cm long, 0.2–0.7 cm thick, cylindrical with an enlarged base or more often flattened with a longitudinal groove. Surface dry, smooth, at base often with a large "boot" of cottony white mycelium. Concolorous but usually lighter than the cap color. **Flesh:** Thin, fibrous, stipe often hollow and slightly cartilaginous. **Odor:** Indistinct to slightly spicy, cheesy or unpleasant. **Taste:** Indistinct to slightly farinaceous or unpleasant. **Spore deposit:** Pinkish to dingy salmon. **Microscopy:** Spores variable, generally ranging from 6–12 × 5.5–8.5 µm, 5 to 6 sided in side view, moderately angled. Clamp connections absent in cells of cap cuticle.

Ecology: Solitary, scattered, or in clumps of in humus and needle litter; typically in disturbed forest settings, especially with western redcedar. Occasional, widespread. Fruiting in late fall into early winter, occasionally well into spring.

Edibility: Unknown.

Comments: This is one of the most challenging groups to identify within an already-difficult family. Although there are dozens of presumably-valid "*Cyanula*" names based on specimens described as *Leptonia* from northern California and Cascadia, given current knowledge, understanding their original species concepts and differentiating them from one another is an exercise in frustration. The domed to bell-shaped or slightly umbilicate caps, colors ranging from black to blue to gray to pale beige-brown, smooth stipes, and dull pinkish spores are helpful features for recognizing this genus. We encourage collectors who take an interest in this group to make vouchered collections paired with careful notes on morphology and habitat. Little by little, DNA sequence benchmarks for older names are becoming available and progress is being made on updating our understanding of their diversity.

"Cyanula" serrulata

sensu PNW

Cap: 1.5–6 cm across, domed to broadly convex when young, plane or uplifted and slightly wavy in age, often with a broad, round umbo that is slightly depressed at the center. Surface dry, finely tomentose-scruffy when young, becoming appressed-tomentose with fine radially arranged scales that become uplifted as cap expands. Translucent-striate when wet. Most often deep navy-blue but ranging from royal blue in bright forms to grayish steel blue in dull forms. **Gills:** Attached, often notched with a decurrent tooth, close to moderately spaced. Grayish white to bluish when young, becoming bluer especially near margin, sometimes vivid blue overall; edges distinctly dark, with contrasting navy-blue to royal blue clusters of cheilocystidia. **Stipe:** 2–8 cm long, 0.3–0.5 cm thick, cylindrical. Surface finely punctate with dark clusters of cystidia at least at apex, otherwise smooth to obscurely vertically striate. Grayish to navy-blue, steel-blue to gray in dull forms, brighter blue in others. **Flesh:** Thin; stipe hollow; pale grayish to dark bluish. **Odor:** Indistinct. **Taste:** Not sampled. **Spore deposit:** Dull salmon to pale brownish pink. **Microscopy:** Spores 8–12 × 5.5–8.5 µm, angled in side view. Cheilocystidia abundant, with bluish to grayish brown contents.

Ecology: Solitary or scattered in small groups in duff, moss, or bare soil of road cuts, often under western redcedar but also in other forest types. Widespread but uncommon at low elevations and in the western foothills of the Cascades. Fruiting from fall to early winter.

Edibility: Unknown.

Comments: The dark royal blue to navy-blue colors, striate cap, and blue-edged gills are a distinctive suite of features. However, this taxon provides a good example of a bad habit —indiscriminate use of an existing name based on a single highly distinctive feature; in this case, the dark-marginate gills. Genetic data indicate that the species illustrated here ranges south to central California but is not conspecific with either the European or eastern North American collections going by this name. Moreover, there are likely 3–4 cryptic species that share the blue-edged gills even in western North America.

Misapplied name: *Leptonia serrulata* (Fr.) P. Kumm. *Entoloma serrulatum* (Fr.) Hesler

"Cyanula" formosa group

Cap: 1–5 cm across, domed when young with a narrowly incurved margin, becoming broadly convex to plane or shallowly uplifted with a slightly wavy margin. Surface dry, covered in many tiny scales overall when young, but soon appressed-fibrillose outward, with a streaky appearance, sometimes smooth except at center. Warm ocher to orange-brown, sometimes darker brown at center, becoming golden-yellow to yellowish beige but covered with fine brownish scales. Margin usually bright to pale golden-yellow, distinctly translucent-striate with brownish tan lines. **Gills:** Narrowly attached, close; edge even to strongly serrate or eroded. White to pale cream, soon with dingy pinkish tones; edges at times darker brown. **Stipe:** 3–10 cm long, 0.3–0.7 (1) cm thick, cylindrical or compressed and with a central groove, with a slightly enlarged base. Creamy yellow to ocher, dingy pale tan, or light brownish. Base usually with a large, bright white "boot" of fluffy-cottony mycelium. **Flesh:** Thin, fibrillose, hollow in stipe; whitish to beige. **Odor:** Indistinct. **Taste:** Not sampled. **Spore deposit:** Dingy salmon-pink. **Microscopy:** Spores 9–12 × 6–10.5 µm, 5 or 6 sided in side view, distinctly angular.

Ecology: Scattered in small groups or occasionally clusters of two to four fruitbodies in moss, duff, or on bare soil on road cuts and trail sides. Common in conifer forest habitats throughout Cascadia, especially under western redcedar. Fruiting from fall through early spring.

Edibility: Unknown.

Comments: The ocher to yellowish cap with minute brownish scales, smooth yellowish stipe, and whitish gills are characteristic; but there are at least three or four cryptic species in this group. *Leptonia formosa* var. *microspora* is described as having spores averaging less than 10 µm long and has mostly 2-spored basidia. *L. badissima* has a darker chocolate-brown cap and brownish stipe. *Rhodophana nitellina* has a smooth, carrot-orange to reddish orange cap and stipe, and smells like cucumber. Similar species of *Nolanea* are usually duller beige-brown, distinctly hygrophanous, and often have more conical caps. *Inocephalus* have smoother, duller creamy to pale brownish caps and have cube-like spores.

Synonyms: *Leptonia formosa* (Fr.) Gillet, *Entoloma formosum* (Fr.) Noordel.

Entolomataceae

Rhodophana nitellina group

Cucumber Pinkgills

Cap: 2–6 cm across, broadly convex to plane, often with a depressed center; margin wavy and sometimes uplifted in age. Surface dry, smooth, with small uplifted scales. Carrot-orange to orangish red, dull orange or dark yellowish orange, often with irregularly distributed reddish orange blotches; translucent-striate when wet. Hygrophanous, fading to pale beige when dry. **Gills:** Broadly attached, fairly widely spaced; edges even to ragged; partial gills numerous. Dingy pallid orange, sometimes cream-orange or even paler to dull beige or whitish. **Stipe:** 4–12 cm long, 0.3–0.8 cm thick, cylindrical, typically slender, straight or sinuous. Surface smooth or sometimes bumpy. Dull, often somewhat translucent, bright orange to orangish beige, apex with pale beige or whitish scurfy scales. Base with streaks of white mycelium. **Flesh:** Thin, very fragile. Dingy pale orange or yellowish orange. **Odor:** Strongly cucumber-like or farinaceous. **Taste:** Farinaceous. **KOH:** No reaction. **Spore deposit:** Dull pinkish tan. **Microscopy:** 6.5–7.5 × 4.5–5 µm, teardrop-shaped, distinctly roughened.

Ecology: Solitary or scattered in small groups in duff under conifers, especially western redcedar. Common throughout our range, apparently more common in the Coast Range and Cascade foothills than along the coast. Fruiting in fall, with stragglers into winter, and often a spring montane flush.

Edibility: Unknown.

Comments: Although fairly nondescript, the rather bright orange color is distinctive—especially with the combination of a translucent-striate margin and cucumber-farinaceous odor. However, it should be noted that fruitbodies generally matching this description exhibit substantial variation in appearance and genetic data have corroborated the idea that there are probably at least three species currently going by this name group in our area (none of which are likely to be the "true" *R. nitellina*). One such species seems to be consistently darker brick-red and tends to grow at higher elevations. *Lactarius rubidus*, which produces milky-watery droplets emerging from broken gills and has white spores and *Tubaria* species, which have brown spore deposits, a flabbier texture, and tendency to grow in disturbed and urban areas rather than in forest duff.

Rhodocybe aureicystidiata

Lennox ex T.J. Baroni

Goldcell Pinkgill

Cap: 1–4 cm across, convex with an inrolled and sometimes finely ribbed margin, expanding to broadly convex, often weakly lobed. Surface dry, smooth. Grayish tan to dull brown, often with a few concentric pale bands near margin, opaque, sometimes weakly translucent-striate in age. **Gills:** Broadly attached or slightly decurrent, widely spaced, thick-edged, a few often forked near stipe. Grayish to tan-brown. **Stipe:** 2–6 cm long, 0.2–0.4 cm thick, straight to slightly curved or sinuous, narrowly club-shaped at base with a whitish "boot" of mycelium. Dingy beige-tan to grayish tan or brownish with pale scurfs near the apex. **Flesh:** Fairly thin, stipe solid or becoming hollow in age, slightly cartilaginous. pale grayish to tan-brown. **Odor:** Indistinct or slightly farinaceous. **Taste:** Indistinct or weakly bitter-acrid. **Spore deposit:** Dull pinkish tan. **Microscopy:** 7–9.5 × 3.5–4.5 µm, shaped like elongated teardrops, covered with obscure small warts or bumps. Cheilocystidia filled with yellowish contents.

Ecology: Solitary, or in small groups in duff or humus, or on mossy soil in moist mixed forests. Widespread in Cascadia but generally uncommon. Fruiting in fall and winter.

Edibility: Unknown.

Comments: The convex, brownish-gray, mostly opaque cap, thick cartilaginous stipe, and pinkish spore deposit are characteristic. Microscopically, the bumpy, teardrop-shaped spores and cystidia with yellowish contents are important diagnostic traits. This taxon belongs to a group of species which mostly remain undescribed in our area and which may be placed in an expanded concept of the genus *Clitopilus*. There are a few similar species in our area. *R.* "*caelata*" (likely a misapplied name) is similar in stature but with bluish-gray, often chalky-looking caps. *Clitopilopsis hirneola* is a small, slender species with a gray-brown to chalky-gray cap that is often depressed at the center. "*Entoloma*" *laceratum* is an atypical-looking species with a deeply umbilicate, grayish cap with that splits radially, giving it a "lacerated" appearance. *Rhodocybe nuciolens* is a nondescript peachy-beige member of this group that often shows white rhizomorphs.

Entoloma pleopodium

(Bull.) Noordel.

Olive Pinkgill

Cap: 1–5 cm across, broadly convex to plane, with a slightly depressed center; margin uplifted or wavy in age. Surface smooth to finely scaly, slightly greasy to dry. Olive-brown to dingy yellowish green, becoming beige-tan with a pale yellowish cream margin, becoming translucent-striate. **Gills:** Narrowly attached, widely spaced, often inter-veined. Dull beige to tan, becoming pinkish in age. **Stipe:** 2–7 cm long, 0.3–0.6 cm thick, straight to slightly curved, cylindrical or somewhat club-shaped at base. Dry, olive-brown to dull beige or greenish tan with a paler silky-fibrillose surface. Base with a boot of whitish mycelium. **Flesh:** Thin, fragile in cap; stipe fibrous, becoming hollow in age. Dingy olive-beige. **Odor:** Indistinct, to pungently sweet-fruity and sharp. **Taste:** Indistinct. **KOH:** No reaction. **Spore deposit:** Pinkish to dingy salmon. **Microscopy:** Spores 8–12.5 × 6.5–8 µm, 5 to 6 sided in side view, moderately angled.

Ecology: Solitary or in small groups and clusters, often on decaying woody debris or rich humus. We have seen it under trees in urban settings, and under rhododendrons at the edges of spruce forest. Fruiting in fall and winter. Apparently rare in Cascadia but hard to recognize and thus probably more common than current records suggest.

Edibility: Unknown.

Comments: The widely-spaced gills, olive-yellow to beige-tan colors, dingy pinkish spore deposit, and tendency to grow near rotting wood are helpful identification clues for this unusual pinkgill. It will certainly not remain with the "core *Entoloma*" after the genus is split and may, in fact, end up in a genus practically all on its own. *Callistosporium luteo-olivaceum* is remarkably similar but has white spore deposit, a reddish KOH reaction, and is more consistently associated with rotting wood. *Nolanea cetrata* has a honey-tan hue and grows in forested areas, while a species closely related to *E. borgenii* has been found in a fen habitat in Cascadia; it has a browner cap and whitish gills, with spores measuring 8–10 × 7–8 µm.

"Nolanea" vindobonense group

Silky Grass Pinkgills

Cap: 2–6 cm across, domed, bell-shaped with a rounded umbo or round-conical at first, becoming plane or slightly depressed at center in maturity; margin often slightly wavy in age. Surface dry, smooth. Dull, dark brown at first, soon fading to grayish brown, pale grayish tan or brownish beige, with a mottled or radially streaked, slightly metallic appearance. Often concentrically zonate, with bands of dark and pale colors. Margin obscurely translucent-striate when moist. **Gills:** Usually notched or broadly attached with a decurrent tooth, moderately spaced; edges smooth to irregularly ragged. Dingy pale grayish beige, becoming brownish tan and then pinkish as spores mature. **Stipe:** 2–7 cm long, 0.3–0.8 (1) cm thick, cylindrical, often tapered toward base, sometimes weakly club shaped. Surface dry, silky-fibrillose or finely scurfy, soon becoming smoother. Dingy beige-tan to light grayish tan or silvery. **Flesh:** Thin, fibrous or slightly rubbery, whitish gray; in stipe watery-fibrous, becoming hollow in age, tan to brown. **Odor:** Mild to slightly farinaceous. **Taste:** Mild to slightly nutty or weakly farinaceous. **Spore deposit:** Pinkish tan to dull pinkish brown. **Microscopy:** Spores 7–9) × 5.5–8 µm, isodiametric, 5 to 6 sided, distinctly angled in side view. Cheilocystidia absent. Cells of cap surface coarsely incrusted with pigment.

Ecology: Solitary or more often scattered in troops on soil, often in lawns, or amid small herbaceous plants in exposed settings along trails or roads and in urban areas. Fairly common throughout our range in appropriate habitats. Scattered fruitings occur from late fall into winter, with larger flushes in spring.

Edibility: Likely non-toxic.

Comments: For years, these drab-brown, vaguely concentrically, hygrophanous, grass-dwelling *Nolanea* have gone under the name *N. sericea*. However, there are multiple cryptic species which match the description above. One such species, *N. fusciceps*, was named from Grey's Harbor, Washington. It differs from other members of this group by having intracellular pigmentation in the cells of the cap, rather than coarse, externally-incrusted pigments.

Synonym: *Entoloma vindobonense* Noordel. & Hauskn.

Nolanea subcapitata

Largent

Cap: 3–7 (10) cm across, conical with a distinct umbo, expanding to broadly conical. Surface moist to dry, smooth. Dark brown to warm tan-brown, hygrophanous and thus paler to beige-tan in age, translucent-striate when moist, becoming opaque when dry. **Gills:** Attached, close, broad, soft; edges often slightly irregular. Pale grayish beige to tan, becoming dingy salmon as spores mature. **Stipe:** 4–10 cm long, 0.4–1 (1.5) cm thick, cylindrical or often compressed with a longitudinal groove; base slightly enlarged to club shaped. Surface nearly smooth, but vertically striate and often twisted. Light brownish to tan, grayish tan, often with a pale sheen over upper half. **Flesh:** Fairly thin, in cap fragile, pale grayish tan to whitish; in stipe fibrillose-stringy, becoming hollow in age. **Odor:** Indistinct. **Taste:** Indistinct. **Spore deposit:** Dull pinkish to salmon tan. **Microscopy: Spores** 8.5–11 × 8–10 µm, 5 or 6 sided in side view, distinctly angled. Cheilocystidia capitate (with enlarged, rounded heads). Cap cuticle cells with internal globules of brown pigment. Stipe surface with distinctive cylindrical to capitate caulocystidia, often emerging from center of hyphae (rather than from end).

Ecology: Solitary or in scattered troops in moss and conifer duff. Widespread and common throughout Cascadia. Fruiting in fall and winter.

Edibility: Unknown.

Comments: The distinctly umbonate-conical, hygrophanous, dark brownish to beige-brown cap, relatively stout stipe covered in a whitish bloom, and capitate cystidia help identify this species. Many other *Nolanea* superficially match the description above but, keep in mind, that identifying *Nolanea* is difficult (bordering on impossible without microscopy and DNA) given the current state of knowledge. Although your mushroom may match one of the descriptions below fairly well, proving its identity in a definitive way will take a great deal of work. *N. holoconiota* is more yellow-brown to dingy honey-colored or pale grayish-yellow when dry and fruits in spring. *N. clandestina* tends to be paler, more slender, with a taller stipe. *N. cetrata* has light golden brown to tan cap which is nearly flat rather than distinctly conical and a farinaceous odor; it appears to be common throughout the lowlands of Cascadia, fruiting in winter and spring. *N. obscurata* has clamped basidia and was described from mossy conifer forest on Mt. Rainier. The small, slender, deeply-striate *N.* "*oculobrunnea*" is currently treated as a variety of *N. clandestina*, but deserving full species status. The tall, straight-statured, strongly-striate *N. hirtipes* and the grayish, umbonate-campanulate, obscurely-striate *N. latifolia* (=*Entoloma fuligineo-cinereum*) are also similar.

Pouzarella alissae

Largent & Bergemann

Alissa's Pouz

Cap: 0.8–3 cm across, convex to plane; margin sometimes slightly wavy in age; occasionally with a low or broad umbo. Surface dry, silky-fibrillose when young, soon shaggy looking with small, uplifted scales made of clumped fibrils. Edge of cap fringed with small tufts of pale hairs. Silvery when young, soon grayish tan to gray, occasionally with an olive cast, streaked brownish to blackish in age. **Gills:** Broadly attached with a decurrent tooth, rather widely spaced with fairly thick and even edges. Pale grayish tan when young, becoming dingy grayish brown with smoky grayish blotches, with a pinkish cast from spores in age. **Stipe:** 4–8 cm long, 0.2–0.3 cm thick, cylindrical with a slightly enlarged base. Surface entirely covered with whitish fibrils and scurfs when young, slightly smoother in age but still noticeably silky-striate. Silvery whitish when young, becoming dingy tan to grayish brown, darker brown where handled. **Flesh:** Thin, pale grayish. **Odor:** Indistinct. **Taste:** Not sampled. **Spore deposit:** Pinkish to dingy salmon tan. **Microscopy:** Spores 8.5–11.5 × 6–8.5 µm, with 5 to 6 distinct angles in side view or slightly bumpy or warty. Cheilocystidia scattered; pleurocystidia occasional. Cells of cap cuticle with coarsely incrusted pigment. Clamp connections absent.

Ecology: Solitary, in clumps or troops in duff of a variety of broadleaf trees including bigleaf maple and bay-laurel, but also under alder. Seems to prefer damp, cool microhabitats with ferns. Rare, known from lowlands throughout Cascadia. Fruiting in fall and early winter.

Edibility: Unknown.

Comments: *Pouzarella* are distinguished from other members of the Entolomataceae by their silky-fibrillose or hairy-scaly cap and stipe surfaces, which give them an overall shaggy look. Additionally, the colors are often dingy or dark with a silvery sheen but they are not usually as brown as in many *Nolanea*, nor are they blue or purple as in *Leptonia*. Because of their small size, dull colors, and hairy caps, *Pouzarella* can resemble some *Inocybe*. Species in the latter genus are told apart by their brown spore deposits and smooth or knobby (but not angular) spores. It appears that most *Pouzarella* species tend to be rare and most are known from fewer than a dozen collections. Other species in Cascadia include an apparently undescribed species with small, slender gray fruitbodies with strikingly shaggy-hairy cap and stipe surfaces. A species going by *P. fulvostrigosa* has reddish hairs around the stipe base. *P. versatilis* has greenish olive hues. Both of the latter were described based on specimens from elsewhere and it is likely that these names are being misapplied in Cascadia. With its silvery brownish colors and radially-silky texture, *Nolanea fernandae* (=*Pouzarella fernandae*) resembles a *Pouzarella* but has a more convex cap; it is rare in Cascadia.

11 - Pluteus and Allies

This is an easy-to-recognize group of mushrooms thanks to their distinctly free gills, dingy-pink to salmon-colored spores, and tendency to grow on wood or coarse plant debris. Most *Pluteus* species grow directly on logs, rotting stumps, or wood chips. For the smaller, sometimes brightly colored members of this genus, field identifiers broadly separate them into two groups based on size, color, and cap texture: *Celluloderma* have wrinkled caps and *Hispidodermini* have silky-hairy, velvety, or granular caps. Keep in mind that this is an imperfect rule of thumb and that additional microscopic study is needed to achieve greater certainty.

Volvopluteus have smooth viscid when wet, and are often found in grass or nutrient-rich soil in urban or disturbed areas. The rarer *Volvariella* are found in a variety of habitats including standing trees, decaying wood, in rich humus or, in some case, on the distorted fruitbodies of other gilled mushrooms! These genera are easily distinguished from *Pluteus* by their universal veil tissue that leaves a sheathing volva around the base of the stipe. This is especially interesting because it provides a clue to the evolutionary history of these mushrooms: *Pluteus*, *Volvopluteus*, and *Volvariella* are quite closely related to *Amanita*, which is the only other group of macrofungi in our area that has such prominent universal veil tissue!

Important features to note when identifying members of this group include the textures of the cap and stipe surfaces, overall coloration, and identity of the rotting wood substrate. Confirmation to species often requires microscopy, including the structure of the cells of the cap surface, presence or absence of clamp connections, presence or absence of pleuro- and/or cheilocystidia (as well as their size and shape), and the size and shape of the spores.

Pluteus leoninus

(Schaeff.) P. Kumm.

Lion Pluteus

Cap: 2–6.5 (8) cm across, rounded-conical to bell shaped, becoming broadly convex to plane. Surface dry, evenly velvety when young with a pale sheen, smoother in age. Yellowish to golden-brown, or dull orange-brown in age. **Gills:** Free (sometimes obscurely), close, broad. Pinkish white when young, sometimes very light yellow, salmon as spores mature. **Stipe:** 3–8 (10) cm long, 0.5–1 cm thick, cylindrical, usually quite tall and straight, base slightly enlarged. Surface dry, smooth. Creamy whitish to light creamy yellow at first, yellowish to light tan with twisting, silvery vertical striations in age, often orangey pink near very base. **Flesh:** Thin, soft; stipe fibrous; whitish. **Odor:** Indistinct. **Taste:** Mild. **Spore deposit:** Dull pink. **Microscopy:** Spores 7–8.5 × 6–7 µm, broadly ellipsoid to subglobose, smooth. Cystidia on gills large, bottle-shaped, thin-walled, tips rounded or with a few odd warts and nipples. Cells of cap surface long, upright, cylindrical, with yellow contents.

Ecology: Solitary or in small groups on branches, logs, and woody debris of hardwoods and conifers. Uncommon throughout our range. Fruiting from fall into early winter.

Edibility: Unknown, likely non-toxic.

Comments: The beautiful warm yellow-gold to orange-brown colors, velvety cap surface with a pale sheen, free gills that tend to be pinkish white even when young, and growth on wood are good identifying features for this species. Like so many *Pluteus*, an all-white form of this species has been confirmed by sequencing. *P. fulvibadius* can be similar in color and habitat but has a smooth to wrinkled or veiny cap surface (made up of round to broadly club-shaped cells), is less tall, more dull, with an olive to tan-brown cap. The recently described *P. vellingae* is nearly identical to *P. fulvibadius* but may be distinguishable by paler gills, slightly smaller spores, and may be more partial to conifer wood. The *P. chrysophlebius*/*P. aurantiorugosus* complex are brilliant yellow to orange to red, respectively. All of the similar species mentioned in the comments belong to the Celluloderma group of *Pluteus*.

Synonym: *Pluteus flavofuligineus* G.F. Atk.

Pluteus cervinus

(Schaeff.) P. Kumm.

Deer Mushroom

Cap: 4–12 (15+) cm across, ovoid or domed when young, soon expanding to broadly convex to plane; margin sometimes slightly wavy. Surface greasy to moist, soon dry, wrinkled when young, soon smooth and innately radially fibrillose, sometimes slightly tomentose at center. Dark brown, soon medium brown to tan-gray or light brown, often more metallic gray in dry weather. Occasional "albino" forms can be nearly pure white. **Gills:** Distinctly free, close to crowded, broad, soft to mushy. White at first, becoming light pink and then salmon to dingy pinkish tan in age, with whitish gill edge. **Stipe:** 5–15 cm long, 0.7–1.5 cm thick at apex, cylindrical, straight to curved, enlarged toward base. White to creamy but covered with a sparse (sometimes dense) layer of grayish to dark blackish brown vertically arranged fibrillose patches. These scales can wear off or become flattened in age, then stipe appears extensively creamy or with dingy light tan tones. **Flesh:** Thick, moderately firm when young, soon soft; white. **Odor:** Radishy. **Taste:** Mild to radishy. **Spore deposit:** Salmon-buff, more brownish when dry. **Microscopy:** Spores 7–8 × 5–5.5 µm, broadly ellipsoid to subglobose, smooth. Pleurocystidia abundant, thick-walled, bottle shaped, with a crown of 2–4 hooks or "horns." Cheilocystidia club-shaped, often with a narrow stalk. Cap cuticle a cutis without clamp connections.

Ecology: Solitary, in pairs, or in small groups on stumps and logs of hardwoods and conifers, causing a white rot. Also found on wood chips or landscaped areas in urban settings. Common in the Pacific Northwest but exact distribution is uncertain due to confusion with similar species (see comments). Fruiting throughout the wet season, with peaks of fruiting in early fall and spring

Edibility: Edible but soft-fleshed.

Comments: This species is recognized by its fairly large size, dark brown to grayish brown cap, free, white gills that turn pink in age, brownish fibrils on the stipe, and growth on wood. Another common *Pluteus* in Cascadia is *P. exilis.* Although it may, on average, fruit earlier in the season than *P. cervinus*, telling the two species apart is extremely difficult due to conflicting reports about morphology. As such, large, dull brown "Deer Mushrooms" may not be identifiable to species without a genetic sequence. As with most *Pluteus*, "albino" forms of both occur and can be confused with many other species. *P. petasatus*, has large, pale fruitbodies, and is usually found in urban areas – it lacks clamp connections in the cap cuticle and has spores measuring 6–7 × 4–5 µm. *P. laricinus* (possibly a synonym of the more commonly applied name *P. atromarginatus*) is easily told apart by its distinctly dark gill edg-

Pluteus exilis

Pluteus atromarginatus

es. As a group, *P. primus*, *P. orestes*, *P. "parilis"* nom. prov., and *P. pouzarianus* can be distinguished by the clamps connections in their cap cuticle. Average differences in darker or lighter cap color may eventually prove useful in distinguishing these species but are unreliable given current knowledge. *P. brunneidiscus* is another species which may have smaller fruitbodies on average. Two medium-sized species with pale gray-brown to whitish caps are best differentiated by microscopy: *P. leucoborealis* lacks clamp connections in the cap and has spores measuring 6–8 × 5–6 µm and *P. orestes* has clamps in the cap cuticle and spores longer than 8 µm.

Pluteus plautus group

Flatfoot Deer Mushrooms

Cap: 3–7 (10) cm across, domed at first, expanding to broadly convex to plane, sometimes a bit wavy in age. Surface dry to moist, often slightly bumpy or wrinkly, densely tomentose at first, soon breaking up into small granular-hairy scales (often flattening or wearing off), then extensively smooth near margin. Dark reddish brown to brown, or ocher-brown to tan, distinctly hygrophanous and fading to beige, margin translucent-striate (especially in age). **Gills:** Free, close, broad. Light creamy, becoming pinkish beige and then dingy salmon tan. **Stipe:** 3–6 cm long, 0.2–0.5 cm thick, cylindrical, slightly enlarged near base. Surface mostly smooth but usually covered with fine brown hairs near base. Whitish gray to tan, with twisting, silvery striations, **Flesh:** This, fibrous, often quite soft and waterlogged (especially in cap). Whitish, bruising slowly olive brownish where injured. **Odor:** Indistinct or earthy. **Taste:** Mild. **Spore deposit:** Dingy salmon. **Microscopy:** Spores 6–7 × 5–6 µm, broadly ellipsoid to subglobose. Pileipellis made of upright, long, cylindrical brown cells. Pleurocystidia smooth, pear-shaped, or club-shaped.

Ecology: Solitary or in small groups on decaying hardwood logs and branches, with some variants also on conifers. Uncommon throughout Cascadia, more frequent in lowland bigleaf-maple habitat, as well as inland oak-dominated and riparian areas. Fruiting throughout wet season.

Edibility: Unknown, likely non-toxic.

Comments: Several species in this group are present in our area and some are quite distinct from the others. Features that unify and distinguish the group are a finely fuzzy to distinctly granular-hairy cap surface (not usually strongly netted or wrinkled), lack of yellow tones, striate, hygrophanous cap (typically fading to beige), and a tendency for the base of the stipe to be covered in dark fibrils. *P. phaeocyanopus* has a dark gray-brown cap and develops a whitish to yellowish stipe which soon develops bluish gray tones in age. *P. umbrosus* is typically larger with a darker, granular-scaly or wrinkled-potholed cap surface, a stipe covered in dark granular ornamentation, and dark gill edges.

Pluteus podospileus group

Cap: 1–4 cm across, broadly convex to nearly plane with a low, broad umbo. Surface dry, distinctly granular-velvety or finely tomentose. Margin of cap smooth to distinctly sulcate or ribbed, often deeply wrinkled to center of cap; obscurely hygrophanous or opaque. Dark, warm brown to grayish-brown. **Gills:** Free, close to subdistant, narrow. White to very pale grayish, soon becoming pale pink to salmon-tinged. **Stipe:** 1–5 cm long, 0.2–0.5 cm thick, cylindrical, equal or slightly enlarged at base. Surface dry, smooth or slightly vertically fibrillose. White to pale grayish, somewhat translucent. **Flesh:** Thin, fragile; stipe fibrous; pale grayish to white. **Odor:** Indistinct or faintly metallic-earthy. **Taste:** Indistinct. **Spore deposit:** Cinnamon-salmon to dingy pinkish. **Microscopy:** Spores 4–7 × 3.5–5 µm, broadly ellipsoid, smooth. Cystidia swollen, club-shaped or urn-shaped. Cap surface composed of vertically-arranged club-shaped cells (a hymeniderm).

Ecology: In small groups or gregarious, on woody debris, sometimes abundant, and in disturbed areas. Members of this group are widespread across North America but we don't yet have a detailed picture of their biogeography. They often fruit opportunistically, sometimes outside of the main rainy season in appropriate microhabitats (e.g. decaying logs in shady drainages).

Edibility: Tiny, thin-fleshed, flavorless.

Comments: Recognized by the small size, granular-velvety brown caps, pinkish gills, and pink spores—although a number of small *Pluteus* found in Cascadia will roughly match this description (though they are not all closely related). Much remains to be learned about diversity of our smaller *Pluteus* species. Members of the *P. plautus* group are granular or slightly hairy and often with dull or pale gray-brown to tan-brown caps. Members of the *P. nanus* group have slightly wrinkled to smooth cap texture and whitish stipes which may have faint bluish tones. The smaller, slender *P. longistriatus* has a grayish, deeply sulcate cap; it is usually found on rotting woody debris in disturbed or urban areas, even inside houses!

Volvariella bombycina

(Schaeff.) Singer

Tree Silksheath

Cap: 5–20 cm across, egg-shaped to round-conical at first, expanding to bell-shaped, convex, or nearly plane in age. Surface dry, finely silky, with small but distinct radiating fibrils; opaque; margin often with a narrow band of shaggy-hairy overhanging tissue. White overall, sometimes washed yellowish to pale tan at center. **Gills:** Free, close, broad. White when young, soon pale pinkish and then dull salmon-colored in age. **Stipe:** 2–15 cm long, 0.5–2 cm thick, cylindrical, often strongly curved, slightly enlarged towards base. Surface smooth, often finely longitudinally striate; white. **Partial veil:** Absent. **Volva:** A thin, spacious, egg-shaped membrane. Outer surface whitish to grayish or tan-brownish, interior pale. **Flesh:** Relatively thin in cap, soft and fragile; stipe solid, fibrous; whitish. **Odor:** Indistinct to radishy-musty. **Taste:** Indistinct. **Spore deposit:** Dull salmon. **Microscopy:** Spores 6–9 × 3.5–5 µm, ellipsoid, smooth. Pleurocystidia scattered to abundant, large, club-shaped.

Ecology: Almost always solitary. Fruiting on the trunks of a variety of hardwoods, including willow, cottonwood, and non-native ornamental trees. Usually near a broken branch or other wound on live, standing trees. Widespread but rare in inland areas of Cascadia, especially in urban areas and riparian corridors. Fruiting mostly in summer.

Edibility: Edible but soft-fleshed.

Comments: This large, elegant mushroom is identified by its whitish, silky cap texture, large membranous volva, pink mature gills and spore deposit, and habit of solitary fruiting directly on standing trees or dead wood. Other Silksheaths (*Volvariella*) in Cascadia are considerably smaller. These include *V. hypopithys*, which is similar in color and has a volva surrounding the stipe base,\ but with much smaller caps (only 2–5 cm across). It grows directly on duff and soil, and has spores 5–7 µm long. Another small species, *V. smithii*, has narrower spores (4.5–7 × 3–4 µm), a shorter stature, and a grayish cap. *V. terrea* is a somewhat stouter species with a gray cap and volva and, most interestingly, it may be a parasite on *Agaricus xanthodermus*! *V. surrecta* is another member of this genus with a myco-parasitic ecology as it grows on the fruitbodies of *Clitocybe nebularis*, deforming and digesting them until it eventually fruits directly out of the host mushroom. *Pluteus tomentosulus* has a thicker stipe and lacks a volva but is otherwise similar with a white, silky cap and pinkish gills at maturity.

Volvopluteus gloiocephalus

(DC.) Vizzini, Contu & Justo

Stubble Rosegill

Cap: 5–15 cm across, egg-shaped to round-conical or domed at first, expanding to broadly convex and often with a low, round umbo; margin occasionally uplifted and wavy in age,opaque. Surface smooth, viscid to moist when young, remaining so in wet weather, otherwise soon becoming dry and shiny. Extensive white to cool pale to medium gray with a brownish or olive wash, darkest at center, often radially streaked with appressed dark grayish fibrils, overall paler and metallic in age. **Gills:** Distinctly free, close to crowded, broad. White with a gray wash to creamy when young, soon pinkish cream and then salmon-pink as spores mature. **Stipe:** 5–20 cm long, 1–4 cm thick, Change: enlarged towards base.. Surface dry, silky or smooth. White to creamy, becoming dingy tan in age, or covered partially with pinkish spores. **Volva:** Fairly thick and membranous (but fragile), copiously sheathing stipe base when young, occasionally breaking up and becoming indistinct in age. White to grayish. **Flesh:** Moderately thick, firm at first, becoming rather soft. White or marbled with grayish. **Odor:** Indistinct to slightly musty-mealy. **Taste:** Mild. **Spore deposit:** Salmon-tan to pinkish tan. **Microscopy**: Spores 13–15 (20) × 8–9.5 (12) µm, ellipsoid, smooth. Pleurocystidia and cheilocystidia swollen, club-shaped or tapered at both ends, smooth.

Ecology: Solitary, scattered in groups or clusters, occasionally occurring in large troops in lignin-rich mulch, wood chips, grass, nutrient-rich gardens, and even on agricultural soil. Occasional, mostly found in gardens and landscaped areas in western Cascadia. Fruiting throughout year but more abundant in spring and early fall.

Edibility: Not recommended. Although edible, it is soft-fleshed, musty-tasting, and often grows where pesticides and other forms of pollution are a concern. Be careful to avoid confusion with the Death Cap (*Amanita phalloides)*.

Comments: The fairly large size, pinkish gills, and presence of a volva distinguish this mushroom. If the volva is destroyed or overlooked, the fruitbodies are similar to some larger species of *Pluteus* but those are usually found on more intact wood and usually have cystidia with hooks at the tip. Some *Amanita* in section *Vaginatae* (the Grisettes) are similar but have a less bulbous stipe base, have prominently grooved (sulcate) cap margins, as well as white spores. Greenish specimens with a silvery appearance have been confused with the deadly poisonous *Amanita phalloides* but that species has a skirt on the stipe, whitish or creamy gills at maturity, and produces a white spore deposit. *Volvariella bombycina* is a spectacular species with rather large, whitish fruitbodies and silky, fibrillose caps that emerge from an elongate, egg-like volva. It typically fruits high off the ground on rotten parts of standing trees.

Synonyms: *Volvariella gloiocephala* (D.C.) Boekhout & Enderle, *Volvariella speciosa* (Fr.) Singer.

12 - Russula and Lactarius

Russula and *Lactarius* are some of the most familiar and abundant of our woodland mushrooms. Fruitbodies of both genera are recognizable by their generally stout statures (with a stipe that is shorter than the width of the cap), convex to concave caps, and chalky-granular or hard-brittle textures. The flesh breaks cleanly and readily due to the presence of spherocysts (non-binding globose cells). Those that exude milk or 'latex' when cut pertain to *Lactarius*; while *Russula* lack latex. A few *Russula* and *Lactarius* produce hypogeous, truffle-like fruitbodies (some of the latter even exude latex!), but these are less common in our area and are most reliably identified by microscopy. Both *Lactarius* and *Russula* are ectomycorrhizal and play important roles in the ecology of temperate forests around the world

Although *Lactarius* are relatively diverse in Cascadia, and can be a bit tricky to identify, our concepts for the local species are fairly strong. Be sure to note the color of the latex (and any changes in color after exposure to air), taste (mild or peppery?), any distinctive odor, and the coloration of the cap (especially the presence of concentric zones). Microscopically, the structure of the amyloid spore ornamentation (which requires the use of Melzer's reagent) can be an important identifying feature.

Amateur mycologist Ben Woo systematically documented *Russula* in Cascadia for over 30 years. His work has been the foundation for subsequent genetic studies (Bazzicalupo et al., 2017) which found 72 genetically distinct species among his collections. These are summarized at Danny Miller's website https://www.alpental.com/psms/ddd/Russula/index.htm

Despite these heroic efforts, identification of *Russula* in the Pacific Northwest remains difficult and often impossible. There are close to a hundred species in Cascadia that we know of, but many of these remain poorly characterized and/or nameless. The species we highlight here are just some of the more common and recognizable species. Take note of overall coloration (but keep in mind that this feature can be extremely variable even within a single species), size, stature, presence or absence of a grooved cap margin, taste (acrid, bitter or mild?), presence or absence of partial gills, and any bruising reactions and spore color (bright white, creamy, yellowish or ocher?). Microscopically, the structure of the amyloid spore ornamentation (using Melzer's reagent) and the structure of the cap cells are the most useful features needed for identification. Keep in mind that *Russula* are notoriously resistant to categorization.

Russula emetica

(Schaeff.) Pers.

Emetic Brittlegill

Cap: 3–6 (8) cm across, convex when young, becoming broadly convex to plane, often with a depressed center; margin often grooved in age. Surface smooth, thinly viscid and shiny when wet, dull if dry. Highly variable in color: extensively bright red, pale red to pink, at times with a yellow to peachy orange center, or nearly overall creamy white to pale yellow. **Gills:** Broadly attached, but sometimes becoming free in age, close to subdistant, narrow, fragile; partial gills rare or absent. White to cream. **Stipe:** 3.5–6 cm long, 0.8–1.8 cm thick, equal or enlarged slightly toward base. Surface smooth, dry. White. **Flesh:** Thin, fragile, white. **Odor:** Indistinct. **Taste:** Quickly and strongly acrid, leaving a lingering peppery sensation. **Spore deposit:** White. **Microscopy:** Spores 8–11 × 7–9 µm, subglobose to broadly ellipsoid, ornamented with isolated warts 0.5–1.5 µm high, occasionally connected with ridges, forming a partial reticulum.

Ecology: Solitary or scattered, often on well-decayed wood or moss-covered stumps and logs, occasionally in wet moss or soil under conifers. Common and widespread across western Cascadia, occasional on the east side. Fruiting from late summer into early winter.

Edibility: Reportedly toxic.

Comments: *R. emetica* is one of several variably red, pink, yellow to cream-capped species with white gills and stipe, white spores, and a strongly acrid taste. The name *R. bicolor* has been misapplied to pink-capped specimens with a yellow to peachy orange center and *R. silvicola* for bright red-capped specimens. *R. montana* is very similar but is more likely to have yellow or creamy cap colors (at least in Cascadia). It is best distinguished microscopically, with smaller spores (7–10 × 6–8 µm) and fine short warts < 0.5 µm high. Additionally, a couple of species in the *R. crenulata* group could be mistaken for *R. emetica*; they may be larger and stockier and firmer, but that may be wishful thinking. *R. cremoricolor* is a larger and stockier and firmer species which fruits with tanoak in late fall and winter in the southern portion of Cascadia. It has two distinct color forms: bright cherry-red and yellowish cream to creamy white—often growing intermingled in the same patches.

Russula rhodocephala

Bazzic., D. Mill. & Buyck

Redhead Brittlegill

Cap: (2) 4–12 cm across, rounded to convex when young, becoming broadly convex to plane, often with a depressed center; margin occasionally uplifted in age. Surface smooth, thinly viscid to tacky, often shiny when wet, dull when dry. Bright scarlet, to rosy red, often slightly darker on disc; can fade to pink to dingy white after heavy rain or prolonged exposure to sun. **Gills:** Broadly attached to slightly decurrent, close to subdistant, narrow, fragile, occasionally forking near stipe; partial gills rare. White to creamy white at first, becoming cream to pale yellow in age. **Stipe:** 2.5–7 (10) cm long, 1–2.5 cm thick, equal or enlarged slightly toward base. Surface smooth, dry. Rosy pink to pale pink, often with a whitish bloom, discoloring when handled. **Flesh:** Thin to moderately thick, firm, fragile. White. **Odor:** Indistinct. **Taste** Quickly and strongly acrid, leaving a lingering peppery taste. **Spore deposit:** Cream to pale yellow. **Microscopy:** Spores 7.5–8 (10) × 6–7 μm, ovoid to broadly ellipsoid, ornamented with fine isolated warts 0.4–0.7 μm high.

Ecology: Scattered or in troops, rarely solitary, on ground or in duff under pine. Especially common and abundant in coastal shore pine habitat, rare to occasional in montane areas with lodgepole pine or in urban settings with planted pines. Fruiting from early fall into winter.

Edibility: Extremely acrid; to be avoided.

Comments: The scarlet-red cap, pink to reddish stipe, white to creamy gills, acrid taste, and growth with pine make the this an easy species to recognize. *R. americana* is quite similar but the caps are a slightly different shade of red, often has a more slender stature, larger spores, and grows with true fir and hemlock. Members of the *R. queletii* group have dark wine-red to reddish purple caps and rosy or purple-blushed stipes. Members of the *R. xerampelina/R. olympiana* group can have red caps and pinkish blushed stipes but have a mild taste, fishy odor, and slow yellowish brown staining on the stipe when scratched or handled. There are a large number of poorly understood red-capped *Russula* with white stipes; see under *R. emetica.*

Misapplied name: *Russula sanguinea* Fr.

Russula americana

(Singer) Singer

Cap: 3–6 cm across, convex at first, becoming broadly convex to plane, often with a depressed center. Surface slightly viscid when wet, to dry, smooth. Rosy red pinkish red to bright red. Gills Broadly attached or with a short decurrent tooth, close, narrow, fragile; partial gills absent. White to cream. **Stipe**: 4–8 cm long, 0.7–1.5 cm thick, equal or tapered slightly toward base. Surface dry, with a fine powdery bloom to smooth. Rosy red to pinkish red. **Flesh**: Thin, firm when young, soon soft, fragile. White. **Odor**: Indistinct. **Taste**: Quickly and strongly acrid. **Spore deposit**: Pale cream. **Microscopy**: Spores 9–11 × 8–9.5 µm, subglobose to ovoid; ornamented with warts 0.5–1 µm high, with sparse connecting lines or ridges.

Ecology: Solitary or scattered on ground, duff, and moss under hemlock and fir. Occasional, more frequently found in mid-elevation mature and old-growth forests. Fruiting in fall.

Edibility: Strongly acrid, best avoided.

Comments: The rosy red to pinkish red cap and stipe, white to cream gills and growth with fir and hemlock distinguish this species. *R. rhodocephala* can look quite similar but has a darker scarlet red cap, creamy to pale yellowish gills, and grows with pine. Red-capped members of the *R. xerampelina*/*R. olympiana* group are often larger, have yellowish ocher to brown-staining stipes, and fishy odors. *R. emetica* can have a similar cap color but has a white stipe and gills.

Russula queletii group

Cap: 3–10 cm across, rounded to convex when young, becoming broadly convex to plane, occasionally with a depressed center in age. Surface smooth, thinly viscid to tacky, often shiny when wet, dull when dry. Dark wine-red to purplish when young, staying dark wine-red on disc, but becoming paler purple-red toward margin. Can fade to pale purple or almost white after heavy rain or prolonged exposure to sun. **Gills**: Broadly attached to slightly decurrent, close, narrow, fragile, occasionally forking near stipe; partial gills rare. White at first, becoming cream in age. **Stipe**: 3–7 cm long, 0.8–2 cm thick, equal or slightly enlarged toward base. Surface smooth, dry. Blushed rosy purple to pale pinkish purple, darkening when handled. **Flesh**: Thin to moderately thick, firm, fragile. White. **Odor**: Indistinct to mildly fruity. **Taste**: Quickly and strongly acrid, at least when young, leaving a lingering peppery taste in mouth. **Spore deposit**: Cream to pale yellow. **Microscopy**: Spores 6.5–9.5 × 6–8 µm, subglobose to broadly ellipsoid, ornamented with well-developed isolated warts 0.7–1.2 µm high.

Ecology: Solitary, scattered, or in troops on ground under pine and spruce. Abundant in coastal forests, occasional to rare elsewhere. Fruiting in fall.

Edibility: Strongly acrid; to be avoided.

Comments: The dark wine-red to purple cap, white to creamy gills, rosy purple stipe, and acrid taste are distinctive. *R. rhodocephala* is similar and often grows intermingled with *R. queletii* but has a scarlet to red cap and a pink to red-blushed stipe. *R. salishensis* (with a purplish, purplish red to pinkish red cap), and *R. pseudopelargonia* (with a deep red, purplish red or brownish red cap) are two similar, difficult to distinguish, small species with pinkish-blushed stipes, and white to pale cream spores. They can be found across Cascadia with Douglas-fir, fir and hemlock. Members of the *R. xerampelina*/*R. olympiana* group can have wine-red caps and blushed stipes, but have a fishy odor, a mild taste, and yellowish to orangey brown staining on the stipe when handled.

Russula mustelina

Fr.

Dense Brittlegill

Cap: 7–12 (15) cm across, convex to broadly convex at first, becoming plane; disc often depressed in age. Surface smooth, dull, dry to thinly viscid if wet. Color highly variable; typically pale cream, yellowish to pinkish when young, developing pink to vinaceous patches, becoming mottled with these colors, often vinaceous to red-brown overall in age. **Gills**: Broadly attached, close to subdistant, broad, fragile; partial gills rare. White when young, becoming creamy in age. **Stipe**: 4–9 cm long, 1.5–3 cm thick, equal or enlarged slightly towards base. White, at times staining brownish when handled. **Flesh**: Thick, very firm, stipe solid. White, at times staining brownish in stipe. **Odor**: Indistinct. **Taste**: Mild. **Spore deposit**: Pale yellow. **Microscopy**: Spores 7.5–10.5 × 6.5–9 µm, subglobose to broadly ellipsoid, ornamented with isolated warts 0.5 µm high, with connecting lines and heavy ridges forming a partial to complete network.

Ecology: Solitary or scattered in duff or soil, at times remaining hidden under duff well into maturity. Especially in drier, high elevation, and east-side true fir forests in the Cascade Range. Occasional to locally common, more likely in the southern portion of our range. Fruiting in summer and fall, rarely in late spring.

Edibility: Edible.

Comments: Recognized by its firm, medium-sized to large fruitbodies, highly variable cap color ranging from creamy white to pale yellow, pink-blushed and mottled with vinaceous red, or completely vinaceous to red-brown. Additional identifying features include the creamy gills, pale yellow spores, white stipe (which occasionally stains brownish), and growth in late summer and fall under true firs. *R. xerampelina/R. olympiana* can have a similarly colored cap but often shows a rosy blush to the stipe, yellow to brown staining, and a softer, pith-like core in the stipe. *R. olivacea* is another large, variably colored species but typically has a rosy-blushed, unstaining stipe. *R. cyanoxantha* has a violet, rosy to purplish gray cap, and forking white gills. It's rare in Cascadia, mostly growing with tanoak in the southwest portion of our range.

Russula mordax

Burl.

Biting Brittlegill

Cap: 4–10 (13) cm across, convex when young, becoming broadly convex to plane, often with a depressed center. Surface moist to dry, smooth. Color highly variable: Reddish brown, pinkish red, orange-red to orangish-brown, typically developing yellow to orange tones on disc, occasionally more extensively yellowish. **Gills**: Narrowly attached, close to subdistant, broad, fragile, often forking at stipe; partial gills rare or absent. Creamy white when young, soon yellowish. **Stipe**: 4–8 cm long, 1.5–3 cm thick, equal or enlarged slightly toward base. Surface smooth, dry. White, often with a reddish or pinkish blush. **Flesh**: Thick, firm, but fragile, white. **Odor**: Indistinct to floral. **Taste:** Quickly and strongly acrid, leaving a lingering peppery, burning sensation in mouth. **Spore deposit**: Yellow-ocher. **Microscopy**: Spores 7.5–9 (10) × 6.5–7.5 µm, ovoid to broadly ellipsoid, rarely subglobose, ornamented with warts 0.3–0.7 µm high, with heavy ridges, not forming a reticulum.

Ecology: Solitary or scattered on ground, in duff or moss in conifer forests with Douglas-fir, hemlock and possibly true fir. Common and at times abundant. Fruiting from late summer through fall.

Edibility: Strongly acrid, best avoided.

Comments: The reddish, orangish to evenly brownish cap which often develops an orangish to yellowish disc, yellowish gills, yellow-ocher spores and strongly acrid taste identify this common *Russula*. Fruitbodies of the *R. xerampelina/R. olympiana* group have a mild to slightly acrid taste, yellowish to brownish staining stipes, and often have a fishy odor. The variably-colored *R. benwooi* has creamy-colored spores and a mild taste. *R. firmula* has a deep purple cap, a whitish or violet-blushed stipe, yellow spores, and a strongly acrid taste. It grows in coastal spruce forests.

Russula olympiana

Noffsinger & Buyck

Western Shrimp Brittlegill

Cap: 5–15 (20) cm across, convex when young, becoming broadly convex to plane, with a depressed center. Surface smooth, thinly viscid when wet, finely tomentose and often with debris stuck to it when dry. Color extremely variable, even within the same patch. Generally some shade of deep purple, reddish purple, reddish brown to rosy purple, but can have any shade of these colors, as well as brown, gray, greenish olive, olive-brown, or yellowish orange. **Gills:** Broadly attached to notched, close to subdistant, broad, fragile, rarely forking; with occasional partial gills. Creamy white when young, becoming cream to ocher-yellow in age. Stipe 4–10 (15) cm long, 1–4 cm thick, equal or club-shaped, with a slightly enlarged base, or narrower in middle. Surface smooth, dry. White, generally with a rosy pink blush, which is more pronounced in age, occasionally entirely white or rosy pink, staining pale ocher-yellow, then orange-brown when handled. **Flesh:** Thick, very firm when young, often brittle in age. White, discoloring pale ocher-yellow to pale orangish brown, or light brown overall in age. Stipe with soft pith-like core. **Odor:** Mild when young, soon developing a fishy odor, often described as shrimpy or crab-like. **Taste:** Mild to sweet. Spore deposit: Yellow. **Microscopy:** Spores 7.5–9.5 × 6.5–7.5 µm, ovoid to broadly ellipsoid, ornamented with isolated warts 0.5–1 µm high, with a few or no connecting lines or ridges.

Ecology: Scattered, in small troops, or occasionally solitary on ground, in duff or moss with a wide range of conifers, especially Douglas-fir. Widespread and abundant, from wet coastal forests well into dry east-side habitats. Scattered fruiting will occur in summer at higher elevations and in coastal fog zones, with the primary flush in fall.

Edibility: Edible and excellent. The firm flesh, sweet-savory flavor, and suite of reliable field marks make this a good species to seek out for the table.

Russula olympiana

Russula serissima

Comments: The variable dark purple to reddish cap (in its most common form), the white to creamy gills that darken to ocher-yellow in age, a rosy-blushed stipe— with yellow to light ocher-brown staining when scratched—helps define this common species. This staining reaction is most pronounced when the stipes are scraped but often develops after normal handling as well. It is not a quick reaction but after a few minutes (or by the time you get home with a full basket), the stipe will often be covered in ocher brownish fingerprints. The odor can be mild in young fruitbodies or cold weather but is usually quite strong and sometimes even repulsive in mature fruitbodies. Other Cascadian species in this complex include *R. viridofusca*, with more pallid cap (tan, pinkish yellowish or olive) and a white or faintly blushed stipe. *R. serissima (= R. favrei* or *R. elaeodes* sensu PNW), which typically has a dark brown cap. *R. brunneola* also has a dark brown to reddish brown cap, and a often a grayish blushed stipe. *R. olivacea* is a large, variably colored species with an olive, red to purplish cap with a band of concentric wrinkles near the margin, creamy gills, a rosy-blushed, unstaining stipe, and mild taste. Although edible for most, some people have an adverse reaction to it; cook thoroughly and sample sparingly at first.

Misapplied name: *Russula xerampelina*. The Western North American species is distinct from the European *R. xerampelina*, and was recently described as *R. olympiana*.

Russula benwooi

Bazzic., D. Mill. & Buyck

Ben's Brittlegill

Cap: (4) 8–12 (16) cm across, convex when young, becoming broadly convex to plane, often with a shallowly depressed center. Surface smooth, viscid when wet, shiny when dry. Color highly variable: mottled or extensively brownish tan, olive, reddish, pink, vinaceous or purple. **Gills**: Attached, close to subdistant, broad, fragile, with occasional partial gills. Creamy white to pale yellowish. **Stipe**: 5–10 cm long, 1–3 cm thick, equal or enlarged toward base. Surface smooth, dry. Typically white, but can be grayish tinged, reddish or purple-blushed to extensively reddish purple. Occasionally staining brownish. **Flesh**: Moderately thick, firm, fragile. White. **Odor**: Indistinct. **Taste**: Mild, or occasionally acrid. **Spore deposit**: Cream. **Microscopy**: Spores 8.5–9.5 × 7–7.5 µm, broadly ellipsoid, ornamented with scattered warts ~1 µm high, with occasional ridges forming a partial reticulum.

Ecology: Solitary, scattered, or in troops on ground, in moss or duff of hemlock and possibly other conifers. Common and occasionally abundant across northern Cascadia. Distribution in the southern portion of our range still poorly known. Fruiting in fall.

Edibility: Edible.

Comments: This species was named for Ben Woo, a *Russula* enthusiast who collected and made detailed notes on over a thousand collections from the Pacific Northwest. Like many russulas, the highly variable cap colors make this a difficult species to recognize. Caps can show extensive brown, red, purple, or olive tones, or any combination or variation thereof—this species doesn't seem to have a 'typical' color form. Features to aid in identification are the medium to large fruitbodies, creamy to pale yellow gills, white to reddish purple blushed stipe—which does not stain or occasionally browns slightly—and cream spores. Shrimp Russula (*R. olympiana* group) has darker ocher-yellow gills at maturity, usually a distinctly rosy-blushed stipe that stains yellowish to orangish brown when handled, a mild taste, and a fishy odor. *R. vinosa* has a variably olive to purple cap and a fleeting pinkish red staining reaction which slowly turns gray on the stipe and flesh.

Russula vinosa

Lindblad

Graying Olive Brittlegill

Cap: 5–15 cm across, rounded to convex when young, becoming broadly convex, to plane, often with a depressed center in age. Surface smooth, dull and often with a powdery bloom when young and dry, moist to thinly viscid when wet; margin often grooved in age. Color variable, olive, olive-gray, olive-purple to all purple at first, often variegated with a combination of these colors as it expands, usually darker on disc, becoming paler, developing yellowish olive to grayish lilac colors in age. **Gills**: Broadly attached to notched, close to subdistant, broad, fragile; often with short, irregular partial gills near margin. Surface smooth, dry. White when young, cream to pale creamy yellow in age. Occasionally slightly blackening near cap margin, or along edges. **Stipe**: 4–8 cm long, 1.5–3 cm thick, equal or enlarged slightly toward base. White when young, discoloring to grayish in age. Slowly staining pinkish red, then gray when handled or scratched. **Flesh**: Thick, firm when young, brittle in age. White to gray, slowly staining pinkish red, then gray. **Odor**: Indistinct. **Taste**: Mild. Spore deposit: Pale yellow to cream colored. **Microscopy**: Spores 8.5–11.5 × 6.5–8 μm, broadly ellipsoid to ovoid, ornamented with pointed isolated spines or warts up to 2 μm high, occasionally with short ridges or lines, but not forming a reticulum.

Ecology: Solitary, scattered, or in troops in moss or duff under a wide range of conifers, especially Sitka spruce. Common and widespread across Cascadia. Fruiting from late summer through fall and occasionally in spring on the coast.

Edibility: Reported to be edible.

Comments: This species can be separated from other large, variably olive- to purple-capped *Russula* by the staining reaction on the stipe and flesh: a fleeting pinkish red followed by gray. The staining is best observed by scratching the stipes of young fruitbodies. Other distinguishing features are the extensively gray stipe and flesh of older specimens, mild taste, and pale yellow spores. Iron sulfate produces a faint rusty stain on the stipe of this species, as opposed to the blue-green colors it produces on the stipes of the similar, but darker-spored *R. viridofusca* and *R. serissima* (members of the broader Shrimp Russula group – section *Xerampelinae*). The gray staining helps distinguish *R. vinosa* from the more variably colored but otherwise similar *R. benwooi. R. olivacea* sensu PNW can share similar colors but is generally larger, has a rosy-blushed stipe, yellow spores, and does not stain. *R. decolorans* has a coppery orange cap, and a distinct gray staining (or discoloring) stipe and flesh.

Synonym: *Russula occidentalis* (Singer) Singer

Russula obscurozelleri

Bazzic., D. Mill. & Buyck

Cap: 3–8 (12) cm across, convex when young, becoming broadly convex to plane, often with a depressed center. Surface viscid when wet, shiny, smooth. Color highly variable, often darkest at disc, ranging from purple, to reddish purple, rosy purple, or pinkish red, occasionally bright red. Typically developing yellowish to tan spots and blotches with age. **Gills**: Broadly attached, close, broad, fragile. White at first, becoming creamy to yellow, or even ocher-yellow in age; partial gills absent. **Stipe**: 4–9 cm long, 1–2.5 cm thick, equal or enlarged slightly toward base. Surface dry, smooth. White or with rusty stains near base. **Flesh**: Moderately thin, firm, fragile; white. **Odor**: Indistinct. **Taste**: Mild. **Spore deposit**: Cream to yellow. **Microscopy**: Spores 6–10.5 × 5.5–8 µm, broadly ellipsoid to ovoid, ornamented with scattered isolated warts 0.2–1 µm high, with some connecting ridges.

Ecology: Scattered in troops on ground, occasionally solitary, in duff or moss under conifers, especially frequent with Douglas-fir. Common and widespread, occurring at low and mid elevations across Cascadia. Fruiting in fall.

Edibility: Edible.

Comments: This species is one of five in the *R. zelleri* group in Cascadia, two of which are undescribed. As a group, they tend to have purple to red viscid to shiny caps which develop yellowish spots and splotches, creamy to yellow gills, and a mild taste. They are just about impossible to distinguish without an ITS sequence. *R. obscurozelleri* and *R. pseudotsugarum* (which may be more likely to have olive tones to the cap) are both quite common with Douglas-fir and hemlock but can also grow with spruce, fir, and pine. *R. zelleri* appears to be restricted to spruce and pine and has more brownish, olive to yellowish tones mixed with the reds and purples. Microscopically, the three described species have overlapping spore ranges but the average length of a large sample of spores may be helpful—for those who want to torture themselves trying to identify *Russula* based on morphology. *R. obscurozelleri* has the shortest average spore length at 7.7 µm, *R. pseudotsugarum* spores average 8.1 µm long, while those of *R. zelleri* average 8.9 µm. The poorly known (yet supposedly common) *R. sierrensis* could be mistaken for species in this group and comes in every color imaginable. It appears to have duller (not shiny) caps and spores with isolated warts without connecting ridges.

Russula turci

Bres.

Purple-plush Brittlegill

Cap: 3–9 cm across, convex when young, becoming broadly convex to plane, often with a depressed center. Surface viscid when wet, often finely roughened and wrinkled, dull, pruinose, often finely cracking when dry. Dark purple to royal purple, occasionally more reddish purple, typically with a whitish bloom when dry. At times developing yellowish spots, rarely extensively yellowish or cream. **Gills**: Broadly attached to notched, close to subdistant, narrow, fragile; partial gills absent. White at first, becoming cream-colored to pale yellow-ocher as spores mature. **Stipe**: 3–7 cm long, 1–2.5 cm thick, equal or slightly enlarged towards base. Surface pruinose to smooth, dry. White to off-white. **Flesh**: Thin, firm but fragile; white. **Odor**: Iodine-like in stipe base. **Taste**: Mild. **Spore deposit**: Ocher-yellow. **Microscopy**: Spores 7–9 × 6–7.5 µm, subglobose to broadly ellipsoid, ornamented with warts ~0.5 µm high, connected with ridges, forming an almost complete reticulum.

Ecology: Solitary, scattered, or in troops on ground under conifers, especially Douglas-fir. Common across western Cascadia. Fruiting in fall.

Edibility: Edible.

Comments: Distinctive features include the purple cap that is viscid when wet but pruinose (covered in a powdery bloom) when dry, mild taste, creamy to pale yellow gills, and yellow-ocher spores. There is often an iodine odor in the flesh of the stipe base, usually noticed after crushing the flesh. *R. murrillii* is similar in many regards but lacks an iodine odor and its caps are often reddish purple to rosy purple. *R. obscurozelleri* (and other members of the *R. zelleri* complex) have a smooth, shiny cap.

Russula phoenicea

Bazzic., D. Mill. & Buyck

Pastel Brittlegill

Cap: 3–7 cm across, convex when young, becoming broadly convex to plane, often with a depressed center; margin occasionally translucent-striate or grooved in age. Surface smooth, thinly viscid and shiny when wet, dull when dry. Color extremely variable – generally with a dark purple, gray, reddish brown or pinkish center, transitioning to a zone that can be any of these shades, but often with olive tones, to a paler margin, rarely extensivly any one of these colors. **Gills:** Broadly attached, sometimes becoming free in age, close to subdistant, narrow, fragile; partial gills rare or absent. White to pale creamy white. **Stipe:** 2–5 cm long, 0.5–1.5 cm thick, equal. Surface smooth, dry. White to off-white. **Flesh:** Thin, very fragile, white. **Odor:** Indistinct. **Taste:** Mild or slightly acrid. **Spore deposit:** White. **Microscopy:** Spores 7–9 × 5.5–8 µm, subglobose to broadly ellipsoid, ornamented with isolated warts, 0.5–1.5 µm high, connected with thin ridges, forming a partial to complete reticulum.

Ecology: Solitary, scattered or in troops on well-decayed wood, moss-covered logs, or stumps, or on ground under Douglas-fir (and possibly other conifers). Common and widespread. Fruiting in fall.

Edibility: As with most acrid *Russula*, not recommended.

Comments: This species can exhibit a wide range of cap colors, sometimes all on the same fruitbody at once! The small fruitbodies, variable cap color (usually darkest at the disc and with some olive or purple tones), and white gills, stipe, and spores help identify it. *R. hypofragilis* is very similar and can be difficult to reliably distinguish without an ITS sequence; it tends to have more pronounced purple to reddish purple cap colors, is less likely to show olive tones, and often grows with true firs. Both of these are recently described members of the *R. fragilis* group and would have been identified as such in the past. *R. versicolor* grows with introduced birch trees and has yellow spores.

Russula aeruginea group

Green Brittlegill

Cap: 4–10 cm across, convex when young, becoming broadly convex to plane, with a depressed center and/or an uplifted margin in age. Surface smooth, dull and often with a powdery bloom when young and dry, moist to thinly viscid when wet; margin often grooved in age. Deep green, olive-green or grayish green at first and usually developing yellowish brown stains on disc in age. **Gills**: Broadly attached to notched, close to subdistant, broad, fragile, often forking at stipe; partial gills rare. White when young, cream to pale yellow in age. Occasionally discoloring brownish when old. **Stipe**: 3–8 cm long, 1.5–3 cm thick, equal. Surface smooth, dry. White when young, often developing yellowish to dingy yellowish orange stains in age. **Flesh**: Moderately thick, firm when young, brittle in age. White. **Odor**: Indistinct. **Taste**: Mild. **Spore deposit**: Pale yellow. **Microscopy**: Spores 6.5–9 (11) × 5.5–7 (9) µm, broadly ellipsoid to nearly round, ornamented with isolated low warts 0.2-0.5 µm high, with short ridges or lines, but not forming a reticulum.

Ecology: Solitary, scattered, or in small troops on ground, in moss, or duff in a wide variety of forest types. Common and widespread, occurring throughout Cascadia. Fruiting from mid-summer until freeze.

Edibility: Edible and good; especially the young firm buttons.

Comments: *R. aeruginea* has a deep green, greenish gray or yellowish green cap, pale yellow gills that often fork near the stipe, pale yellow spores, and a mild taste. There are several other green Russula in Cascadia. *Russula graminea* also has a green to greenish yellow cap (although it can also have purple tones), darker yellow spores, and a mild to acrid taste. It is more common in coastal spruce-hemlock forests. A common, undescribed species with a beautiful grass-green cap (*R.* 'Woo sp.13') fruits in high elevation fir-hemlock forests in summer and early fall. *Russula* 'Woo sp. 15' has a grayish green to olive brown cap. *Russula* sp. 'Green Madrone' is an undescribed species found under madrone and manzanita. It has large fruitbodies with yellow-green caps that develop orangish blotches in age and pale orangish staining on the stipe when handled. Although generally rare in Cascadia, it is often found in the southern part. A locally common species from coastal spruce forests in the *R. heterophylla* group has a dark olive-green, greenish brown, or greenish purple cap that often ages more purplish brown. It develops distinctive lightning-bolt-like cracks and areolate patches on the cap, has whitish gills that develop orangey stains, and has a mild taste and white spores.

Russula versicolor

Jul. Schäff.

Birch Brittlegill

Cap: 2–6 cm across, convex when young, becoming broadly convex to plane, often with a depressed center in age; margin grooved. Surface thinly viscid and shiny when wet, dull if dry. Color variable, often some shade of, or a mix of, purple, wine-red, red, pink, brown, gray, olive, green, yellow, or whitish. Generally with a darker disc and paler margin. **Gills:** Broadly attached, but sometimes becoming free in age, close to subdistant, narrow, fragile. White at first, becoming cream colored to pale yellow in age. **Stipe:** 2–5 cm long, 0.5–1.5 cm thick, equal or enlarged slightly toward base. Surface smooth, dry. White at first, developing yellowish stains. **Flesh:** Thin, fragile; white, slowly discoloring pale yellow when cut or in age. **Odor:** Indistinct. **Taste:** Quickly and strongly acrid. **Spore deposit:** Yellow to ocher-yellow. **Microscopy:** Spores 5.5–8.5 × 4.5–6.5 µm, broadly ellipsoid to nearly round, ornamented with coarse warts with scattered ridges, forming a partial to complete reticulum

Ecology: Solitary, scattered, or in grass under planted birch. Often first fruiting in late summer around irrigated trees, continuing through fall.

Edibility: Not recommended.

Comments: Even though this mushroom is highly variable in color, it is easily recognized by its growth under non-native birch trees, small size, yellowish spores, and acrid taste. A sister species discovered by ITS sequencing also occurs in Cascadia, typically with cottonwood (but also with birch); it may show more olive tones but is not well known. There are three similar-colored, conifer-associated, mild-tasting *Russula* in Cascadia which have yellowish discolorations on the stipe. *R. puellaris* sensu PNW has a dark purple to brick red cap that fades and develops yellow to orange colors in age and a white stipe that stains yellow to pale orange when handled. It is locally common in coastal spruce-hemlock forests. *R. sapinea* and *R. abietina* typically have purple to reddish purple caps, which develop yellowish to brownish tones over the disc. These two are practically indistinguishable and may require an ITS sequence to separate them. Also see *R. phoenicea;* an acrid-tasting species with white spores that grows under conifers.

Russula stuntzii

Grund

Gray Brittlegill

Cap: 3.5–8 cm across, rounded to convex when young, becoming broadly convex to plane, often with a depressed center in age. Surface smooth, viscid to tacky when wet, shiny when dry. Whitish, grayish to pale lilac-gray, often darker at disc, paler toward margin or in age. **Gills:** Broadly to narrowly attached, close, narrow, fragile; partial gills rare. White to pale creamy white. **Stipe:** 3–8 cm long, 1–2 cm thick, equal or enlarged slightly toward base. Surface smooth, dry. White to pale grayish white, often discoloring pale grayish when handled. **Flesh:** Thin to moderately thick, firm, fragile, white. **Odor:** Indistinct. **Taste:** Quickly and strongly acrid, rarely faintly acrid. Spore deposit: White to pale creamy white. **Microscopy:** Spores 6–9 × 6–7.5 µm, globose to ovoid, occasionally broadly ellipsoid, ornamented with scattered warts 0.5–1 µm high, with rare lines or ridges

Ecology: Solitarily or scattered, usually on well-decayed moss-covered logs or stumps, occasionally on ground. Common in west-side mixed conifer forests, especially with hemlock and fir. Fruiting in fall

Edibility: Unknown; the strongly acrid taste is a deterrent

Comments: The whitish cap with grayish or lilac tones, white gills, whitish stipe that becomes grayish, acrid taste, and habit of growing on well-decayed logs are distinctive. Most members of the *R. emetica* group can have pallid forms but they all have "warmer" creamy tones to the cap and bright white gills. *R. crassotunicata* has a whitish to creamy cap when young, which develops brown stains.

Russula cerolens

Shaffer

Pocket-stalked Brittlegill

Cap: 3–10 cm across, rounded when young, becoming broadly convex to plane, with a depressed center at all ages. Margin often becoming wavy and uplifted in age. Surface smooth, with grooves near margin, viscid and shiny when wet, dull and often cracking when dry. Dark brown, light brown to tan, occasionally yellow-brown, often darker on disc. Developing orangish stains and occasionally becoming dingy grayish brown, darkening in age. **Gills**: Broadly attached to notched, close to subdistant, broad, fragile; partial gills rare or absent. White to creamy at first, developing orangish brown spots and stains, becoming dingy cream colored with brownish stains in age. **Stipe**: 3–8 cm long, 1–2.5 cm thick, equal or enlarged slightly toward base. Surface smooth, dry. White but often with an orange base, developing rusty orange or brownish stains in age. **Flesh**: Thin, firm, fragile when young, mealy and brittle in age; stipe often with 2 or 3 hollow, diamond-shaped pockets. White, occasionally with brownish stains. **Odor**: Mild at first, soon unpleasant, like burnt hair, or more chemical, like plastic or rubber. **Taste**: Disagreeable, bitter at first, soon acrid. **Spore deposit**: White. **Microscopy**: Spores 6.5–9 × 5.5–6.5 μm, subglobose to broadly ellipsoid, ornamented with scattered warts 0.5–0.7 μm high, occasionally with ridges, but not forming a reticulum.

Ecology: Solitary, scattered, or in troops on ground in a wide variety of forest types, associating with hardwoods and conifers. Common and locally abundant, widespread across Cascadia. Fruiting from late summer into winter, with the primary flush in late fall.

Edibility: Not recommended.

Comments: This ubiquitous *Russula* can fruit by the hundreds under shore pine along the coast. However, it is not limited to this forest type and fruits in many kinds of hardwood and conifer forests throughout our range. It is easily recognizable by its light to dark brown cap with a grooved margin, white gills, and white stipe that develops reddish orange stains. The distinctive and unpleasant odor is also a good identification feature. There are a few similar, undescribed species in Cascadia and we currently lack criteria to distinguish them. One of these is close to *R. recondita*, which tends to have a paler yellow-brown cap; it may be restricted to oaks and other hardwoods. The Almond-scented Russula (*R. fragrantissima* group) looks similar but has an almond or maraschino cherry odor when young and tends to be substantially larger.

Russula fragrantissima group

Almond-scented Russulas

Cap: 5–15 (25) cm across, round-convex, convex to plane, often with a depressed center in age. Surface smooth to bumpy; margin grooved; viscid when wet; skin fairly thick and rubbery, easily peeled, often cracking when dry. Honey-brown to beige-brown, often with reddish brown to orange-brown blotches and stains. **Gills**: Broadly attached to notched, close to subdistant, broad. Pale tan to yellowish beige or clay-colored, often with ocher-brown stains. Edges often decorated with clear droplets, especially when young. **Stipe**: 6–20 cm long, 3–6 cm thick, club-shaped or swollen in middle. Surface moist to dry, roughened to smooth. Whitish to clay tan; with dingy brownish stains in age. **Flesh**: Moderately thick to thin, firm when young, soon softer and brittle or mealy in age. Stipe chambered at first, soon hollow throughout. **Odor**: Very strong, sweetish like almond extract or maraschino cherries, foul in age. **Taste**: Bitter or unpleasant at first, soon very acrid. **Spore deposit**: Cream to pale yellow. **Microscopy**: Spores 8.5–10.5 × 6.5–9.5 µm, subglobose to broadly ellipsoid, ornamented with scattered warts 0.8–1.2 µm high, with heavy ridges or lines, forming a partial to full reticulum.

Ecology: Solitary or scattered, often in large troops on moss and duff; typically under conifers (true fir and hemlocks) but some members of this group grow with oak and tanoak. Collectively common and widespread across Cascadia, fruiting in summer and early fall, with stragglers later into the season.

Edibility: Likely toxic.

Comments: The pungent sweet odor, yellow-brown cap, and chalk-like or brittle flesh distinguish this complex of *Russula*. The name *R. laurocerasi* has also been applied to these almond-scented species but recent work has shown there are at least four species in Cascadia, which all need a name. A few *Hygrophorus* have a similar odor but are more fibrous-fleshy, taste mild, and have thick, waxy gills. *R. cerolens* is similar but smaller, with a dark brown to tan cap, and a distinct odor (like burnt hair or plastic). *R. pallescens* is a small pale yellow to yellowish brown-capped species with a grooved margin, and white gills and stipe; it typically lacks an odor and has a strongly acrid taste.

Russula crassotunicata

Singer

Rubbery-capped Russula

Cap: 3–8 (10) cm across, rounded to convex when young, becoming broadly convex to plane, often with a depressed center, occasionally concave in age. Surface viscid, covered with a thick slime layer when wet, often dull and cracked when dry. Cap skin thick, gelatinous and rubbery when wet, almost leathery when dry. Pale creamy yellow when young, soon developing yellowish to ocher-brown stains and discoloration, occasionally becoming completely light brown in age. **Gills**: Broadly attached to notched, close to subdistant, narrow, occasional forking near stipe, fragile; partial gills rare. White to creamy white at first, developing brownish spots and blotches, occasionally becoming all brown in age. **Stipe**: 3–5 cm long, 0.8–2 cm thick, often short, stout, equal or enlarged slightly toward base. Surface smooth, dry. White at first, developing brownish spots and stains. **Flesh**: Thin to moderately thick, rubbery-gelatinous near cap skin, very firm when dry. White, slowly staining brownish when cut. **Odor**: Indistinct. **Taste**: Usually quickly and strongly acrid and bitter but can be mildly acrid or mostly bitter. **Spore deposit**: White. **Microscopy**: Spores 9–11 × 7–9 µm, broadly ellipsoid to subglobose, ornamented with fine isolated warts 0.8–1.2 µm high.

Ecology: Solitary, scattered, or in small troops on ground or in duff under hemlock, spruce, and fir. Locally common and widespread across western Cascadia, becoming scarce inland and towards the southern portion of our range. Fruiting in summer and fall, with stragglers continuing into winter.

Edibility: Unknown; not recommended.

Comments: The thick, gelatinous, and rubbery cap skin (when wet), creamy yellow colors with brown stains, and growth with conifers help identify this *Russula*. Members of the *R. emetica* group are typically fragile, have bright white gills and don't stain brown. *R. stuntzii* has a whitish to pale grayish cap and a white stipe that discolors grayish. *R. cerolens* has a light to dark brown cap, a burnt-hair or plasticky odor, and develops orangish stains on the gills and stipe.

Russula brevipes group

Short-stalked Russulas

Cap: (4) 7–30 (35) cm across, convex with an incurved margin and a depressed center when young, vase-shaped with an uplifted or wavy margin in age. Surface dry, finely tomentose to smooth. White at first, soon dingy white to whitish buff with yellowish brown spots or stains. *Gills*: Broadly attached to slightly decurrent; close, broad, fragile, occasionally forking or fusing near the stipe; partial gills numerous. White when young, sometimes with a bluish cast (see comments), often developing brownish spots and stains in age; edges often covered with clear droplets when young. **Stipe**: 2–8 cm long, 2–5 cm thick, short, stout, more or less equal. Surface dry, smooth. White, with brownish spots or stains. Occasionally with a bluish band at apex. **Flesh**: Thick, very firm when young, granular in age. White, but often with brownish stains around larva tunnels. **Odor**: Mild when young, strong and unpleasantly fishy in age. **Taste**: Mild to slightly acrid. **Spore deposit**: White to cream. **Microscopy**: Spores 8–10.5 × 6.5–9 µm,broadly ellipsoid to ovoid, occasionally nearly round, ornamented with scattered warts 0.3-1.3 µm high, and heavy ridges forming a partial reticulum.

Ecology: Solitary, in small clusters, or scattered in large troops in a wide variety of forest types. Widespread and very common throughout our range. Fruiting from late summer into early winter, with the primary flush coming in early to mid-fall.

Edibility: Edible but while some forms are good, others are downright awful (it's a species complex). Firm young buttons are best, especially those growing with hardwoods.

Comments: Species in this group of *Russula* can produce extremely large fruitbodies, with caps measuring 25–30 cm across regularly encountered. The white colors overall (often with yellowish to brown stains), cap that becomes funnel-shaped in age, lack of latex, and absence of gray to black staining are helpful identifying features. Genetic data indicate that there are at least six species in the *R. brevipes* complex in Cascadia but features to distinguish them haven't been established. The name *R. brevipes* var. *acrior* has been applied to forms with a bluish green band at the stipe apex but it appears multiple lineages can show this feature. Those growing with hardwoods are generally smaller and brighter white in color. *R. cascadensis* has a white cap and gills but is more likely to develop orangish stains as it ages, has a strongly acrid taste, and smaller spores (6.5–8 × 5.5–6.5 µm) which are cream to yellowish. Members of the blackening Russula group all show grayish to black staining (as the name suggests). Some large *Lactarius* species can look similar but bleed white or colored latex when the gills are broken.

Hypomyces lactifluorum

(Schwein.) Tul.

Lobster Mushroom

Fruitbody: 4–20 cm across, often highly irregular and distorted, vaguely taking the shape of the host mushroom Russula brevipes. Upper surface dry to moist, often cracked or pitted. Underside dry, covered with small pimples (openings of perithecia). Gills completely covered, appearing as obscure ridges. Usually bright orange but can be whitish with pale orange blotches when young and buried in duff; developing reddish purple stains and blotches as it starts to decay. **Stipe**: Often short, stout, and distorted, but can be more regularly formed. **Flesh**: Very firm and white with a thin orange crust at first, becoming mottled with orange and softer as it ages. **Odor**: Indistinct when young, disagreeable with age. **Taste**: Mild. **KOH**: Bright magenta-purple. **Spore deposit**: White. **Microscopy**: Spores 35–40 (45) × 4.5–7 µm, spindle-shaped, distinctly warted. Asci long-cylindrical.

Ecology: Solitary, scattered, or in large troops, often buried in duff before erupting at maturity, growing on fruitbodies of the Short-stalked Russula, *R. brevipes*. Non-parasitized specimens of this Russula are usually present within patches, at least later in the season. Common and at times abundant in western Cascadia, rare on the east side, and largely absent from the southern Cascade and Klamath ranges. First fruiting in mid-summer, even when dry, continuing into fall and with stragglers lasting into winter.

Edibility: Edible and good but only firm specimens should be eaten. A fun trick is to sauté Lobster Mushrooms in a lot of butter, then use the butter to cook eggs – they'll turn purple!

Comments: A fungus on a fungus–what could be better? The Lobster Mushroom is a parasite that forms a bright orange crust which engulfs and distorts its host mushroom (*Russula brevipes* in most, if not all cases). It is easily recognized by its large size, bright orange color, and convoluted shapes. One should start looking for this species in mid to late summer, as it typically starts fruiting in the middle of the dry season. It is highly sought after as a source of fabric dye, yielding peach, pink to magenta colors on wool and silk. *H. aurantius* forms a bright orange layer on the undersides of polypores, especially older fruitbodies of *Fomitopsis mounceae* and occasionally on Oyster Mushrooms (*Pleurotus* spp.); it is common in western Cascadia.

Hypomyces luteovirens

(Fr.) Tul.

Green Langoustine

Fruitbody: A thin crust extensively covering the gills and often the stipe of normal or slightly deformed Russula species, soon appearing on the cap margin, rarely covering the cap. **Underside**: Dry, completely covering the gills, which will appearing as nondescript shallow ridges. Gills and upper stipe portion covered with imbedded small pimples (perithecia). Sulfur-yellow to lime-green when young, developing olive tones; perithecia becoming dark forest-green to almost black. **Stipe**: Typically fully formed, lower portion occasionally bare. **Flesh**: Very firm and white with a thin yellow crust, softer and dingier in age. **Odor**: Indistinct when young, disagreeable with age. **Taste**: Mild. **Spore deposit**: White. **Microscopy**: Spores 30–35 × 4.5–6 µm, spindle-shaped, finely warted. Asci thread-like to long-cylindrical.

Ecology: Solitary or scattered, rarely in large numbers on the underside of *Russula*. It seems to prefer denser, stockier *Russula* (e.g., *R. mustelina*) but the identity of the host species is mostly unknown. Widespread but rare; more likely to be found in mid to high elevations in the Cascade Range. Fruiting in late summer and fall.

Edibility: Reportedly edible but due to unknown host edibility, not recommended.

Comments: This species is often overlooked, unless one is flipping over random *Russula* specimens. It first becomes apparent on the gills of the host, before covering much of the stipe, and then creeping over the cap in age. The yellow to greenish colors, darker pimple-like dots, and growth on the undersides of *Russula* readily identify it. Unlike its edible orange cousin *H. lactifluorum*, this species does not seem to be restricted to a specific group of *Russula* as its host. There are a number of other *Hypomyces* species, many of which are specific to a single species or genus of host mushrooms. *H. lateritius* grows on *Lactarius* (most often those in the *L. deliciosus* group) and covers only the gills of its host with a whitish layer; it is rare in Cascadia. *H. rosellus* is a rare, bright pink to rosy, pimpled crust that forms a thin layer on the decaying fruitbodies of many kinds of mushrooms.

Russula adusta group

Viscid Blackening Russulas

Cap: (5) 7–12 cm across, convex when young, becoming broadly convex to plane, occasionally with a depressed center. Surface thinly viscid when wet, shiny if dry, smooth. Dingy white at first, darkening in patches, grayish brown to dingy brown, to dark gray or blackish in age. **Gills:** Broadly attached to subdecurrent, close, thick, broad, fragile, occasionally forking or inter-veined; partial gills interspersed near margin. Whitish when young, becoming creamy white to pale grayish yellow, slowly staining pinkish red, then dark gray when damaged. **Stipe:** 3–6 cm long, 1.5–3 cm thick, more or less equal. Surface smooth, dry. White to pale gray, slowly staining pinkish red at first to dark gray. **Flesh:** Thick, very firm. White when young, often grayish in age, staining pinkish, then gray. **Odor:** Indistinct. **Taste:** Mild. **Spore deposit:** White. **Microscopy:** Spores 7–9 × 6–8 µm, broadly ellipsoid to subglobose, finely ornamented with short warts and spines 0.1–0.4 µm high, interconnected with ridges or lines, forming a partial to complete reticulum.

Ecology: Scattered or in troops, rarely solitary from soil, duff, and moss under conifers. Common and widespread throughout Cascadia but more abundant in west-side forests. Fruiting from late summer until freeze.

Edibility: To be avoided; some blackening *Russula* are poisonous.

Comments: Members of the *R. adusta* group are distinguished from other blackening *Russula* by their thinly viscid or shiny caps, mild taste, and pinkish or reddish staining that develops over about five minutes after being cut or damaged. ITS sequence data has shown three species in Cascadia but reliable morphological features to separate them have not been established. *R. acrifolia* sensu PNW (often called *R. densifolia)* is quite similar but has an acrid taste. *R. nigricans* has a dull dry cap and darker red staining before darkening to black. *R. atrata* also has a dry cap and stains black without the red in-between phase.

Russula nigricans

Fr.

Blackening Brittlegill

Cap: 5–15 (20) cm across, convex when young, becoming broadly convex with a depressed center, to funnel-shaped and often irregular, with a wavy or uplifted margin in age. Surface dull, finely velvety to smooth, often cracking into areolate patches in age, slightly viscid if wet when young, soon dry. Dingy white when young, darkening in patches, gray to grayish brown, to eventually black. **Gills**: Broadly attached, subdistant to distant, thick, broad, fragile; partial gills numerous. Creamy to pale grayish yellow, when broken slowly staining red and turning black, becoming dark gray to all black in age. **Stipe**: 2–7 cm long, 1.5–5 cm thick, stout, more or less equal. Surface smooth, dry. Whitish at first, soon pale gray to grayish brown, to dark gray or black in age, slowly staining red, then black when handled. **Flesh**: Thick, very firm. Whitish when young, staining red, with an intermediate reddish-brown phase, then dark gray to black; often grayish to all black in older fruitbodies. **Odor**: Mild to musty or earthy when young, generally unpleasant in age. **Taste**: Mild to slightly acrid. **Spore deposit**: White. **Microscopy**: Spores 6.5–8 × 5–7 µm, nearly round to broadly ellipsoid, ornamented with short isolated warts and spines 0.1–0.3 µm high, with short ridges or lines, forming a partial to full reticulum.

Ecology: Solitary, scattered, or in troops on ground under pine and Sitka spruce in coastal forests and Douglas-fir and hemlock in west-side forests. Common and often abundant in western Cascadia, occasional on the east side. Fruiting throughout the year on the coast, from late summer until the first freeze; the primary flush is during early fall everywhere.

Edibility: Some blackening *Russula* are poisonous; to be avoided.

Comments: This *Russula* is the most common member of the blackening *Russula* group in Cascadia. It is identified by its dry, often finely velvety, whitish cap that becomes grayish brown and then black in age, pale grayish yellow gills, and slow red staining when damaged. The red staining reaction usually takes five or more minutes, then fades to reddish brown, grayish, and then black after about twenty minutes. The acrid-tasting *R. acrifolia* sensu PNW (often called *R. densifolia*) has a thinly viscid or shiny whitish cap that becomes gray-brown, close to crowded, creamy to pale yellowish gray gills, and a staining reaction progressing from pinkish red to dark gray. *R. adusta* has a thinly viscid to shiny whitish cap that becomes brown to brownish gray and stains dingy grayish brown to dark gray (sometimes with a fleeting pinkish stage). *R. atrata* (and other members of the *R. albonigra* group) slowly stain black, with no red phase. The whole group is in need of modern taxonomic work to establish solid species concepts.

Lactarius rubidus

(Hesler & A.H. Sm.) Methven

Candy Cap

Cap: 1.5–8 (12) cm across, broadly convex to plane, disc often slightly sunken. Margin downcurved at first, becoming uplifted, wavy and ruffled in age. Surface dry to moist, wrinkled, slightly roughened (like an orange peel). Orange-brown, warm orange to pale buff-orange, occasionally reddish orange in age. **Gills:** Broadly attached to subdecurrent, close to subdistant, narrow. Buff to pale orange. **Latex** Scant or absent, watery white at first, slowly becoming clear. **Stipe:** 2–10 cm long, 0.3–1.5 cm thick, equal or tapered slightly toward base, slender, fragile. Surface smooth, dry; base covered with small whitish hairs. Pale orangish buff to pale orange, with a whitish sheen that is easily rubbed off. **Flesh:** Thin, brittle. Buff to pale orange; stipe fragile, hollow (sometimes stuffed when young). **Odor:** Distinctive, sweet and sharp, like fresh fenugreek; becoming much more pronounced, sweeter, like brown sugar or maple syrup when dried. **Taste:** Mild. **Spore deposit:** White to pale creamy. **Microscopy:** Spores 6–8 × 6–7.5 µm, globose to subglobose, ornamented with scattered warts and irregular ridges; forming a partial reticulum 0.5–1 µm high.

Ecology: Solitary or in small clusters and troops on ground, or on well-decayed stumps and logs in a wide range of forest types. More common in south-west Cascadia with tanoaks but can be found in lower elevation interior valleys with Douglas-fir across western Cascadia. Fruiting from late fall through winter.

Edibility: This sweet-flavored morsel of a mushroom is a real treat. The mapley aroma is at its best after slow (but thorough) drying. Powdered or crumbled; small amounts will flavor ice cream, cookies, cheesecake, and even vodka! Also used fresh in savory meat dishes and curries

Comments: The dry to moist (never viscid), warm orange, wrinkled-roughened cap (often with a ruffly margin), watery latex and a hollow stipe help identify this mushroom. Although the fenugreek odor when fresh is distinctive, it is different from the maple sugar odor that develops when the mushroom is dried. The dried mushrooms retain their odor for years; even traces amount in vacuum cleaners or car defrost vents. It has many orangish Milk Cap look-alikes, but all lack one or more of the features mentioned above. The poisonous *L. rufus* has a dull pinkish to reddish brown cap, creamy white latex, and a very acrid taste. The poisonous *L. xanthogalactus* has a pinkish to peachy buff cap and white latex that quickly turns yellow. Also compare with *L. substriatus*, *L. subviscidus*, and *L. luculentus*. From above, *Paralepista gilva/flaccida* looks very similar to the Candy Cap, but has fibrous flesh, no latex, and a slight odor of black pepper. Be especially cautious of collecting specimens growing on or near decayed stumps because the wood-dwelling Deadly Galerina (*G. marginata*) can have a vaguely similar warm color but has a smooth, hygrophanous, honey-brown cap, pale tan to brown gills, no latex, a partial veil that leaves a faint ring on the stipe, and brown spores.

Little Orange Milk Caps

Cascadia has several small, orange Milk Caps. The description below covers the features shared by members of this group, while the respective distinguishing features of the individual species are noted in the short accounts.

Cap: 2–8 cm across, convex to plane, slightly depressed at disc; margin down-curved when young, slightly uplifted in age. Gills: Broadly attached to subdecurrent, narrow, fragile, close to crowded; partial gills numerous. **Latex**: White to creamy, changing differs by species. **Stipe**: 2–10 cm long, 0.5–1 (1.5) cm thick, equal or enlarged toward base, firm to fragile, stuffed with tan pith when young, some becoming hollow. **Flesh**: Thin, fragile, often pale orange. Odor: Indistinct. **Microscopy**: Spores broadly ellipsoid to ellipsoid, ornamented with scattered warts and irregular ridges and lines, forming a partial reticulum 0.5–1 µm high. **Edibility**: Apparently non-toxic.

Lactarius luculentus var. *laetus*

Hesler & A.H. Sm.

Cap: Bright orange to orange-brown or slightly paler in age. Surface smooth, viscid to subviscid, greasy in dry conditions. **Gills:** Pale buff to orangish buff, with orangish brown spots or stains in age. **Latex** Copious, thick, white, unchanging, staining gills pale brownish orange**. Taste:** Mild to slightly bitter. **Spore deposit:** White. **Microscopy:** Spores 8–10 × 7–8.5 µm.

Ecology: Solitary or scattered on ground or in moss in conifer forests, typically under spruce. Common in Cascadia, mostly in coastal and west-side forests. Fruiting in fall and occasionally spring.

Comments: The orange to orange-brown cap, white unchanging latex, and mild to slightly bitter taste help identify this milk cap. *L. luculentus* var. *luculentus* is often slightly larger and has more orange-brown tones. It is found in western Cascadia with Douglas-fir and possible other conifers.

Lactarius substriatus

A.H. Sm.

Cap: Dark orange, orange-red to orange-brown, slightly paler toward margin and in age. Surface slightly viscid when wet, greasy when dry**. Latex:** Copious, thick, white at first, slowly discoloring creamy yellow, drying pale yellow, staining gills orangish**. Taste:** Slowly acrid, usually taking 10–15 seconds. **Spore deposit:** White. **Microscopy:** Spores 8–9.5 × 7–8 µm.

Ecology: Solitary or scattered in small groups on ground or in mossy areas in conifer forests. Occasional to locally common in western Cascadia. Fruiting from late fall into spring.

Comments: This species corresponds to our concept of *L. subviscidus* in MotRC. The orange opaque caps, pale buff gills, white latex that slowly dries yellowish, slowly-developing acrid taste help distinguish this species. The name *L. subflammeus* has been applied to a more reddish orange species with white latex that doesn't change to yellow. It is otherwise nearly identical and preliminary DNA work suggests that they may, in fact, be the same. *Lactarius sp.* 'Acrid Orange' is widespread pale orange to yellowish orange species with white latex and a slightly to strongly acrid taste.

Lactarius subviscidus

Hesler & A.H. Sm.

Cap: Reddish orange, orange to brownish orange, paler toward a faintly translucent-striate margin. Surface viscid to moist. **Latex:** Copious to scant, white, thin, whey-like, slowly and erratically turning pale yellowish or creamy. **Taste:** Slightly acrid. **Spore deposit:** White to creamy. **Microscopy:** Spores 7.5–9 × 5.5–7 µm.

Ecology: Solitary or scattered on ground or in moss in conifer forests, typically under spruce. Locally common but rarely abundant, occurring in western Cascadia. Fruiting in fall and early winter.

Comments: This species corresponds to our concept of *L. substriatus* in MotRC. The small fruitbodies, deep orangish red to orange-brown caps that are translucent-striate at the margin, white latex that erratically turns pale yellow (often overlooked), and slightly acrid taste help identify this species.

Lactarius alpinus

Peck

Cap: pale orangish when young, soon apricot orange. Surface dry, finely pubescent when young, finely scaly to smooth in age. **Gills**: Creamy to pale orange. **Latex**: Copious to scant; white, unchanging. Taste: Slowly acrid. **Spore deposit**: White. **Microscopy**: Spores 7–9 × 6–7 µm.

Ecology: Solitary or scattered on ground or in moss under alder. Uncommon in Cascadia, mostly in coastal and west-side forests. Fruiting in fall

Comments: The pale orange to apricot colors, dry cap and growth with alder help define this group. *L. cuspidoaurantiacus* tends to be slightly browner toned and has a mild taste. *Lactarius* sp. 'Acrid Orange' has a light orange to orange viscid cap and slight to strongly acrid taste.

Lactarius rufus

(Scop.) Fr.

Peppery Milk Cap

Cap: 3–10 (12) cm across, convex with an incurved margin when young, becoming broadly convex to plane or slightly uplifted, often with a depressed center, with or without an umbo. Surface dull, smooth to slightly wrinkled, dry to moist. Pinkish brown, dull brick-red to orange-brown. **Gills:** Broadly attached to subdecurrent, close to crowded, often forking near stipe, narrow, fragile. Pale orangish white, buff to pale orange, sometimes developing reddish spots or discoloration in age. **Latex:** Copious, white, usually rather thick, unchanging and not staining tissue. **Stipe:** 3–12 cm long, 0.5–2 cm thick, equal or nearly so, but often pinched or misshapen at base. Surface dry, often covered with a whitish sheen of fine powder. Reddish brown, orange-brown to reddish tan, often paler toward base. **Flesh:** Thin, fragile, off-white, tan to pale orange. Stipe stuffed at first, becoming hollow. **Odor:** Indistinct. **Taste:** Slowly very acrid, sometimes taking up to 30 seconds, can be mild to slightly bitter at first, but soon hot and peppery. **KOH:** Olive-green on cap and stipe. **Spore deposit:** White to pale creamy white. **Microscopy:** Spores 7–9.5 × 6–7 µm broadly ellipsoid, ornamented with scattered warts and irregular ridges and lines, forming a partial reticulum, up to 0.5 µm high.

Ecology: Solitary, scattered, or in small clusters; often forming large troops in duff and moss under pine, spruce, and possibly other conifers. Widespread and locally abundant across Cascadia. Fruiting from late summer though fall.

Edibility: Poisonous.

Comments: The dry, drab, brick-red to pinkish brown cap, pale buff to orangish white gills, unchanging white latex, and strongly acrid taste (sometimes slow to develop) help distinguish this species from many other (small), reddish orange Milk Caps. The smaller *L. rubidus* has a paler orange-brown cap with a rough surface, a mild taste, and a sweet (most pronounced after drying) odor. Other small reddish orange Milk Caps have moist to viscid caps.

Lactarius badiosanguineus

Kühner & Romagn.

Dark Alder Milk Cap

Cap: 2.5–6 cm across, broadly convex with a down-curved margin when young, becoming plane or uplifted, often with a depressed center and sometimes an umbo. Surface dry to slightly greasy, smooth to wrinkled or rough. Dark purple-red, liver-red to dark orange-red; margin slightly lighter in age; center remaining dark. **Gills:** Broadly attached to subdecurrent, narrow, close to crowded; partial gills present. Light pinkish buff to tan, becoming orange-buff, often with dark red-brown stains. **Latex:** Scant and white, like skim milk. **Stipe:** 4–9 cm long, 0.7–1.3 cm thick, enlarged slightly toward base, creased or pinched at times, brittle; base covered with whitish orange mycelial hairs. Surface smooth, dry to moist. Reddish orange, paler toward apex. **Odor:** Indistinct. **Taste:** Mild. **KOH:** Slight olive-orange on all parts. **Spore deposit:** Creamy white. **Microscopy:** Spores 8–10 × 6–8 µm, broadly ellipsoid to ellipsoid, ornamented with scattered warts and a broken or partial reticulum up to 1 µm high.

Ecology: Solitary or scattered on ground in wet areas or along streams near alder but possibly associating with conifers. Common at lower elevations across western Cascadia. Fruiting in fall and early winter.

Edibility: Unknown, probably non-toxic.

Comments: The dark liver red to dark orange-red cap, scant white latex, mild taste, and growth under alder help distinguish this species. Other Milk Caps with alder include: *L. occidentalis,* a small species with an olive-brown to orange-brown cap that is faintly striate in age, scant white latex, and a mild taste; and *L. cuspidoaurantiacus* (=*L. alpinus* var. *mitis*) with a brownish orange to brown, finely textured cap that fades to beige as it dries. *L. tabidus* has an orange-brown to reddish brown cap, and a slightly bitter-acrid taste. It appears to grow with alder, birch and a host of conifers.

Synonym: *Lactarius atrobadius* Hesler & A.H. Sm.

Lactarius occidentalis

A.H. Sm.

Cap: 1–3 cm across, convex to broadly convex when young, becoming plane and then centrally depressed. Margin even, inrolled when young, finely translucent-striate in age. Surface moist to dry, smooth, wrinkled or bumpy. Variable in color, olive to olive-brown when young, losing the olive tones and becoming orange-brown to cinnamon in age. **Gills:** Broadly attached to slightly decurrent, narrow, close, fragile. Warm tan to pinkish orange when young, darkening and developing orange-brown stains in age. **Latex**: Scant, white, becoming creamy to pale yellow upon drying. **Stipe:** 2–6 cm long, 0.4–0.7 cm thick, equal or enlarged at base. Surface dry, smooth, with a pale bloom over a pale olive-orange ground color. **Flesh:** Thin, fragile, orangish to tan; stipe solid. **Odor:** Indistinct. **Taste:** Mild. **KOH:** Unknown. **Spore deposit:** White. **Microscopy: Spores** 8–10 × 6.5–8 µm, ellipsoid, ornamented with scattered warts and a broken reticulum.

Ecology: Solitary, scattered, or in troops from soil and moss under alder. Especially common in red alder stands at lower elevation but also found at higher elevations in shrubby alder patches. Fruiting in summer and fall.

Edibility: Unknown, likely non-toxic.

Comments: The small size, olive tones when young, translucent-striate cap margin at maturity, white latex, mild taste, and growth with alder help identify this species. *L. tabidus* has an orange-brown to reddish brown cap and a slightly bitter-acrid taste. It has been reported as an alder associate in Cascadia but grows with a wide range of hardwoods and conifers in Europe. *L. cuspidoaurantiacus* (=*L. alpinus* var. *mitis*) is an uncommon but widespread species with alder, with a dry, light brown to orange-brown cap, scant watery white, unchanging latex and a mild taste. *L. subviscidus* is a conifer-associated species with a more reddish orange cap and an acrid taste.

Lactarius deliciosus group

Saffron Milk Caps

Cap: 3–10 cm across, convex with an inrolled margin when young, becoming plane with a depressed center; margin often wavy and uplifted in age. Surface smooth, moist to dry. Variable in color, often mottled, obscurely zonate or spotted, orange, brownish orange to orangish buff, often with patchy greenish turquoise discoloration to entirely turquoise-green in age or when very young. **Gills:** Broadly attached to subdecurrent, close to crowded, narrow, fragile. Bright orange to orangish buff, staining slowly turquoise-green when damaged or with a greenish cast, to extensively green in age. **Latex:** Scant, orange, slowly staining tissue turquoise-green. **Stipe:** 2–6 cm long, 1–3 cm thick, equal, slightly enlarged or tapered toward base, fragile, hollow in age. Surface smooth, dry to moist, not scrobiculate, or rarely so. Orange to orangish buff, with a whitish sheen or bloom when young, often paler at apex, developing greenish to turquoise-green stains, especially when damaged or in age. **Flesh:** Thin to thick, firm, fragile and grainy. Orange to buff, discoloring slowly green, paler orangish white in stipe center up into cap. **Odor:** Indistinct. **Taste:** Slowly slightly acrid. **KOH:** Unknown. **Spore deposit:** Creamy to pale yellow. **Microscopy:** Spores 7–9 × 5.5–6.5 µm, ellipsoid, ornamented with isolated warts and ridges, forming a partial reticulum up to 0.5 µm high.

Ecology: Scattered in troops or small clusters (rarely solitary) from soil and duff under lodgepole and shore pine. Similar species in this complex occur with different hosts (see comments). Fruiting in fall.

Edibility: Edible but most don't find it particularly delicious. The grainy texture is off-putting and it is somewhat lacking in flavor. Slow cooking helps a bit, and firm, younger fruitbodies are less grainy.

Comments: This common Milk Cap is distinguished from other species in this group by the growth with pines, orangish cap, scant, unchanging orange latex, and turquoise-green stains. *L. deliciosus* is a large species complex with species best recognized by tree associates, green staining (or lack thereof), and changes in latex color. The latex is often scant and orange to start, but some species turn red or vinaceous after exposure. *L. deliciosus* var. *piceus* (sensu PNW) grows with spruce, has an orange cap that readily discolors green, and orange latex that slowly turns vinaceous purple. *L. aurantiosordidus* also grows with spruce but has an orange-olive cap and dingy orange, unchanging latex. *L. aestivus* has a faintly zonate, orange cap that rarely discolors green, unchanging bright orange latex, and associates with true fir and hemlock *L. rubrilacteus* can look similar but maintains wine-red to brownish red latex from the start. Although all are edible, be sure not to confuse them with the poisonous *L. xanthogalactus* with a vaguely similar cap color (usually more pinkish) but white latex that quickly turns bright yellow.

Lactarius aestivus

Nuytinck & Ammirati

Orange Fir Milk Cap

Cap: 3–12 cm across, convex with an incurved margin when young, becoming plane with a depressed center or broadly vase-shaped in age. Surface moist to dry, smooth. Bright orange to orange or peachy orange, paler in age or when dry, often zonate with alternating light and dark rings; sometimes developing slight greenish stains or discolorations in age. **Gills:** Broadly attached to subdecurrent, close to crowded, broad; partial gills numerous. Bright orange to peachy orange or orangish buff. **Latex** Scant, bright neon-orange, not discoloring or slowly staining tissue pale greenish to greenish gray. **Stipe:** 3–6 cm long, 1–2.5 cm thick, more or less equal or tapered slightly toward the base. Marbled orange to pale orangish buff; often scrobiculate, with bright orange spots. **Flesh:** Thin to moderately thick, firm, fragile. Bright orange to peachy orange, sometimes slowly staining pale greenish to greenish gray when damaged. Stipe stuffed with pale orange pith, becoming hollow in age. **Odor:** Indistinct. **Taste:** Mild or slowly slightly acrid. **KOH:** Orange on stipe, little change elsewhere. **Spore deposit:** Creamy to pale yellow. **Microscopy:** Spores 8.5–10 × 6.5–8 µm, ellipsoid, ornamented with irregular ridges, forming a partial reticulum up to 0.5 µm high.

Ecology: Solitary or scattered in moss, duff, or soil under true fir and hemlock. Locally common but rarely abundant. Fruiting in summer and early fall in the mountains and into late fall coastally.

Edibility: The flesh is firmer and less grainy than the related species found under pine. Slow baking brings out a sweet flavor and firms up the texture.

Comments: The bright orange zonate cap, orange gills, scrobiculate stipe, neon-orange latex, sparse greenish gray staining, and growth with fir and hemlock help identify this species. An undescribed species in Cascadia matching most of these features has latex that turns red shortly after exposure to air. *L. 'deliciosus'* grows with two-needle pines and exhibits pronounced greenish staining. *L. xanthogalactus* has a pinkish brown cap and white latex that becomes bright yellow.

Lactarius aurantiosordidus

Nuytinck & S.L. Mill.

Sordid Orange Milk Cap

Cap: 2.5–5 (7) cm across, convex with an incurved margin when young, becoming plane with a depressed center or broadly vase-shaped. Surface moist to dry, smooth. Variable in color, orange to dingy orangish brown to olive-orange, often with dark turquoise to olive-green discoloration, especially at the center of the cap and in age, zonate at times, with darker concentric bands and/or mottled with darker areas. **Gills:** Broadly attached to subdecurrent, close to crowded, narrow, fragile. Orange, ocher to orangish buff, developing turquoise to greenish olive discolorations, especially when damaged or in age. **Latex:** Scant, dark dingy orange, not discoloring, slowly staining tissues turquoise-green. **Stipe:** 2–6 cm long, 0.7–2 cm thick, equal, enlarged slightly or tapering slightly at base, fragile, stuffed when young, becoming hollow in age. Orange to mottled orangish brown, discoloring turquoise-green to olive green. **Flesh:** Thin, firm, fragile, orangish to orange-buff or greenish, slowly discoloring turquoise-green. **Odor:** Indistinct. **Taste:** Mild or slowly slightly acrid. **KOH:** Dingy orange on stipe, no reaction elsewhere. **Spore deposit:** Creamy to pale yellow. **Microscopy:** Spores 8.5–10 × 6.5–7.5 µm, broadly ellipsoid to ellipsoid, ornamented with isolated warts and ridges, forming a partial reticulum.

Ecology: Solitary, scattered, or in groups on ground in moss or duff under Sitka spruce. Locally common but limited to coastal spruce forest. Fruiting abundantly in fall and early winter.

Edibility: Edible.

Comments: This species is recognized by its small to medium size, dingy orange-green stained cap, unchanging orange latex, and growth with Sitka spruce. It often grows alongside the similar-looking *L. deliciosus* var. *piceus* which is usually larger, with brighter orange colors, and bright orange latex that slowly turns deep red to vinaceous purple.

Lactarius rubrilacteus

Hesler & A.H. Sm.

Bleeding Milk Cap

Cap: 4–12 (20) cm across, convex with an inrolled margin when young, becoming broadly convex to plane with a depressed center, margin even or uplifted in age. Surface smooth, slightly viscid to dry. Variable in color, dingy salmon, pinkish orange, orangish buff to buff-brown, often mottled, zonate to obscurely zonate or spotted, staining or discoloring greenish turquoise when very young or in age. **Gills:** Broadly attached to subdecurrent, close to crowded, narrow, fragile; often forking near stipe. Pinkish orange, orangish buff to buff-brown, often with greenish stains, staining reddish brown and then slowly turquoise-green when damaged. **Latex:** Scant, vinaceous red to reddish brown, discoloring and staining tissue reddish brown and then slowly turquoise-green, orange in stipe base only. **Stipe:** 2–8 cm long, 1–3 cm thick, equal or slightly tapered toward base. Surface dry to moist, smooth, occasionally scrobiculate, often with orangish hairs at base. Pinkish orange to orange to buff, often with a pale bloom when young, developing greenish stains in age. **Flesh:** Thin to thick, firm and grainy. Orangish buff to reddish, discoloring reddish brown and then slowly greenish, whitish to pale buff in stipe center up into cap, stipe becoming hollow in age. **Odor:** Indistinct. **Taste:** Mild. **Spore deposit:** Pale yellow. **Microscopy:** Spores 7.5–9 × 6–7 µm, broadly ellipsoid to ellipsoid, ornamented with isolated warts and short ridges forming a partial reticulum.

Ecology: Solitary, scattered, or in small clusters on ground or in duff under Douglas-fir. Common throughout Cascadia. The primary flush is in late fall but occurs in summer at higher elevations and persists later into early winter at lower elevations.

Edibility: The young fruitbodies are good edibles but mature specimens have a disagreeable grainy texture. Slow baking or roasting helps keep the flesh texture palatable.

Comments: This common component of Douglas-fir forests is identified reliably by its dingy salmon to pinkish orange cap, greenish staining, and vinaceous red to reddish brown latex. Some members of the *L. deliciosus* group have orange latex that slowly turns to the same color as the fresh latex of *L. rubrilacteus* but are associated with different conifer species. The poisonous *L. xanthogalactus* has white latex that quickly changes to bright yellow. *L. barrowsii* is similar but has a more pallid cap and gills, and reddish latex, and is a southwestern species associated with ponderosa pine, but may occur in eastern Washington.

Lactarius xanthogalactus

Peck

Yellow-staining Milk Cap

Cap: 3–10 (12) cm across, convex with a down-curved margin when young, becoming broadly convex to plane, or slightly uplifted in age, often with a depressed center. Surface smooth, slightly viscid to dry. Variable in color, pinkish orange, peachy orange, orange-red to orangish brown, often clearly to obscurely zonate, mottled or water-spotted, discoloring orange-brown in age. **Gills:** Broadly attached to slightly decurrent, close to crowded, narrow, fragile. Creamy, pinkish yellow to orangish buff, occasionally with orangish brown spots or stains in age. **Latex:** Copious to scant, creamy white at first, quickly or slowly becoming neon-yellow to pale yellow. **Stipe:** 2–7 (10) cm long, 0.6–2.5 cm thick, equal, slightly tapered or enlarged toward base. Surface dry, smooth, often with whitish hairs at base. Creamy tan to pale orange, darkening in age. **Flesh:** Thin to thick, firm, stipe stuffed when young, hollow in age. Creamy to creamy orange, discoloring yellow from latex. **Odor:** Indistinct. **Taste:** Slightly bitter-acrid. **Spore deposit:** Pale yellow. **Microscopy:** Spores 7–9 × 5.5–7 µm, broadly ellipsoid to ellipsoid, ornamented with scattered warts and irregular ridges and lines, forming a partial reticulum up to 0.5 µm high.

Ecology: Solitary, scattered, or in small clusters on ground in moss or duff in mixed forests. With a range of hardwoods and conifers, especially Douglas-fir and Oregon white oak. Common in southern Cascadia and rare north of central Oregon. Fruiting in fall.

Edibility: Poisonous.

Comments: The pinkish orange, zonate and often spotted cap, creamy to pale orange gills, and white latex that turns bright yellow are a distinctive set of features. The color change of the latex usually takes 5–10 seconds but varies depending on the age and dryness of the fruitbody. *L. rubrilacteus* can look similar but has vinaceous red to reddish brown latex and develops greenish stains. Members of the *L. deliciosus* group have scant orange latex when first damaged and stain greenish. *L. alnicola* has a creamy to yellowish cap, a strongly acrid taste, and white latex that sometimes changes to pale yellow.

Lactarius olympianus

Hesler & A.H. Sm.

Olympic Milk Cap

Cap: 5–10 (15) cm across, broadly convex and centrally depressed when young, to broadly funnel-shaped with an uplifted margin in age. Surface smooth, thinly viscid to moist. Orange, often with conspicuous concentric zonations, occasionally more evenly colored. **Gills:** Broadly attached to subdecurrent, somewhat broad, fragile, close to subdistant. Creamy to yellowish, staining orangish brown. **Latex** Copious, thick, white, unchanging, staining tissue orange-brown. **Stipe:** 2–6 cm long, 1–3 cm thick, stout, equal or tapering toward base. Surface moist to dry, Creamy white with a whitish bloom when young, to more orangish in age. **Flesh:** Firm, but brittle, white; stipe hollow. **Odor:** Indistinct. **Taste:** Acrid. **Spore deposit:** White to creamy white. **Microscopy:** Spores 8–11.5 × 7.5–9.5 µm, ellipsoid, ornamented with isolated warts and ridges, forming a partial to nearly complete reticulum, 0.5–1 µm high.

Ecology: Scattered in troops from duff and soil. Occasional but more often in the eastern Cascades with Engelmann spruce and west-side with true fir. Fruiting in late summer and fall.

Edibility: Very acrid, possibly poisonous.

Comments: The smooth, zonate orange cap, creamy white gills, white latex, and acrid taste are characteristic. *L. alnicola* has a finely pubescent margin when young and a creamy yellow to yellow-brown cap. It grows with montane conifers, especially in the eastern portion of Cascadia. *L.*"*alnicola*" sensu CA has a creamy to yellowish, often zonate, cap and abundant white latex (across the southern portions of Cascadia with oak). *L. subvillosus* has a zonate, orangish buff to peachy orange cap that a bearded margin (when young) and pale pinkish buff gills; uncommon in Cascadia, more likely in west-side forests with Douglas-fir from central Oregon south. *L. affinis* has a non-zonate buff to dull orange cap and grows with spruce (and possibly other conifers in northeastern Cascadia.

Taxonomic note: Based on ITS sequence, the European *L. zonarioides* is closely related and may be conspecific. If they are synonymized, *L. zonarioides* would take precedent.

Lactarius pubescens var. *betulae*

(A.H. Sm.) Hesler & A.H. Sm.

Fuzzy Birch Milk Cap

Cap: 3–10 (13) cm across, convex, broadly convex to plane, disc centrally depressed. Margin inrolled tight to the stipe when young, staying inrolled until cap fully expands, sometimes becoming ruffled or wavy in age. Surface slightly viscid to dry, smooth to slightly hairy on disc, coarsely bearded at margin until expanded, then more matted-tomentose to nearly smooth in age. Pale pinkish buff, pinkish cinnamon to tan on disc, lighter toward margin, sometimes with inconspicuous concentric zonations. **Gills:** Broadly attached to subdecurrent, narrow, fragile, close to crowded. Creamy white, pale pinkish to pale pinkish orange. **Latex:** White, often scant, sometimes drying yellowish. **Stipe:** 2–7 cm long, 1–2.5 cm thick, stout, equal. Surface smooth, moist to dry. Creamy white, pinkish to pinkish yellow, often pinker at apex. **Flesh:** Firm, fragile, moderately thin. Pale pinkish white to creamy, stipe stuffed when young, soon hollow. **Odor:** Indistinct. **Taste:** Strongly acrid, usually taking 15–30 seconds to develop. **Spore deposit:** Creamy yellow. **Microscopy:** Spores 6–7.5 × 4.5–6.0 µm, broadly ellipsoid to ellipsoid, ornamented with scattered warts and irregular ridges and lines, forming a partial reticulum, 0.5–1 µm high.

Ecology: Scattered in groups or small clusters under introduced birch trees. Locally common in urban areas with planted birch and abundant northward in native birch forests. Fruiting from late summer through fall but can appear in spring as well.

Edibility: Very spicy, possibly poisonous.

Comments: The pale pinkish white cap with a coarsely bearded margin, strongly acrid taste, and growth under birch help identify this species. *L. torminosus* is another bearded Milk Cap occurring with birch but has darker pinkish colors. The name *L. torminosus* var. *nordmanensis* has been applied to a species occurring with willow, aspen, or cottonwood in southern Cascadia and California. It has a slightly more pinkish-orange cap and larger spores (7.5–9.5 × 6–7.5 µm). *L. subvillosus* has a peachy orange cap, scrobiculate stipe, and a finely bearded margin. *L. resimus* has a coarsely bearded, white to pale yellow cap, white latex that turns yellow, and a mild to slightly peppery taste. Members of the *L. scrobiculatus* group have yellowish, centrally depressed to funnel-shaped caps, strongly scrobiculate stipes, white to pale yellow latex, and a strongly acrid taste.

Lactarius scrobiculatus group

Potholed Milk Caps

Cap: 6–15 (20) cm across, broadly convex and centrally depressed when young, becoming plane to funnel-shaped. Margin tightly inrolled at first, expanding and uplifting in age. Surface viscid when wet, matted-tomentose to appressed-fibrillose, giving it a roughened feel; margin finely bearded and often covered with water droplets when young. Yellowish, yellowish orange to brownish orange, sometimes with inconspicuous concentric zones. **Gills**: Attached, subdecurrent to decurrent, close, narrow, fragile. Pale yellow to pale yellowish orange, discoloring dingy orangish brown. **Latex**: Creamy white, drying yellowish. Stipe: 3–8 cm long, 1–4 cm thick, stout, equal. Surface rough, pitted, moist to dry. Pale yellowish to yellowish orange, heavily scrobiculate, with darker orangish spots and pock-marks. **Flesh**: Firm, off-white to pale yellowish; stipe hollow, usually with orangish discolorations. Odor: Indistinct. **Taste**: Quickly and strongly acrid. **KOH**: No reaction. **Spore deposit**: Creamy white. **Microscopy**: Spores 7–9.5 × 5.5–7.5 µm, broadly ellipsoid, ornamented with scattered amyloid warts and irregular ridges and lines forming a partial reticulum up to 0.5 µm high.

Ecology: Scattered in groups or loose clusters on ground under conifers; especially spruce. Common and widespread in Cascadia. Fruiting in fall.

Edibility: Not edible, possibly poisonous.

Comments: *L.* "*scrobiculatus*" appears to be a species complex in western North America and there much uncertainty regarding which names to use. The description above pertains to the common form of *L* "*scrobiculatus*" found in Cascadia. It is told apart from similar Milk Caps by its yellowish, funnel-shaped cap, coarsely scrobiculate stipe, white latex, and strongly acrid taste. There is a slightly smaller and paler-capped species with white latex that quickly changes to yellow with a mild to slightly acrid taste that is found with fir and hemlock. *L. alnicola* has a finely pubescent to nearly smooth margin and a creamy yellow to yellow-brown cap. It occurs with spruce (and possibly other conifers) in the eastern Cascades. *L.* "*alnicola*" sensu CA has a creamy to yellowish, often zonate cap and abundant white latex. It is found across the southern portions of Cascadia with oaks. *L. resimus* has a white to pale yellow cap with a finely bearded margin, white latex that turns yellow, and a mild to slightly peppery taste. *L. repraesentaneus* can look almost identical to *L. scrobiculatus* but has latex that slowly stains lilac-purple.

Lactarius controversus

Pers.

Poplar Milk Cap

Cap: 6–15 (20) cm across, broadly convex, centrally depressed, becoming broadly funnel-shaped in age; margin inrolled when young, becoming uplifted in age. Surface slightly viscid to dry, margin finely pubescent when young, matted-fibrillose toward disc, becoming smooth or nearly so with age. White, dingy white to pale pinkish tan, often with vinaceous to lavender-brown stains. **Gills:** Broadly attached to subdecurrent, narrow, fragile, crowded; partial gills numerous. Pale pinkish, creamy pink to pale pinkish buff, slightly darker in age. **Latex:** Scant to copious, white, unchanging and not staining, or slightly yellowing. **Stipe:** 2.5–7 cm long, 1.5–3 cm thick, equal or tapering slightly toward base, fragile, hollow. Surface smooth, moist to dry. Dingy white to pale pinkish white. **Flesh:** Firm, whitish, not staining. **Odor:** Indistinct. **Taste:** Slowly but strongly acrid. **Spore deposit:** Pale yellow. **Microscopy:** Spores 6–7.5 × 4.5–5.5 µm, ellipsoid, ornamented with isolated warts and ridges, forming a partial reticulum up to 0.5 µm high.

Ecology: Scattered or in small clusters on ground, often in riparian zones under aspen or cottonwood (*Populus* spp.). Occasional across Cascadia. Fruiting in fall.

Edibility: Not edible, possibly poisonous.

Comments: The whitish cap, pinkish gills, white latex, slowly acrid taste, and growth under cottonwoods help identify this milk cap. *L. resimus* has a white to pale yellow cap with a bearded margin, creamy white gills that can become pinkish, white latex that quickly turns yellow, and a mild to slightly peppery taste. *L. pseudodeceptivus* has a white to pinkish-buff cap with a cottony margin and creamy gills; it grows with conifers in eastern Cascadia. *L. scrobiculatus* has a yellowish cap and heavily scrobiculate stipe and is found with conifers.

Lactarius repraesentaneus

Britzelm.

Purple-staining Bearded Milk Cap

Cap: 5–15 (20) cm across, broadly convex and centrally deeply depressed when young, becoming broadly funnel shaped. Surface dry to slightly viscid, tomentose to fibrillose; margin coarsely bearded when young, becoming matted-fibrillose; disk nearly smooth. Creamy to yellowish, with slightly darker, yellow-ocher fibrils when young, slightly dingier in age. **Gills:** Broadly attached to decurrent, narrow, fragile, close to crowded, often forking near stipe. Creamy to pale ocher, staining lilac to purple when damaged. **Latex:** Copious, white at first, staining the tissue lilac to purple after ~5 minutes. **Stipe:** 4–7 (10) cm long, 1–3.5 cm thick, stout, equal or tapering toward the base. Surface slightly viscid, Creamy to yellowish, scrobiculate, with yellow-orange spots and pockmarks. Staining purplish when damaged. **Flesh:** Firm, thick, off-white, stipe hollow. Off-white to creamy, slowly discoloring lilac to purple. **Odor:** Indistinct. **Taste:** Mild to slowly slightly bitter or acrid. **Spore deposit:** Pale yellowish. **Microscopy:** Spores 8.5–10.5 × 6–8.5 µm, broadly ellipsoid, ornamented with scattered warts and irregular ridges and lines, forming a partial reticulum, up to 1 µm high.

Ecology: Solitary, scattered, or in small groups on ground under spruce and high elevation firs. Rare in Cascadia, found mostly in high elevation east-side forests in the Cascade Range, but also on the coast with Sitka spruce. Fruiting in late summer and fall.

Edibility: Unknown.

Comments: The yellowish cap with a bearded margin, white latex that stains the tissues lilac to purple, and the scrobiculate stipe help distinguish this species. *L. resimus* has a white to creamy yellow cap with a bearded margin and white latex that quickly turns yellow; it is locally common across western Cascadia. Members of the *L. scrobiculatus* group have finely bearded cap margins, and white latex that turns yellow. The birch-associate *L. pubescens* var. *betulae* has a pinkish white cap with a bearded margin, strongly acrid taste, and white unchanging latex.

Lactarius pallescens

Hesler & A.H. Sm.

Pallid Purple-staining Milk Cap

Cap: 3–10 cm across, convex with an incurved margin when young, becoming broadly convex to plane, often with a depressed center. Surface smooth, viscid, slimy when wet, moist to shiny when dry. Bright white when young, becoming dingy white to pale grayish white or pale gray in age, occasionally with faint concentric zones or "water spots". Staining lilac-purple when damaged and often with orange to orange-brown stains in age. **Gills:** Broadly attached to subdecurrent, close to subdistant, narrow, fragile; partial gills numerous. White to off-white, becoming creamy white, often with orangish spots in age, slowly staining lilac-purple when damaged and then slowly turning purplish brown to brownish orange. **Latex:** Copious, bright white, staining the tissues lilac-purple. **Stipe:** 3–10 cm long, 0.8–2 cm thick, equal or slightly enlarged toward base. Surface smooth, viscid or slimy. White to off-white, often developing orange stains in age, slowly staining lilac-purple when handled or bruised. **Flesh:** Thick to thin, firm, fragile, white to dingy white; stipe stuffed when young, sometimes becoming hollow in age; all parts slowly staining lilac-purple. **Odor:** Indistinct. **Taste:** Mild to slowly, slightly acrid. **KOH:** Pale yellow on cap. **Spore deposit:** White to creamy. **Microscopy:** Spores 9–10.5 × 7–9 µm, ellipsoid, ornamented with scattered warts and irregular ridges and lines, forming a partial reticulum 0.5–1 µm high.

Ecology: Solitary or scattered in moss or duff under conifers, especially true fir and hemlock. Occasional to locally common across Cascadia. Fruiting in fall.

Edibility: Unknown, but probably poisonous.

Comments: The combination of a viscid white cap and stipe, pale gills, slow lilac-purple staining of all tissues when damaged, and growth with conifers is diagnostic for this species. Similar violet-staining Milk Caps in Cascadia include: *L. californiensis*, with a darker gray to grayish white cap and often with orangish stains on the stipe—it grows with tanoak and possibly conifers in the southwestern portion of Cascadia. *L. montanus* has a vinaceous gray cap that turns green in KOH and whitish gills that stain purple; it is more common at higher elevations in Cascadia, often with cottonwood and aspen. *L. cordovaensis* (=*L. cascadensis*) is a large mushroom with a gray-brown, zonate cap, tan to pale ocher gills, and white latex which stains the tissues pinkish violet; although reported from alder habitats, it is likely a conifer associate. There is also a pale yellow-capped species with creamy gills and stipe in Cascadia in the *L. aspideoides* group.

Lactarius kauffmanii

Hesler & A.H. Sm.

Cap: 4–15 cm across, convex to plane, occasionally with a slightly depressed center; margin inrolled when young. Surface smooth, viscid when wet, becoming shiny and often with stuck-on debris when dry. Dark grayish brown to grayish tan, often fading and becoming more brownish as it ages. **Gills:** Broadly attached to subdecurrent, narrow when young, becoming broad in age, fragile, close to crowded, often forking near stipe. Creamy white to pale ocher-white, developing dingy tan, orangish or ocher discoloration as it ages. **Latex:** Copious, white to dingy white, discoloring grayish as it dries, staining tissues orangish tan. **Stipe:** 4–12 cm long, 1–3 cm thick, equal, tapering toward base or with a swollen middle. Surface smooth, moist to tacky but not viscid. Dingy white to creamy white, becoming orangish tan and discoloring dingy tan, orangish or ocher in age. **Flesh:** Firm, brittle, creamy white, slowly staining orange to brownish orange; stipe hollow in age. **Odor:** Indistinct. **Taste:** Slowly acrid. **KOH:** Bleaching cap color to pale yellowish. **Spore deposit:** White to creamy. **Microscopy:** Spores 8–9.5 × 6–7.5 µm, broadly ellipsoid to ellipsoid, ornamented with irregular ridges and lines forming a partial reticulum, 0.5–1 µm high.

Ecology: Solitary, scattered, or in troops in soil, moss, or duff under hemlock and true fir. Common and at times abundant across Cascadia. Fruiting from summer at higher elevations into late fall or early winter on the coast.

Edibility: Unknown.

Comments: The viscid, grayish brown cap, creamy white gills, and stipe that becomes stained dingy ocher-tan in age, and growth under hemlock and fir are the best identifying features for this species. *L. glutigriseus* is generally smaller and has a viscid, dark to pale gray or grayish tan cap and stipe, and white gills. *L. caespitosus* has a paler grayish brown to brown cap, pallid gills, a white stipe, white latex that slowly turns yellowish, and a strongly acrid taste. It often fruits in clusters under aspen. There is a similar, apparently unnamed summer or early fall fruiting species in the mountains with a viscid gray cap and often scrobiculate stipe. *L. trivialis* has a paler grayish, buff or tan, obscurely zonate, thinly viscid cap, but is rare in Cascadia and more likely to be found to the north, growing with spruce. *L. argillaceifolius* var. *megacarpus* is often very large (caps to 30 cm), with a violet-gray to dingy gray cap, and grows with oak.

Lactarius glutigriseus

V.L. Wells & Kempton

Slimy Gray Milk Cap

Cap: 2–8 (10) cm across, convex with an incurved margin when young, becoming broadly convex to plane, often with a depressed center and an uplifted margin in age. Surface smooth, viscid when wet, moist to shiny when dry. Dark gray to dark grayish brown, often paler toward margin in age. **Gills:** Broadly attached to subdecurrent, close to crowded, narrow, fragile. White, becoming creamy white to pale yellowish buff, staining brownish when damaged. **Latex:** Copious to scant, bright white to watery white, drying pale yellow and staining gills brownish. **Stipe:** 3–10 cm long, 0.5–1.5 cm thick, equal or enlarged slightly toward base, at times with a crease or fold. Surface smooth, viscid or tacky. Dark gray to brownish gray, often concolorous with or slightly lighter than cap, paler toward base and often off-white at extreme base, usually with a distinct pale zone at apex. **Flesh:** Thin, fragile, off-white to gray, stipe stuffed when young, becoming hollow in age. **Odor:** Indistinct. **Taste:** Acrid. **Spore deposit:** Creamy white to pale yellow. **Microscopy:** Spores 7–9 × 6–7 µm, broadly ellipsoid to ellipsoid, ornamented with scattered warts and irregular ridges and lines forming a partial to nearly complete reticulum, 0.5–1 µm high.

Ecology: Solitary, scattered, or in large troops in moss or duff of conifer forests where it associates with hemlock and fir. Common and abundant across western Cascadia. Fruiting in summer in the mountains and through fall into early winter at lower elevations.

Edibility: Unknown.

Comments: The viscid, dark gray to light gray cap, white to creamy white gills, slimy gray stipe, white latex, acrid taste, and growth under hemlock and fir help distinguish this species from other gray-capped Milk Caps. Older specimens can look like *L. kauffmanii,* which is often found growing in the same habitat. However, *L. kauffmanii* has a dry to moist (never viscid), creamy white to pale buff stipe that is concolorous with the gills. *L. californiensis* has a viscid gray cap and white latex that stains its tissues lilac-purple.

Synonym: *Lactarius pseudomucidus* Hesler & A.H. Sm.

Lactarius glyciosmus

(Fr.) Fr.

Coconut Milk Cap

Cap: 2–8 cm across, convex to plane, centrally depressed; margin inrolled when young, even, wavy to uplifted in age. Surface dry, often with a pallid bloom when young, becoming smooth or with scattered scales in age. Gray to lilac-gray, developing some brownish tones and often fading with age. Occasionally with pale zonations. **Gills:** Broadly attached to decurrent, close to crowded, narrow, fragile. Creamy to pale pinkish buff or yellowish buff. **Latex:** Scant, white, unchanging. **Stipe:** 2–6 (10) cm long, 1–1.5 cm thick, equal, enlarged or tapering slightly toward base. Surface dry, finely pruinose to smooth. Pale gray to whitish. **Flesh:** Thin, firm, fragile, stipe stuffed when young, hollow in age. Off-white to creamy. **Odor:** Sweet, like coconut. **Taste:** Indistinct. **Spore deposit:** Creamy. **Microscopy:** Spores 7–9 × 6.5–7 µm, broadly ellipsoid, ornamented with scattered warts and irregular ridges and lines, forming a partial or nearly complete reticulum up to 0.5 µm high.

Ecology: Scattered or in small clusters, often in groups from soil, moss, or grassy lawns under birch. Occasional with planted birch trees in urban areas. Common northwards in native birch forests. Fruiting in summer and fall.

Edibility: Reportedly edible, we have no firsthand experience.

Comments: The dry gray cap, creamy to yellowish buff gills, scant white latex, mild taste, growth with birch, and distinctive coconut odor readily identify this species. *L. vietus* is a similar small, gray, northern birch associate that differs in having white latex that dries to olive-gray, no odor, and an acrid taste. *L. flexuosus (=L. circellatus* var. *borealis)* is larger, has a distinctly zonate cap, and an acrid taste; it appears to prefer cottonwood and willow habitat in Cascadia. *L. montanus* is larger and has white latex which stains the tissues violet.

Lactarius fallax

A.H. Sm. & Hesler

Velvety Milk Cap

Cap: 2.5–6 cm across, convex to plane, at times with a depressed center and/or a sharp umbo. Margin incurved when young, often finely ribbed, becoming ruffled, wavy, or uplifted in age. Surface velvety, dry to slightly moist, finely to coarsely wrinkled or roughened. Dark brown to brown, or deep reddish brown. **Gills:** Broadly attached to subdecurrent, close to crowded, narrow, fragile. White, creamy, to creamy yellow or pale buff-yellow, at times with a brown marginate edge, discoloring pale orange-brown when damaged. **Latex:** Copious to scant, white, slowly staining tissues orange-brown. **Stipe:** 3–10 cm long, 0.5–1.5 cm thick, equal or enlarged slightly and often curved toward base, fragile. Dark brown to brown or grayish brown, paler toward base where it can be off-white to pale buff. **Flesh:** Thin, fragile, off-white to buff-brown, stipe stuffed with buff pith, can become hollow in age. **Odor:** Indistinct. **Taste:** Mild. **KOH:** No reaction. **Spore deposit:** Creamy white to pale yellowish. **Microscopy: Spores** 7–9 × 7–9 µm, subglobose to globose, ornamented with irregular ridges and lines, forming a partial reticulum 1–2 µm high.

Ecology: Solitary, in fused pairs, or scattered on ground or well-rotted logs and stumps in mossy conifer forests where it likely associates with spruce, hemlock, and true fir. Common but rarely abundant; occurring across Cascadia, from the coast to high elevations. Fruiting in late summer and fall.

Edibility: Edible and good, but often thin-fleshed and rarely found in large numbers.

Comments: The dark brown, velvety-wrinkled cap and the creamy white gills that contrast strongly with the brown stipe stand out for this beautiful species. The prounced spore ornamentation of this species is spectacular. ITS sequence data shows two species in Cascadia; at this time no distinguishing morphological features have been established. *L. pallidiolivaceus* is typically stockier, with an olive-gray, grayish to beige, often blotchy, finely velvety cap, and creamy latex staining the tissues pale orange. It is rare in Cascadia, more likely to be found in the southern portion of our range.

Lactarius olivaceoumbrinus

Hesler & A.H. Sm.

Toadskin Milk Cap

Cap: 3–10 (12) cm across, convex with an inrolled margin when young, becoming plane with a depressed center; margin lobed or wavy in age. Surface slightly scaly-fibrillose, matted into a slimy layer. Variable in color, often mottled or spotted, dingy dark olive to deep olive-green when young, becoming olive-brown and paler, olive-buff toward margin and in age. **Gills:** Broadly attached to subdecurrent, close to crowded, often forking near stipe, narrow and fragile. Dingy cream to pale gray, spotted and stained olive, becoming dirty olive-brown in age. **Latex:** Copious, thick, white, staining gills olive-brown. **Stipe:** 4–8 (10) cm long, 1–3 cm thick, equal, enlarged or tapered toward base. Surface slightly viscid to moist, fibrous, pockmarked to smooth. Olive to olive-brown, scrobiculate, with dark spots, often colored similar to cap. **Flesh:** Thin to thick, firm, fragile, stipe stuffed when young, becoming hollow. Ivory to pale tan or grayish tan, unchanging or slowly staining olive-tan. **Odor** Indistinct. **Taste:** Strongly acrid. **KOH:** Magenta to red on cap. **Spore deposit:** Creamy to pale yellow. **Microscopy:** Spores 7–9 × 6–8 µm, broadly ellipsoid to ellipsoid, ornamented with scattered warts and irregular ridges and lines, forming a partial reticulum 0.5–1.5 µm high.

Ecology: Solitary or scattered on ground, in duff or moss under conifers, especially Sitka spruce. Uncommon, mostly in coastal forests and west side foothills. Fruiting in summer and fall.

Edibility: Not edible, possibly poisonous.

Comments: The green to olive-brown colors, pitted and pockmarked stipe, pallid gills, and growth under Sitka spruce help identify this mushroom. Young specimens of this milk cap are rather beautiful but the forest green colors quickly fade with mature specimens looking rather dingy. The similar *L. necator* is generally paler and stouter and only grows under birch in urban areas *L. sordidus* is slightly smaller, has a paler olive cap, and whitish gills. It is more common in conifer forests in Alaska (but may range into northern Cascadia). *L. fallax* has a deep dark brown velvety cap, brown stipe, white to buff gills, and a mild taste.

Lactarius necator

(Bull.) Pers.

Green Birch Milk Cap

Cap: 4–10 (15) cm across, convex with an inrolled margin when young, often staying incurved well into maturity, becoming plane with a depressed center or slightly uplifted in age. Surface slightly viscid to dry, wrinkled to pitted at times, scaly-fibrillose to radially fibrillose; margin tomentose when young. Dark dingy olive-green, olive-brown to olive-orange, often mottled, spotted or streaked. **Gills:** Broadly attached to subdecurrent, close to crowded, often forking near stipe. Creamy white to pale orangish when young, becoming dingy, dirty cream to greenish gray, developing olive-brown spots and stains in age. **Latex** Copious, thick, white, drying pale greenish, staining gills olive-brown. **Stipe:** 2–6 cm long, 1–4 cm thick, equal or tapered and pinched toward base. Olive, olive-brown to dingy olive-yellow, often paler toward apex, sometimes scrobiculate, with darker spots. Surface smooth, subviscid to dry. **Flesh:** Thin to thick, firm, fragile. Whitish, ivory-gray to pale dingy yellow, slowly darkening olive-tan; stipe stuffed with whitish pith when young, becoming hollow and dingy olive-brown. **Odor:** Indistinct. **Taste:** Slowly acrid. **KOH:** Violet-purple on cap. **Spore deposit:** Creamy to pale yellow. **Microscopy:** Spores 7–9 × 4.5–6.0 µm, broadly ellipsoid to ellipsoid, ornamented with scattered warts and irregular ridges and lines, forming a partial reticulum, 0.5 µm high.

Ecology: Solitary, scattered, in troops or clusters on ground under birch. Common and often abundant in Cascadia, especially with planted birches in urban settings. Fruiting from late summer through fall.

Edibility: Not edible, likely toxic.

Comments: The dark, dingy olive-green to olive-brown cap and stipe, stout stature, and growth under birch help identify this Milk Cap. *L. olivaceoumbrinus* looks quite similar but typically has darker colors when young, an extensively scrobiculate stipe, and is found with Sitka spruce. The northern *L. sordidus* is often smaller, has a paler olive cap, and whitish gills and grows in conifer forests.

Synonyms: *Lactarius turpis* (Weinm.) Fr., *Lactarius plumbeus* (Bull.) Gray

13 - Waxy Caps

Waxy caps belong to the Family Hygrophoraceae, which includes a variety of mushroom-forming fungi, and even a few basidiolichens, including the leafy *Dictyonema*!

There are two major groups in this section. The first are medium-sized to large, ectomycorrhizal *Hygrophorus*, which resemble *Tricholoma* in stature but have broadly attached to decurrent, thick, waxy or greasy gills. The second are smaller, slender, often brightly colored *Hygrocybe*-like taxa. Some familiar waxy caps from this latter group are now placed in smaller genera, including *Gliophorus*, *Cuphophyllus* and *Humidicutis.* Microscopic features are fairly plain: all have smooth, inamyloid, often ellipsoid and pigmentless spores, and many have long basidia.

Hygrocybe are mostly brightly colored with lots of red, orange, and yellow species. In addition, they often have dry to thinly viscid caps and dry to slightly viscid stipes. *Gliophorus,* on the other hand, have extremely viscid caps and stipes which often makes them tough to handle because they are so slippery. *Cuphophyllus* are recognized by their decurrent, widely spaced gills and frequently drab colors. These genera are most common in forests with western redcedar and are especially common in coast redwood forests in California (and are more extensively covered in *Mushrooms of the Redwood Coast*). The ecological roles of these genera are unclear. Research by Seitzman et al. (2010) suggests many Hygrophoraceae formerly thought to be saprotrophic are likely in biotrophic relationships with understory plants and/or bryophytes. In addition to the genera mentioned above, the wood-dwelling omphalinoid mushrooms *Chrysomphalina* and *Chromosera* are actually closely related to *Gliophorus.* The bryophyte-associated *Arrhenia* and the basido-lichen *Lichenomphalia* are also in *Hygrophoraceae* but are treated with the multitudes in the White-spored chapter.

Important identification characters to note for all waxy caps include the colors of all surfaces (subtle differences in color can be very important, as well as young versus mature gills), stature, and texture of the cap and stipe (especially degree of viscidity). For the mycorrhizal *Hygrophorus*, be sure to note the identity of nearby trees. Also pay attention to fruiting times, as some species are spring-fruiters, while others fruit in fall.

Hygrophorus camarophyllus

(Alb. & Schwein.) Dumée

Cap: 3–10 cm across, rounded to convex with an inrolled margin when young, often broadly umbonate, becoming broadly convex to plane or with uplifted margin. Surface appressed-fibrillose, dry or slightly tacky if wet. Dark gray to dark grayish brown, usually with dark brown or blackish radial streaking. **Gills:** Broadly attached to decurrent, well spaced, thick, broad, greasy. Contrastingly white at first, becoming slightly creamy or grayish with a pale edge. **Stipe:** 2–10 cm long, 1–3 cm thick, more or less equal or tapering toward base. Surface appressed-fibrillose to smooth, dry. Dark to light beige or grayish, usually with longitudinal brownish streaks, often with darker blotches at gill attachment. **Flesh:** Moderately thick, firm, fibrous; white. **Odor:** Indistinct. **Taste:** Indistinct. **Spore deposit:** White. **Microscopy:** Spores 7–10 × 4–6 µm, ellipsoid, smooth, colorless.

Ecology: Solitary or scattered on ground, in duff or moss under conifers, especially hemlock. Uncommon, mostly at middle to high elevation forests in the Cascade Range. Fruiting in fall.

Edibility: Edible.

Comments: The robust stature, dry, streaked grayish brown cap, widely spaced, decurrent white gills, and dry, gray stipe are distinctive. *H. marzuolus* (sensu PNW) is occasional during springtime in the mountains, often fruiting at the edges of melting snowbanks; it is typically slightly paler and often has a disagreeable odor. *H. calophyllus* has a viscid, dark grayish brown cap and stipe, and pinkish gills when young (although some collections lack the pink tones). *Gerronema atrialbum* looks quite similar, but has a tall, slender stipe covered in scurfy gray dots; it grows among decaying hardwood debris.

Hygrophorus fuscoalboides

Hesler & A.H. Sm.

Sheathed Waxy Cap

Cap: 3–10 cm across, rounded-conical with an incurved margin at first, becoming broadly convex to plane or with a broad umbo, occasionally with uplifted margin in age. Surface viscid, with a layer of glutinous slime over appressed fibrils. Gray to dark brownish gray, often with darker streaks, fading slightly in age. **Gills:** Broadly attached to slightly decurrent, subdistant to distant, narrow, waxy to greasy. White. **Stipe:** 3–9 cm long, 0.8–2.5 cm thick, tall, slender, often thicker in the middle with a tapered base. Surface viscid to tacky, silky-fibrillose. White with irregular patterns of sheathing gray fibrils or chevrons below a glutinous-fibrous, upward-flaring ring; white above ring. **Partial veil:** Composed of slime-covered fibers, leaving a dry, smooth annular zone near apex. **Flesh:** Thin, soft, fibrous in stipe; white. **Odor:** Indistinct. **Taste:** Indistinct. **Spore deposit:** White. **Microscopy:** Spores 9–14 × 5.5–8 µm, ellipsoid, smooth, colorless.

Ecology: Solitary, scattered, or in small clusters on ground, in moss or thick duff under conifers, especially spruce. Uncommon to rare in Cascadia. Fruiting in late fall and winter.

Edibility: Edible.

Comments: There are three similar species in Cascadia, which until recently were all under the name *H. olivaceoalbus*. As a group, they are recognized by their viscid gray to brown caps, pale gills, and viscid white stipes ornamented with brown fibers. *H. olivaceoalbus* does occur in western North America but may be more northerly in distribution. Although criteria to distinguish it from *H. fuscoalboides* are lacking, *H. olivaceoalbus* may have darker colors (at least the stipe fibrils). Both of these species tend to be larger with more gray tones than *H. whitei*, which has caps 3–6 cm across, a slender stipe, and a gray-brown to dark brown cap. *H. whitei* appears to be restricted to coastal Sitka spruce forests and is more common in the southern portion of Cascadia. *H. inocybiformis* has a dry cap covered with gray scaly fibrils and a gray, scaly stipe with a silky partial veil; it is rare in Cascadia.

Hygrophorus siccipes

A.H. Sm. & Hesler

Herald of Winter

Cap: 2–8 (10) cm across, broadly conical to convex with an inrolled margin at first, becoming broadly convex to plane, sometimes umbonate; margin often uplifted and wavy in age. Surface smooth or finely appressed-fibrillose, viscid to tacky when wet, often with adhering debris, shiny when dry. Color variable, typically dark brown or reddish brown at first; margin often paler; yellow-brown, yellow-orange to orange-red colors developing overall with age. **Gills:** Broadly attached to decurrent, close to subdistant, broad, greasy. Creamy to pale yellowish buff at first, yellowish to orange in age. **Stipe:** 3–12 cm long, 0.5–3 cm thick, tapering toward base. Surface silky-fibrillose on upper portions, more or less smooth below, viscid to moist. Whitish, creamy yellow, dark yellow or pale orange. **Partial veil:** Thin, consisting of fine glutinous fibers forming a vague ring when young, disappearing in age. **Flesh:** Thin to thick, soft, fibrous in stipe; yellow to whitish. **Odor:** Indistinct. **Taste:** Indistinct. **Spore deposit:** White. **Microscopy:** Spores 5.5–7.5 × 4–5.5 µm, ellipsoid, smooth, thin-walled, colorless.

Ecology: Solitary, scattered, or clustered on ground in duff under two- and three-needle pines. Occasional to locally common throughout Cascadia. Fruiting in fall, into winter on the coast.

Edibility: Edible.

Comments: The viscid, variably colored cap with olive-brown to bright orange-red or yellow tones, whitish to yellowish ocher gills, viscid stipe, and growth with pines are helpful identifying characters. *H. boyeri* is very similar and often difficult to distinguish; it is typically more slender in stature and has paler olive, brownish, or orangish yellow colors. *H. speciosus* has brighter, more pure orange-red colors and grows with larch. *H. flavodiscus* has a glutinous, creamy cap with a darker yellow disc and grows with five-needle pines. Members of the *H. fuscoalboides/olivaceoalbus/whitei* complex have dark to light gray-brown caps, white gills, stipes with brownish bands, and they grow with spruce.

Misapplied name: *Hygrophorus hypothejus* (Fr.) Fr.

Hygrophorus speciosus

Peck

Larch Waxy Cap

Cap: 2–5 cm across, broadly conical to convex with an inrolled margin at first, becoming broadly convex to plane, sometimes with a low broad umbo. Surface smooth, viscid to tacky when wet, shiny when dry. Red, orange-red to orange, occasionally fading to golden orange in age. **Gills:** Broadly attached to subdecurrent, subdistant, narrow, greasy. White when young, to yellowish in age. **Stipe:** 4–10 cm long, 0.3–1 cm thick, equal or enlarged toward base. Surface pruinose above annular zone, glutinous below. Whitish to yellowish when young, with pale orangish slime, which darkens and dries orange. **Partial veil:** Thin, consisting of fine glutinous fibers that leave an annular zone on stipe when young. **Flesh:** Thin, soft, fibrous in stipe; whitish. **Odor:** Indistinct. **Taste:** Indistinct. **Spore deposit:** White. **Microscopy:** Spores 8–10 × 4.5–6 µm, ellipsoid, smooth, colorless.

Ecology: Solitary, scattered, or in small clusters on ground under larch. Locally common in the eastern Cascade Range but can show up anywhere larch is planted. Fruiting in fall.

Edibility: Edible.

Comments: The viscid red-orange cap, white to yellowish gills, and growth with larch make this colorful mushroom easy to identify. *H. siccipes* has a variably colored cap with olive-brown to bright orange-red or yellow tones and creamy to pale to yellowish buff when young, becoming yellowish to orange in age. It is best distinguished by its habitat, growing with pines. Another pine associate, *H. boyeri*, is quite similar to *H. siccipes*, but is typically more slender in stature and has paler olive, brownish, or orangish yellow colors. Although there are some viscid, orangey *Cortinarius* species, they typically lack such intense colors and have brown spores.

Hygrophorus subalpinus

A.H. Sm.

Snowbank Waxy Cap

Cap: 5–15 (20) cm across, convex with an inrolled margin at first, becoming broadly convex to plane; margin even, wavy or slightly uplifted in age. Surface thinly viscid to tacky when wet, often with adhering debris when dry, smooth or with veil fibrils toward the margin. Bright white. **Gills:** Broadly attached to decurrent, close to subdistant, narrow, soft, waxy. White. **Stipe:** 4–10 cm long, 2–5 (7) cm thick, stout, often bulbous when young, more or less equal in age. Surface smooth and dry at the apex, with fibrils over lower portions. White. **Partial veil:** A thin, membranous-fibrillose sheath extending from the stipe bulb to the cap margin, leaving an annular zone low on the stipe or disappearing in age. **Flesh:** Thick, firm, fleshy, stipe fibrous-fleshy; white. **Odor:** Indistinct. **Taste:** Mild to sweet. **KOH:** No reaction. **Spore deposit:** White. **Microscopy:** Spores 8–10 × 4.5–6 µm, ellipsoid, smooth, colorless.

Ecology: Solitary or scattered on ground under conifers, likely restricted to true firs. Especially common at higher elevations near melting snowbanks. Fruiting in spring and early summer, occasionally in fall.

Edibility: A great edible! This is one of the better spring mushrooms for the table —it's often abundant, sweet, and has a great meaty texture.

Comments: The stocky stature, decurrent, waxy gills, white colors, and growth under conifers, often near snowbanks, make this a distinctive mushroom. Another snowmelt *Hygrophorus*, *H. marzuolus*, is smaller, typically has a gray cap (some specimens can be nearly white), and lacks a partial veil. Fall fruiting specimens of *H. subalpinus* can be confused with a few similar species. *H. ponderatus* is a smaller, compact species in Coast Range forests in southern Cascadia. *H. subellenae* is medium-sized with a whitish waxy cap that lacks a partial veil and often develops a yellowish cap center (especially as it dries out). Another large white species we are tentatively calling *H. perfumus* also lacks a veil and has a bright golden KOH reaction on the stipe.

Hygrophorus piceae group

White Waxy Caps

Cap: 1–5 cm across, bell-shaped to convex at first, becoming broadly convex to plane; margin often uplifted, wavy and irregular in age. Surface viscid, smooth or with fibrils imbedded in a glutinous layer. Bright white. **Gills:** Broadly attached to weakly decurrent, fairly widely spaced, thick, broad, waxy or greasy. White, occasionally tinged very pale yellowish in age. **Stipe:** 2–7 cm long, 0.3–1 cm thick, more or less equal or tapering toward base. Surface moist to dry (not viscid), finely scurfy over upper portion when young, smooth overall in age. White, discoloring pale dingy yellowish white in age. **Flesh:** Thin, soft, fibrous in stipe; white. **Odor:** Indistinct. **Taste:** Indistinct. **Spore deposit:** White. **Microscopy:** Spores 6–8 × 4–5 µm, ellipsoid, smooth, colorless.

Ecology: Solitary or scattered in moss or duff under spruce. Locally common in coastal spruce forests, occasional at higher elevations. Fruiting in fall and early winter

Edibility: Edible but slimy and lacking flavor.

Comments: The slender stature, white colors, slimy cap, dry stipe, and growth with spruce are distinctive. There are at least two similar species in Cascadia hiding under this name. One may average slightly larger and stockier but that remains to be reliably established. *H. eburneus* sensu PNW is larger and stockier, has a glutinous cap and stipe, and grows with oak, tanoak, and madrone; it is more common in southern Cascadia. *H. sitchensis* is a poorly-known Sitka spruce associate with a viscid white cap and stipe that develops yellow tones from the base up with an odor of almond extract. *H. gliocyclus* sensu PNW is a stocky species with an off-white to yellowish beige cap and a darker center; it grows with two and three-needle pines and is more commonly found in southern Cascadia. *H. flavodiscus* has a glutinous cap with a creamy yellow disc and paler margin, and a viscid stipe. It is found with five-needle pines. *H. pusillus* is a small species with a viscid, white to pale pinkish cap with darker splashes of pinkish red, widely spaced pale gills, and dry white stipe. *Cuphophyllus virgineus* has a moist to dry, white to creamy cap which is translucent-striate when wet, and decurrent, widely-spaced gills.

Hygrophorus chrysodon

sensu PNW

Gold Dust Waxy Cap

Cap: 2.5–8 cm across, convex with an inrolled and often scalloped margin at first, soon broadly convex to plane. Surface fibrillose to smooth, viscid to dry. Whitish to creamy, but covered with golden yellow fibrils and patches when young (especially near the margin), often bright whitish with scattered lemon-yellow splotches in wet weather. **Gills:** Subdecurrent to decurrent, close to distant, thick edged, often fairly narrow, greasy. Whitish to creamy or slightly pinkish. **Stipe:** 3–8 cm long, 0.5–1.5 cm thick, equal or tapered slightly toward base. Surface granular to smooth, dry to slightly viscid. When young whitish with scattered yellow flakes, sometimes nearly entirely white in age. **Flesh:** Thin, soft, fibrous in stipe; white. **Odor:** Indistinct. **Taste:** Indistinct. **KOH:** Yellow to orange on stipe and young cap. **Spore deposit:** Whitish. **Microscopy:** Spores 7–9 × 3.–4.5 µm, ellipsoid to oblong, smooth, thin-walled, colorless.

Ecology: Solitary, clustered, or in troops on ground, in duff, or in moss in a wide variety of habitats; growing with both conifers (especially fir and pine) and hardwoods (especially oak). Widespread across Cascadia, more common in the eastern Cascade Range and the southern portion of our range, and rare in coastal areas. Typically fruiting in late summer at elevation, or from mid-fall into winter at lower elevations, and occasionally in spring in the eastern Cascades.

Edibility: Edible.

Comments: The gold flakes on the cap and stipe and the white to creamy gills make this one of the more lovely and distinctive waxy caps. Although the yellow flakes can wash off in wet weather, the center and/or margin of the cap and stipe apex will usually retain some adhering yellow fibrils. This western North American species is distinct from the true European *H. chrysodon* and needs its own name. *H. eburneus* (with hardwoods) and members of the *H. piceae* group (with conifers) are pure white when young but may discolor very slightly yellowish as they age. The former is dramatically slimy, unlike *H. chrysodon*.

Hygrophorus bakerensis

A.H. Sm. & Hesler

Brown Almond Waxy Cap

Cap: 4–12 (15) cm across, conical to convex with an inrolled margin at first, becoming broadly convex to plane, or wavy and irregular in age. Surface covered with glutinous slime when wet, tacky in dry weather, smooth with appressed fibrils under the glutin. Yellowish brown to orangey brown to straw-brown at center, contrastingly whitish at margin. **Gills:** Broadly attached to decurrent, subdistant to distant, thick, broad, with a greasy feel. White when young, often with creamy yellow or pinkish tones in age. **Stipe:** 4–10 (15) cm long, 0.7–2.5 cm thick, more or less equal, but often tapering or rooting toward base. Surface dry to moist, with scurfy white scales near apex when young, smooth overall in age. White, discoloring pale yellowish in age. **Flesh:** Moderately thick, soft; stipe fibrous; white. **Odor:** Strongly of almond extract when young, weaker in age. **Taste:** Mild to slightly disagreeable. **Spore deposit:** White. **Microscopy:** Spores 7–9 × 4–6 µm, ellipsoid, smooth, colorless.

Ecology: Often in small clusters or in scattered troops, rarely solitary, on ground, in duff or moss under conifers, especially hemlock and fir. Common and widespread across western Cascadia. Fruiting in fall.

Edibility: Edible but slimy and bland (the odor doesn't remain after cooking).

Comments: This attractive waxy cap is recognized by its slimy, warm brownish to yellowish brown cap with a white margin, white gills and stipe, and odor of almond extract. The cap, gills, and stipe of young fruitbodies are often adorned with clear water droplets. *Rhodocollybia oregonensis* also has a strong odor of almond extract but usually has a darker reddish brown cap with a warm beige margin and a taller, more slender, usually rooting stipe. Its gills usually have serrate edges at maturity and the stipe often develops reddish brown blotches. Other waxy caps with a similar odor include: *H. secretanii* (=*H. monticola* and probably *H. pacificus*) with a cinnamon to buff cap and widely spaced gills; *H. agathosmoides* and *H. albofloccosus* with gray to gray-brown caps; and *H. odoratus* with a yellowish gray cap.

Taxonomic note: The names *Hygrophorus variicolor* and *H. amarus* likely pertain to the same taxon as this species. Pending studies of the type specimens, we have chosen to continue using *H. bakerensis* because it is the most well-known name. If these taxa are found to be to the same, *H. variicolor* is the oldest name, and would have priority.

Hygrophorus agathosmoides

Lebeuf, E. Larss. & Bellanger

Gray Almond Waxy Cap

Cap: 2–8 cm across, convex to plane, often with a low broad umbo; margin inrolled at first, becoming plane, to slightly uplifted in age. Surface smooth, viscid and shiny when wet but often drying out with stuck-on bits of debris. Gray-brown to gray, fading slightly with age. **Gills:** Broadly attached to slightly decurrent, close to fairly well spaced, narrow to broad, greasy. Bright white when young, staying white or becoming pale grayish white. **Stipe:** 3–7 cm long, 0.3–1 cm thick, equal or enlarged toward base. Surface finely scurfy or pruinose at first, more smooth in age, thinly viscid to dry. White to off-white when young, developing grayish brown tones in age. **Flesh:** Thin to moderately thick, soft, fibrous in stipe; white to pale gray. **Odor:** Mildly to strongly sweet, like almond extract. **Taste:** Indistinct. **Spore deposit:** White. **Microscopy:** Spores 8.5–9.5 × 5–6 µm, ellipsoid, smooth, colorless.

Ecology: Solitary, scattered, or in small clusters on ground or in thick duff, especially under Sitka spruce. Locally common in western Cascadia, occasional at higher elevations and under Engelmann spruce on the east side. Fruiting in fall and early winter.

Edibility: Edible but sweet odor disappears when cooked.

Comments: Until recently, all of Cascadia's gray-capped almond-scented waxy caps were called *H. agathosmus*. We now recognize two common species. The first is *H. agathosmoides*, a small to medium-sized species with a gray-brown to gray cap, white stipe which develops grayish brown tones, and growth with spruce. *H. alboflocossus* is larger, tends to have a more grayish cap and stipe which remains bright white, and is more general in its habitat preferences—occurring with conifers (especially Douglas-fir) and madrone. The rare almond-scented *H. odoratus* is small, often slender, with a creamy gray to grayish cap and large spores (11–14 × 6.5–8 µm). *H. pustulatoides* has grayish scales on the stipe and lacks any distinctive odor. Other species with an almond extract odor include *H. bakerensis* with a yellowish brown cap and whitish margin and *H. secretanii* (=*H. monticola* and possibly =*H. pacificus*) with a cinnamon to buff cap and widely spaced gills.

Hygrophorus pustulatoides

Lebeuf, E. Larss. & Bellanger

Cap: 1.5–5 cm across, convex to plane, at times with an obtuse umbo when young. Surface glutinous when wet, appressed-fibrillose under a layer of glutin, but often appearing smooth. Margin with fine white tomentum when young. Pale gray-brown to gray, often darker at disc. **Gills:** Broadly attached to slightly decurrent, well-spaced, thick, waxy. White to off-white. **Stipe:** 2–6 cm long, 0.3–1 cm thick, equal or tapered towards base. Surface dry, white, sparsely to densely covered with white scales which soon turn grayish to brownish. **Flesh:** Thin, soft, fibrous in stipe; white. **Odor:** Indistinct. **Taste:** Mild. **Spore deposit:** White. **Microscopy:** Spores 8–11.0 × 4.5–6 µm, ellipsoid, smooth, colorless.

Ecology: Solitary or scattered in duff and moss under spruce. Generally uncommon, occurring in both coastal forests and in east-side forests of the Cascade Range with both Sitka and Engelmann spruce. Fruiting in fall.

Edibility: Edible.

Comments: The small size, gray cap, white waxy gills, white stipe covered with small scales that darken in age, and lack of an odor help identify this species. *H. agathosmoides* can look quite similar but lacks the stipe scales and has an almond extract odor. *H. odoratus* also has the almond odor, a creamy gray to grayish cap, and large spores (11–14 × 6.5–8 µm). *H. inocybiformis* has a dry, gray, fibrillose-scaly cap and lower stipe, and a contrasting white stipe apex and gills. *Gerronema atrialbum* has a dry, convex to funnel-shaped cap, a tall gray stipe with gray fibrils, and grows from woody debris.

Misapplied name: *Hygrophorus pustulatus* (Pers.) Fr.

Hygrophorus capreolarius

sensu PNW

Cap: 3–8 cm across, convex with an inrolled margin at first, soon broadly convex to plane; margin becoming wavy and occasionally uplifted in age. Surface appressed-fibrillose to finely scaly, thinly viscid when wet, often with stuck-on debris when dry. Reddish purple with a grayish hue and darker wine-red scales at first, soon extensively patchy with dark, sordid purplish colors. Often with darker concentric rings or a ring of dark spots around the margin. **Gills:** Broadly attached to decurrent, subdistant to distant, thick-edged, greasy. Pinkish gray to vinaceous buff at first, soon developing darker wine-colored stains, becoming dark, dingy reddish purple in age. **Stipe:** 3–8 cm long, 0.5–1.2 cm thick, equal or enlarged toward base. Surface smooth to finely scurfy, dry. Whitish but streaked with vinaceous pink, darkening to wine-red from the base up as it ages. **Flesh:** Thin to moderately thick, firm, fibrous in stipe; pinkish. **Odor:** Indistinct. **Taste:** Indistinct. **Spore deposit:** Whitish. **Microscopy:** Spores 6–9 × 4–5 µm, ellipsoid, smooth, colorless.

Ecology: Solitary or scattered on ground or in moss under true fir and hemlock. Widespread and locally common but rarely abundant. Fruiting from mid fall into winter.

Edibility: Edible.

Comments: This small to medium-sized waxy cap is identified by the vinaceous gray cap and gills that darken as it ages, darker ring or spots around the cap margin, and growth with fir and hemlock. One could easily mistake it for *H. russula* but that species is usually larger, has an overall paler color scheme, and grows with hardwoods. *H. erubescens* has taller fruitbodies with pale pinkish caps that develop vinaceous pink blushes and yellowish stains in age; it grows primarily with spruce. Members of the *H. purpurascens* group are also similar but have a cortina-like partial veil, are usually larger, and have a rosy purple cap.

Hygrophorus purpurascens group

Purple Waxy Caps

Cap: 4–12 (18) cm across, convex with an inrolled margin at first, soon broadly convex to plane; margin becoming wavy and occasionally uplifted in age. Surface appressed-fibrillose to nearly smooth, thinly viscid when wet. Whitish with purplish to reddish streaks, spots, and stains at first, more extensively reddish purple in age. **Gills:** Broadly attached to decurrent, close to subdistant, broad, greasy. White at first, soon developing reddish purple spots, edges and blotches, occasionally evenly reddish purple in age. **Stipe:** 3–10 cm long, 1–3 cm thick, equal, tapered downward or rarely club-shaped. Surface smooth to finely fibrillose, dry. Whitish at first, soon extensively streaked reddish purple, darkening and becoming dingier in age. **Partial veil:** A scant cortina, typically only visible when young, at times leaving a thin ring on stipe. **Flesh:** Thick, firm, fibrous in stipe; whitish with vinaceous stains around larval tunnels. **Odor:** Indistinct. **Taste:** Mild or slightly bitter. **KOH:** No reaction or slight yellowish on all parts. **Spore deposit:** White. **Microscopy:** Spores 5.5–7 × 3–4.5 µm, ellipsoid, smooth, colorless.

Ecology: In small clusters or scattered troops, sometimes abundant in arcs, rarely solitary. Typically fruiting under the duff layer (or at least partially buried) and associated with true fir. Common at higher elevation and east-side forests. Fruiting in spring, beginning as soon as the snow melts and continuing into summer (see comments for fall-fruiting specimens). At times, fruiting right at the edges of melting snowbanks but not restricted to this habitat.

Edibility: Edible but the taste is poor.

Comments: The stout stature, streaked and stained reddish purple cap, white waxy gills which develop reddish purple spots, and half-buried growth habit under conifers help identify this species. The above description is for the spring fruiting species in the *H. purpurascens* group. There is at least one other summer and fall fruiting species that tends to be larger, with a more pronounced partial veil, and slightly paler colors. *H. russula* has a pinkish cap with dark wine-red spots around the margin and grows with hardwoods. *H. capreolarius* is a smaller, darker vinaceous purple species that grows with fir and hemlock. *H. neoerubescens* (=*H. erubescens*) has a taller stature, a paler, pinkish-streaked cap, and often develops yellowish stains as it ages. *H. fragrans* and *H. pudorinus* have more pinkish cap tones and have a bright orange KOH reaction on the white stipe.

Hygrophorus fragrans

(Murrill) Hesler & A.H. Sm.

Cap: 3–15 (25) cm across, rounded to convex with an inrolled margin at first, becoming broadly convex to plane, occasionally wavy and with uplifted margin in age. Surface smooth, viscid to tacky when wet, shiny and often with adhering debris when dry. Pink, pinkish orange or salmon at first, fading slightly in age. **Gills:** Broadly attached to subdecurrent, close to subdistant, broad, waxy to greasy. Whitish, creamy to pinkish cream. **Stipe:** 4–12 cm long, 1.5–5 cm thick, more or less equal, often with a slightly rooting base. Surface punctate at apex, fibrous to smooth over lower portion, moist to dry; upper stipe covered with clear droplets when young. Whitish, pinkish white to orangish white, staining orange at base when handled. **Flesh:** Moderately thick, firm, fibrous in stipe; whitish to pale pinkish white. **Odor:** Mild to fragrant, chemical sweet. **Taste:** Indistinct. **KOH:** Bright orange on stipe apex. **Spore deposit:** White. **Microscopy:** Spores 7–9.5 × 4.5–5.5 µm, ellipsoid, smooth, colorless.

Ecology: Scattered or in small clusters in duff under conifers, especially true fir and hemlock. Locally common but rarely abundant, widely distributed across Cascadia. Fruiting in fall.

Edibility: Edible.

Comments: This beautiful waxy cap is one of two species which have long been referred to as *H. pudorinus*. They are easily recognized by their pink to pinkish orange caps, whitish to pinkish white stipes, and the orange KOH reactions at the stipe apex. *H. fragrans*, the more common west-side species, tends to exhibit brighter pink tones, orange staining on the lower stipe, and a brilliant KOH reaction. *H. pudorinus* is more common on the east-side of the Cascade Range, with a rosy pink to purplish pink cap, tendency to lack the orange staining of the lower stipe, and a more muted orange KOH reaction. *H. neoerubescens* (=*H. erubescens*) has a pale rosy-pink-splashed cap, which darkens and yellows in age, and lack of reaction with KOH on the stipe.

Hygrophorus saxatilis

A.H. Sm. & Hesler

Cap: 3–10 cm across, obtuse to convex with an inrolled margin at first, soon broadly convex to plane; margin becoming wavy and occasionally uplifted in age. Surface finely appressed-fibrillose to smooth, thinly viscid to dry. Pinkish buff, fading to creamy buff, occasionally with cinnamon specks. **Gills:** Broadly attached to decurrent, subdistant to distant, thick-edged, greasy. Salmon-pink when young, ocher-pink to ocher in age. **Stipe:** 4–9 cm long, 0.8–2 cm thick, equal or tapered towards base. Surface smooth to pruinose near apex, moist to dry. Whitish to pinkish buff, darkening in age or when handled. **Flesh:** Thin to moderately thick, firm, fibrous in stipe, salmon-pink to buff. **Odor:** Pleasant. **Taste:** Indistinct. **Spore deposit:** Whitish. **Microscopy:** Spores 7–9.5 × 4–5 µm, ellipsoid, smooth, colorless.

Ecology: Solitary or scattered on ground in moss or duff under true firs. Rare. Most collections come from higher elevation, old-growth forests in the Cascade Range. Fruiting in late summer and fall.

Edibility: Unknown.

Comments: This rare waxy cap is readily identified by the pinkish buff cap, rich salmon-pink gills, pale stipe, and habit of fruiting in late summer and fall. *H. goetzei* has a pink cap that fades as it ages, pale gills, and fruits near melting snow around mountain hemlocks in late spring and summer. *Cuphophyllus pratensis* is quite similar but has a salmon-orange to pinkish buff cap, paler creamy apricot-colored gills, and often grows at lower elevations in mixed forests or in grassy areas.

Taxonomic note: More work is needed to see if *H. saxatilis* is conspecific with the European *H. karstenii*. If they are, the latter name would take precedence.

Hygrophorus goetzei

Hesler & A.H. Sm.

Rosy Snowbank Waxy Cap, Snowrose

Cap: 2–5 cm across, obtuse to convex at first, becoming broadly convex to plane or with slightly uplifted margin in age. Surface smooth, viscid when wet. Porcelain-pink to rosy pink at first, often with darker streaks, fading to creamy-pink or cream-colored in age. **Gills**: Broadly attached to subdecurrent, moderately to widely spaced, thick and waxy. Pale pinkish to pinkish cream at first, becoming cream to pallid. **Stipe**: 3–7 cm long, 0.3–0.8 cm thick, equal or enlarged slightly lower, occasionally rooting at base. Surface moist to dry, smooth; base with fine strigose hairs. Whitish to cream-colored, at times with pinkish streaks when young. **Flesh**: Thin, fibrous in stipe; whitish. **Odor**: Indistinct. **Taste**: Indistinct. **Spore deposit**: White. **Microscopy**: Spores 12–15 × 7–9 µm, ellipsoid, smooth, colorless.

Ecology: Solitary or scattered near melting snow where it often melts a hole through the edge of the snow as it grows or along the immediate edge of the snowbank. Occasional at higher elevations in the Cascade Range, in forests with mountain hemlock. Fruiting in late spring or summer, depending on stage of snowmelt. Because it has a relatively small range and is restricted to a narrow band of elevation where mountain hemlock forests experience favorable snow accumulation and melt-off dynamics, this species is probably threatened by climate change.

Edibility: Unknown.

Comments: This beautiful species is easily recognized by the pink cap that fades as it ages, thick, widely spaced gills, and growth at or near the edges of melting snowbanks. *H. vernalis* is another rare snowbank waxy cap with a similar stature. It has a yellowish-tan to vinaceous fawn cap and whitish gills and is also restricted to higher elevations across Cascadia. The late summer and fall fruiting *H. saxatilis* has a pinkish buff cap and brighter salmon-pink gills. *Pseudoomphalina angelesiana* (=*Neohygrophorus angelesianus*) has smaller, more slender fruitbodies with darker brown to grayish brown caps, purplish to lilac-pink gills, and amyloid spores. *Clitocybe albirhiza* is a common snowbank species that can have pinkish buff to pale buff caps that fade in age, but it has crowded gills, thick white hairs, and rhizomorphs at the stipe base.

Cuphophyllus pratensis

(Fr.) Bon

Salmon Waxy Cap

Cap: 2.5–8 (10) cm across, convex to plane, often with a broad umbo; margin inrolled and sometimes ribbed at first, becoming uplifted in age. Surface smooth, moist to dry. Ochraceous orange to salmon-orange or paler buff-apricot in age. **Gills:** Slightly to deeply decurrent, well-spaced; often with cross veins, thick, soft and greasy. Ochraceous orange to salmon-orange at first, then pale pinkish beige. **Stipe:** 4–10 cm long, 0.7–2 cm thick, equal, club-shaped, or with a tapered base. Surface smooth, moist to dry. Pale ochraceous orange to whitish. **Flesh:** Moderately thick to thin, fibrous in stipe; pale orangish buff to whitish. **Odor:** Indistinct to faintly fragrant. **Taste:** Mild to sweet. **Spore deposit:** White. **Microscopy:** Spores 5–8 × 4–5.5 µm, ellipsoid, smooth, colorless.

Ecology: Solitary, scattered, or in small clusters in duff or moss, or on ground in mixed forest, especially under western redcedar, and also in unimproved grassy areas. Locally common throughout western Cascadia. Fruiting in fall and early winter.

Edibility: Edible and good.

Comments: The salmon-orange to pinkish buff cap, thick, widely spaced, decurrent gills, and lack of a strong odor distinguish this species. *C. graveolens* is similar but has a paler cap and often a strong, pungent-fragrant odor. *Hygrophorus saxatilis* has a pinkish buff cap and more richly colored contrasting salmon-pink gills. *Chroogomphus tomentosus* is similar in color (although more pale ocher than pink) but has a slightly matted-shaggy cap and olive-black spores. *Cantharellus formosus* has brighter, yellowish orange colors and shallow, thick, often irregularly veined gills.

Synonyms: *Camarophyllus pratensis* (Fr.) P. Kumm.

"*Cuphophyllus*" *graveolens*

(A.H. Sm. & Hesler)

Cap 2.5–6 cm across, convex to plane with a depressed center in age; margin inrolled and faintly ribbed at first, often ragged in age. Surface silky-fibrillose to smooth, moist to dry. Pale peachy orange, pinkish cinnamon to creamy buff, with a silky, whitish sheen, becoming paler overall in age. **Gills:** Broadly attached to decurrent, fairly widely spaced, occasionally with cross veins, rather thick, soft and greasy. Pale peachy orange to creamy buff. **Stipe:** 4–10 cm long, 0.7–1.5 cm thick, equal, club-shaped with an enlarged base, or enlarged in middle. Surface smooth, longitudinally striate, moist to dry. Pale creamy peach to pale creamy buff. **Flesh:** Thin in cap, pale peachy orange to orangish white. **Odor:** Strong, fragrant, almost sickeningly sweet with a fruity-chemical component (weaker in age or cold weather, or even in some young collections). **Taste:** Indistinct. **Spore deposit:** White. **Microscopy:** Spores 6–9 × 4–6 µm, ellipsoid, smooth, colorless.

Ecology: Solitary or scattered in small troops in duff or moss, or on ground in mixed forest. Widespread but locally common in western Cascadia, rare elsewhere. Fruiting in fall and early winter.

Edibility: Unknown, likely non-toxic.

Comments: The combination of a silky-looking pale peachy cap, thick, widely spaced, decurrent gills, and fragrant odor are distinctive. Often the first few sniffs are pleasant, after which it often becomes unpleasantly overpowering (although not everyone can detect this odor). *C. pratensis* is very similar but is usually larger and stockier, with a darker peachy orange cap lacking the silky sheen, with little odor.

Synonym: *Hygrophorus graveolens* A.H. Sm. & Hesler

Cuphophyllus russocoriaceus

(Berk. & T.K. Mill.) Bon

Cedar-scented Waxy Cap

Cap 0.7–4.5 cm across, convex when young, becoming plane in age; margin incurved, translucent-striate and often ribbed, occasionally uplifted in age. Surface smooth, moist to dry. White to creamy white, sometimes tinged pale yellowish, hygrophanous, brighter white when dry. **Gills:** Decurrent, relatively widely spaced, often veined with narrow to broad ridges, soft and greasy. White to creamy white. **Stipe:** 4–10 cm long, 0.2–1 cm thick, equal or enlarged slightly toward base. Surface smooth to vertically fibrillose, moist to dry. White to cream at first, often pinkish at base in age. **Flesh:** Thin, fibrous in stipe, white. **Odor:** Strongly fragrant, like cedar wood. **Taste:** Usually disagreeable, metallic or medicinal, sometimes mild. **KOH:** No reaction. **Spore deposit:** White. **Microscopy:** Spores 7–10 × 4.5–6 µm, ellipsoid, smooth, colorless.

Ecology: Solitary or scattered in troops from moss and duff. Occasional in western Cascadia, typically in lower elevation forests, with a preference for western redcedar duff. Fruiting from late fall through winter.

Edibility: Unknown.

Comments: One sniff of this mushroom is practically enough to identify it! The strong odor of cedar wood is shared by only a few other species, none of which visually resemble this species. Occasionally the odor will be faint or absent in older specimens. In these cases, the white to cream colors, decurrent gills, and disagreeable taste help distinguish it from other white *Cuphophyllus*. "*Hygrophorus*" *lawrencei* is a very similar species described from Oregon. It may be slightly larger and dingier overall, but more sequenced collections are needed to delimit the two. Members of the *C. virgineus* group are similar but lack any distinctive odor, have a mild taste, and are typically larger. *Marasmius calhouniae* has more widely spaced, wavy-edged gills and lacks any odor; it grows in troops on conifer duff (especially pine).

Synonyms: *Hygrocybe russocoriacea* (Berk. & T.K. Mill.) P.D. Orton & Watling, *Camarophyllus russocoriaceus* (Berk. & T.K. Mill.) J.E. Lange.

Cuphophyllus rainierensis

(Hesler & A.H. Sm.) Lebeuf

Cap: 1–3.5 cm across, rounded to obtuse at first, becoming broadly convex to plane, often with an umbo. Surface thinly viscid to moist, smooth. Lilac gray to dull purplish gray, at times light gray with a lilac wash. **Gills:** Decurrent, subdistant, thick, waxy, sometimes forking, often with many inter-gill veins. Whitish, light gray to purplish gray. **Stipe:** 4–8.5 cm long, 0.3–0.7 cm thick, equal to a pointed base. Surface moist to dry, smooth. White to off-white. **Flesh:** Thin, fleshy, translucent grayish in cap. Stipe fibrous; white. **Odor:** Strongly of green corn. **Taste:** Indistinct. **Spore deposit:** White. **Microscopy:** Spores 5–7 × 3.5–5.5 µm, broadly ellipsoid. smooth, colorless.

Ecology: Solitary or scattered in duff and soil, typically under redcedar. Rare, most collections come from mid-elevation forests in western Cascadia. Fruiting in fall.

Edibility: Unknown.

Comments: The small to medium-sized fruitbodies, purplish gray, lilac to pale gray cap, decurrent gills, whitish stipe, and growth from duff and soil help identify this species. The green corn odor has been strong in all the collections we have found, but supposedly is not always present. It may be the same as the European *C. lacmus*, although ours lacks the yellowing of the stipe base and has a distinctive odor. Collections with larger spores (7–9 × 6–7 µm) from eastern Washington and Idaho have been called *C. nordmanensis*. *C. cinerellus* is a slightly duller purplish, grayish to gray-brown species which grows in *Sphagnum* bogs. *C. colemannianus* has a dull brown to ruddy brown cap, thick, widely spaced decurrent gills, and dry stipe. *C. recurvatus* is a small species (1–3 cm across) with a dark brown to olive-brown cap and white gills; it typically grows in nutrient-poor grassy areas.

Synonym: *Hygrophorus rainierensis* Hesler & A.H. Sm.

Gliophorus psittacinus group

Parrot Mushrooms

Cap: 1–4 cm across, rounded-conical to egg-shaped or convex when young, becoming broadly convex to plane, often with an umbo. Surface smooth, extremely viscid, with a thick layer of slime. Strongly hygrophanous, translucent-striate when wet. Colors extraordinarily variable and often mixed: bright or dark green to bluish green, olive-green, olive-brown or orange, fading to light yellowish green, peachy orange, pinkish or lilac, duller overall when dry. **Gills:** Attached, with distinct notch and decurrent tooth, thick, greasy, subdistant to distant. Colors variable: creamy white to pale yellow or brilliant yellow, yellowish green or greenish white, sometimes pinkish cream to peachy. **Stipe:** 2–6 cm long, 0.3–0. 7 cm thick, equal, often curved. Surface smooth or with a wavy-bumpy appearance like mother-of-pearl, viscid and shiny. Yellow to greenish yellow to pale yellow or creamy white, often brighter yellow toward base, at times with a bluish green or lilac band at apex. **Flesh:** Thin, fragile, pale yellowish. **Odor:** Indistinct. **Taste:** Indistinct. **KOH:** Bleaching out cap pigment. **Spore deposit:** White. **Microscopy:** Spores 8–10 × 5–5.5 µm, elongate-ellipsoid, smooth, colorless.

Ecology: Solitary or scattered, sometimes clustered in duff, moss, or soil. Locally common in lower elevations across western Cascadia, typically in redcedar or bigleaf maple forests. Fruiting in late fall and winter.

Edibility: Non-toxic.

Comments: Chameleon Waxy Cap would be a fitting name for this species because the range of colors (even in a single fruitbody as it ages!) is extraordinary. Although the typical sequence is from greenish to orangish or pinkish yellow, there are also "Backward Parrots" that start off orange and become green! Other forms skip the green colors entirely. As you might expect, this is a species complex with at least five species in Cascadia. The overlap between their ranges of variation makes stable distinctions very difficult to establish. The colors of *G. laetus* are also bright and variable (yellowish green, orange, pink, lilac) but it can be distinguished by its decurrent gills and rough, finely bumpy cap texture. *G. irrigatus* has a slimy grayish brown to brown cap and stipe and white gills. *"Hygrophorus" (=Gliophorus) flavifolius* has a milky white cap, brilliant chrome-yellow gills, and a glutinous, porcelain-white stipe. *Hygrocybe virescens* has lime-green colors, is larger, and has a moist to dry cap and stipe.

Synonym: *Hygrocybe psittacina* (Schaeff.) P. Kumm

Humidicutis marginata var. *olivacea*

(A.H. Sm. & Hesler) Singer

Cap: 2–4 cm across, conical to bell-shaped, with a pointed to obtuse umbo; margin becoming uneven, often splitting in age. Surface moist to dry, smooth. Color variable: olive-green to green overall when young (or restricted to disc) and olive-orange to orange-brown or red-brown outward or extensively orange-brown to red-brown in age. Occasionally fading to whitish in age when dry. **Gills:** Notched, with a decurrent tooth, moderately to widely spaced, broad. Bright orange to ocher-orange when young, fading with age; edges often darker. **Stipe:** 3–7 cm long, 0.2–0.5 cm thick, more or less equal, usually developing a longitudinal groove in age. Splitting easily; base often breaking or decaying. Surface moist to dry, smooth. Olive-yellow to orangish (sometimes extensively grass-greenish), often somewhat translucent. **Flesh:** Thin, fragile, greenish to orangish; stipe hollow, fragile, pallid. **Odor:** Indistinct. **Taste:** Indistinct. **Spore deposit:** Whitish. **Microscopy:** Spores 6.5–8 × 4–5 µm, ellipsoid to broadly ellipsoid, smooth, colorless.

Ecology: Solitary, scattered, or in small clusters in humus and duff, most often under western redcedar. Rare, mostly at lower elevations in the northern portion of Cascadia. Fruiting in summer and early fall.

Edibility: Unknown.

Comments: The umbonate, variably colored cap which typically shows some olive-green tones when young, striking orange gills, fragile stipe, and tendency to fruit during the drier part of the season help identify this rare waxy cap. *Mycena aurantiomarginata* is similar in color but is considerably smaller, with a thin stipe (<0.2 cm). *Hygrocybe virescens* is similar in shape and texture but has a lime-green to yellowish green cap and pale yellow-green gills. Some forms of *Gliophorus psittacinus* are also greenish to orange but have viscid to glutinous caps and stipes. Members of the *Lactarius deliciosus* group have convex to concave caps, are stockier, and have chalky to grainy flesh.

Hygrocybe flavescens

(Kauffman) Singer

Yellow Waxy Cap

Cap: 2–8 cm across, convex to plane with a small umbo; margin becoming uneven, often wavy in age. Surface moist to viscid when wet, often with adhering debris in older specimens or when dry. Pale lemon-yellow to bright yellow or yellowish orange to orange, fading in age. **Gills:** Notched and narrowly attached, moderately to widely spaced, narrow to broad, soft, greasy. Off-white to pale yellow when young, remaining pale yellow or becoming darker yellow to yellowish orange in age. **Stipe:** 3–5 cm long, 0.3–1 cm thick, more or less equal, usually developing a longitudinal groove in age. Surface fibrous to smooth, moist to slippery when wet, quickly drying out. Yellow, pale lemon-yellow or yellowish orange. **Flesh:** Thin, watery, creamy white to pale yellow in cap; stipe stuffed when young, becoming hollow, yellowish. **Odor:** Indistinct. **Taste:** Indistinct. **KOH:** Bleaching cap pigment. **Spore deposit:** Whitish. **Microscopy:** Spores 7.5–9 × 4–5 µm, ellipsoid, smooth, colorless.

Ecology: Solitary, scattered, or in small clusters on humus and duff, most often under western redcedar and bigleaf maple. This appears to be the most abundant waxy cap in Cascadia, particularly throughout the lower elevations. Fruiting from late fall into early spring.

Edibility: Reportedly edible.

Comments: This species is identified by its thinly viscid, convex to plane, yellow to yellow-orange cap (without any red tones) and whitish to yellow gills and stipe. *H. acutoconica* is similar but has a more peaked-conical, yellow-orange to reddish cap, and a fibrous stipe. Faded fruitbodies of the much rarer *H. virescens* can be similar but have dry caps and usually at least faint traces of greenish hues. *H. constrictospora* sensu PNW is a small species with a pale yellow cap and decurrent gills. *H. ceracea* has a translucent-striate, viscid, orange cap that fades to yellow, weakly decurrent gills, small to medium-sized fruitbodies, and grows in mossy and/or grassy areas.

Hygrocybe singeri group

Western Witch's Hats

Cap: 3–10 cm across, bell shaped to narrowly conical, becoming irregularly broadly conical, umbonate, often pointed. Surface smooth, viscid and shiny when wet, dull when dry. Yellow-orange to reddish orange, occasionally extensively scarlet, sometimes yellowish black to olive without orange-red tones, developing extensive black streaking, often entirely black in age. **Gills:** Attached to nearly free, close to subdistant. Off-white when young, becoming pallid gray, irregularly developing black edges and blotches, sometimes entirely black in age. **Stipe:** 5–20 cm long, 0.7–2 cm thick, equal or enlarged toward base. Surface longitudinally fibrillose-striate, smooth, viscid to dry. Yellow, yellow-orange, or olive-grayish, often whitish at extreme base, slowly developing black stains. **Flesh:** Thin, fragile; stipe fibrous, hollow; off-white to yellowish, slowly staining black where damaged. **Odor:** Indistinct. **Taste:** Indistinct. **KOH:** No reaction. **Spore deposit:** White. **Microscopy:** Spores 9–13 × 5–7 µm, elongate-ellipsoid, smooth, colorless.

Ecology: Solitary, scattered or in troops from soil and duff. The most common representative of this species group is widespread in mixed forests in western Cascadia, while others in the group prefer wet moss or bogs, lawns or other grassy areas, high elevation meadows, or alpine heaths. Fruiting in fall, early winter, at lower elevations, or spring and summer at high elevations.

Edibility: Not recommended. There are (unfounded?) reports of this and relatives being poisonous.

Comments: The bright colors, peaked-conical cap, and blackening on all parts make this an easy mushroom to identify, at least in a general sense. *H. singeri* is a species complex (also referred to as the *H. conica* group) with at least five species in Cascadia. Molecular work has shown that even in Europe the name *H. conica* has been applied to a complex of ten or more species, many of which remain unnamed. The name *H. olivaceoniger* has been applied to a smaller, yellow to greenish form. *H. nigrescens* has been used for a taxon with larger, consistently redder caps that grows in hardwood duff. *H. acutoconica* is similar, with a slimy, orange-yellow, conical cap, but does not discolor gray and black.

Hygrocybe constans

Murrill

Cap: 1–3 cm across, convex to plane, often with a slightly depressed center; margin downcurved and often ruffled when young. Surface dry to moist (never viscid), at center with small tufts or scales which become more noticeable in age, giving it a felty appearance. Bright red to orange-red, fading to reddish orange or yellow-orange. **Gills:** Broadly attached, thick, waxy, widely spaced. Creamy yellow to pale yellow when young, becoming yellow to yellow orange in age. **Stipe:** 2–7 cm long, 0.2–0.4 cm thick, equal. Surface smooth, moist to dry, often with lighter longitudinal striations as it dries. Reddish orange to orange, becoming yellow-orange to yellow, often paler toward base. **Flesh:** Thin, fragile, yellow-orange. **Odor:** Indistinct. **Taste:** Indistinct. **KOH:** No reaction. **Spore deposit:** White. **Microscopy:** Spores 7–9 × 4–6 µm, variably shaped, generally broadly ellipsoid to ellipsoid, pear-shaped or cylindrical and constricted (like peanuts in a shell), or like corn kernels, smooth, colorless.

Ecology: Scattered, in small clusters, or solitary, often in duff or on ground. Locally common in lower elevation forests across western Cascadia, with a preference for western redcedar stands. Fruiting in fall and winter.

Edibility: Edible.

Comments: The small size, reddish orange colors, and minutely scruffy cap help distinguish this species. There are other small, similarly-colored, scurfy-capped species. *H. phaeococcinea* has a darker red to flame-red cap and stipe, and slightly decurrent, orange gills. *H. substrangulata* has dark red to orange-red colors, warty-scurfy cap textures, and growth in bogs or wet mossy areas. Members of the *H. cantharellus* group have red, orange to yellow caps, with or without dark scales, decurrent gills and long, slender stipes.

Misapplied name: *Hygrocybe miniata* (Fr.) P. Kumm.

Hygrocybe ceracea

(Sowerby) P. Kumm.

Cap: 1–4 cm across, convex to broadly convex, at times with a slight umbo; margin sometimes uplifted and splitting in age. Surface smooth, viscid when young, moist in age. Bright golden orange when young, paler and translucent-striate at margin, hygrophanous, often patchily fading to pale orange or yellowish orange to yellow. **Gills:** Broadly attached to subdecurrent, close to subdistant, thick, waxy. Golden yellow with a paler edge at first, soon fading to pale yellowish in age. **Stipe:** 2–4 cm long, 0.2–0.5 cm thick, equal or tapering slightly toward base. Surface smooth, moist to dry. Orange to gold, fading to pale yellow. **Flesh:** Thin, fragile, stipe hollow in age; concolorous with cap and stipe. **Odor:** Indistinct. **Taste:** Indistinct. **KOH:** Bleaching pigment on all parts. **Spore deposit:** White. **Microscopy:** Spores 7–9 × 5–6.5 µm, ellipsoid, smooth, colorless.

Ecology: Solitary, scattered, or in groups on grassy lawns or moss. Locally common in western Cascadia, mostly in urban areas at lower elevations. Fruiting in late fall and winter.

Edibility: Unknown.

Comments: The translucent-striate, viscid, orange cap that fades to yellow, weakly decurrent gills, small to medium-sized fruitbodies, and growth in grassy and mossy areas help set this species apart from other small, colorful waxy caps. "*Gliophorus fenestratus*" (a provisionally named species) has a paler cap with a whitish to clear center, and a viscid stipe. It is rare in lower elevations of western Cascadia but more common in redwood and cypress duff in California. *H. mucronella* (sensu PNW) is a small species with a red, orange to yellow cap, and a bitter taste. *H. parvula* is also variably colored but tends to have decurrent gills and an orange stipe. There are a number of other small to tiny, red to orange-red *Hygrocybe* in Cascadia which remain unnamed, so don't be too frustrated if your find doesn't match any of these descriptions—you may well have an undescribed species on your hands!

Hygrocybe sp. 'Scaly Bog'

Cap: 1.5–4 cm across, convex to plane, typically with a depressed disc at all ages; margin incurved at first, becoming downcurved to plane, often ribbed. Surface moist to dry, covered with fine scales, densest at disc. Color variable, when young dark with blackish scales, soon more orange, yellow, brown to olive, with sparser scales in age. **Gills:** Decurrent, well-spaced, broad, soft, greasy. Creamy to yellowish ocher. **Stipe:** 3–7 (10) cm long, 0.3–0.8 cm thick, equal or enlarged slightly toward base. Surface moist to dry, smooth or with scattered scales or fibrils. Orangish to yellowish, at times with a grayish to blackish wash over middle portion, with a paler base. **Flesh:** Thin, fragile, stipe hollow; concolorous with exterior. **Odor:** Indistinct. **Taste:** Indistinct. **KOH:** No reaction. **Spore deposit:** White. **Microscopy:** Spores 8–14 × 5–7.5 µm, ellipsoid, smooth, colorless.

Ecology: Solitary or scattered in troops in moss. Typically in bogs or wet areas among sphagnum moss. Uncommon, but can be abundant in suitable microhabitats throughout Cascadia. Fruiting in summer or early fall.

Edibility: Unknown.

Comments: The variably orange to yellow cap with blackish scales, decurrent gills and growth in wet mossy areas help distinguish this species. It is similar to *H. turunda*, but the Cascadian species is distinct, and appears to be undescribed. *H. cantharellus* is quite similar, but usually has more red to orange colors, and typically lacks the dark scales (although scaly forms do occur). It also tends to grow in drier habitats. *H. phaeococcinea* has a dark red to flame-red cap and stipe, and slightly decurrent orange gills.

Hygrocybe coccinea group

Scarlet Waxy Caps

Cap: 2–5 (7) cm across, rounded-conical to convex, becoming broadly convex to plane, occasionally with a low umbo; margin incurved at first, becoming wavy and irregular, sometimes uplifted and splitting in age. Surface smooth to very slightly roughened, thinly viscid to greasy or dry. Deep scarlet-red to blood-red, sometimes becoming patchy orangish red in age. **Gills:** Attached, sometimes notched with a decurrent tooth, close to well-spaced, broad, soft, greasy. Pinkish red to red, with a paler whitish yellow to pinkish edge at first, becoming orange-red overall in age. **Stipe:** 2–7 cm long, 0.5–1.5 cm thick, equal or enlarged slightly toward base, occasionally with a longitudinal groove in age. Surface smooth, only faintly longitudinally striate when old and dry. Red, often with a whitish to orangish base. **Flesh:** Thin, fragile; stipe hollow; red in cap, orangish in stipe. **Odor:** Indistinct. **Taste:** Indistinct. **KOH:** No reaction. **Spore deposit:** White. **Microscopy:** Spores 7–10 × 4–5.5 µm, ellipsoid, smooth, colorless.

Ecology: Solitary, in small clusters, or scattered in small to large troops in duff, moss, or soil, usually under western redcedar. Locally common at lower elevations in western Cascadia. Fruiting in late fall and winter.

Edibility: Reportedly edible but not recommended. Closely related species (*H. punicea* group) have caused problems for some people.

Comments: Red *Hygrocybe* in Cascadia are a taxonomic jumble with most going by misapplied European names. That includes the *H. coccinea* group, which appears to be comprised of two distinct species in Cascadia, neither of which are conspecific with the European species. They are recognized by their medium-sized fruitbodies, deep scarlet-red to blood-red caps, red gills, and smooth stipes. The name *H. marchii* has been applied to smaller red-capped species but it is another dubious name to use in Cascadia. *H. laetissima* is a medium to large species (2–8 cm across) with cherry-red to scarlet young caps, gills that start out pale and develop peachy tones, and longitudinally fibrillose stipes that are extensively white when young (developing yellow to orange or reddish tones from the base upward). A still larger species, *H. punicea* (sensu western NA), has a dark red to blood-red cap, gills that become reddish in age, and a longitudinally striate stipe. *H. aurantiosplendens* has a reddish cap that quickly and blotchily fades to orange, and a smooth stipe.

14 - The White-Spored Multitude

Although this group includes many interesting mushrooms, it also contains many of the plainest-looking, most featureless fungi in our area. The list of genera here is quite long, and most were at one time placed in the Tricholomataceae – a family which was long used as a "taxonomic waste bin" for white-spored mushrooms whose evolutionary relationships were difficult to determine. Little by little, many of the genera represented have been being sorted into named lineages of their own; likewise, individual species are being more critically examined to determine which genus they "truly" belong in. Many new genera have been created and many existing species have been reshuffled in recent years, and this is likely to continue for the forseeable future.

Ecologically, most of the genera represented here are fairly generalist saprobes and litter decayers, although a few have rather odd and/or specific substrates (for example, only on spruce cones or on rotting fruitbodies of certain mushrooms). *Catathelasma, Tricholoma* and *Laccaria* are ectomycorrhizal; the latter two genera are important symbiotic partners of trees in temperate forests in many places around the globe.

We've made up a number of artificial subgroups here (mostly based on stature type) to help you navigate the White-Spored Multitude. Microscopic examination paired with good technical references will likely be necessary to identify some of the little white-spored mystery mushrooms you come across

Group A (p. 285-292) – The membranous to flocculent partial veil (present at least when young) is the uniting feature for this group (except *Ossicaulis*). *Armillaria, Neolentinus* and *Pleurotus* grow on or near wood, as does the similar but veil-less *Ossicaulis. Floccularia, Catathelasma* and the four veiled *Tricholoma* included here grow on ground.

Group B (p. 292-305) – Medium-sized to large mushrooms, with white or pale yellowish, typically close to crowded, notched (sinuate) gills, and fibrous flesh, often with a farinaceous odor. Cap textures range from smooth to (more often) fibrillose or scaly: *Tricholoma.*

Group C (p. 306-315) – Medium-sized to large mushrooms with notched, broadly attached, or decurrent gills. Often in rings, arcs, or dense clusters: *Lyophyllum, Leucocybe, Aspropaxillus, Leucopaxillus, Ampulloclitocybe, Gerronema, Clitocybe, Cleistocybe, Melanoleuca* and "*Hygrophorus*" *caeruleus.* Also see Groups D and E.

Group D (p. 316-317) – Fruitbodies tough, with fibrous stipes and thick, often widely spaced gills. Fruitbody coloration usually in warm pastels, from pinkish buff to orangey, or in one species with intense purple colors: *Laccaria.*

Group E (p. 318-321) – Small to medium-sized mushrooms with broadly attached to decurrent gills, stipe fleshy-fibrous (not cartilaginous or tough): *Paralepista, Infundibulicybe* and *Clitocybe.*

Group F (p. 321-324) – Miscellaneous medium-sized to fairly large mushrooms growing from wood (can be quite decayed with some species), or brightly colored (pink in grass, purplish with yellow gills from duff). If you can't find your species here, check other white-spored sections, especially Groups A–E: *Tricholomopsis, Calocybe* and *Flammulina.*

Group G (p. 325-334) – Stipe usually cartilaginous or tough but not wiry, sometimes tapering and rooting, frequently hollow or stuffed with a stringy pith, fruitbodies small to medium sized (a few species occasionally have caps 8+ cm wide): *Baeospora, Collybia, Clitocybula, Neolentinus, Rhodocollybia, Gymnopus, Collybiopsis, Connopus, Marasmius, Gymnopus, Macrocystidia, Sagaranella* and *Tephrocybe*. Also check the larger *Mycena* in Group N.

Group H (p. 335-336) – Mushrooms small, caps typically less than 2.5 cm across. Growing on cones of various conifers, or on woody debris and duff. Cap convex to plane, stipe slender, often with spiky hairs at base: *Pseudobaeospora, Baeospora* and *Strobilurus.*

Group I (p. 337-339) – Growing on mushrooms or decaying remains of mushrooms, gills usually present (absent or rudimentary in two species), most with a sclerotium (small hard mass of tissue) at base of stipe (but often embedded in substrate of decaying mushroom tissue), one species with branches coming off stipe: *Collybia, Dendrocollybia* and *Asterophora.*

Group J (p. 337-339) – Small mushrooms with convex caps that have a distinctly depressed center and broadly attached to decurrent gills. This distinctive stature type is called omphalinoid. Growing on the ground, on moss or duff: *Cantharellula, Pseudolaccaria, Spodocybe, Myxomphalia* and *Arrhenia* (also see *Lichenomphalia ericetorum* in Group K)

Group K (p. 343-346) – Same as Group J, but growing on wood. *Arrhenia, Lichenomphalia, Chromosera, Pseudoarmillariella, Chrysomphalina* and *Xeromphalina.*

Group L (p. 347-349) – Stipes thin, tough, and often wiry, many with fruitbodies that can dry out and revive when wet: *Xeromphalina, Mycetinis, Paragymnopus* and *Marasmiellus*. Also compare with Groups G and H.

Group M (p. 350-352) – Tiny mushrooms with a slender stature, attaches to decurrent gills, a fragile, dry stipe, and conical, convex, or slightly umbilicate caps: *Resinomycena, Phloeomana, Mycena* and *Rickenella*.

Group N (p. 353-368) – Fruitbodies small, stipe fragile, cap broadly conical, peaked, or convex, usually translucent-striate when fresh and moist, fruitbodies dull to brightly colored: *Atheniella* and *Mycena.*

Armillaria solidipes

Armillaria nabsnona

Armillaria sinapina

Armillaria cf. *gallica*

White-spored (A)

Armillaria species

Honey Mushrooms

Cap: 3–15 cm across, convex or bell-shaped when young, expanding to broadly convex, plane or concave, often retaining a broad umbo. Surface with pointed tufts of grayish-brown to black hairs when young, these becoming sparse as it expands. Color varies widely in the different species (and even within species): Beige, pinkish-tan, and ocher-brown predominate, but darker browns and yellow tones are common in some collections, to paler cream in age or dry weather. **Gills:** Broadly attached, often with a decurrent tooth, close, broad. Whitish to cream, beige, or dingy pinkish tan. At times brownish (which leads to confusing with brown-spored genera). **Stipe:** 5–20 (30) cm long, 0.5–5 cm thick, usually cylindrical and curved. Surface dry, covered in a yellowish fuzz, or in cottony white chevrons adhering with grayish brown scales, or nearly smooth overall. Whitish to beige, tan or yellow, often with dusky olive-brown to grayish tones. **Partial veil:** Variable; in some forms silky to cottony-fibrous, in others more membranous, leaving an annular zone to membranous veil on stipe. **Flesh:** Thick, fleshy-fibrous in cap; stipe cottony-stringy with a cartilaginous rind. Beige to pinkish cream, staining darker pinkish buff. **Odor:** Indistinct. **Taste:** Mild. **Spore deposit:** White. **Microscopy:** Spores 6–10 × 5–8 µm, ellipsoid, smooth, colorless.

Ecology: As a group, honey mushrooms are found in practically every habitat in Cascadia. Some species are virulent pathogens, while others are weakly so. Fruiting primarily in fall.

Edibility: Edible and popular. However, the fruitbodies seem to be somewhat indigestible for many people. Cook thoroughly and sample cautiously.

Comments: The medium to large size, presence of a partial veil, growth on wood, and white spores are helpful clues for recognizing this genus. However, fruitbodies are remarkably variable even within species and the benchmarks for the different species are subtle and still not resolved. The taxonomy and nomenclature of honey mushrooms is one of the most complicated examples among all Cascadia's macrofungi (which is really saying something). *A. solidipes* (sister to *A. ostoyae*) is a robust conifer-associated species that often has floccose stipe and white membranous veil. *A. gallica* and *A. cepistipes* typically show a pinkish-brown to reddish-brown cap, a cobwebby cortina-like partial veil, bulbous stipe base, and tendency to grow more solitary. *A. sinipina* has a similar stature but usually has yellowish fuzz on the cap and yellow floccose stipe when young. *A. altimontana* is more common in the eastern Cascades and Rocky Mountains. *A. mellea* (sensu western NA) are typically in dense clusters on hardwoods with tapering stipes. *A. nabsnona* has a nearly smooth cap, and a thin, tissue-like partial veil. It grows with a wide range of trees but prefers alder and bigleaf maple.

Neolentinus ponderosus

(O.K. Mill.) Redhead and Ginns

Giant Sawgill

Cap: 5–30 (40) cm across, hemispheric when young, becoming broadly convex to plane, becoming more wavy in age. Surface dry, covered in broad, flat, patches and scales, cracks between scales becoming deep and pronounced when dry. Whitish to beige to yellowish-beige in age. **Gills:** Broadly attached to slightly decurrent, fairly close, edges distinctly serrated when young, becoming more ragged in age. Whitish to creamy to pale yellow in age. **Stipe:** 5–15 cm long, 2–6 cm thick, central to off-center, rarely lateral, equal or swollen in the middle, usually curved and often tapered downwards and rooting into the substrate. Dry, covered with recurved scales that become appressed or obscure in age. Whitish to pale creamy-ivory, yellower in age, often with rusty spots and stains near base. **Flesh:** Thick, firm-rubbery, fibrous and very tough, especially in stipe. Whitish. **Odor:** Indistinct to fragrant, fruity when young, musty-grainy in age. **Taste:** Mild. **Spore deposit:** White. **Microscopy:** Spores 8.5–10 × 3.5–4.5 µm, long-ellipsoid, smooth.

Ecology: Solitary or in small clusters on well-decayed stumps and bases of standing dead trunks of various conifers; especially partial to lodgepole pine. Most common on the eastern slopes of the Cascade Range, but is scattered in lowlands, especially in urban areas. Fruiting primarily from late spring through summer.

Edibility: Edible and fairly popular when young and fresh, but needs special cooking attention—the flesh is tough and rubbery.

Comments: The size of the fruitbodies alone set this species apart from almost all other mushrooms in our area; combined with the serrate gill edges, pale colors, and growth on conifer debris, it is nearly unmistakable. This species is not closely related to other gilled mushrooms, but rather to *Gloeophyllum*. *Cantharocybe gruberi* is similar in shape and habitat but grows from duff, has whitish to yellow colors, and lacks cap scales and serrate gills. Some *Pholiota* have very scaly caps and grow on wood but are more richly colored, have brown spores, and are never as large.

Pleurotus dryinus

(Pers.) P. Kumm.

Veiled Oyster

Cap: 5–13 cm across, convex with an inrolled edge, becoming broadly convex to plane, and often slightly wavy and uplifted in age. Surface dry, covered with pale, broad appressed-fibrillose scales when young, nearly smooth in age, margin often with thicker whitish remnants of partial veil. Gray to grayish tan, becoming grayish white to creamy white overall. **Gills:** Deeply decurrent, close. Pale creamy to white. **Stipe:** 3–8 cm long, 1.5–4 cm thick, attached asymmetrically to cap, tapered toward base and often deeply embedded in substrate. White. **Partial veil:** White, ragged membranous tissue forming a weak zone on stipe, or more often remaining attached to margin of cap. **Flesh:** Thick, firm, rubbery, white. **Odor:** Slightly fragrant (anise-like) to musty. **Taste:** Mild. **Spore deposit:** White. **Microscopy:** Spores 9.5–14 × 3.5–4.5 µm, cylindrical to elongate-ellipsoid, inamyloid, smooth.

Ecology: Solitary, in fused pairs or in small groups on dead or partially dead hardwoods, often on bigleaf maple and alder. Rare in most of Cascadia, fruiting from late summer through fall.

Edibility: Edible and good, but hard to find and not as prolific as other *Pleurotus*.

Comments: The growth on wood, off-center stipe, decurrent white gills, and partial veil (often as remnants at cap edge or in a weak zone on stipe) are distinctive. Although confusion with other *Pleurotus* species is possible when the partial veil is not apparent, the distinctively textured cap (with appressed-fibrillose patches) helps to set this species apart. *Ossicaulis lignatilis* is typically smaller, lacks a partial veil, and does not have deeply decurrent gills. *Hypsizygus tessulatus* has a more central stipe, broadly attached (not decurrent) gills, lacks a partial veil, and has a smooth cap, often with a spotted pattern.

Ossicaulis lignatilis

(Pers.) Redhead & Ginns

Bonestem

Cap: 3–10 cm, circular or asymmetric, rounded to convex when young, becoming plane in age, margin finely inrolled and sometimes slightly ribbed. Surface dry, opaque, smooth or somewhat rough-textured. Chalky-white when young, often appearing 'water-spotted' and developing dull yellowy or rusty tones in age. **Gills:** Broadly attached, or with a narrow notch, crowded, narrow. White to creamy. **Stipe:** 1–6 cm long, 0.5–2 cm thick, usually tapered downwards and fused into clusters; on some specimens the stipe is very reduced. Surface dry, smooth, base often with white mycelial rhizomorphs. Whitish to pale beige, discoloring brownish at base in age. **Flesh:** White, solid, firm-rubbery when young but becoming very hard upon drying. **Odor:** Farinaceous. **Taste:** Indistinct to farinaceous. **Spore deposit:** White. **Microscopy:** Spores 4–5.5 × 2.5–3.5 µm, ellipsoid, smooth, inamyloid, colorless in KOH.

Ecology: Usually found in small clusters in crevices and hollows of large-diameter decaying wood of both hardwood and conifers. Uncommon throughout Cascadia, mostly fruiting in fall and early winter.

Edibility: Likely non-toxic, but tough, rare, and not sought as food.

Comments: The crowded white gills, habit of fruiting hidden inside hollow trunks and stumps, and tough flesh help distinguish this otherwise nondescript mushroom. *Pleurotus* usually have a fleshy texture, more asymmetrical caps, less well-defined central stipes (or no stipes at all), and more deeply decurrent gills. Various *Clitocybe* and other veil-less white-spored genera can be superficially similar, but have slightly different surface and flesh textures, and are mostly terrestrial. The Elm Oyster (*Hypsizygus tessulatus*) has a white to creamy cap—which often shows a blotchy pattern of watery spots. It grows on the exposed outer surfaces of standing, dead, or decaying hardwood trunks—especially in riparian areas. It is rarely reported from Cascadia but may be overlooked.

Floccularia albolanaripes

(G.F. Atk.) Redhead

Golden Robes

Cap: 4–12 cm across, ovoid to convex with an umbo when young, becoming broadly convex; margin at times uplifted and wavy in age. Surface moist to dry, often retaining remnants of partial veil as a cottony band around margin, or as thin, triangular flaps of tissue. Dark brown at center, ocher-brown to golden brownish to bright yellow towards margin, often with appressed-fibrillose brown scales and streaks, creamy-yellow to whitish at extreme margin. **Gills:** Narrowly attached, notched, close; edge eroded in age. Whitish to creamy or buttery-yellow, often with a dull apricot wash in age. **Stipe:** 5–10 cm long, 1–4 cm thick, enlarged towards base when young, equal in age. Surface dry, smooth and creamy or yellowish to light apricot above annular zone, lower stipe pale creamy and abundantly covered in shaggy, cottony white veil with scattered bright lemon-yellow to golden-brown plaques and chevrons. **Partial veil:** Thick, cottony, and membranous, leaving an annular zone on stipe and remnants on cap margin. **Flesh:** Thick, soft, fleshy in cap, stipe fibrous, solid. Creamy white. **Odor:** Indistinct to slightly mealy. **Taste:** Mild or faintly like corn meal. **Spore deposit:** White. **Microscopy:** Spores 6–9 × 4–5 µm, ellipsoid, smooth, amyloid

Ecology: Solitary, in pairs, or scattered in small troops, in thick duff of conifers, especially pine and Douglas-fir. Occasional to locally common in mid-elevation west-side forests (occasional coastal), and under a range of conifers—especially east side pine forests. Fruiting from mid-fall through winter (lower elevations) and spring (mountains).

Edibility: Edible and very good when young, older collections are often slightly musty.

Comments: This attractive mushroom is found throughout Cascadia but rarely is abundant. It is recognized by the yellow to golden-brown cap, cream-white gills, and creamy stipe with shaggy scales. *Amanita augusta* is similar in color but has discrete yellowish warts on the cap surface, a skirt-like partial veil, a distinctly bulbous stipe base, and typically taller stature. *Tricholoma focale* is vaguely similar but has orange to greenish cap colors, white gills which develop orange-brown stains, and a strong farinaceous odor and taste

Tricholoma murrillianum

Singer

Western Matsutake

Cap: 6–20 cm across, peg-shaped to convex in button-stage, expanding to broadly convex and eventually plane; margin inrolled and covered with veil when young, to down-curved with veil remnants, eventually flattening or flaring slightly upward. Surface dry to slightly viscid when wet, often with appressed cottony patches or fibrils. White to creamy buff at first, with fibrils becoming orangish to pale brown. **Gills:** Narrowly to broadly attached, close to really crowded, narrow; partial gills numerous. Whitish, creamy white to pale creamy pink when young, developing orangish spots or stains in age. **Stipe:** 5–15 cm long, 1.5–4.5 cm thick, equal or tapering to a pointed base. Surface dry to slightly viscid, with fine chevrons of powdery scales above veil, with appressed flocculose scales below veil that can gelatinize slightly in wet weather. Whitish to creamy with orangish to brownish scales or fibrils below ring, whitish above. **Partial veil**: Thick, cottony, flaring upward at first, becoming appressed in age. **Flesh:** Thick, very firm to rubbery, whitish to creamy white at first, slowly discoloring orangish or with orangish stains around worm holes. **Odor:** A mix of sweet cinnamon spiciness and odd mustiness. Some people only smell the "good" spicy odor, while others recoil with disgust. **Taste:** Indistinct. **Spore deposit:** White. **Microscopy:** Spores 6–8 × 4.5–6 µm, broadly ellipsoid, smooth.

Ecology: Solitary, in small "nests," or scattered on ground, often buried in duff. Found in a wide variety of forest types, but largely absent from wetter west-side forests, except along the immediate coast. Common in sandy soil under coastal shore pine and in drier Douglas-fir, true fir, and hemlock forests, or with lodgepole pine in the eastern Cascades. Also fond of tanoak in the southern portion of our range. Fruiting in fall.

Edibility: Edible and highly prized. The strong flavor goes a long way, and a small amount can season a whole dish. Most people recommend thinly slicing the mushrooms into a stir-fry or using them to flavor rice or other grains. They also make a great addition to clear soups.

Tricholoma murrillianum

Tricholoma murrillianum

Comments: Other common names for this mushroom include Pine Mushroom or White Matsutake. The whitish cap, crowded, creamy white gills, thick, cottony veil, firm, rubbery flesh, and distinctive cinnamon odor make the Matsutake quite distinctive. *T. dulciolens* shares the odor but is usually smaller and has brown scales on the cap and stipe; it has a bitter taste. The often-massive *Catathelasma ventricosum* has a whitish to gray cap, decurrent gills, a thick partial veil that forms a double ring, an often deeply rooting stipe, and a strong, farinaceous odor and taste. Care should be taken not to mistake Matsutake for the potentially deadly *Amanita smithiana* and *A. silvicola*; both have relatively fragile flesh and an unpleasant "wet dog" or chlorine-like odor, but occasionally young specimens of *A. smithiana* can smell just like Matsutake.

Misapplied Name: *Tricholoma magnivelare* (Peck) Redhead, which is a distinct, eastern North American species.

Tricholoma dulciolens

Kytöv.

Bitter Matsutake

Cap: 4–9 cm across, rounded to convex with an inrolled margin at first, becoming broadly convex to plane, or with a low umbo. Surface dry, covered with radially arranged appressed to slightly recurved, dark brown to blackish fibrils or floccules over a creamy white to buff base color. **Gills:** Narrowly attached with a short notch, close to crowded, narrow. White, developing orangish brown spots and stains in age. **Stipe:** 7–12 (18) cm long 1–2.5 cm thick, tall and cylindrical, equal or tapering towards an often-rounded base. Surface dry, with belts or scattered patches of appressed to floccose, dark brown scales over a creamy white to pale tan base color below the veil, whitish to cream with faint belts of recurved fibrils above the veil. **Partial veil:** Forming an upward flaring membranous ring when young, then hanging and ragged in age. Whitish to pale brown. **Flesh:** Very firm, somewhat rubbery, stipe solid and fibrous. White, unchanging. **Odor:** Cinnamon-spicy, with musty undertones. **Taste:** Mildly to strongly bitter. **Spore deposit:** White. **Microscopy:** Spores 5.3–7.2 × 4.3–6.2 µm, broadly ellipsoid to subglobose, smooth.

Ecology: Solitary or scattered from soil in mixed forests dominated by Douglas-fir, western hemlock, and fir. Rather uncommon, with most records from east-side Cascade forests. Fruiting in fall.

Edibility: Edible, but often bitter tasting. However, the bitterness seems to vary considerably between collections.

Comments: *T. dulciolens* is recognized by its tall, often slender stature, and dark brown fibrils or scales over a creamy white to buff cap and stipe, the membranous partial veil, and the spicy-musty odor. *T. murrillianum* has a similar distinctive odor but typically has a whitish cap, and a thick, creamy white cottony veil. Although it can show orangish to orange-brown scales, they are more warmly colored (rarely as dark as in *T. dulciolens*). Additionally, *T. murrillianum* lacks the bitter taste of *T. dulciolens* and tends to be shorter and much stouter.

Misapplied name: *Tricholoma caligatum* (Viv.) Ricken

Catathelasma ventricosum group

Large Cats

Cap: 8–25 (35) cm across, tightly rounded and peg-like when young, with an inrolled margin, becoming convex to broadly convex, eventually becoming plane to slightly wavy and uplifted in age. Surface dry to slightly tacky, smooth or with flocculent patches. Color variable within a range; pale to dark gray or grayish brown, often with whitish blotches when young, becoming paler, to dingy gray, pale gray to almost white, with darker and paler areas, and often with whitish, flocculent patches, then typically becoming grayish brown in age. Occasionally with orangish stains. **Gills:** Decurrent, often in a curving manner, close to crowded. Creamy white to pale creamy buff. **Stipe:** 10–30 (40) cm long, 2–7 (10) cm thick, equal aboveground, then with a long tapering taproot. Occasionally swollen at ground level, then tapering below. Grayish to whitish, staining or discoloring orangish. **Partial veil:** 2-layered, persistent, set rather high on the stipe. Outer layer grayish, membranous, leaving bands on the stipe. Inner layer thick, white, flaring to ragged, often leaving patches on the cap margin. **Flesh:** Very thick, firm, rubbery–hard. White. **Odor:** Strongly farinaceous; older specimens can be more rancid. **Taste:** Farinaceous. **Spore deposit:** White. **Microscopy:** Spores 8–12 (14) × 4–5.5 µm, ellipsoid, smooth, amyloid.

Ecology: Solitary or scattered, more rarely in small clusters of 2–4 fruitbodies in thick humus, needle duff, or mossy conifer forest. Common along the coast under Sitka spruce and on the west slope of the Cascades, scarce east of the crest. Fruiting from summer (earliest at high elevations) into late fall, most abundant in early fall.

Edibility: Edible but tough.

Comments: This massive fruitbodies of this species are among the largest and heaviest in all of western north America! Besides the large size, the white to grayish colors, decurrent gills, thick, flaring double veil, tapered, rooting stipe, and farinaceous odor help clinch the identification. The name *C. ventricosum* is problematic, as it was originally described from Alabama, referring to a mushroom with a "shining white" cap growing with hardwoods; it is undoubtedly distinct from the (as yet unnamed) western species described here. *C. evanescens* is a bit smaller, with a shorter stipe, and cinnamon-brown to dingy brown colors; it is currently known from east of the Cascade crest with Engelmann spruce (but should be looked for in the Olympic Mountains). The Western Matsutake (*Tricholoma murrillianum*) has a musty cinnamon-like odor and notched (not decurrent) gills, and is overall whiter, with ocher-brown blotches in age. *Amanita smithiana* has a powdery partial veil that breaks up and leaves hanging remnants around the cap margin.

Tricholoma focale

(Fr.) Ricken

Veiled Orange Trich

Cap: 3–12 (15) cm across, convex with an inrolled margin when young, becoming broadly convex to nearly plane. Surface viscid to moist, with small appressed scales or matted fibrils, to nearly smooth. Color variable, orange but often with deep olive-green to forest green colorations in button stage, with reddish brown appressed scales or fibers, becoming orange-brown to reddish orange in age. **Gills:** Narrowly attached with a distinct notch, or even appearing free at times, close to crowded, relatively narrow; partial gills numerous. Dingy white to pale buff-white, developing orange-brown stains in age and slowly bruising orange-brown when damaged. **Stipe:** 4–11 cm long, 1–3 cm thick, more or less equal, but tapering or pointed at base. Surface dry, covered with orangish belts of scales over a paler base color, whitish above veil, discoloring orange-brown in age below veil. **Partial veil:** Thick, membranous, somewhat sheathing, leaving a flaring ring and often orangish scaly belts. **Flesh:** Moderately thick to thin, firm, fibrous; stipe solid. Whitish, slowly staining orange to orange-brown. **Odor:** Strongly farinaceous. **Taste:** Unpleasant, strongly bitter-farinaceous. **KOH:** Bright red on cap, reddish on stipe, little reaction elsewhere. **Spore deposit:** White. **Microscopy:** Spores 5–6 × 3–4 µm, broadly ellipsoid, smooth.

Ecology: Solitary or scattered on ground, in moss or duff in a wide variety of forest types. Common in sandy coastal forests with shore pine and spruce, and in the eastern Cascades with mixed conifers. Fruiting in late summer and fall.

Edibility: Inedible due to unpleasant taste.

Comments: The orange to greenish-blushed cap, whitish gills which stain orangish, membranous partial veil, and farinaceous odor and taste help distinguish *T. focale*. *T. badicephalum* is similar in many regards but has predominantly brown colors. *T. aurantium* has the orange to green-stained cap but lacks a partial veil; its stipe has orange chevrons or belts of scales covering the lower three-quarters, terminating abruptly below a smooth white band near the apex. *Floccularia albolanaripes* has a yellow to yellow-brown cap, creamy white to pale yellow gills, a flocculose white partial veil, and a creamy whitish shaggy-scaly stipe with yellow belts or bands.

Synonym: *Tricholoma zelleri* (D.E. Stuntz & A.H. Sm.) Ovrebo &Tylutki

Tricholoma badicephalum

(Zeller) N. Siegel, S.A. Trudell & M.J. Gordon

Bay-headed Trich

Cap: 3–10 cm across, convex with an inrolled margin when young, becoming broadly convex to nearly plane. Surface moist to dry, often with distinct appressed scales or matted fibrils, to nearly smooth. Tan, light brown to orange-brown, at times dark brown, or with more orange fibrils or scales. Occasionally with greenish discolorations. **Gills**: Narrowly attached with a distinct notch, or even appearing free at times, close to crowded, relatively narrow; partial gills numerous. Dingy white to pale buff-white, developing orange-brown stains, at times extensively dull orange. **Stipe**: 4–8 (10) cm long, 1–3 cm thick, equal with a tapering or pointed base. Surface dry, covered with irregular bands of scales below the annular zone. Base color whitish; scales brownish to orangish, darkening overall with age. **Partial veil**: Thick, membranous, somewhat sheathing, leaving a flaring ring at first, sparse in age. **Flesh**: Moderately thick to thin, firm, fibrous; stipe solid. Whitish, slowly staining orange to orange-brown. **Odor**: Strongly farinaceous. Taste: Farinaceous. **Spore deposi**t: White. **Microscopy**: Spores 4.5–7 × 3.3–4.5 µm, broadly ellipsoid to ovoid, smooth.

Ecology: Solitary or scattered on ground, in moss or duff under conifers, typically in sandy soil. Locally common in coastal forests, rare elsewhere. Fruiting in fall.

Edibility: Unknown.

Comments: This *Tricholoma* is recognized by its brown cap, whitish gills which develop orangish stains, membranous partial veil, and scaly stipe. *T. focale* is very similar but typically has brighter orange colors and is more likely to show greenish discolorations. *T. vaccinum* has a dry, scaly, brown to orange-brown cap, and a cottony partial veil. *T. aurantio-olivaceum* is a small, slender species which lacks a partial veil; it has a creamy orange to dingy olive or orange-brown cap and white gills that become dingy ocher-brown in age. Older specimens resemble *Inocybe* but can be distinguished by the white spores.

Tricholoma vaccinum

(Schaeff.) P. Kumm.

Cap: 2–8 cm across, conical to rounded when young, becoming convex to plane; margin tightly inrolled and cottony in the button stage, often ribbed and tufted with veil remnants when young. Surface dry, matted-fibrillose to appressed-fibrillose, covered with small, upturned tufts or scales and often radially cracking in age. Reddish brown to rusty brown, lighter toward margin. **Gills:** Narrowly to broadly attached, often with a distinct notch, close to subdistant, moderately broad; partial gills numerous. Creamy to pale buff, with orangish brown spots or stains in age. **Stipe:** 3–9 cm long, 0.5–2 cm thick, more or less equal or tapered toward base. Surface dry, matted-fibrillose, covered with tufts and scales and pubescent veil remnants when young. Whitish to dingy buff base color with darker orangish buff scales or fibers, darkening to orangey brown in age. **Partial veil:** Cottony, often clinging to cap margin and not leaving a ring on stipe. **Flesh:** Thin, fleshy-fibrous, whitish to buff; stipe tough, fibrous, stuffed when young, becoming hollow, all parts slowly staining orangish brown. **Odor:** Farinaceous. **Taste:** Farinaceous-bitter, although mild at times. **Spore deposit:** White. **Microscopy:** Spores 6–7 × 4–5.5 µm, ellipsoid, smooth.

Ecology: Often scattered in troops or in small clusters on ground under conifers, especially spruce or fir. Common in coastal forests, and in the high cascades. Fruiting in fall.

Edibility: Besides conflicting reports on the edibility of this species, the tough flesh and bitter-farinaceous taste discourage experimentation.

Comments: The dry, scaly, reddish brown to orange-brown cap, whitish to buff gills that discolor orangish, cottony partial veil when young, finely scaly stipe, and growth with spruce help identify this pretty Trich. *T. imbricatum* is larger, has a dark brown cap with small scales, whitish to buff gills, a light brown stipe, and often grows with pine. *T. aurantio-olivaceum* is a small, slender species that has an ocher-beige to buff-brown cap and grows in mixed forests. A poorly known, and apparently undescribed species closely related to the European *T. psammopus* is known from the Oregon Cascades. It has dark brown finely-scaly conical-convex cap and scaly stipe.

Tricholoma ammophilum

A.D. Parker, Grubisha & S.A. Trudell

Sand-loving Trich

Cap: 4–12 cm across, convex with an inrolled margin at first, becoming broadly convex to plane or lobed and wavy in age. Surface smooth, viscid, typically with adhering debris. Light brown, orange-brown to reddish brown, becoming quite dingy in age. **Gills:** Narrowly to broadly attached, with a distinct notch and often a decurrent tooth, close to crowded, broad; partial gills numerous. Creamy white to orangish white, or pale orangish buff, developing orangish brown spots or stains in age. **Stipe:** 4–10 cm long, 1–3 cm thick, more or less equal. Surface dry, smooth. Whitish, discoloring and becoming dingy in age. **Flesh:** Moderately thick, firm; stipe fibrous. Whitish with buff to brownish discolorations in stipe. **Odor:** Strongly farinaceous. **Taste:** Strongly farinaceous. **Spore deposit:** White. **Microscopy:** Spores 4.5–6.5 × 3–4.5 µm, ellipsoid, smooth.

Ecology: Often in troops or clusters in sandy or muddy soil under cottonwood and aspen. Common across Cascadia, fruiting in fall.

Edibility: Edible but not generally recommended. Easily confused with a number of similar brown *Tricholoma*, of which many are toxic.

Comments: The brown colors which become dingier with age, white spores and growth with cottonwood and aspen are helpful in identifying this Trich. Older specimens with brown gills often resemble *Hebeloma*; white vs. brown spore deposits (respectively) separate them. Other thinly viscid brownish *Tricholoma* are best separated by tree association and stature: With oaks we have the tall-statured *T. dryophilum* and the stockier *T. subannulatum.* Taller statured specimens with conifers (and sometimes birch) are likely in the *T. fulvum* group; *T. albobrunneum* (= *T. muricatum*) is typically stockier and grows with pine. All these look-alikes are toxic.

Misapplied name: *Tricholoma populinum* J.E. Lange

Tricholoma subannulatum

(Peck) Zeller

Cap: 4–15 cm across, convex to broadly convex, margin slightly inrolled and often ribbed when young, becoming wavy or lobed in age. Surface with scattered appressed fibrils or smooth; thinly viscid when wet. Color variable, grayish orange, orangish brown to brown, often darker at disc, darkening overall in age. **Gills:** Narrowly attached with a distinct notch, close to crowded, narrow to relatively broad; partial gills numerous. Creamy white to yellowish, developing orange-brown spots or stains in age and slowly bruising rusty orange-brown when damaged. **Stipe:** 4–9 cm long, 1.5–3 cm thick, more or less equal, or slightly tapered towards base. Surface dry, slightly fibrous or finely scaly or scruffy at apex. White to creamy white at apex, becoming orange-brown to brown from base up in age. **Flesh:** Firm, thick, tough, whitish, slowly staining orange-brown in cap; stipe tough, fibrous, solid or sometimes hollow in age, whitish. **Odor:** Indistinct or slightly farinaceous. **Taste:** Bitter. **Spore deposit:** White. **Microscopy:** Spores 5.5–7 × 3.5–5 µm, broadly ellipsoid, smooth.

Ecology: Solitary or scattered under oaks or other hardwoods. Locally common in suitable habitat but rather localized in Cascadia. Fruiting in fall.

Edibility: Likely toxic.

Comments: The grayish orange to brown cap, whitish to yellowish gills, lack of a distinct odor, strongly bitter taste, and growth with oaks are distinctive. Many Cascadia collections are misidentified as *T. dryophilum*, which is typically taller in stature and often starts off paler in color, developing orangish streaks and stains. It also has a strongly farinaceous odor and taste (although the viscid cap layer is bitter). *T. albobrunneum* (= *T. muricatum*?) has a pinkish brown to reddish brown cap, buff to orangish brown gills, a bitter-farinaceous taste, and grows with pine. *T. manzanitae* has a creamy white to pinkish buff cap when young but darkens as it ages and could be mistaken for *T. subannulatum*; however, it has a creamy stipe with yellowish granules or fine scales at the apex and grows with manzanita or madrone. Also compare with the conifer-associated *T. fulvum* and with *T. ammophilum* which grows with cottonwood and aspen.

Tricholoma fulvum

(DC.) Bigeard & H. Guill.

Cap: 3.5–12 (15) cm across, conical to convex when young, becoming broadly convex to plane with a low broad umbo, margin inrolled and ribbed when young, staying downcurved well into maturity. Surface viscid, appressed-fibrillose, rusty brown, orange-brown to brown, often with darker reddish brown appressed fibrils or fine scales at disc. **Gills:** Narrowly to broadly attached, with a distinct notch, close to subdistant, broad; partial gills numerous. Creamy white, pale buff or yellowish young, developing reddish brown spots or stains in age. **Stipe:** 5–15 cm long, 1–2 cm thick, tall, often twisted, equal, enlarged lower, but tapering or slightly rooting at base. Surface dry, matted-fibrillose, often with fine silky hairs, base often with matted whitish mycelium. Dingy buff to orange-brown when young, darkening to brown from base up in age. **Flesh:** Relatively thin, firm, fibrous, whitish to creamy buff; stipe stuffed when young, quickly becoming hollow, fibrous, whitish when young, soon buff to brownish. **Odor:** Farinaceous. **Taste:** Farinaceous. **Spore deposit:** White. **Microscopy:** Spores 5–8 × 3.5–6 µm, ellipsoid, smooth.

Ecology: Solitary, scattered, or in small clusters on ground or in moss or duff. Common under conifers across much of western Cascadia, occasionally with birch to the north. Fruiting in fall.

Edibility: Likely toxic.

Comments: This Trich is recognized by its tall stature, rusty brown viscid cap, creamy to pale buff gills that are often spotted, orangish brown, hollow stipe, and growth with conifers. It belongs to a confusing group of toxic, brown-capped *Tricholoma* which include *T. albobrunneum* (= *T. muricatum?*) which typically grows with pine, has a streaked, dull reddish brown to pinkish brown cap and creamy orange gills. *T. stans* (sensu PNW) is also very similar but seems to have paler gills and a punctate stipe with a contrasting apex. *T. ammophilum* grows in troops under cottonwood and aspen. *T. subannulatum* and *T. dryophilum* with oaks. *T. batschii* (= *T. fracticum)* has an orange-brown to red-brown cap, and a viscid, cortina-like partial veil when young that leaves a distinct line separating the white upper and orange-brown lower stipe; it is found with pines in urban areas in California but should be expected in Cascadia.

Taxonomic note: We called this species *T. nictitans* sensu CA in MotRC, but recent work suggests that *T. nictitans* and *T. transmutans* are synonyms of *T. fulvum*; it has also been called *T. pessundatum* by some authors. More work is needed to delimit cryptic species in this complex.

White-spored (B)

Tricholoma terreum var. *cystidiotum*
(Shanks) Blanco-Dios

Mouse Trich

Cap: 1.5–6 cm across, conical to convex when young, becoming broadly convex to nearly plane and often with a low broad umbo; margin tomentose, inrolled at first, often slightly uplifted and splitting in age. Surface dry, matted-tomentose to felty when young, becoming more appressed-fibrillose to finely scaly in age. Dark grayish black to mouse-gray with a paler margin, becoming pale gray to grayish tan in age. **Gills:** Narrowly attached with a distinct notch, or appearing free at times, close to subdistant, broad; partial gills numerous. Pale gray when young, becoming grayish white to whitish, occasionally with darker spots or stains. **Stipe:** 2–6 cm long, 0.4–1 cm thick, equal or slightly bulbous at base. Surface dry, silky-fibrillose. Whitish to pale gray, becoming yellowish gray, often discoloring yellowish when handled. **Partial veil:** Faint, silky cortina leaving gray hairs on stipe, usually only visible when young. **Flesh:** Thin, fragile in cap, whitish to pale gray; stipe fibrous, solid to hollow, whitish. **Odor:** Indistinct. **Taste:** Indistinct. **Spore deposit:** White. **Microscopy:** Spores 6–8.5 × 3.5–5.5 µm, ellipsoid, smooth. Cheilocystidia scattered to abundant, club-to spoon-shaped.

Ecology: Solitary, scattered, or in small clusters under conifers (Douglas-fir or pine). Widespread and common, especially at lower elevations and in urban areas. Fruiting from mid-fall into winter.

Edibility: Edible but not recommended. Be careful with your identification, as the edibility of many gray *Tricholoma* is still unknown and a few are poisonous.

Comments: The dark gray to mouse-gray, felty cap, pale gray to whitish gills, small size, and growth with conifers help identify this dainty Trich. *T. triste* fruits in the mountains shortly after snowmelt. It has an appressed-fibrillose, gray cap (often paler than that of the Mouse Trich), whitish to pale gray gills, a white to pale grayish stipe, and no partial veil. *T. atrosquamosum* sensu PNW has a gray cap with blackish fibrils or scales, white gills, and a pale gray stipe ornamented with appressed blackish fibrils or scales; it is rare in Cascadia. *T. argyraceum* (= *T. sp. 'Stanford Gray'*) has slightly larger fruitbodies, with a strong farinaceous odor and taste; grows with oak.

Tricholoma triste
(Scop.) Quél.

Cap: 1–4 cm across, convex to nearly plane, margin fibrillose, undulating in age. Surface dry, tomentose to appressed-fibrillose to felty when young, becoming more appressed-fibrillose to finely scaly, and often radially splitting in age. Dark gray, brownish gray to pale gray, with paler colors between fibrils and cracks. **Gills:** Narrowly attached with a distinct notch, or appearing free at times, close to subdistant, broad; partial gills numerous. White to pale gray. **Stipe:** 2–6 cm long, 0.4–1 cm thick, equal or club-shaped. Surface dry, fibrillose, often pruinose at apex. Whitish to pale gray. **Flesh:** Very thin, fragile in cap, whitish to pale gray; stipe fibrous, solid to hollow, whitish. **Odor:** Indistinct. **Taste:** Indistinct. **Spore deposit:** White. **Microscopy:** Spores 7–10.5 × 4–6 µm, ellipsoid to oblong, smooth.

Ecology: Scattered or in small clusters, more rarely solitary, from ground or duff under conifers, especially true firs. Often forming "shrumps" that push up duff and debris. Common in the eastern Cascades, fruiting immediately following snowmelt in spring and early summer.

Comments: The small size, gray, fibrillose cap, white stipe and growth in spring help identify this dainty Trich. Although not strictly a "snowbank fungus," it will fruit within days of the winter snowpack melting. *T. terreum* var. *cystidiotum* is very similar, but often has a fuzzier appearance and fruits in fall. *T. argyraceum* is a larger fall-fruiting, oak-associated species with a fine cortina-like partial veil. ITS sequences have shown a very close relationship between *T. triste* and *T. bonii* but more work is needed to sort out this group in Cascadia.

Synonym: *Tricholoma moseri* Singer

Tricholoma vernaticum

Shanks

Cucumber Trich

Cap: 5–15 cm across, convex to plane; margin often becoming wavy and occasionally slightly uplifted in age. Surface dry, with whitish cottony tissue pasted to disc when young, appressed-fibrillose outward. Color variably whitish to creamy gray, with a white bloom of veil over the disc when young, blotchy light to dark gray in age; at times developing brownish tones. **Gills:** Attached with a distinct notch, close to somewhat crowded, broad. White to dingy white, developing grayish or yellowish tones in age. **Stipe:** 5–15 cm long, 1–3.5 cm thick, equal or club-shaped. Surface dry, fibrillose-scaly below annular zone, scurfy above. White to off-white, staining pale orangish brown when handled. **Partial veil:** Membranous to cottony, leaving a ring on upper stipe, disappearing in age. **Flesh:** Thick, firm; stipe solid, fibrous. White. **Odor:** Strongly of cucumber. **Taste:** Like cucumber or watermelon rind. **Spore deposit:** White. **Microscopy:** Spores 8–12 × 5–6 µm, ellipsoid, smooth.

Ecology: Solitary or scattered from soil, often in deep duff and forming "shrumps" in east-slope Cascade forests. Fruiting in spring and early summer.

Edibility: Reportedly edible, but only with limited sampling. Caution is warranted.

Comments: Few spring-fruiting mushrooms rival the size of *Tricholoma vernaticum* in the eastern Cascades; this, along with the gray colors, white veil patch on the cap, partial veil, and cucumber odor clinch the identification. There are two similar undescribed *Tricholoma* we have seen in the southern Cascades: A large species with pale citrine-yellow to creamy colors and a fine partial veil, as well as an all-white species with few other distinguishing features; both lack any distinct odor. Some *Melanoleuca* are similar but are more soft and thin-fleshed, have more slender stipes and lack the partial veil. A few similar *Lyophyllum* species grow in clusters.

Tricholoma pardinum

sensu PNW

Leopard Trich

Cap: 4–12 (15) cm across, convex, broadly convex to plane; margin down-curved when young, becoming wavy and sometimes splitting in age. Surface dry, appressed-fibrillose when young, becoming scaly. Whitish to pale gray with gray to blackish tufted scales on the disc or appressed fibers that break up into dark gray scales. **Gills:** Narrowly attached, with a distinct notch, close to crowded, broad; partial gills numerous; edge becoming ragged in age. Whitish, creamy to pale creamy buff in age. **Stipe:** 7–11 cm long, 1–3 cm thick, equal to club-shaped with an enlarged base. Surface dry, smooth to appressed-fibrillose. White to dingy white, often bruising pale orangish brown near base. **Flesh:** Thick, firm, fibrous and solid in stipe. Whitish to pale gray; stipe base often with orangish stains. **Odor:** Farinaceous. **Taste:** Farinaceous. **Spore deposit:** White. **Microscopy:** Spores 6–10.5 × 4–7 µm, ellipsoid, smooth.

Ecology: Solitary or scattered on ground or moss with both conifers and hardwoods. Uncommon in most years. Fruiting in fall.

Edibility: Poisonous, causing severe gastrointestinal distress.

Comments: The dry, whitish to gray cap with gray scales, white gills and stipe, and large, stocky stature are the best features to identify this mushroom. Younger or wetter caps sometimes lack the gray color and can be whitish and without the gray scales or be extensively gray with dense scales. *T. venenatoides* typically has tan to gray-brown scales; there is some question as to whether it is conspecific with *T. pardinum* as presented here. *T. smithii* is a very similar, perhaps slightly paler species from the Rocky Mountains. *T. vernaticum* is a large, gray, montane species that fruits in the spring or summer with a smooth, whitish to gray cap that often has a whitish patch on the disc, a faint partial veil that leaves a slight ring on the stipe, and a strong cucumber odor.

Tricholoma subacutum

Peck

Cap: 4–10 (12) cm across, conical when young, becoming convex to plane or with a slightly uplifted margin when mature, with a well-defined umbo at all stages. Surface dry to moist, smooth when young, but made up of appressed streaked fibers that can form small, uplifted scales when old or dry. Dark gray to gray or brownish gray with black or dark gray streaks or appressed virgate fibers. **Gills:** Narrowly attached with a distinct notch or appearing free at times, close, broad, often ragged in appearance. Off-white, dingy white to pale gray; edge often darker gray. **Stipe:** 7–15 cm long, 1–2 cm thick, equal or slightly swollen in middle, base with an abrupt bulb at times. Surface dry, smooth. White to dingy white, sometimes discoloring yellowish orange on lower portions. **Flesh:** Fibrous, relatively thin. Off-white to pale gray in cap, stipe stuffed with white pith when young, becoming hollow in age. **Odor:** Faint, farinaceous. **Taste:** Slightly bitter-farinaceous or disagreeable at first, slowly (20–30 seconds) becoming strongly acrid. **Spore deposit:** White. **Microscopy:** Spores 6–8.5 × 4.5–6.5 µm, broadly ellipsoid, smooth.

Ecology: Solitary or scattered in duff or moss in conifer forest. Locally common in western Cascadia, occasional elsewhere in our range, fruiting in fall.

Edibility: Likely toxic.

Comments: The umbonate, gray, streaked cap that is never truly viscid, dingy whitish gills, and acrid taste help identify this species. *T. subumbrinum* has a similarly shaped dingy yellow-brown to gray-brown cap and a strongly farinaceous odor and taste, but lacks the streaks. *T. portentosum* differs by its dark gray, viscid cap that is never as distinctly umbonate or peaked, whitish gills that become yellowish in age, and mild to slightly farinaceous taste. *T. nigrum* has a moist to tacky cap that is dark gray on the disc with a paler margin, whitish gills, a white stipe that often has small blackish scales on the upper portions, and a strong farinaceous odor and taste.

Misapplied name: *Tricholoma virgatum* (Fr.) P. Kumm., a European species not known from North America.

Tricholoma nigrum

Shanks & Ovrebo

Cap: 3–10 cm across, broadly conical when young, becoming convex to plane with a broad umbo, margin undulating in age. Surface thinly viscid to dry, appressed-fibrillose over disc, to nearly smooth. Extensively dark gray when young, soon dark gray over the disc, to light gray towards margin; at times developing brownish gray tones in age. **Gills:** Narrowly attached with a distinct notch or appearing free at times, close, broad, often ragged in appearance. White to pale gray; edge often darker gray, and eroding in age. **Stipe:** 7–15 cm long, 1–3 cm thick, equal or club-shaped, with a bulbous or tapered base. Surface dry, with sparse fibrils and small blackish scales near apex. White to dingy white. **Flesh:** Relatively thin, fibrous. Off-white to pale gray in cap; stipe solid. **Odor:** Strongly farinaceous. **Taste:** Strongly farinaceous. **Spore deposit:** White. **Microscopy:** Spores 6.5–8.5 × 4.5–6 µm, broadly ellipsoid, smooth.

Ecology: Solitary or scattered in duff or moss with a wide range of conifers. Widespread in Cascadia, but uncommon. Fruiting in fall.

Edibility: Unknown.

Comments: The thinly viscid to dry, often two-toned gray cap, white stipe with small blackish scales near the apex, and strongly farinaceous odor and taste help identify *T. nigrum*. The similarly colored *T. mutabile* has a viscid cap, brighter, white gills, and lacks black scales on the stipe. *T. subacutum* has a gray, umbonate cap with blackish streaks, and a slowly acrid taste.

Tricholoma portentosum

sensu PNW

Gray Streaked Trich

Cap: 4–12 (15) cm across, broadly convex to plane, margin downcurved when young, becoming uplifted, wavy and often splitting in age. Surface smooth to appressed fibrous, viscid to moist, or shiny if dry; skin often peels easily. Dark gray to grayish brown over a paler gray base, with dark streaks or appressed fibers, sometimes developing yellowish gray colors in age. **Gills:** Narrowly attached with a distinct notch or appearing free at times, close to subdistant, broad and often ragged in appearance. Creamy white when young, becoming creamy to pale yellowish, especially around margin. **Stipe:** 5–15 cm long, 1–3 cm thick, often curved or twisted, equal, tapering downward or bulbous at base. Surface dry, smooth to slightly fibrous. White to creamy when young, becoming dingy white to pale yellow. **Flesh:** Fibrous, relatively thin, off-white to pale gray or yellowish gray, stipe fibrous, solid, off-white to pale gray or yellowish near surface. **Odor:** Faint, mildly farinaceous to indistinct. **Taste:** Mildly farinaceous. **Spore deposit:** White. **Microscopy:** Spores 5–8.5 × 3–6 µm, broadly ellipsoid, smooth.

Ecology: Solitary, scattered, or in small clusters on ground, in moss or duff under conifers. Common with fir and hemlock, but more likely to be found with pine in the southern Cascades. Occasional to locally common, fruiting in fall, or early winter.

Edibility: Edible and very good, but caution is warranted because it has a number of toxic look-alikes.

Comments: The dark gray to gray-brown, streaked viscid cap, whitish gills that become yellowish, mild to slightly farinaceous taste, and growth under conifers help identify this species. There is some genetic variation between the PNW concept of *T. portentosum* and European collections; more work is needed to delimit species concepts in this group. *T. mutabile* is similar, with a viscid, pale gray to medium gray cap (paler near the margin), white gills and stipe, and strong farinaceous odor; it is reportedly toxic. *T. griseoviolaceum* can be told apart by its growth under hardwoods (oak and tanoak), variably violet-gray cap, and stronger farinaceous odor. *T. subacutum* has a peaked or umbonate, gray, streaked cap that is dry to moist, dingy whitish gills, and a delayed strongly acrid taste. Members of the *T. saponaceum* group can look similar, but have dry, dull-colored, unstreaked caps and often stain pink in the lower stipe. Similar *Entoloma* species (some of which are poisonous) usually have pinkish gills at maturity and pinkish spores.

Tricholoma atrofibrillosum

S.A. Trudell, A.D. Parker & E.T. Cline

Green Streaked Trich

Cap: 3–8 cm across, conical to convex when young, becoming broadly convex to plane or uplifted with a low broad to pronounced umbo; margin often crimped when young, becoming wavy, slightly uplifted, and often splitting in age. Surface slightly fibrous or finely scaly to smooth, viscid when wet, but often drying out. Dark greenish black disc over a yellow to yellowish green base color, with dark streaks or appressed fibers, becoming dark greenish yellow with radially arranged blackish streaks. **Gills:** Narrowly attached with a distinct notch or appearing free at times, close to subdistant, broad; partial gills numerous. Off-white, pale yellowish white to pale yellowish green, often darker toward cap margin. **Stipe:** 3–10 cm long, 1–2 cm thick, often twisted and curved, enlarged or slightly bulbous toward base. Surface dry, slightly fibrous to smooth. Color variable and often patchy, whitish, yellowish to greenish yellow. **Flesh:** Fragile to fibrous, thin. Off-white to pale gray in cap, fibrous, whitish to yellowish near surface in stipe. **Odor:** Faint, mildly farinaceous or indistinct. **Taste:** Mildly bitter-farinaceous to strongly bitter or unpleasant. **Spore deposit:** White. **Microscopy:** Spores 5.5–8 × 4.5–6.5 µm, broadly ellipsoid to subglobose, smooth.

Ecology: Solitary or scattered on ground or in moss under conifers. It is especially fond of wetter fir-hemlock forests in western Cascadia, or occasionally in coastal spruce forests. Common, fruiting from fall into early winter.

Edibility: Unknown, possibly poisonous.

Comments: The yellow to yellow-green cap with radial blackish streaks or fibrils, whitish to yellowish gills and stipe, and growth under conifers help identify this species. Long referred to as *T. subsejunctum* or *T. sejunctum*, the PNW species was recently recognized as distinct. *T. arvernense* can look similar but lacks the dark streaks and is generally more yellow-brown when young, fading to whitish yellow as it ages. The edible *T. portentosum* has a more evenly colored dark gray to gray-brown, streaked, viscid cap, whitish gills that become yellowish, and a mild to slightly farinaceous taste. *T. equestre* has a golden yellow to yellow-brown cap that can have small brownish scales (but never streaked), bright yellow gills, and a farinaceous taste and odor.

Tricholoma atroviolaceum

A.H. Sm.

Cap: 4–12 (15) cm across, rounded to conical in the button stage, soon broadly convex to plane, margin downcurved and pinched in when young, often uplifted, wavy or lobed in age. Surface dry, covered with fine scales. Dark grayish black to dark violet-gray on disc with a paler margin that often has darker 'water spots' when young, becoming grayish violet to grayish brown. **Gills:** Narrowly attached with a distinct broad notch or appearing free at times. Close to subdistant, thick-edged, broad, becoming ragged in age; partial gills numerous. Pale gray to pinkish gray, becoming dingy grayish buff; edges darker; often with dark grayish black to purplish black spots or stains. **Stipe:** 5–12 cm long, 1–3.5 cm thick, equal or enlarged at base. Surface dry, fibrillose, often with fine dark scales on upper portions. Whitish, grayish to pale violet-gray when young, becoming grayish buff and often developing pale orangish tan discolorations at base when handled. **Flesh:** Thin, fibrous to fragile in cap; stipe solid, fibrous. Whitish to pale gray, staining reddish gray and becoming dingy grayish brown, occasionally orangish tan in stipe base. **Odor:** Farinaceous. **Taste:** Strongly farinaceous. **Spore deposit:** White. **Microscopy:** Spores 7–10 × 5–7 µm, broadly ellipsoid, smooth.

Ecology: Solitary or scattered in duff or moss under conifers, especially true firs and hemlock. Widespread across Cascadia, but uncommon. Fruiting in fall.

Edibility: Unknown.

Comments: The dry, dark grayish black to violet-gray, finely scaly cap, dingy pinkish gray to grayish buff gills (often with darker edges), and reddish gray staining of the flesh are the distinguishing features of this species. *T. pardinum* has a whitish to gray cap with small dark scales, whitish gills, and usually has a club-shaped stipe. *T. atrosquamosum* sensu PNW has gray to blackish tufts and scales covering the cap (contrasting with the white background flesh) and a pale gray stipe with contrasting blackish fibrils and scales. A few *Entoloma* species can look similar but have smoother caps and pink spores.

Tricholoma megalophaeum

N. Siegel, S.A. Trudell & A.D. Parker

Big Dark Trich

Cap: 6–15 cm across, conical when young, becoming broadly conical to convex, often with a broad umbo; margin inrolled to incurved when young, often undulating and uplifted in age. Surface dry, appressed-fibrillose, at times with fine recurved scales. Dark olivaceous gray when young, developing yellowish color between fibrils and scales as cap expands. **Gills:** Narrowly attached with a distinct notch, close, broad. Partial gills numerous, edges often becoming ragged in age; creamy with a grayish cast when young, developing dingy yellowish tones in age. **Stipe:** 7–15 cm long, 1–5 cm thick, equal or club-shaped with an abrupt rounded to bulbous base. Surface dry, fibrillose-scaly to nearly smooth. Whitish to pale yellowish, at times with slight orangish stains in age. **Flesh:** Thick, very firm, grayish in the cap, stipe solid, fibrous, whitish to grayish. **Odor:** Slightly farinaceous or like green corn. **Taste:** Mild to slightly farinaceous. **Spore deposit:** White. **Microscopy:** Spores 6.5–10.5 × 4.5–7 µm, subglobose to ellipsoid, smooth.

Ecology: Solitary or scattered, often deeply 'rooted' in ground or thick duff, or in moss under conifers, especially Sitka spruce. So far, only known from northern Humboldt County, California, and the Olympic Peninsula in Washington, but likely more widespread. Fruiting in fall.

Edibility: Unknown, do not experiment.

Comments: The large size, dark olivaceous gray cap which develops yellowish colors, creamy to dingy grayish yellow gills, and a slightly farinaceous odor and taste help distinguish this species. Little is known about habitat preferences and abundance, though it appears to be rare and restricted to coastal spruce forests. *T. arvernense* can have olive tones but is smaller and more typically golden to yellow-colored. *T. atrofibrillosum* has a dark olive-green cap with black fibrils over the disc. *T. portentosum* has a thinly viscid gray cap which develops yellowish tones with age.

Tricholoma aestuans

(Fr.) Gillet

Cap: 4–7 (10) cm across, convex to plane, often wavy or lobed in age. Surface thinly viscid when wet, smooth to finely appressed-fibrillose. Greenish gray with dark gray to blackish streaks and appressed fibrils, dingier in age. **Gills:** Narrowly attached with a distinct notch or appearing free at times, close to crowded, broad; partial gills numerous. Citrine green to grayish green, fading and becoming dingy in age. **Stipe:** 5–10 cm long, 1–2 cm thick, equal or tapered downward. Surface dry, fibrillose to smooth. Off-white to dingy grayish. **Flesh:** Thin, firm, fibrous; stipe solid. Whitish to dingy grayish green. **Odor:** Sour-farinaceous. **Taste:** Bitter-farinaceous with a slight acrid aftertaste. **Spore deposit:** White. **Microscopy:** Spores 6.5–8 × 4–5 µm, broadly ellipsoid, smooth.

Ecology: Solitary or scattered in troops in sandy soil, often buried in needle duff under shore pine. Locally common coastally, but under-reported. Fruiting in late fall and early winter.

Edibility: Unknown, do not experiment.

Comments: The dry, grayish green cap with darker streaks, greenish gills, and disagreeable odor and taste are distinctive. The citrine greenish colors often fade, and can become quite dingy with age. *T. equestre* has a warmer-toned cap and bright yellow gills. *T. atrofibrillosum* has a yellow to yellow-green cap with radial blackish streaks and occurs under a wide range of conifers, but we have not seen it with pine.

Tricholoma intermedium group

Cap: 3–10 cm across, convex when young, becoming broadly convex to plane, often wavy or lobed in age. Surface slightly viscid at first, soon dry, finely appressed-fibrillose to matted-tomentose, becoming fine-scaly in age. Golden yellow, yellowish tan, olive-yellow to yellow-brown on disc, lighter yellow toward margin. **Gills:** Narrowly attached with a distinct notch or appearing free at times, close to crowded, broad; partial gills numerous. White, creamy white to pale pinkish white, not staining. **Stipe:** 4–12 cm long, 0.7–3 cm thick, equal to club-shaped with an enlarged or rounded base. Surface dry, fibrillose, often with fine scales on upper portions. White to pale yellowish white, with pale yellowish to brownish scales at apex, often developing orange-brown discolorations on base, or when handled. **Flesh:** Firm, fibrous; stipe solid. White, sometimes staining orange-brown. **Odor:** Farinaceous. **Taste:** Farinaceous. **Spore deposit:** White. **Microscopy:** Spores 5–9.5 × 3.5–6 µm, broadly ellipsoid, smooth.

Ecology: Solitary or scattered on ground under conifers, especially pines, often buried in needle duff. Widespread but uncommon. Fruiting in fall.

Edibility: Edible and good, similar in taste to the Canary Trich (*T. equestre*); see comments regarding edibility under that species. Care should be taken not to mistake it for the possibly poisonous Soapy Trich (*T. saponaceum* group).

Comments: Until you flip this mushroom over, it looks just like the Canary Trich (*T. equestre*). However, it has white gills and in general small dark scales on the upper stipe. Like *T. equestre*, it is a species complex, with 2–3 species in Cascadia going under this name. One of the forms of *T. saponaceum* can look very similar to *T. intermedium*, but has a smooth, dull yellow cap and pink-staining flesh in the stipe.

Taxonomic note: *Tricholoma leucophyllum*, described from Idaho, may be the correct name for one of the species in this complex.

Tricholoma arvernense

Bon

Cap: 5–12 (15) cm across, conical to convex when young, becoming broadly convex to plane or uplifted with a low broad to pronounced umbo, margin often crimped when young, becoming wavy, uplifted and often splitting in age. Surface dry, covered with fine scale or tufts, to appressed-fibrillose, often radially cracking in age. Color highly variable, generally olive-yellow to olive-brown with blackish olive scales when in button stage, fading as it expands, to brownish yellow, golden yellow to grayish yellow and eventually pale yellow-tan to yellowish white. **Gills:** Narrowly attached with a distinct notch, close to crowded, broad; partial gills numerous. Orangish yellow to creamy yellow when very young, fading to whitish with yellow around the margin. **Stipe:** 5–15 cm long, 0.8–3 cm thick, often twisted and curved, more or less equal but often pointed toward base. Surface dry, pruinose at apex, appressed-fibrillose, silky to smooth. Whitish to pale yellowish white, pinkish white to pale buff. **Flesh:** Thin to moderately thick, fibrous to fragile, whitish to pale grayish yellow in cap, stipe tough, fibrous, whitish to pale grayish white. **Odor:** Farinaceous. **Taste:** Farinaceous. **Spore deposit:** White. **Microscopy:** Spores 4.5–7 × 3.5–5.5 µm, ellipsoid, smooth.

Ecology: Solitary, scattered, or in small to large groups or clusters in duff under conifers. Widespread, occasional to locally common, occurring in a wide variety of habitat types. Fruiting in fall.

Edibility: Unknown.

Comments: The dry olive-yellow, brownish yellow, or yellowish white cap with darker scales, and yellowish to whitish gills help distinguish this species. The highly variable cap can resemble a number of other *Tricholoma* depending on the stage of development. Young specimens could be mistaken for *T. megalophaeum*, which is usually larger and darker. *T. atrofibrillosum* differs by having a greenish yellow cap with blackish streaks. *T. intermedium* has a yellow to yellow-brown cap with small scales over the disc and clean white gills. *T. equestre* also has a yellow to yellow-brown cap and bright yellow gills. *T. davisiae* has an acutely umbonate cap that is more greenish yellow when young and fades to pinkish yellow; it also tends to discolor pinkish in the stipe base. *T. aestuans* has a light greenish gray cap with dark appressed fibrils and dingy, pallid greenish gray gills; it grows in sandy shore pine habitat. *Floccularia albolanaripes* has very similar colors, but usually shows appressed brownish scales on the cap, and has a distinctly scaly-shaggy stipe.

Tricholoma equestre group

Canary Trichs, Yellow-gilled Trichs

Cap: 3.5–12 cm across, rounded to convex with a downcurved margin when young, becoming broadly convex to nearly plane, margin often wavy or lobed and sometimes uplifted in age. Surface thinly viscid at first, becoming dry and often with adhering debris. Mostly smooth, but often appressed-fibrillose over disc, breaking up into fine scales. Golden yellow to pale yellow with brownish scales on disc when young, becoming dingy yellow to yellowish brown or even all-brown. **Gills:** Narrowly attached with a distinct notch, or appearing free at times, close to crowded, broad; partial gills numerous. Pale to bright yellow to greenish yellow when young, becoming slightly paler in age. **Stipe:** 2–10 cm long, 0.7–2 cm thick, equal or enlarged toward or even bulbous at base. Surface dry, fibrillose. Whitish to pale yellow, darkening slightly in age. **Flesh:** Thin to moderately thick, firm, white to pale yellowish near cap surface, stipe fibrous, solid to hollow in age, whitish. **Odor:** Farinaceous. **Taste:** Mild to sweet-farinaceous. **Spore deposit:** White. **Microscopy:** Spores 5–8.5 × 3.5–6 µm, broadly ellipsoid, smooth.

Ecology: Solitary or scattered on ground under conifers or poplars usually buried in duff. Especially common along the coast in sandy soil with pine, or in the eastern Cascades; occasional elsewhere. Fruiting in fall and into early winter.

Edibility: Considered by many to be a great edible. However, it apparently contains a rhabdomyolitic compound implicated in several deaths in Europe (although in that case, large amounts were consumed daily for a week). There haven't been any poisonings reported from North America but, if you choose to eat this mushroom, you should be cautious never to consume it in large amounts or for many days in a row as it may contain a cumulative toxin.

Comments: Species in the *T. equestre* group can be recognized by their golden yellow caps with brownish scales, yellow gills, and whitish to pale yellow stipes. Based on genetic studies, there are at least five species in this complex in Cascadia. At this time, we are unable to distinguish them based on morphological features. One of these (*T. aff. frondosae*) appears to be brighter yellow and prefers aspen and cottonwood. Other, similar species include: *T. intermedium*, which has white gills; *T. aestuans* with a light greenish gray cap with dark appressed fibrils, and dingy greenish gray gills, and *T. sulphureum* also has a yellow cap and gills but has smaller, slender fruitbodies with an awful coal-tar or mothball odor. *T. atrofibrillosum* has a greenish yellow cap with olive and blackish streaks or fibrils and whitish gills that become blotchy with creamy and yellow tones.

Tricholoma saponaceum group

Soapy Trichs

Cap: 1.5–15 cm across, conical to convex with an inrolled margin when young, becoming broadly convex, plane, with or without an umbo; margin often slightly uplifted, lobed and wavy in age. Surface dry to slightly moist, smooth or with scattered appressed scales on disc, radially splitting and cracking into large scales in dry weather. Greenish gray, gray, bluish gray, grayish brown, brown, yellowish brown to yellow. **Gills:** Narrowly to broadly attached, often with a distinct notch and/or a decurrent tooth, or even appearing free at times, close to subdistant, broad; partial gills numerous. Whitish, creamy white to pale yellowish tan, sometimes staining pinkish. **Stipe:** 3–15 (20) cm long, 0.5–3 cm thick, more or less equal, bulbous, tapering or rooting at base. Surface smooth to slightly fibrous, moist to dry. Whitish, creamy white to pale yellowish tan, staining or discoloring pinkish at base. **Flesh:** Moderately thick, firm to fragile, whitish to pale gray; stipe solid, fibrous, whitish to pale gray with a pinkish base. **Odor:** Distinctive, like old-fashioned unscented soap. **Taste:** Mild to disagreeable, often described as "soapy." **Spore deposit:** White. **Microscopy:** Spores 5–7 × 3.5–5 µm, ellipsoid, smooth.

Ecology: Solitary, scattered, or in small clusters on ground in a wide variety of forests. One member is especially common with pines, while others seem to prefer tanoak or mixed conifer forests. Mostly fruiting in fall but at least one member of this group fruits in spring in the eastern Cascades.

Edibility: Unknown, possibly poisonous.

Comments: *T. saponaceum* is a complex, with 4–5 species in the PNW. As a group, they share the following features: caps smooth and variably colored, whitish gills, stipes with pinkish flesh in the base, and a soapy odor. The pink in the stipe is often slow to develop and not everyone detects the soapy odor. The grayish green "form" appears to be the most widespread, occurring in mixed conifer forests across Cascadia. The more yellow-toned species is fairly common with pine. Another form, found in shore pine habitats, has a broadly umbonate, two-toned cap (brown on the disc, gray to grayish yellow at the margin) and a long, tapering stipe with a bright pink base. *T. saponaceum* var. *squamosum* (in the PNW sense) has a dark grayish green cap with black scales on the disc and matted dark gray fibrils on the stipe; it appears to prefer wet fir forests.

Tricholoma sulphureum group

Yellow Stinky Trichs

Cap: 1.5–5 cm across, conv ex to plane or wavy in age. Surface dry, smooth or with scattered fine scales. Creamy yellow, dingy yellow to yellow-brown, rarely extensively citrine yellow or pinkish yellow. **Gills:** Narrowly attached with a distinct notch or appearing free at times, subdistant to widely spaced; partial gills numerous. Yellowish. **Stipe:** 4–10 cm long, 0.4–1 cm thick, often tall and spindly, equal or club-shaped with a bulbous or tapered base. Surface dry, silky-fibrillose to smooth. Yellow or creamy yellow, darkening slightly when handled, at times developing brown tones from base up in age. **Flesh:** Thin, fibrous, yellow. **Odor:** Nauseatingly strong; like coal tar, mothballs or paperwhite narcissus. **Taste:** Repulsive, burnt-tar-like. **Spore deposit:** White. **Microscopy:** Spores 8.5–12 × 5.5–7 µm, almond-shaped, spindle-shaped or ellipsoid, smooth.

Ecology: Solitary or scattered on ground or in moss under conifers. Uncommon in our area and rarely in large numbers when found. Fruiting in fall.

Edibility: Possibly poisonous. In any case, the odor and taste are major deterrents.

Comments: Like *T. platyphyllum*, members of the *T. sulphureum* group are slender with a stinky mothball or coal tar odor but are easily distinguished by their yellowish colors. This is a species complex in Cascadia, with two to three species showing varying degrees of yellow, ranging from pale creamy yellow to more extensively bright yellow. *T. lutescens* is a larger white species which stains yellow when damaged and it lacks the stinky odor.

Tricholoma platyphyllum

(Murrill) Murrill

White Stinky Trich

Cap: 1.5–5 cm across, convex to plane or wavy in age. Surface dry, smooth. White, creamy, pale pinkish white to pale tan, slightly darker at disc. **Gills:** Narrowly attached with a distinct notch or appearing free at times; widely spaced, broad; partial gills numerous, often becoming ragged in age. White, creamy white to pale tan in age. **Stipe:** 4–12 cm long, 0.3–0.8 cm thick, tall, spindly, equal, enlarged toward, or pointed at base. Surface dry, silky-fibrillose to smooth. White, creamy white to pinkish white, with pale orangish tan stains and often with a rosy pink base. **Flesh:** White, very thin, fragile in cap, stipe solid, fibrous. **Odor:** Nauseatingly strong; like coal tar, mothballs or paperwhite narcissus. **Taste:** Repulsive, like burnt tar. **Spore deposit:** White. **Microscopy:** Spores 9.5–14.5 × 5–8 µm, almond-shaped, spindle-shaped or ellipsoid, smooth

Ecology: Solitary or scattered on ground or in moss under conifers. Locally common across western Cascadia. Fruiting in fall.

Edibility: Possibly poisonous. In any case, the odor and taste are major deterrents.

Comments: The slender stature, whitish colors, widely spaced gills, and peculiar odor make this a distinctive mushroom. The strong odor is reminiscent of coal tar or mothballs; some people get a pleasant (albeit heavy) floral scent at first, others are nauseated by the faintest whiff. *T. inamoenum* is practically indistinguishable but appears to be more common in spruce forests of British Columbia and Alaska. *T. lutescens* is larger and stains yellow when damaged. Members of the *T. sulphureum* group have a similar odor and stature but are pale yellow to yellowish buff overall and generally slightly stockier.

Misapplied name: *Tricholoma inamoenum* (Fr.) Gillet

Tricholoma apium

Jul. Schäff.

Cap: 4–10 cm across, convex to broadly umbonate with an inrolled margin when young, becoming broadly convex to plane; margin wavy and often uplifted in age. Surface dry, dull, finely tomentose to slightly flocculose, becoming areolate in patches with age. Whitish to ocher-buff when young, often patchy, ocher brownish to olive brownish with age. **Gills:** Narrowly attached, often with a broad notch, close to crowded, narrow, with ragged edge in age. White to creamy white, with yellowish stains in age. **Stipe:** 3–8 cm long, 1–2 cm thick, equal or narrowing towards base. Surface dry, fibrillose-scaly; base with white mycelium. Whitish to cream-colored, with ocher brownish scales, darkening overall with age. **Flesh:** Thin to moderately thick, firm, fragile in cap, fibrous in stipe, white to cream-colored. **Odor:** Pleasant, faint to strong, reminiscent of fennel or celery. **Taste:** Mild to slightly farinaceous. **Spore deposit:** White. **Microscopy:** Spores 4–6 × 4–5 µm, ovoid to nearly globose, smooth.

Ecology: Scattered in troops or arcs in soil, duff, or moss; typically in mature and old-growth forests under conifers. Rare in Cascadia; most collections are from lower elevation coastal and Coast Range forests. Fruiting in fall.

Comments: Recognized by its dry, often areolate, whitish to ocher-buff cap which develops brownish to olive brownish tones in age, a whitish stipe with ocher-brown scales, and white mycelium at the base. Although quite rare in our area, it can locally occur in large numbers. *T. manzanitae* has a creamy white cap when young, soon becoming creamy buff to pinkish buff, and eventually orangish brown to buff-brown in age. The cap is distinctly viscid to sticky when wet, and it occurs with madrone and manzanita. Similar to *T. manzanitae* is a rare species with conifers, *T.* aff. *japonicum.* It has a viscid pinkish tan cap, whitish gills and stipe. It is reminiscent of *Leucopaxillus* due to the persistent clump of white mycelium at the stipe base. However, *L. albissimus* is typically larger, has a chalky white cap which yellows slightly in age, and broadly attached to decurrent gills. *L. gentianeus* has a smoother, brown to tan cap, broadly attached to slightly decurrent gills, and a bitter taste. *Leucopaxillus* have amyloid, warted spores.

Lyophyllum decastes group

Fried Chicken Mushrooms

Cap: 3–10 cm across, rounded with an inrolled margin when young, expanding to broadly convex to plane, or wavy in age. Surface greasy to dry, smooth or with fine fibrils near center. Color quite variable: dark brown, gray-brown, mottled grayish beige, ocher-tan or honey colored. Occasionally much paler, nearly whitish. Often with spots, blotches, and streaks of darker color. Young caps of darker variants often have a pale grayish bloom overall. **Gills:** Narrowly attached, fairly close, soft and waxy. White to creamy, dingy in age. **Stipe:** 4–12 cm long, 0.7–2 cm thick, equal or enlarged and often fused in large clusters. Surface smooth, moist to dry. Whitish cream at first, developing pale yellowish tones, then becoming tan. **Flesh:** Moderately thick, fibrous-rubbery; stipe solid or with a small hollow. Grayish to whitish. **Odor:** Indistinct. **Taste:** Indistinct. **Spore deposit:** White. **Microscopy:** Spores 4–6 µm, round, smooth, inamyloid.

Ecology: In dense clusters and clumps, often on nutrient-poor soil or in thin humus in disturbed areas. Members of this group can be found throughout the year, but peaking in late summer and fall.

Edibility: Edible and fairly popular due to its firm texture and abundance, but the flavor is fairly bland and often found growing in 'unsavory' habitats.

Comments: The smooth, greasy to dry caps fruiting in large to huge clusters with fused stipe bases, lack of any veil tissue, and rubbery white flesh are characteristic. Due to the variable and overlapping cap colors, distinguishing species is difficult. Caps of *L. decastes* tend to be light brown, gray-brown to tan, whereas those of *L. shimeji* (which has been called *L. loricatum* in the past) are typically dark brown to gray-brown. Caps of *L. fumosum* are also dark brown to gray-brown caps, but have fibrillose scales over the surface. There is an additional unnamed white to yellowish beige species which fruits in spring in the eastern Cascades. Similarly colored (and toxic) *Entoloma* species can look similar, but don't grow in dense clusters and have pinkish spores. *Lepista subconnexa* grows in less dense clusters, has grayish to buff or whitish caps, dingy buff gills, and pinkish buff spores. *Armillaria* species have a partial veil and usually show small dark hairs on the cap.

Leucocybe connata

(Schumacher) Vizzini, P. Alvarado, et al.

Tufted Paledome

Cap: 2–8 cm across, rounded to broadly convex to wavy; margin finely inrolled when young, often lobed. Surface moist to dry, smooth to finely powdered with a chalky bloom; often bumpy and concentric-zonate. Bright white when young, becoming duller ivory white to grayish-white, cream or even pale beige as the bloom weathers. **Gills:** Narrowly attached to slightly decurrent, close; edges even to slightly irregular. White to cream, often washed with pinkish, becoming beige in age. **Stipe:** 4–10 cm long, 0.5–1 cm thick, cylindrical, often slightly sinuous, slightly enlarged towards base. Surface smooth to finely fibrillose or faintly scurfy near apex. Whitish to pale beige or cream-tan in age. **Flesh:** Moderately thick, fibrous-fleshy; whitish to pale beige-tan. **Odor:** Indistinct. **Taste:** Mild. **Iron sulfate:** Turning blue-black to violet on cap and gills. **Spore deposit:** White. **Microscopy:** Spores 5.5–7 × 3–4 µm, ellipsoid, smooth, inamyloid.

Ecology: In small to large clusters, rarely solitary. On soil and gravel in disturbed areas such as roadsides, trail edges, or skid trails around logging operations. Common in the Cascade Range and the Olympic Mountains, occasional in inland valleys, rarer towards the coast and mostly absent from southern Cascadia. Fruiting in fall and early winter.

Edibility: Not recommended. Although it has long been considered suspect at best and poisonous at worst (due to conflation with *C. dilatata*), it's been slightly exonerated by its recognition as a separate entity.

Comments: This species is recognizable by its powdered- or chalky-looking white caps, moderate size, and tendency to grow in dense clusters; the purplish iron sulfate (FeSO4) reaction can be helpful in atypical or aged fruitbodies. *L. candicans* tends to fruit singly (rather than in clusters). There is a common, unnamed, spring-fruiting *Lyophyllum* with white to cream colors; it grows in dry forests in the eastern Cascades. *Clitocybe sclerotoidea* can be fairly similar but is smaller, grayer, and usually nearly fused at the base of the stipes, which attach to a dense "nugget" of marbled beige-yellow tissue – the remains of the *Helvella* fruitbodies which it parasitizes! *Clitocybe rivulosa* can be quite similar but often has pinkish-beige spots and streaks on the cap, is much smaller, and usually grows in grassy areas or along gravelly trails and roadsides.

Misapplied Name: *Clitocybe dilatata* P. Karst.

Synonyms: *Clitocybe connata* (Schumacher) Gillet, *Lyophyllum connatum* (Schumacher) Singer

Aspropaxillus septentrionalis

(Singer & A.H.Sm.) Vizzini

Dusky Aspro

Cap: 7–30 (40) cm across, broadly convex with an inrolled margin when young, becoming plane with a depressed center and then shallowly funnel-shaped; margin thin and undulating in age. Surface dry, smooth, suede-like to finely appressed-scaly, with a fine, streaky whitish bloom when young. Streaky beige-tan to dingy olive-tan, paler creamy-beige outwards to whitish near margin. **Gills:** Narrowly attached or weakly decurrent, close to crowded, broad. White to pale pinkish. **Stipe:** 6–15 cm long, 1.5–3 cm thick, more or less equal. Surface dry, smooth to finely vertically fibrillose. Whitish to tan, or with darker vertical striations in age. **Flesh:** Fairly thick, fleshy-fibrous; whitish. **Odor:** Musty-farinaceous to skunky, unpleasant. **Taste:** Indistinct or musty. **Spore deposit:** White. **Microscopy:** Spores 6–8 × 3–4.5 µm, ellipsoid to ovoid, smooth, amyloid.

Ecology: Usually in arcs and troops numbering from dozens to hundreds, occasionally scattered or in small groups, rarely solitary, in needle duff and debris of conifers. Uncommon; mostly found in northern Cascadia in inland and mountainous areas. Fruiting in fall.

Edibility: Unpleasant tasting and tough. Not recommended.

Comments: This species is recognized by its large size, brownish-gray cap that fades and becomes depressed at the center in age, and growth in troops. The smooth amyloid spores further help distinguish it. *A. gigantea* is more whitish overall and may grow in a different habitat (more open, grassy areas); its presence has not yet been confirmed in Cascadia. *Leucopaxillus* species at times rival the size, with dry, dull caps but their caps tend not to become funnel-shaped in age and their spores have amyloid spines. *Clitocybe nebularis* can be quite similar but tends to have a skunky aroma, a smooth gray cap, and smooth, inamyloid spores. A somewhat similar-looking species of *Clitopaxillus* has smaller fruitbodies, a finely velvety-felty, grayish cap with dark spots and blotches, and has inamyloid spores. *Harmajaea harperi* is very similar to the latter species but has a smooth cap, a slightly musty odor, and lacks clamp connections.

Leucopaxillus gentianeus

(Quél.) Kotl.

Pancake-of-the-Woods

Cap: 7–15 cm across, rounded to convex with a strongly inrolled, lobed, and usually ribbed margin when young, expanding to broadly convex or plane, sometimes uplifted and slightly wavy. Surface finely velvety, becoming smoother and often finely cracked into plaques. Reddish brown to medium brown, ocher-brown or tan, or fading to dull beige, margin often nearly whitish. **Gills:** Broadly attached to slightly decurrent, close to crowded, broad. White to cream. **Stipe:** 5–10 (15) cm long, 1–3 cm thick, cylindrical with an enlarged, club-shaped base. Surface dry, fibrillose to smooth. White to cream, discoloring beige to yellowish. Base usually attached to a thick mat of binding white mycelium. **Flesh:** Thick, firm, and rubbery at first, often becoming riddled with maggot larvae and becoming crumbly. White. **Odor:** A bit musty, distinct, slightly unpleasant. **Taste:** Extremely bitter. **Spore deposit:** White. **Microscopy:** Spores 4–6 × 3–5 µm, broadly ellipsoid to subglobose, ornamented with amyloid warts and ridges.

Ecology: Solitary, scattered, or clustered, often in large troops, rings, or arcs in heavy duff of many different forest types and under trees in parks and other urban settings. Fairly common throughout Cascadia. Fruiting from late fall into spring, peaking in late winter.

Edibility: Inedible, extremely bitter.

Comments: The finely velvety, brownish cap that is often ribbed around the margin, white gills, bitter taste, and stout, white stipe attached to a mat of pale mycelium make this species fairly easy to recognize. Mature caps look remarkably like pancakes. A few *Tricholoma* in our area are vaguely similar but have notched gills that are often not as white (beige to tan, or blotchy orange brown) and usually have less robust stipes that are not attached to extensive mats of mycelium.

Synonym: *Leucopaxillus amarus* (Alb. & Schwein.) Kühner

Leucopaxillus albissimus

(Peck) Singer

Large White Leucopax

Cap: 6–30 cm across, rounded with an inrolled margin when young, becoming broadly convex to plane, often wavy and lobed in age. Surface dry, finely velvety to smooth, cracking into small uplifted scales or more deeply cracked in age or dry weather. White to cream or dingy yellowish tan at center in age. **Gills:** Subdecurrent to decurrent, close to crowded. White to creamy, becoming light yellowish in age. **Stipe:** 9–30 cm long, 2–6 (9!) cm thick, usually swollen downward, then tapered at base, sometimes cylindrical or irregularly swollen above. Smooth to vertically fibrillose, often rough or with small, pointed scales in age. White to creamy, developing yellowish beige tones in age. Base usually with adhering pad of white mycelium and rhizomorphs bound to duff and needle litter. **Flesh:** Very thick and firm, solid-rubbery, solid; tough and slow to decay. Whitish to cream. **Odor:** Distinctive, musty, usually unpleasant. **Taste:** Mild to bitter. **Spore deposit:** White. **Microscopy:** Spores 5.5–8.5 × 4–6 µm, ellipsoid, with amyloid warts and spines.

Ecology: Solitary or more often forming arcs and rings in duff. Fairly common in forested settings of inland areas and lower elevations of the Cascades, including urban areas with thick duff layers. Fruiting in fall and early winter, but fruitbodies are persistent, sometimes lasting for months.

Edibility: Inedible; tough, odd smelling, and often bitter.

Comments: The large to giant fruitbodies, white to dingy creamy colors, decurrent gills, and habit of growing in arcs and troops make this species fairly easy to recognize. *Clitocybe nebularis* can be quite similar in size, appearance, and fruiting habits but has narrowly attached or only slightly decurrent gills, a more grayish-toned (and often more neatly circular) cap, and usually doesn't have a thick pad of mycelium and litter adhering to the stipe base. *Aspropaxillus septentrionalis* is more fleshy (less tough), with a more grayish, broadly funnel-shaped cap, and lacks a large pad of mycelium adhering to leaf litter around the stipe base.

Ampulloclitocybe avellaneialba

(Murrill) Harmaja

Tan Clubfoot

Cap: 2–7 (10) cm across, convex with a low, round umbo when young, soon concave, to funnel-shaped in age. Margin slightly inrolled when young, sometimes with pronounced, dark radial ribs, thin and uplifted in age. Surface smooth or finely felty when young, occasionally sparsely radially fibrillose near margin in age; moist to dry or slightly greasy. Brownish to dark gray to olive-tan, fading to beige or creamy towards margin. **Gills:** Subdecurrent to decurrent, subdistant to distant, narrow; edges even. White to cream. **Stipe:** 4–15 cm long, 0.6–1.5 cm thick, cylindrical overall, with a enlarged base, usually tall and relatively slender. Surface dry, smooth to finely longitudinally roughened. Beige-tan to dingy pale gray-brown, apex often darker and contrasting with gills. **Flesh:** Moderately thick, fleshy, fibrillose in stipe; whitish. **Odor:** Indistinct. **Taste:** Indistinct. **Spore deposit:** White. **Microscopy:** Spores 6–8 × 3–5 µm, almond-shaped with elongated, tapering ends, smooth, inamyloid.

Ecology: Scattered in small troops, occasionally solitary, from mossy duff, often near rotting wood, usually in mature conifer forest. Uncommon, most often encountered at higher elevations on the west slope of the Cascades, but actual distribution unclear (see comments). Fruiting in fall, with scattered fruitings into winter.

Edibility: Reportedly non-toxic but bland. The similar *A. clavipes* reportedly contains compounds which interfere with the function of aldehyde dehydrogenase in the liver, causing toxic interactions when consumed with alcohol. Best avoided.

Comments: This species is recognizable by its decurrent whitish gills, tall, slender brownish stipe, and brownish-gray to olive-tan cap that becomes funnel-shaped at maturity and which often shows dark "ribbing" near the margin. *A. clavipes* sensu PNW is smaller, with a paler gray cap that lacks the dark ribbing, has a more bulbous stipe base, and often (but not always) smells sweet (compared to grape soda). It is more common coastally and at lower elevations, growing in moss and duff. *Gerronema atrialba* is taller and more slender, with more gray-colored gills and a gray-dotted stipe.

Gerronema atrialbum

(Murrill) Borovicka & Kolarík

Dot-stipe

Cap: 2.5–7 (10) cm across, convex when young, becoming plane with a depressed center, to broadly funnel shaped in age. Surface finely appressed-fibrillose to smooth, moist to dry. Dark gray to dark gray-brown at first, becoming pale grayish brown to gray. **Gills:** Subdecurrent to decurrent, well-spaced, broad. White to pale gray. **Stipe:** 5–15 cm long, 0.4–1.5 cm thick, tall and slender, equal or enlarged slightly toward base. Surface covered with small blackish scurfs and scales, especially pronounced near apex. Gray to grayish brown, paler toward base and in age. **Flesh:** Thin, fibrous in stipe; whitish to pale gray. **Odor:** Indistinct. **Taste:** Indistinct. **Spore deposit:** White **Microscopy:** Spores 6–9 × 7–8 µm, subglobose to globose from 4-spored basidia, up to 13 × 9 µm and broadly ellipsoid to ovoid from 1- and 2-spored basidia, smooth, amyloid.

Ecology: Solitary or scattered on well-decayed branches and logs of hardwoods, especially red alder and bigleaf maple. Common and often abundant at low elevations in western Cascadia, from Eugene northward. Fruiting in fall.

Edibility: Unknown.

Comments: The grayish brown cap, whitish to gray, decurrent gills, long, slender, scurfy stipe, and growth on wood provide a distinctive set of features. The weedy growth in urban areas is suggestive of an introduced species (likely from California). Most similar are *Hygrophorus* species like *H. camarophyllus* and *H. pustulatoides*, both of which grow from the ground rather than wood and have thicker gills with a waxy feel. *Ampulloclitocybe avellaneialba* and *A. clavipes* have smooth stipes and creamy white gills that are typically more crowded; the former grows from well decayed conifer wood, the latter in moss and duff. *Pseudoclitocybe cyathiformis* has close to crowded gray gills, a strongly hygrophanous dark gray to brown cap, and grows from both decaying wood and duff.

Synonym: *Clitocybula atrialba* (Murrill) Singer

Clitocybe nebularis

(Batsch) P. Kumm.

Cloudy Clitocybe

Cap: 4–16 cm across, convex with an inrolled margin, becoming plane, uplifted, and often wavy in age. Surface moist to dry, smooth or with scattered patches of moldy-looking whitish fuzz. Grayish white to darker gray with a fine whitish bloom, mottled grayish tan with a whitish margin in age. Often with darker streaks, and frequently with dark grayish "water spots" around margin. **Gills:** Broadly attached or slightly decurrent, close to subdistant, broad. Whitish cream to dingy beige, or grayish tan in age. **Stipe:** 6–15 cm long, 1.5–4 cm thick, equal to slightly enlarged toward base. Surface smooth or with darker scales near apex, base often with irregularly scattered moldy-looking, fuzzy whitish patches. Whitish to grayish or tan-gray. **Flesh:** Thick, fleshy-fibrous; whitish. **Odor:** Strong and distinctive, a musty-musky mixture often compared to skunk, swamp gas, rancid flour, or marijuana. **Taste:** Difficult to assess separately from odor. **Spore deposit:** Creamy. **Microscopy:** Spores 5.5–8 × 3–4 µm, ellipsoid, smooth, inamyloid.

Ecology: Scattered in small groups or large rings and arcs, occasionally solitary in forest duff. Common and widespread, occurring in most forest types. Most abundant in low elevation bigleaf maple-alder-cedar forests. Fruiting in fall and winter.

Edibility: Although some consume, there are reports of it causing mild gastrointestinal upsets. The off-putting odor deters most.

Comments: The grayish cap, large size, slightly decurrent white to cream gills, and distinctive musky odor distinguish this species. *Harmajaea harperi* is quite similar in all respects except for its darker grayish brown gills, lack of a skunky odor, and lack of clamp connections in all tissues; it is widespread but rare. *Leucopaxillus gentianeus* has a brown cap and strong bitter taste, and *L. albissimus* has whitish fruitbodies (becoming yellowish) lacking grayish tones and does not have the strong unpleasant odor. The less common *Ampulloclitocybe clavipes* produces smaller fruitbodies with more decurrent gills, strongly swollen stipe bases, and often with a pleasant sweet odor.

Cleistocybe vernalis

Ammirati, Parker & Matheny

Spring Cloistercap

Cap: 2–6 cm across, convex with a tightly inrolled margin when young, expanding to broadly convex with a slightly depressed center; margin slightly lobed and wavy but often remaining inrolled into age (often appearing weakly ribbed). Surface dry, with extensive fine, flat scales, often with patches of appressed whitish veil tissue, smoother and slightly metallic in age. Scales dingy brown to dull vinaceous brown, rest of surface dull brown to tan fading to beige, margin pale pinkish-gray to buff when young. **Gills:** Broadly attached to decurrent, occasionally forking, close to subdistant, narrow; edges even. Dingy pinkish-brown to buff-tan or dirty beige in age. **Stipe:** 3–6 cm long, 0.7–1.5 cm thick, stout, cylindrical, rather straight to slightly curved. Surface dry, fibrous, with ragged scraps and chevrons of beige to dull ocher tissue. Pale creamy-beige to tan, often with a pinkish-gray band near apex. **Partial veil:** Weakly membranous or cortina-like, soon rupturing and leaving a ragged zone on the upper stipe and adhering patches near edge of cap. **Flesh:** Thick, fibrous, pallid. **Odor:** Farinaceous. **Taste:** Unpleasant-farinaceous. **KOH:** Green on cap. **Spore deposit:** White. **Microscopy:** Spores 7–10.5 × 4–5 µm, elongate-ellipsoid to almond-shaped, smooth, inamyloid.

Ecology: Scattered, or in small clusters from duff and debris in conifer forest. Rare in Cascadia, only known from east of the Cascade crest at mid to high elevations. Fruiting almost exclusively in spring, with rare fall fruitings.

Edibility: Unknown.

Comments: This rather extraordinary mushroom is recognized by its dull colors, stout stature, spring fruiting, partial veil zone and ragged bands of tissue on the stipe, and finely scaly cap with an inrolled margin which turns green in KOH. *C. gomphidioides* is similar, with more irregular gills, fruits in fall, and has more elongate, slightly larger spores measuring up to 12 (15) µm long. It is known from mixed forests including cedar and hemlock from the Cascades as well as coastal locations on the Olympic Peninsula. At least one rather similar-looking spring-fruiting species of *Lulesia* occurs in montane conifer forest in the Cascades, but it lacks a partial veil and has a smoother, beige-tan cap (although it also turns green in KOH).

Clitocybe glacialis

Redhead, Ammirati, Norvell & Seidl

Snowbank Cly

Cap: 2–9 cm across, round-convex with an inrolled margin when young, expanding to nearly plane, often slightly wavy around margin. Surface smooth, moist to dry, opaque, with a fine bloom of white powder or faintly streaky when young. Pale silvery gray when young, soon dark gray to gray-brown in age, drying to whitish beige, often shiny-metallic in dry conditions. **Gills:** Narrowly to broadly attached, close, broad. Pale gray to dusky tan-gray or pinkish-gray. **Stipe:** 3–9 cm long, 0.5–1.5 cm thick, cylindrical, stout, equal to narrowly club-shaped. Grayish to tan or brown. Extreme base with whitish mycelium often binding a large chunk of duff and forming a strange, watery, gel-like layer. **Flesh:** Thick, pale grayish to light gray-tan, fibrous, fleshy. **Odor:** Indistinct. **Taste:** Indistinct. **Spore deposit:** White. **Microscopy:** Spores 5.5–7 × 3.5–4.5 µm, ellipsoid, smooth, inamyloid.

Ecology: Scattered in troops or small clusters from needle duff around the edges of melting snowbanks (or even melting through the edge of the snow!) in higher-elevation conifer forests, especially around true fir. Widely distributed on both slopes of the Cascades, the highest elevations of the Olympic Peninsula, as well on the higher peaks in the coastal mountains (where accumulating snow occurs annually). Fruiting in spring and early summer.

Edibility: Unknown.

Comments: This common species is an iconic representative of the assemblage of fungi associated with melting snowbanks, sometimes called the "niveomycota." Although identification of older fruitbodies can be confusing, young fruitbodies are reliably recognized by their silvery-gray caps with a white bloom that streakily wears off to reveal a brownish gray color. At all stages, the watery whitish gel-like mycelium around the stipe base and association with melting snow are extremely helpful clues. *Melanoleuca angelesiana* has a browner cap even when young, brighter white, more crowded gills, and lacks the distinctive pad of gel-like basal mycelium. The group of *Clitocybe* with dark gray-brown gills (including this species) will likely be separated into the genus *Pseudolyophyllum*; most of the other species in this group have slender, thin-fleshed fruitbodies and hygrophanous caps. Other snowbank fungi include *Pseudoomphalina* (*Neohygrophorus*) *angelesiana* with dark vinaceous brown to grayish brown colors and thick-edged, decurrent gills and *Hygrophorus marzuolus* (sensu PNW), with a dark gray to blotchy gray cap and waxy, white to grayish gills.

Melanoleuca angelesiana

A.H. Smith

Angel's Cavalier

Cap: 3–10 cm across, convex to broadly convex with a low, round umbo, margin often slightly inrolled to incurved and weakly ribbed, expanding to plane or concave and wavy in age. Surface smooth, moist to greasy, dry to metallic-looking in dry weather. Dark brown to dark grayish brown when young (often dusted with a faint pale bloom, becoming medium brown to tan or yellowy-brown overall. **Gills:** Narrowly attached, notched, and appearing almost free, close to crowded, broad; edge slightly irregular to weakly serrate. White to pale gray. **Stipe:** 3–10 cm long, 0.7–2 cm thick, cylindrical, straight, often swollen to club-shaped at base. Surface dry, finely vertically roughened to smooth, sometimes with long, slightly twisting striations. Pale grayish to tan-brown or dingy yellowish, to dark gray-brown towards base; often appearing "frosted" when young. **Flesh:** Thin, soft, fibrous; whitish. **Odor:** Indistinct. **Taste:** Indistinct. **Spore deposit:** White. **Microscopy:** Spores 7–9 × 5–6.5 µm, ellipsoid, finely-warted with amyloid spines. Cheilocystidia and pleurocystidia absent.

Ecology: Solitary or scattered in troops in thick duff of montane conifer forests; most often with true fir and pine near melting snowbanks. Common at high elevations throughout Cascadia. Fruiting in spring.

Edibility: Edible but not recommended.

Comments: The smooth, dark gray-brown cap, crowded gills and straight stipe, and springtime fruiting (often near melting snowbanks) help identify this species. A number of other *Melanoleuca* are quite similar but don't tend to be as closely associated with melting snow as this species (although some may fruit in springtime more generally). One species currently going under the name *M. evenosa* has a paler gray cap and cystidia on the gill edges. *M. melaleuca* and a few other dark-colored *Melanoleuca* are found in fall in west side forests. *Melanoleuca verrucipes* is an uncommon mostly white species with dark scales on the stipe. It occurs in wood chips or rich soil in urban areas. *Clitocybe glacialis* also grows on duff at the edges of melting snowbanks, but has darker gray colors when young. Older fruitbodies can be similar but have a pale pad of gel-like mycelium around the base.

Clitocybe odora var. *pacifica*

Kauffman

Pacific Anise Cly

Cap: 2–8 cm across, broadly convex with an inrolled margin at first, becoming plane to concave. Surface dry or greasy, smooth. Sky blue with teal hues to grayish turquoise with innate silvery streaks and thus appearing pale and frosty, sometimes with "water spots" and darker blue-green blotches. Becoming dingy olive-gray with bluish hues, or fading to paler beige with hints of blue in age. **Gills:** Broadly attached, close, narrow. Pale grayish blue to dingy turquoise-blue and then sordid olive grayish. **Stipe:** 2–6 cm long, 0.5–2 cm thick, equal or enlarged toward base, sometimes tapering (or bulbous and tapering). Surface dry, smooth, base with binding mycelium, and elastic whitish rhizomorphs. Whitish with sky blue and grayish tones, beige-tan toward base, irregularly developing dingy olive and orange blotches in age. **Flesh:** Thin, fibrous; white, dingy tan to variably pallid bluish gray. **Odor:** Strong and sweet like aniseed or fennel. **Taste:** Mild. **Spore deposit:** Off-white. **Microscopy:** Spores 6–8 × 3–4 µm, ellipsoid, smooth, inamyloid.

Ecology: Often in small clusters or scattered in arcs, in thicker duff of conifer forests. Widespread and common throughout Cascadia. Fruiting in fall.

Edibility: Edible but not commonly eaten. The odor is not reflected in the flavor when cooked, but some people have tried infusing the aroma of the mushrooms into vodka, with varying success.

Comments: This mushroom is quite distinct by its blue-green hues and strong, sweet anise odor. The variety *pacifica* was described as having grayish-beige to bluish gills, unlike the white-gilled species going under the name *C. odora* in eastern and boreal North America. Thus, the common version in Cascadia should almost certainly be elevated to full species status. *Clitocybe deceptiva* shares the strong anise odor but is smaller and has a paler pinkish to tan or creamy cap. "*Marasmius*" *calhouniae* occasionally ages to bluish green (possibly due to viral infection) but is much smaller, has very widely spaced gills, and lacks any odor. "*Hygrophorus*" *caeruleus* is a similarly-colored, stout, spring-fruiting species with a foul odor.

"Hygrophorus" caeruleus

O.K. Mill.

Blue Stinker

Cap: 2.5–9 cm across, rounded to convex with an inrolled margin when young, becoming broadly convex to plane; margin often wavy and uplifted in age. Surface moist to dry, occasionally felty to appressed-fibrillose, at times with small scales in age, typically cracked when dry. Silvery blue, grayish blue to creamy with a bluish tint, or more rarely entirely sky blue when young, becoming bluish gray, and developing dingy stains, to gray in age; occasionally with brownish orange tones. **Gills:** Broadly attached to slightly decurrent, close to well-spaced, broad, rather thick. Pale blue, bluish gray to bluish green when young, becoming grayish in age. **Stipe:** 2–5 cm long, 1–3 cm thick, typically short and stout, irregular, tapering abruptly at base. Surface dry, finely powdery at apex when young, with twisted, longitudinally appressed fibrils, base with white rhizomorphs. Concolorous with the cap. **Flesh:** Moderately thick, firm; fibrous in stipe. Dingy white, with blue to bluish gray tones, brownish in stipe base, more grayish overall in age. **Odor:** Farinaceous when young, becoming strongly rancid-farinaceous and foul in age. **Taste:** Mild when young, rather unpleasant, rancid in age. **Spore deposit:** White. **Microscopy:** Spores 6.5–9 × 4–5 µm, ellipsoid, smooth, inamyloid.

Ecology: Solitary, scattered, or in small clusters; typically partly buried in duff under conifers. Uncommon, occurring in east-side forests. Fruiting in spring soon after snowmelt, continuing into summer.

Edibility: Unknown.

Comments: This species is recognized by the silvery blue to bluish gray colors, short, stocky stature, and often strong, rancid-farinaceous odor. Unless it is growing in an area with hard packed soil and thin duff, it is rarely exposed. Despite its stout-statured fruitbodies, springtime fruiting in montane habitats, and foul rather than sweet aroma, its genetic ITS barcod e is virtually identical to that of *C. odora* var. *pacifica* — a mysterious, unresolved finding. *Clitocybe odora* var. *pacifica* has a similar colored cap and gills, but is more slender in stature and has a sweet, pleasant odor of anise or fennel; it typically fruits in the fall. *Hygrophorus marzuolus* has a grayish to blackish cap, and whitish to grayish gills, both of which can have a bluish tone (but never as pronounced as in '*H*'. *caeruleus*); it typically lacks an odor when young but can smell like swamp gas or a mouse nest in age. It grows at the edges of melting snowbanks in spring. *Clitocybe nuda* typically has more lilac to purplish colors in young fruitbodies (fading to gray or buff in age), crowded gills, often-bulbous stipe base, and a faint sweet odor.

Clitocybe nuda

(Bull.) H.E. Bigelow & A.H. Sm.

Blewit

Cap: 3–15 cm across, rounded with an inrolled margin when young, soon broadly convex to nearly plane, sometimes with an umbo, often wavy or uplifted in age. Surface smooth, slightly greasy when wet, often shiny when dry. Color highly variable – amethyst purple or pinkish purple when young with a paler margin, but may be light lavender, pale pinkish beige with a lilac hue, buff, beige, or even entirely brownish in age. **Gills:** Attached, often notched, close to crowded. Deep lavender or amethyst when young, becoming duller in age, developing ocher-brownish hues, sometimes extensively pale gray. **Stipe:** 4.5–10 cm long, 2–4 cm thick, equal to club-shaped, often with a bulbous base. Surface dry, scurfy (especially near apex), otherwise smooth to fibrillose. Base with fluffy lilac-white or purple mycelium, often binding to humus, and with pale lilac rhizomorphs. Pale silvery lilac to evenly light purple, sometimes with ocher or brownish tones. **Flesh:** Fleshy-fibrous, firm. marbled amethyst or pallid, when fresh often strikingly full of water. **Odor:** Often slightly sweet or floral. David Arora describes it as "frozen orange juice." **Taste** Mild to slightly spicy. **Spore deposit:** Pinkish buff. **Microscopy:** Spores 5.5–8 × 3.5–5 µm, ellipsoid, finely warted, inamyloid.

Ecology: Solitary, scattered, or clustered in small loose troops, sometimes in large fairy rings or arcs. An aggressive decomposer of organic matter, this species is found almost anywhere accumulated plant debris is present. As you might expect, it is easily cultivated. Common throughout urban areas and lower elevations, largely absent from undisturbed higher elevation forests. Fruiting from late fall into early spring.

Edibility: Edible and popular, but not everyone appreciates the strong, somewhat "earthy" flavor.

Comments: This species is quite variable in color within a consistent range: purple to brown with varying shades of beige, pink, and silvery gray. The gills and stipe base tend to stay purple after the cap fades, and this can be a good clue when examining browned-out older fruitbodies. The stout stipe and smooth cap with an inrolled margin give it a distinctive gestalt, and the pale pinkish buff spore print further help identify this species. The sister species, *Clitocybe violaceifolia* is very similar in all respects but may show deeper amethyst colors on average when fresh, and more of a tendency to fade to dingy or metallic silvery-gray in age; it may have a preference for the duff of cypress, cedar, redwood, and juniper. The Brownit (*Collybia brunneocephala)* is beige to brownish with no purple tones (some forms are even creamy to nearly milky-white overall). It is found in warmer, drier microclimates of inland valleys and short-grass meadows east of the Cascade crest. Some *Cortinarius* can look strikingly similar, but have cobwebby fibers on the stipe and rusty brown spores.

Synonym: *Lepista nuda* (Bull.) Cooke, *Collybia nuda* (Bull.) Z.M. He & Zhu L. Yang

Laccaria amethysteo-occidentalis

G.M. Muell.

Western Amethyst Deceiver

Cap: 1.5–10 cm across, convex to plane, often with a depressed center, or wavy to highly irregular. Surface dry to moist, smooth to finely fibrillose-scaly. Color extremely variable, depending on age and moisture content. Often distinctly royal purple when young, but can be purplish, lilac, pinkish buff to gray. Typically developing tan tones in age or fading to beige or grayish buff when dry. **Gills:** Broadly attached, notched with a decurrent tooth or slightly decurrent, well-spaced, often cross veined. Royal purple to lilac when young, and retaining purple tones longer than cap, before fading to pinkish buff. **Stipe:** 3–12 (16) cm long, 0.4–2 cm thick, equal, enlarged toward base, or quite twisted and distorted. Surface dry, fibrillose, purple, purplish gray, purple-brown, pinkish buff to grayish beige, typically with numerous whitish or pale buff loose fibrils and scurfs. Basal mycelium distinctly violet, often well after rest of fruitbody has faded. **Flesh:** Thin to moderately thick, rubbery-tough, fibrous. Violet, grayish lilac to whitish beige. **Odor:** Indistinct. **Taste:** Mild. **Spore deposit:** White. **Microscopy:** Spores (6.5) 7.5–10.5 × 6.5–9 µm, subglobose to ellipsoid, ornamented with sparse to crowded spines, 0.5–1.5 µm long. Basidia 4-spored.

Ecology: Solitary, scattered, or in small clusters on ground under conifers and hardwoods, and even in urban settings. Especially fond of Douglas-fir, oak, and pine. Very common, found at all elevations and on both slopes of the Cascades. Fruiting primarily in fall, with stragglers into early winter.

Edibility: Edible and good, at least the caps (the stipes are rather tough). Avoid eating collections from urban areas, along roads, or potentially polluted areas as *Laccaria* concentrate heavy metals in their fruitbodies.

Comments: When specimens are fresh, the vibrant purple colors, widely spaced gills, fibrous stipe, and white spores make this species practically unmistakable. When faded, it is mistaken for other *Laccaria.* The larger size, markedly scurfy, fibrous stipe, violet basal mycelium, and lack of warm pinkish buff tones help distinguish it. *L. bicolor* is generally smaller, and has a warmer pinkish brown to buff cap and paler gills (but often with a violet cast when young). *Clitocybe nuda* has a smooth cap, softer flesh, and buff spores. Some *Cortinarius* and *Inocybe* can look similar but have brown spores.

Laccaria nobilis

Laccaria bicolor

Laccaria proxima

Laccaria laccata

Laccaria nobilis

A.H. Sm.

Noble Deceiver

Cap: 3–10 cm across, convex with an incurved margin, expanding to plane or slightly uplifted and wavy in age. Surface dry, finely scaly or cracked. Pinkish buff to ocher-pink or orangish; hygrophanous, fading to paler orange-beige. **Gills:** Notched with a small decurrent tooth, widely spaced, broad. Pale lavender-white to pallid pinkish, often developing cinnamon-pinkish blotches in age. **Stipe:** 7–15 (20) cm long, 0.4–1 cm thick; tall, slender, upright, cylindrical with a slightly enlarged base. Surface vertically roughened, finely hairy-shaggy with recurved fibrils, Pinkish-tan to dull, pale orange-brown, sometimes with obscure lavender tints. Base with a coating of whitish (rarely violet) mycelium. **Flesh:** Cartilaginous to slightly rubbery-fibrous; stipe tough, fibrous. Pinkish buff. **Odor:** Indistinct. **Taste:** Indistinct. **Spore deposit:** White. **Microscopy:** Spores 7–10 × 6–9 µm, globose to broadly ellipsoid, spiny, spines up to 1.5 µm tall, inamyloid.

Ecology: Solitary or scattered in troops under conifers, perhaps more common in higher-elevation forests (although true extent of habitat and range obscured by confusion with other *Laccaria* species). Fruiting from summer at higher elevations into early winter at lower elevations.

Edibility: Edible.

Comments: As a genus, *Laccaria* are relatively easy to recognize by their widely spaced gills, tough, fibrous stipes, lack of veil tissue, and colors which range from purple to pink to dingy orange. However, trying to pin down exactly which species you have at hand will quickly reveal why they're called "Deceivers!" There's a wide range of variation in color, size, and stature *within* species, and substantial overlap *between* species. Thus, the following tentative generalizations are humbly offered: *L. proxima* is difficult to distinguish from *L. nobilis* macroscopically, although it tends to be shorter in stature, has a finely scaly cap, white basal mycelium, and is more common with pine. *L. bicolor* is usually a bit smaller, smoother, and pinkish-orange overall except for the contrasting lilac or lavender-whitish mycelium at the stipe base. Members of the *L. laccata* group are smaller still, light pinkish to orangish overall, and mostly grow with oaks. *L. montana*, *L. pumila*, and *L. tortilis* are miniature Deceivers (rarely over 2 cm) that prefer alpine habitats, often with willow. *L. amethysteo-occidentalis* has a scaly cap and a fibrous-hairy stipe surface but shows striking amethyst colors when young, and usually retains some lavender color on the gills into age.

Paralepista gilva

(Sowerby) Vizzini

Cap: 3–9 cm across, broadly convex with a depressed disc, and distinctly inrolled margin at first, becoming broadly funnel-shaped, with a wavy margin. Surface greasy or dry, smooth, but often with a pale sheen when fresh and with small moldy-looking patches in age. Rich pinkish to ocher, fading to dingy tan when dry or dingy brick red in age, lighter variants yellowish buff to pinkish beige and creamy at margin; often with darker spots and streaks. **Gills:** Decurrent, close to crowded, often forked near stipe. Whitish to cream, becoming beige, often dingy pinkish tan in age. **Stipe:** 5–12 cm long, 0.4–1 cm thick, cylindrical, upright but often curvy, base slightly enlarged. Smooth to fibrillose, base often covered in conspicuous pale mycelium. Creamy to pinkish tan or ocher brown (darker in age). **Flesh:** Fibrous, thin, pale. **Odor:** Instinct. **Taste:** Indistinct. **Spore deposit:** Off-white. **Microscopy:** Spores 4–5 × 3.5–4 µm, globose to broadly ellipsoid, finely warted, inamyloid.

Ecology: Scattered in small clumps, clusters, or in denser troops forming arcs and rings. Found in rich humus and needle duff of a wide range of forest types throughout Cascadia. Fruiting in fall and winter.

Edibility: Unknown.

Comments: The medium size, yellowy-tan to pinkish-ocher, funnel-shaped caps, and close, pale, decurrent gills are diagnostic. *P. flaccida* differs by having slightly darker cinnamon-orange to ocher and a black pepper odor; unsure if both species occur in Cascadia. A number of *Clitocybe* are somewhat similar but are usually more whitish to brown or grayish (less warm-toned). At least one similar *Bonomyces* species occurs in Cascadia but has a slightly richer pinkish-brown to orange-brown cap and smooth spores. *Infundibulicybe* can be quite similar but usually have whitish to creamy gills. *Rhizocybe* have pronounced spiky white rhizomorphs at the base of the stipe. Paler species of *Rhodocybe* can be similar but are usually smaller, don't show distinctly depressed caps, and have more elongated-ridged spores. *Hertzogia martiorum* looks somewhat similar but is more brownish, with a slightly chalky-looking cap, and lacks an odor.

Synonyms: *Clitocybe gilva* (Pers.) P. Kumm., *Lepista gilva* (Pers.) Roze.

"*Infundibulicybe*" *squamulosa* var. *montana*

comb. prov.

Mountain Cly

Cap: 2–7 cm across, broadly convex with a low umbo, soon uplifted to funnel-shaped in age. Surface dry, finely scaly to nearly smooth, opaque. Ocher-orange, bright ocher or reddish brown at first, fading to tan, beige or nearly white in age in dry conditions. **Gills:** Subdecurrent to decurrent, moderately to widely spaced, narrow; edges even. White to creamy. **Stipe:** 3–7 cm long, 0.3–0.7 cm thick, cylindrical, equal. Surface dry, faintly vertically-roughened. Ocher-tan to brownish with whitish fibrils. **Flesh:** Thin, fleshy, fibrous in stipe. Whitish. **Odor:** Indistinct to faintly fungal. **Taste:** Indistinct. **KOH:** No reaction. **Spore deposit:** White. **Microscopy:** Spores 7–11 × 3–4.5 µm, variably long-ellipsoid to elongate-teardrop or banana-shaped, sometimes irregular in outline; smooth, inamyloid.

Ecology: Solitary or in small clusters, often in troops in duff of conifers, usually at mid to high elevations. Widely distributed on both slopes of the Cascades but more common on the east side. Fruiting in spring and summer.

Edibility: Unknown.

Comments: The finely scaly to nearly smooth ocher-orange cap, white to cream, decurrent gills, contrasting stipe, and growth in spring identify this common species. This species needs to be formally combined into *Infundibulicybe* and elevated to species rank. *Bonomyces squamulosus* is quite similar but has a slightly scalier cap, a strongly farinaceous odor, and spores that are more broadly ellipsoid. *I. gibba* (sensu PNW) has a smooth, buff to grayish cap, and crowded decurrent gills. *Paralepista* are similar in shape but usually taller, with more pinkish-beige to ocher-beige colors (lacking such contrastingly pale gills) and tougher-textured stipes.

Synonym: *Clitocybe squamulosa* var. *montana* H.E. Bigelow

Clitocybe sp.

Lyophyllum sp.

Clitocybe (Atractosporocybe) sclerotoidea

Pseudoclitocybe cyathiformis

Gray-brown Clitocyboids

The diversity of small to medium-sized, dull-colored, veil-less, white-spored mushrooms in Cascadia is enough to frustrate even the most patient and persistent taxonomist. These species are, by definition, lacking in distinctive morphological features. The various species, genera, and even family-level groups seem to blend into one another and recent accumulation of DNA sequences has shed light on just how much work remains to be done. Broadly speaking, the "Gray Group" contains representatives from many different families and genera, including: Clitocybaceae (*Clitocybe*, *Collybia*, *Macrosporocybe*, *Pseudolyophyllum*) Lyophyllaceae (*Atractosporocybe*, *Gerhardtia*, *Leucocybe*, *Lyophyllum*, *Myochromella*, *Sagaranella*, *Sphagnurus*, *Tephrocybe*, *Tephroderma*, *Tricholyophyllum*), Porotheleaceae (*Clitocybula*, *Gerronema*, *Hydropus*, and *Megacollybia*), Pseudoclitocybaceae (*Harmajaea*, *Clitopaxillus*, *Pseudoclitocybe*), and even some far-flung waxcap relatives! With enough practice (and years of additional data), it's likely that we'll eventually be able to reliably distinguish a number of these species in the field. However, it is just as likely that some of this diversity will remain insurmountably subtle and variable, and that confident identification will hinge on genetic data. Although irritating to our desire for certainty and concrete names, this is a common pattern across the tree of life — our five human-scaled senses may very well fail to detect the variations in features that are relevant to the organisms themselves.

"Clitocybe" albirhiza

H.E. Bigelow & A.H. Sm.

Snowy Roothead, Whiteroot Cly

Cap: 2–8 cm across, convex with an inrolled margin when young, soon broadly convex or slightly wavy in age. Surface smooth, slightly greasy, moist to dry. Dingy buff-beige or whitish with pinkish-tan concentric zones and peachy orange blotches, dingy tan in age or irregularly fading to whitish when dry. **Gills:** Narrowly attached, thin, moderately spaced. Pale tan to pinkish beige or buff. **Stipe:** 3–6 cm long, 0.5–1 cm thick, cylindrical. Surface dry, smooth to vertically fibrillose with a pale-scurfy apex. Whitish-tan to dingy beige. Base with abundant white rhizomorphs, some pointed and spiky, others sinuous and elastic clinging to surrounding substrate. **Flesh:** Fibrous-fleshy; pale buff. **Odor:** Indistinct. **Taste:** Mild to slightly bitter. **Spore deposit:** White. **Microscopy:** Spores 5–6 × 2.5–3.5 µm, ellipsoid, smooth, inamyloid.

Ecology: Scattered in troops, often in clumps and clusters in duff under conifers in montane habitats, often near melting snowbanks. Widespread at higher elevations (determined by persistence of accumulating snow) on both sides of the Cascade crest. Fruiting in spring and early summer.

Edibility: Unknown.

Comments: This species is barely rescued from bland anonymity by the slightly peachy orange blotches and streaks of the pale young caps, the abundant spiky white rhizomorphs around the base of the stipe, and habit of spring fruiting usually near melting snow. Genetic data indicate that this species should be placed in *Rhizocybe*. There are a few other members of this genus in Cascadia, all of which show whitish rhizomorphs around the stipe base. *R. vermicularis* subsp. *americana* has a more depressed or funnel-shaped orange-tan to dingy ocher cap. Although it often fruits in spring, it is not strongly associated with snowmelt, and is more common in lower elevation forests. *R. rhizoides* has a weakly depressed to funnel-shaped, mottled grayish brown cap (often with chalky-looking blotches) and a bitter taste. It is also a spring-fruiting species.

Clitocybe deceptiva

H.E. Bigelow

Anise Cly

Cap: 1–5 cm across, broadly convex when young, becoming plane, usually with a shallow depression at disc; margin ruffled or wavy in age. Surface smooth, slightly greasy to dry. Pinkish-buff to pale tan, or creamy white to nearly white. Hygrophanous, margin translucent-striate when wet. **Gills:** Broadly attached to subdecurrent, or with a decurrent tooth, close to moderately spaced. Pallid whitish to watery beige. **Stipe:** 2.5–7.5 cm long, 0.2–0.8 cm thick, cylindrical, sometimes slightly curved. Surface smooth to slightly fibrillose, dry. Pallid whitish to beige-tan. **Flesh:** Very thin and watery. **Odor:** Sweet or anise-like when fresh. **Taste:** Slightly sweet or mild. **Spore deposit:** Pale pinkish-white. **Microscopy:** Spores 6.5–8.5 × 4–4.5 µm, ellipsoid, smooth, inamyloid.

Ecology: Solitary or scattered in troops in duff under conifers. Widespread and common but most abundant in low and mid elevation, west side forests. Fruiting from mid fall into spring.

Edibility: Edible but easy to confuse with other species of unknown edibility. The odor is not retained after cooking.

Comments: This species would be one of the most featureless of local gilled mushrooms were it not for its strong fragrance. Although not everyone detects the anise character, almost everyone can sense the strength and sweetness of the odor. *C. fragrans* is very similar but tends to be slightly larger, with a whitish to cream, often wavy cap; it is more common in redwood duff in California. *C. odora var. pacifica* shares the sweet-anise odor but is larger and entirely bluish to dingy bluish-olive gray in age. Many other "*Clitocybe*" are structurally similar and drab colored but smell mild or farinaceous. *Rhodocybe nuciolens* is not as pale (darker pinkish tan), is thicker-fleshed, has a pinkish spore print, with an unremarkable odor.

Clitocybe rivulosa

(Pers.) P. Kumm.

Sweating Cly

Cap: 1–5 cm across, broadly convex with a finely inrolled margin when young, becoming nearly plane and slightly depressed at center, and often uplifted and ruffled in age. Surface dry, smooth or wrinkled and cracking in age. White to whitish gray with grayish tan "water spots," concentric rings, and radial streaks, often becoming pinkish brown in blotches or in cracks. **Gills:** Broadly attached to decurrent, close to somewhat crowded, narrow. Dingy pale gray to creamy gray or gray. Often with irregular, contrasting lighter orange zones or bands. **Stipe:** 2–6 cm long, 0.5–2 cm thick, cylindrical or more often flattened and grooved and irregularly curved or twisted, usually slightly or distinctly enlarged near base. Surface dry, often vertically lined and roughened. Pale whitish gray to creamy grayish, becoming dingy grayish brown or ocher-brown where handled or in age. **Flesh:** Thin to moderately thick, rather firm-stringy or rubbery; pale grayish to grayish or dingy brownish. **Odor:** Indistinct. **Taste:** Indistinct. **Spore deposit:** Whitish. **Microscopy:** Spores 5–5.5 × 3–35 µm, ellipsoid, smooth, inamyloid.

Ecology: Solitary, scattered, or in small clusters, often forming arcs or rings in grassy areas such as pastures and lawns. Common and widespread, especially in urban areas. Fruiting primarily in fall, less often in spring.

Edibility: Highly toxic, containing muscarine. Sweating, salivation, and gastrointestinal distress are typical symptoms. Although small, the fruitbodies are frequently abundant in urban areas, and if many are eaten, they are potentially deadly for small children or pets.

Comments: This species is identified by its growth in clusters and arcs in grassy areas, drab whitish gray colors accented with pinkish to ocher-tan hues, often flattened and grooved stipe, and frequently concentrically zoned and "water spotted" caps. *Clitopilus prunulus* can be very similar but has a slightly velvety whitish to gray cap when young, pinkish spores, and a strongly farinaceous odor and taste.

Tricholomopsis flavissima

(A.H. Sm.) Singer

Yellow Mop

Cap: 1.5–5 cm across, rounded at first, soon broadly convex to plane, margin often narrowly inrolled when young. Surface moist to dry, appressed-fibrillose to finely scaly. Yellow to light golden yellow, often with slightly darker fibrils. **Gills:** Broadly attached to notched, close to subdistant, narrow; edge with a narrow gelatinous band (use hand lens). Yellow to golden; edges often paler. **Stipe:** 2–6 cm long, 0.3–0.7 cm thick, equal, often curved and rooted into substrate. Surface finely fibrillose to nearly smooth. Whitish to pale yellow, at times darker in age. **Partial veil:** A sparse, fleeting cortina of whitish to pale yellowish fibrils. **Flesh:** Thin, fibrous; light yellow. **KOH:** Not tested. **Odor:** Indistinct. **Taste:** Indistinct to slightly acrid. **Spore deposit:** White. **Microscopy:** Spores 7–9 × 6–8 µm, globose to subglobose, smooth, inamyloid. Cheilocystidia abundant, threadlike, somewhat gelatinized.

Ecology: Solitary or in small groups on decaying stumps and logs of conifers. Rare, mostly in old growth forests in the Cascade Range and higher elevations in the Coast Range and Olympic Mountains. Fruiting in late summer and fall.

Edibility: Unknown.

Comments: This rare species is recognized by its appressed-fibrillose to finely scaly yellow to golden cap, yellow gills and pallid stipe. *T. fulvescens* is similar, but tends to be darker yellow to ocher-yellow, stains rusty brown when damaged, and has larger, broadly ellipsoid spores (8–10 × 6–7 µm) and club-shaped to nearly spindle-shaped cheilocystidia. It is also rare, and occurs in similar habitats. *T. sulfureoides* is duller yellow, and has a smooth cap with darker appressed streaks, and small, ellipsoid spores (5.5–6.5 × 4.5–5 µm). It is more common and widespread in Cascadia. *Calocybe naucoria* (=*C. fallax*) often has a powdery stipe apex, and grows in litter and duff (not on wood).

Tricholomopsis decora
(Fr.) Singer

Decorated Mop

Cap: 2–6 cm across, rounded at first, soon broadly convex to plane, margin often finely inrolled. Surface moist to dry, densely granular to tufted-scaly when young, sparser in age. Bright lemon yellow to dingy yellow, covered in dark gray, brown to nearly black fibrils and scales. Hygrophanous, fading to yellowish creamy as it loses moisture, edge of cap sometimes translucent-striate. **Gills:** Broadly attached to notched, broad. Bright canary yellow to duller yellow, often with peachy tones in age or after drying out. **Stipe:** 3–7 cm long, 0.4–1.2 cm thick, equal, often curved, and rooted into substrate. Surface smooth or finely scurfy. Yellow to dingy light yellow or pale creamy beige. **Flesh:** Thin, fibrous; stipe hollow. Yellow. **KOH:** Orange on cap. **Odor:** Indistinct. **Taste:** Indistinct. **Spore deposit:** White. **Microscopy:** Spores 6–7.5 × 4.5–5 µm, ellipsoid, smooth, inamyloid.

Ecology: Solitary or in small groups on decaying conifer wood. Widespread, and locally common. Fruiting in fall.

Edibility: Unknown.

Comments: This wood dweller is recognized by the yellow cap with dark granular scales, bright yellow gills and stipe, and white spores. *T. floccosa* looks quite similar, but has olive-brown to brown scales on the cap. *T. sulfureoides* is typically duller yellow, and lacks the cap scales (but often has darker streaks). It also has a sparse, silky partial veil when young and has smaller spores (5.5–6.5 × 4.5–5 µm). Although similarly golden yellow in color, *Calocybe naucoria* (=*C. fallax*) often has a powdery stipe apex, and grows in litter and duff (not on wood). It is widespread in montane forests, but rare. *Chrysomphalina chrysophylla* has a slender stature, a golden orange cap with dark scales that becomes brighter and smoother in age, and bright golden gills. *Callistosporium luteo-olivaceum* is fairly similar, but has a smooth cap and is usually much duller, with olive to brown tones on the cap.

Tricholomopsis rutilans
(Schaeff.) Singer

Plums and Custard

Cap: 5–11 cm across, conical-convex to rounded when young, soon broadly convex to plane, often with a low, broad umbo. Surface densely hairy-velvety, more appressed-fibrillose in age. Rosy pink, magenta to wine red, fading in age to pale pinkish over a pale creamy yellow background, more extensively yellowish in age. **Gills:** Attached, notched, close, broad. Creamy when young, soon buttery yellow. **Stipe:** 5–16 cm long, 1–2.5 cm thick, equal, often curved at base. Surface dry, covered with chevrons of fibrils, very dense when young, becoming flattened or disappearing in age. Purplish red to pinkish red fibrils over a whitish to creamy yellow background. **Flesh:** Thin, fleshy-fibrous. Creamy yellow, slowly bruising ocher-brown where damaged. **KOH:** Reddish orange on gills and stipe. **Odor:** Indistinct to slightly farinaceous. **Taste:** Slightly nutty or indistinct. **Spore deposit:** Whitish. **Microscopy:** Spores 5–7 (9) × 3.5–5 (6.5) µm, ellipsoid, smooth, inamyloid.

Ecology: Solitary or scattered in troops on rotting conifer wood, wood chips, or lignin-rich humus. Widespread and fairly common. Fruiting in fall.

Edibility: Edible, but not to everyone's liking.

Comments: The reddish pink to purplish fibrillose cap, cream to yellowish gills, and contrasting reddish chevrons on the pale creamy stipe help distinguish this species. *T. flammula* has a similar cap, but is more slender, and lacks the red stipe fibrils. It also grows on decaying conifers but is rare. *Gymnopilus luteofolius* can look quite similar and also grows on wood, but has a silky-cobwebby partial veil and orange-brown spores. *Calocybe onychina* has yellowish gills and a purplish stipe, but has a more amethyst-purple cap that lacks hairy fibrils; it grows on duff and soil.

Calocybe onychina

(Fr.) Donk

Purple Mountain–dome

Cap: 2–6 cm across, rounded with an inrolled and sometimes finely ribbed margin, becoming broadly convex to wavy in age. Surface dry, opaque, finely velvety to smooth, often with a pale bloom, especially near margin. Burgundy to reddish-amethyst, paler toward margin, fading to duller purplish-brown overall in age. **Gills:** Narrowly attached to notched and appearing nearly free, close to subdistant, broad. Rich yellow to pale cream with a yellowish wash in age. **Stipe:** 3–7 cm long, 0.4–1 cm thick, cylindrical, straightish to curved, often slightly irregularly flattened. Surface dry, whitish to pallid cream with a dusky pinkish or rosy purplish wash, often frosty-looking when young, apex often dotted with purplish-red scruffiness. **Flesh:** Thin, fibrous; whitish to cream. **Odor:** Indistinct. **Taste:** Mild. **Spore deposit:** White. **Microscopy:** Spores 3–4 × 2–3.5 µm, subglobose to broadly ellipsoid, smooth, inamyloid.

Ecology: Solitary or in small troops or clusters in duff under conifers. Uncommon, throughout the mid to higher elevations on both slopes of the Cascades, but more often encountered on the east side; rare at lower elevations. Fruiting primarily in spring with rare fruitings in fall.

Edibility: Unknown.

Comments: The deep wine-purple cap and contrasting yellow gills make this a distinctive and beautiful species; the montane conifer habitat and habit of spring fruiting clinch the identification. *Tricholomopsis rutilans* can be similar in color, but tends to be a bit larger, has a finely hairy-scaly cap and stipe, the cap tends to be more reddish than deep purple, develops more extensive yellow tones showing between the rosy fibrils, and grows on rotting wood. *C. naucoria* (=*C. fallax*) is also found in high elevation forest, usually among moss or aspen litter. It is very differently colored, with a bright golden orange cap and yellowish-orange gills, and tends to be smaller and more slender.

Synonyms: *Rugosomyces onychinus* (Fr.) Raithelh.

Calocybe carnea

(Bull.) Donk

Pink Grass–dome

Cap: 1–4 cm across, convex to broadly convex with a finely inrolled margin, soon becoming depressed at center, concave and often wavy in age. Surface dry, finely velvety to smooth, opaque. Pastel pink to ocher pink or dusky pink, often with pallid chalky blotches. Sometimes fading to watery-tan or grayish-pink when weathered. **Gills:** Narrowly attached, often notched, or appearing vaguely free, close to somewhat crowded, thin, narrow. White to cream. **Stipe:** 2–6 cm long, 0.3–0.5 cm thick, cylindrical overall, but tapering towards stipe base; sometimes flattened or grooved. Surface finely scurfy with pink fibrils near apex, smooth to fibrillose and whitish to dull dingy pink below. Base often with whitish fuzz or hairs. **Flesh:** Thin, fibrous; whitish. **Odor:** Indistinct. **Taste:** Indistinct. **Spore deposit:** White. **Microscopy:** Spores 4.5–6.5 × 2–3 µm, ellipsoid, smooth, inamyloid.

Ecology: Scattered in troops or small clusters in grass lawns in urban areas. Rare to locally regular; most frequently reported around Seattle but could occur anywhere with suitable lawn habitat—more likely in warmer inland areas. Fruiting in summer and early fall.

Edibility: Likely non-toxic but little information available. Too small and infrequent to be of much interest.

Comments: The small size, pink colors, and gregarious growth on grass lawns make this species virtually unmistakable. Its urban habitat preference raises questions about its native range; most probably it is an introduced species from Europe. *Calocybe* cf. *ionides* is similar in habitat and stature (usually a bit larger and stouter) but has an amethyst to pale lavender cap and stipe. The brightly-colored Lyophyllaceae (including this species and its relatives) have been variously treated in the genera *Calocybe* and *Rugosomyces*, but this distinction is based on weakly defined genetic differences.

Synonyms: *Rugosomyces carneus* (Bull.) Bon

Flammulina populicola

Redhead & R.H. Petersen

Cottonwood Velvet-foot

Cap: 1.5–6 (8) cm across, convex to plane, occasionally with a low umbo. Surface smooth, thinly viscid and faintly translucent-striate when wet, opaque when dry. Yellow-brown to orange-brown, with a slightly paler margin when wet; hygrophanous, fading to yellowish to pale ocher-buff when dry. **Gills:** Narrowly attached with a distinct notch, close to crowded, narrow. Whitish to pale buff. **Stipe:** 5–12 cm long, 0.5–1.5 cm thick, equal or enlarged lower, with a tapered and often rooting base. Surface dry, distinctly velvety, especially over portion. Cream to pale yellowish at apex, gradually darker, from orange to orange-brown, with a dark brown base in age. **Flesh:** Thin, fibrous to cartilaginous in stipe. Concolorous. **Odor:** Indistinct. **Taste:** Indistinct. **Spore deposit:** White. **Microscopy:** Spores 6.5–7.5 × 4–4.5 um elliptic, ellipsoid to ovoid, smooth, inamyloid.

Ecology: Typically in loose to dense clusters, or scattered in troops, emerging from dead roots, or around the bases of aspen or cottonwood, more rarely on the lower trunks of snags, or on other hardwoods. Widespread but uncommon; more frequent on the eastern slopes of the Cascade Range. Fruiting from fall to spring.

Edibility: Edible.

Comments: The clustered growth from aspen roots or lower trunks, yellow-brown, thinly viscid cap, whitish gills, velvety, rooting stipe, and white spores readily identify this species. Members of the *F. velutipes* group tend to be smaller, often have darker, ocher-brown to reddish brown caps, and grow on a wide variety of hardwoods. *Flammula alnicola* has a similar clustered growth but more orange to ocher-brown colors and brown spores. The toxic *Omphalotus olivascens* is also orange and grows in clusters on wood, but is much larger, with a short, off-center stipe and grayish-olive flesh. We expect it will turn up in oak forests of inland southern Oregon. The large bright orange *O. illudens* has been introduced to the Seattle area.

Flammulina velutipes group

Velvet-foot

Cap: 1–5 (7) cm across, convex to plane, occasionally with a low umbo. Surface smooth, thinly viscid and translucent-striate when wet, opaque when dry. Ocher-brown to orange-brown to dark reddish brown, with a slightly paler margin when young, or wet; hygrophanous, fading to light orange-brown to orange-buff when dry or in age. **Gills:** Broadly to narrowly attached with a distinct notch, close, narrow. Whitish to pale ocher-buff. **Stipe:** 2–10 cm long, 0.2–0.7 cm thick, more or less equal. Surface dry, distinctly velvety, especially over lower portion. Whitish at apex, gradually darker, from orange to orange-brown, with a dark brown base. **Flesh:** Thin, fibrous to cartilaginous in stipe. Pallid. **Odor:** Indistinct. **Taste:** Indistinct. **Spore deposit:** White. **Microscopy:** Spores 6.5–9 × 3–4 µm, ellipsoid, cylindrical to teardrop-shaped, smooth, inamyloid.

Ecology: In clusters or scattered troops (rarely solitary), on hardwood snags, stumps, or logs. Common and widespread across Cascadia. Fruiting from fall to spring but more abundant in cold weather.

Edibility: Edible, with good flavor. Be especially sure of your identification as the deadly toxic *Galerina* have similar colors and also grow on wood.

Comments: This species is recognized by the ocher-brown to reddish brown, thinly viscid caps, whitish to pale buff gills, velvety stipe, and growth on hardwoods. It appears that the common species in Cascadia is *F. filiformis* but distinguishing it from *F. velutipes* is difficult without an ITS sequence. Pigment-less specimens of *F. filiformis* are cultivated and marketed under the name Enoki. *F. rossica* is often smaller and may be paler overall, with longer spores (7.5–11 × 3.5–4.5 µm). *F. lupinicola* is restricted to growth on bush lupine; it is rare in Cascadia, restricted to coastal dunes. *F. populicola* is larger and stockier, with a yellow- to orange-brown cap; it grows off the roots and lower trunks of aspen and cottonwood. All these species are edible but be aware of similar, brown-spored mushrooms, some of which are seriously toxic.

Baeospora myriadophylla

(Peck) Singer

Lavender Baeo

Cap: 1–2.5 cm across, convex when young, becoming broadly convex to plane; margin incurved at first, often wavy in age. Surface moist to dry, covered with a whitish bloom when young, finely floccose to nearly smooth. Lavender to lilac-gray when young, becoming lilac-brown to grayish brown in age, fading to grayish tan to buff if dry. **Gills:** Attached, sometimes with a small notch, very crowded, narrow, partial gills numerous. Purplish, lilac to lilac-gray when young, becoming purplish brown to purplish gray in age. **Stipe:** 1.5–6 cm long, 0.2–0.4 cm thick, equal, often rather long and thin. Surface dry, powdery when young, becoming smooth, base covered with coarse whitish to lilac hairs. Purplish, lilac to lilac-gray and covered with a powdery white bloom when young, becoming purplish brown to grayish. **Flesh:** Very thin, cartilaginous. **Odor:** Strongly fungal. **Taste:** Indistinct. **KOH**: No reaction. **Spore deposit:** Whitish. **Microscopy:** Spores 2.5–4.5 × 2–3 µm, rounded-ellipsoid, smooth, weakly amyloid. Cystidia large and encrusted.

Ecology: In small, scattered groups on rotting wood of both conifers and hardwoods. We have seen it on fir, alder, and cottonwood. Generally uncommon; more often encountered in mature forests on the west slope of the Cascades in Washington, mid to higher elevations of the Olympic Peninsula, and low elevation cottonwood forests. Rarer in Oregon. Fruiting from fall through spring.

Edibility: Unknown.

Comments: Though infrequently encountered, this lovely species is readily identified by its crowded purple gills, lavender to lilac-gray to grayish brown cap, and growth on wood. *Chromosera cyanophylla* is smaller, viscid, tends to have a yellowish cap, and has distinctly decurrent, more well-space gills. Other small purplish mushrooms include species in the *Mycena pura* group and some *Pseudobaeospora* species, neither of which grow on wood.

"Collybia" bakerensis

A.H. Smith

Spring Fir-coin

Cap: 1–4 cm across, convex with a slightly inrolled margin when young, expanding to plane, sometimes slightly wavy. Surface dry, finely hairy to silky, nearly smooth in age, opaque. White, pinkish to beige with whitish hairs. **Gills:** Narrowly attached or with a decurrent tooth, close, narrow; edges finely serrated or beaded to smooth. Whitish. **Stipe:** 1–6 cm long, 0.3–0.4 cm thick, slender, thin, cylindrical to slightly curving. Surface dry, innately fibrillose or scurfy over most of length. White to creamy, base often pinkish. **Flesh:** Thin, fibrous; white to pale pinkish in lower stipe. **Odor:** Indistinct. **Taste:** Indistinct. **Spore deposit:** White. **Microscopy:** Spores 5.5–7.5 × 3–4.5 µm, ellipsoid, smooth, inamyloid.

Ecology: Solitary, in small tufts, or scattered in small groups, directly on rotting wood, usually inside sheltered nooks and crannies and inner surfaces of bark sheets of true fir. Occurring in mid- to high-elevation conifer forests. Uncommon on both slopes of the Cascades, to be expected in the Olympic Mountains. Fruiting in spring and summer.

Edibility: Unknown. Tiny and thin-fleshed.

Comments: Although rather anonymous, the small size, overall pale colors, finely hairy to silky cap, and springtime fruiting in the nooks and crannies of conifer bark help distinguish this species. Genetic data place it in *Collybiopsis* but it has not yet been transferred. Other small wood-dwelling species have different textures or have partial veil tissue, differently-colored fruitbodies, or different mature gill/spore colors. Most similar in appearance is a *Trichocybe* that grows on decomposing hardwoods at low elevations in fall and winter. It has a more tan-beige cap and is rare in Cascadia. *Ripartites tricholoma* has a white, finely silky cap with a hairy-fringed margin, slightly decurrent gills, and pale brown spores. It is rare in Cascadia, fruiting from duff or well decayed wood.

Clitocybula abundans

(Peck) Singer

Cap: 1–4 cm across, bell-shaped to convex with slightly incurved margin, often with a depressed center, more concave in age. Surface dry, smooth to innately fibrillose. Whitish to pale grayish white, with a tan to grayish brown disc when young, paler in age or when dry. **Gills:** Broadly attached to subdecurrent, close to moderately spaced, narrow. White. **Stipe:** 3–6 cm long, 0.2–0.5 cm thick, equal or enlarged slightly towards base, often curved. Surface dry, with a faint bloom to smooth; base with whitish tomentum. White to whitish gray, slightly translucent. **Flesh:** Thin, fibrous; stipe hollow in age. Whitish to pale grayish. **Odor:** Indistinct. **Taste:** Indistinct. **Spore deposit:** White. **Microscopy:** Spores 4.5–6 × 3.5–5 µm, subglobose to broadly ellipsoid, smooth, amyloid.

Ecology: In loose clusters or troops on stumps, logs and other large woody debris of both conifers and hardwoods. Widespread but rare. Fruiting primarily in spring and early summer, rarely in fall.

Edibility: Unknown.

Comments: The loosely clustered to scattered growth on wood, whitish to grayish tan colors, and subglobose, amyloid spores help identify this species. *C. familia* tends to grow in denser, fused clusters but is otherwise similar; more study is needed to see if they differ. *Connopus acervatus* has a lilac-tinged beige to tan or brown cap, a darker lavender-toned brownish stipe, and usually fruits in dense clusters from wood. *Collybiopsis confluens* has a tough stipe with a faint coating of whitish scurfs, more tan-brown colors, and fruits from duff. *Mycena galericulata* doesn't usually fruit in such dense groups and has a rather tough stipe. *Mycena overholtsii* is quite similar in appearance and habit but has a darker, translucent-striate cap (at least when fresh), grayish gills, a very fuzzy stipe base, and fruits in spring near melting snowbanks.

Neolentinus kauffmanii

(A.H. Sm.) Redhead & Ginns

Kauffman's Sawgill

Cap: 1.5–5 (8) cm across, broadly convex to plane, often with a depressed center. Surface dry, finely pruinose or hairy, becoming smooth. Whitish beige to pale pinkish beige at first, darkening to pinkish tan to tan in age. **Gills:** Broadly attached to decurrent, close to crowded, partial gills numerous. Edges very finely serrate when young, becoming more ragged in age. Whitish to pale pinkish white at first, darkening slightly, pale creamy buff to pale pinkish buff in age. **Stipe:** 2–6 cm long, 0.3–1 cm thick, central to off-center, rarely lateral, equal or slightly rooting, often curved. Surface dry, finely pruinose, covered with a white sheen when young. Whitish, pale pinkish buff to pale tan. **Flesh:** Thin, fibrous and tough, especially in stipe, whitish to pale buff. **Odor:** Indistinct. **Taste:** Slightly acrid. **Spore deposit:** White **Microscopy:** Spores 5–6 × 2–2.5 µm, oblong to ellipsoid, inamyloid, smooth.

Ecology: Solitary or scattered in troops on well-decayed logs, stumps, and branches of Sitka spruce; occasional on other conifers. Common in coastal forests, rare elsewhere. Fruiting primarily from late fall into spring.

Edibility: Unknown.

Comments: The whitish-buff to pinkish colors, serrate gills, small size, and growth on decaying spruce wood are key identifying features. *Lentinellus micheneri* is often slightly larger, has more brown tones to the cap, and grows on hardwood branches and logs. The Cockleshell Sawgill *L. cochleatus* is larger, often grows in clusters on hardwoods, and has a shoe-horn shaped cap. *L. flabelliformis* sensu PNW (which appears to represent two species, *L. occidentalis* and *L. subargillaceus)* is smaller and has a short, lateral stipe and more widely-spaced serrate gills. A few *Clitocybe* are vaguely similar but have more closely-spaced gills with even edges and less tough-textured flesh.

Rhodocollybia oregonensis

(A. H. Sm.) Lennox

Almond Coin

Cap: 3–10 cm across, convex to plane, often with a low, broad umbo; margin often slightly uplifted and wavy in age. Surface smooth, moist to dry. Dark reddish brown to dark brown overall when young, soon fading slightly, to more orange-brown and becoming distantly two toned, with a darker center, and a paler, buff to orange-brown margin. **Gills:** Broadly to narrowly attached, often with a distinct notch, close to subdistant; edges finely serrated when young, becoming more ragged in age. Creamy white to pale yellowish. **Stipe:** 6–20 (30) cm long, 0.5–2 cm thick, equal, enlarged lower, then tapering to a rooting base. Surface smooth, dry, longitudinally striate or fibrous, at times with a powdery apex. Whitish, often faintly translucent, upper stipe often cool gray. Developing reddish brown stains in age. **Flesh:** Thin, tough, fibrous in stipe. **Odor:** Strongly like almond extract. **Taste:** Mild to slightly bitter. **Spore deposit:** Creamy pink to pale pinkish orange. **Microscopy**: Spores 6–8 × 4–5 µm, ovoid, smooth, dextrinoid.

Ecology: Solitary or scattered in thick duff, or around decaying stumps and logs in conifer forests. Rather scarce along the coast, somewhat more frequently encountered at mid- to high elevations of the Cascades. Fruiting in fall.

Edibility: Inedible.

Comments: The reddish brown, often two-toned cap, serrated to ragged gill edge, long rooting stipe that develops reddish spots and stains, and, most importantly, strong almond extract odor are distinctive. *R. prolixa (=R. badiialba)* generally has a dark reddish-brown cap that evenly fades and does not become two toned; it also lacks an odor. *R. maculata* has a paler cap, crowded gills, and a rooting stipe. It lacks an odor, and all parts become red-blotched in age. *Hygrophorus bakerensis* also has a distinct almondy odor but differs by having a two-toned cap that is more of a yellow-brown with a pale, whitish margin, and rather thick, waxy white gills.

Rhodocollybia maculata

(Alb. & Schwein.: Fr.) Singer

Spotted Coin

Cap: 2–10 cm across, convex with an incurved to weakly inrolled margin, expanding to broadly convex, or wavy in age. Surface smooth, greasy to dry. Whitish to cream, developing ocher to reddish-brown spots, in age often extensively washed reddish-brown over disc. **Gills:** Notched to finely attached or seceding; crowded, margins distinctly serrated. Whitish to cream. **Stipe:** 5–10 (15) cm long, 0.5–1 cm thick, equal or tapered and rooting into substrate. Surface smooth to longitudinally striate. Whitish to pale grayish-cream, often developing pinkish to red-brown blotches. **Flesh:** Moderately thick, fleshy in cap, fibrous-cartilaginous in stipe Whitish to cream. **Odor:** Indistinct. **Taste:** Bitter. **Spore deposit:** White or pale buff. **Microscopy:** Spores 5–7 × 4.5–6 µm, broadly ellipsoid to nearly round, smooth, at least some dextrinoid at maturity.

Ecology: Solitary or in small groups, usually on well decayed wood, typically at the stage of reddish-brown, crumbly debris. Prefers mature conifer forest, at mid to high elevations. Fruitings can appear nearly any time of year, concentrated in spring and fall.

Edibility: Inedible, very bitter.

Comments: The medium-sized fruitbodies, crowded gills, bitter taste, and tendency to develop reddish-brown blotches on all surfaces help identify this mushroom. Fruitbodies with yellow colors and less-crowded gills have gone by the name *R. maculata* var. *scorzonerea*; these likely represent a distinct species. *R. oregonensis* has a darker, evenly reddish-brown cap, and smells strongly of almond extract. Other *Rhodocollybia* mostly have evenly-colored darker caps and less of a tendency to show long, rooting stipes. Some *Hygrophorus* develop pinkish blotches and can be somewhat similar in appearance (e.g. *H. pusillus*, *H. neoerubescens*) but with more widely spaced gills and a mild taste. *Megacollybia fallax* is larger, with a radially-splitting gray cap, broad, widely-spaced gills, lacks rusty blotches, and usually grows around more obvious (less decomposed) stumps and other woody debris.

Rhodocollybia butyracea

(Bull.) Lennox

Buttery Coin

Cap: 2–7 cm across, convex to plane, with a low, broad umbo, often slightly uplifted and wavy in age. Surface smooth, lubricious and slippery when wet, to dry. Dark brown, reddish brown to dark cinnamon brown at first, soon fading to light brown or cinnamon brown, to buff if dry. **Gills:** Narrowly attached to free, close to crowded; edges finely serrated when young, becoming more ragged in age. White to pale pinkish. **Stipe:** 2–7 cm long, 0.4–1 cm thick, equal to club shaped with a swollen base. Smooth, dry to greasy, often translucent and longitudinally striate, upper portion smooth, lower portion covered with whitish hairs. Pale pinkish buff when young, becoming cinnamon brown in age. **Flesh:** Thin, stipe fibrous, often hollow in age. Concolorous with, or slightly paler than exterior. **Odor:** Indistinct. **Taste:** Mild. **Spore deposit:** Creamy buff to pale pinkish buff. **Microscopy**: Spores 5.5–7.5 × 2.5–4 µm, ellipsoid, smooth, some dextrinoid.

Ecology: Solitary, in small clusters, or in scattered troops in duff and needles of both hardwood and conifers. Widespread throughout Cascadia. Fruiting in fall, or occasionally in spring and summer in montane areas.

Edibility: Edible.

Comments: The greasy to soapy, dark brown to cinnamon brown cap, serrated to ragged gill edges, and creamy buff to pinkish buff spores are the best identifying features of this modest-sized mushroom. *R. asema* is nearly identical, with a light to medium brown cap. *R. prolixa (=R. badiialba)* is generally larger and has a darker reddish-brown cap and a stockier, often rooting stipe. *Gerhardtia foliicola* looks somewhat similar, with a reddish brown to cinnamon-brown cap, narrow, rather crowded white gills with even edges, and a bit stouter stature; it has been found from spring through fall in disturbed grassy areas as well forest settings. *Gymnopus dryophilus* is smaller, with paler ocher-brown to orangish caps, gills with even edges, and white spores. *R. subsulcatipes* has a tall, tough-cartilaginous stipe and dingy lavender-brown to ocher-tan colors. It is apparently rare, occurring under conifers at higher elevations, fruiting in fall.

Gymnopus dryophilus

(Bull.) Murrill

Oak-loving Coin

Cap: 2–8 cm across, convex with a narrow inrolled margin and often with a low, round umbo at first, becoming plane, to uplifted and wavy and occasionally centrally depressed in age. Surface smooth, greasy when young, soon dry. Warm brick to ocher-brown at first, then honey colored to beige, finally pale tan, hygrophanous, fading to pale beige or whitish when dry. **Gills:** Variably attached, ranging from broadly attached with a decurrent tooth to nearly free or seceding, close to crowded, soft, greasy feeling; partial gills numerous. White to cream. **Stipe:** 3–10 cm long, 0.3–1 cm thick, usually curved, cylindrical at first but often becoming irregularly flattened and grooved. Surface smooth to fibrillose-wrinkled. Base covered in blotches of hairy-spiky, pale contrasting whitish tomentum. Dull pale beige to tan, in some collections brighter honey tan to ocher. **Flesh:** Thin, somewhat tough in cap, cartilaginous in stipe. **Odor:** Pleasant. **Taste:** Indistinct. **KOH:** No reaction. **Spore deposit:** White. **Microscopy:** Spores 5–7.5 × 3–4 µm, ellipsoid, smooth, inamyloid.

Ecology: Usually in arcing troops or clusters, occasionally solitary, in forests as well as shrubland, with a preference for oaks. Widespread at lower elevations in Cascadia, more common in warmer, drier inland valleys. Fruiting in early fall, with scattered fruitings in spring.

Edibility: Edible but not widely collected for food.

Comments: The cheery ocher cap with a smooth greasy surface, smallish (but not tiny) size, and growth in troops help distinguish this relatively featureless fungus. The very similar *Rhodocollybia butyracea* has ragged-edged gills and is larger. *R. prolixa* is larger and darker reddish brown. *Melanoleuca* can be similarly colored but have taller, straighter stipes, more crowded gills, and amyloid ornamentation on the spores. *Clitocybe* generally have broadly attached or decurrent gills and paler or "cooler" colors.

Collybiopsis peronata

(Bolton) R.H. Petersen

Woollyfoot Coin

Cap: 2–7 cm across, convex or bell-shaped with a finely inrolled margin when young, expanding to plane, then uplifted and wavy in age. Surface dry, smooth to slightly radially fibrillose, margin sometimes with tightly appressed scraps of fibrillose white tissue. Dull grayish-pink, reddish-brown, or dingy yellowish, often streaky looking, fading to pale grayish, tan or whitish when dry. **Gills:** Narrowly attached or free with a distinct gap around the top of the stipe; narrow to moderately broad, rather widely spaced; edges even. Dull yellowish to dingy pale brownish, edges paler. **Stipe:** 3–9 cm long, 0.2–0.7 cm thick, cylindrical, slightly enlarged to club-shaped at base. Surface finely scurfy over upper surface, becoming woolly, cottony, or remarkably tangled-shaggy over the lower stipe. Pale grayish to yellowish, but densely covered overall with pale fibrils, thus appearing pallid. **Flesh:** Thin but tough, pallid to yellowish-tan. **Odor:** Indistinct or faintly fragrant. **Taste:** Peppery-spicy or indistinct in some cases. **KOH:** Indistinct to faintly green. **Spore deposit:** White. **Microscopy:** Spores 8.5–10 × 3–4 µm, ellipsoid to elongate, smooth, inamyloid.

Ecology: Solitary, occasionally in tufted clumps, in scattered troops, usually attached to bits of small woody debris, or on rich organic matter including needle duff and leaf litter. Abundant in a variety of lowland habitats, but especially partial to alder and bigleaf maple duff and disturbed urban "forests" in park settings. Currently abundant from British Columbia south to Portland, and likely continue to expand its range southwards. Fruiting from summer through early winter.

Edibility: Edible, but not highly sought-after.

Comments: When young, these mushrooms are recognizable by their pinkish to tan to pale grayish-white cap color, yellowish to dingy pinkish brown gills, and tough stipe covered in whitish fuzz or long-shaggy hairs (like a teased-apart cotton ball). In age, the fruitbodies become so browned-out and nondescript that the stipe texture is the only feature left to look for. This species appears to have arrived in Washington from Europe (perhaps as early as the 1920s), became common by the 1990s, and is now spreading rapidly—making it one of the more clear-cut cases of an invasive macrofungus in our area. *C. confluens* has a more short-fuzzy stipe surface, lacks the yellowish and pinkish colors, and tends to routinely fruit in tight clusters. *Gymnopus ocior* can also have yellowish gills but has a more brownish to reddish-brown cap and lacks the fuzzy to shaggy stipe ornamentation.

Synonyms: *Gymnopus peronatus* (Bolton) Gray, *Collybia peronata* (Bolton) P. Kumm.

Gymnopus erythropus

(Pers.) Antonin, Halling & Noordel.

Red-footed Coin

Cap: 1–6 cm across, broadly convex to plane, or slightly uplifted and wavy in age. Surface moist to dry, smooth, translucent-striate. Brown to honey-colored at center, beige to cream outward and with a pale margin. **Gills:** Narrowly attached, close to subdistant, narrow; edges even. Whitish to cream. **Stipe:** 2–9 cm long, 0.3–0.5 cm thick, cylindrical, at times longitudinally compressed, bases often fused. Surface smooth, shiny. Lower half strikingly reddish-orange, fading in a gradient upwards to beige at apex. **Flesh:** Thin, tough, pallid in cap; stipe hollow, concolorous. **Odor:** Indistinct or faintly like rotting cabbage. **Taste:** Indistinct. **KOH:** No reaction. **Spore deposit:** White. **Microscopy:** Spores 6–8.5 × 3.5–4 µm, ellipsoid, smooth, inamyloid. Spores and cap tissue green in KOH.

Ecology: Fruiting in dense, tight clusters in wood chips and lawns in human-modified and disturbed areas but may be spreading into native habitats. Widely distributed through the lowlands of Cascadia but primarily found in the immediate vicinity of urban areas.

Edibility: Unknown.

Comments: The pale caps, tough, polished stipes with extensive orangey-red coloration, and growth in clusters in disturbed areas help to distinguish this species. This species was likely introduced to the Pacific Northwest, from Europe, with the first records appearing in the 1960's in the Seattle Arboretum. Another introduced species (which is becoming 'weedy' in California), *Gymnopus brassicolens*, has a darker brown cap, a stipe that is brownish to black over the lower portion, and smells strongly of rotting cabbage. *Mycetinis applanatipes* is a rare species found in duff of higher-elevation fir forests; although it often has a similarly dusky reddish-brown lower stipe, the stipe is minutely scurfy over most of its length. *Flammulina* species also have pale caps, grow in clusters, and have velvety stipes that show a color gradient through yellow-orange to cream near the apex. The lower stipe tends to be much darker brown to nearly black (sometimes dusky brown overall) and they grow directly on wood.

Connopus acervatus

(Fr.) K. W. Hughes, Mather & R. H. Petersen

Dusky Clustercoin

Cap: (0.7) 1.5–4 cm across, convex with a thinly inrolled margin at first, becoming broadly convex, plane or centrally depressed in age. Surface smooth, moist to dry. Reddish purple when young, soon reddish brown, fading to brownish in age, strongly hygrophanous; pinkish brown to buff when dry. **Gills:** Narrowly attached, with a curved notch, to free in age, close to crowded, narrow, partial gills numerous. Whitish to pale pinkish buff. **Stipe:** 3–10 cm long, 0.2–0.5 cm thick, equal, fragile. Surface smooth, base covered with a whitish tomentum when young. Reddish purple when young, soon reddish brown, darkening in age. **Flesh:** Concolorous, thin, fragile in cap, brittle, but fibrous in stipe. **Odor:** Indistinct. **Taste:** Indistinct. **KOH:** No reaction. **Spore deposit:** White. **Microscopy:** Spores 5.5–7 × 2.5–3 µm, cylindrical to ellipsoid, smooth, inamyloid.

Ecology: In fused clusters of overlapping fruitbodies (sometimes with dozens of caps), often growing from well-decayed stumps, logs or buried roots, or thick duff. Common throughout Cascadia in forests from sea level to the highest parts of the Cascades, including mesic areas of the east slope. Fruiting in fall or early winter.

Edibility: Not edible, possibly mildly poisonous.

Comments: The clustered growth, reddish brown to lilac-gray, hygrophanous caps, and smooth stipes are distinctive. Faded specimens could be mistaken for *Collybiopsis confluens* but that species is usually taller and less tightly clustered and has a minutely fuzzy stipe. *G. erythropus* has an ocher-buff to tan cap and an orange to reddish orange stipe. It often grows in large clusters in wood chips or lawns. *G. brassicolens* has a brown cap that fades to ocher-buff to tan when dry, but has a black lower stipe and a strong disagreeable odor.

Synonyms: *Gymnopus acervatus* (Fr.) Murrill

Collybiopsis subpruinosa

(Murrill) R.H. Petersen

Walnut Coin

Cap: 1–7 cm across, broadly convex to plane, often wavy in age. Surface dry, with pallid fibrils or a powdery bloom when young; smooth when young but soon bumpy and extensively grooved nearly all the way to disc. Dull beige to tan with brownish grooves, disc darker brown, usually paler toward margin. **Gills:** Attached, well-spaced, occasionally inter-veined; partial gills numerous. Distinctively clay-tan to grayish beige. **Stipe:** 3–10 cm long, 0.2–0.4 cm thick, more or less equal, thin, tall. Surface dry, floccose-pruinose. Whitish to tan. Base often with white, binding mycelial rhizomorphs. **Flesh:** Very thin in cap, tough and slightly wiry in stipe. **Odor:** Indistinct. **Taste:** Indistinct. **KOH:** Darkening slightly. **Spore deposit:** White. **Microscopy**: Spores 7.5–9 × 3.5–4 µm, ellipsoid to teardrop-shaped, smooth, thin-walled, inamyloid, colorless in KOH.

Ecology: Often clustered, sometimes in large arcs or rings in lignin-rich humus, wood chips, or thick duff, usually in urban areas. Fruiting throughout the year but most abundant in late summer and fall.

Edibility: Unknown.

Comments: The deeply sulcate, pallid beige cap (reminiscent in color and texture of a walnut shell) in combination with the tough, finely pale-fuzzy stipe, and urban milieu distinguish this species. It is most likely to be confused with *C. villosipes*, which is a darker brown to vinaceous brown overall with a more distinctly umbilicate cap.

Synonym: *Gymnopus subpruinosus* (Murrill) Desjardin, Hailing & Hemmes

Collybiopsis villosipes

(Cleland) R.H. Petersen

Scurfy-legged Coin

Cap: 1–3.5 cm across, convex to plane with a sunken center, often wrinkled to irregular in age. Surface dry to moist, smooth to wrinkled, often leathery in age. Dark reddish brown, vinaceous brown to dark brown when young and wet, dull brown when dry, becoming dingy brown to beige in age. Strongly hygrophanous, translucent-striate when wet, dull, much paler when dry. **Gills:** Attached to a gelatinous collar, which can dissolve, making them appear free in age, close to subdistant, narrow, partial gills numerous. Beige to light brown. **Stipe:** 1.5–8 cm long, 0.1–0.3 cm thick, more or less equal, tough, wiry. Surface dry, covered with fine, stiff hairs, base with binding mycelium and rhizomorphs. Reddish brown to dark brown, often paler toward apex. **Flesh:** Very thin, leathery, stipe tough, fibrous, hollow. **Odor:** Indistinct. **Taste:** Indistinct. **KOH:** No reaction. **Spore deposit:** White. **Microscopy:** Spores 6.5–10.5 × 3.5–4.5 µm, ellipsoid, smooth, thin-walled, inamyloid.

Ecology: Scattered in troops or clusters, occasionally solitary in duff under conifers, lawns, or in wood chips. Most often found in disturbed areas, such as trail-sides or in urban landscaping. Common throughout Cascadia. Fruiting soon after first rains in fall, with stragglers through the winter, and a second flush in spring; or summer in irrigated areas.. The fruitbodies are tough and long-lived and seem to be able to dry out and rehydrate to some degree.

Edibility: Unknown.

Comments: This common and widespread mushroom has a small size, hygrophanous cap that varies from reddish or vinaceous brown to light brown or beige with a thin, wiry stipe covered in a fine pale coating of tiny hairs. *Gymnopus fuscopurpureus* is often larger with a darker vinaceous cap, pale vinaceous gills, a smooth stipe, and a bright green KOH reaction. *C. confluens* is larger, with pale beige to yellowish beige colors, and grows in fused tufts and clusters. *C. subpruinosus* is a common wood chip species that differs by its light tan to beige and deeply translucent-striate to sulcate cap.

Synonyms: *Gymnopus villosipes* (Cleland) Desjardin, Halling & B. A. Perry

Marasmius oreades

(Bolton) Fr.

Fairy Ring Mushroom

Cap: 1–6 cm across, rounded to bell-shaped when young, becoming broadly convex to plane; margin sometimes ruffled, and often wavy in age. Surface smooth, greasy when young, soon dry. Pinkish tan to ocher-beige with a darker edge at first, soon paler to beige, strongly hygrophanous; fading when dry, or in age, to nearly white. **Gills:** Narrowly attached or free, moderately close to well-spaced, broad. Creamy whitish to pale beige, becoming ocher blotched or overall warm tan in age. **Stipe:** 3–7 cm long, 0.3–0.5 cm thick, cylindrical and often quite straight. Surface dry, finely pruinose near apex, finely velvety to obscurely matted-tomentose over lower half when young, or smooth in age. Base attached to a corky mycelial mat (although often inconspicuous). Ivory to creamy or beige-tan near apex, ocher-tan downward. **Flesh:** Thin, somewhat tough, buff to whitish in cap; tough, fibrous, whitish in stipe. **Odor:** Mild to slightly almondy or sweet, chemical-like. **Taste:** Mild. **Spore deposit:** Whitish. **Microscopy:** Spores 6.5–9 × 4–6 µm, broadly ellipsoid, almond- to lemon-shaped, smooth, inamyloid.

Ecology: In clumps, arcs, and rings. Occurring almost exclusively in grassy lawns, thus mostly often encountered in the low elevation central valleys of Cascadia but also in grass lawns, short-grass meadows at higher elevations, and in grassy steppe habitat on the east side. Fruitings can be found throughout the year with peaks in spring and early summer, and again in fall.

Edibility: Edible and good but not recommended, as it often grows in areas treated with herbicides or polluted by urban emissions.

Comments: Few other mushrooms that regularly occur on grass lawns could be confused with this species. The warm buff-beige caps, tough stipes with a fuzzy lower surface, and habit of forming rings or arcs on lawns are a distinctive combination of characters. As with many mushrooms, care should be taken when eating it from treated lawns or roadsides, two places where it commonly grows, as the fruitbodies of some mushrooms concentrate heavy metals. *Gymnopus dryophilus* is somewhat similar but usually a bit larger, has a fleshy cap and cartilaginous stipe, and grows in duff and liter in forested habitats. *M. albogriseus* is quite similar but generally paler or with dingier grayer tones. It grows in duff in forested habitats but is rare in Cascadia. *M. oreades* shares the same habitat as the poisonous *Clitocybe rivulosa* but is unlikely to be mistaken for this species.

Marasmius plicatulus

Peck

Red Pinwheel

Cap: 1–4 (5.5) cm across, conical to bell-shaped when young, becoming broadly conical to convex, often umbonate in age. Surface dull, dry, finely velvety and often with a whitish sheen. Water droplets often bead up on the cap in wet weather. Deep wine red, burgundy to reddish brown, but more orange-brown to reddish tan colors are possible. Often paler toward margin, at least in age. **Gills:** Narrowly attached, subdistant to distant, broad, often with 1 or 2 partial gills in between each full gill. Pale pinkish white to creamy white when young, becoming pale buff to pale yellowish buff. **Stipe:** 6–12 cm long, 0.2–0.4 cm thick, more or less equal, tough, wiry, fibrous. Surface smooth, shiny, base covered with a "pom-pom" of spiky whitish mycelial hairs. Colors quite variable, white at extreme apex when young, then neon or pastel magenta, to reddish or purplish, transitioning through dark reddish brown or purplish brown to black toward base; often extensively polished reddish brown or blackish in age. **Odor:** Indistinct. **Taste:** Indistinct. **Spore deposit:** White. **Microscopy:** Spores 12–16.5 × 4.5–6.5 µm, almond-shaped to spindle-shaped, smooth, inamyloid.

Ecology: Solitary, scattered, or in small groups on duff or leaves. Common throughout lower elevations of Cascadia. A true generalist, this species can grow in a wide range of forest types; it seems especially fond of cedar, Douglas-fir, oak, and other trees which produce thick, accumulating layers of duff. Fruiting primarily in fall, with scattered fruitings into winter.

Edibility: Edible.

Comments: This beautiful mushroom is nearly unmistakable thanks to the shiny, polished stipe that grades from pink to black over its length, the velvety cap whose color ranges from burgundy, rosy pink to orange-brown, and widely spaced creamy gills. *M. curreyi* is a tiny species with a brighter orange cap; it is found in short-grass habitats.

Macrocystidia cucumis

(Pers.) Joss.

Cucumber Cap

Cap: 1–4 cm across, hemispherical when young, becoming broadly convex in age, occasionally with a wavy margin. Surface moist or dry, very finely pruinose to smooth. Blackish brown to dark mahogany with a whitish sheen, paler, reddish brown outward, to an abruptly pale, nearly white extreme margin, and translucent-striate when wet. Hygrophanous, pale mahogany buff when dry, **Gills:** Narrowly attached to free, close, broad. Pale creamy to beige, becoming cinnamon to pinkish brown in age. **Stipe:** 3–8 cm long, 0.3–0.5 cm thick, equal and rather straight. Surface dry, irregularly wavy, covered with a dusting of pruinose cystidia, becoming nearly smooth in age. Dark reddish brown to nearly black over lower portion, variably pale pinkish tan to brownish over upper portion; covered with a whitish pruinose sheen, darkening from base up in age. **Flesh:** Thin, tough, springy. Dark in stipe, pallid in cap. **Odor:** Usually distinctive, like cucumber or sometimes unpleasant. Described, rather aptly, as smelling like sushi. **Taste:** Slightly farinaceous to mild. **KOH**: No reaction. **Spore deposit:** Pinkish to dingy salmon tan. **Microscopy** Spores 7.5–10 × 4–5 µm, ellipsoid, smooth. Cystidia quite large with pointed tips, abundantly scattered over faces and edges of gills, clamp connections abundant.

Ecology: Scattered in clumps and clusters, or troops in wood chips and other disturbed landscaped areas. Common around major urban areas in Cascadia. Fruiting in fall and winter.

Edibility: Unknown.

Comments: The dark cap with a narrow pale edge, dark, tough stipe, and cucumber odor make this a very distinctive species with only one real lookalike. *Gymnopus brassicolens* is similar in some respects but has a cap that often becomes plane to uplifted in age, a stipe that is often grooved and less upright, and a much more disagreeable odor — like rotting cabbage.

Tephrocybe rancida

(Fr.) Donk

Rancid Grayling

Cap: 1–6 cm across, round-conical to bell-shaped when young, expanding to nearly plane, but usually retaining a distinct, obtuse umbo; margin finely inrolled when young. Surface, moist to dry margin chalky-pruinose when young, weakly translucent-striate in age. Dark cool gray to grayish-tan, developing beige-brown tones at maturity or when dry. **Gills:** Narrowly attached or notched, pale gray to dark gray, fairly broad; edges even. **Stipe:** 4–12 cm long 0.2–0.6 cm thick, tall, straight, more or less cylindrical, often with a long, rooting base. Surface smooth, occasionally vertically grooved; base with white, cottony patches of mycelium. Cool medium to dark gray, whitish over lower quarter, in age fading overall to beige, tan or brownish-gray. **Flesh:** Thin, fibrous, slightly fragile; hollow in stipe. Pale gray to whitish. **Odor:** Unpleasantly rancid farinaceous. **Taste:** Farinaceous. **Spore deposit:** White. **Microscopy:** Spores 6–8.5 × 3–4.5 µm, smooth, ellipsoid, inamyloid.

Ecology: Solitary or scattered in small troops, never abundant. Widely distributed in mature conifers throughout Cascadia but often overlooked, probably more common than reported. Fruiting in fall and early winter.

Edibility: Unknown.

Comments: The dark, cool-gray cap and stipe, umbonate cap, and foul scent help distinguish this odd species. *Sphagnurus paluster* is smaller, paler beige to tan-brown, and grows in boggy or mossy habitats (e.g. on *Sphagnum*). *Sagaranella* and *Myochromella* species are typically smaller, often more deeply translucent-striate, and usually show distinct pale scurfiness over the upper stipe. *Gerronema atrialba* grows on wood, has a depressed cap center, and dark gray dots over the stipe surface.

Sagaranella tylicolor

(Fr.) V. Hofstetter, Clémençon, et al.

Brown Sag

Cap: 0.5–2.5 cm across, bell-shaped to conical-convex, usually with a small, acute umbo. Surface finely fibrillose to smooth, moist to dry, translucent-striate. Light, dull honey-brown to pinkish-brown, darker to grayish-brown at disc, often with a small sheen of white fibrils. **Gills:** Narrowly attached to slightly notched, moderately spaced, broad; edges even to slightly eroded. Pale creamy to pale whitish-tan, slightly translucent. **Stipe:** 2.5–6 cm long, 0.2–0.3 cm thick, tall, slender, cylindrical, fairly straight. Silvery grayish-white to pinkish-tan or dull honey-brown, covered with small white scurfs over most of length, denser over lower half. **Flesh:** Thin, fragile, watery-fibrous, slightly translucent in stipe. **Odor:** Mild to radishy. **Taste:** Mild to earthy. **Spore deposit:** White. **Microscopy:** 6–8 × 5–6 µm, round-ellipsoid, surface spiny.

Ecology: In troops in nitrogen-enriched soil and mossy forest duff. The nitrogen source is commonly from a decomposing carcass, urine, or dung. Widespread in Cascadia, probably more common than currently known. Fruiting in fall.

Edibility: Unknown.

Comments: This elegant little mushroom can be identified by its pointed, yellow-brown to pinkish-brown translucent-striate cap, pale, scattered-scurfy stipe, and tendency to grow in troops on suspicious substrates. Although vaguely resembling a *Mycena*, this species is more stout-statured and a bit larger and browner than most. *Myochromella* are closely related and quite similar but usually more gray-toned and less markedly translucent-striate. *Sphagnurus paluster* has smooth, pale gray cap and stipe; it is a parasite on sphagnum moss. A few *Nolanea* are similar but usually larger, with a more strongly hygrophanous cap, less of tendency to fruit in dense troops, and have pink spore deposits. This species belongs to a fascinating group of "ammonia" fungi that fruit in response to inputs of nitrogen to the soil – these include a number of *Coprinopsis*, *Hebeloma*, *Laccaria*, and few different kinds of ascomycetes.

Pseudobaeospora 'Dark'

Cap: 0.3–1.7 cm across, rounded to convex with a narrowly inrolled margin when young, soon broadly convex, plane to uplifted and wavy in age. Surface dry, with a fine powdery bloom when young. Dark gray over disc, lighter gray outward, dingy grayish brown in age. **Gills:** Narrowly attached or deeply notched, moderately spaced, broad; thick-edged. Grayish yellow, gray to tan. **Stipe:** 1–3.5 cm long, 0.05–0.2 cm thick, equal or slightly enlarged at base. Surface with scattered scurfy scales over upper portion; base covered with coarse, matted hairs. Gray at apex, darker gray to black below, developing brownish tones in age. **Flesh:** Thin, cartilaginous-fibrillose; concolorous with exterior. **Odor:** Indistinct to unpleasant. **Taste:** Not sampled. **KOH:** Slightly bluish gray-green. **Spore deposit:** Whitish, often scant. **Microscopy:** Spores 3.5–4 × 3–3.5 µm, subglobose, smooth.

Ecology: Solitary or scattered on ground or in humus under bigleaf maple and western redcedar, often interspersed in areas with the more noticeable waxy caps (*Hygrocybe*). Locally common; known from the Olympia area, but likely more widespread. Fruiting in late fall and early winter.

Edibility: Unknown.

Comments: *Pseudobaeospora* are recognized by their opaque caps with finely inrolled (and sometimes ribbed) margins; rather thick-edged, moderately- to well-spaced gills; slender stipes which often have a scurfy apex and variable amount of coarse, spiky hairs at their base. Many species have distinct KOH reactions. They are fairly common and diverse in California's redwood and cypress forests, but appear to be rare in Cascadia (although this may be due collector bias, as most mushroomers don't spend much time in cedar forests). We know of three (likely unnamed) species in Cascadia: the one described above, a beautiful cherry-red species from mid-elevations in the Washington Cascades, and a vinaceous purple species (similar to *P. deckeri*) from the Seattle area.

Baeospora myosura

(Fries) Singer

Conesprout Baeo

Cap: 0.5–3 cm across, broadly convex, plane or weakly wavy, usually with a small umbo. Surface opaque, smooth, moist to dry. Light brown, brownish tan to butterscotch tan or yellowish, usually paler around margin. **Gills:** Narrowly attached to free, very crowded, narrow. Off-white to pale grayish white. **Stipe:** 2.5–6 cm long, 0.05–0.3 cm thick, equal, often rooting and curved. Pinkish beige to tan-brown with fine, contrasting pale scruffiness over most of length, especially near apex. Base covered with fuzzy whitish hairs. **Flesh:** Very thin, cap leathery, stipe fibrous. **Odor:** Indistinct. **Taste:** Indistinct. **KOH:** No Reaction. **Spore deposit:** Off-white. **Microscopy:** Spores 3–4 × 2–2.5 µm, ellipsoid, smooth, amyloid.

Ecology: Solitary, scattered, or in small troops fruiting directly out of conifer cones, most often those of Sitka spruce or Douglas-Fir, occasionally pine. Common throughout Cascadia west of the Cascade Crest. Fruiting from the first rains of the fall season into spring, but most abundant in early fall.

Edibility: Unknown.

Comments: The broadly convex and often slightly umbonate, light caramel-colored cap, crowded whitish gills, small size, and growth on conifer cones easily distinguish this species. This species frequently grows intermixed with other cone-dwelling species, especially *Strobilurus trullistaus* and *S. occidentalis*, which have pinkish-white to whitish-yellow or grayish caps, less densely-packed gills, yellow lower stipes, and inamyloid spores.

Strobilurus trullisatus

(Murrill) Lennox

Doug-fir Conesprout

Cap: 0.5–2 cm across, convex to plane, occasionally wavy and uplifted in age. Surface smooth, moist to dry. Whitish with a pinkish orange to buff cast to center, occasionally pale grayish white when young, to all white. Slightly hygrophanous, faintly translucent-striate when moist, opaque, dull when dry. **Gills:** Broadly attached, close to subdistant. White or with a pale pinkish buff cast. **Stipe:** 2–6 cm long, 0.1–0.2 cm thick, thin, tough, with a deep rotting base if cone is buried. Surface dry, appearing powdery, with a whitish sheen, often with whitish to pale orange hairs at base. Yellowish to yellow-brown over lower portions with a white apex, darkening slightly, to orangish brown overall in age. **Flesh:** Very thin, tough; concolorous. **Odor:** Indistinct. **Taste:** Mild. **KOH:** No reaction. **Spore deposit:** White. **Microscopy:** Spores 3.5–6 × 2–3.5 µm, ellipsoid to narrowly almond-shaped, smooth, inamyloid. Pleurocystidia capitate and with thickened walls.

Ecology: Solitary or scattered in small troops, growing directly out of Douglas-fir cones, rarely those of other conifers. Common and abundant throughout most of Cascadia (wherever its host is present). Fruiting soon after first rains in fall, occasionally with a few persisting into winter.

Edibility: Unknown.

Comments: The whitish cap (often with a yellowish or buff blush at the disc), white gills, thin stipe with a yellowish to yellow-brown, fuzzy base, and growth on Douglas-fir cones are distinctive. *Baeospora myosura* has a light grayish-brown cap and very crowded gills; it can grow on the cones of Douglas-fir but is most common on spruce cones. *S. occidentalis* is similar but mostly grows on cones of Sitka spruce and occasionally on loose woody debris (but apparently not on Douglas-fir cones); it has a whitish or grayish cap and a darker orange lower stipe and lacks capitate cystidia. See comments under *S. lignitilis* for other species of *Strobilurus.*

Strobilurus lignitilis

Wells & Kempt.

Wood Conesprout

Cap: 1–5 cm across, convex to broadly convex when young, expanding to nearly plane, sometimes with a diffuse, low umbo. Surface smooth, dry; margin translucent-striate. Dark brown to gray-brown with a silvery-hoary sheen when young, fading to ocher-tan, pale gray or creamy-beige; margin always paler, often with a narrow white band; rarely extensively whitish. **Gills:** Narrowly attached, close to subdistant, broad. White. **Stipe:** 3–10 cm long, 0.2–0.4 cm thick, equal, often rooting, and with fuzzy white hairs at base; smooth above. Whitish near apex, gradually becoming yellow lower. **Flesh:** Thin, fibrous; whitish in cap, stipe concolorous. **Odor:** Indistinct. **Taste:** Indistinct. **Spore deposit:** White. **Microscopy:** Spores 5–7.5 × 3–4 µm, smooth, ellipsoid to broadly teardrop-shaped, inamyloid. Cystidia thin-walled, cylindrical or narrowed to a blunt tip.

Ecology: Solitary or scattered in small, loose troops near decomposing logs, stumps, other woody debris or thick duff, occasionally on cone scales of fir. Common at higher elevations of the Cascade Range, occasional elsewhere. Fruiting in late summer and fall.

Edibility: Unknown. Tiny and thin-fleshed.

Comments: Typical forms are recognized by their relatively large size (for a *Strobilurus*), dark grayish-brown caps, stipes with yellow lower portions and hairy, slightly rooting bases, and growth near decomposing woody debris (rather than directly on cones). *S. occidentalis* is usually paler and typically grows on spruce cones. However, *S. lignitilis* can be very pale-capped, and *S. occidentalis* is rarely found on conifer litter/woody debris rather than on cones—in these cases, microscopy may be useful. The gill cystidia of the latter species have thicker walls (1-2 µm) and may have a blob of yellowish "resin" at the tip. *S. diminutivus* is tiny, with caps less than 0.5 cm across; it grows on the cone scales of sugar pine and occasionally ponderosa pine. *Baeospora myosura* also occurs on cones (mostly spruce and Douglas-fir) but has a grayish-tan cap and very crowded gills.

Misapplied name: *Strobilurus albipilatus* (Peck) V.L. Wells & Kempton

Dendrocollybia racemosa

(Pers.) R. H. Petersen & Redhead

Long-branched Dendro

Cap: 0.5–1.5 cm across, bell-shaped to convex, often with a low umbo, remaining convex or becoming nearly plane in age. Surface moist to dry, smooth. Opaque, light gray to grayish beige, sometimes dark gray or tan to brownish. **Gills:** Obscurely attached, close to subdistant. Dull, pale gray to dingy yellowish tan. **Stipe:** 1–6.5 (9) cm long, 0.1–0.2 cm thick, cylindrical for much of length, but rather sinuous and tapering downward, mostly below level of duff. Surface irregularly covered in straight to curved branches 0.3–0.5 cm long with enlarged, rounded tips. These branches, called conidial pegs, can be lost in age, and stipe can appear mostly smooth. Base of stipe emerging directly from a small sclerotia. **Sclerotium:** 0.2–0.7 cm wide, rounded to oval, surface smooth. Matte black, hard. **Flesh:** Very thin, grayish to tan. **Odor:** Indistinct. **Taste:** Indistinct. **KOH:** No reaction. **Spore deposit:** Whitish, scant. **Microscopy:** Spores 4–5 × 2–3 µm, narrowly ellipsoid, smooth. Asexual spores from stipe 7–12 × 3–5 µm, highly irregular, cylindrical, peanut-shaped, or bent.

Ecology: Often in clusters, rarely solitary, on remains of rotting mushrooms, particularly *Lactarius* and *Russula* (especially *R. crassotunicata* and species in the blackening group). Fairly common but often overlooked, found from sea level to the high elevations of the Cascades as well as the east slope. Fruiting in fall but the sclerotia can be found on the surfaces and hollowed inner stipes of rotting mushrooms anytime during the year before the fruitbodies emerge.

Edibility: Unknown.

Comments: With the opaque gray to grayish beige cap, grayish gills, small size, and long, sinuous stipe with irregular branches, this is an easy mushroom to identify. This species often fruits in areas rich with other small-capped trooping mushrooms (*Mycena* and others), making it difficult to initially detect. Once you become very familiar with the mushrooms in your area, you can key in on the unique gestalt impression made by the small, clustered, grayish brown, non-striate caps. Specimens that have lost their side branches can be mistaken for *Pseudobaeospora*, but the smooth, black sclerotia set them apart. However, these sclerotia are easily overlooked unless carefully excavated. The only truly similar species is *D. "pycnoramella" nom. prov.*, which has much shorter, more densely-packed branches, and shorter, more straight-stemmed fruitbodies which arise directly from knobby sclerotia (rather than long, trailing stipes emerging from smooth, rounded sclerotia).

Synonym: *Collybia racemosa* (Pers.) Quél.

Collybia tuberosa

Collybia cirrhata

Collybia cookei

Collybia tuberosa and *Dendrocollybia racemosa*

Collybia tuberosa

(Bull.) P. Kumm.

Apple-seed Coin

Cap: 0.5–2.5 cm across, convex with a narrowly inrolled margin when young, expanding to nearly plane in age, often slightly irregular. Surface smooth, dry, with a very faint bloom. White to creamy or beige or yellowish over disc. **Gills:** Narrowly attached, moderately spaced. White to creamy-beige, becoming dingy in age. **Stipe:** 1–4 cm long, 0.01–0.02 cm thick, cylindrical, often sinuous and irregular. Surface smooth, dry. White to cream, or orangish blushed below. **Sclerotium:** Found at the base of the stipe; a hard, smooth, elongate, reddish-brown structure that resembles an oversized apple seed. **Flesh:** Thin, and fragile in cap; slightly tough in stipe. White. **Odor:** Indistinct. **Taste:** Indistinct. **Spore deposit:** White. **Microscopy:** Spores 5–5.5 × 3–3.5 µm, ellipsoid, smooth, inamyloid.

Ecology: Fruiting in loose troops, rarely solitary on the remains of old, decayed fruitbodies of *Russula*, *Lactarius*, and other mushrooms. These fruitbodies are usually "mummified" (e.g. partly dried). Common and widely distributed across Cascadia. Fruiting from late summer through fall.

Edibility: Unknown. Tiny and unappealing due to the substrate.

Comments: This is one of three similar-looking, closely-related species of small, slender, whitish mushrooms that grow in troops on the decayed remains of macrofungi. The reddish-brown, apple-seed like sclerotia of *C. tuberosa* set it apart from *C. cookei*, which has smaller, rounder, yellowish sclerotia that resemble withered corn kernels. *C. cirrhata* is differentiated by its lack of any sclerotium and tendency to fruit in denser troop. *C. cookei* is more often found on decaying polypores (e.g. *Bondarzewia*) or honey mushrooms (*Armillaria* spp.). All three of these occur in Cascadia and often grow alongside each other — *C. cirrhata* is probably the most common of the three. *Strobilurus* species can be quite similar but have yellow to orangey lower stipes and grow in troops on the fallen cones of conifers and decayed conifer wood rather than on fungal remains. *Asterophora* species grow on decaying mushrooms but *A. parasitica* is grayish with strange blunt-edged gills, and lacks a sclerotium, while *A. lycoperdoides* has stouter stipes and globose caps that disintegrate into brownish powder. Two *Dendrocollybia* species occur in Cascadia and also grow on the decayed remains of mushrooms; both of these have elaborate side-branches covering their stipes with small, blackish, smooth, or knobby sclerotia.

Asterophora parasitica

(Bull.) Singer

Silky Piggyback

Cap: 0.3–2 cm across, rounded, bell-shaped to conical with an inrolled and crimped margin when young, becoming broadly conical, convex, to plane with a low, broad umbo. Surface dry, silky when young, to smooth in age. Pale gray to grayish beige base color, covered with silvery white fibrils; becoming gray to gray-brown in age. **Gills:** Attached, distant, very thick, irregular, often with a fuzzy appearance. Whitish, light gray to grayish brown. **Stipe:** 0.5–3 cm long, 0.1–0.4 cm thick, equal or tapered toward base. Pale gray to grayish brown base color, covered with silvery white, silky fibrils, becoming darker, gray to grayish brown when handled or in age. **Flesh:** Very thin, soft, fragile in cap, fibrous in stipe. Grayish brown to watery brown. **Odor:** Unpleasant, rancid farinaceous. **Taste:** Strongly farinaceous. **Spore deposit:** Whitish. **Microscopy:** Basidiospores 5–7 × 3–4 µm, ellipsoid, smooth. Asexual chlamydospores 12–17 (20) × 9–11 µm, spindle-shaped to oval, thick walled, with large droplets.

Ecology: In small to large clusters, rarely solitary, on decaying *Russula* (especially black-staining species and species in the *R. brevipes* group). Usually growing from the cap disc or near the stipe base of the host. Uncommon in coastal forests but can be locally regular in locations with large populations of its preferred host species. Fruiting mostly from late summer through early winter but can appear whenever mummified, decaying fruitbodies of its hosts become saturated.

Edibility: Unknown.

Comments: The silky, whitish to gray cap, irregular gills, and growth on decaying *Russula* are distinctive. It is most likely to be mistaken for *A. lycoperdoides*, which has a white to light brown, rounded cap that becomes powdery as it ages, and usually lacks gills. *Volvariella surrecta* is a silky white species that grows on large *Clitocybe* species. It is larger, with crowded, white to light pink gills, and a volva forming a sock-like cup around the stipe base. It is extremely rare, known in Cascadia from only a handful of records. Also see the *Collybia* species, all of which colonize and grow from rotting mushrooms.

Asterophora lycoperdoides

(Bull.) Ditmar

Powder Cap, Powdery Piggyback

Cap: 0.5–2 cm across, rounded to convex with an inrolled margin. White at first, soon becoming a light brown powdery mass. **Gills:** Often absent, or only present as rudimentary veins or ridges. Grayish to watery brown. **Stipe:** 3–5 cm long, 0.3–0.6 cm thick, equal or tapered toward base. Surface dry, pubescent to silky. White when young, becoming grayish to dingy gray in age. **Flesh:** Thin, soft and white at first in cap, soon light brown and powdery. Stipe firm, fibrous, becoming hollow in age, gray to watery brown. **Odor:** Unpleasant, rancid farinaceous. **Taste:** Strongly farinaceous. **KOH:** No reaction. **Spore deposit:** Basidiospores rare, whitish, chlamydospores light brown. **Microscopy:** Basidiospores 3–6 × 2–4 µm, ellipsoid, smooth. Asexual chlamydospores 13–20 × 10–20 µm, ovoid to nearly round, covered with blunt spines or warts.

Ecology: In small groups or clusters on decaying *Russula*, especially the Blackening Russula (*R. nigricans, R. adusta,* etc.) and the *R. brevipes* group, more rarely on *Lactarius* species. Fairly common in coastal forests, as well as around Puget Sound, and on the west slope of the Cascades. Often fruits in late summer or early fall. In wet years, excess moisture rots the *Russula* faster than the *Asterophora* can establish itself and fruit.

Edibility: Unknown.

Comments: This oddball is one of a handful of mushrooms that grow on decaying mushrooms. It is easily identified by the white to light brown cap that becomes powdery as it ages, lack of defined gills, and growth on rotting mushrooms. *A. parasitica* shares this habitat but has a silky cap and thick, irregular gills. The powdery cap of this *A. lycoperdoides* is mostly made up of asexual chlamydospores, which are dispersed by wind and splashing raindrops. Upon reaching a suitable host, these spores can grow much faster than the sexual basidiospores, allowing them to colonize and fruit from the host before it is fully decayed. The decaying mushroom colonizers, *Dendrocollybia* also produce asexual spores, but they occur on conidial pegs (short side-branches on the stipe).

Cantharellula umbonata

(J.F. Gmel.) Singer

Grayling

Cap: 1–4 cm across, broadly convex with a distinct round umbo, expanding to nearly plane, becoming depressed at center, but typically retaining a low umbo; margin inrolled and finely fuzzy when young, unfurling in age, then often finely scalloped. Surface dry, finely felty to smooth, often concentrically cracking or zonate in age. Dull grayish with a bluish cast when young, becoming paler brownish-gray, often with a pale bloom and dusky blotches and streaks. **Gills:** Decurrent, close to crowded, narrow, repeatedly forking towards margin. Whitish to pale cream, developing blotches of ocher to rusty-red stains. **Stipe:** 2–6 cm long, 0.2–0.5 cm thick, cylindrical, slender. Surface dry, finely fibrillose. Pale silvery-gray, often more bluish-gray near apex. **Flesh:** Thin, fibrous; whitish, staining dingy rusty-red when cut or handled. **Odor:** Indistinct to faintly farinaceous. **Taste:** Indistinct. **Spore deposit:** White. **Microscopy:** Spores 8–11 × 2.5–4 µm, long-spindle-shaped to cylindrical, smooth, amyloid.

Ecology: Scattered in troops, always amongst patches of haircap moss. Uncommon at mid- to high elevations in northern Cascadia. Fruiting from late summer through fall.

Edibility: Edible and good, but not particularly easy to recognize for beginners — take care with identification.

Comments: This species is recognized by its small size, gray, often umbonate cap, crowded gills which fork repeatedly, and growth in beds of haircap moss. *Spodocybe trulliformis* can be quite similar but tends to be slightly warmer gray-brown in color, grows in duff of many forest types (not tightly associated with moss), and, most importantly, does not have the thick-edged, repeatedly forking gills. There are many similar mushrooms which grow in moss but those are either differently colored (many are darker brownish), have a differently-colored spore deposit (many are pinkish), have smoother, more translucent-striate caps, or lack the distinctive forking gills and rusty blotches.

Pseudolaccaria pachyphylla

(Fr.) Vizzini & Contu

Cap: 0.5–2.5 cm across, convex to plane, with a depressed center; margin downcurved and often ribbed when young, becoming ragged in age. Surface dry, covered with appressed fibrils, which often break up into small, uplifted scales in age. Buff to pale straw colored, with paler appressed fibrils. **Gills:** Broadly attached, close to subdistant, broad, thick, and often slightly irregular. Off-white to pale buff. **Stipe:** 1–3 (5) cm long, 0.1–0.3 cm thick, more or less equal, often curved at base. Surface dry, covered with fine silken fibrils to smooth. Off-white to pale buff. **Flesh:** Very thin, fibrous; off-white to pale buff. **Odor:** Indistinct. **Taste:** Mild. **KOH:** No reaction. **Spore deposit:** White. **Microscopy:** Spores 7.5–10 × 5–6 µm, nearly round to broadly ellipsoid, smooth, amyloid. Basidia 4-spored. Pleurocystidia and cheilocystidia abundant, cylindrical to spindle-shaped, clamp connections present.

Ecology: Solitary or scattered on nutrient-poor, moss-covered soil, clay, or sandy areas, especially along old logging roads or in disturbed areas. Widespread throughout Cascadia, but usually overlooked because of their small size and anonymous appearance. Fruiting from spring into fall.

Edibility: Unknown.

Comments: The small size, umbilicate cap, buff to pale straw colors, slightly scurfy textures, and growth in nutrient-poor soil are distinctive. *Lichenomphalia ericetorum* has a ribbed cap, widely spaced gills, and grows on green algal mats. Small *Clitocybe* generally have smooth caps and stipes and often grow in duff or moss. *Pseudoomphalina intermedia* is quite similar but tends to be slightly larger, with a smooth cap, brittle flesh, and a farinaceous odor and taste. *Laccaria tortilis* and *L. pumila* are vaguely similar but the flesh texture is usually a bit more fleshy-cartilaginous with brighter pinkish buff colors when young.

Synonym: *Pseudoomphalina pachyphylla* (Fr.) Knudsen

Spodocybe trulliformis

sensu PNW

Dapper Grayhead

Cap: 0.7–4 cm across, convex with a narrow inrolled margin and often a small umbo when young, expanding to broadly convex, becoming depressed at center and often with a ribbed or crinkled margin in age. Surface dry, slightly fuzzy when young, smooth in age. Dark grayish at center (blackish in some specimens), to gray overall with a whitish bloom, or grayish tan to grayish beige outward with scattered gray-tan spots, extreme margin often white. **Gills:** Broadly attached to slightly decurrent, quite widely spaced. White to creamy. **Stipe:** 1–5 cm long, 0.2–0.5 cm thick, cylindrical, usually upright, base slightly enlarged. Surface smooth to innately fibrillose. Grayish to beige or whitish gray in age, usually markedly contrasting with bright whitish gills. **Flesh:** Thin, fibrous; white. **Odor:** Indistinct to farinaceous. **Taste:** Mild to farinaceous. **Spore deposit:** White. **Microscopy:** Spores 4.5–6.5 × 2.5–4 µm, ellipsoid, smooth, inamyloid.

Ecology: Scattered in troops and small clumps in leaf litter and needle duff of many forest types, especially common under Douglas-fir. Common throughout Cascadia. Fruiting from late summer into early winter, peaking in late fall.

Edibility: Unknown.

Comments: The small size, mottled to dark grayish, finely suede-textured cap, and widely spaced, slightly decurrent whitish gills that often contrast with the gray to tan upper stipe are distinguishing characteristics for this species. Many other small-to-medium, drab species are superficially similar but most are slightly larger, smoother-capped, paler, and/or have more closely spaced gills. *Melanoleuca* species are usually larger and have crowded gills that are not decurrent. A very similar-looking species from Cascadia has recently been determined to be a member of the genus *Hygrophorocybe*; it may have more deeply decurrent gills and a tendency to show an entirely pale cap but is very difficult to recognize.

Synonym: *Clitocybe trulliformis* sensu PNW

Myxomphalia maura

(Fr.) Hora

Coal Umble

Cap: 0.5–3 (5) cm across, convex to plane, with a depressed center; margin incurved when young, becoming plane to uplifted in age. Surface slightly viscid when wet, to dry, smooth. Dark blackish brown, dark gray to grayish brown when wet or young. Strongly hygrophanous, fading to pale gray when dry, translucent-striate when wet. **Gills:** Broadly attached to subdecurrent, close to somewhat crowded, narrow, partial gills numerous. Bright white at first, contrasting with cap and stipe color, becoming off-white to pale grayish in age. **Stipe:** 1–5 (6) cm long, 0.2–0.5 cm thick, equal or enlarged slightly toward base. Surface dry, apex with whitish powder when young, soon smooth overall. Dark blackish brown, dark gray to grayish brown. **Flesh:** Very thin, grayish. Stipe stuffed with whitish pith, occasionally hollow in age. **Odor:** Indistinct to slightly farinaceous. **Taste:** Mild to slightly farinaceous. **KOH:** No reaction. **Spore deposit:** White. **Microscopy:** Spores 4.5–6.5 × 3.5–4.5 µm, nearly round to broadly ellipsoid, smooth, amyloid. Pleurocystidia and cystidia abundant, cylindrical to spindle-shaped.

Ecology: Solitary, scattered, or in small clusters on burnt ground around ash. Found throughout Cascadia; it is more common in wildfire burn scars in montane habitats on the east side of the Cascade Range in burn areas. Fruiting in fall and spring.

Edibility: Unknown.

Comments: The umbilicate cap, decurrent whitish gills, dark blackish brown to grayish brown hygrophanous cap, and growth in burnt areas are distinctive. Could be mistaken for *Lyophyllum anthracophilum* (=*Tephrocybe anthracophila*), another small dark gray-brown mushroom at burn sites. It differs by its more conical-convex or plane cap (often umbonate) and grayer gills. *Arrhenia obscurata* is similar in cap color and stature but has darker, thicker, more widely spaced gills; it often grows on bare soil or in moss. *A. epichysium* grows on decaying wood and has darker gills. *Pseudoclitocybe cyathiformis* is larger, has grayish gills, and grows on unburnt wood or duff.

Synonyms: *Mycena maura* (Fr.) Kühner, *Omphalina maura* (Fr.) Quél.

White-spored (J)

Arrhenia chlorocyanea

(Pat.) Redhead, Lutzoni, Moncalvo & Vilgalys

Blue-green Umble, Verdigris Navel

Cap: 0.3–2 cm across, convex to plane, with a depressed center, margin downcurved and often ribbed when young, becoming plane to slightly uplifted in age. Surface dry to moist, slightly roughened when young, becoming smooth. Dark turquoise-green to dark bluish green when young, fading to grayish blue to grayish green with yellowish undertones in age. **Gills:** Subdecurrent to decurrent, widely spaced, fairly thick, broad, partial gills often irregular, inter-veined and numerous. Pale bluish at first, becoming bluish white to dingy white in age. **Stipe:** 1–4 cm long, 0.1–0.3 cm thick, equal or enlarged slightly toward base. Surface dry to moist, smooth. Dark turquoise-green to dark bluish green when young, fading slightly in age. **Flesh:** Very thin, turquoise-green to grayish green, stipe stuffed when young, often hollow in age. **Odor:** Indistinct. **Taste:** Indistinct. **KOH:** No reaction. **Spore deposit:** White. **Microscopy:** Spores (7.5) 8.5–11 × 4.5–6 µm, broadly ellipsoid to oval, smooth, inamyloid.

Ecology: Solitary, scattered, or occasionally in small clusters in moss, often in areas with nutrient-poor or disturbed soil such as old logging roads or areas of sandy or clay soils. Widespread in Cascadia but far more common at lower elevations. Fruiting in winter and spring.

Edibility: Unknown.

Comments: The umbilicate, dark teal to olive-gray cap, decurrent bluish green gills, and small size are distinctive; it is unlikely to be mistaken for anything else. Other *Arrhenia* have dark brown to grayish black colors. *Lichenomphalia ericetorum* often occurs in similar settings (among moss and lichens) and has the same stature but is usually a bit larger and has a light brownish to pinkish beige cap.

Synonyms: *Clitocybe atroviridis* H.E. Bigelow, *Omphalina chlorocyanea* (Pat.) Singer, *O. viridis* (Hornem.) Kuyper

Arrhenia obscurata group

Shadow Umbles

Cap: 1–3 cm across, convex at first, but soon plane with a depressed center, at times uplifted and funnel-shaped in age. Surface moist to dry, smooth to slightly wrinkled. Dark gray-brown to beige-brown near margin, paler when dry. Hygrophanous, translucent-striate in age when wet. **Gills:** Broadly attached to decurrent, widely spaced, thick-edged. Beige to gray-tan. **Stipe:** 1–2 (3.5) cm long, 0.1–0.4 cm thick, cylindrical. Surface dry to moist, smooth. Dark gray to brownish gray. **Flesh:** Thin, fibrous. Grayish-brown. **Odor:** Indistinct. **Taste:** Indistinct. **KOH:** No reaction. **Spore deposit:** White. **Microscopy:** Spores 6.5–9 × 3.5–5 µm, ellipsoid, smooth, inamyloid.

Ecology: Solitary or scattered in small troops or clusters from soil, needle duff, or moss in conifer-dominated habitats. Common and widespread in Cascadia, more often at higher-elevations and in east side forests. Easily overlooked and misidentified, thus likely underreported. Fruiting throughout the year, but most abundant in spring and early summer.

Edibility: Unknown.

Comments: This species is difficult to distinguish from other *Arrhenia* and also superficially resembles members of a number of other genera. The small, dark fruitbodies with decurrent gills and growth from soil and duff help distinguish it. DNA sequences have suggested that there are a number of similar species going under this name. *A. onisca* and *A. sphagnicola* are similar in color and stature, but are slightly larger and grow in boggy areas with *Sphagnum* and other mosses. The latter may show appressed, scaly tufts of hairs on the cap in age. Although poorly understood, *A. hohensis* supposedly has paler gills and may show a finely hairy stipe surface. *A. rainierensis* is supposedly more grayish than brown-capped.

Lichenomphalia ericetorum

(Pers.) Voitk, Thorn & I. Saar

Lichen Agaric

Cap: 0.5–2.5 (3) cm across, convex to plane, with a depressed center, margin downcurved, ribbed and often scalloped, becoming plane to uplifted in age. Surface dry to moist, pleated or grooved, otherwise smooth. Light buff, cinnamon buff, to light brown when young, quickly fading to creamy beige, with darker lines or groves. **Gills:** Slightly to deeply decurrent, well-spaced, thick, broad, partial gills often short and inter-veined. Buff to beige at first, becoming creamy beige to dingy whitish beige, edges usually concolorous, but occasionally darker. **Stipe:** 1–3 (4) cm long, 0.1–0.3 cm thick, equal or enlarged slightly toward base. Surface dry, smooth. Buff, beige to creamy beige, often darker, with a vinaceous tint at apex. **Flesh:** Very thin, off-white to beige. **Odor:** Indistinct. **Taste:** Indistinct. **KOH:** No reaction. **Spore deposit:** White. **Microscopy:** Spores 7–9 × 4–6 µm (but larger, 12 × 8 µm from 1-spored basidia), ellipsoid, smooth, inamyloid. Basidia often 2-spored but can be 1-, 3-, or 4-spored, clamp connections absent.

Ecology: Solitary or scattered in varied habitats. Often on logs, stumps, or bark of living and dead trees but also in areas with nutrient-poor sandy or clay soil. Associated with green algae (*Coccomyxa* spp.), which forms a visible bluish-green to hunter green scummy layer on the surface of the logs or soil from which the fruitbody grows. Common and widespread from the Cascade crest down to sea level; uncommon to rare in the drier habitats of the east slope. Fruitings can be found throughout the year but most abundant from late fall into spring.

Edibility: Unknown.

Comments: The omphalinoid stature, buff to light brown cap that fades creamy beige, thick, widely spaced, creamy beige gills, and growth from a green algal mat help identify this common lichenized agaric. It is one of a handful of basidio-lichens (nearly all other lichens are ascomycetes). *Lichenomphalia* have specialized modified hyphae that loosely encase the cells of *Coccomyxa* algae and siphon off some of their photosynthetic products. Older *Contumyces rosellus* can look similar but generally have some pinkish coloration and grow in sandy areas or on bare soil.

Synonyms: Although it had passed through a number of different genera, this species had long been known by the epithet "*umbellifera*" (e.g. *L. umbellifera* (L.) Redhead, Lutzoni, Moncalvo & Vilgalys). Recently it was discovered that that name referred to a mushroom of an entirely different genus, leaving ericetorum as the earliest appropriate epithet.

Arrhenia epichysium

(Pers.) Redhead, et al.

Dark Wood-Umble

Cap: 1.5–6 cm across, convex to plane, with a distinct depression at disc; margin downcurved, becoming slightly uplifted in age; sometimes appearing like a small goblet or vase. Surface smooth, dry to moist, translucent-striate when wet. Dark, dull brownish black when young, hygrophanous, becoming gray to smoky brown to light grayish tan when dry. **Gills:** Broadly attached to decurrent, arching, close to subdistant, narrow. Smoky gray to pale gray, or dingy grayish tan. **Stipe:** 2.5–6 cm long, 0.1–0.4 cm thick, equal or with a slightly enlarged base, straight or curved. Surface smooth or slightly bumpy, dry, base often with a tuft of fuzzy whitish mycelium. Gray to tan, at times darker, to brownish black. **Flesh:** Very thin, cartilaginous, stipe hollow. **Odor:** Indistinct. **Taste:** Mild. **KOH:** No reaction. **Spore deposit:** White. **Microscopy:** Spores 7–8.5 × 4–4.5 µm, ellipsoid, smooth, colorless, inamyloid. Pileipellis strongly encrusted, with clamp connections.

Ecology: Solitary or scattered in small groups on well-decayed wood. Fairly common throughout Cascadia. Fruiting primarily from fall through spring but can fruit whenever its substrate is wet enough.

Edibility: Non-toxic.

Comments: This species is recognized its dark blackish-brown to gray or tan cap, arching gray gills, and growth on wood. *A. obscurata* (and others) are smaller, and grow in moss, duff, or soil. *Myxomphalia maura* grows in recently burnt areas (campfire pits or ashy soil after forest fires); it has a dark gray to blackish brown cap (fading to pale gray) and strongly contrasting white gills when young. *Pseudoclitocybe cyathiformis* is superficially similar but is larger and can grow on wood or duff.

Synonyms: *Omphalina epichysium* (Pers.) Quél., *Clitocybe epichysium* (Pers.) H.E. Bigelow

Chromosera cyanophylla

sensu PNW

Lilagolds

Cap: 0.4–2.5 cm across, bell-shaped to convex when young, becoming broadly convex to plane, with a depressed center. Surface viscid to tacky, smooth. Bright to dull yellow to peachy yellow, fading in age to yellowish beige, translucent-striate with darker lines when wet. **Gills:** Broadly attached to decurrent, moderately spaced, broad, partial gills short and numerous. Lilac-purple when young, fading to grayish lilac, pinkish beige and occasionally becoming pale yellowish in age. **Stipe:** 1–4 cm long, 0.1–0.3 cm thick, equal or club shaped with an enlarged base. Surface viscid, smooth, often slightly fuzzy at base. Lilac-purple at apex and yellowish lower when young, fading and becoming more yellow in age. Base with bright lilac-purple disc when young. **Flesh:** Very thin, fragile, yellowish, stipe stuffed with watery pith or hollow in age. **Odor:** Indistinct. **Taste:** Mild. **KOH:** No reaction. **Spore deposit:** White. **Microscopy:** Spores 6–7 × 3–3.5 µm, broadly ellipsoid to almond-shaped, smooth.

Ecology: Solitary, scattered, or in small clusters on decaying conifer logs and branches, especially those of grand fir and western hemlock. Rare in fall and winter in coastal forests, becoming more common inland around Puget Sound, and quite common in spring at middle to higher elevations of the Cascades on both west- and east-side forests.

Edibility: Unknown.

Comments: The translucent-striate, yellow, viscid cap, lilac-purple young gills, viscid stipe, and growth on conifers are distinctive. *C. citrinopallida* is quite similar, also with a viscid cap and stipe but has overall bright yellow colors. It is rare, mostly fruiting in summer in the high Cascades. Older *Chromosera* specimens often lose their vivid colors, but most other look-alikes lack the viscid cap and stipe. Other brightly colored omphalinoid species on wood include: *Chrysomphalina aurantiaca*, with a dry, bright orange cap and stipe; *C. chrysophylla*, with a yellow-brown, scaly cap when young (becoming orange-yellow in age), and bright golden gills; and *C. grossula*, with pale yellowish green colors overall. *Gliophorus* species have thick "waxy" gills and grow on the ground.

Pseudoarmillariella ectypoides

(Peck) Singer

Bristly Woodfunnel

Cap: 1–5 cm across, convex with a depressed disc at first, becoming uplifted and deeply funnel-shaped in age. Margin sometimes finely inrolled and slightly ribbed. Surface dry to moist, covered in fine pointed tufts or scales of dark hairs, becoming smoother in age, but retaining some scaly hairs. Dingy beige to honey-brown or yellowish-tan, sometimes developing reddish-brown blotches. Hygrophanous, becoming pale beige to cream or sometimes nearly whitish when dry. **Gills:** Decurrent, arched, occasionally forked, well-spaced, narrow; edges even. Dingy tan to beige, sometimes pale to medium-rich yellow, sometimes blotched reddish-brown. **Stipe:** 2–7 cm long, 0.2–0.5 cm thick, fairly short, slightly curved, cylindrical or slightly enlarged at base. Surface smooth to pale-scurfy. Creamy-beige to tan, base with pale mycelium. **Flesh:** Thin, fibrous; pallid. **Odor:** Indistinct. **Taste:** Indistinct. **Spore deposit:** White. **Microscopy:** Spores 6.5–9 × 3.5–5 µm, ellipsoid, smooth, amyloid.

Ecology: Solitary or in troops or loose clusters on wood of decomposing conifers logs. Found in mid to high elevation forests in northern Cascadia; including both slopes of the Cascades. Fruiting from late summer through fall.

Edibility: Unknown.

Comments: The uplifted to funnel-shaped, beige to honey-brown cap with fine scales, decurrent tan to yellowish gills, and growth on decaying conifer wood help identify this species. *Chrysomphalina* are typically a bit smaller and show brighter yellow to orange colors. Dingy specimens could be confused with *Gerronema atrialba* which is usually taller, with grayer colors, and a dark-dotted stipe surface. *Armillaria* species can be yellowish, have scaly caps, and grow in clusters on wood but are usually substantially larger and have a partial veil. The remarkably similar-looking *Gerronema strombodes* is an eastern species with brighter yellow gills and a smoother cap.

Synonyms: *Clitocybe ectypoides* (Peck) Sacc., *Omphalina ectypoides* (Peck) H.E. Bigelow

Chrysomphalina grossula

(Pers.) Norvell, Redhead & Ammirati

Greenish Wood-Umble

Cap: 0.5–3.5 (5) cm across, convex to broadly convex with an incurved margin when young, becoming plane with a depressed center; margin often distinctly scalloped. Surface moist to dry, smooth or finely scaly-hairy at disc. Greenish, greenish yellow to yellow, at times with brownish yellow tones when young, soon fading to yellow or citron yellow, to pale yellow or whitish in age. **Gills:** Subdecurrent to decurrent, moderate to widely spaced, arched, broad. Yellowish to greenish yellow, becoming pale yellow to whitish in age. **Stipe:** 0.5–4 cm long, 0.1–0.4 cm thick, slender, cylindrical to slightly curved, tapered to base. Surface moist to dry, smooth. Greenish yellow to yellow, paler towards base and in age. **Flesh:** Thin, watery-fibrous, pale. **Odor:** Indistinct. **Taste:** Indistinct. **Spore deposit:** White. **Microscopy:** Spores 6–9.5 × 4–6 µm, ellipsoid, smooth, inamyloid.

Ecology: Solitary or scattered in small troops on woody debris. Most frequently encountered in urban areas on wood chips, but also found in old growth forests. Curiously, it may be the case that this species was fairly common in old-growth forests before industrial-area logging, after which it became rare. It appears to have "discovered" a new niche on wood chips and has subsequently become more common again. Widely distributed in Cascadia but generally uncommon, with records concentrated around inland urban areas. Fruiting primarily in late fall and winter but can show up almost any time of year.

Edibility: Unknown.

Comments: This species is recognized by the combination of its striking yellow-green color, small size, umbilicate cap, and growth on woody debris. Most likely to be confused with *Lichenomphalia ericetorum*, which also grows on rotting wood (on an algal mat), but that species is more pinkish-tan to creamy-capped. *Hypholoma fasciculare* can show a similar chartreuse color and grows on decaying wood but is usually much larger, fruits in tufts and clusters, has dark purplish-gray spores, and has a bitter taste. Species of *Rickenella* and *Loreleia* are similar in stature but are smaller, more slender, and orange in color.

Chrysomphalina aurantiaca

(Peck) Redhead

Peachfuzz Wood-Umble

Cap: 0.5–3 cm across, convex to plane, often with a depressed center in age. Surface dry, covered with fine stiff hairs when young, becoming smooth or nearly so, often translucent-striate in age. Bright orange at first, fading to pale orange to yellowish orange or sometime creamy orange in age. **Gills:** Broadly attached to slightly decurrent, moderately spaced, narrow, fragile, partial gills short and numerous. Creamy orange to pale orange at first, becoming yellowish orange in age. **Stipe:** 1–3 cm long, 0.1–0.3 cm thick, equal or enlarged slightly at base. Surface dry, smooth, base often with fuzzy white hairs. Orange to yellowish orange. **Flesh:** Very thin, fragile, orange, stipe stuffed with orangish pith, occasionally hollow in age. **Odor:** Indistinct. **Taste:** Mild. **KOH:** No reaction. **Spore deposit:** Whitish. **Microscopy:** Spores 6.5–9 (11.5) × 4–5.5 µm, ellipsoid, smooth. Basidia usually 4-spored, rarely 1-, 2-, or 3-spored.

Ecology: Scattered in small clusters or troops on decaying conifer stumps and logs, most commonly on grand fir. Uncommon in coastal forests, common to abundant at middle and high elevations of the Cascades. Fruiting primarily in fall and winter, with stragglers into spring or early summer.

Edibility: Unknown.

Comments: The orange cap with a depressed disc, and fine hairs around the margin (obscure in age), pale orange gills, and growth on wood help identify this colorful mushroom. Brightly colored, dry-capped *Hygrocybe* species (such as *H. constans*) are similar but don't usually grow on wood. *C. chrysophylla* is often mistaken for *C. aurantiaca* but has yellow to golden gills and a scaly yellow-brown cap, it also lacks the fringe of hairs around the cap margin. *C. grossula* has the same stature but has pale yellowish green colors that fade to yellowish beige in age; it is often found on wood chips as well as on conifer forest debris. *Xeromphalina* are superficially similar but much smaller, with translucent-striate caps, and dark lower stipes.

Synonym: *Omphalia aurantiaca* Peck

Xeromphalina campanella

(Fr.) Kuehner & Maire

Fuzzy Foot

Cap: 0.3–2.5 (3) cm across, convex to plane, with a depressed center, margin downcurved, occasionally splitting in age. Surface dry to moist, smooth, translucent-striate when wet. Dull orange, orange-brown to yellowish brown. **Gills:** Sub-decurrent to decurrent, moderately spaced, broad, partial gills numerous, strongly inter-veined. Yellow-beige to pale orange. **Stipe:** 1–3 (5) cm long, 0.1–0.3 cm thick, equal or enlarged slightly toward base, thin, somewhat wiry, fibrous. Surface smooth, dry, with fine hairs at base. Yellow to orangish at apex, transitioning to orange, to a dark brown or reddish brown at lower portion and often with a fuzzy yellowish-orange base. **Flesh:** Very thin, concolorous. **Odor:** Indistinct. **Taste:** Indistinct. **KOH:** Orangish brown. **Spore deposit:** White. **Microscopy:** Spores 5–8 × 3–3.5 µm, ellipsoid to cylindrical, smooth, amyloid.

Ecology: Scattered or clustered in small to large troops on decaying conifer logs and stumps. Widespread and abundant throughout Cascadia. Fruiting from early fall well into spring, most abundant in mid-fall.

Edibility: Unknown, probably non-toxic.

Comments: The decurrent gills, orange cap, dark lower stipe, and growth on wood help identify this species. The well-named *X. enigmatica* is practically identical except for its DNA sequence. *X. campanelloides* is a poorly-understood wood-dwelling species that supposedly differs primarily by its bitter taste. *X. brunneola* is also quite similar but has a darker brown to orange-brown cap, an unpleasant rancid-oily taste, and may average a bit stouter in stature. It appears to be uncommon, but large fruitings are occasionally found and it may be generally underreported. See the *X. cornui/cirris* group for other species in this genus that grow on needles, twigs, duff, or moss. Some small wood-dwelling *Gymnopilus* can superficially resemble *Xeromphalina* but have rusty-brown spores.

Xeromphalina fulvipes

(Murrill) A.H. Sm.

Domed Gingertail

Cap: 1–4 cm across, rounded at first, sometimes with low umbo, expanding to broadly convex to nearly plane in age, occasionally weakly sunken over center but not distinctly umbilicate. Surface smooth, moist to dry, finely powdered when young, shiny when wet, becoming slightly radially wrinkled in age, sometimes obscurely translucent-striate in age. Dingy ocher to brighter orange-brown, brighter yellowish outwards, pale beige-yellow around margin. **Gills:** Narrowly to broadly attached or notched, somewhat widely spaced, moderately broad. Dingy pale gray-yellow, edges slightly paler. **Stipe:** 2–5 cm long, 0.1–0.3 cm thick, cylindrical, fairly straight, tall. Surface dry, finely pale-scurfy to powdery, with orangey dots over most of length contrasting with darker tones beneath. Creamy-yellow at apex and over upper half, otherwise orangey brown to dark reddish brown overall, nearly blackish brown at base. **Flesh:** Thin, yellowish and somewhat fragile in cap, tough-wiry and dark in stipe. **Odor:** Indistinct. **Taste:** Bitter. **KOH:** Reddish to brown. **Spore deposit:** White. **Microscopy:** Spores 4.5–6 × 1.5–2 µm, sausage-shaped, weakly amyloid.

Ecology: Solitary or scattered in small troops on woody debris of both conifers and hardwoods, common in most forested habitats across Cascadia. Can be found fruiting almost any time of year, although more common in fall and spring.

Edibility: Tiny and inconsequential.

Comments: Whereas most *Xeromphalina* have umbilicate caps and decurrent gills, this species is readily distinguished by its convex cap and narrowly to broadly attached gills — it has a subtly more upright, stout stature. These differences have led some taxonomists to place it in a separate genus, *Heimiomyces*, but we consider this split unwarranted. Species in the *X. campanella* group have decurrent gills and grow in clusters or troops on conifer wood.

Synonyms: *Heimiomyces fulvipes* (Murrill) Singer

Xeromphalina cornui/cirris group

Moss Gingertails

Cap: 0.5–3 cm across, broadly convex to nearly plane. with a depressed or umbilicate disc. Surface smooth to slightly powdery, dry, wrinkled in age. Orange to ocher-brown, becoming yellowy-orange, paler around margin. **Gills:** Broadly attached to slightly decurrent, close to subdistant; margins even. Pale cream to dusky yellow. **Stipe:** 2–6 cm long, 0.1–0.2 cm thick, cylindrical, slender, tall. Surface dry, finely powdery to smooth, base with yellowish or orange hairs. Yellowish at apex, darker reddish-brown to dusky blackish below. **Flesh:** Thin, pale and somewhat tough in cap, tough and wiry, dark in stipe. **Odor:** Indistinct. **Taste:** Mild to bitter. **KOH:** Reddish-brown on cap. **Spore deposit:** White. **Microscopy:** Spores 5–7.5 × 3–4.5 µm, ovoid to ellipsoid or nearly round, weakly amyloid.

Ecology: Solitary or in scattered troops on conifer needle litter. Common across Cascadia from sea level to montane forest. Fruiting from late summer into early winter.

Edibility: Inconsequential.

Comments: The names *X. cornui* and *X. cirris* have been applied somewhat indiscriminately to species in this group, members of which are recognized by their tall, slender stature, slightly decurrent gills, and mild taste. Studies based on morphology (in the pre-DNA age) distinguished these species by subtle differences in gill attachment, taste, and substrate preferences. Species which have been reported from Cascadia include: *X. cirris*, with broadly attached to subdecurrent gills and a mild taste; *X. parvibulbosa*, with broadly attached gills that may show a decurrent tooth, and with a slightly bitter to astringent taste; *X. cornui*, with decurrent gills and a mild taste; and *X. cauticinalis*, with decurrent gills and a strongly bitter taste. *X. fulvipes* has broadly attached to notched gills, a distinctly powdery stipe surface, and a bitter taste; it typically grows on small woody debris of conifers. Species in the *X. campanella* group have deeply decurrent gills and grow in clusters or troops on conifer wood. *X. cirris* and *X. fulvipes* are the only names described from western North America, but even these names should be taken with a grain of salt...the whole genus needs modern taxonomic work.

Mycetinis salalis

(Desjardin & Redhead) Redhead

Salal-wire

Cap: 0.5–2 cm across, convex to plane. Surface smooth to radially grooved, dry. Pale beige to tan; slightly depressed and darker ocher to brownish at disc. Obscurely hygrophanous, becoming concentrically paler in dry weather. **Gills:** Broadly attached, widely-spaced. Creamy to beige or pale tan. **Stipe:** 2–5 cm long, 0.1–0.2 cm thick, slender, upright, cylindrical. Surface dry, finely scurfy over most of length, sometimes partly smooth in age. Pale beige to tan over upper part, lower half fading through orangey-brown to dark brown-blackish at base. **Flesh:** Thin, tough; pallid. **Odor:** Garlicky. **Taste:** Indistinct or slightly bitter-garlicky. **Spore deposit:** White. **Microscopy:** Spores 15–19 × 3–5 µm, elongate-ellipsoid to long-teardrop or crescent-shaped, smooth, inamyloid.

Ecology: Solitary or in small groups on fallen leaves of the salal (*Gaultheria shallon*) and Oregon grape (*Mahonia* spp.). Found throughout Cascadia, primarily in mature conifer forests. Fruiting from late summer through fall.

Edibility: Although some people find the aroma attractive for use as a spice, the few available reports from people who have tried it aren't particularly enthusiastic – the aroma and flavor are somewhat harsh and bitter.

Comments: This scent of this species often permeates the air of the forest understory. The small (but not tiny) fruitbodies with scurfy stipes that are dark over their lower parts and growth on dead leaves of salal and Oregon grape make this species readily recognizable. However, its status with relation to *M. copelandii* is unclear – that species is virtually identical in every respect but is found on the leaves of tanoak. There is little difference in genetic barcode between these two species, and some sources consider them synonymous. *M. scorodonius* also has a garlicky odor but with a smooth stipe. There are a few poorly known *Paragymnopus* species in the *P. perforans* group with similar statures and often putrid odors. *Rhizomarasmius undatus* is a bit taller, with a dark stipe, and a finely frosted appearance on all surfaces (often extensively adorned with dewy droplets).

Paragymnopus ponderosae

(R.H. Petersen) J.S. Oliveira

Ponderosa Threadstem

Cap: 0.3–1.5 cm across, convex, expanding to plane. Surface rough to wrinkly, radially grooved, dry. Beige to greyish-tan with a darker brown center, extreme margin with a narrow whitish band. **Gills:** Obscurely attached to notched or seceding, fairly broad, widely spaced with a few short gills. Dingy beige to tan. **Stipe:** 2.5–6 cm long, 0.05–0.1 cm thick, thin, tough, wiry, often irregularly flattened and grooved. Apex pale, dingy tan-beige, shiny dark reddish-brown to nearly blackish at base. Base of stipe and surrounding duff covered with fine, hairlike black rhizomorphs. **Flesh:** Very thin, tough; dark. **Odor:** Indistinct. **Taste:** Indistinct. **Spore deposit:** White. **Microscopy:** Spores 5.5–7 × 3.5–5 µm, smooth, ellipsoid to teardrop-shaped, inamyloid.

Ecology: Scattered in troops, sometimes in larger swarms on duff of pines, especially ponderosa pine. Common on the east side, largely absent from most of western Cascadia. Fruiting in fall.

Edibility: Inconsequentially small.

Comments: This inconspicuous species is recognizable by its dark, wiry stipes, dingy beige-tan to dull brownish cap and gills, and growth in pine needle litter. There are a number of other "gymnopoids" with dark, wiry stems in Cascadia, but differ in their aroma (cabbage-like or garlicky) and/or substrate. *Gymnopus androsaceus* has a darker brown cap but with the dark wiry stipe and surrounding black rhizomorphs. It occurs in a wide range of conifer needle duff. *Pseudomarasmius pallidocephalus* has a brighter white cap with a tan disc, more pronounced "pinwheel" grooves, and a thin, shiny, wiry stipe that is extensive dark reddish brown to nearly black over the lower part. It grows in a wide variety of needle duff of conifers. *Crinipellis piceae* has brown, radiating, matted fibrils on the cap, which overhang the margin, with a fringed appearance, has a tough, wiry stipe, and grows on spruce needles. *Paragymnopus* (=*Micromphale*) *sequoiae* grows on redwood and cedar needles and has a pungent odor.

Marasmiellus filopes

(Peck) Redhead

Threadfoot Parachute

Cap: 0.2–0.5 cm across, at first appearing as a tiny "pinhead", ludicrously disproportionate to the stipe. Expanding to convex and then plane, slightly radially grooved and often finely crinkled around margin. Surface glabrous, dry. Whitish to creamy, or pale yellowish-tan at disc. **Gills:** Broadly attached, distant, sparse, narrow. White to cream. **Stipe:** 2–6 (10) cm long, 0.04–0.06 cm thick, thread-like and sinuous, often partially lying on the ground (decumbent) or tangled; sometimes more upright. Surface smooth to minutely and obscurely scurfy. Creamy to pale beige or tan. **Flesh:** Extremely thin. **Odor:** Garlicky; faint individually, but stronger *en masse* or when the mycelium-bound duff is stepped on. **Taste:** Indistinct. **Spore deposit:** Difficult to obtain. **Microscopy:** Spores 6.5–10 × 3–4.5 µm, teardrop- or almond-shaped, smooth, inamyloid. Cheilocystidia scattered, with many short, knobby projections.

Ecology: Fruiting in enormous swarms of hundreds or thousands on thick duff of mixed conifers. Fruitbodies emerging from individual needles of true fir or Douglas-fir. Prefers damp, open understory of conifer forest at mid to high elevations, and especially east side forests in Cascadia. Fruiting in fall.

Edibility: Inconsequential.

Comments: The swarms of tiny white fruitbodies with a tangled mess of thin white stipes from duff make this an easy species to recognize. *Rhizomarasmius setosus* is quite similar but doesn't fruit in such tremendous quantities and tends to have a dark brown stipe base, fading through yellow to pale cream upwards; it appears to be rare or overlooked. A few similar-looking species of *Owingsia* (including a species long referred to as *Marasmius epiphyllus*) occur in Cascadia. They are usually slightly larger with shorter, less sinuous stipes, slightly more visible cross-veins between the gills, and often two-toned stipes (whitish above with yellowish-brown lower parts). They often grow in scattered troops on rotting vegetation in damp riparian areas, especially on cottonwood leaves.

Marasmiellus candidus

(Bolton) Singer

Fairy Parachute

Cap: 0.5–3.5 (6) cm across, round to convex when young, becoming broadly convex to nearly plane, disc often depressed in age. Margin translucent-striate when wet, grooved or pleated; often wavy when dry. Surface finely powdery to smooth, moist to dry. White to translucent white at first, staying white, or developing creamy yellowish, pinkish to brick red blotches in age. **Gills:** Broadly attached to decurrent, sparse, very widely spaced, with few to many intergill veins. White, or developing creamy yellowish or pink to reddish spots, splotches, or overall color in age. **Stipe:** 0.5–4 cm long, 0.1–0.4 cm thick, central to off-center, curved away from center of cap into substrate; more or less equal. Surface dry, smooth or with fine, silky fibrils. Whitish to pale grayish, darkening to brown or ashy black upward from base. **Flesh:** Very thin, tough-leathery; concolorous. **Odor:** Indistinct. **Taste:** Indistinct. **Spore deposit:** White. **Microscopy:** Spores 11.5–16 × 3.5–5.5 µm, elongated teardrop- to spindle-shaped, smooth, inamyloid.

Ecology: Scattered in small to large troops and clusters, occasionally solitary, on twigs, branches, and mossy logs of conifers, hardwoods, shrubs, and herbaceous plants, especially *Rubus* spp. and red alder. Very common throughout the west slope of Cascadia. Generally fruiting in a large flush in early to mid–fall with scattered fruitings through winter and spring at lower elevations

Edibility: Unknown.

Comments: The white cap, widely spaced gills with radiating cross-veins, short, often curved stipe with a dark base, and growth on wood make this mushroom easy to identify. Occasionally it will develop yellowish to pink or reddish colors. *Metacampanella dendrophora* f. *washingtonensis* (aka *Campanella* sp. 'Veiny' in MotRC) differs by the more vein-like "pseudogills" and tiny or absent stipe. Although whitish when young, it is never bright white like *Marasmius candidus* and often develops bluish gray colors overall. *Metacampanella subdendrophora* is much smaller and grows primarily on the stems of dune grass. *Marasmius calhouniae* has a taller stature, more fragile flesh, and a white stipe; it grows in troops on conifer duff.

Resinomycena saccharifera
(Berk. & Broome) Redhead

Sugary Resin Bonnet

Cap: 0.2–0.8 cm across, convex to broadly convex; margin down-curved, often scalloped. Surface dry, finely powdery, pleated. White to pale ivory. **Gills:** Broadly attached, distant, narrow; edges fringed. Whitish to pale creamy-tan. **Stipe:** 0.2–0.5 cm long, 0.05 cm thick, central to off-center, equal or slightly enlarged at attachment point. Surface dry, powdery-granular. Whitish to pale ivory-tan. **Flesh:** Very thin, somewhat pliable. **Odor:** Indistinct. **Taste:** Indistinct. **Spore deposit:** Presumably white, difficult to obtain. **Microscopy:** Spores 8–11 × 5–6 µm, narrowly ellipsoid, smooth, thin-walled, amyloid. Cheilocystidia narrowly spindle-shaped, cylindrical to thread-like. Pleurocystidia embedded, spindle-shaped.

Ecology: Gregarious on dead stems of sedges and grasses, herbaceous plant debris, hardwood twigs or other small debris, especially dead *Rubus* canes. Common at lower elevations across western Cascadia. Fruiting in fall.

Edibility: Unknown.

Comments: The tiny fruitbodies, whitish to pale ivory colors, frosted-powdery appearance of the cap, gill edges and stipe surfaces, and the central to off-center stipe (typically shorter than the cap is wide) help define this species. *R. montana* is quite similar but typically has a stipe longer than the cap is wide, grows on conifer twigs and small branches and has shorter, narrower spores (8–9 × 3–4.5 µm). *Mycena culmigena* is another tiny species on grass and sedge stems with a smooth, grayish pink to nearly white cap and stipe. A number of *Hemimycena* species also resemble *Resinomycena* but typically have smooth or hairy caps and stipes, bright white colors, and inamyloid spores.

Phloeomana speirea
(Fr.) Redhead

Bark Bonnet

Cap 0.3–1 cm across, rounded, convex and often sunken at disc, often ribbed, or with broad sulcate grooves. Surface very finely powdery at first, soon smooth, moist to dry; translucent-striate when wet. Disc light brown, grayish brown to grayish yellow at first, fading, buff, yellowish to grayish; margin paler, grayish buff, pale gray to whitish, and often with darker lines in grooves. **Gills:** Broadly attached to decurrent, close to subdistant, narrow. White to pale grayish white. **Stipe:** 1–3.5 (5) cm long, 0.05–0.1 cm thick, fragile, slender, equal or slightly enlarged toward base. Surface finely powdery to smooth, dry to slightly moist; base with fine whitish hairs. Watery whitish, pale grayish white to pale yellowish white, occasionally with pale yellowish to brownish colors at base in age. **Flesh:** Very thin, fragile, whitish. **Odor:** Indistinct. **Taste:** Indistinct. **Spore deposit:** White. **Microscopy:** Spores 8.5–11.5 µm × 4.5–6.5 µm, ellipsoid to almond-shaped, with a prominent hilar appendage, smooth, thin-walled, inamyloid. Basidia 2-spored. Cheilocystidia elongate to narrowly club-shaped, often forked at tip with one to several knob-like projections. **Ecology:** Solitary, scattered, or in small clusters or troops on moss-covered pieces of bark (often on the inside of the bark) or small woody debris. Mostly found on bigleaf maple and alder but also on a number of other hardwoods and occasionally conifers. Common but easily overlooked. Fruiting soon after first rains in fall, continuing through winter and into wet springs.

Edibility: Unknown.

Comments: The small size, grayish brown convex cap with a paler margin broadly attached to decurrent gills, thin stipe, and growth on tree bark help identify this mushroom. Microscopically, the inamyloid spores and 2-spored basidia are useful diagnostic characters. *Mycena clavata* has the same stature but with a dark grayish brown, cinnamon-brown to buff cap; it grows on the bark of living and dead western red cedar. There are many other bark dwelling species of *Mycena* in Cascadia but most have a conical, bell-shaped to hemispherical cap and amyloid spores.

Synonym: *Mycena speirea* (Fr.) Gillet.

Mycena pterigena

(Fr.) P. Kumm.

Pink Fern Bonnet

Cap: 0.1–0.3 (0.5) cm across, egg-shaped to conical at first, becoming broadly conical to bell-shaped. Surface dry, finely powdery to smooth, pleated. Apricot orange to pinkish orange at first, fading to pale coral pink or pinkish white. **Gills:** Broadly attached to subdecurrent, distant, narrow. Whitish with pale pinkish orange edges. **Stipe:** 0.2–2 cm long, <0.05 cm thick, thread-like with a slight enlargement at basal attachment. Surface dry, smooth with short tufts of hairs at base. Pinkish orange at first, fading to pale pinkish gray, often retaining color near base. **Flesh:** Very thin, pliable. **Odor:** Indistinct. **Taste:** Indistinct. **Spore deposit:** Presumably white. **Microscopy:** Spores 7–9 × 3.5–5 µm, narrowly to broadly ellipsoid, smooth, thin-walled, amyloid. Cheilocystidia club- to sack-shaped, covered with finger-like projections. Pleurocystidia absent.

Ecology: Solitary, scattered or in troops on the dead stalks of lady fern *(Athyrium filix-femina)*. Common in lower and mid elevations across western Cascadia, typically on lady ferns in wetter areas. Fruiting in fall.

Edibility: Unknown.

Comments: The tiny size, apricot-pink to pinkish white colors, and growth on the dead stalks of lady fern make this species easy to identify. This species is common but requires a dedicated effort to find! The instructions are simple: just crawl into any clump of damp lady-ferns in the fall and carefully examine the dead stalks. *M. culmigena* is a similarly sized species on grass and sedge stems. It has a grayish pink cap when young, soon fading to pinkish white to nearly white, and is common in wet sedge thickets on the coast.

Mycena picta

(Fr.) Harmaja

Square Bonnet

Cap: 0.1–0.7 cm across, cylindrical to bell-shaped, with a sunken disc and distinct darker grooves. Surface smooth, moist to dry, hygrophanous, translucent-striate when wet, dull when dry. Dark brown to ocher-brown or grayish brown on disc, paler, grayish to yellowish toward margin. **Gills:** Broadly attached to decurrent, moderately spaced, unusual in appearance, extending straight across from stipe to cap margin. Creamy white to pale gray at first, to grayish buff in age. **Stipe:** 1–6 cm long, 0.1 cm thick, more or less equal, or slightly wider toward apex. Surface dry to moist, smooth above; base covered in fine yellowish brown hairs. Yellowish brown to darker reddish brown at base. **Flesh:** Very thin, fragile, pale grayish beige. **Odor:** Indistinct. **Taste:** Mild. **Spore deposit:** White. **Microscopy:** Spores 8–10 µm × 4–5 µm, narrowly ellipsoid to elongate, smooth, thin-walled, amyloid. Cheilocystidia club-shaped to cylindrical, with short rod-like projections at tip. Pleurocystidia absent.

Ecology: Solitary or scattered in moss or conifer duff across lower elevations in western Cascadia. Uncommon but, due to the small size, likely underreported. Fruiting in fall.

Edibility: Unknown.

Comments: The brown, boxy or bell-shaped and often umbilicate cap, unusual gills (forming a straight edge from the stipe to the cap margin), and tiny size are distinctive. It could be mistaken for *Rickenella swartzii* which has a purplish to gray convex cap and a dusky purple or gray stipe apex.

Rickenella swartzii

(Fr.) Kuyper

Dusky Moss Agaric

Cap: 0.3–0.8 (1.2) cm across, broadly convex to plane with a sunken center, margin down-curved when young, becoming slightly uplifted or wavy. Surface dry, appearing slightly powdered (pruinose) when young to smooth. Dark grayish purple to purplish black on young disc, fading to orangish purple, paler toward margin. **Gills:** Slightly to deeply decurrent, close to subdistant. Whitish to pale vinaceous gray. **Stipe:** 2–4 (6) cm long, 0.05–0.15 cm thick, equal, thin. Surface dry, powdery to smooth. Dark purple to purplish gray at apex, transitioning to orangish at base. **Flesh:** Very thin, pallid. **Odor:** Indistinct. **Taste:** Mild. **Spore deposit:** White. **Microscopy:** Spores 4–5 (7) × 2–3 µm, narrowly ellipsoid, smooth, inamyloid. Cheilocystidia and pleurocystidia abundant, with swollen middles and rounded heads (like a wide bowling pin). Cap and stipe surface with numerous cystidia, spindle-shaped or swollen at base, with elongated necks.

Ecology: Scattered in mossy woods or in mossy lawns across western Cascadia. Locally common; fruiting in fall and early winter.

Edibility: Unknown.

Comments: This small, pretty, and often overlooked mushroom is recognized by its plane, umbilicate cap with a purplish center (that fades to purplish orange), decurrent gills, and contrasting dusky purplish stipe apex (orangish toward the base). *R. fibula* often grows intermingled with this species in moss beds and has the same stature but is orange overall. There are lots of small, grayish, moss-dwelling *Mycena* but they rarely have the same stature or the dark purple colors.

Rickenella fibula

(Bull.) Raithelh.

Orange Moss Agaric

Cap: 0.3–1.2 cm across, broadly convex to plane with a sunken center; margin down-curved and often ribbed when young, becoming slightly uplifted or wavy in age. Surface dry, appearing slightly powdered (pruinose) when young to smooth in age. Translucent-striate when wet. Orange at first, soon fading to pale orange or orangish buff in age, often retaining orange color on disc. **Gills:** Slightly to deeply decurrent, close to subdistant. Whitish to pale creamy orange. **Stipe:** 1–4 (6) cm long, 0.05–0.2 cm thick, equal. Surface dry, pruinose to smooth. Bright to pale orange, often paler at the base. **Flesh:** Very thin, fragile, orangish; stipe often hollow. **Odor:** Indistinct. **Taste:** Mild. **Spore deposit:** White. **Microscopy:** Spores 4.5–6 × 2–3 µm, narrowly ellipsoid, smooth, inamyloid. Cheilocystidia and pleurocystidia abundant, subcylindrical to spindle-shaped. Cap and stipe surface covered with long cystidia.

Ecology: Solitary, scattered in troops, or small clusters in moss with a preference for open mossy areas, often in yards and roadsides. Widespread and common at low and mid elevations in western Cascadia, occasional at higher elevations and on the east side. Fruiting throughout the year, most abundant in fall and early winter.

Edibility: Unknown.

Comments: This small moss inhabitant is recognized by the orange, convex to plane cap with a depressed disc, pale, decurrent gills, and long, thin, orangish stipe. *R. mellea* is similar but has a paler creamy orangish colors, less prominent cystidia, and appears to have a preference for growing among sphagnum moss. Two similar but slightly larger species are in the genus *Loreleia*: *L. marchantiae* has a bright orange cap, and *L. postii* has a dull orange stipe and deeply decurrent gills. Both of these are rare in Cascadia. *L. marchantiae* grows in association with liverworts and *L. postii* on wet soil with mosses and liverworts. Many species of orangish brown to buff *Galerina* often grow alongside *R. fibula* in moss beds; however, they have conical to convex caps and have light rusty brown spores. *Xeromphalina* species are often larger, have orange-brown caps, and tough stipes.

Atheniella delectabilis

(Peck) Lüderitz & H. Lehmann

White Fairy Bonnet

Cap: 0.2–1 cm across, conical to convex at first, expanding to broadly convex or plane, often with a small, rounded umbo. Surface smooth, moist to dry. Watery white to pale creamy white with a translucent-striate margin when wet; dull, opaque, chalk white when dry. **Gills:** Broadly attached to decurrent, moderately spaced, often with short veins between gills; edges fringed when young. White. **Stipe:** 2–4 cm long, 0.05–0.2 cm thin, fragile. Surface smooth, dry to moist, base with sparse whitish hairs. Watery white to white. **Flesh:** Concolorous, very thin, fragile. **Odor:** Bleachy, distinct when crushed. **Taste:** Indistinct. **Spore deposit:** White. **Microscopy:** Spores 5–7 × 3.5–4 µm, ellipsoid, smooth, inamyloid. Cheilocystidia abundant, spindle-shaped to cylindrical. Pleurocystidia similar, scattered to abundant.

Ecology: Scattered or in troops, rarely solitary; in duff and needles, especially of Douglas-fir and true firs. Common and widespread, occurring throughout Cascadia but more abundant on the west side. Fruiting soon after first rain in fall, with stragglers into early winter.

Edibility: Inedible.

Comments: This small (but not necessarily tiny) white mushroom often grows in troops on needle duff soon after the first fall rains. Its most distinct feature is the relatively larger size (if something 0.2–1 cm across can be considered large) and the bleach-like odor when crushed. There are numerous white *Hemimycena* in Cascadia which look similar but most are even smaller, and most lack the bleachy odor. Members of the *Hemimycena lactea* group are quite similar, though they tend to be slightly smaller and have more convex caps, as well as longer, narrower spores (7–12 × 2.5–3.5 µm). *H. tortuosa* is smaller (caps 0.1– 0.5 cm across), with fruitbodies often attractively adorned with dew droplets, and commonly encountered growing on bark of both hardwood and conifers. Microscopically, it has very distinctive, corkscrew-shaped cystidia on the cap and stipe. Another member of this genus is *H. hirsuta*, with a minutely fuzzy cap and poorly formed gills. This group needs considerable work, and despite the diversity in Cascadia (around 20 species can be recognized based on DNA sequence data), species names are lacking or misapplied.

Synonym: *Hemimycena delectabilis* (Peck) Singer

White-spored (N)

Atheniella aurantiidisca

(Murrill) Redhead, et al.

Orange-disc Fairy Bonnet, Tangerine Bonnet

Cap: 0.3–1.5 cm across, conical at first, becoming broadly conical to bell-shaped. Surface smooth, moist to dry; translucent-striate when wet. Color variable: Bright orange over disc, to yellow-orange or yellow outwards, with a whitish margin, sometimes pale yellowish to creamy white overall in age. **Gills:** Narrowly attached, notched, occasionally with a decurrent tooth, close to subdistant, narrow. Whitish, pale yellowish to yellowish orange. **Stipe:** 1.5–5 cm long, 0.1–0.2 cm thick, fragile, equal, or enlarged slightly toward base. Surface finely powdered at first, becoming smooth, moist to dry. Base with coarse whitish hairs. Pale yellowish, translucent yellow to whitish. **Flesh:** Very thin, fragile, concolorous with cap; stipe hollow, translucent yellow to whitish. **Odor:** Indistinct. **Taste:** Mild. **Spore deposit:** White. **Microscopy:** Spores 6–9 × 3–4.5 µm, narrowly ellipsoid, smooth, thin-walled, inamyloid. Cheilocystidia and pleurocystidia spindle-shaped with long, narrow necks.

Ecology: Solitary, scattered, or in small to large troops in moss, duff, or conifer debris. Widespread across Cascadia and abundant in wetter areas. Fruiting primarily in fall and early winter but can be found throughout the year.

Edibility: Unknown.

Comments: The three-toned (orange, yellow, whitish), conical cap, pallid to orangish gills, thin, translucent yellow to whitish stipe, and growth in duff make this an easy mushroom to recognize. Faded specimens could be mistaken for the pale yellow to creamy white *A. flavoalba*. *A. adonis* has a bright pink cap when young that fades to pale pink or pinkish white in age. *Mycena strobilinoidea* has a more rounded to bell-shaped cap that is bright reddish orange to orange and has pale orangish yellow gills with a dark reddish orange edge. *M. oregonensis* is bright yellow overall and has a bell-shaped or plane cap and a very thin stipe. *M. acicula* is a tiny species with a reddish orange to coral-colored cap that fades to light orange and a long, thin, yellowish stipe.

Synonym: *Mycena aurantiidisca* (Murrill) Murrill

Atheniella adonis

sensu PNW

Pink Fairy Bonnet

Cap: 0.4–2 cm across, conic, convex to uneven-convex, usually with a small umbo, margin often splitting in age. Surface moist to dry, smooth, translucent-striate when wet. Bright pink to salmon pink when young, fading to paler pinkish yellow or pinkish white. **Gills:** Attached, close to subdistant, often inter-veined. Pale pink, off-white to creamy or pale yellowish. **Stipe:** 2–4 cm long, 0.05–0.3 cm thick, very fragile, equal or slightly thinner and curved toward base. Surface often dusted with whitish powder when young, becoming smooth, moist to dry; base with whitish fibrils. Pale pink to pinkish white, creamy white or translucent, often lighter to white toward base. **Flesh:** Very thin, fragile, pinkish. **Odor:** Indistinct. **Taste:** Mild. **Spore deposit:** White. **Microscopy:** Spores 7–8 × 3–5 µm, narrowly ellipsoid, smooth, inamyloid. Cheilocystidia and pleurocystidia spindle-shaped.

Ecology: Solitary, scattered, or in small groups or clusters in humus and duff, especially Douglas-fir needle litter across Cascadia. Fairly common. Fruiting in fall.

Edibility: Unknown.

Comments: The small size and pink conical-convex caps of young specimens are usually enough to separate this mushroom from all but the similarly colored *Mycena rosella*, which differs in having pink gills with darker pink edges. Faded older specimens of *A. adonis* can be quite pale, these resemble both *A. flavoalba* which starts off pale yellow and fades to creamy white and *A. delectabilis* which is brighter white. *Mycena monticola* has a darker pinkish red to rosy red conical to bell-shaped cap and a slightly cartilaginous pink stipe which develops gray-brown colors from the base up. The tiny *Mycena acicula* has a bright orange to orange-red cap when young that can fade to pinkish in age and a thin, long yellowish stipe; it grows from hardwood bark and debris.

Synonym: *Mycena adonis* (Bull.) Gray

Mycena monticola

A.H. Sm.

Pink Mountain Bonnet

Cap: 1–3 cm across, conical to bell-shaped at first, occasionally convex with a low, broad umbo. Margin downcurved when young, wavy, flaring to uplifted in age. Surface moist to dry, hygrophanous; translucent-striate when wet. Pinkish red to rosy red, often with a pale bloom when young, fading from the margin to pale pinkish. **Gills:** Attached with a shallow notch and short decurrent tooth, close to subdistant, narrow. Pale creamy pink to pink; edges concolorous. **Stipe:** 3–7 cm long, 0.1–0.3 cm thick, equal or enlarged slightly toward base. Surface moist to dry, smooth; base with sparse, stiff, white to yellowish hairs. Pinkish, pinkish gray to pinkish brown, darkening to gray-brown from base up in age. **Flesh:** Very thin, fragile in cap, stipe somewhat cartilaginous; concolorous. **Odor:** Indistinct. **Taste:** Mild. **Spore deposit:** White. **Microscopy:** Spores 7–10 × 4–5 µm, elongate-ellipsoid, smooth, thin-walled, amyloid. Cheilocystidia club-shaped with short finger-like projections. Pleurocystidia absent.

Ecology: Scattered in troops, tufts, or small clusters from duff under conifers in higher elevation forests in the Cascade Range and on the Olympic Peninsula. Uncommon to rare. Fruiting in fall.

Edibility: Unknown.

Comments: The pinkish red to rosy red conical to bell-shaped cap, creamy pink gills with concolorous edges, and a pink stipe which develops gray-brown colors from the base up as it ages are the best features to identify this pretty *Mycena*. *M. rosella* is quite similar but tends to fade more drastically as it ages and has pink gills with darker pink to reddish edges. *Atheniella adonis* has a bright pink to salmon pink cap and a fragile pinkish to almost translucent white stipe.

Mycena rosella

(Fr.) P. Kumm.

Rosy Pink Mycena, Pink Bonnet

Cap: 0.3–1.5 (2) cm across, rounded to bell-shaped at first, becoming convex, broadly convex to plane, occasionally with a small umbo, or depressed on disc. Surface smooth, dry to moist, translucent-striate when wet. Deep pink to rosy red at first, soon fading to pinkish or pinkish gray, often with yellowish over the disc in age. **Gills:** Broadly attached, often with a decurrent tooth, close to subdistant, narrow. Pinkish at first, becoming pale pinkish peach to pale pinkish white. Edges darker pink to rosy pink. **Stipe:** 1–5 (7) cm long, 0.05–0.2 cm thick, fragile, equal or slightly enlarged toward base. Surface dry, smooth, base with sparse whitish fibrils. Watery pink, pinkish gray to light pinkish brown at first, becoming dingy pinkish gray to light grayish brown. **Flesh:** Very thin, fragile, pinkish to pinkish gray. **Odor:** Indistinct to faintly radish-like. **Taste:** Mild. **Spore deposit:** White. **Microscopy:** Spores 8–9.5 (10.5) × 4–5 µm, narrowly ellipsoid, smooth, thin-walled, amyloid. Cheilocystidia abundant, club-shaped, covered short projections, with dark reddish content. Pleurocystidia abundant, narrowly spindle-shaped, with reddish content.

Ecology: Solitary, scattered, or in small troops on conifer needle duff or in moss. Common in west-side forests, rare on the east side of the Cascade Range, and largely absent from the southeast portion of Cascadia. Fruiting in fall and early winter.

Edibility: Unknown.

Comments: The rosy cap that fades to pinkish gray, pinkish gills with darker rosy pink edges, and pinkish gray stipe help identify this *Mycena*. *M. monticola* has a darker pinkish red to pinkish gray cap, gills with concolorous or paler edges, and a pink stipe which develops gray-brown colors from the base up. *Atheniella adonis* shares the same pink colors but generally has a conical to convex cap and lacks the darker pink gill edges; it also has inamyloid spores. Older specimens could be mistaken for *M. capillaripes*, which has a dark to light gray cap, pinkish gill edges, and an odor of chlorine or bean sprouts when crushed. *M. subsanguinolenta* has a reddish brown cap, pallid gills with a reddish edge, and “bleeds” a reddish brown juice from the stipe.

Mycena strobilinoidea

Peck

Flame Bonnet

Cap: 1–2 cm across, conical at first, soon bell-shaped to convex; margin often scalloped or finely pleated. Surface thinly viscid to dry, smooth, slightly hygrophanous; margin translucent-striate when wet. Deep red-orange to bright orange at first, fading to paler orange or yellow-orange in age. **Gills:** Broadly attached or with a shallow notch, close, subdistant, narrow. Creamy to pale yellowish orange with dark red-orange to orange edges. **Stipe:** 2–5 cm long, 0.1–0.2 cm thick, equal or slightly enlarged toward base. Surface moist to dry, finely powdery over upper portion when young, becoming smooth; base with short, stiff hairs. Bright orange to yellow-orange. **Flesh:** Very thin, pliable; stipe somewhat cartilaginous, yellow-orange. **Odor:** Indistinct. **Taste:** Mild. **Spore deposit:** White. **Microscopy:** Spores 7–9 × 4–5 µm, ellipsoid, smooth, amyloid. Cheilocystidia abundant, club-shaped to broadly spindle-shaped with finger-like projections on one side, with bright to pale orange content. Pleurocystidia scattered to abundant, similar in shape to cheilocystidia.

Ecology: Scattered in small to large troops from conifer duff and moss in mid- or high-elevation forests. Common at higher elevations across the Cascade Range, rare to occasional in higher elevation forests in the Coast Range, largely absent from lower elevations. Fruiting in late summer and fall.

Edibility: Unknown.

Comments: The bright orange colors, bell-shaped to convex cap, pale yellowish orange gills with a dark red-orange edge, and growth at higher elevations make this an easy *Mycena* to recognize. *M. aurantiomarginata* has orangish gills with dark orange edges and a translucent-striate olive-brown to yellowish brown cap. *Atheniella aurantiidisca* often has a three-toned (orange, yellow, whitish) conical cap, pallid to orangish gills with concolorous edges, and a translucent yellow to whitish stipe. *M. oregonensis* is a tiny species with bright yellow to yellow-orange cap and thin, brightly colored stipe.

Mycena oregonensis

A.H. Sm.

Western Yellow Bonnet

Cap: 0.2–1 cm across, egg-shaped to bell-shaped when young, expanding to convex, occasionally with a small umbo. Surface dry, smooth or sometimes dusted with fine yellowish granules. Bright yellow to bright yellow-orange, fading to yellow or pale yellow, but often remaining darker on disc. **Gills:** Attached, close to subdistant. Pale yellow, yellowish orange or creamy white. **Stipe:** 1.5–5 cm long, 0.05–0.1 cm thick, fragile, equal, often curved, sometimes rooting. Surface dry, finely powdered, covered with tiny orange granules; base with yellowish orange hairs. Bright yellow to yellow-orange, becoming paler yellow to orangish yellow. **Flesh:** Very thin, fragile, yellowish. **Odor:** Indistinct. **Taste:** Mild. **Spore deposit:** White. **Microscopy:** Spores 7–10 × 2.5–4 µm, broadly cylindrical with a pointed end, smooth, inamyloid. Cheilocystidia abundant, club- to spindle-shaped, with a rounded head, and with yellow contents. Pleurocystidia similar, infrequent.

Ecology: In small groups, scattered troops, or solitary in duff and debris of all forest types across Cascadia but especially common under Douglas-fir. Fruiting in fall with occasional stragglers into winter.

Edibility: Unknown. It would take hundreds to make a bite

Comments: The tiny size, bright yellow to yellow-orange cap, and thin, brightly colored stipe help distinguish this species from other small, colorful *Mycena*. As the inamyloid spores suggest, it does not belong in *Mycena* and is closely related to *Phloeomana*. *M. acicula* has a reddish orange cap when young and a pale yellowish stipe and grows on hardwood debris. *Atheniella aurantiidisca* has a more conical orange cap when young that fades to yellow and a pale stipe; it is stockier overall. *M. strobilinoidea* has larger caps (1–2 cm), darker reddish orange colors when young, orange gill edges, and grows in duff and moss at higher elevation.

Mycena epipterygia group
Yellowleg Bonnets

Cap: 0.7–2 cm across, egg-shaped at first, soon conical to bell-shaped, becoming broadly conical to convex in age. Surface sticky to viscid when wet, smooth, or dusted with fine whitish powder. Grayish olive to grayish green, grayish yellow to yellowish white. **Gills:** Attached with a decurrent tooth, close to subdistant. Whitish to pale gray, sometimes developing pale pinkish or yellowish colors in age. **Stipe:** 3–7 cm long, 0.1–0.2 cm thick, equal or slightly enlarged toward base. Surface viscid to sticky, smooth; base with fine off-white to grayish hairs. Bright yellowish green when young, becoming pale yellowish green to pale yellowish or yellowish white. **Flesh:** Very thin, fragile, grayish white. **Odor:** Faint, pleasant to unpleasant, often like cucumbers or sometimes rancid-farinaceous in age. **Taste:** Farinaceous. **Spore deposit:** White. **Microscopy:** Spores 9–12 (13) × 5–7 µm, narrowly ellipsoid, smooth, thin-walled, amyloid. Cheilocystidia club-shaped, with scattered finger- to peg-like projections. Pleurocystidia absent.

Ecology: Solitary or in small groups on duff, in moss, or on rotting conifer logs and branches. Common across western Cascadia, occasional on the east side. Primarily fruiting in fall with occasional appearances in winter or spring.

Edibility: Unknown.

Comments: The viscid yellow stipe and some combination of grayish olive to grayish yellow cap colors make this an easy *Mycena* to identify. There are several described varieties of *M. epipterygia*, including var. *viscosa*, which has a strong farinaceous odor and taste, and a tendency to develop dirty brownish stains; and var. *lignicola*, which has yellower cap colors and grows on wood. ITS sequence data has shown multiple species in this group but morphological features corresponding to these differences have yet to be defined. *M. nivicola* has a yellow-brown, olivaceous brown to dark brown cap, and a viscid to dry, yellow stipe; it is restricted to higher elevations and fruits on woody debris or duff at the edges of melting snowbanks.

Mycena nivicola
B.A. Perry & Desjardin

Olive Snowbank Bonnet

Cap: 1–3 cm across, conical to bell-shaped at first, becoming broadly bell-shaped with a flaring margin in age. Surface viscid when wet, shiny when dry, smooth, translucent-striate. Cap color variable; ranging from nearly blackish brown to reddish brown, olivaceous to yellowish olive with a darker brown disc or paler overall in age (although in some specimens darkening in age). **Gills:** Attached, with a broad notch and decurrent tooth, subdistant, moderately broad. Whitish, grayish to yellow. **Stipe:** 4–14 cm long, 0.1–0.5 cm thick, equal, but often with a slightly enlarged and curved base. Surface viscid to dry, smooth; base covered with white hairs. Pale to bright yellow, developing reddish brown tones from the base up in age. **Flesh:** Thin, pliable, cap skin a separable gelatinous cuticle. **Odor:** Indistinct. **Taste:** Mild or slightly farinaceous. **Spore deposit:** White. **Microscopy:** Spores 9.5–12 × 5.5–7 µm, ellipsoid, smooth, thin-walled, amyloid. Cheilocystidia irregularly club-shaped with short to long finger-like projections. Pleurocystidia absent.

Ecology: Solitary or scattered from duff and woody debris at the edges of melting snowbanks in higher elevation forests across Cascadia. Common on the east side and southern portion of the Cascade Range, less often on the west side. Fruiting in spring and early summer.

Edibility: Unknown.

Comments: Easily recognized by the yellowish to olive-brown cap, yellow stipe which develops reddish brown colors, and habitat near melting snowbanks. This is one of our more common members of the snowbank funga but easily overlooked due to its small size and colors which blend into the duff. *M. epipterygia* is more common in low to mid-elevation west side forests, has a variable grayish olive to yellow cap, a viscid yellow stipe, and fruits in fall and winter from duff, moss, or on small woody debris.

Mycena haematopus

(Pers.) P. Kumm

Fringed Bleeding Bonnet

Cap: 0.8–3.5 cm across, rounded to bell-shaped at first, becoming convex to broadly convex. Surface very finely powdered, giving it a sheen at first, becoming smooth, dry to moist, translucent-striate in age. Margin with overhanging flaps of sterile tissue, crimped at first, becoming ragged in age. Pinkish brown, reddish brown to vinaceous brown at center, lighter reddish brown to pinkish buff toward margin, often developing reddish brown to purplish brown stains throughout. **Gills:** Attached, close to subdistant. Pale pinkish white to pale pink when young, becoming pinkish and developing dark reddish to purplish brown stains, spots, and blotches. Edges concolorous. **Stipe:** 2–10 cm long, 0.15–0.5 cm thick, fragile, equal or slightly enlarged toward base. Surface smooth, dry to moist; base covered with off-white to grayish hairs. Reddish, reddish brown, reddish gray to violet-gray. **Flesh:** Thin, fragile, reddish brown. Stipe hollow, exuding dark red, reddish brown to wine-colored juice when fresh specimens are broken. **Odor:** Indistinct. **Taste:** Mild. **Spore deposit:** White. **Microscopy:** Spores 9–11 × 5.5–7 µm, ellipsoid to almond-shaped, smooth, thin-walled, amyloid. Cheilocystidia abundant, club-shaped, or more swollen in middle, occasionally forked, or covered with numerous short peg-like projections. Pleurocystidia rare to abundant, similar to cheilocystidia.

Ecology: Generally fruiting in clusters, occasionally in scattered troops or solitary on rotting wood, with a preference for hardwoods (especially bigleaf maple and alder) but also on conifers. Very common at low to mid elevations of western Cascadia, locally common in higher elevation forests and on the east side of the Cascade Range. Fruiting throughout the year but more abundant in fall and spring.

Edibility: Unknown.

Comments: This *Mycena* is easily identified by the reddish brown cap with ragged overhanging tissue at margin, growth on wood, and the dark red juice it bleeds when broken (especially from the stipe). Another "bleeder" in Cascadia is *M. subsanguinolenta* which has a small, conical to bell-shaped cap, colored gill edges, with growth on conifer duff. Similar non-bleeding species include *M. purpureofusca* which has a purplish to violet-gray cap and purplish gill edges. *M. bulliformis* looks quite similar to *M. haematopus* and often grows in clusters on rotting wood, but does not bleed when broken, nor does it have the overhanging tissue on the cap margin. *M. maculata* has a gray cap that develops reddish stains and blotches in age, it grows on decaying conifer wood.

Mycena subsanguinolenta

A.H. Sm.

Small Bleeding Bonnet

Cap: 0.3–1 (2) cm across, conical to bell-shaped at first, becoming broadly bell-shaped, more rarely convex. Surface dry, often with finely powdery bloom when young, soon smooth; translucent-striate. Dark red-brown when young, retaining the red-brown on disc, but becoming orange-brown to yellow-brown outward with darker lines, developing dingy blotches and stains with age. **Gills:** Broadly attached, at times with a narrow notch, close to subdistant, narrow. Pinkish white to creamy pink, often developing brownish blotches in age. Edges dark red to red-brown. **Stipe:** 2–8 cm long, 0.1–0.2 cm thick, fragile, equal or slightly enlarged toward base. Surface moist to dry, very finely powdery near apex, soon smooth, base with whitish to brownish hairs. Pinkish gray when young, darkening to grayish red to reddish brown with age. **Flesh:** Very thin, fragile. Stipe hollow, exuding dark red-brown juice when broken. **Odor:** Indistinct. **Taste:** Mild. **Spore deposit:** White. **Microscopy:** Spores 7–8.5 × 3.5–4 µm, narrowly ellipsoid, smooth, thin-walled, amyloid. Cheilocystidia abundant, somewhat spindle-shaped, with a pointed tip, reddish brown. Pleurocystidia similar.

Ecology: Solitary, scattered, or in small troops on duff under conifers across Cascadia. Occasional but likely underreported due to the small size. Fruiting during wet periods, most often in winter and spring in western Cascadia, and during late summer and fall on the east side of the Cascade Range.

Edibility: Unknown.

Comments: This "bleeding" *Mycena* is identified by the red-brown to orange-brown colors, red-brown gill edges, and growth in conifer duff. When the stipe is broken, it often produces a drop of dark red-brown liquid. There is some question whether the European *M. sanguinolenta* also occurs in Cascadia, or only *M. subsanguinolenta*, originally described from Oregon. The more common *M. haematopus* grows on wood, has overhanging flaps of sterile tissue on the cap margin, bleeds redder juice, and does not have colored gill edges. See *M. purpureofusca* for similar "non-bleeders."

Mycena purpureofusca

(Peck) Sacc.

Purple-edged Bonnet

Cap: 0.8–2.5 cm across, conical to bell-shaped at first, becoming broadly bell-shaped, convex to nearly plane. Surface dull, dry to moist; distinctly translucent-striate when young. Dark purple to reddish purple when young, becoming purple to purplish pink over disc, paler toward margin, becoming purplish brown to purplish gray in age. **Gills:** Attached, close to subdistant. White to off-white at first, becoming pale pinkish white to pale purplish white. Edges purple to purplish red. **Stipe:** 2–8 cm long, 0.1–0.3 cm thick, fragile, slightly enlarged toward base. Surface finely pubescent to smooth, dry to moist, base covered with purplish, off-white to grayish hairs, hollow, not exuding juice when broken. Dark purplish gray to purple at first, becoming purplish, reddish purple, purplish brown to purplish gray. **Flesh:** Very thin, fragile, purplish. **Odor:** Indistinct. **Taste:** Mild. **Spore deposit:** White. **Microscopy:** Spores 9–12 × 6–8 µm, ellipsoid, smooth, thin-walled, amyloid. Cheilocystidia abundant, spindle-shaped with a distinctly swollen middle, forking at tip in age, with purplish content.

Ecology: Solitary, scattered, or in small clusters on cones, twigs, branches, bark, or other debris of conifers, especially pines. Common across Cascadia, especially in pine forests. Fruiting in fall and winter.

Edibility: Unknown.

Comments: This *Mycena* is readily identified by the purple coloration, pale gills with dark purple edges, and growth on woody debris. *M. bulliformis* is similar but is generally paler, grows in clusters on decaying hardwood logs, and has white gills when young with pale violet-brown edges. *M. subsanguinolenta* grows on conifer duff, has reddish brown gill edges, and bleeds a dark reddish juice from the stipe when broken. *M. rubromarginata* has a vinaceous gray to grayish white, faintly translucent-striate cap and vinaceous brown gill edges.

Mycena pura group

Lilac Bonnets

Cap: 1–4 cm across, bell-shaped to convex at first, becoming broadly convex to plane, usually with a broad umbo. Margin incurved when young, often wavy, ragged and occasionally slightly uplifted in age. Surface smooth, moist to dry; strongly hygrophanous, translucent-striate and shiny when wet, dull, opaque when dry. Color extremely variable, generally some shade of purple, purple-gray, lilac, pinkish lilac to pinkish gray at first, becoming grayer as it ages, or becoming dull grayish lilac to whitish when dry. **Gills:** Broadly attached or notched with a slightly decurrent tooth, close to subdistant, broad; partial gills numerous, occasionally strongly inter-veined. Lilac-gray, pinkish lilac to purplish when young, becoming grayish, whitish to brownish gray in age. **Stipe:** 3–10 cm long, 0.2–0.8 cm thick, equal or enlarged toward base. Surface dry, slightly scurfy at apex when young, becoming smooth; base with whitish to yellowish hairs. Lilac-gray, purplish, gray to grayish brown, often developing yellowish to brownish discoloration at base in age. **Flesh:** Thin, fragile; stipe hollow, cartilaginous. Purplish, grayish to whitish. **Odor:** Cucumber- or radish-like. **Taste:** Indistinct, mildly earthy, or radish-like. **Spore deposit:** Whitish. **Microscopy:** Spores 7–9.5 × 3–5 µm, narrowly ellipsoid, cylindrical to oblong, smooth, thin-walled, weakly amyloid (or occasionally inamyloid). Cheilocystidia abundant, spindle to club-shaped with a rounded to pointed tip. Pleurocystidia similar.

Ecology: Solitary, scattered, or in small clusters in duff, moss, or occasionally on well-decayed wood. Common in all forested habitats across Cascadia. Fruiting throughout the year with the primary flush in fall.

Edibility: Possibly poisonous.

Comments: The strongly hygrophanous, translucent-striate cap with purple, lilac-gray or pinkish colors (when young), lilac-gray to pinkish lilac gills and stipe, and radishy odor are distinctive. This is a variable species that appears drastically different when wet versus dry or young versus old. There are many described varieties and forms, and recent genetic work has shown that it is part of a species complex. As with most groups of mushrooms, more work is needed. *Mycena* cf. *pelianthina* shares a similar stature but has a dark purple to violet-gray cap and purple-edged gills and is rare in Cascadia. Violet-capped *Inocybe* differ in having a cobwebby partial veil, a silky, opaque cap, and brown spores. *Laccaria amethysteo-occidentalis* is larger, has a tough, fibrous stipe, and thick, widely spaced gills. *Clitocybe tarda* grows in grass and has pinkish buff spores.

Synonym: *Prunulus purus* (Pers.) Murrill

Mycena rutilantiformis

(Murrill) Murrill

Ruddy Bonnet

Cap: 2–6 cm across, convex at first, becoming broadly convex to plane or concave; margin down-curved when young, often ragged and uplifted in age. Surface moist to dry, roughened to smooth. Dull vinaceous red when young, fading to vinaceous pink to pinkish gray; occasionally darker red-brown. **Gills:** Narrowly attached with a pronounced notch, subdistant, broad; partial gills numerous, occasionally inter-veined. Rosy-pinkish when young, soon pinkish buff to pinkish gray; edges dark wine-red at all stages. **Stipe:** 4–8 cm long, 0.3–0.8 cm thick, equal or enlarged downwards; base often dog-legged. Moist to dry, appressed-fibrillose to scurfy at apex when young, becoming smoother with age; base with whitish hairs. Vinaceous red at apex, pinkish gray lower, at times developing yellow-brown tones. **Flesh:** Thin, fragile in cap; stipe hollow, cartilaginous. Pinkish gray in cap, yellowish in upper stipe, pallid lower. **Odor:** Faint, radish-like. **Taste:** Radish-like. **Spore deposit:** White. **Microscopy:** Spores 7–8 (9) × (3.5) 4–4.5 µm, ellipsoid, smooth, thin-walled, amyloid. Cheilocystidia abundant, narrowly ovoid to spindle-shaped, dark reddish brown. Pleurocystidia abundant, similar in shape, but longer than cheilocystidia.

Ecology: Solitary, scattered, or in small clusters in needle duff, especially Douglas-fir, at low to mid elevation in western Cascadia. Rare. Fruiting in fall.

Edibility: Unknown.

Comments: The wine-red colors which fade with age, dark red gill edges, and amyloid spores are distinctive. *Mycena* cf. *pelianthina* is similar but has a dark purple to violet-gray cap, purple-edged gills, and smaller spores (5.5–7 × 3–3.5 µm). Both of these species do not have the "typical" *Mycena* stature and may not be initially recognized as belonging to the genus. *M. pura* has a purple, lilac-gray, or pinkish cap and concolorous gill edges. Some *Rhodocollybia* are vaguely similar but have white to pinkish gills with concolorous edges.

Mycena maculata

P. Karst

Red-spotted Bonnet

Cap: 1–3 (4) cm across, conical to bell-shaped at first, becoming broadly bell-shaped to convex, usually with a distinct, low broad umbo. Surface smooth, moist to dry, hygrophanous, translucent-striate when wet, dull when dry. Dark gray, dark grayish brown to blackish brown at first, remaining so on disc, but fading outward to light gray or grayish buff. Developing reddish to vinaceous spots and stains in age. **Gills:** Narrowly attached with a broad notch, close to subdistant. Off-white to pale gray, developing dingy reddish brown to vinaceous spots and stains in age. **Stipe:** 3–10 (13) cm long, 0.1–0.4 cm thick, more or less equal, occasionally with a rooting base. Surface dry to moist, smooth above; base covered in coarse white hairs. Gray to whitish gray, paler toward base, developing reddish brown to vinaceous spots and stains from base up in age. **Flesh:** Thin, fragile, watery gray to grayish white, slowly staining reddish brown. Stipe hollow, not bleeding when broken. **Odor:** Indistinct to faintly farinaceous. **Taste:** Mild to slightly farinaceous. **Spore deposit:** White. **Microscopy:** Spores 9–10.5 × 5–6.5 µm, narrowly ellipsoid, smooth, thin-walled, amyloid. Cheilocystidia club-shaped, covered with a few simple to much-branched projections. Pleurocystidia absent.

Ecology: In small to large clusters, scattered in troops, or rarely solitary. Fruiting on well-decayed logs, stumps, and other woody debris. Prefers conifer debris but will occasionally fruit from hardwoods. Locally common in western Cascadia but rare on the east side. Fruiting in fall, with stragglers into winter.

Edibility: Unknown.

Comments: The gray cap, often broadly notched gray gills, growth in clusters on decaying wood, and especially the development of reddish brown to vinaceous spots and stains in age make this *Mycena* easy to identify. Young specimens often lack the reddish staining or show it only at the stipe base. *M. haematopus* has more reddish brown colors, bits of sterile tissue hanging from the cap margin, and bleeds a reddish juice when broken. *M. galericulata* generally has a taller stature, a tough stipe, and white to pinkish gills, lacks reddish stains and generally on hardwood stumps and logs.

Mycena galericulata

(Scop.) Gray

Rooting Bonnet

Cap: 1.5–6 cm across, conical to bell-shaped at first, becoming broadly convex to plane, usually with a distinct, broad umbo at all stages. Margin incurved when young, often wavy, splitting and occasionally slightly uplifted in age. Surface often wrinkled or finely sulcate to smooth, slightly viscid to dry, translucent-striate when wet. Dark grayish brown, brown to gray at center, paler toward margin at first, fading to lighter grayish brown, light brown to whitish. **Gills:** Broadly attached to decurrent, sometimes with a broad notch and decurrent tooth, close to subdistant, broad; partial gills numerous; occasionally strongly inter-veined. White to pale whitish gray, developing pinkish tones in age. **Stipe:** 3–15 cm long, 0.2–0.6 cm thick, equal or enlarged below, slightly compressed or with a twisting groove down middle, and often with rooting base. Surface dry, smooth; base tomentose, covered in dense white hairs. Dark to pale grayish brown, paler toward apex, often developing yellowish discoloration in age. **Flesh:** Thin, fragile, watery brown to grayish. Stipe hollow, somewhat tough, cartilaginous. **Odor:** Mild to farinaceous or radish-like. **Taste:** Indistinct, mildly farinaceous or earthy. **Spore deposit:** Whitish to creamy. **Microscopy:** Spores (8) 9–11 (12) × 6–8 µm, ellipsoid, smooth, thin-walled, containing oil droplets, amyloid. Cheilocystidia club-shaped to irregularly shaped, with irregular finger-like projections. Pleurocystidia absent.

Ecology: Scattered or solitary, occasionally in small clusters, on stumps, logs, or other coarse woody debris. Prefers hardwoods, especially oak, bigleaf maple, and alder but occurs on conifers as well. Common and widespread across Cascadia. Fruiting from early fall well into winter with rare spring and summer fruitings .

Edibility: Unknown.

Comments: This *Mycena* is recognized by its brownish gray cap, whitish gills that often develop pinkish tones in age, tall and often slightly rooting cartilaginous stipe, and growth on wood. The large size and growth on wood alone help separate it from most other *Mycena*. *M. megaspora* is very similar macroscopically but has larger spores (12–17 × 6–8 µm from 2-spored basidia, 10–12 × 5–7 µm from 4-spored basidia) and grows in sphagnum moss. *M. maculata* is a similar large species that grows on wood but generally fruits in clusters, has grayish gills, and develops reddish brown to vinaceous spots and stains in age. *M. robusta* is a common, large, gray, winter-fruiting species in moss and duff. *Mycopan scabripes* has a slightly darker, brownish to grayish brown, often wrinkled cap, whitish gills that become dingy pinkish white to grayish white in age, and a shorter stipe with fine fibrils.

Mycena overholtsii

A.H. Sm. & Solheim

Clustered Snowbonnet

Cap: 2–5 cm across, obtusely conical, bell-shaped to convex when young, becoming plane or with a broad umbo. Margin even, occasionally ragged or uplifted in age. Surface moist to slightly lubricous when young, soon dry, smooth. Hygrophanous, translucent-striate and dark gray, grayish black to grayish brown when moist, fading greatly in age or when dry to pale gray or whitish. **Gills:** Broadly attached, often with a decurrent tooth, more rarely notched or decurrent, close to subdistant, broad, at times inter-veined. Whitish to pale gray, staining or discoloring yellowish to grayish in age. **Stipe:** 4–10 (14) cm long, 0.2–0-6 cm thick at apex, widening towards base, cylindrical to flattened. Surface dry, smooth at apex; base densely covered with silky hairs. Whitish to brownish, darkening from base up, becoming reddish brown in age. Basal hairs bright white when young, discoloring with age. **Flesh:** Thin, cartilaginous; stipe hollow. Watery gray. **Odor:** Mild to pungent, often yeast-like. **Taste:** Mild to sharp. **Spore deposit:** White. **Microscopy:** Spores 6–8 (8.5) × 3.5–4 µm, narrowly ellipsoid, teardrop-shaped to narrowly ovoid, smooth, thin-walled, amyloid. Cheilocystidia scattered, mixed with basidia, somewhat cylindrical to narrowly spindle-shaped. Pleurocystidia rare to absent.

Ecology: Typically in large clusters, more rarely in small tufts, or solitary from conifer stumps or large logs at the edges of melting snowbanks. It prefers old-growth forests, or legacy stumps of logged large old-growth trees. Locally common in higher elevation forests on the east side of the Cascade Range in the north portion of Cascadia, and across higher elevations in the southern portion of Cascadia. Fruiting as soon as snowmelt exposes stumps and logs in spring into early summer.

Edibility: Unknown.

Comments: Recognized by the often-clustered growth on large pieces of wood near snowbanks, the dark gray to grayish brown to whitish cap, white to yellowish gills, and stipe with a prominently fuzzy white base. The large size and cap shape is not reminiscent of *Mycena* at first glance. *M. hudsoniana* is another snowbank *Mycena* in Cascadia which grows on duff or woody debris. It has a conical to convex, blackish to dark gray cap, grayish gills, and club-shaped to obpyriform (like an upside-down pear) cheilocystidia covered with bumpy to spiny outgrowths. *M. maculata* fruits in clusters on decaying conifer logs and stumps in fall, a gray cap, gills, and stipe when young, and soon develops red stains as it ages. The stipe has sparse to dense white hairs at the base; however, it lacks the woolly-to-mentose lower stipe of *M. overholtsii*. *Clitocybula abundans* fruits in clusters, has a broadly convex to plane, whitish to pale grayish tan cap, and broadly ellipsoid, smooth, amyloid spores measuring 4.5–6 × 3.5–5 µm. Although it can fruit in spring, these usually appear well after snowmelt.

Mycena capillaripes

Peck

Bean Sprout Bonnet

Cap: 0.5–2.5 cm across, conical at first, becoming broadly conical to bell-shaped, occasionally umbonate; margin sometimes uplifted in age. Surface moist to dry, smooth; translucent-striate and strongly hygrophanous, with darker lines when wet, duller and much paler when dry. When wet, dark gray to dark grayish brown, fading to pale gray to pale grayish brown; grayish buff to pale pinkish gray or grayish white when dry. **Gills:** Attached, notched, often with a decurrent tooth, close to subdistant, narrow. Whitish, pale grayish white to pale pinkish buff. Marginate, edges darker, although sometimes only faintly so, pinkish brown to pale reddish brown. **Stipe:** 2–8 cm long, 0.05–0.25 cm thick, fragile, equal or slightly enlarged toward base. Surface smooth, dry, base with whitish to grayish fibrils or hairs. Light gray to grayish brown and often developing pinkish tones on lower portions, paler, whitish to pale gray toward apex. **Flesh:** Very thin, fragile, pallid, grayish. **Odor:** Like chlorine or bean sprouts, especially when crushed. **Taste:** Radishy to unpleasant. **Spore deposit:** White. **Microscopy:** Spores 8.5–11 (12.5) × 4.5–6.5 (7) µm, ellipsoid, smooth, thin-walled, amyloid. Cheilocystidia and pleurocystidia typically abundant, broadly to narrowly spindle-shaped, with pinkish brown content.

Ecology: Scattered or in small clusters forming large troops in duff and needle litter under pines and other conifers. Very common, especially in lower elevation and coastal pine forests, occasional in east-side pine forests. Often fruiting in huge numbers soon after the first fall rains in fall, with scattered fruitings into winter.

Edibility: Unknown.

Comments: The gray to grayish brown, translucent-striate cap that fades in age or when dry, whitish to pale gray gills that have a faint darker pinkish brown edge (use a hand lens to check), chlorine or bean sprout–like odor, and growth in conifer duff are distinctive. There are many similar gray *Mycena* in Cascadia but few have darker gill edges. Older specimens could be mistaken for faded *M. rosella*, which has pinkish gills with darker rosy-pink edges. *M. rubromarginata* also has a gray cap but is faintly translucent-striate, has darker vinaceous brown gill edges, and lacks an odor. *M. citrinomarginata* has an extremely variable cap color (sometimes overlapping with the range shown by *M. capillaripes*) but is usually more yellowish olive and has gills that are whitish to pale grayish with darker yellow to yellowish brown edge.

Mycena citrinomarginata

Gillet

Yellow-edged Bonnet

Cap: 0.5–3 cm across, conical to bell-shaped, margin often flaring in age. Surface moist to dry, finely powdery when young but soon smooth, hygrophanous, translucent-striate when wet, paler and opaque when dry. Color variable: Typically pale yellow, yellow-gray, olivaceous yellow to olive-brown, but at time extensively brighter yellow, dark to light gray or almost white. **Gills:** Narrowly attached, notched, subdistant, narrow. Whitish, yellowish gray to pale gray; edges pale to bright yellow or yellow-brown. **Stipe:** 2–8 cm long, 0.1–0.2 cm thick, equal or with slightly enlarged base. Surface moist to dry, smooth. Grayish, yellowish gray to olive-gray; apex paler and often more yellow. **Flesh:** Very thin, fragile, grayish. **Odor:** Indistinct. **Taste:** Mild. **Spore deposit:** White. **Microscopy:** Spores 8–14 × 4–6 µm, narrowly ellipsoid, smooth, thin-walled, amyloid. Cheilocystidia abundant, tapered, spindle- to club-shaped, smooth or with finger-like projections on one side, yellow. Pleurocystidia absent or occasional, similar in shape to cheilocystidia.

Ecology: Solitary, scattered, or in troops from duff or grassy lawns. Common and often abundantly, fruiting by the thousands from pine duff in east-side forests but seems to prefer grassy areas on the west side. Fruiting in fall.

Edibility: Unknown.

Comments: The variably colored cap, often with pale yellow to olive-gray tones, yellowish gray gills, with faintly to distinctly colored edges, and growth in grass or duff help identify this species. If the coloration of the gill edge is overlooked, it can blend into the plethora of other grayish *Mycena.* The name *M. olivaceobrunnea* has been applied to darker capped collections with smaller spores (7.5–10 × 4.5–5.5 µm) but many consider these to be synonymous. *M. capillaripes* has a dark to light gray cap that sometimes shows pink tones, a pinkish gill edge, and an odor of chlorine or bean sprouts when crushed.

Mycena filopes

(Bull.) P. Kumm.

Iodine Bonnet

Cap: 0.5–2 cm across, egg-shaped to rounded at first, becoming conical to bell-shaped. Margin with overlapping sterile tissue, often resembling a partial veil when young, becoming jagged as cap expands or disappearing altogether in age. Surface with a powdery whitish to pale grayish bloom when young, becoming smooth, moist to dry, translucent-striate near margin when old and wet. Grayish brown to gray, at margin paler when young, fading to light gray, pale grayish brown or grayish white. **Gills:** Narrowly attached, notched, occasionally with a decurrent tooth, close, narrow. Whitish to pale grayish white. **Stipe:** 2–12 (15) cm long, 0.1–0.2 cm thick, equal or enlarged slightly toward base, often tall and slender. Surface finely powdery at first, becoming smooth, moist to dry. Base covered with fine to coarse, long whitish hairs. Whitish to pale gray at apex, darker, gray, dark gray to grayish brown lower. **Flesh:** Very thin, fragile, concolorous with cap; stipe hollow, gray. **Odor:** Iodine-like. **Taste:** Mild. **Spore deposit:** White. **Microscopy:** Spores 10.5–11 (12) × 5.5–6.5 µm from 2-spored basidia, 8–10.5 × 4.5–6 from 4-spored basidia, ellipsoid, smooth, thin-walled, amyloid. Cheilocystidia abundant, irregularly cylindrical, club-shaped to swollen, with infrequent finger-like projections. Pleurocystidia rare to absent.

Ecology: Scattered or in loose clusters or troops on branches, twigs, woody debris, and thick duff of both conifers and hardwoods. Common across western Cascadia. Fruiting in fall and early winter.

Edibility: Unknown.

Comments: The grayish brown cap that often fades to whitish, overhanging tissue at the cap margin, gray stipe which becomes quite long, coarse whitish hairs at the stipe base, and iodine odor are distinctive. *M. metata* looks very similar to *M. filopes* but lacks the sterile tissue around the margin of the cap, its gills often turn pinkish in age, and it typically grows on conifer duff. *M. vitilis* has a clay-brown cap which fades to white and has a similar leggy stature, but lacks any odor, doesn't have sterile tissue around the cap margin, and is common on alder debris.

Mycena clavicularis

(Fr.) Gillet

Slimy Needle Bonnet

Cap: 1–2 cm across, rounded to convex at first, soon broadly convex to plane; margin sometimes uplifted in age. Surface dry to moist, faintly translucent-striate when wet. Dark to light gray to gray-brown; margin slightly paler. **Gills:** Broadly attached or notched with a decurrent tooth, close to subdistant, narrow. White to pale grayish. **Stipe:** 3–6 cm long, 0.1–0.2 cm thick, equal or enlarged toward base. Surface glutinous to viscid, smooth; base with fine whitish to grayish hairs. Gray to gray-brown, paler with age. **Flesh:** Very thin, fragile, grayish; stipe hollow, dark grayish brown. **Odor:** Indistinct. **Taste:** Mild. **Spore deposit:** White. **Microscopy:** Spores 7–8.5 × 3–4 µm, narrowly ellipsoid, smooth, thin-walled, amyloid. Cheilocystidia and pleurocystidia scattered to abundant, broadly club-shaped with short irregular projections.

Ecology: Solitary, scattered, or in small clusters. Often forming large troops carpeting the ground in moss, duff, and needle litter. Locally common in western Cascadia, especially in Sitka Spruce and Douglas-fir forests, not known from the east side. Fruiting soon after the first rains, with stragglers into late fall.

Edibility: Unknown.

Comments: The dry to moist gray cap, white gills, viscid stipe, and indistinct odor and taste help identify this from amongst a difficult to distinguish group of gray *Mycena*. *M. tenax* is quite similar and will often grow in the same coastal habitats—it differs in having a distinctly viscid to moist cap, whitish gills that often develop brownish to reddish brown spots in age, and a farinaceous odor and taste. *M. vulgaris* is more common at mid to higher elevations and in east-side forests, it differs by having decurrent gills and lacking pleurocystidia. *Roridomyces roridus* is generally smaller, has a rounded-convex grayish to whitish cap, white decurrent gills, and a white stipe coated with a thick layer of slime. '*Mycopan*' *quiniaultensis* is typically larger, a conical to bell-shaped cap with a darker center, and a viscid gray stipe.

Mycena abramsii

(Murrill) Murrill

Cap: 1–3 cm across, rounded to somewhat conical at first, becoming broadly conical to bell-shaped, occasionally umbonate. Surface moist to dry, smooth; translucent-striate, with darker lines when wet, strongly hygrophanous, dull, much paler when dry. Dark to light gray. **Gills:** Attached, notched, often with a decurrent tooth, close to subdistant, narrow. Whitish to pale grayish. **Stipe:** 4–9 cm long, 0.1–0.25 cm thick, very fragile, equal or slightly enlarged toward base. Surface smooth, dry, base with whitish to grayish fibrils or hairs. Whitish to light gray over upper portion, to grayish brown below. **Flesh:** Very thin, fragile; stipe hollow, watery. Grayish. **Odor:** Indistinct or slightly bleachy. **Taste:** Indistinct to slightly. **Spore deposit:** White. **Microscopy:** Spores 9–12 × 4.5–6 µm, teardrop-shaped to elongate, smooth, amyloid.

Ecology: Scattered or in pairs or tufts on decaying wood; typically on alder and bigleaf maple, but also on conifers. Common on the coast and in the Coast Range (at least in Oregon), occasional at lower elevations in Washington. Fruiting soon after the first soaking rains in fall.

Edibility: Unknown.

Comments: One of many gray *Mycena* in Cascadia, we use this name loosely to define this wood dwelling bonnet. It is recognized by the smooth gray cap and stipe, pallid gills, and an indistinct or faint bleachy odor. The name *M. alnicola* likely refers to the same species. *M. leptocephala* tends to be darker, has a distinct bleach-like odor, and is as likely to be found in duff as it is on woody debris. Also compare with *M. filopes* (and kin), which tends to be taller-statured, and has an iodine odor. *M. stylobates* is a small, grayish to whitish species with a basal disc on the stipe, and fine hairs on the cap; it grows on branches. *M. tenerrima* is even smaller, with a slightly bulbous stipe base and a granular cap texture.

Mycena leptocephala

(Pers.) Gillet

Dark Bleachy Bonnet

Cap: 0.5–3.5 cm across, egg-shaped to conical at first, becoming broadly conical to bell-shaped. Surface with a powdery bloom when young and dry, becoming smooth; translucent-striate when wet. Dark gray, dark grayish brown to nearly black at first, fading slightly, gray to light grayish brown but retaining the dark striations. Extreme margin often paler, whitish to pale gray. **Gills:** Narrowly attached, notched, occasionally with a decurrent tooth, close to subdistant. Gray, grayish to whitish; edges often paler. **Stipe:** 2–8 cm long, 0.1–0.3 cm thick, fragile, equal, or enlarged slightly toward base. Surface finely powdery at first, becoming smooth, moist to dry. Base tomentose, covered with fine to coarse whitish hairs. Gray to dark grayish brown, fading slightly in age. **Flesh:** Very thin, fragile, concolorous with cap; stipe hollow, dark gray. **Odor:** Usually strongly chlorine-like when crushed, becoming fainter in age. **Taste:** Indistinct. **Spore deposit:** White. **Microscopy:** Spores 7.5–12 × 5–6 µm, ellipsoid to nearly teardrop-shaped, smooth, thin-walled, amyloid. Cheilocystidia abundant, spindle- to club-shaped, or with a swollen base and elongated neck. Pleurocystidia similar, scattered to absent.

Ecology: Scattered or clustered in small to large troops in an impressive array of habitats including wood chips, soil, duff, woody debris, or on bark of both living and dead trees. Common throughout Cascadia, especially in urban areas. Primarily fruiting in fall and early winter but can be found during wet periods throughout the year.

Edibility: Unknown.

Comments: This species is distinguished by the dark blackish to grayish brown cap (retaining dark striations even after fading), grayish gills, and strong chlorine odor. We have seen large fruitings in urban areas landscaped with wood chips and in garden soil. Other species with a chlorine odor in Cascadia include *M. alnicola* with a gray to grayish brown cap which fades to pale gray in age, white to pale gray gills, and growth on woody debris, especially alder. *M. abramsii* is very similar but generally has a pale bluish to bluish gray powdery bloom when young and is more often found on conifer branches. *M. silvae-nigrae* has a gray-brown cap, larger spores (10.5–15 × 7–10 µm), and cheilocystidia that are variably club-shaped (often with very broad heads) to spindle-shaped. Also compare the usually paler *M. capillaripes.*

Mycena amicta

(Fr.) Quél.

Blue Bonnet

Cap: 0.4–1.5 cm across, egg-shaped to conical at first, becoming broadly conical to bell-shaped, often with a rounded translucent or brownish gray umbo, margin occasionally upturned in age. Surface slightly powdered at first, becoming smooth, moist to slightly viscid. Color variable, extensively bluish, turquoise-blue to bluish gray, soon fading to gray or grayish brown, or even pale gray or whitish in age. **Gills:** Narrowly attached, notched, occasionally with a decurrent tooth, often breaking away from stipe in age, close to subdistant, narrow. Whitish, pale grayish to pale grayish brown. **Stipe:** 3–10 cm long, 0.1–0.2 cm thick, equal, often curved and rooting at base. Surface often covered with fine, short whitish hairs, giving it a powdered appearance, becoming smooth, dry. Pale gray to grayish brown, base blue to blue-gray and often covered with fine to coarse whitish to bluish hairs. **Flesh:** Very thin, fragile, concolorous with cap; cap skin often separable as a gelatinous layer. Stipe hollow, grayish. **Odor:** Indistinct. **Taste:** Mild. **Spore deposit:** White. **Microscopy:** Spores 7–9 × 4–5 µm, ellipsoid, smooth, thin-walled, amyloid. Cheilocystidia narrowly spindle-shaped to somewhat cylindrical. Pleurocystidia absent.

Ecology: Solitary or scattered in small groups in duff, on well-rotted branches, or other woody debris of both conifers and hardwoods. Widespread and fairly common, rarely fruiting in large numbers, and easily overlooked or confused with any of the other myriad gray *Mycena* after its bluish colors fade. Fruiting during any wet period throughout the year.

Edibility: Unknown.

Comments: When this *Mycena* is young, the blue coloration renders it practically unmistakable, however, it fades quickly as it matures and the blue often remains only at the very base of the stipe in age. Other distinguishing features include the easily separable gelatinous cap cuticle, often translucent umbo, and fine whitish fibrils covering the young stipes. Genetic data suggest that multiple species go by this name in Cascadia, including one which grows on the edges of melting snowbanks in the east-side/southern Cascades.

'Mycopan' quiniaultensis

comb. prov.

Cap: 1–4 cm across, conical to bell-shaped when young, becoming broadly bell-shaped to convex, umbonate at all stages. Margin down-curved when young, uplifted, wavy and often splitting in age. Surface glutinous to viscid, smooth, translucent-striate from margin to disc. Dark gray to gray-brown with paler grayish margin when young, slightly lighter and developing brownish gray tones in age. **Gills:** Broadly attached or notched with a decurrent tooth, subdistant, narrow, with short partial gills and inter-veined. White to grayish white. **Stipe:** 4–7 cm long, 0.1–0.4 cm thick, equal or slightly enlarged toward base. Surface glutinous to slightly viscid, smooth; base with whitish to yellowish orange hairs. Watery gray to brownish gray. **Odor:** Indistinct. **Taste:** Mild. **Spore deposit:** White. **Microscopy:** Spores 6.5–8 × 3–3.5 µm, narrowly ellipsoid, smooth, amyloid. Cheilocystidia and pleurocystidia abundant, long, cylindrical or with a slightly wider base.

Ecology: Solitary or scattered in small troops in duff or moss under conifers, mostly in old growth forests. Uncommon, known from lower elevations in western Cascadia. Fruiting in fall.

Edibility: Unknown.

Comments: The viscid, deeply translucent-striate, gray-brown, bell-shaped cap, whitish gills, and viscid gray stipe are helpful identifying features of this uncommon inhabitant of low-elevation old growth forests. An undescribed species of *Mycopan* looks quite similar but differs in having a slightly viscid to moist cap, moist to dry stipe, and growth on small hardwood branches and other debris. *Gliophorus irrigatus* differs by its stockier stature, broad, thick-edged waxy gills, and inamyloid spores. *Mycena tenax* shares the gray-brown colors but lacks a distinct umbo, has white gills which develop reddish brown stains in age, and has a rancid-farinaceous odor and taste.

Synonym: *Mycena quiniaultensis* Kauffman

15 - Pleurotoid Mushrooms

These mushrooms all share a similar stature—growing sideways from decaying wood or, in a few cases, moss or soil. They tend to have short, off-center stipes (or no stipe at all) and all are saprobic (although *Hohenbuehelia* and *Pleurotus* gain additional nutrients by preying on nematodes!). Despite similarities in stature, they show a wide range of sizes, textures, and spore colors. The genera with white spore deposits are *Arrhenia, Cheimonophyllum, Hohenbuehelia, Lentinellus, Metacampanella, Muscinupta, Panellus, Phyllotopsis, Pleurocybella, Pleurotus, Rimbachia, Sarcomyxa, Schizophyllum,* and *Scytinotus*. As expected for members of Entolomataceae, *Claudopus* and *Clitopilus* have pinkish spore deposits, their gills can remain pale for a long time, leading to confusion with the previously mentioned white-spored genera. The only pleurotoid *Deconica* included here has purplish-black spores. *Crepidotus* and *Tapinella* have dingy brown spore deposits and usually have brownish gills to match.

Important identifying features include substrate (what kind of wood and what state of decay), presence or absence of a stipe, structure and attachment of the gills, gill color, and texture of the fruitbody. The color of the spore deposit goes a long way in sorting out the groups in this chapter!

Pleurocybella porrigens

sensu PNW

Western Angel Wings

Cap: 2–10 cm wide, kidney-, shoehorn- or fan-shaped when young, becoming irregularly lobed, wavy, and uplifted in age. Margin thin, slightly inrolled when young, undulating and ruffly in age. Surface very smooth, moist to dry. Usually distinctly white to pearly-ivory, sometimes with light creamy or pale pinkish tones; often so thin as to be translucent. **Gills:** Deeply decurrent, often to attachment point, close to crowded, narrow. White to creamy. **Stipe:** Absent. **Flesh:** Very thin, fleshy-rubbery. Whitish. **Odor:** Indistinct. **Taste:** Indistinct. **Spore deposit:** White. **Microscopy:** Spores 5–7.5 × 4–6 µm, subglobose, smooth, inamyloid, colorless in KOH.

Ecology: Fruiting in small clusters or large troops on well-decayed conifer wood, perhaps most often on western hemlock. Common throughout Cascadia, fruiting in fall.

Edibility: Not recommended. Although many people have eaten this mushroom with no ill effects, a spate of deaths in Japan were attributed to ingestion of a closely related species. The presence of high levels of cytotoxic amino acids and cyanide-rich compounds in that species suggest our local Angel Wings are best avoided, although it probably does not pose an acute toxicity risk for most people.

Comments: The thin, often wavy, delicate-looking fruitbodies, shining white coloration, and growth on conifer wood distinguish this species. Distinguishing Angel Wings from the 'true' Oyster Mushrooms (*Pleurotus* spp.) is tricky, but the latter are usually larger, have thicker flesh, often show beige, brown, tan, or gray colors (not as strikingly white), and are found on a wide variety of wood—more often hardwoods than conifers. *Scytinotus longinquus* usually shows some pink coloration, and has smaller, less wavy fruitbodies. Some *Crepidotus* are also white but are less wavy and uplifted, and have brown spore deposits.

Taxonomic note: *Pleurocybella porrigens* was described from Europe; western North American Angel Wings are a distinct species and need a new name.

Pleurotoid

Pleurotus ostreatus group

Oyster Mushrooms

Cap: 3–15 (25) cm across, semicircular to fan-shaped with an inrolled margin when young, expanding to plane or uplifted in age. Surface moist to dry, smooth. White to pearly gray when fresh, at times buff to dark bluish gray or evenly brown. Usually becoming paler, off-white to chalky white when dry, and with yellow discoloration in age. **Gills:** Decurrent, close to subdistant, broad. White, cream, to pinkish buff, yellowing in age. **Stipe:** 0.5–3 (6) cm long, 0.3–2 (4) cm thick, laterally attached, often stout, more or less equal. Surface dry, smooth, with a hairy base. White to creamy, yellowing in age. **Flesh:** Thin to thick, stringy-fibrous, whitish. **Odor:** Distinct, an anise-like sweetness and mushroomy-earthy tones. **Taste:** Mild. **Spore deposit:** White, buff to lilac-gray. **Microscopy:** Spores 7.5–9.5 × 3–4 µm, narrowly ellipsoid, smooth, inamyloid.

Ecology: In overlapping clusters and troops on dead wood. Most are found on hardwoods, especially alder—but are also found on conifers. Common throughout Cascadia, with two main fruitings (spring and fall), but smaller numbers occur throughout the year whenever there is sufficient moisture.

Edibility: Edible and very good, with a robust texture. Some collections are more fragrant and flavorful than others.

Comments: These ubiquitous mushrooms are easily recognized by their fan-shaped, white to gray to brown caps, whitish decurrent gills, short, off-center stipes, thick rubbery flesh, and growth on wood. However, the common name "Oyster Mushroom" obscures the fact that there are multiple rather similar-looking species in our area. Although recently the use of *P. pulmonarius* has become popular, it is unclear how that species can be reliably distinguished from *P. ostreatus*, or whether either name is appropriate for the commonly encountered taxa in Cascadia. The larger, browner Oyster Mushrooms that often form 'pancake stacks' on decaying cottonwoods are usually referred to as *P. populinus*. Another distinctive but unnamed species occurs on conifer wood in wet and montane areas of Cascadia; its fruitbodies are often quite large (>20 cm wide), have brown caps with patches of dusty gray bloom, and a fine spiky texture covering the often-ridged lower stipe. Ultimately, a clear and comprehensive understanding of Oyster Mushrooms in the Western USA will require broad sequencing of carefully documented collections, mating studies, and phylogenies including specimens from around the globe. *P. dryinus* has a partial veil, a central to off-center stipe, and a grayish, often patchy or felt-textured cap surface. *Hohenbuehelia petaloides* is smaller and rubbery-fleshed, with ocher-brown caps and uplifted, wavy, shoehorn-shaped fruitbodies. *Pleurocybella porrigens* is smaller, with thin flesh, and wavy or lobed, uplifted pure white caps; they grow on decaying conifer wood.

Hohenbuehelia petaloides group

Shoehorn Oysters

Cap: 2–7 (10) cm across, in fused clusters up to 50 cm across, shoehorn to funnel shaped, becoming more fan shaped in age, margin tightly inrolled when young, often staying downcurved in age. Surface smooth, slightly viscid when wet, otherwise lubricious or moist, often with a white bloom when young. Brown, grayish brown to gray at first, fading, whitish gray to whitish buff in age. **Gills:** Decurrent, very narrow near attachment, close to crowded, becoming wavy. White, pale grayish white to creamy, at times with a yellowish cast in age. **Stipe:** 1–4 cm long, 0.5–3 cm thick, off-centered to lateral, stout, narrowing toward base. Surface dry, covered with whitish hairs. Whitish to grayish buff. **Flesh:** Moderately thin to thick, rather tough, rubbery and with a gelatinized layer when wet. Watery whitish to gray. **Odor:** Indistinct to farinaceous. **Taste:** Mild to slightly farinaceous. **KOH:** No reaction. **Spore deposit:** Whitish. **Microscopy:** Spores 5–6.5 × 3.5–5.5 µm, ellipsoid, smooth, inamyloid, colorless in KOH. Pleurocystidia spindle shaped, thick walled, with incrusted crystals over tip. Cheilocystidia similar, or more cylindrical and thin-walled.

Ecology: In dense, overlapping clusters when growing on wood chips; solitary or in small clusters on decaying hardwood logs and stumps. Generally scarce, but can be locally common in urban areas. Fruiting mostly in fall, but also opportunistically whenever there is enough moisture.

Edibility: Edible, but chewy and bland.

Comments: Fruitbodies are easily recognized by their uplifted and grooved shape, brownish color, and rubbery texture. Although the name *H. petaloides* has been widely applied in western North America, recent sequence data has shown *H. thornii* and *H. tremula* in Cascadia. They are virtually identical to *H. petaloides*, and may require a DNA sequence to distinguish. *Pleurotus ostreatus* are larger, thicker-fleshed, and have less gelatinized cap cuticles.

Schizophyllum commune

Fr.

Split Gill

Cap: 1–4 (5) cm across, fan shaped, margin inrolled when young, becoming lobed and wavy and often ragged in age. Surface coarsely hairy, dry. Pinkish buff, light grayish brown, grayish to whitish. **Gills:** Radiating out from attachment point, consisting of closely spaced, "split" or grooved gills that open up in wet weather. Grayish to pale pinkish-tan or beige. **Stipe:** Absent; fruitbodies are laterally attached to substrate. **Flesh:** Thin, tough, leathery. **Odor:** Indistinct. **Taste:** Mild. **Spore deposit:** Whitish. **Microscopy:** Spores 3–4 × 1–1.5 µm, cylindrical, smooth, colorless in KOH.

Ecology: In overlapping, shelving clusters, in small groups, or occasionally solitary on a wide variety of woody debris. Very common throughout Cascadia. Fruiting in wet weather but the fruitbodies can dry out and rehydrate multiple times and, thus, persist throughout the year.

Edibility: Edible but small and rather tough.

Comments: The small size, pale color, and fuzzy margin are similar to many pleurotoid wood rotters but the blunt-edged, "split" gills are diagnostic. These grooved gills fold closed during dry weather and open back up to produce spores when conditions are favorable. This species is remarkable for many reasons, not least of which is that it known to have more than 23,000 mating types (the fungal equivalent of sexes)!

Lentinellus montanus

(Bolton) S. Ito

Snowmelt Sawgill

Cap: 2–8 cm wide, broadly kidney- to shell- or fan-shaped, becoming flatter and somewhat wavy in age. Surface finely hairy, dry. Brown to beige or tan towards margin, paler overall in age or when dry. Center of cap frosty white, with pale hairs concentrated near attachment to substrate. **Gills:** Fairly widely spaced; pale beige to creamy tan, edges paler to whitish and strongly serrate. **Stipe:** Absent, attachment point often with a tan tomentum, and thick rhizomorphs. **Flesh:** Somewhat rubbery, fleshy. Pale beige to tan. **Odor:** Not distinctive. **Taste:** Spicy-acrid. **Spore deposit:** White. **Microscopy:** Spores 5.0–6.5 × 4.0–5.0 µm, ovoid to subglobose, slightly spiny, amyloid.

Ecology: Solitary or in small groups on well-decayed conifer wood at higher elevations at the edges of melting snowbanks. Widespread and common in areas with dense winter snowpack, fruiting in spring and summer.

Edibility: Not considered edible due to the acrid taste.

Comments: The fan shaped ocher-tan cap, serrated gill edges, lack of a stipe, medium size, and habit of fruiting near melting snow identify this species. *L. ursinus* also lacks a stipe but has tougher flesh and grows in more coastal settings. *L. vulpinus* grows in dense clusters of overlapping caps. Members of the *L. flabelliformis* group are smaller and grow in fall in lower elevations. *Neolentinus kauffmanii* has a central to slightly off-centered stipe and grows on spruce logs and stumps.

Lentinellus ursinus group

(Fr.) Kühner

Bear Sawgill

Cap: 2–10 cm across, fan shaped, semicircular, to kidney shaped, soon irregular lobed. Surface moist to dry, coarsely hairy to velvety over much of cap with a smooth margin. Often brown to reddish brown over disc, paler light brown to cinnamon buff on margin, darkening in age; hygrophanous, fading overall when dry. **Gills:** Decurrent, close to crowded, edges saw-toothed and often ragged. Whitish to buff, developing cinnamon color in age. **Stipe:** Laterally attached to wood without a stipe, or with a short indistinct stipe. **Flesh:** Concolorous, thin, moderately tough, and often fibrous. **Odor:** Mild, like black pepper, or slightly fragrant. **Taste:** Acrid and often bitter. **KOH:** No reaction. **Spore deposit:** White. **Microscopy:** Spores 4–4.5 × 3–3.5 µm, subglobose to broadly ovoid, roughened with small warts, amyloid.

Ecology: Often clustered in overlapping shelves, scattered, or rarely solitary on decaying logs or dead trees. Often on hardwoods, also occurring on conifers. Occasional across Cascadia, fruiting in fall.

Edibility: Inedible.

Comments: The pleurotoid shape, velvety to coarsely hairy cap, strongly serrate gills, white spores, and growth on wood are distinctive. *L. castoreus* is generally slightly smaller, paler, typically mild tasting, and has more crowded gills. (Note that both are European names and may be misapplied here; it is also likely that we have additional undescribed taxa.) *L. vulpinus* forms fused clusters from a common base, grows on hardwoods, and has smaller spores. *L. montanus* grows on conifer branches near melting snowbanks in the mountains and has a light brown to pinkish buff cap with a felty disc. *L. flabelliformis* (group) is smaller and has a smooth cap. Other similarly-shaped mushrooms growing on wood lack the jagged gill edges.

Sarcomyxa serotina

(Pers.) P. Karst.

Green Oyster

Cap: 3–12 (15) cm across, fan shaped, semicircular, to kidney shaped, margin inrolled, becoming wavy and often lobed in age. Surface viscid to moist, shiny when dry, more or less smooth, but often with a velvety patch near attachment point, occasionally finely hairy and with scales near disc when dry. Olive-green to bluish gray or olivaceous gray, often developing yellow, ocher or brownish tones as it ages. Rarely ocher even when young. **Gills:** Radiating out from a broad, curb-like stipe, close to crowded, occasionally forking, partial gills numerous. Orange-buff to orange at first, fading; yellowish buff to beige in age. **Stipe:** Laterally attached to wood, indistinct from top, and a short, curb-like attachment on underside. **Flesh:** Moderately thick to thin near margin, firm, rubbery. Watery whitish. **Odor:** Often indistinct, or slightly earthy. **Taste:** Mild to bitter-metallic. **Spore deposit:** Creamy yellowish. **Microscopy:** Spores 3.5–6.5 × 1–2 µm, cylindrical to sausage shaped, smooth, amyloid. Cheilocystidia abundant, club to spindle shaped, pleurocystidia similar, sparse.

Ecology: Scattered in overlapping or solitary shelves, occasionally in clusters, on decaying logs or dead trees, almost always on red alder or bigleaf maple. Seems to fruit more abundantly in cold weather but scattered fruitings happen throughout the rainy season.

Edibility: Edible but not highly regarded.

Comments: The pleurotoid-shape, olive cap, and crowded orangish gills with a curb-like stipe are distinctive. *Phyllotopsis nidulans* has a fuzzy cap, brighter orange gills, and often a disagreeable odor. *Tapinella panuoides* has a dry cap, crimped, often cross-veined gills, and brown spores. *Hohenbuehelia petaloides* is smaller and thinner, with a gray cap and white gills.

Synonym: *Panellus serotinus* (Pers.) Kühner. It is more closely related to the Hygrophoraceae than to true *Panellus* (*P. stipticus* and allies), which are related to *Mycena*.

Phyllotopsis nidulans

(Pers.) Singer

Orange Mock Oyster

Cap: 2–8 cm across, fan shaped, semicircular, to kidney shaped, margin inrolled, often into maturity, often wavy and irregular lobed. Surface moist to dry, covered with finely pubescent hairs, which can become matted or occasionally disappear in age. Orange, but covered with whitish hairs, and appearing paler, occasionally all orange, or orangish buff in age. **Gills:** Radiating out from attachment point, close to crowded, partial gills numerous. Orange-buff to orange at first, fading slightly, to yellow-orange. **Stipe:** Laterally attached to wood without a stipe, often with white velvety fuzz at attachment point. **Flesh:** Concolorous, thin, moderately tough and rubbery. **Odor:** From mild and indistinct to strong, rancid and disagreeable; like rotting cabbage, swamp gas, or mouse urine. **Taste:** Pungent, unpleasant, occasionally bitter. **KOH:** No reaction. **Spore deposit:** Pinkish orange to creamy peach. **Microscopy:** Spores 5–7 × 2–3.5 µm, cylindrical to bead shaped, smooth, inamyloid.

Ecology: Often clustered in overlapping shelves, scattered, or solitary on decaying logs or standing snags of hardwoods and conifers. Occasional throughout Cascadia, with pronounced fruiting pulses in fall and spring.

Edibility: Inedible.

Comments: The pleurotoid shape, velvety cap (at least when young), and orange colors are distinctive of this common "mock oyster." The odor is often downright repulsive (but can be nearly absent in older specimens). *Crepidotus crocophyllus* also has a hairy cap and orangey gills (from light orangish beige to rather bright orange), but differs by its lack of a distinct odor, smaller fruitbodies, and brown spores. *Sarcomyxa serotina* has a mostly smooth cap with olive tones and yellowish to orange gills with a short, wide, wedge-like stipe. *Tapinella panuoides* has crimped, often cross-veined gills, and brown spores. *Lentinellus* has jagged, sawtooth gill edges and generally lacks the orange colors. *Gloeophyllum sepiarium* is tough and leathery, with thick-edged distant gills.

Scytinotus longinquus

(Berk.) Thorn

Pink Alder Oyster

Cap: 1–5 cm across, kidney-shaped or fan-shaped, flat in age, margin thin, even when young but often finely scalloped, slightly irregular or wavy in age. Surface smooth and viscid with a gelatinous layer, translucent-striate, hygrophanous, more opaque when dry. Color variable: Pastel pink to coral pink to white, often paler towards near the attachment point. Fading pinkish-white, dingy white, or with tan tones in age. **Gills:** Broadly attached or slightly decurrent, moderate to widely spaced, with a few short gills present, narrow, edges even, Whitish to pale pink or pinkish-cream. **Stipe:** 0.2–1 (1.5) cm long, 0.1–0.5 cm thick. Short to nearly absent, often wider near attachment to gills and slightly tapered into substrate. **Flesh:** Thin, slightly gelatinous-rubbery, pale to buff. **Odor:** Indistinct. **Taste:** Indistinct. **Spore deposit:** Whitish to cream. **Microscopy:** Spores 6-11.5 × 3-5 µm, quite variable in shape, ranging from ellipsoid to elongate and sometimes constricted near midpoint, smooth, amyloid.

Ecology: In troops on dead wood, primarily of alder and bigleaf maple. Common throughout northern Cascadia from the coast to the slopes of the Cascades. Less common southward and apparently absent south of Coos Bay. Fruiting from fall into spring.

Edibility: Non-toxic; not sought out for food.

Comments: The pink colors, viscid surface, and small stipe make this species easy to recognize. *Panellus mitis* is very similar but brighter white and grows on conifer branches. *Crepidotus* are typically laterally attached, and stipe-less, dry, often fibrillose, lack pink coloration, and have brown spores. *Sarcomyxa serotina* is quite similar in some respects but is substantially larger, fleshier, not translucent-striate, and comes in a range of greens, oranges, and blue-gray colors, —but not pink. *Pleurotus* often have more decurrent gills and more well-developed stipes. Those which are stipeless are larger, fleshier, and white, gray or brownish. *Scytinotus ringens* is quite similar but darker brick red when young with dry, opaque caps. It grows on decaying wood (particularly birch).

Panellus mitis

(Berk. & M. A. Curtis) Singer

Elastic Oysterling

Cap: 0.3–2 cm across, fan-shaped with a distinctly inrolled margin at first, expanding and becoming flatter. Surface dry to gelatinous, smooth. White to grayish-white or creamy, sometimes developing tan-brownish tones in age. **Gills:** Radiating out from stipe or attachment point, close, fairly thick-edged. White at first, becoming cream colored. **Stipe:** 0.1–0.5 cm long, 0.1–0.3 cm wide; a small lateral nub with whitish scurfs, sometimes entirely absent. **Flesh:** Thin, slightly rubbery, whitish to pale grayish. **Odor:** Indistinct **Taste:** Indistinct. **KOH:** No reaction. **Spore deposit:** White. **Microscopy:** Spores 3.5–6.5 × 0.9–1.2 µm, sausage-shaped, smooth, amyloid. Cheilocystidia highly irregular in outline, constricted and knobbed.

Ecology: Usually in troops, rarely solitary on decaying logs and branches of conifers. Fairly common but rarely reported from Cascadia. Fruiting in fall and winter, well into cold-weather periods after frost.

Edibility: Unknown.

Comments: This small pleurotoid mushroom is recognized by its white to grayish colors, inrolled margin when young, rubbery-gelatinized cap flesh, and reduced stipe. *Cheimonophyllum candidissimum* has thinner flesh, a dry texture, and more crowded gills. *Clitopilus hobsonii* is somewhat similar but has thinner flesh and pinkish spores. Several *Crepidotus* resemble *P. mitis* when young but soon develop buff to brown colors on the gills and have yellowish brown to brown spores: *C. versutus* often grows on deciduous bark, *C. herbarum* grows on grass stems or leaves, *C. variabilis* grows on stems and small branches (and has more pinkish buff spores), and C. *subverrucisporus* has more clay-colored gills.

Arrhenia acerosa

sensu PNW

Moss Oysterling

Cap: 0.5–3 cm wide, fan-shaped to irregularly lobed, rarely more evenly circular, in age ruffly and wavy, sometimes uplifted. Surface dry to moist, smooth to faintly wrinkled, translucent-striate when wet, opaque when dry. Light gray to gray-brown. **Gills**: Widely spaced, thick-edged, slightly veiny or wrinkly, short gills present. Gray to tan. **Stipe**: 0.1–0.2 cm long, 0.1–0.2 cm wide, short, reduced to a small, off-center 'stub' often with fuzzy whitish mycelium near attachment to substrate. **Flesh**: Very thin, weakly gelatinous. **Odor**: Indistinct. **Taste**: Indistinct. **KOH**: No reaction. **Spore deposit**: White, but difficult to obtain. **Microscopy**: Spores 6.5–9 × 3.5–5 µm, narrow-ellipsoid.

Ecology: Fruiting singly or in small groups on woody debris, damp soil, and moss. Uncommonly reported but widespread and probably mostly overlooked or misidentified. Fruiting mostly in fall.

Edibility: Too small and thin-fleshed to consider.

Comments: This non-descript mushroom is recognized by its small size, gray colors, and white spores. Recent taxonomic work on the A. *acerosa* complex in Europe and eastern Canada has revealed substantial cryptic diversity. Additional work is needed in western North America to delimit our species. *Claudopus byssisedus* is readily distinguished by its more opaque cap and pinkish-buff spores. *Arrhenia lobata* has gills that are reduced to a shallow network of veins. It often grows in clusters on moss in wet alpine areas. *A. retiruga* is a small, light gray species with a veiny underside (sometimes with a few reduced gills) and grows on moss. The tiny, brownish *Arrhenia aursicalpium* has a narrow, eccentric stipe and reduced vein-like gills—it is rare in Cascadia, but more common northwards into Alaska.

Panellus stipticus

(Bull.) P. Karst.

Luminescent Oysterling

Cap: 2–6 cm across, convex, soon fan shaped, then plane to irregularly lobed and wavy in age, margin tightly inrolled at first, remaining so well into maturely. Surface moist to dry, suede-like to smooth, soon finely areolate with fine, flat fibrillose patches. Beige-tan to dull, pale ocher-tan, dusty whitish near margin, sometimes zonate with a grayish area near attachment. **Gills:** Broadly attached to shallowly decurrent, often crisped and wrinkled, forking and sometimes anastomosing, with many cross veins. Beige-tan to brown. **Stipe:** 0.5–2 cm long, 0.05–0.2 cm thick, very short and rudimentary, laterally attached to cap, thickest near gills and tapered to attachment to wood. Surface dry, covered with fine or coarse spiky whitish tomentum, breaking into fine, obscure chevrons near the gills. Orange-tan to beige or brownish. **Flesh:** Thin, rubbery-tough, pliant. **Odor:** Indistinct. **Taste:** Mild to bitter. **KOH:** No reaction. **Spore deposit:** Whitish. **Microscopy:** Spores 3.5–5.5 × 2–3.5 µm, narrowly elliptic, smooth, amyloid, colorless in KOH.

Ecology: Often in dense clusters or scattered in troops and rows on rotting branches and logs of hardwood trees. Uncommon throughout Cascadia, fruiting even in cold weather. Fruitbodies are quite persistent and half-dried specimens can be encountered throughout season.

Edibility: Inedible.

Comments: This species is recognized by the dry, opaque, beige-tan caps, short off-center stipes, and growth in troops on wood. The narrow, often finely forked and wrinkled gills are another good clue. *Crepidotus* generally have a flabbier and moister texture, more regular gills, and a brown spore deposit. *Panellus mitis* has a smooth whitish cap and are found on conifer branches. *Phyllotopsis nidulans* produces larger fruitbodies with obvious orange tones, and a strong unpleasant aroma when fresh. Eastern North American specimens of *P. stipticus* glow green in the dark while specimens from western North America appear to lack the bioluminescent chemistry.

Rimbachia bryophila

(Pers.) Redhead

Veined Mossear

Cap: (0.1) 0.3–1 (1.5) cm across, at first a downward-facing cup with an incurved margin, becoming convex, semicircular to fan shaped, wavy and occasionally lobed in age. Surface dry, very finely hairy to nearly smooth. Bright white, chalky white, or pale cream. **Gills:** Irregular, shallow, radiating out from attachment, often with numerous cross-gill veins. White. **Stipe:** Absent. **Flesh:** Very thin, concolorous. **Odor:** Indistinct. **Taste:** Mild. **KOH:** No reaction. **Spore deposit:** White. **Microscopy:** Spores 5–7 × 4.5–7 µm, subglobose to apple seed shaped with a pointed apiculus, smooth, inamyloid, colorless in KOH.

Ecology: Solitary or in scattered troops on mosses, often laterally attached to their main stems. Common in mossy habitats throughout western Cascadia. Fruitings can occur whenever moss beds are sufficiently wet and temperatures are above freezing.

Edibility: Unknown.

Comments: This common moss parasite is identified by its small size, white colors, and shallow gills. *Arrhenia retiruga* is similar in appearance and habitat but is told apart by its smooth to veined underside and grayish colors. *Muscinupta laevis* also has a smooth underside and is bright white in color. *Cheimonophyllum candidissimum* has a bright chalky-white cap and grows on dead wood.

Muscinupta laevis

(Fr.) Redhead, Lücking & Lawrey

Minute Mossmaid

Cap: 0.1–1 cm across, fan-shaped to nearly circular, often irregularly lobed and wavy, margin finely inrolled in young specimens. Surface dry, opaque to translucent, obscurely silky, finely wrinkled to nearly smooth. Bright white, chalky white to pale grayish. **Underside:** Smooth to finely wrinkled, with scattered veinlike ridges in larger fruitbodies. White. **Stipe:** Absent, but fruitbodies often asymmetrical and elongate towards the point attached to the substrate and thus appearing weakly stipe-like. **Flesh:** Practically non-existent. **Odor:** Indistinct. **Taste:** Indistinct. **Spore deposit:** White, but almost impossible to obtain a visible deposit. **Microscopy:** Spores 3-4 × 2-2.5 µm, ellipsoid, smooth, inamyloid. Pleurocystidia abundant.

Ecology: Solitary or in small groups directly on the fronds of living mosses, especially haircap moss (*Polytrichum* spp.). Occurs in wetter areas across western Cascadia but is largely absent south of Coos Bay. Fruiting throughout the wet season, including winter wherever microhabitats insulate from freezing temperatures.

Edibility: Tiny, fleshless.

Comments: This tiny, often overlooked species is common in the northwest part of Cascadia. Once you become familiar with the look of these "white specks on moss", their prevalence becomes apparent. The lack of gills helps distinguish this species from *Rimbachia bryophila*, while the chalky, bright white caps set it apart from those of the similar, but more gray-tan *Arrhenia retiruga*. *Calyptella* often have a tiny, distinctly narrowed stipe and more convex-cupped fruitbodies which tend to grow in swarms on wood or small-diameter plant debris. *Cheimonophyllum candidissimum* has a bright chalky-white cap, crowded gills and grows on dead wood.

Synonym: *Cyphellostereum laeve* (Fr.) D.A. Reid

Metacampanella subdendrophora

(Redhead) R.H. Petersen

Cap: 0.1–1 cm across, convex to semicircular. Surface dry, wrinkled, very finely hairy to smooth. White to creamy white, developing bluish black spots in age. Gills: 3–5 primary anastomosing veins radiating out from stipe, with shallow irregular cross veils. Whitish to cream colored, developing bluish black spots in age. **Stipe**: 0.1–0.3 cm long, 0.03–0.07 cm thick, off-center, appearing lateral, but cap is creased around it. Grayish with a white sheen when young, to bluish black in age. **Flesh**: Extremely thin, somewhat tough. **Odor**: Indistinct. **Taste**: Mild. **KOH**: No reaction. **Spore deposit**: White. **Microscopy**: Spores 8–10 × 6–7 µm, triangular to pyramidal, smooth, inamyloid.

Ecology: Solitary or scattered on stems of dune grass (*Ammophila* spp.), less often on other grasses, sedges, or *Rubus* canes. Locally common in coastal dunes habitats but also in inland forests with appropriate substrates. Fruiting from fall well into winter.

Edibility: Too small and thin-fleshed to seriously consider.

Comments: This tiny mushroom is distinguished by the whitish cap that develops bluish black spots, irregularly veined gills, short, off-center stipe, and preference for dune grass *Metacampanella dendrophora f. washingtonensis* (=*Campanella* sp. 'Veiny' in MotRC) also develops bluish gray colors in age and has highly irregular cross-branched gills, but is usually larger and often lacks a stipe completely. It grows on small branch litter, especially that of Scotch Broom (*Cytisus scoparius)* and blackberries *(Rubus* spp*). Calyptella capula* produces downward-facing bell-shaped or pipe-shaped fruitbodies with smooth or faintly veined fertile surfaces and grows on small branches or herbaceous stems (we have seen it on debris of *Rubus*, grand fir, and western hemlock). The much more common *Marasmiellus candidus* is a bit larger and has a more well-developed, central to off-center stipe.

Synonym: *Tetrapyrgos subdendrophora* (Redhead) E. Horak

Claudopus byssisedus

(Pers.) Gille

Cap: 1–4.5 cm across, fan-shaped or plane, margin inrolled when young, often wavy in age. Grayish brown to mouse gray or pallid grayish tan, often with a sheen of white or silvery fibrils, with cobwebby white mycelium near attachment point. **Gills:** Radiating out from attachment point, subdistant to distant, narrow. Grayish white at first, becoming pinkish gray to dingy salmon in age. **Stipe:** Absent or rudimentary, laterally attached to substrate. Whitish to gray with a pale bloom. Base usually with cords of white mycelium. **Flesh:** Thin, fragile. **Odor:** Indistinct to farinaceous. **Taste:** Indistinct. **Spore deposit:** Pinkish to salmon. **Microscopy:** Spores 7.5–11 × 5–8.5 µm, elongate, multifaceted with 5 or 6 sides. Cheilocystidia rare to frequent, capitate.

Ecology: Often in small troops or clusters, rarely solitary. Typically in small hollows in soil, or sheltered areas with overhanging debris. Generally uncommon but can be locally abundant throughout western Cascadia. Fruiting in fall and early winter.

Edibility: Unknown.

Comments: The small, fan-shaped grayish caps, rudimentary stipe, pinkish gills, and white, rhizomorphic mycelium are helpful identifying features. The name *C. byssisedus* is based on European specimens, and is likely misapplied in Cascadia. *C. graveolens* is a poorly-known species with tiny caps (<1 cm), spores averaging < 9 µm, and a skunky odor. The European name *C. parasiticus* has been applied to a species with small, dark brownish-gray fruitbodies with decurrent gills and tiny, rudimentary stipes that fruits directly on the underside of chanterelle fruitbodies! It has been found on *Cantharellus subalbidus* and *C. formosus* in forested habitats throughout Cascadia, but is rare. *Clitopilus hobsonii* is a small, whitish species with a finely fuzzy cap surface; it grows on decaying wood. There are a number of similar small, pale pleurotoid mushrooms, but these all have differently-colored spore deposits; usually white (e.g. *Arrhenia, Cheimonophyllum*), or dull brown (*Crepidotus*).

Deconica horizontalis

(Bull.) Noordel.

Cap: 0.2–2.5 cm across, convex, plane to fan shaped, margin finely inrolled and often ribbed when young, sometimes wavy in age. Surface moist to dry, finely velvety or silky. Dark purplish brownish to reddish brown or ocher-brown, fading to beige or tan as it loses moisture. **Gills:** Broadly attached, often with a decurrent tooth. Cinnamon brown when young, mottled with lilac gray, and dark gray-brown to brown in age, edges fringed whitish. **Stipe:** 0.2–0.7 cm long, 0.1–0.2 cm thick, off-center (but not lateral), short, curved. Surface dry, covered with a fine pale bloom to smooth. Light to dark brown, but often covered with a white tomentose when young, becoming darker, to blackish brown in age. **Flesh:** Very thin, dark. **Odor:** Indistinct. **Taste:** Indistinct. **Spore deposit:** Dark purplish gray. **Microscopy:** Spores 6–7.5 × 4–5 µm, ellipsoid to almond shaped, with a germ pore, thick walled, smooth.

Ecology: Solitary, scattered, or in troops on small branches and sticks, canes of blackberry (*Rubus* spp.) and Scotch Broom (*Cytisus scoparius*), rotting wood, almost any cellulose-rich debris (even discarded denim jeans!). Common, but often overlooked. Fruiting in fall and spring.

Edibility: Unknown, probably non-toxic.

Comments: The brown cap, small, off-center stipe, growth on wood, and dark purplish brown spore print distinguish this species. *D. phillipsii* is generally smaller and grows from grass and sedge stems and debris. *Crepidotus* and *Panellus* can look similar but have dull brown and white spore deposits, respectively. *Phaeomarasmius* have hairy cap and stipe surfaces and ocher-brown spores. Some *Simocybe* can look similar when stunted but are usually more olive-brown and have brown spores. *Tectella patellaris* has a partial veil when young and fruitbodies that are asymmetrical, funnel-shaped, or like stalked suction cups, becoming flatter and more fan-shaped in age. They have beige-tan caps and whitish spores.

Synonym: *Melanotus horizontalis* (Bull.) P. D. Orton.

Crepidotus applanatus group

White Crep

Cap: 1-6 cm wide, fan-shaped to oyster-shaped. Surface dry to moist, sparsely covered with small fibrils. Pallid grayish-tan to milk white, hygrophanous and often two-toned, becoming paler upon drying. **Gills**: Radiating out from central attachment to substrate, close, narrow, edges even or finely pale-beaded. Whitish to pale gray, becoming dull brownish as spores mature. **Stipe**: Absent. **Flesh**: Very thin, fragile, pale beige. **Odor**: Indistinct. Taste: Indistinct. **KOH**: No reaction. **Spore deposit**: Dull brown. **Microscopy**: Spores 4–5µm, rounded, with small, low warts. Cheilocystidia present, fairly large and variable in shape.

Ecology: Growing in small troops on decaying trunks, logs, and branches of hardwoods. Fairly common in Cascadia, fruiting in fall.

Edibility: Unknown.

Comments: The fan-shaped fruitbodies that lack stipes, pallid, sparsely hairy to nearly smooth caps, and greyish to dull brown gills suggest this species, but don't suffice to definitively rule out other *Crepidotus*. *Clitopilus* have pinkish spores, while *Cheimonophyllum* has white spores. *Pleurotus* are larger, with firmer rubbery-fleshy texture and white spores. From the perspective of a taxonomist, *Crepidotus* is one of the most problematic genera in North America—an unfortunate situation given how diverse and abundant they are! A source of this difficulty is that fruitbodies are relatively feature-poor: Even the more distinctive species are macroscopically separated by rather subtle differences in size, shape, color, and substrate. Microscopy adds many additional features (size and shape of spores and cystidia, details of pileipellis structures) but runs into our frustratingly incomplete state of knowledge. Early workers often provided sparse descriptions, leaving us with weak and overlapping species concepts. Slowly but surely, genetic data is helping to clarify the picture, although not always in favor of field workers—cryptic species (those that are practically identical macroscopically but phylogenetically distinct) appear to be common.

Crepidotus calolepis

(Fr.) P. Karst.

Hairy Creps

Cap: 2–6 cm across, fan shaped to semicircular, margin slightly inrolled when very young but soon expanding to plane or slightly scalloped or wavy at margin. Surface dry to moist to waterlogged, in age slightly viscid or flabby, sometimes obscurely striate when wet. Finely hairy when very young, becoming nearly smooth with scattered brownish fibrils in age. Dull tan-brown to beige-brown, strongly hygrophanous and becoming watery grayish buff to light beige as it dries out. **Gills:** Radiating out from attachment point, many partial gills present. Grayish white to dingy cream, browner in age. **Stipe:** Absent, but with distinct tuft of whitish mycelium at attachment to substrate. **Flesh:** Moderately thin, soft, often waterlogged, dingy whitish gray to yellowish tan. **Odor:** Indistinct. **Taste:** Indistinct. **KOH:** No reaction. **Spore deposit:** Dull brownish. **Microscopy:** Spores 7–10 × 4.5–6 µm, smooth, ellipsoid.

Ecology: Solitary or in small groups on dead wood, especially hardwoods. Very common in Cascadia, especially on red alder, fruiting primarily in fall and spring.

Edibility: Unknown and best avoided.

Comments: The semicircular, hygrophanous, dingy grayish beige to light brownish cap with appressed brown hairs helps identify this species. *C. crocophyllus* has richer orange coloration to the cap and around the reduced stipe/attachment point. *C. ehrendorferi* is much smaller, with a smoother, pale cap, and orange tones restricted to the tiny stipe; it appears to be rare in Cascadia. *C. applanatus* is similar but has a nearly smooth or weakly fibrillose whitish or dingy grayish white cap (which can become covered with brownish spores in age) and small (4–5.5 µm), globose spores. Although the name *C. mollis* has long been used for specimens matching this description from the Pacific Coast, recent sequence data indicate a closer relationship to *C. calolepis*. Given the difficulty of circumscribing clear species concepts in this genus, readers should expect our interpretation of these names to continue to change. We encourage readers to consult Hesler and Smith's monograph on this genus, which is freely available at MykoWeb (see Resources).

Tapinella panuoides

(Batsch) E.-J. Gilbert

Cap: 2–12 cm across, oblong to fan shaped, with an inrolled margin when young, expanding to plane or uplifted, and often lobed in age. Surface moist to dry, finely velvety to nearly smooth in age. Light brown, yellow-brown to olive-brown, often paler toward margin. **Gills:** Radiating out from attachment point, often wrinkled and sometimes anastomosing near attachment, with many cross veins, narrow. Creamy buff, yellowish beige to tan, at times with an olive cast. **Stipe:** Absent or rudimentary, laterally attached to substrate. **Flesh:** Moderately thin, soft, pale whitish to pale yellow-brown. **Odor:** Indistinct. **Taste:** Mild to slightly bitter. **KOH:** Olive-yellow on all parts. **Spore deposit:** Pale brown to yellowish brown. **Microscopy:** Spores 4.5–6 × 3–4 µm, ellipsoid to ovoid, smooth, yellowish in KOH.

Ecology: Solitary or in overlapping clusters on rotting conifer logs, or in clumps and clusters on woody debris. Generally uncommon but large fruitings can occur in urban areas on wood chips and along skid roads following logging. Fruiting throughout the wet season, more common in late winter and spring.

Edibility: Unknown.

Comments: The pleurotoid shape, mustardy yellow-brown colors, wrinkled gills, brown spores, and growth on woody debris are distinctive. *Crepidotus crocophyllus* and *C. calolepis* have fragile flesh, more regular gills, and darker brown spores. *Tapinella atrotomentosa* is larger and has a fuzzy, dark brown, stipe. *Hygrophoropsis* have forking gills, central stipes, and white spores.

16 - Boletes

The term *'bolete'* refers to those mushrooms with a central stipe, cap, a tube layer with pores, and a soft or fleshy texture. Until recently, the family Boletaceae included only a few genera, with most of our familiar boletes placed in the genus *Boletus*. However, the family has undergone dramatic revision in the past decade and now includes many smaller genera. There are a few species with partially hypogeous and often semisequestrate or contorted growth forms (species described in the genus *Gastroboletus*) but for the typical boletoid species, it's useful to learn these major subgroups.

Boletus – "Core" boletes, including the King Bolete and its equally delicious relatives. Members of this group are commonly called *porcini (por-chee-nee)* – Italian for 'little pig.' These species produce robust fruitbodies with white flesh, reticulate, club-shaped stipes, and small, densely packed tubes. The pores are white or light creamy yellowish when young.

Butyriboletus – "Butter Boletes" have large, dense fruitbodies and reticulate stipes like *Boletus*, but differ in having yellow flesh and yellow pores that bruise blue.

Caloboletus – "Bitter Boletes" are sometimes confused with Butter Boletes due to their large-statured fruitbodies and blue bruising reactions. However, they often have intensely blue-staining whitish to grayish flesh, and usually have a sour-bitter flavor (nibble a small piece of the cap, then spit).

Aureoboletus – Fruitbodies with intensely colored citrine yellow pores, often tapered stipe bases, and lemony-tasting cap surfaces. Not staining blue.

Xerocomellus – Boletes with medium brown to dark gray to pinkish red or dark purplish black, velvety, often cracking caps, yellow pores that stain blue in some species, and red to reddish yellow stipes. There is extensive variation in cap color, degree of cracking, stipe color and ornamentation, and staining reactions even within species. Part of the 'xerocomoid' bolete group, along with *Hortiboletus, Phylloporus, Pulchroboletus*, and *Xerocomus*.

Hortiboletus **and** ***Pulchroboletus*** – Small to medium boletes with a red cap, yellow pores, yellow to red stipe, and yellow flesh. Some species stain blue, while others only sporadically blue, and others don't stain at all. *Pulchroboletus* tend to have a red band at stipe apex.

Xerocomus – Small to medium sized, with dry, velvety, brownish caps and rather coarse yellowish pores that sometimes stain blue. Unlike *Xerocomellus*, *Xerocomus* never have extensively red-colored stipes.

Hemileccinum – The only representative in our area is a medium-sized bolete with yellowish-brown cap, bright yellow pores, and, most distinctively, a stipe covered in pale scurfy sca ers.

Chalciporus – Small to tiny boletes with coppery or pinkish tan pores, yellow mycelium at the base of the stipe, and a peppery taste. Usually found near *Amanita muscaria*.

Porphyrellus – Medium to large boletes with brown coloration overall, including dark chocolate brown pores. Some stain obscurely dark bluish to olive-green.

Pulveroboletus **and** ***Buchwaldoboletus*** – Uncommon boletes with medium to large, brilliant-yellow caps. The pores are bright yellow to reddish and stain blue. The former genus is mycorrhizal; members of the latter are apparently saprotrophic (or possibly parasitic) and often found near rotting wood.

Rubroboletus, Neoboletus* and *Suillellus – Boletes with bright red to dark red pore mouths that stain blue when bruised. *Rubroboletus* are the only dangerously toxic boletes in our area.

Gyroporus – In Cascadia, the only species are small to medium-sized boletes with orange-brown caps and stipes, white to yellow pore mouths, and straw-yellow spores. The stipes are distinctively hollow or pith-stuffed and rather fragile.

Leccinum – Easily recognized by cylindrical to clavate stipes covered in orangey-brownish to dark grayish or blackish tufts or scabers. The caps are usually orangey to reddish, brown, gray to white, and the pores range from grayish to pale. Species are typically associated with a narrow range of host plants (aspen, birch, madrone, manzanita, or conifers). Taxonomy very complicated.

Suillus – A very distinct group, commonly known as 'Slippery Jacks' (although some are entirely dry). Fruitbodies tend to have short statures, and most have either a partial veil or sticky "glandular dots" on the stipe. The pore mouths tend to be coarse and irregular and vaguely radially arranged. All are mycorrhizal with trees in the pine family and are particularly common with Douglas-fir and pine in our area. Many *Suillus* have associated with *Chroogomphus* or *Gomphidius* on the mycelium, likely as parasites

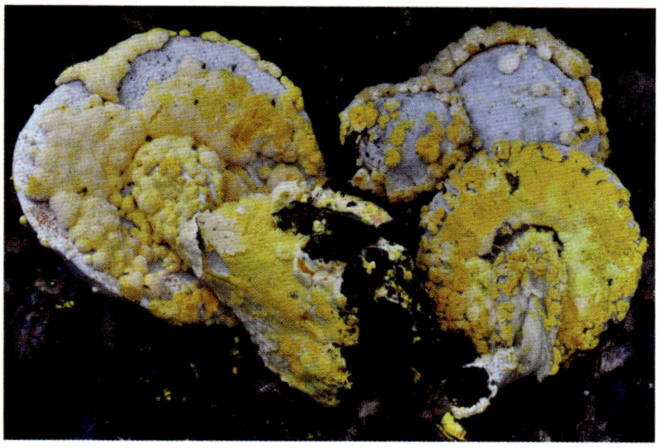

Hypomyces microspermus

Rogerson & Samuels

Bolete Eater

Fruitbody: First appearing as a white mold on boletes, soon covering whole fruitbody, occasionally blistering. The host is generally still firm in this stage. Soon starts turning bright yellow to golden colored, with the host often collapsing and becoming quite decayed. Final (sexual) stage is rarely observed in the wild; a reddish brown, pimply crust. **Odor:** Indistinct to putrid. **Taste:** Not sampled. **Spore deposit:** White in first stage, golden yellow in second stage. **Microscopy:** Spores in white stage, 5–17 × 3–7 µm, oblong to broadly elliptic. In yellow stage, 8.5–15 µm, round, distinctly spiny, very thick walled. In sexual stage, 8–15 × 2.5–4.5 µm, spindle to lance shaped.

Ecology: Very common on 'xerocomoid' boletes, especially *Xerocomellus diffractus*. Very common throughout our range, possible whenever boletes are fruiting, which is usually late summer through fall.

Edibility: Unknown, likely poisonous.

Comments: This common bolete-parasitic mold goes through a sequence of two asexual-sporulating stages, as well as a less-frequently observed sexual stage (the host is often buried in duff by this point). *H. chrysospermus* also occurs in Cascadia but is more likely to be found on non-xerocomoid boletes. It is only reliably distinguished by its much larger spores measuring (10) 12–25 µm (in yellow stage). *Dicranophora fulva* forms tiny, translucent white stalks holding up bright yellow, globular heads. These completely engulf *Suillus* (and occasionally *Gomphidius*) fruitbodies. Unlike *Hypomyces*, it is a zygomycete fungus.

Boletus barrowsii

Thiers & A. H. Sm.

White King

Cap: 5–15 cm across, bun shaped to convex at first, becoming broadly convex or unevenly flat and sometimes wavy in age. Surface dry, smooth to slightly velvety with a bloom of whitish powder when young. Dusty grayish white at first, then pale beige-white to beige-tan or pale tan-brown. **Tubes:** Broadly attached or sunken around stipe, 1–3.5 cm long, separating easily from cap. **Pores:** Rather small, round to slightly angular, stuffed with white tissue at first. White when young, becoming pale yellowish, yellowish tan or orangish tan, to olive brown in age. **Stipe:** 7.5–20 cm long, 4–9 cm thick, swollen to club shaped or bulbous toward base. Surface dry, reticulated with a raised fishnet pattern; usually more pronounced toward apex, but sometimes covering entire stipe. Pale whitish beige or grayish white at first, becoming pale whitish tan, often developing peach tan or orangish tan tones. **Flesh:** Thick, firm when young, soft in age. Whitish; not staining. **Odor:** Indistinct or pleasant. **Taste:** Mild to sweet, or like carrots. **KOH**: No reaction. **Spore deposit** Olive-brown. **Microscopy:** Spores 13–15 × 4–5 µm, long-ellipsoid to somewhat spindle-shaped, smooth.

Ecology: Often in small clusters or scattered troops. Typically in urban areas with introduced oaks, or occasionally other exotic hardwoods (hornbeam, linden). More rarely with Oregon white oak and may also occur with ponderosa pine in the eastern Cascades. Fruiting in summer and fall, less commonly in spring.

Edibility Edible and excellent! Very firm, with a lovely, sweet odor and nutty taste. As for most boletes, drying concentrates flavor and emphasizes rich earthy tones (at the expense of the delicate, sweet odor and flavors when fresh).

Comments The hefty stature, dry, pale cap, white flesh, lack of bruising reactions, mild to nutty taste, and strong reticulation on the stipe help distinguish this species. *Caloboletus marshii* also has a pale cap with a whitish bloom, but has a yellow stipe without any fishnet pattern, turns blue rapidly when bruised, and tastes very bitter. *B. mottiae* (=*B. regineus*) has a whitish bloom on the cap when young but soon shows a dark mahogany brown cap with tubes more greenish olive in age. Pale-capped *B. edulis* generally have a taller stature and a viscid cap when moist. *B. rex-veris* can have a pale cap when young or buried in duff but develops reddish-brown colors and fruits in the spring and early summer under montane conifers.

Boletus subalpinus

(Trappe & Thiers) M. Nuhn et al.

Humble King

Fruitbody: Semi-sequestrate, with an often convoluted or irregular cap, irregularly exposed or contorted tubes and a variably short, often contorted stipe. **Cap:** (3) 5–12 cm across, rounded to convex or irregularly convex when young, becoming centrally depressed, margin typically uplifted and lobed in age. Surface dry or slightly lubricious if wet, finely velvety or somewhat appressed-fibrillose near margin to nearly smooth, often with adhering debris. White, whitish buff to pale tan. **Tubes:** 0.5–4 cm long, typically vertically oriented when young, more convoluted and in various angles at maturity. Whitish when young, to yellowish, then olive at maturity. **Pores:** Often covered with a thin white membrane when young, and stuffed with white hyphae, irregularly opening in age. Concolorous with tubes. **Stipe:** 1–5 cm long, (1) 2–5 cm thick at apex, often short, stout, typically swollen in middle, to irregularly club-shaped. Surface dry, finely velvety to smooth, occasionally reticulate at apex. White, beige to tan, occasionally yellowish at base. **Flesh:** Thick, fleshy, firm when young, soft in age. Whitish, at times with yellowish stains, slowly staining pinkish to pale vinaceous in cap when cut. **Odor:** Indistinct. **Taste:** Mild, pleasant, often nutty. **Microscopy:** Spores 10–16 (18) × 4.5–6 (8) µm, ellipsoid, oblong to subovoid, smooth.

Ecology: Solitary or scattered, completely buried to partially exposed in duff under high elevation conifers—especially true firs. Best found by investigating "mushrumps" or the truffle digs of animals. Apparently restricted to higher elevations. Locally common in the southern Cascade Range, rarer to the north. Fruiting in summer and fall.

Edibility: Edible but often difficult to find clean specimens.

Comments: Fruitbodies of *B. subalpinus* are easily recognized by an irregular, lobed, whitish to tan cap, white to olive pores, a short, stout white stipe, and white flesh with slow pinkish staining. The lack of bright colors or blue staining on the fruitbody distinguish it from other "Gastroboletus" in our area.

Synonym: *Gastroboletus subalpinus* Trappe & Thiers

Boletus rex-veris

D. Arora & Simonini

Spring King

Cap: (7) 10–30 cm across, rounded to domed when young, becoming broadly convex, plane, or wavy. Surface dry, covered with a fine white bloom when young, soon smooth, occasionally thinly viscid in age. Typically pinkish brown to reddish brown, buried or young specimens can be paler, more ocher to orange-brown in age. **Tubes:** Broadly attached to sunken around stipe, 0.5–4 cm long. **Pores:** Stuffed when young, small, round to slightly irregular. Creamy to pale yellowish at first, becoming yellow to yellowish olive in age. At times brownish flushed. **Stipe:** 5–12 (20) cm long, 2.5–6 cm thick at apex, equal or becoming more enlarged or bulbous lower, but often with a pointed base. Surface dry, covered with fine reticulation. Whitish at first, soon developing tan, pinkish brown to reddish brown color. **Flesh:** Thick, firm, fleshy. White, not staining. **Odor:** Indistinct. **Taste:** Mild to sweet. **KOH:** Unknown. **Spore deposit:** Olive-brown. **Microscopy:** Spores 15–17 × 4.5–5 µm, elongate-ellipsoid to spindle-shaped, smooth.

Ecology: Solitary, scattered or in small clusters. Most often under duff at first, forming 'mushrumps', and pushing free of duff when mature. More exposed in mossy areas, or areas with thin duff. Common on the eastside of the Cascade Range in fir and pine forests. Largely absent from western Cascadia except in rain shadow pockets. First fruiting in spring occurs a few weeks after snow melt, continuing into summer at higher elevations.

Edibility: Edible and great. The thick firm flesh and sweet, nutty flavor stand out.

Comments: The reddish brown cap that often shows a pallid 'bloom' when young, white to yellow pores, and a whitish to reddish brown stipe are helpful in identifying this species. The spring fruiting or early summer fruiting under eastside conifers is the clincher. *B. mottiae* (=*B. regineus*) is similar in stature and by its whitish bloom when young. However, the caps mature darker brown, it grows with hardwoods, and fruits in the fall in westside forests. *B. edulis* lacks the bloom when young with a creamy tan to orange-brown cap—which is thinly viscid when wet. It also tends to have a taller stature, and fruits in summer and fall.

Boletus edulis var. *edulis*

Bull.

King Bolete, Porcini

Cap: 5-20 (35) cm across, rounded to bun shaped at first, expanding to broadly convex, plane, or wavy in age. Surface thinly viscid or more often greasy to dry, smooth or wrinkled. Typically creamy beige or tan to light brown, more rarely dark brown or orange-brown. Occasionally pale; white to whitish beige. **Tubes:** Sunken around stipe, fairly thin at first, becoming quite long in age. **Pores:** Stuffed with whitish tissue at first, opening as they mature, round and rather small. White when young, becoming creamy to yellow and then yellowish olive or discolored brownish in age. **Stipe:** (5) 8–17 (20+) cm long, 2–5 (8) cm thick at apex, club shaped at first, but often becoming quite slender and elongate, at time short and stocky. Surface dry, covered in a fine to coarse reticulum, especially pronounced at the apex. Whitish to cream, developing tan tones in age. **Odor:** Indistinct. **Taste:** Mild to sweet or nutty. **Spore deposit:** Olive-brown. **Microscopy:** Spores 15–17 × 6–7 µm, spindle-shaped to long-ellipsoid, smooth.

Ecology: Solitary or scattered in troops in a wide variety of habitats. Commonly fruiting in a synchronized flush in early to mid-fall with Sitka spruce in western Cascadia. *Clitopilus prunulus* is often in the immediate vicinity with many patches conspicuously accompanied by *Amanita muscaria*. Summer and early fall fruitings are found at higher elevations with firs and on the eastern slopes with spruce. It has a preference for forest edges, or disturbed areas (meadow edges, road or trail sides, or forest openings). Can be abundant in the mountains in years with summer or warm fall rains. Although *B. edulis* var. *edulis* occurs in urban areas with a range of non-native trees, *B. edulis* var. *grandedulis* is more prolific in these habitats.

Edibility: Edible and wonderful. This is one of the most sought-after mushrooms with a rich flavor, large size, and firm texture. Thin slices of young fruitbodies can be eaten raw or marinated. Tubes of older specimens can be soft and mushy but are full of flavor. Many dry them separately for use as a rub or as a base for soup stock or gravy.

Boletus edulis 'Coastal Spruce'

Boletus edulis var. *grandedulis*

Comments: Easily recognized by their large bun-shaped caps, white pores when young, club-shaped stipes with white reticulation, and white, non-staining flesh. *B. edulis* var. *grandedulis* has a more richly colored cap at maturity (warmer reddish, ocher, to orange-brown) and the pores become reddish to bronze-red in age. In the coastal dunes both varieties grow together—*grandedulis* with shore pine and *edulis* with spruce. Although *B. edulis* var. *grandedulis* associates with a wider range of trees overall (including both hardwoods and conifers), it seems to be absent from spruce forests. *B. barrowsii* has a dry, whitish cap, typically grows with hardwoods (at least in Cascadia), and has a stout stature. *B. mottiae* (=*B. regineus*) usually has a much darker cap with a prominent whitish bloom, a shorter stipe, and a preference for hardwoods.

Boletus fibrillosus

Thiers

Fib King

Cap: 5–15 (20) cm across, rounded to domed when young, becoming broadly convex. Surface dry, finely velvety at first, becoming matted-tomentose, breaking up into appressed-fibrillose tufts arranged in radial wrinkles, occasionally cracking when dry. Dark brown when young, becoming brown to yellow-brown. Completely yellow forms occur rarely. **Tubes:** Broadly attached to sunken around stipe, 0.5–2.5 cm long. **Pores:** Stuffed when young, small, round to slightly irregular. Creamy to pale yellowish at first, becoming yellow to yellowish olive in age. Generally not staining, but can discolor slightly greenish in age. **Stipe:** 7–15 (20+) cm long, 1.5–3 cm thick at apex, equal, or club shaped, becoming more enlarged or bulbous lower, but often with a rooting or pointed base. Covered with coarse to fine reticulation; off-white at first, aging brown. Base color shades of brown, often with a paler band at apex, and a whitish base. **Flesh:** Thick, firm when young, becoming soft, stipe sometimes becoming hollow in age. White to beige in cap, white in stipe, not staining. **Odor:** Indistinct. **Taste:** Mild to sweet. **KOH:** Darkening, reddish brown on cap, no reaction elsewhere. **Spore deposit:** Dark olive-brown. **Microscopy:** Spores 13–16 × 5.5–6.5 µm, elongate to subellipsoid, smooth.

Ecology: Solitary or scattered on ground, in duff or moss under conifers—especially Douglas-fir and true firs. Widespread across Cascadia, but rarely abundant. Fruiting from late summer through fall or into early winter.

Edibility: Edible and very good, with a sweet, nutty flavor. Older specimens can become very soft and are better suited to being dried and used as a powdered seasoning.

Comments: The dark brown to honey brown matted-tomentose to fibrillose cap brownish, reticulate stipe with a pale, often rooting base, are distinctive. *B. mottiae* (=*B. regineus*) also has brown cap but has a powdery bloom when young and is typically smooth in age; it grows with hardwoods. *B. edulis* tends to be paler with a smoother, thinly viscid cap.

Caloboletus frustosus

(Snell & E.A. Dick) D. Arora & J.L. Frank

Dark Bitter Bolete

Cap: 7–25 (30) cm across, rounded when young, becoming broadly convex to bun shaped in age, margin inrolled at first, then downcurved well into maturely, to rounded or slightly uplifted in age. Surface dry, finely velvety-tomentose, giving it a suede-leather feel, often becoming areolate cracked, with minute tomentose tufts in age, or with deep cracks when dry. Dark olivaceous gray to dark gray at first, fading as it ages to grayish brown or dingy brown, especially in drier habitats, **Tubes:** Broadly attached or with a slight notch, 0.3–3 cm long**. Pores:** Small, angular to round. Dull, cool yellow when young, soon pale olive-yellow to dingy yellow-olive in age. Quickly staining blue when bruised, eventually fading to dingy orangish brown. **Stipe:** 5–15 cm long, 2–5 cm thick at apex, often club shaped, with a thicker base when young, to more or less equal in age. Surface dry, covered with fine reticulation, at least at apex. Yellow overall at first, developing dingy grayish colors from base up as it matures, and sometimes a reddish base in age, staining blue when handled. **Flesh:** Very thick and firm when young, soft in age. Yellow to whitish at stipe apex, quickly staining blue when cut, slowly fading to grayish. Stipe often with reddish in base and around larva tunnels, at least when old. **Odor:** Indistinct or sour. **Taste:** Extremely bitter. **KOH:** Brown on cap, orange on flesh, tubes, and upper stipe, brown on lower stipe. **Spore deposit:** Olive-brown. **Microscopy:** Spores (12) 14–17 × 4–6 µm, subcylindric to subellipsoid, smooth, ochraceous in KOH.

Ecology: Solitary or scattered on ground, in moss or duff under conifers—especially true firs and hemlock. Common and widespread, fruiting from summer through fall.

Edibility: Unknown, extremely bitter tasting.

Comments: The dark olivaceous-gray cap which becomes areolate-cracked in age, yellow pores, flesh that bruises blue when damaged, yellow, finely reticulated stipe, and extremely bitter taste are the key identifying features of this large, stately bolete. It is quite abundant in years with summer rains. The spring and fall fruiting *C. rubripes* have non-reticulate stipes with a reddish base and paler colors when young. In older fruitbodies of *C. frustosus* the cap color becomes lighter (darker in *C. rubripes*) and the stipe might develop a dingy red color at the base. The rosy sterile margin on the mature caps of *C. rubripes* can help separate the two species at this stage.

Synonyms: *Caloboletus conifericola* Vizzini, *Boletus coniferarum* E.A. Dick & Snell

Caloboletus rubripes

(Thiers) Vizzini

Red-footed Bitter Bolete

Cap: 5–18 (28) cm across, rounded when young, becoming broadly convex to bun shaped in age, margin inrolled at first, downcurved, plane or slightly uplifted in age; often with a narrow band of sterile tissue which becomes rosy pink in age. Surface dry, finely velvety-tomentose, giving it a suede-leather feel, often becoming areolate cracked, with minute tomentose tufts in age. Buff, or olivaceous buff when young, browner in age, bruising brown when rubbed. **Tubes:** Broadly attached to stipe, 0.3–2 cm long. **Pores:** Very small, angular to nearly round. Dull to bright yellow or golden yellow when young; dingy yellow-olive in age. Quickly staining blue when bruised, eventually fading to orangish brown. **Stipe:** 5–15 (20) cm long, 2–4.5 cm thick at apex, often club shaped when young, to more or less equal in age. Surface dry, smooth. Yellow at apex, with a rosy red to red-blushed base when young, then redder from base up as it matures, becoming dingy red with a blackish red base in age. Staining blue when handled. **Flesh:** Very thick and firm when young, soft in age. Yellow to whitish at stipe apex, quickly staining blue when cut, slowly fading grayish. Stipe base and larva tunnels reddish. **Odor:** Indistinct. **Taste:** Slightly bitter or sour. **KOH:** Brown on cap, orange on flesh, tubes, and upper stipe, brown on lower stipe. **Spore deposit:** Olive-brown. **Microscopy:** Spores 12.5–17.5 × 4–5 µm, subcylindric to subellipsoid, smooth, ochraceous in KOH.

Ecology: Solitary or scattered on ground, in moss or duff under conifers, especially fir and hemlock. Common and widespread, fruiting mostly in fall, with scattered montane fruitings in the spring.

Edibility: Unknown, not recommended. Although only mildly bitter, there are reports of minor gastrointestinal distress from other *Caloboletus*.

Comments: The buff, suede-textured cap (when dry), yellow pores, blue-staining flesh, non-reticulate stipe with a reddish base, and slightly bitter taste are distinctive. When young, *C. frustosus* is told apart by its dark olivaceous gray cap, finely reticulated stipe that does not have a red base, and extremely bitter taste. *C. marshii* has a beige to grayish buff cap and grows under oaks. The stocky *Butyriboletus persolidus* has a cinnamon to light brown cap, yellow pores, yellow flesh which sporadically stains blue and a mild taste; it grows with oaks. *Pulchroboletus smithii* has a rosy red to pinkish velvety cap that fades to gray, yellow pores that become reddish blushed, a stipe with a red apex and yellow base, and a mild taste.

Synonym: *Boletus rubripes* Thiers

Butyriboletus primiregius

D. Arora & J. L. Frank

Spring Butter Bolete

Cap: 7–20 cm across, bun shaped to broadly convex, sometimes slightly wavy or lobed at margin. Surface tacky to dry, often covered with dust or soil. Dark burgundy red to vinaceous red at first, developing brown tones in age. **Tubes:** Short and densely packed when young, 1–2.5 cm long when mature. **Pores:** Small, round. Pale yellow at first, becoming bright yellow to olive-yellow. Weakly staining bluish gray when young, quickly bluing in age. **Stipe:** 5–15 cm long, 2–6 cm thick, typically club-shaped when young, or swollen lower, then pointed towards base, more equal in age. Surface dry, covered with a fine reticulation over upper part portion. Lemon yellow to butter yellow, occasionally with a reddish blush over the lower part, staining bluish when handled. **Flesh:** Thick, very firm when young, distinctly yellow; especially in the stipe flesh, not staining, or erratically staining bluish when cut. **Odor:** Indistinct. **Taste:** Mild. **Spore deposit:** Olive-brown. **Microscopy:** Spores 13–17 × 3.5–5 µm; spindle shaped, smooth.

Ecology: Solitary or scattered, often in groups or arcs in montane areas with fir. Typically remaining under duff well into maturity or forming "mushrumps." Scattered in drier, high elevation and east-side forests in the Cascades, more common in the southern portion of our range. Fruiting in spring and early summer.

Edibility: Edible and very good, especially favored for its dense texture, but lacking the rich flavor of the porcini (*Boletus* spp.).

Comments: The dark red to vinaceous cap, yellow pores and reticulated yellow stipe are helpful identification clues. The habit of fruiting in spring with montane conifers clinches it. *B. abieticola* grows in similar habitat but is more likely to fruits in summer and early fall. It has a pale pinkish beige cap, often covered with small scales. *B. autumniregius* tends to be more rosy or purplish pink, and fruits in fall in mixed forests at lower elevations in western Cascadia. *B. querciregius* has a pinkish cap and grows with oaks. *B. persolidus* has a brown to cinnamon brown cap and fruits in fall with hardwoods. All butter boletes are edible but should be cooked well. None are common in Cascadia and are more likely to be found on the southern portion of our range.

Misapplied name: *Boletus regius* (a name for a European species) had long been applied to three different red-capped butter boletes in western North America.

'Pulchroboletus' smithii

(Thiers) comb. prov.

Smith's Bolete

Cap: 5–15 (25) cm across, convex at first, becoming broadly convex to nearly plane, and often wavy and slightly uplifted in age. Surface dry, finely velvety, or suede-like. Dark to pale rosy pink when young, fading patchily to yellow and then gray; often with a combination of pink, yellow, and gray at the same time. Older caps can be extensively gray to olivaceous gray. **Tubes:** Often deeply sunken near stipe, 0.5–2 cm long. **Pores:** Small, round to irregular. Yellow at first, occasionally with a reddish blush to dingy yellow to reddish in age. Often staining blue slowly and erratically when bruised (quickly in some specimens, while others don't stain at all). **Stipe:** 5–18 cm long, 1–3.5 cm thick, equal or enlarged lower, often tapering at base. Rosy pink over upper portion, yellow to yellowish white at base. Surface dry, finely punctate with pinkish dots when young, mostly smooth in age. Base often with fine whitish to yellow rhizomorphs. **Flesh:** Thick to thin, firm when young, soft in age. Yellow, but retaining thin pinkish zone under skin of cap, even in age. **Odor:** Indistinct. **Taste:** Mild. **Spore deposit:** Olive-brown. **Microscopy:** Spores 14.5–19 × 4–6 µm, spindle shaped to cylindrical, smooth.

Ecology: Solitary, scattered on in small clusters on ground, in moss or duff under conifers, especially hemlock, fir, and spruce. Common in conifer forests from the coast through the Cascades. Fruiting in summer and fall.

Edibility: Unknown. A closely related eastern counterpart is mildly toxic.

Comments: This common bolete is recognized by the rosy pink cap that fades irregularly to gray, with an intermediate yellow phase, yellow pores that become reddish blushed, and rosy pink stipe apex. *Butyriboletus autumniregius* has a rosy red, smooth (not velvety) cap and a yellow, finely reticulated stipe. Faded specimens could be mistaken for *Caloboletus rubripes,* which has a gray to grayish brown soft suede-like cap, pale yellow flesh that quickly stains blue, a bitter taste, and red that is restricted to the base of the stipe when young (spreading upward in age). *Xerocomellus* are generally smaller and most have cracking caps.

Synonym: *Boletus smithii* Thiers

'Boletus' mirabilis

(Murrill) Murrill

Admirable Bolete

Cap: 4–15 (22) cm across, domed at first, soon becoming broadly convex, occasionally slightly uplifted in age. Margin with a narrow band of overhanging tissue. Surface moist to dry, distinctly roughened-tomentose to coarsely hairy, with stiff short hairs. Dark reddish brown, maroon-brown to brown, often with a paler margin, and pale pinkish brown to whitish spots, especially in age. **Tubes:** Attached or sunken around stipe, 1–4 cm long. **Pores:** Small to moderately large, round to irregular. Pale yellow to greenish yellow at first, becoming dingy olivaceous yellow in age. **Stipe:** 7–20 (25) cm long, 1–3 cm thick at apex, up to 7 cm thick at base, club shaped with a large, swollen base. Surface with sparse reticulation over apex, smooth below, base often viscid. Dark brown to reddish brown, spotted and streaked with paler pinkish buff to beige color. **Flesh:** Thin (cap is mostly tubes), soft in cap; firm, fibrous in stipe. Whitish, buff to pale yellow, often with vinaceous stains in age. **Odor:** Indistinct. **Taste:** Lemony. **Spore deposit:** Olive-brown. **Microscopy:** Spores 18–22 × 7–9 µm, spindle-shaped to elongate-ellipsoid, smooth.

Ecology: Solitary or scattered on well-decayed, moss-covered logs and stumps, or occasionally on ground under hemlock. Often found near *Chroogomphus tomentosus*, which is likely a parasite of this species. Common and abundant in western Cascadia, occasional on the eastern slopes, fruiting in fall.

Edibility: Edible and good. Older caps can be quite mushy and waterlogged but the stipes usually stay solid.

Comments: One of our most distinctive boletes. The roughened-tomentose reddish-brown cap with pale spots, long club-shaped stipe, and growth from rotten stumps and logs readily distinguish this species. Although it has been classified as a *Boletus*, *Boletellus*, and *Xerocomus* in recent years, genetic data suggests that it is close to *Aureoboletus*, but the case can be made that it belongs in a genus of its own. *Boletus fibrillosus* has an appressed-fibrillose dark brown to yellow-brown cap and white pores when young. *Xerocomellus atropurpureus* lacks the tomentose cap and has an equal to clavate red-punctate stipe.

Hemileccinum subglabripes

(Peck) Halling

Butterfoot Bolete

Cap: 4–10 cm across, convex to nearly flat. Surface moist to dry, smooth. Ocher-orange to pale reddish-brown, occasionally more yellow-brown. **Tubes:** Broadly attached to stipe or with a narrow notch, 1-2 cm long. **Pores:** Small, round or slightly irregular, yellow, unchanging. **Stipe:** 5–10 cm long, 1–2 cm thick, cylindrical, equal. Surface dry, extensively finely scurfy. Butter yellow to brighter yellow, occasionally with a reddish blush over lower portion. **Flesh:** Firm when young, mushy in age. Pale yellow in cap, brighter yellow in stipe. **Odor:** Indistinct. **Taste:** Mild. **Spore deposit:** Olive-brown. **Microscopy:** Spores 11–14 × 3–5 µm, narrowly spindle shaped, smooth.

Ecology: Solitary or scattered from duff or moss. Most common under conifers, especially spruce in Cascadia, with birch northward. Typically fruiting in mid-summer in areas with coastal fog, into early fall at lower elevations across Cascadia.

Edibility: Edible.

Comments The smooth cap, non-staining yellow pores, and the cylindrical, finely scurfy stipe are distinctive. This widespread species occurs with birch across much of North America but is found with conifers in the Pacific Northwest. *Aureoboletus citriniporus* has a dark reddish brown to dark brown cap and brilliant yellow pores. It is rare in Cascadia, occurring with oaks. *A. flaviporus* also has bright yellow pores but has a slimy or sticky cap that is paler pinkish tan to orange-brown. Common in California with oak, it has been reported (but not confirmed) from Oregon. *Xerocomus subtomentosus* has a dry, velvety cap surface, and yellow pores that slowly bruise blue. *Buchwaldoboletus sphaerocephalus* has a bright yellow tomentose cap and yellow to reddish pores. *B. lignicola* has a similarly-textured cap but is ocher-brown and grows among decaying wood. *Pulveroboletus ravenelii* has a sulphur yellow to red-blushed cap and a bright yellow, powdery partial veil. It is rare in Cascadia and more frequently found southwards.

Xerocomus subtomentosus group

Boring Brown Bolete

Cap: 4–15 cm across, convex at first, soon broadly convex to plane, occasionally uplifted, lobed or wavy in age. Surface dry, velvety to suede-like, often with cracks near edge of cap, becoming matted or smoother with age and sometimes nearly metallic in dry weather. Brown to olive-brown or reddish brown but often with a pale sheen, becoming tan-brown with dark blotches, to paler overall in age. **Tubes:** Sunken around stipe, or occasionally broadly attached, usually with a small zone of elongated pores running slightly down stipe. **Pores:** Round, irregular to angular, small to rather large, even on same fruitbody. Bright yellow to straw yellow, or becoming ocher-brown in age. Slowly bruising blue when damaged. **Stipe:** 4–10 cm long, 1–3 cm thick, cylindrical and often curving, with a tapered base attached to a tuft of yellow rhizomorphs. Surface smooth or with weak to prominent elongated reticulum. Whitish to yellowish at first, often extensively brown streaked in age. **Flesh:** Whitish to pale yellowish, occasionally bluing slightly. **Odor:** Indistinct. **Taste:** Mild. **Spore deposit:** Dull brown. **Microscopy:** Spores 11.5–16 × 3.5–5 µm somewhat spindle shaped to broadly cylindrical, smooth.

Ecology: Solitary or scattered in troops on ground or in duff with both conifers and hardwoods. Common throughout western Cascadia, rare to occasional on the eastern slopes. Fruiting in fall.

Edibility: Edible but bland.

Comments: The velvety brown cap, slender stipe, irregular and rather large pore mouths that slowly bruise blue, and yellow basal rhizomorphs distinguish this species. It could be confused with an *Aureoboletus* but the cap is soft and velvety, the taste is mild, and the pores are duller yellow. *X. spadiceus* is similar but has whitish mycelium at the stipe base and a sterile band of tissue on the cap margin, and usually grows is coastal spruce-pine forests. *Ceriomyces oregonensis* Murrill is a western North American name for a species in the *X. subtomentosus* group but more work is needed to sort out these species.

Xerocomellus diffractus

N. Siegel, C.F. Schwarz, & J.L. Frank

Cracked-cap Bolete

Cap: 3–10 (14) cm across, rounded, bun shaped to broadly convex at first, becoming flat, occasionally slightly wavy in age. Surface dry, finely velvety-tomentose, becoming extensively cracked. Olive, olive-gray to leather-brown or tan, becoming paler in age, more rarely reddish to pinkish brown. At first showing whitish to pale yellow flesh in cracks, becoming pinkish in age. **Tubes:** Sunken around stipe, 0.5–1.5 cm long. **Pores:** Small and round at first, enlarging and becoming slightly angular, pale yellow to dingy greenish yellow in age, staining blue to bluish gray, slowly to moderately when bruised. **Stipe:** 4–10 cm long, 0.8–2 cm thick, equal or clavate, Surface dry, yellow to golden ground color, streaked with fine red punctations with red pigment more concentrated toward base, often sparse or absent towards apex. Slowly staining bluish to bluish gray when handled. **Flesh:** Firm when young, becoming soft, whitish to pale yellow in cap, yellow in stipe, bluing slowly or erratically. **Odor:** Indistinct. **Taste:** Mild. **Spore deposit:** Dull olive-brown. **Microscopy:** Spores 13–16 × 5–6 µm, subfusoid to subcylindrical, smooth, inamyloid.

Ecology: Solitary, scattered, or in troops occurring in moss, duff, or soil. Common and often abundant across Cascadia, with a wide variety of trees and habitat types. It is often abundant in urban areas under Douglas-fir. Fruiting from late summer in higher elevations, through fall or early winter in lower elevations.

Edibility: Edible, with good flavor. Older specimens become soft-fleshed, and mushy.

Comments: Long called by the European epithet "*chrysenteron,*" our common western North American Cracked-capped Bolete was recently described as a distinct species. The olive to gray, cracking cap, pale yellow pores which slowly stain bluish, and stipe with red tones restricted to the lower portion are helpful identification clues. Stature and size are quite variable in this species. The speed and intensity of staining and the red ornamentation are better features to key in on. *X. salicicola* has a brown to reddish cap, an obscurely reticulate stipe that is typically extensively red, and grows with willow, birch, and cottonwood. *X. rainisiae* has a light red blush to the lower stipe, and stains intensely blue-green when handled. *X. amylosporus* and *X. mendocinensis* have darker caps with scattered and irregular cracks, stipes with red tones up to the apex, quick blue staining, and truncate spores.

Xerocomellus mendocinensis

(Thiers) N. Siegel, C.F. Schwarz & J.L. Frank

Cap: 3–10 cm across domed or bun shaped when young, expanding to broadly convex, eventually nearly plane, occasionally wavy. Surface dry, finely velvety, cracking near the margin, often only cracking on outer quarter of cap, but occasionally extensively cracked across entire cap. Typically dark olive-brown when young (rarely almost blackish), soon olive-tan to medium brown, sometimes lighter gray in age. Flesh in cracks dull tan-brown or light olive to pallid creamy (rarely pinkish). **Tubes**: Sunken around stipe, 0.5–1.5 cm long. **Pores**: Small and irregularly round at first, soon moderately large and distinctly angular. Bright yellow to dull yellow, soon straw-colored, often with ocher or reddish brown blush in age, sometimes distinctly orange-red from start. Quickly and intensely staining dark blue when bruised. **Stipe**: 4–10 cm long, 1–2.5 cm thick, equal to clavate. Surface dry, yellow when very young, but soon extensively covered with red punctation over a yellow ground color; uppermost part of stipe often with strongest red, staining blue when handled. **Flesh**: Yellow, marbled with creamy white in some areas, reddish to tan near base, usually quickly bruising blue. **Odor**: Indistinct. **Taste**: Mild. **Spore deposit**: Olive-brown. **Microscopy**: Spores 12–15 × 4.5–5.5 µm, spindle-shaped, smooth, many with a strongly truncate (squared-off) end, smooth, inamyloid.

Ecology: Solitary, scattered in troops, or in small clumps. Fruiting in fall under both conifers and hardwoods. Widespread in western Cascadia but only common in the southern portion of our range.

Edibility: Edible but soft-textured.

Comments: In comparison to the other *Xerocomellus* in our area, the relatively dark, moderately cracked cap, quick blue staining, and extensively red, punctate stipe are good clues. The stipe of *X. amylosporus* typically discolors brownish in age and the pores stain slightly more slowly (usually darkest within 5-10 seconds). *X. diffractus* typically has the red restricted to the lower stipe and stains blue still slowly (usually to darkest tone in ~20 seconds). *X. rainisii* has a thick velvety-tomentose cap skin that only cracks in age, a golden yellow stipe with a reddish blush at the base, and stains intensely dark blue-green on the stipe when handled. *X. salicicola* typically has pinkish or reddish tones to the cap and a smeary red-toned stipe without distinct punctation. Additionally, its associations with willow, poplar, and birch (often in swampy habitats) help to distinguish it.

Misapplied name: *Xerocomellus (Boletus) truncatus* represents an eastern North American species.

Xerocomellus amylosporus

(A.H. Sm.) J.L. Frank & N. Siegel

Dark Cracked-cap Bolete

Cap: 3–10 cm across, rounded, bun shaped to broadly convex at first, becoming flat or slightly wavy in age. Surface dry, very finely velvety, becoming cracked, at first showing whitish to pale yellow flesh in these cracks, becoming pinkish in age. Vinaceous-brown, dark gray-brown to olive-brown. **Tubes:** Sunken around stipe, 0.5–1.5 cm long. **Pores:** Small and round at first, enlarging and becoming slightly angular. Pale yellow to dingy greenish yellow in age. Staining inky blue moderately quickly when bruised. **Stipe:** 4–10 cm long, 0.8–2 cm thick, equal or club shaped. Surface dry, covered with red punctation or longitudinal striations over a yellowish base color when young, soon becoming brownish to reddish brown over much of the stipe to extensively brown with a red band at apex. **Flesh:** Firm, fleshy. Whitish to pale yellow in cap, yellow in stipe, erratically and slowly bluing. **Odor** Indistinct. **Taste:** Mild. **Spore deposit:** Dark reddish olive to olive-brown. **Microscopy:** Spores 13–16 × 5–6.5 µm, fusoid to subcylindrical, most with truncate end, smooth, weakly to distinctly amyloid.

Ecology: Solitary or scattered from soil and duff under conifers. Occasional across Cascadia with most collections from mature coastal spruce forests and old-growth forests in the Cascade foothills. Fruiting in fall.

Edibility: Edible.

Comments: *X. amylosporus* is best identified by its red-punctate stipe (which develops brown tones in age) and amyloid, truncate spores. Distinguishing this species from other *Xerocomellus* can be difficult in the field, as their cap color and cracking is highly variable. The speed of staining, stipe color, and ornamentation are better features to focus on. *X. mendocinensis* is most similar but has a more extensively red stipe that does not discolor brownish and the pores stain blue more quickly. *X. diffractus* typically has red tones restricted to the lower stipe and pores which stain more slowly.

Xerocomellus salicicola

C.F. Schwarz, N. Siegel & J.L Frank

Swamp Cracked-cap Bolete

Cap: 3–10 cm across, rounded-convex at first, becoming broadly convex to plane., Surface dry, finely velvety-tomentose or suede-like when young, soon extensively cracked. Dull red, tan to brown, pinkish yellow in cracks. **Tubes:** Slightly sunken around stipe, 0.5-1.5 cm long. **Pores:** Slightly angular or irregular. Pale yellow to dingy yellow, staining dark dingy blue quickly when bruised. **Stipe:** 3–9 cm long, 1–2 (4) cm thick, equal or with a tapered base. Surface dry, at times coarsely reticulated over upper portion. Reddish overall or yellowish with red streaks and a reddish band near the apex. **Flesh:** Moderately thick, firm, fleshy, stipe fibrous. Light yellow, staining blue when cut. **Odor:** Indistinct. **Taste:** Mild. **KOH:** Yellow on flesh. **Spore deposit:** Dull olive brown. **Microscopy:** Spores 12–14.2 × 4.5–6 µm, elongate ellipsoid to spindle-shaped, smooth.

Ecology: Solitary or scattered in small groups often from wet, muddy, or hard-packed soil—typically associated with willow or cottonwood. Also in lawns and other landscaped areas with birch. Typically fruiting in fall but can fruit in spring and summer, especially in irrigated areas.

Edibility: Edible.

Comments: Although quite similar to the other *Xerocomellus*, the easiest way to recognized *X. salicicola* is by its growth with willow, cottonwood, and birch. *X. diffractus* can have similar cap colors but the red on the stipe is typically concentrated around the base. *X. dryophilus* has a felty red cap and grows with oaks. Both *X. mendocinensis* and *X. amylosporus* have extensively red stipes which are distinctly punctate and have truncate spores. *Hortiboletus coccyginus* can occur in similar habitat but occurs with a wider variety of trees overall. It has a red cap and stipe and yellow pores which do not stain blue. There are two other *Hortiboletus* species introduced to the region with hardwoods. Both have red caps, reddish-blushed to red stipes, and weak staining reactions.

Xerocomellus rainisiae

(Bessette & O.K. Mill.) N. Siegel, et al.

Cap: 4–12 cm across, rounded to convex when young, becoming broadly convex to nearly plane. Surface dry, velvety to velvety-tomentose when young, more matted-tomentose and becoming cracked in age. Dark olive, olive-brown to blackish brown when young, becoming paler olive-brown, olive-gray to yellowish brown with yellowish flesh showing between cracks. **Tubes:** Broadly attached, or with a narrow notch at stipe, 0.5–1.5 cm long. **Pores:** Small, round to angular. Yellow to golden yellow when young, becoming dark golden to yellow-olive, and often developing a reddish blush with age. Quickly staining dark blue-green when bruised, then slowly brownish. **Stipe:** 3–9 cm long, 1–3.5 cm thick, club-shaped, enlarged lower when young, becoming more equal with age, typically with a pinched base at all stages. Surface dry, smooth or with scattered punctations, base with white mycelium. Bright yellow to golden yellow, often tinged with reddish longitudinally streaks, and red blotches near base. Staining dark blue-green to dark teal-blue to greenish black when handled. **Flesh:** Moderately thick, firm when young. Yellow with red in stipe base and around larva tunnels, staining blue when cut. **Odor:** Indistinct. **Taste:** Indistinct. **Spore deposit:** Olive-brown. **Microscopy:** Spores 13–16.5 × 5–6.5 µm; elongate ellipsoid to spindle-shaped, smooth, ochraceous in KOH.

Ecology: Solitary, scattered in troops, or small clusters under conifers across Cascadia. Uncommon to rare with most collections from old-growth forests. Fruiting along the coast from mid to late fall and into winter, and at higher elevations in summer and fall.

Edibility: Edible.

Comments: *X. rainisiae* is distinguished by its velvety cap, lack of extensive red punctations on the stipe, and intense greenish black to bluish-black staining on the lower stipe. This species has a thicker velvety-tomentose layer on the cap than other Cascadia *Xerocomellus*, which can be helpful in identification of older specimens. Both *X. mendocinensis* and *X. amylosporus* look similar but have an extensively red-punctate stipe and truncate spores. *X. diffractus* stains slowly bluish to bluish gray on the stipe.

Xerocomellus zelleri

(Murrill) Klofac

Zeller's Bolete

Cap: 2–6 (9) cm across, convex to plane, occasionally uplifted and wavy in age. Surface moist to dry, smooth or finely roughened, wrinkled or pitted. Dark vinaceous black, occasionally dark olivaceous black to brownish black or olive-brown to gray, with a paler, whitish beige to yellowish tan band around the extreme margin. **Tubes:** Sunken around stipe to slightly decurrent, 0.5-1.5 cm long. **Pores:** Small, round to slightly angular or irregular. Pale creamy yellow to dingy yellow when young, to dingy yellow-olive, occasionally developing reddish blushes in age, not bruising blue, or occasionally erratically so in older, waterlogged specimens. **Stipe** 2–7 cm long, 0.5–2 cm thick, equal or with a tapered base. Surface dry, covered extensively with fine rosy red punctuations over a yellow ground color when young, becoming evenly rosy red to dark red in age. **Flesh:** Thin, firm, light creamy yellow, not bruising blue, or slightly so in older specimens. Stipe fibrous, light dingy yellow, reddish at base in age. **Odor:** Indistinct. **Taste:** Mild to lemony. **Spore deposit:** Olive-brown. **Microscopy:** Spores 12.2–16.7 × 4.6–6.2 µm, elongate ellipsoid to spindle-shaped, inamyloid, smooth. Pileipellis a trichoderm with subglobose to ellipsoid underlying cells and clavate to cylindrical awl-shaped terminal cells.

Ecology: Scattered or in troops, on or around well-decayed stumps, logs and in moss-covered debris under conifers. Rare to locally common, typically found in mature and old-growth forests in coastal habitats, especially on the Olympic Peninsula and in the Cascade Foothills. Fruiting in fall.

Edibility: Edible.

Comments: The dark cap with a pale band around the margin, pallid yellow pores that rarely bruise blue, and slender red stipe distinguish *X. zelleri*. Although it can be locally common in appropriate habitat, (e.g. old-growth forests on the Olympic Peninsula) it is quite rare across Cascadia. The abundant *X. atropurpureus* can be nearly indistinguishable. It is best distinguished by the stockier stature, more variable cap color (often with more purple, reddish, or vinaceous tones), typically without the paler margin, and often with a pale sheen when young.

Xerocomellus atropurpureus

J.L. Frank, N. Siegel & C.F. Schwarz

The Imposter

Cap: (3) 5–10 (14) cm across, rounded to broadly convex at first, becoming flat occasionally slightly wavy in age. Surface dry to moist, often with a faint pale bloom, sometimes finely pruinose when young, rugulose to extensively bumpy-wrinkled, occasionally smoother. Irregular cracking can occur in age or under dry conditions. dark blackish purple, deep reddish purple to dark wine-red or reddish brown, sometimes paler red, rarely with extensive olivaceous tones. **Tubes:** Sunken around stipe, 0.5–1.5 cm long. **Pores:** Moderately small, round to slightly angular. Pale yellow to dingy greenish yellow in age, occasionally reddish-blushed. Unchanging, or rarely bruising blue. **Stipe:** 4–12 cm long, 0.8–3 cm thick, equal or club shaped with an enlarged base. Surface dry, streaked, or with sparse punctations. Red to rosy red, over a yellowish ground color, base whitish to pale yellow. **Flesh:** Firm, fleshy, yellow, unchanging or bluing slightly. **Odor:** Indistinct. **Taste:** Mild to lemony. **Spore deposit:** Dull olive-brown. **Microscopy:** Spores 12–16 × 4.8–5.8 µm, elongate to ellipsoid. Pileipellis a trichoderm with cylindrical to elongated cells.

Ecology: Solitary or scattered in troops from duff, soil, or well-decayed wood. Growing with a wide range of trees, both conifers and hardwoods. Very common and widespread. Mostly fruiting in fall and winter, with scattered fruitings in spring.

Edibility: Edible and good.

Comments: The dark, mostly uncracked caps, dull yellow pores, red stipe, and lack of blue staining (in most cases) help set this species apart. Although very similar to *X. zelleri*, *X. atropurpurpureus* is far more common and widespread, occurring across Cascadia in a wide variety of habitats. *X. zelleri* appears to be restricted to wet, west-side, coastal forests—typically with an old-growth component. Morphological criteria to distinguish these two species are subtle – *X. atropurpureus* is typically stockier, with a reddish purple to wine purple cap, rather than the dark vinaceous black to dark olivaceous black cap of the more slender-statured *X. zelleri*. Additionally, *X. zelleri* almost always shows a narrow, pale band around the cap margin, whereas this feature is more variable in *X. atropurpureus*. With atypical specimens of either species, a quick check of the pileipellis will be helpful: the trichoderm of *X. atropurpureus* is made up of only of cylindrical cells, whereas *X. zelleri has* globose to ellipsoid cells.

Rubroboletus pulcherrimus

(Thiers & Halling) D. Arora, et al.

Beautiful Satan's Bolete

Cap: 8–20 (25) cm across, rounded when young, becoming broadly convex, margin with a narrow sterile band, incurved when young, rounded to slightly irregular in age. Surface dry, finely velvety to smooth, occasionally with flattened warts or patches, becoming finely cracked in age. Color variable, often deep rosy red, burgundy to reddish brown when young, but can be olive to gray with a rosy pink blush on margin. Becoming more grayish olive overall in age. **Tubes**: Sunken around stipe, 0.5–2 cm long. **Pores**: Small, round to slightly irregular. Deep red at first, occasionally paler toward margin, becoming orange-red to reddish brown in age. Staining blue-black intensely and immediately when bruised. **Stipe**: 7–15 (20) cm long, 3–7 cm thick at apex, 5–12 thick at base, club shaped with an enlarged or bulbous base, occasionally equal or nearly so. Red, pink to orangish red, covered with deep red reticulation. Staining deep bluish to bluish black when handled, discoloring from base up in age. **Flesh**: Thick, firm. Pale yellow, staining blue and fading to grayish yellow, often reddish in lower stipe and around larva tunnels. **Odor**: Indistinct. **Taste**: Mild. **KOH**: Brown on cap, tubes, and stipe, orangish brown on flesh. Spore deposit: Dark olive-brown. **Microscopy**: Spores 13–16 × 5.5–6.5 µm, elongate to subellipsoid, smooth.

Ecology: Solitary or scattered in small patches on ground under conifers, especially grand fir and western hemlock. Rare in Cascadia; mostly in coastal forests as well as the Cascade Foothills. Fruiting in late summer and fall.

Edibility: Extremely toxic, causing severe gastric distress.

Comments: This beautiful, highly toxic bolete is an uncommon resident of lowland conifer forests. It is recognized by the large size, rosy pink to red cap that fades in age, red pores, large, red reticulated stipe, and intense blue staining on all parts of the fruitbody. *R. eastwoodiae* has a pinkish cap that fades to gray,a stout stipe with a large swollen base and narrow apex, and it grows with oak. *R. haematinus* is a similar conifer-associated species but it differs by having a pinkish cap that quickly goes brown as it ages and paler, orangish red pores when young. It occurs in high elevation fir forests in the Cascade Range.

Synonym: *Boletus pulcherrimus* Thiers & Halling

Suillellus sp. 'Pacific Red'

Cap: 7–15 (20) cm across, convex to broadly convex, occasionally wavy in age. Surface dry to slightly tacky, smooth. Orange-brown to reddish brown, darkening with age. Quickly staining dark blue when bruised. **Tubes:** Sunken around stipe, 1-3 cm long, yellow, quickly staining blue. **Pores:** Small, round to slightly angular. Reddish-orange to red at first, becoming orange to ocher-orange, quickly bruising bluish-black when damaged. **Stipe:** 8–18 cm long, 2–4 (6) cm thick, equal to narrowly club-shaped. Surface finely punctate or appearing smooth. Punctuation red over a yellow ground color, to more extensively yellow with a reddish wash in age. Quickly staining bluish black when handled. **Flesh:** Thick, firm, fleshy. Pale yellow, red in stipe base, quickly staining blue when cut. **Odor:** Indistinct. **Taste:** mild. **KOH:** Unknown. **Spore deposit:** Olive-brown. **Microscopy:** Spores 11–14 × 5–6.5 µm, Spindle-shaped to elongate, smooth.

Ecology: Solitary or scattered, occasionally in pairs; fruiting from soil, moss, and duff under conifers—especially Douglas-fir. Uncommon, found in wetter forests in the Coast Range and Cascade Foothills. Fruiting in fall.

Edibility: Unknown.

Comments: This robust species is recognized by its orange-brown to reddish brown cap, reddish orange pores, non-reticulate stipe, and deep blue staining. Fruitbodies of *S. amygdalinus* are typically more stout and grow with oaks. *Neoboletus luridiformis* sensu western North America has a deep red velvety cap and red pores. It fruits in summer and early fall in coastal spruce forests. *Rubroboletus* have finely reticulated club-shaped or bulbous stipes.

Misapplied name: *Boletus erythropus* sensu Thiers

Neoboletus flammeus

(A.H. Sm.) D. Arora & J.L. Frank nom. prov.

Cap: 3–8 cm across, irregular domed to convex at first, generally with a sunken center, with an uplifted and highly irregular margin in age. Surface dry, velvety at first, matted-tomentose to appressed-fibrillose in age, bruising navy blue when damaged. Reddish orange, reddish brown to yellowish brown when young, becoming dingy brown to dark brown, but often with a yellowish margin in age. **Tubes**: Distorted, often anastomosing, and pulling away from stipe, quite long. **Pores**: Highly irregular, fused, and facing many directions, mouths often sealed with tissue when young. Yellowish reddish to orange-red, quickly staining dark blue to bluish black, becoming quite dingy in age. **Stipe**: 1–3.5 (7) cm long, 0.7–2 cm thick, central to off-centered, equal or tapering toward base, often short. Yellowish to orange with a reddish base, darkening from base up in age. Quickly staining deep bluish black when handled. **Flesh**: Thin, firm, yellow, with red in stipe base and around larva tunnels, quickly bluing when cut. **Odor**: Indistinct. **Taste**: Mild. **KOH**: Orangish on cap, flesh, and tubes. **Spore deposit**: None. **Microscopy**: Spores 13.5–17 × 5–8.5 µm, ellipsoid to somewhat spindle shaped, smooth.

Ecology: Solitary, scattered, or in smaller clusters. Often partially buried in duff under conifers, especially fir. Common in montane fir forests and occasional in coastal spruce forests. Fruiting in summer and early fall.

Edibility: Unknown.

Comments: The distorted, reddish-orange fruitbodies, highly irregular tubes and pores, quick blue staining, and partially buried growth habit distinguish this species. It can be abundant in years with rainy summers in the mountains. Gastroid fruitbody morphologies have evolved within many different genera of boletes (in the past, these were lumped into their own genus, *Gastroboletus*). Most of these species occur in the drier, east-side, high-elevation forests of the Cascade Range. Among these are gastroid porcini (*Boletus subalpinus*), a gastroid from of *Xerocomellus amylosporus* (=*Gastroboletus xerocomoides*), a close relative of *'Pulchroboletus' smithii* (*Gastroboletus vividus*), gastroid *Suillus* and others.

Synonym: *Gastroboletus turbinatus* var. *flammeus* A.H. Sm.

Chalciporus piperatus

(Bull.) Bataille

Peppery Bolete

Cap: 2–9 cm across, rounded to weakly conical or bun shaped when young, becoming broadly convex to nearly plane and wavy in age. Margin with a narrow band of overhanging tissue. Surface viscid to moist, innately fibrillose to smooth. Color variably reddish brown, ocher-brown, brick red, light brown to pinkish tan with darker streaks; usually paler toward margin, and often pinkish ocher to beige in age. **Tubes**: Sunken around stipe, relatively thin. **Pores**: Rather large and angular. Coppery pinkish tan to pinkish or ocher-brown. Unchanging, or staining slightly reddish brown. **Stipe**: 4–10 cm long, 0.6–1.5 cm thick, cylindrical with an enlarged base. Pinkish beige to reddish tan or ocher brown, with brilliant yellow mycelium at base. **Flesh**: Thin, soft. Pinkish buff to light brown, to pale ocher-brown in stipe. **Odor**: Indistinct. **Taste**: Acrid, peppery. **KOH**: Dark reddish brown on cap. Spore deposit: Olive-brown. **Microscopy**: Spores 8.5–12 × 3–4 μm, somewhat spindle shaped to ellipsoid, smooth, thick walled.

Ecology: In scattered troops, sometimes in small clusters on ground, in duff, or moss. Most often in conifer forest, almost always near *Amanita muscaria* (and sometimes members of the *A. gemmata* and *A. pantherinoides* groups). Common throughout our range. Fruiting from late summer through fall.

Edibility: Best to avoid, possibly toxic (at least when raw).

Comments: The fruitbodies of *Chalciporus* species are the smallest of any of the boletes in our area. In combination with their coppery pinkish tan to ocher pore mouths, peppery taste, and bright yellow mycelium at the stipe base, identification of the genus is straightforward. Those *Chalciporus* with blue-staining pores are usually called *C. piperatoides* but a number of different species likely go by each of these names. All these *Chalciporus* usually fruit in close proximity to *Amanita* (those in the *Amanita* section, especially *A. muscaria*) and probably parasitize their mycelia.

Gyroporus borealis

Davoodian, O. Asher, et al.

Chestnut Bolete

Cap: 4–10 cm across, rounded to broadly convex young, to nearly plane, slightly lobed or with uplifted areas in age. Surface dry, velvety or suede-like, to smoother in age. Warm brown to orange-brown, sometimes with rather bright orange areas, fading to brownish tan or ocher-tan. **Tubes:** Broadly to narrowly joined to stipe, shallow. **Pores:** Round, quite small, white to creamy or pale straw-colored, developing yellowish ocher and brown tones in age. **Stipe:** 4–8 cm long, 1–2.5 cm thick, equal, or often squat with a swollen, club-shaped base. Surface finely pruinose to smooth, often with multiple cracks around circumference in dry weather. Orange to ocher-brown, strongly contrasting with pores. **Flesh:** Thin, mealy and brittle, white. Stipe stuffed with white pith when young, often hollow at maturity. **Odor:** Pleasant. **Taste:** Mild to sweet and nutty. **KOH:** Orange on cap. **Spore deposit:** Creamy yellow to straw colored. **Microscopy:** Spores 10–14 × 5–6 μm, ellipsoid, smooth, colorless to pale yellow in KOH.

Ecology: Solitary, in groups or sometimes in arcs under oaks, and very rarely pine. Rare, typically fruiting in urban areas in summer and fall.

Edibility: Edible and good.

Comments: This bolete is identified by its small size; orange-brown cap and stipe with contrasting creamy white pores; hollow stipe; and brittle, mealy texture. Although rarely found in the Pacific Northwest, patches can be quite prolific.

Misapplied name: *Gyroporus castaneus* (Bull.) Quél.

Porphyrellus porphyrosporus

(Fr. & Hök) E.-J. Gilbert

Dark Bolete

Cap: 5–15 (18) cm across, rounded to bun shaped at first, becoming broadly convex to plane. Surface dry, finely velvety to suede-like. Dark grayish black to dark vinaceous brown at first, fading to grayish brown, or dark gray. Often getting parasitized by *Hypomyces*, which starts as a whitish bloom over cap, before covering whole fruitbody with a white mold. **Tubes:** Broadly attached to deeply grooved around stipe, short when young, up to 1.5 cm long in age. **Pores:** Small and angular at first, becoming moderately large and irregular in age. Color variable, typically deep reddish brown to reddish black at first, fading to grayish brown, reddish gray to gray. Staining dingy greenish blue to bluish black, then maroon-brown when bruised (sometimes without blue phase). **Stipe:** 6–15 cm long, 1–3 cm thick, often club-shaped with a swollen lower portion, but tapering at extreme base, to more or less equal in age. Surface dry, not reticulated or with extensive reticulum over upper portion. Dark grayish black, deep reddish brown, grayish brown to gray with a whitish base, often staining reddish when scraped. **Flesh:** Thick, firm, fleshy. Whitish, often staining bluish in cap, then reddish brown, typically unchanging, or reddening in stipe. **Odor:** Indistinct. **Taste:** Mild. **Spore deposit:** Reddish brown. **Microscopy:** Spores 12–18 × 6–7.5 µm, cylindrical to narrowly ellipsoid, smooth.

Ecology: Solitary or scattered in troops on ground, moss, or duff, occasionally on decaying logs and stumps in conifer forest—especially in coastal areas with Sitka spruce. Common in coastal forests, rare elsewhere. Fruiting from late summer into early winter.

Edibility: Reportedly edible but local experience is lacking.

Comments: The dark colors alone set this bolete apart. Other distinguishing features are the dry velvety cap, contrasting white flesh, and bluish to reddish staining reaction. More work is needed to determine whether the western North American taxon is the same as the European *P. porphyrosporus.* If the western USA taxon turns out to be distinct, *P. olivaceobrunneus* would be the correct name our species.

Leccinum ponderosum group

Copper Tops

Cap: 7–20 (25) cm across, rounded when young, becoming broadly convex to bun shaped and then nearly plane and sometimes wavy. Margin with a thin flap of overhanging tissue. Surface viscid to dry, appressed-fibrillose to nearly smooth. Dark red-orange, red-brown to orange. **Tubes:** Broadly attached or sunken around stipe, 1-4 com long. **Pores:** Round, fairly small, whitish cream (in some collections dusky gray-brown at first but becoming paler as cap expands), then darkening and becoming dingy in age. **Stipe:** 9–20 cm long, 2–5 (8) cm thick, club-shaped, or swollen and bulbous at midpoint. Surface dry, covered with whitish to pinkish brown scabers over a whitish ground color, scabers darkening with age. Base with blotches of blue to blue-green stains. **Flesh:** Thick, firm, stipe tough, fleshy-fibrous. Whitish, slowly staining grayish, with or without a pinkish stage. **Odor:** Indistinct. **Taste:** Mild. **Spore deposit:** Dull brown. **Microscopy:** Spores 14–17 × 4.5–6 µm, spindle shaped to cylindrical, smooth.

Ecology: Solitary or scattered from duff, moss, and soil under conifers—especially two-needle pines. Common across Cascadia, fruiting in late summer and fall.

Edibility: Edible with caution. Some people have adverse reactions to orange-capped *Leccinum.* If you do choose to eat it, always cook it well.

Comments: The orange-red capped *Leccinum* are a confusing complex of species. Although there are a number of available names for western North American species, there are no clear concepts for applying them. The species described above roughly matches the current concept of *L. ponderosum.* A darker-capped species often found in fir-hemlock forests with vinaceous to grayish black-staining flesh appears to be *L. fibrillosum. L. manzanitae* is a large species that grows with manzanita and madrone. *L. insigne* has a paler orange cap and grows with aspen.

Boletes

Leccinum scabrum

(Bull.) Gray

Birch Bolete

Cap: 4–12 cm across, hemispheric at first, soon bun shaped, to broadly convex, sometimes uplifted with a wavy margin in age. Surface thinly viscid to dry, suede-like to finely tomentose to nearly smooth. Dark brown to dark gray at first, generally fading to gray to tan in age. **Tubes:** Deeply depressed at stipe, relatively shallow at first, becoming long. **Pores:** Round, tiny at first, staying small in age. Creamy white to beige when young, darkening and becoming dingy with age, slowly staining light gray-brown when bruised. **Stipe:** 5–15 cm long, 1.5–4 cm thick, typically club-shaped, to more equal in age. Surface dry, densely covered with blackish tufts or scabers over a whitish to pale gray ground color, scabers lightning, and ground color becoming dingier with age. Base often with whitish mycelium. **Flesh:** Thick, firm, fleshy in cap. Stipe fibrous. Whitish marbled with beige; unchanging when cut. **Odor:** Indistinct. **Taste:** Mild. **KOH:** Reddish brown on cap. **Spore deposit:** Dusky brown. **Microscopy:** Spores 15–19 × 5–7 µm, spindle shaped, smooth.

Ecology: Solitary or scattered in troops, often on lawns, always with birch. This is one of the most common mushrooms found with planted birch trees, along with *Paxillus cuprinus* and *Lactarius necator*. Common in fall but can occur in summer in irrigated areas.

Edibility: Edible and with a good flavor, but often quite slimy.

Comments: The association with birch, dark brown to gay cap, pallid pores, and whitish stipe covered with dark scabers help with identification. Habitat, grayish cap, pale stipe with grayish scabers, and blue-green stains at the base distinguish this species. *L. schistophilum* and *L. cyaneobasileucum* are two similar urban birch associates which stain blue on the lower stipe and typically have paler brown to gray caps. *L. brunneum* has a dark brown cap with a thin band of overhanging sterile tissue, and grows with aspen. Identification of *Leccinum* is difficult, as they have many overlapping features. Tree association, staining of flesh, and the presence or absence of a sterile band on the cap margin are important features to note. Diversity of *Leccinum* greatly increases as you go east into the Rocky Mountains or north into boreal forests.

Suillus punctatipes

(Snell & E.A. Dick) Singer

Cap: 4–15 (20) cm across, round to convex at first, becoming broadly convex to plane, margin incurved when young, undulating in age. Surface smooth to appressed-fibrillose, viscid to glutinous when wet, tacky when dry. Color extremely variable and often blotchy, typically dark brown to orange-brown, but often grayish brown, or even pallid, creamy, orange or tan (or any combination of these). **Tubes:** Broadly attached to decurrent, shallow, 0.2–0.7 cm long. **Pores:** Radially arranged, small and slightly irregular when young, more elongate in age. Creamy, at times with milky droplets when young, becoming warm yellow to ocher-yellow, dingier olive-yellow when old. **Stipe:** 3–10 cm long, 1–3 (4) cm thick, variable, stout to elongate, tapered towards base, equal or clavate with a pinched base. Surface greasy to sticky, covered with glandular dots. Whitish to yellowish, developing orange to brown stains, dots pinkish tan at first, darkening in age. **Flesh:** Firm and whitish when young, soft and becoming yellowish in age, showing brownish stains around larva tunnels. **Odor:** Indistinct. **Taste:** Mild. **KOH:** Unknown. **Spore deposit:** Ocher-brown. **Microscopy:** Spores 7.5–9 × 3–3.5 µm, oblong to cylindric, smooth.

Ecology: Solitary, scattered, or clustered from soil, duff, and moss in mid- to high elevation Cascade forests with Douglas-fir, true firs, and hemlock—rarer at lower elevations. It can be quite common in high elevation forests of hemlock and true fir. It's one of the few *Suillus* species which made the "host jump" from the core pines to the 'Abietoid' conifers. Fruiting in late summer and fall.

Edibility: Edible. Young specimens have firmer flesh. Many people prefer to peel the slimy cap skin before cooking. A very small percentage of people have adverse reactions to the glutinous Slippery Jacks, so they should be tried in moderation at first.

Comments: This common montane *Suillus* is highly variable in appearance, with a wide range of cap colors and statures. It is best identified by its habitat (Douglas-fir, fir, and hemlock forests), radially arranged pores, and dotted stipe. *S. subalpinus* is often slightly paler and only grows with five-needle pines. *S. pungens* has a whitish to greenish cap when young, becoming orange in age, and has dark dots on the stipe. It is a common Slippery Jack in California, where it grows with Monterey and bishop pine. Although not confirmed from the PNW, it may show up in urban areas. *S. quiescens* has a short stipe with fine glandular dots in age. It grows with two- and three-needle pines.

Misapplied name: *Suillus granulatus.*

Suillus subalpinus

M.M. Moser

Cap: 3–7.5 cm across, convex with an incurved margin when young, becoming broadly convex to plane; margin even, occasionally wavy or uplifted in age. Surface smooth, glutinous when young, viscid to tacky in age. Color variable; whitish near margin, gray-brown, reddish brown to vinaceous brown towards disc, usually variegated or blotchy, becoming more orange-brown to warm brown in age. **Tubes:** Broadly attached or slightly decurrent, 0.3–0.5 cm long. **Pores:** Small, round to angular, occasionally radially arranged. White to ivory when young, developing creamy tones, and maturing yellow. **Stipe:** 2–5.5 cm long, 1–2.5 cm thick, equal, club-shaped or tapered downwards, extreme base pointed. White when young, covered with white to reddish glandular dots that become brown in age, base developing yellowish tones. **Flesh:** Moderately thick, firm when young, soft in age, white to pale yellowish in age. **Odor:** Indistinct. **Taste:** Mild. Spore deposit: Olive to brownish olive. **Microscopy:** Spores 6.5–7.7 (9) × 2.5–3 (4) µm, slender cylindric to subfusoid, smooth, thin-walled.

Ecology: Solitary or scattered in soil under five-needle pines, especially whitebark pine. Uncommon across higher elevations in the Cascade Range, fruiting in summer or early fall.

Edibility: Edible.

Comments: Recognized by the patchy whitish to brownish cap; short, glandular-dotted stipe; lack of any partial veil tissue; and growth with five-needle pines. We include the populations occurring with western white pine in our concept of this species. *S. punctatipes* is very similar but typically has a darker cap and grows with Douglas-fir, true firs, and hemlock. *S. brunnescens* has a similarly colored cap but has a white to vinaceous gray membranous veil that leaves remnants on the cap margin and sometimes an annular zone on the stipe (often disappearing completely in age). It also grows with five-needle pines. *S. glandulosipes* has a glutinous, cinnamon pink to yellowish or reddish brown cap that darkens in age; a dotted white stipe; and a partial veil that clings to the cap margin and does not form a ring on the stipe. It grows with two- and three-needle pines.

Suillus pseudobrevipes

A.H. Sm. & Thiers

Cap: 4–12 (15) cm across, round to convex at first, becoming broadly convex to plane, margin incurved and covered with partial veil tissue when young, expanding in age. Surface smooth, but often with appressed-fibrillose streaks, viscid when wet. Color variable, often dingy yellowish to honey brown at first, darkening to reddish brown to brown in age. **Tubes:** Sunken around stipe to decurrent, shallow, 0.5–1 cm long. **Pores:** Round to irregular, tiny at first, enlarging slightly in age. Yellow at first, becoming golden, then dingy dull ocher-yellow in age. **Stipe:** 2–8 cm long, 1–3 cm thick, equal, enlarged or tapered toward base. Surface greasy to sticky, lacking glandular dots. Whitish when young, becoming yellowish and often developing dingy yellowish brown discolorations or stains. **Partial veil:** Thin, whitish (sometimes tinged purplish), floccose to thinly membranous, leaving hanging remnants on cap margin and a simple, thin ring on stipe, which can disappear altogether in age **Flesh:** Firm when young, soft in age. Whitish to pale yellow, discoloring pinkish around larva tunnels, often orangish brown in stipe base in age. **Odor:** Indistinct. **Taste:** Mild to sour; lemony. **KOH:** Brownish on cap. **Spore deposit:** Cinnamon buff to pale brown. **Microscopy:** Spores 7–9 × 2.5–3 µm, oblong, smooth, inamyloid.

Ecology: Solitary, scattered, in small clusters or troops under two- and three-needle pines. Common in coastal dune forest under shore pine and often abundant in wet years under ponderosa pine. Primarily fruiting in fall but fairly common in spring in the eastern cascades.

Edibility: Edible but soft-fleshed, slimy, and sour.

Comments: The combination of a viscid, yellowish to honey brown cap that darkens in age, pale yellow pores, stipe lacking glandular dots, and evanescent veil are distinctive. *S. glandulosipes* is similar but has dark glandular dots all over the stipe. *S. luteus* has a viscid membranous veil that forms a thick purplish skirt or ring in age. *S. brevipes* and *S. quiescens* are generally darker when young, with short, white or pale yellow stipes lacking glandular dots and veil tissue. Also compare with the Douglas-fir–associated *S. caerulescens* and *S. ponderosus.*

Suillus brevipes

(Peck) Kuntze

Short-stalked Slippery Jack

Cap: 3–10 cm across, round to convex at first, becoming broadly convex to plane. Surface smooth, viscid, covered with glutinous slime when wet, shiny if dry. Color variable, generally dark brown, but can be reddish brown, vinaceous brown to grayish brown when young, fading to cinnamon brown or even yellowish brown in age. **Tubes:** Broadly attached to slightly decurrent, relatively shallow, 0.4–1 cm long. **Pores:** Small, round to slightly irregular. Creamy when young, soon yellow, becoming golden and developing an olivaceous tint as spores mature. **Stipe:** 2–6 cm long, 1–3 cm thick, short, stout, equal or swollen towards base. Surface dry, smooth to slightly pruinose, lacking glandular dots when young, occasionally scattered in age. White, developing yellowish colorations in age. **Flesh:** Firm when young, soft in age. White, but often developing yellow color in stipe in age. **Odor:** Indistinct. **Taste:** Mild. **KOH:** Dark gray on cap, grayish brown on tubes, lilac on flesh. **Spore deposit:** Cinnamon brown. **Microscopy:** Spores 7–9 (10) × 2.5–3.5 µm, elliptic to oblong, smooth, inamyloid.

Ecology: Solitary, scattered, or in small clusters under two- and three-needle pines. Common across Cascadia; especially in coastal dune forests, and the eastern Cascades. Often fruits in large flushes in fall, but scattered specimens can pop up anytime.

Edibility: Edible, one of the better Slippery Jacks. The slimy cap "skin" peels off easily. Older specimens become very soft-fleshed and cook up rather slimy; these can be dried and powdered for use as flavoring.

Comments: This is one of the more ubiquitous Slippery Jacks in pine forests across Cascadia. It is recognized by its variably-colored, viscid cap that fades in age; yellow pores; and short, white stipe. *S. quiescens* is more cool-brown in color when young, but otherwise very similar and it can be tough to distinguish from *S. brevipes* in age. *S. pseudobrevipes* is often more robust, has a paler cap, and an evanescent partial veil – it is primarily known from ponderosa pine forests. *S. subalpinus* has a glandular-dotted stipe and grows with five-needle pines.

Suillus quiescens

T.D. Bruns & Vellinga

Quiet Jack

Cap: 4–10 (15+) cm across, round to convex at first, becoming broadly convex to plane or wavy. Surface viscid, glutinous when wet, if dry glutin drying into streaks. Color variable, extensively dark brown, or with pallid blotches or occasionally a whitish margin when young, becoming more evenly dark brown, brown to orange-brown in age. At time with greenish blue patches. **Tubes:** Broadly attached to slightly decurrent, shallow, 0.3–1 cm long. **Pores:** Small, round to slightly irregular. Creamy when young, soon yellow, becoming golden and developing an olivaceous tint as spores mature. **Stipe:** 2–5 (8) cm long, 1–3.5 cm thick, short, stout, equal or swollen towards base. Surface dry, slightly pruinose, developing fine yellow glandular dots near apex. Whitish to pale yellowish, darkening slightly, and developing dingy stains on lower portion in age. **Flesh:** Firm when young, soft in age. Whitish to pale yellow, orangish around larva tunnels in stipe. **Odor:** Indistinct. **Taste:** Mild. **KOH:** Unknown. **Spore deposit:** Cinnamon brown. **Microscopy:** Spores 6–10 (14.5) × 2.5–3.7 µm, elliptic to oblong, smooth, inamyloid.

Ecology: Solitary, scattered, or in small clusters under pines two- and three-needle pines. Widespread in Cascadia, common in coastal dune forests with shore pine, occasional elsewhere. Fruiting in fall.

Edibility: Edible. The slimy cap "skin" peels off easily.

Comments: For years *Suillus quiescens* hid in plain sight, lumped withing the range of variation of *S. brevipes*. The caps of *S. quiescens* are evenly brown when young (often with bluish-gray blotches) and darken in age. Conversely, caps of *S. brevipes* usually fade in age. The stipe of *S. quiescens* is generally pale yellowish and develops fine glandular dots in age, while the stipes of *S. brevipes* are smooth. Additionally, the pores of *S. quiescens* are typically brighter yellow than in *S. brevipes*. All of these features are more easily judged after seeing good examples of both species. S. pseudobrevipes has a pale, often blotchy cap, and an evanescent partial veil.

Suillus tomentosus

(Kauffman) Singer

Bluing Slippery Jack

Cap: 4–12 (15) cm across, round to convex at first, becoming broadly convex to plane, margin incurved and often fuzzy when young, becoming irregular and occasionally slightly uplifted in age. Surface often becoming completely smooth in age and especially in wet weather, viscid but not thick-glutinous. Yellowish to golden at first, darkening to golden-orange. Fibrils and scales grayish to yellow-brown at first, becoming reddish brown in age. **Tubes:** Broadly attached to slightly decurrent at first, becoming sunken in age, 0.5–1.5 cm long. **Pores:** Tiny at first, enlarging in age, angular. Cinnamon brown to ocher at first, becoming ocher-yellow to yellow, to dingy olive-yellow in age. Slowly staining blue to bluish gray when bruised. **Stipe:** 2–12 cm long, 1–3 cm thick, equal, enlarged lower, but often tapered at base. Yellowish to ocher-orange, surface resinous, sticky, covered whitish to pale gray glandular dots when young, dots darkening, grayish brown to reddish in age. **Flesh:** Firm when young, soft in age, whitish, yellow to ocher-orange, slowly staining bluish, especially in cap. Stipe flesh soft, sometimes becoming hollow in age. **Odor:** Indistinct. **Taste:** Mild to sour; lemony. **KOH:** Vinaceous brown on flesh. **Spore deposit:** Olive-brown to cinnamon brown. **Microscopy:** Spores 7–10 (12) × 3–4 (5) µm, spindle shaped to elongate-ovoid, smooth, inamyloid, pale yellowish in KOH

Ecology: Solitary or scattered on ground or in duff under two- and three-needle pines. Especially abundant in coastal shore pine forests, in montane lodgepole pine, and occasionally with ponderosa pine. Fruiting soon after first rains in fall into early winter on the coast.

Edibility: Edible but really soft-fleshed and sometimes sour tasting.

Comments: The yellow cap covered in grayish yellow fibrils or scales when young, ocher to yellow pores, stipe covered with glandular dots, and blue-staining flesh make this Slippery Jack distinctive. The staining can be faint or slow at times, but is best observed by pinching a bit of flesh out of the center of the cap. Darker specimens could be mistaken for *S. fuscotomentosus* but that has a dark brown cap when young and does not stain blue (or does so only slightly in the stipe base). We have yet to see it in Cascadia, but it is common in California. *S. discolor* is often dingier colored, but without experience with both species, it's tough to compare. Tree association is the best way to distinguish *S. discolor* from *S. tomentosus*, as *S. discolor* only occurs with five-needle pines. *S. caerulescens* has a whitish partial veil, orangish brown-staining pores, and blue-staining flesh in the stipe base. It is common under Douglas-fir. *Xerocomus subtomentosus* has a dry, velvety cap, large angular blue-staining pores, and a dry, not resinous stipe.

Suillus americanus

(Peck) Snell

Chicken Fat Slippery Jack

Cap: 3–7 (10) cm across, rounded to convex at first, becoming broadly convex to plane, occasionally wavy or uplifted in age. Surface glutinous when wet, appressed-fibrillose, smooth or with scattered veil patches. Bright yellow, ocher-yellow to dingy yellow in age, margin with whitish to yellowish veil tissue, which becomes reddish with age. **Tubes:** Broadly attached to subdecurrent, 0.5–1 cm long. **Pores:** Angular, fairly small. Yellow, golden yellow to ocher, staining pinkish to brownish when damaged. **Stipe:** 4–10 cm long, 0.5–1.5 cm thick, equal or tapered lower. Surface sticky, or viscid from veil, covered with glandular dots. Yellow, dots brownish to reddish, overall developing reddish stains and becoming dingy in age. **Partial veil:** Membranous tissue with a gelatinous coating, typically adhering to cap margin, at times with an annular zone on stipe. Dingy whitish to yellowish at first, developing reddish color with age. **Flesh:** Thin, often soft, mushy. Yellow in cap, olive-yellow in stipe, discoloring pinkish brown to vinaceous brown. **Odor:** Indistinct. **Taste:** Indistinct to sour. **KOH:** Blackish on cap. **Spore deposit:** Ocher-brown. **Microscopy:** Spores 8–11 × 3.5–4 µm, spindle shaped to elongate, smooth, inamyloid, pale cinnamon in KOH.

Ecology: Solitary, scattered, or in troops from soil and duff under five-needle pines. Locally common across Cascadia, especially with western white pine, and in urban areas with planted white pines. Fruiting in fall.

Edibility: Edible but slimy.

Comments: The yellow colors, partial veil remnants which discolor reddish, small size, and growth with five-needle pines are helpful in identifying this species. *S. acidus* often grows alongside this species but has an olive-brown cap and dark glandular dots on the stipe. *S. flavidus* doesn't have the vibrant yellow colors, has a remarkably gelatinous sheath- or collar-like partial veil, and grows with two- and three-needle pines.

Synonym: *Suillus sibiricus* (Singer) Singer

Suillus clintonianus

(Peck) Kuntze

Mahogany Larch Jack

Cap: (3) 5–15 cm across, broadly conical to convex when young, becoming broadly convex to plane, occasionally wavy or uplifted in age. Surface glutinous to viscid, smooth, shiny if dry. Deep orange-red, reddish brown to orange-brown, often with a bright yellow margin, rarely more extensively yellow. **Tubes:** Broadly attached to subdecurrent, 0.5–1.5 cm long. **Pores:** Angular, fairly small. Yellow, golden yellow to ocher, with olive tones when spores mature, staining brownish when damaged. **Stipe:** 4–10 cm long, 1–3 cm thick, equal or club-shaped, but often tapered at extreme base. Surface viscid when wet, fibrillose below veil. Orange-brown to reddish brown, occasionally more yellow, dingier in age. **Partial veil:** Pale yellowish membranous tissue covered with a yellow gelatinous coating, leaving a ring on stipe and becoming reddish brown. **Flesh:** Firm when young, very soft and squishy in age. Yellow, staining ocher-brown, occasionally with blue staining in lower stipe. **Odor:** Indistinct. **Taste:** Indistinct to slightly unpleasant. **KOH:** Greenish black. **Spore deposit:** Olive-brown when wet, paler when dry. **Microscopy:** Spores 8–10 × 2.5–3.5 µm, spindle shaped to elongate, smooth, inamyloid, ocher-brown in KOH.

Ecology: Solitary, scattered, or in small clusters, often in troops from soil. Common wherever larch grows in Cascadia, mostly in the eastern Cascades but also in urban areas with planted larch. Fruiting in fall.

Edibility: Edible but not popular.

Comments: The rich reddish to orangish brown glutinous cap, yellow slimy veil, and growth with larch are distinctive. *S. ponderosus* is similar, with a glutinous yellow cap and veil, but associates with Douglas-fir. In areas with mixed larch and Douglas-fir, the stockier stature and slightly paler colors are the best way to distinguish *S. ponderosus*. *S. caerulescens* has an ochraceous buff to yellowish brown, slightly viscid to dry cap, a dry whitish veil when young, and grows with Douglas-fir. *S. americanus* is a slender, bright yellow species found with five-needle pines.

Misapplied name: *Suillus grevillei* (Klotzsch) Singer

Suillus elbensis

(Peck) Kuntze

Gray Larch Jack

Cap: 3–10 cm across, broadly conical to convex when young, becoming broadly convex to plane, occasionally with a broad umbo. Surface viscid to tacky, often roughened, appressed-fibrillose to fibrillose-scaly, at times cracking when dry. Gray, gray-brown to yellowish-brown, at times with greenish tones, dingier with age. **Tubes:** Decurrent, 0.4–0.8 cm long. **Pores:** Angular, radially aligned, moderately large. Whitish to pale brownish at first, darkening to grayish or gray-brown, slowly staining bluish-green when damaged. **Stipe:** 4–10 cm long, 0.7–2.2 cm thick, equal or club-shaped, but often tapered at extreme base. Surface thinly viscid to dry, fibrillose-scaly to matted. Whitish, grayish to brown, darkening with age, staining bluish green when damaged. **Partial veil:** A thin cottony-fibrillose membrane (almost like thin tissue paper) adhering to cap margin and leaving a ring and annular zone on stipe. **Flesh:** Firm when young, soft and squishy in age. Whitish to grayish or yellowish, slowly staining light bluish. Stipe solid, colored like cap, with brownish tones developing on the lower stipe in age; more likely to stain blue. **Odor:** Indistinct. **Taste:** Mild. **KOH:** Unknown. **Spore deposit:** Vinaceous-brown when wet, paler when dry. **Microscopy:** Spores 8.5–12 × 3.5–5 µm, spindle shaped to elongate ellipsoid, smooth, inamyloid, yellowish brown in KOH.

Ecology: Solitary, scattered, or in troops, often in wet soil under larch. Uncommon, restricted to areas with larch, mostly in the eastern Cascades and northern Rocky Mountains. Also in urban areas with planted larch. Fruiting in fall.

Edibility: Edible.

Comments: *S. elbensis* is a distinctive member of Cascadia's larch-associated slippery jacks. It is recognized by its dingy gray to brown colors, bluish staining, and thin partial veil. The slime will stain wax paper a teal green color. Although not as common as the other larch *Suillus* in natural settings, it is more likely to fruit with urban larch than the others. *S. ampliporus* has a darker red-brown cap, yellow pores, and a hollow stipe.

Synonym: *Suillus viscidus* (L.) Roussel

Suillus ampliporus

(Peck) Kuntze

Hollow-stalked Larch Jack

Cap: 3–9 (12) cm across, convex with a slightly inrolled margin at first, becoming broadly convex to plane, occasionally with a broad umbo. Surface dry, tomentose to fibrillose-scaly, more matted with age. Dark brown, orange-brown to red-brown, at times with a paler margin. **Tubes:** Decurrent, 0.2–0.5 cm long. **Pores:** Angular, radially aligned, fairly large. Pale to bright yellow, slightly darker, golden to olive-yellow in age. **Stipe:** 4–10 cm long, 0.7–2 cm thick, typically club-shaped with a swollen base. Surface dry, apex pseudo-reticulate with decurrent pore walls, lower portion fibrillose-tomentose. Yellow above veil, brown, orange-brown to red-brown below. **Partial veil:** Cottony-fibrillose, adhering to cap margin and leaving a ring and annular zone on stipe. **Flesh:** Firm when young, very soft in age, pale yellow not staining. Stipe hollow, fragile, pale yellow with brownish discoloration in age. **Odor:** Indistinct. **Taste:** Mild. **KOH:** Unknown. **Spore deposit:** Olive-brown. **Microscopy:** Spores 7–10 × 3.5–4 µm, spindle shaped to subcylindric, smooth, inamyloid, pale greenish in KOH.

Ecology: Solitary, scattered, or in clusters or large troops on ground. Locally common in the eastern Cascades and the northern Rocky Mountains, but only in areas with larch. Also in urban areas with planted larch. Fruiting in fall.

Edibility: Edible, with a firmer texture than other *Suillus*, but lacking in flavor.

Comments: The hollow stipe is this species' most distinctive feature—shared only with *Gyroporus* among Cascadian boletes. The scaly cap, large angular yellow pores, partial veil, and growth with larch complete the suit of identification features. *S. lakei* is quite similar but grows with Douglas-fir and has a solid stipe. The larch-associated *S. ochraceoroseus* has a rosy pink cap, which can fade in age and a solid stipe.

Misapplied name: The European *Suillus cavipes* (Klotzsch) A.H. Sm. & Thiers, is distinct from the North American species.

Suillus lakei

(Murrill) A.H. Sm. & Thiers

Western Painted Jack

Cap: 4–15 cm across, convex with an inrolled margin enclosed with the partial veil when young, becoming broadly convex to nearly plane. Surface dry to thinly viscid, fibrillose to floccose-scaly when young, more appressed-fibrillose, or nearly smooth in age. Color variable: Reddish brown, orange-brown, dark brown to brick red with a brown to yellowish brown base color in age. **Tubes:** Broadly attached to decurrent, shallow, 0.2–1 cm long. **Pores:** Small, occasionally radially arranged, angular. Yellow at first, becoming ochraceous orange, to dingy brown in age. Staining pinkish brown to brown when bruised. **Stipe:** 3–12 cm long, 1–4 cm thick, equal or with a swollen base. Surface dry, smooth and yellow above veil, often matted-fibrillose and yellowish with reddish brown streaks or discoloration below veil, discoloring to dingy brown in age. **Partial veil:** Thin, fibrillose to felty-tomentose, covering cap margin and forming a sheathing band around stipe when young, leaving ragged hanging remnants on margin and a band-like ring on upper stipe, which can disappear in age. **Flesh:** Thick, firm. Yellowish to pinkish beige when young, darker, dingy yellowish in age, slowly staining pinkish to reddish. Stipe sometimes staining bluish green in base. **Odor:** Indistinct. **Taste:** Mild to slightly acidic. **KOH:** Unknown. **Spore deposit:** Cinnamon brown. **Microscopy:** Spores 7–11 × 3–4 (5) µm, subelliptic to subcylindric, smooth.

Ecology: Solitary, scattered, in small clusters or troops, typically under Douglas-fir. Also found in high elevation hemlock forests.. Common and abundant throughout Cascadia. Fruiting from first rains in fall into early winter, and occasionally in spring in the eastern Cascades. *Gomphidius subroseus* and *G. smithii* are presumed to be parasites on this species and thus are often found fruiting alongside it.

Edibility: Edible.

Comments: This fibrillose- to floccose-scaly capped *Suillus* is distinctive when young, but if it loses the scales, as it sometimes does in age, it can be confused with a few other species. Other distinguishing features are the small angular yellow pores that stain pinkish brown when damaged, white veil, and, typically, growth with Douglas-fir. Two similar scaly-capped larch associates occur in Cascadia: *S. ampliporus* has larger pores, and a hollow stipe, and *S. ochraceoroseus* with a a rosy pink cap (fading in age). Fruitbodies of *S. tomentosus* often have reddish scales on a yellowish to reddish brown cap but lack a veil, and the flesh stains blue (especially in the cap).

Synonym: *Suillus amabilis* probably refers to the same species as *S. lakei* and is an older name (and would thus have priority if the two were ever formally synonymized).

Suillus caerulescens

A.H. Sm. & Thiers

Fat Jack

Cap: 4–15 (20) cm across, convex when young, becoming broadly convex to nearly plane, margin inrolled and enclosed with partial veil when young, often with thin hanging veil remnants as it expands, to bald in age. Surface dry to viscid, nearly smooth or appressed-fibrillose and somewhat streaked, occasionally with small scales. Ochraceous buff, cinnamon buff to yellowish brown, with a paler or yellow margin when young, generally more tawny brown to orangish brown in age. **Tubes:** Broadly attached to decurrent, 0.3–1 cm long. **Pores:** Moderately large, angular, occasionally radially arranged. Pale yellow at first, becoming golden yellow to dingy ochraceous orange in age. Staining cinnamon brown to orangish brown when bruised. **Stipe:** 3–12 cm long, 1–3 (4.5) cm thick, equal, peg shaped (tapering toward base) or with a swollen lower portion but tapering below to base. Surface dry to slightly tacky, often matted-fibrillose below veil, dry and smooth above veil. Yellowish at first, quickly developing orangish brown to brown stains and discoloring to dingy yellowish brown in age. Base slowly staining bluish when scraped. **Partial veil:** Fibrillose, felty-tomentose to membranous, covering cap margin and forming a sheathing band around stipe, leaving hanging remnants on margin and a band-like ring on upper stipe. White, discoloring brownish in age, Not gelatinous or only slightly so. **Flesh:** Thick, firm when young, soft in age. Whitish to creamy at first, becoming pale yellow, discoloring reddish brown around larva tunnels, slowly staining bluish in stipe base. **Odor:** Indistinct. **Taste:** Mild to slightly lemony. **KOH:** Unknown. **Spore deposit:** Cinnamon brown. **Microscopy:** Spores 7.5–11 × 3.5–5 µm, elliptic to nearly oblong, smooth, inamyloid, colorless to ochraceous in KOH.

Ecology: Solitary, scattered, in clusters or troops under Douglas-fir. One of the most abundant mushrooms in Cascadia, fruiting in a large flush soon after the first rains in fall, with stragglers continuing until the first freeze. *Gomphidius oregonensis* often fruits alongside this species, which it is presumed to parasitize.

Edibility: Edible. One of the better-tasting *Suillus*, especially when young.

Comments: The medium to large size, ochraceous buff to yellowish brown, smooth, dry to slightly viscid cap, yellow pores that stain brownish, dry whitish veil when young, bluish staining in the stipe base, and growth with Douglas-fir are the most important identifying characteristics. Very difficult to distinguish from *S. ponderosus* (especially in age). The latter species is distinguished by its richer, more orange-brown to reddish brown cap colors and, when young, by its distinctly viscid yellow partial veil. Could also be mistaken for old or rain-beaten *S. lakei* but that usually has some appressed scales remaining on the cap and has darker, dingy yellow flesh that slowly stains pinkish to reddish brown.

Suillus ponderosus

A.H. Sm. & Thiers

Slimy Fat Jack

Cap: 6–18 (25) cm across, convex when young, becoming broadly convex to nearly plane. Margin inrolled and enclosed with the partial veil when young, often with hanging veil remnants as it expands, soon bald. Surface smooth, glutinous to viscid, appearing streaked or occasionally with small scales in age. Cinnamon brown to orange-brown or occasionally yellow to yellow-brown when young, especially around margin. Becoming paler, cinnamon to ochraceous brown and often with dingy brownish stains in age. Caps sometimes developing forest green to sea-green blotches. **Tubes:** Broadly attached to decurrent, 0.3–1.5 cm long. **Pores:** Moderately large, often radially arranged, angular. Pale yellow at first, becoming yellow to dingy ocher-yellow. Staining cinnamon brown when bruised. **Stipe:** 5–10 (15) cm long, 2–4 (6) cm thick, often widest in middle, and tapering toward base, to more or less equal. Surface viscid from veil, smooth and dry to slightly viscid above and below veil. Yellow at first, quickly developing orangish brown stains and discoloring dingy brown. Base often staining bluish green when scraped. **Partial veil:** Membranous with a yellow to yellow-orange glutinous slime covering underside when young, becoming cinnamon buff to reddish cinnamon and forming a gelatinous band-like ring. **Flesh:** Thick, firm when young, soft in age. Yellow, occasionally discoloring reddish brown around larva tunnels, slowly staining greenish to greenish blue in stipe base. **Odor:** Indistinct. **Taste:** Mild to acidic, sharp, lemony. **KOH:** Blackish on cap, vinaceous on flesh. **Spore deposit:** Brown. **Microscopy:** Spores 6–10 (12) × 3.8–5 µm, elliptic, spindle shaped or nearly oblong, smooth, colorless to yellowish in KOH.

Ecology: Solitary, scattered, in clusters or troops under Douglas-fir. Very common throughout Cascadia. Often fruiting in a large flush soon after the first rains in fall.

Edibility: Edible. Good when young, soft and slimy in age.

Comments: The large size, cinnamon brown to reddish brown cap, yellow pores, viscid yellow veil when young, greenish staining in the stipe base, and growth with Douglas-fir help set this *Suillus* apart from all but the very similar *S. caerulescens.* The latter differs in having a slightly fibrillose, yellowish brown cap when young, a whitish veil that is dry or only slightly viscid, and paler flesh. Older specimens of these two species are nearly indistinguishable. *S. clintonianus* is typically smaller, with a viscid orange-red to reddish brown cap, and grows with larch.

"Jack Fuzz" (*Dicranophora fulva*)

INSET: This is a member of the fungal order Mucorales – the 'pin molds.' It forms a layer of translucent stalks topped with yellow spheres growing on old, rotting Suillus (particularly *S. caerulescens* and *S. ponderosus*). It has also been found on *Gomphidius*, *Chroogomphus*, and even *Paxillus*.

Suillus luteus

(L.) Roussel

Purple-veiled Slippery Jack

Cap: 4–12 (15) cm across, round to convex at first, becoming broadly convex to plane. Surface smooth, viscid, covered with glutinous slime when wet, shiny when dry. Color variable, dark brown, reddish brown to vinaceous brown at first, remaining these colors, or fading slightly, to yellow-brown, and often developing dark streaks and/or olivaceous tones in age. **Tubes:** Sunken around stipe to broadly attached, shallow, 0.2–1 cm long. **Pores:** Round to irregular, tiny at first, enlarging slightly in age. Yellow when young, becoming golden, then dingy olive-yellow to orangish brown in age. **Stipe:** 3–8 cm long, 1–2.5 cm thick, equal or club shaped. Surface greasy to sticky, covered with glandular dots, veil remnants slimy. Whitish to yellowish with darker glandular dots when young, often purplish around veil, developing dingy coloration or stains in age. **Partial veil:** Thick and membranous, white with a purplish, gelatinous underside. Forming a flaring skirt at first, soon collapsing and dissolving in age. **Flesh:** Firm when young, becoming soft. Creamy at first, becoming yellowish in age. **Odor:** Indistinct. **Taste:** Mild. **KOH:** Unknown. **Spore deposit:** Brown to cinnamon. **Microscopy:** Spores 7–9 × 2.5–3 µm, oblong, smooth.

Ecology: Solitary, scattered, clustered, or occasionally in large troops under pine. This is an introduced species, native to northern Europe. Typically found in urban areas, especially with planted Scots pine (*Pinus sylvestris*), but expanding to native pine forests – an ecological-evolutionary trajectory which should be closely watched. Fruiting in fall.

Edibility: Edible. It has a pleasant taste but is rather soft-fleshed and cooks up with a slimy texture. Peeling the glutinous slime off the cap, or drying and powdering the mushrooms and using them as a seasoning (after you have forgotten how slimy they were), are some of the more popular preparations.

Comments: The viscid brown cap, yellow pores, and thick, membranous skirt-like veil are distinctive. The veil often has a purplish gelatinous underside, but occasionally can be all white. *S. brunnescens*, a five-needle pine associate, generally has a pale cap when young (dark in age) and has a purplish partial veil that clings to the cap margin and doesn't form a ring on the stipe.

Suillus acidus

(Peck) Singer

Cap: 3–6 (10) cm across, rounded to conical-convex at first, becoming broadly convex to plane with a low broad umbo, margin even, often with glutinous veil remnants when young. Surface appressed-fibrillose to smooth, glutinous to viscid. Olive-brown, brown to tan, fibrils often darker. **Tubes:** Decurrent to broadly attached, 0.5–1 cm long. **Pores:** Small, angular. Creamy yellow at first, often with clear to milky droplets, becoming yellow to dingy golden in age. **Stipe:** 5–10 cm long, 0.7–1.5 cm thick, equal, or enlarged toward base. Surface viscid to tacky, yellowish ground color, covered with pinkish brown glandular dots when young, darkening to blackish when handled or in age. **Partial veil:** A gelatinous membrane when young, forming a collar-like ring on upper stipe. **Flesh:** Thin to moderately thick, firm when young, soft in age. Pale creamy yellow to dingy yellow, brownish in stipe with age. **Odor:** Indistinct. **Taste:** Often sour, lemony, or slightly unpleasant. **KOH:** Dark brown to blackish on cap. **Spore deposit:** Olive-brown. **Microscopy:** Spores 9–11 × 3–4 µm, narrowly elliptic to nearly oblong, pale ochraceous in KOH.

Ecology: Solitary or scattered on ground, moss, or duff under five-needle pines. Locally common in montane areas across northern Cascadia, fruiting from late summer through fall.

Edibility: Non-toxic. Sour and unpleasant tasting.

Comments: The viscid olive-brown to tan cap, glutinous veil, glandular-dotted stipe, and growth with five-needle pines help set this species apart. *S. flavidus* is often smaller, has a more brightly colored, umbonate cap, and grows with two- and three-needle pines. *S. glandulosipes* has a glutinous, cinnamon pink to yellowish or reddish brown cap at first (darkening to more vinaceous brown), a dotted white stipe, and a thick veil that clings to the cap margin and does not form a ring on the stipe.

Synonym: *Suillus subolivaceus* A.H. Sm. & Thiers

Suillus flavidus

(Fr.) J. Presl

Slim Jack

Cap: 2–6 (10) cm across, conical to conical-convex at first, becoming broadly convex to plane with a low broad umbo, margin even, often with glutinous veil remnants when young. Surface smooth, glutinous to viscid, appearing streaked in age. Yellowish buff, grayish buff, to warm tan at first, covered with pale orangish brown glutinous slime discoloring reddish to orangish brown, often in streaks or stained over disc. **Tubes:** Decurrent to broadly attached, 0.5–1 cm long. **Pores:** Moderately large, radially arranged, angular. Creamy yellow at first, becoming yellow to golden, to dingy golden in age. Often staining pinkish when bruised. **Stipe:** 3–8 cm long, 0.5–1 cm thick, equal, tapering downward or sometimes enlarged toward base. Surface viscid from veil, smooth and dry above, slightly tomentose to smooth below veil. Base often with white rhizomorphs. Whitish at first, becoming yellowish and developing dingy orangish brown stains. **Partial veil:** Gelatinous membrane when young, forming a collar-like ring and dissolving into a viscid, band-like ring in age. **Flesh:** Thin, firm when young, soft in age. Pale creamy yellow to dingy yellow, slowly staining light cinnamon brown. **Odor:** Indistinct. **Taste:** Often sour, lemony, or slightly unpleasant to mild. **KOH:** Reddish brown on cap. **Spore deposit:** Olive-brown. **Microscopy:** Spores 7–11 × 3.5–4.5 µm, narrowly elliptic to nearly oblong, smooth.

Ecology: Solitary or scattered, occasionally in small clusters on ground or in duff under shore and lodgepole pine. Common in coastal dune forests and the eastern Cascades. Fruiting from late summer in the mountains through fall and into early winter on the coast.

Edibility: Non-toxic.

Comments: The small size, often umbonate cap covered with translucent, orangish brown glutinous slime, gelatinous partial veil, and growth with two-needle pines are a distinctive combination of features. Coastal collections tend to be smaller and more slender than montane collections with lodgepole pine. *S. acidus* is more olive-brown colored and grows with five-needle pines.

Synonym: *Suillus umbonatus* E.A. Dick & Snell.

17 - Polypores, Crusts, and Allies

This section includes a wide array of species representing many different evolutionary lineages, some quite distantly related to one another. Despite this evolutionary divergence, they are grouped here because of the shared structure of their fertile surfaces: A layer of tubes and pores on the underside of an often shelf-like cap. A few species in this chapter have a smooth fertile surface rather than pores but are included due to their similar fan- or bracket-shaped fruitbodies.

Some polypores have fruitbodies with distinct stipes and are thus sometimes confused with the boletes. However, these "stalked polypores" tend to have a tough, fibrous, or leathery texture (rather than soft or fleshy). Of these, *Albatrellus* and its allies, *Boletopsis* and *Coltricia* are mycorrhizal. The genus *Albatrellus* has been split up over the past decade, resulting in a number of genera that have not been universally accepted – these include *Xanthoporus*, *Scutiger*, and *Albatrellopsis*. The rest of the stalked polypores appear to be non-mycorrhizal — a few grow on wood, and at least one emerges from a hard tuber-like sclerotium. These include *Polyporus* and its segregate genera (e.g. *Cerioporus, Picipes, Royoporus*) and *Jahnoporus*.

Phaeolus, *Abortiporus*, *Onnia*, and *Bondarzewia* have "rosette-like" or amorphous fruitbodies that frequently develop overlapping tiers or lobes in age, and which often appear terrestrial due to their habit of fruiting on buried wood or roots. Both *Phaeolus* and *Onnia* also form shelves or conks.

Perhaps the most familiar polypores are those that form hoof-shaped to fan-shaped, woody, long-lasting fruitbodies (often called conks). Although the fruitbodies of some are annual, those of the most common and conspicuous species are perennial and add new layers of spore-bearing tissue every year—giving many a layered appearance. Such fruitbodies are produced by species of *Bridgeoporus, Fomitopsis, Ganoderma, Heterobasidion, Laricifomes*, and *Phellinus* and its relatives (including the somewhat smaller *Porodaedalea* and *Pseudoinonotus*).

Other polypores with shelf-like fruitbodies have more watery or fleshy textures and often fruit in large overlapping tiers or groups on wood. These include *Climacocystis*, *Ischnoderma*, and *Laetiporus*, all of which form comparatively short-lived fruitbodies which decay after a few months. The smaller, thinner-fleshed shelving polypores have corky, leathery, or parchment-like textures, and include many common and familiar species in *Trametes*, *Trichaptum*, *Bjerkandera*, and several segregates of the former *Postia (Calcipostia, Fuscopostia, Cyanosporus*, etc.).

A number of oddball polypores with different textures (fuzzy-powdery, chalky-fleshy, flabby-cheesy, ragged-spongy, or like a dry marshmallow) and/or different morphologies (crust- or sheet-forming) are also included here. These include species of *Calcipostia*, *Cyanosporus*, *Phlebia*, and *Pycnoporellus*. The structure of *Cryptoporus* fruitbodies is even more confusing because it has a bubble-like, leathery outer surface that hides a pore layer within —it is polypore with a partial veil!

An assortment of additional species (including a number of "crust fungi") are also included here because of their superficial resemblance to polypores. All of these have smooth or weakly wrinkled fertile surfaces rather than pores. *Stereum* and *Chondrostereum* have thin, leathery, parchment-like fruitbodies forming overlapping shelves on wood. *Thelephora* have rosette-like, vase- or fan-shaped, terrestrial fruitbodies and are ectomycorrhizal.

Albatrellus avellaneus

Pouzar

Hazel Knight

Cap: 3–12 (18) cm across, convex with a downcurved margin at first, often irregular and lobed, becoming plane; margin uplifted and wavy in age. Surface dry, appressed-fibrillose to finely scaly, cracking in age. Whitish, creamy, or with vinaceous buff fibrils and scales at first, generally becoming pale buff to ochraceous, or occasionally darker vinaceous buff or yellowish stained in age. **Pores:** Decurrent, small, round to slightly angular, at times becoming ragged in age. White at first, becoming creamy to pale yellowish, staining yellow when bruised. Tubes shallow, white. **Stipe:** 2–8 cm long, 1–4 cm thick, central to off-center, equal to club shaped. Surface dry, smooth. White to creamy when young, developing yellowish to ocher stains. **Flesh:** Thick and firm, brittle in cap, stipe fibrous. White to cream, staining yellow. **Odor:** Indistinct. **Taste:** Mild. **KOH:** Yellow on all parts, brightest on flesh. **Spore deposit:** White. **Microscopy:** Spores 4.5–6 × 4–4.5 µm, subglobose to ovoid, smooth, inamyloid.

Ecology: Often in clusters, occasionally solitary, usually in large troops and patches on mossy duff under conifers, especially Sitka spruce. Common along the coast, uncommon at mid and higher elevations of the Cascades (but many of these records may pertain to *A. ovinus*). Primarily fruiting from early fall into winter but small numbers can be found any time of year.

Edibility: Edible and good, with a firm, meaty texture and sweet flavor. To top it off, the flesh turns a beautiful golden color when cooked.

Comments: This is one of the most prolific mushrooms of coastal spruce forests, often growing in huge patches comprising hundreds of fruitbodies. It is recognized by its pallid colors, yellowish staining, white pores, and terrestrial growth. *A. ovinus* can look identical but has smaller spores (4–5 × 3–3.5 µm) — it may be more common in the Cascade Range. *A. subrubescens* is also quite similar but often has a sweet odor; it is most reliably distinguished by its distinctly amyloid spores. *Polypus* (=*Albatrellus*) *dispansus* has more complicated rosette-like caps and lacks the yellow staining, while *Scutiger* species have darker colors, and often show green staining.

Albatrellopsis flettii

(Morse ex Pouzar) Audet

Blue Knight

Cap: 5–15 (25) cm across, round and convex with a downcurved margin at first, becoming uplifted with a lobed or wavy margin, often fused when in clumps, wavy-edged rosettes in age. Surface dry, smooth to velvety or finely scaly, cracking in age. Baby- to sky-blue or royal blue when young, often with darker "water spots", becoming mottled blue and pallid grayish or greenish blue, often developing orangish spots, occasionally extensively dingy gray to ocher-gray in age, but usually retaining some blue tones where cap is protected. **Pores:** Slightly to deeply decurrent, small, round to slightly irregular, becoming ragged in age; tubes relatively short. Bright white at first, becoming creamy white, occasionally with orangish stains. **Stipe:** 5–15 cm long, 1–4 cm thick, central or off-center, equal to irregular. White when young, developing orangish stains, and often dingy grayish blue in age. **Flesh:** Thick, firm and rubbery, but somewhat brittle in cap. White to cream. **Odor:** Indistinct to slightly farinaceous. **Taste:** Slightly farinaceous. **KOH:** Slightly yellow on cap. **Spore deposit:** White. **Microscopy:** Spores 4–5 × 3–3.5 µm, ellipsoid to ovoid, smooth, inamyloid.

Ecology: Solitary or scattered, often in fused clusters on ground under conifers, mostly with Sitka spruce and western hemlock, occasionally in mixed Douglas-fir forests. Common along the coast and on both slopes of the Cascades but uncommon or absent from much of the interior low elevations of western Cascadia. Fruiting in late summer and fall; fruitbodies persist into winter.

Edibility: Edible, but not popular.

Comments: This is one of the most beautiful and unusual mushrooms in our region. Besides its striking blue color, this species is easily recognized by its large size, overlapping cap arrangement, and white pores. The blue color may be restricted to the creases or folds of the cap in older fruitbodies. Pale pinkish or brownish specimens (especially from drier habitats) have been called *A. confluens* but there is little genetic difference between the color morphs. *Neoalbatrellus subcaeruleoporus* also has a bluish cap but is much smaller.

Synonym: *Albatrellus flettii* Morse ex Pouzar

Neoalbatrellus subcaeruleoporus

Audet & B. Luther

Little Blue Knight

Cap 1.5–5 cm across, convex to plane, often wavy in age. Surface dry, finely tomentose to smooth. Light sky-blue, baby-blue to grayish blue, more rarely darker sky-blue, fading and developing orangish stains in age. **Pores:** Slightly decurrent, small, round; tubes quite short. Sky-blue to grayish blue. **Stipe:** 1–4 (6) cm long, 0.3–1 cm thick, central to off-center, equal or tapered downward. Surface dry, smooth. Sky-blue to grayish blue, sometimes developing orangish stains in age. **Flesh:** Thin, brittle at first, somewhat corky in age. Whitish to pale bluish gray. **Odor:** Indistinct. **Taste:** Indistinct. **KOH:** Reddish on all parts. **Spore deposit:** White. **Microscopy:** Spores 4–5 × 3.5–4 µm, broadly ellipsoid to subglobose, smooth, inamyloid.

Ecology: Solitary, scattered, or in small clusters on ground or moss under conifers, especially western hemlock. It prefers mossy banks on old road cuts or disturbed trailsides within mature conifer forests. Uncommon to rare throughout Cascadia. Fruiting in fall or early winter.

Edibility: Reportedly edible but rare and small. The counterpart species in eastern North America has a bitter-metallic taste.

Comments: The small size, sky-blue to grayish blue colors, small pores, and often off-center stipe are distinctive. *Albatrellopsis flettii* is another blue-capped species but is larger and has white pores.

Misapplied name: *Neoalbatrellus* (=*Albatrellus*) *caeruleoporus* is an eastern North American species.

Scutiger ellisii

(Berk. ex Cooke & Ellis) Murrill

Greening Goat's Foot

Cap: 8–25 (30+) cm across, circular to lobed or fan-shaped and convex with an inrolled margin at first, becoming plane or undulating with age. Surface dry, hairy to tomentose, covered with vaguely radially-arranged tufts, becoming more matted or scale-like in age. Color variable; golden-yellow to medium brownish; developing green stains and darker brown colors in age. **Pores:** Broadly attached to slightly decurrent, round to irregular, often becoming ragged in age; tubes shallow. Whitish to cream, typically staining greenish to greenish yellow when bruised, discoloring dingy yellow or with greenish stains in age. Shallow to moderately deep; creamy to beige. **Stipe:** 3–10 cm long, 2–4 (6) cm thick, off-center to lateral, more or less equal to irregular. Surface dry, roughened, at times with tufted hairs, often with adhering debris. Yellow-brown to creamy yellow, typically with greenish stains. **Flesh:** Thick, very firm, cream to beige, often greenish stained in age, or slowly staining greenish when cut. **Odor:** Indistinct or slightly unpleasant when young, rancid or strongly unpleasant in age. **Taste:** Mild when young, unpleasant in age. **Spore deposit:** White. **Microscopy:** Spores 8–10 × (5.5) 6–7 µm, ellipsoid to tear-drop shaped, smooth, inamyloid.

Ecology: Solitary or scattered from soil or thick duff under montane conifers. Uncommon, mostly in mid to high elevation and east-side forests in the Cascade Range. Fruiting in late summer and fall.

Edibility: Edible but older fruitbodies develop an unpleasant taste.

Comments: This species is easily recognized by the combination of the hairy to scaly-tufted, yellow to brown cap, whitish to cream pores, off-center to lateral stipe, and a tendency for all parts to discolor greenish. *Scutiger pes-caprae* (=*Albatrellus pes-caprae*) has a reddish brown to brown fibrous-scaly cap and white to creamy pores that have less of a tendency to stain greenish; it is rare in Cascadia, more likely to be found in low to mid-elevation west-side forests. *Jahnoporus hirtus* has a velvety, brown to grayish brown cap, white pores, a strongly bitter taste, and grows on or near decaying stumps and buried roots. *Polyporoletus sylvestris* has a dingy cap, and grayish olive pores that discolor greenish and purplish. *Bondarzewia occidentalis* is typically larger, has a brown to grayish brown cap center, transitioning to warmer and often paler yellowish brown to ocher-brown outward, with a beige to whitish margin. The cap surface is finely velvety and, at times, with scattered felty patches or small hairy scales that become smoother in age and spores that are globose, ornamented, and amyloid.

Synonym: *Albatrellus ellisii* (Berk.) Pouzar

Polyporoletus sylvestris
(Overh. ex Pouzar) Audet

Western Ogre Polypore

Cap: 7–15 (20) cm across, convex with an inrolled, often lobed margin, becoming plane to uplifted. Surface dry, finely velvety to smooth, often finely cracking, sometimes developing broad plaques in age. Chartreuse to beige, tan, or grayish-brown, at times with olive tones. **Pores:** Broadly attached to deeply decurrent, round when young becoming larger, irregularly rounded to slightly maze-like at maturity; tubes rather shallow. Color variable; usually grayish olive to pinkish tan, sometimes entirely dingy lavender, becoming darker brown in age. With a 'shark-skin' sheen, appearing differently colored at different angles of light. Bruising purplish to olive when damaged. **Stipe:** 10–20 cm long, 1–4 cm thick, lateral or off-center, rarely central, more or less equal to irregularly shaped. Surface dry, finely velvety, scaly or smooth, apex often with weak reticulum. Whitish to yellowish when young, darkening to reddish brown in age. **Flesh:** Thick, firm, fleshy-tough. Beige to grayish, with dark, dingy purplish speckles throughout. Staining purple when cut or bruised. **Odor:** Indistinct. **Taste:** Not sampled. **Spore deposit:** White. **Microscopy:** Spores 8–13 × 6.5–10 µm, broadly ellipsoid to subglobose, smooth when young, with a punctate or wrinkled wall when mature, inamyloid.

Ecology: Usually solitary or in small clusters on ground in mature and old-growth conifer forests. Widespread but rare; most frequent in high elevation fir forests in the Cascade Range. Fruiting from late summer through fall.

Edibility: Unknown.

Comments: The combination of an irregular boletoid structure, the pale cap, decurrent gray-olive pore layer, green and purple discolorations, and dark-dotted flesh distinguish this unusual species. *P. bulbosa* is a poorly known species described from a single collection; it is smaller, with a bulbous stipe. The name *P. sublividus* has been misapplied to this species; it refers to a rare hardwood associate in eastern North America. *Scutiger ellisii* is similar in structure and also shows pronounced green staining but has a scaly golden tan cap and whitish to pale creamy pores. *S. pes-caprae* has a scaly, gray-brown to reddish brown cap, less pronounced green staining, and pale pores.

Jahnoporus hirtus
(Cooke) Nuss

Bitter Iodine Polypore

Cap: 4–12 (18) cm across, kidney-shaped to irregularly round, convex to plane; margin wavy, lobed, often uplifted in age. Surface dry, covered with stiff to velvety hairs. Dark brown, warm brown, dull grayish brown or tan. **Pores:** Broadly attached to deeply decurrent, angular, becoming ragged in age; tubes relatively short. Bright white at first, becoming creamy in age. **Stipe:** 2–12 cm long, 1–4 cm thick, often lateral or off-center, rarely central, equal to irregular. Surface dry, velvety. Whitish to gray, becoming brown. **Flesh:** Thick, very firm, tough. White to cream. **Odor:** Mild to fragrant, often iodine-like. **Taste:** Extremely bitter or metallic, persistent. **Spore deposit:** White. **Microscopy:** Spores 12.5–17 × 4.5–5.5 µm, spindle-shaped to cylindrical, smooth, inamyloid.

Ecology: Solitary or scattered, occasionally in small clusters. Often appearing terrestrial but emerging from buried roots around rotting stumps and dead trees, sometimes directly from stumps or logs. Widespread and common throughout Cascadia, occurring across a wide range of elevations and habitat types. Fruiting in fall and early winter, rarely into spring.

Edibility: Inedible.

Comments: The brown, soft-hairy cap, white pores, off-center stipe, and extremely bitter long-lasting taste set this species apart. The rare *Xeroceps skamania* (=*Scutiger* /*Albatrellus skamanius*) has a dark brown cap, pale yellow pores and a mild to astringent taste. *Xanthoporus syringae* (=*Scutiger*/ *Albatrellus syringae*) is similar in shape and stature but has a smoother cap, yellower colors, and lacks a bitter taste. Although absent in much of Cascadia, it has been recorded from the northern Rocky Mountains. *Scutiger pes-caprae* has a fibrous-scaly brown cap and creamy pores and a mild taste, and often stains slightly greenish. *Bondarzewia occidentalis* is often larger, lacks a distinct stipe, has a mildly acrid or only slightly bitter taste, and has (sub)globose spores with amyloid ornamentation.

Boletopsis grisea

(Peck) Bondartsev & Singer

Grey Falsebolete

Cap: 5–15 (25) cm across, convex or lobed with an inrolled margin at first, expanding to plane or uplifted with an irregular, wavy margin in age. Surface dry, smooth to finely fibrillose, often cracking into small scales when dry. Off-white to pale gray at first, developing dark gray streaks and splotches as it ages, occasionally extensively dark gray in age. **Pores:** Slightly to deeply decurrent, small, round to slightly irregular; tubes relatively shallow. Bright white at first, often discoloring grayish to vinaceous gray in age. **Stipe:** 3–10 cm long, 1–5 cm thick, equal, base tapered or rounded. Surface dry, smooth. Off-white to pale gray, developing dark gray streaks. **Flesh:** Thick, firm, becoming brittle in cap, fleshy-fibrous in stipe. White to gray, slowly staining vinaceous gray when cut, occasionally with darker gray or vinaceous spots or stains in age, or with bluish green stains in stipe when waterlogged. **Odor:** Indistinct to slightly farinaceous. **Taste:** Mild to strongly bitter. **KOH:** Forest-green on cap, grayish green on flesh, pores, and stipe. **Spore deposit:** Light brown. **Microscopy:** Spores 5–7 × 4–5 µm, irregular, angular, overall ellipsoid to globose, with irregular warts and projections.

Ecology: Solitary or scattered, often buried in needle duff under conifers, especially pine. Widespread throughout Cascadia along the coast as well as in montane areas (although exact distribution is clouded by taxonomic uncertainty) but largely absent from inland valleys. Fruiting in fall and early winter.

Edibility: Edible but often bitter tasting. Some collections are mild enough to eat without special preparation but not widely consumed in North America.

Comments: Although this mushroom is structurally a polypore, it is more closely related to the toothed genera *Hydnellum* and *Sarcodon* and the smooth to wrinkled fertile tissued *Thelephora* and *Polyozellus*. *Boletopsis* can be easily identified to genus by their chunky stature, firm but somewhat brittle white flesh, shallow white tube layer, and growth on the ground. *B. grisea* appears to be limited to pine forest and has whitish to pale gray colors when young, developing darker streaks and stains in age. There is a common but unnamed species in Cascadia with a lilac-gray to nearly black cap; it grows at higher elevations with fir and hemlock. *B. leucomelaena* has a black cap, white to light gray pores, a gray stipe (often with an orangish base) and grows with spruce. The name *B. smithii* is sometimes applied to collections with an orangish cap but these may be nothing more than color morphs of one (or more) of the other species. Some species of *Albatrellus*, *Scutiger*, and their closely related genera resemble *Boletopsis* but have white spores.

Polyporus tuberaster group

Tuberous Polypores

Cap: 5–12 cm across, broadly convex to plane and often depressed at the center; margin sometimes tightly inrolled at first, slightly wavy or lobed in age. Surface moist to dry, light brown to grayish, with darker brown, upright and pointed tufts and scales, densest at center of cap. **Pores:** Decurrent, quite small, more or less round, sometimes angular-elongate near stipe; tubes shallow. White at first, becoming creamy or beige when dried. **Stipe:** 5–10 cm long, 1–4 cm thick, central and cylindrical; sometimes forked or branched. Surface fibrous, roughened. Creamy whitish to beige, developing ocher or grayish areas upward from base. **Sclerotium:** A nearly round mass of hard, rubbery tissue ranging from size of a golf ball to nearly the size of a soccer ball, buried in ground. Dark grayish, wrinkled surface and a marbled white and ocher-yellow interior. **Flesh:** Thick, solid, firm, rubbery or leathery when dry. Whitish. **Odor:** Indistinct. **Taste:** Indistinct. **Spore deposit:** White. **Microscopy:** Spores 10–16 × 4–7 µm, long-ellipsoid and tapered toward apiculus, smooth, thin-walled.

Ecology: Solitary or in small clusters, fruiting from an underground and often overlooked sclerotium. Uncommon in the interior valleys, lowlands, and foothills. Fruitbodies annual but the sclerotia are longer-lasting. Fruiting especially in spring but can be encountered at any time of year.

Edibility: Inedible.

Comments: The large, scaly cap, fairly long, thick stipe, and growth from an (often overlooked) sclerotium help distinguish this species. A similar species with creamy beige to ocher or orange-brown caps and large, fibrillose scales may pertain to *Polyporus decurrens* (=*P. mcmurphyi*?). Its stipe is off-center to lateral and quite short, emerging directly from wood (usually medium-size logs and thicker branches of hardwoods). *P. alveolaris* (=*Neofavolus alveolaris*) lacks a well-defined stipe (if present, short and off-center, creating a cleft in the cap), has more open angular pore mouths, and grows on hardwoods. Dryad's Saddle (*Cerioporus / Polyporus squamosus*) forms large, annual fruitbodies which appear solitary or in tiers of overlapping shelves on standing hardwoods; it is rare in the inland valleys of Cascadia, fruiting in summer and early fall.

Picipes tubaeformis

(P. Karst.) Zmitr. & Kovalenko

Cap 2–10 (15) cm across, convex to plane with a sunken center and a slightly incurved margin at first, becoming slightly wavy and often uplifted in age. Surface often shiny, dry, smooth to finely scaly or with flattened, matted hairs. Pale gray to ocher or light brown, often concentrically zonate, often with silvery or whitish streaks. **Pores:** Small, round; tubes very short. Creamy white to pale buff, staining slightly darker when fresh. **Stipe:** 1–8 cm long, 0.3–1.5 cm thick, central or slightly off-center. Surface dry, smooth. Ocher brown to dark brown at base and pallid above at first, soon dark brown overall. **Flesh:** Thin in cap, tough and leathery. Older fruitbodies dried in place are hard. **Odor:** Indistinct. **Taste:** Indistinct. **Spore deposit:** White. **Microscopy:** Spores 5.5–9 × 2.5–3.5 µm, cylindrical, smooth, inamyloid.

Ecology: Solitary or scattered on decaying hardwood branches or logs, rarely on conifers. Common throughout Cascadia. New fruitbodies first appear in fall but persist across multiple seasons, long after they've ceased producing spores.

Edibility: Inedible.

Comments: The smooth, pallid grayish to ocher, tan, or brown cap, tall, upwards-darkening stipe, and growth on hardwood are helpful identification features. *P. (= Polyporus / Royoporus) badius* has many overlapping characters but tends to have a darker brown cap and much darker stipe base. *P. melanopus* is quite similar but has a grayer cap, a fuzzy stipe surface, and a tendency to appear terrestrial--fruiting from buried roots and debris. *Cerioporus leptocephalus* (=*Polyporus leptocephalus / P. elegans / P. varius*) is smaller, has a tan to beige cap, and tough black stipe. *Polyporus tuberaster* is larger, has a scaly cap, and grows from a sclerotium.

Synonym: *Polyporus tubaeformis* (P. Karst.) Ryvarden & Gilb.

Coltricia perennis group

Tiger's Eyes

Cap: 2–8 (10) cm across, plane with a depressed center to shallowly funnel-shaped, becoming ruffled, to ragged in age. Surface dry, velvety, often wrinkled or roughened. Dark brown to reddish brown when wet, more orange-brown, ocher to cinnamon when dry, with darker and paler concentric zones and a pale margin; usually with a pale sheen. **Pores:** Often slightly decurrent, small, shallow, round to angular, obscure near margin or present only as a weakly pitted surface; tube layer very thin. Creamy gray at first, soon cinnamon to brown, staining brown when bruised. **Stipe:** 1.5–5 cm long, 0.2–1 cm thick, usually central, sometimes off-center. Orange-brown, cinnamon to dark brown. **Flesh:** Thin, tough, leathery; cinnamon to brown. **Odor:** Indistinct. **Taste:** Unpleasant. **KOH:** Black on all parts. **Spore deposit:** Yellowish brown. **Microscopy:** Spores 6–9 × 3.5–5 µm, ellipsoid to cylindrical, smooth, pale yellowish brown.

Ecology: Solitary, scattered, or occasionally in fused clusters on ground or in moss under a wide range of conifers. Often found in abundance in recently disturbed or burned areas. Common throughout Cascadia. First fruiting in fall but fruitbodies persist for many months.

Edibility: Unknown.

Comments: The brown, velvety, zonate cap, thin, leathery flesh, central stipe, and terrestrial growth help distinguish this polypore; all other genera of small central-stalked polypores grow on wood. *Coltricia* are ectomycorrhizal, unlike most other polypores, which are wood decayers. We have more than one species going by the name *C. perennis* but no modern work has been done on the group in western North America and the name has probably been misapplied. The name *C. cinnamomea* also has been misapplied on the basis of slight differences in cap texture, color, and zonation. *Hydnellum subzonatum* can be nearly indistinguishable from *Coltricia* when seen from above, but its fertile undersurface is composed of spines, not pores — the same goes for the somewhat similar *Phellodon tomentosus. Coltricia montagnei* has a larger fruitbody with a stout stipe and sometimes has remarkable thick-walled "pores" arranged in concentric rings.

Onnia tomentosa

(Fr.) P. Karst.

Tiger's Paw Polypore

Fruitbody: 4–15 cm across, circular or irregularly-lobed, or multi-shelved rosette-shaped, often with bumpy concentric rings, with a depressed center in age. Surface dry, velvety-hairy. Yellowish ocher-brown to caramel-tan. **Pores:** Slightly to deeply decurrent, small, round to angular; tubes short. Whitish becoming buff to brown, flashing different colors when viewed at different angles when young. **Stipe:** 1–4 cm long, 1–3 cm thick, central to off-center, stout. Surface dry, velvety-fuzzy, concolorous with cap. **Flesh:** Thick, tough, fibrous; yellowish tan to brown. **Odor:** Indistinct. **Taste:** Indistinct or sour. **KOH:** Blackish on cap. **Spore deposit:** White. **Microscopy:** Spores 5–6 × 3–4 µm, ellipsoid to ovoid, smooth, inamyloid. Setae abundant in pore layer; tips straight.

Ecology: Solitary or in small groups, sometimes overlapping and fusing. Causes a root-rot of spruce and thus appearing terrestrial. Widespread throughout Cascadia where spruce in present. Fruitbodies are annual, first appearing in late summer or fall and persisting through the wet season.

Edibility: Unknown.

Comments: This polypore is recognized by its velvety, golden tan to caramel-colored cap, whitish to light brown pores, short, central stipe, and growth under spruce. *O. subtriquetra* has hooked rather than straight-tipped setae and grows on the trunk bases of a wider range of conifers, especially pine. It also tends to have more shelf-like fruitbodies. Another species that commonly causes confusion is *Phaeolus schweinitzii*, which has more irregular, maze-like, greenish olive to grayish pores that quickly stain brown when bruised.

Synonym: *Inonotus tomentosus* (Fr.) Teng

Phaeolus schweinitzii group

Dyer's Polypores

Fruitbody: 6–30 (45) cm across, starting as an irregular cushion-shaped blob, soon expanding into a circular disc and then forming a multi-shelved rosette or fused into a nearly circular, centrally depressed mass. Occasionally occurring as a shelf-like bracket on logs or tree trunks. Surface dry, velvety to hairy at first, matted to bumpy in age. Color extremely variable, often correlating with different "ecotypes" growing on different trees: Golden-yellow with a whitish or pinkish margin, sometimes striking neon-orange or with reddish-orange tones; center of cap usually darker, warm brown to dark olive brown with paler concentric bands of mustard-yellow, becoming evenly reddish brown to blackish brown in age. **Pores:** Small, angular to highly irregular, often weakly maze-like; with bands and zones of thicker walls. Pale olive-gray to dingy greenish yellow at first; often with a whitish-gray sheen, becoming olive-brown, quickly staining dark brown to blackish brown when scratched or bruised. Tubes shallow to moderately deep, olive to brown. **Stipe:** Stout but obscure, irregular and slightly rooting, occasionally lateral or absent. **Flesh:** Soft and spongy at first, becoming tough, fibrous, or corky. **Odor:** Indistinct. **Taste:** Unpleasant, rancid or sour. **KOH:** Black on pores, flesh and cap. **Spore deposit:** White. **Microscopy:** Spores 6–9 × 2.5–5 µm, elongate to ellipsoid to ovoid, smooth, inamyloid.

Ecology: Solitary or scattered around base of dead and dying conifer trees and stumps, occasionally on trunks or large downed logs. Causing a cubical butt rot of conifers, this is one of the most common and widespread of all macrofungi in Cascadia. The largest fruitings occur in late summer and early fall, with scattered fruitings into early winter. The spent, dark brown fruitbodies resembling cow patties are persistent and their crumbly remains are found any time of year – sometimes serving as substrate for the firm, bright golden-yellow, pea-shaped fruitbodies of *Endogone* (one of the few Zygomycete fungi to produce a macroscopic structure).

Edibility: Unknown, possibly toxic.

Comments: Genetic data have shown multiple lineages of cryptic species passing under this name, which helps explain the highly variable appearance and habitat preferences observed by mycologists. Despite this variability, species in this group are recognized by their top-shaped or cushion-like fruitbodies that become flat or rosette-like in age, greenish yellow to olive, irregular pores that stain dark brown, and growth on or around the base of conifers. Fresh fruitbodies are highly prized by fiber artists—yielding a bright golden yellow dye when paired with an alum mordant and rich olive greens with iron. This species is frequently confused with *Onnia tomentosa* and *O. subtriquetra*, the former is especially similar in stature but is typically smaller. Both of these have more evenly colored ocher-brown to caramel caps and paler pores that don't stain dark brown when bruised.

Bondarzewia occidentalis

Jia J. Chen, B.K. Cui & Y.C. Dai

Western Bondo

Fruitbody: 5–35 (50) cm across, lumpy and irregular at first, expanding into a single or more often compound uplifted rosette of fan-shaped lobes. Surface moist to dry, finely velvety with scattered felty patches or small hairy scales, smoother in age. Brown to grayish brown at center, warmer and often paler outward; yellowish brown to ocher-brown, usually with a beige to whitish margin, often with light and dark zonations. **Pores:** Small, round at first, soon irregular, becoming rather ragged in age. White to cream, occasionally with pale yellowish buff stains in age. Tubes tough, shallow, white. **Stipe:** Indistinct or central to lateral, often tapered downward and rooting. Brown when exposed, dingy whitish if buried. **Flesh:** Thick, tough, fibrous, whitish to creamy. **Odor:** Pleasant when young, rancid in age. **Taste:** Mild, acrid to slightly bitter. **Spore deposit:** White. **Microscopy:** Spores 6.5–7.5 × 5.5–7 µm, globose to subglobose, with low ornamentation of strongly amyloid ridges and warts.

Ecology: Generally solitary and fruiting at the base of dead or dying conifer trees or stumps. Uncommon, preferring coastal mature forests and in the Cascade Range, perhaps more abundant on the Olympic Peninsula. First emerging in late summer and fall but tough fruitbodies remnants persist.

Edibility: Edible but rather tough and often bitter.

Comments: The fact that this large polypore is actually closely related to *Lactarius* may be difficult to believe given the fruitbody morphology, but the amyloid, ornamented spores tell the tale. The large size, yellow-brown to grayish brown cap color, white pores, and obscure or irregular stipe are the key distinguishing features of this species. *Polypus dispansus* (=*Albatrellus dispansus*) has fruitbodies with many overlapping cap-lobes and brighter golden yellow to tan colors; it is a rare species found across Cascadia in mature conifer forests. *Jahnoporus hirtus* has a finely velvety, brown, a well-defined off-center stipe, an extremely bitter taste, and spindle-shaped to cylindrical spores.

Misapplied name: *Bondarzewia mesenterica* (Schaeff.) Kreisel (European) of which *B. montana* is a synonym.

Abortiporus biennis

(Bull.) Singer

Blushing Rosette

Fruitbody: 4–20 cm across, highly variable in shape: often a simple circular or semicircular blob, often with patches of pores on top; some specimens differentiated into a mass of fused rosettes or shelving clusters, or even peaked turrets. Surface moist to dry, often with red droplets when young, matted-tomentose, often irregular and knobby. Occasionally pores will be over the cap. Whitish to pinkish at first, soon staining dingy reddish, often red-brown overall with a paler margin in age. **Pores:** Often quite irregular, angular to maze-like. Whitish to pale pinkish, staining reddish brown. Tubes shallow, white to beige. **Stipe:** Usually absent, sometimes a short irregular, rooting mass. **Flesh:** Fibrous and watery, becoming tough. Whitish, staining pinkish to reddish. **Odor:** Mild to sweet, sometimes rancid in age. **Taste:** Mild. **Spore deposit:** Whitish to pale yellow. **Microscopy:** Spores 4–6.5 × 3.5–5 µm, broadly ellipsoid to ovoid, smooth, inamyloid. Asexual spores often present, 5–8.5 µm, globose.

Ecology: Solitary or in scattered clusters on buried woody debris; often growing engulfing plant stems and grass. Typically in lawns, gardens, or disturbed areas around dying trees (growing from dead roots). Common in urban areas of interior valleys throughout Cascadia, less common in forested habitats. Fruitbodies annual, mostly fruiting in fall, with occasional late spring or summer fruitings in irrigated areas.

Edibility: Unknown.

Comments: This astonishingly variable mushroom can resemble anything from a honeycombed blob covered in droplets to a shelving polypore to a rosette; its most stable distinguishing features are the pinkish red staining and angular to maze-like white pores. Young specimens with red droplets are frequently mistaken for *Hydnellum peckii*, which has spines on the underside and grows in forest settings. Older fruitbodies have been mistaken for *Bondarzewia occidentalis*, a much larger polypore that lacks the staining reaction and has ornamented, amyloid spores. *Climacocystis borealis* has creamy white, hairy caps and angular pores; it fruits in overlapping shelves on conifers.

Bridgeoporus nobilissimus

(W.B. Cooke) T.J. Volk, Burds., & Ammirati

Noble Polypore

Fruitbody: 30–100 (150!) cm across, large shelf-like, massive hoof-shaped, or top-like with a central stipe when growing on top of a cut stump. Surface coarsely hairy or fuzzy. Off-white, creamy to beige on growing margin, often with green algae, moss, and other debris accumulating on top of it. **Pores:** Small, round. White to creamy beige, occasionally buff when old; tubes multi-layered, moderately short. **Stipe:** Generally absent, except when fruiting on top of a stump. **Flesh:** Very thick, tough, fibrous, whitish. **Odor:** Indistinct. **Taste:** Not sampled. **KOH:** No reaction. **Spore deposit:** White. **Microscopy:** Spores 5.5–7 × 4–5 µm, subglobose to ovoid, smooth, inamyloid.

Ecology: Growing at base of standing snags or large stumps of old-growth true firs. Apparently limited to trees with a trunk diameter of at least 1 m. Widespread but rare; most occurrences are from the west side of the Cascade Range and foothills, extremely rare elsewhere; occurring south to Humboldt County, CA. Fruitbodies perennial, growing for many decades.

Edibility: Inedible.

Comments: The remarkably large fruitbodies, pale colors, deeply fuzzy cap texture, and occurrence on old-growth true firs distinguish this species. Only about a hundred fruitbodies of this species have ever been found, mostly in the Cascade Range in northern Oregon. Sadly, many were growing on stumps of logged old-growth trees. This is one of the only fungi in North America to have received specific legal protection, once being considered a 'sensitive species' by the U.S. Forest Service and Bureau of Land Management. This designation has since been removed due to the apparent prevalence of non-fruiting mycelia (detected by DNA sequencing of wood in fir snags) it is questionable whether this is a meaningful way to characterize the species' population dynamics.

Climacocystis borealis

(Fr.) Kotl. & Pouzar

Boreal Shelfpore

Fruitbody: 5–15 cm across, top-shaped or shelf-like; margin with a moderately thick, rounded edge at first, becoming thin-edged in age. Surface moist to dry, fuzzy at first, becoming soft and lumpy or nearly smooth in age. Apricot-beige when young, becoming pale creamy beige with an apricot-tan margin, overall chalky whitish in age. **Pores:** Slightly decurrent, faintly concentrically zoned; small and irregular when young, becoming larger, more maze-like and ragged to toothy in age. Tube layer shallow. Creamy to pale beige. **Stipe:** Absent to obscure, covered in decurrent pore layer. **Flesh:** Fairly thick, watery-soft to fibrous-tough. Whitish, concentrically zoned. **Odor:** Indistinct to slightly sweet, sometimes faintly like coconut. **Taste:** Indistinct. **KOH:** No reaction. **Spore deposit:** Whitish. **Microscopy:** Spores 4.5–6.5 × 3–4.5 µm, broadly ellipsoid, smooth, inamyloid, colorless.

Ecology: Scattered or in clusters of overlapping shelves on trunks and stumps of dead conifers, causing a mottled white rot. Common throughout higher elevations of Cascadia. Fruitbodies annual, emerging in summer and fall.

Edibility: Inedible.

Comments: Although the individual features of this species are subtle, taken together they make for a fairly distinctive appearance: the tawny to pale, fuzzy to lumpy shelf-like caps, irregular maze-like or toothy pores, soft-fibrous texture, and tendency to fruit in layered clusters on conifers separate it from most others. *Spongipellis spumeus* is fairly similar in appearance but grows on hardwoods, particularly *Populus*. *Trametes gibbosa* often has a simple fan-shaped bracket, lacks the fuzzy-hairy cap surface, with a tougher, more corky flesh texture.

Pseudoinonotus dryadeus group

Giant Weeping Polypores

Fruitbody: 20–70 cm across, semicircular to fan shaped, compressed and shelf-like (2–6 cm thick at rounded margin) or fairly deep (up to 30 cm thick) near attachment to wood. Whitish to creamy or pale beige, mottled with areas of tan-brown to gray or darker brown (especially older parts near attachment to wood). Margin usually palest and covered in many shining droplets, smears, and puddles of ocher-brown to amber liquid, or pitted when dry. Older parts dry, chalky to crumbly or hard and smooth or broken into plaques. **Pores:** Round, tiny; tube layer fairly thick (2–5 cm). White, with brownish tubes. **Flesh:** Thick, corky when fresh, hard and woody in age. Mottled brownish flecked with white. **Odor:** Indistinct to rancid, unpleasant. **Taste:** Mild to sour. **Spore deposit:** Whitish, copious, often covering entire area around mature fruitbodies. **Microscopy:** Spores 6–8 × 5–7 µm, irregularly ellipsoid to globose, smooth, hyaline, inamyloid. Flesh with dark, thick-walled, hooked setae.

Ecology: Solitary or scattered around the bases of living trees or stumps. Occasional throughout Cascadia on both hardwoods and conifers. More common on hardwoods in low-lying areas, inland valleys, and foothills, and on conifers—especially fir along the coast and in montane regions. Fruiting in summer or early fall. Fruitbodies persist for one to two years, at which point they crumble off the tree.

Edibility: Inedible.

Comments: These polypores are impressively large when fresh, rivaled in size only by the largest *Ganoderma* and the much rarer, and even more massive *Bridgeoporus nobilissimus*. Both *Pseudoinonotus* and *Ganoderma* tend to leave heavy deposits of spore dust on their surroundings but they are easily told apart by the color of the spores: White in the former, rusty-brown in the latter. *Bridgeoporus* is restricted to true firs, with a fuzzy hairy cap, and lacks the extensive coating of exudate droplets. Both *Ganoderma* and *Bridgeoporus* produce longer-lasting (perennial) fruitbodies. Genetic studies have shown two species in the *Pseudoinonotus dryadeus* group in Cascadia: The 'real' one on hardwoods and a second, as-yet named species, on conifers.

Synonym: *Inonotus dryadeus* (Pers.) Murrill.

Ischnoderma benzoinum

(Wahlenb.) P. Karst.

Syrup Polypore; Benzoin Bracket

Fruitbody: 4–10 (15) cm across, fan-shaped, slightly lobed or undulating in age; margin rounded. Surface moist to dry, densely velvety. Medium to dark brown, nearly blackish brown in age, often with golden sheen; margin paler to tan, beige, creamy, to white, often dark-streaked. **Pores:** Round, tiny, often decorated with clear golden droplets; tube layer shallow. Whitish to pinkish cinnamon, darkening in age; quickly bruising brown when scratched. **Flesh:** Thick, fleshy when young, tough-fibrous in age. Whitish to pinkish tan, slightly streaked or mottled. **Odor:** Sweet, anise-like or sugary. **Taste:** Indistinct. **KOH:** Olive-black on cap. **Spore deposit:** Whitish. **Microscopy:** Spores 5–7 × 1.5–2 µm, sausage-shaped, smooth, colorless.

Ecology: Solitary or in small groups on fallen and standing trunks of conifers (especially true fir and hemlock), causing a white rot. Fairly common in the west-side foothills and higher elevations of the Cascades and the Olympic Peninsula; scarce throughout lower elevations of Cascadia. Fruiting first in summer or early fall, some appearing later into winter; fruitbodies relatively short-lived.

Edibility: Edible but not recommended.

Comments: This lovely species is distinguished by its dark, densely velvety, fan-shaped caps, soft flesh, anise aroma, and pale pores. The resinous golden droplets, which are often present on the pore surface, are another helpful clue. *I. resinosum* is quite similar but often more evenly fan-shaped, lacking an odor, and grows on hardwoods. *Amylocystis lapponica* has a more shaggy-hairy cap surface, amyloid cystidia in the fertile tissue, and larger spores measuring 8–11 × 2.5–3.5 µm. It favors coastal spruce logs and snags. The Beefsteak Polypore (*Fistulina hepatica*) is rare in Cascadia with a preference for wax myrtle along the immediate coast. It has a minutely pebbly, smooth, or gooey (never hairy) cap surface and has pale-marbled or streaky yellowish flesh that soon turns bloody red. The true *F. hepatica* is a Eurasian taxon; the eastern American species was recently separated as *F. americana* and it is likely that the Beefsteak Polypore along the Pacific coast will get its own name.

Laetiporus conifericola

Burds. & Banik

Conifer Chicken-of-the-Woods

Fruitbody: 5–20 (30) cm across, blob-like or cushion-shaped at first, soon becoming fan-shaped, often with a lobed and wavy margin. Often occurring in clusters of overlapping shelves. Surface dry, spongy-soft and slightly velvety when young, soon bald, bumpy. Sulfur-yellow overall at first, becoming more orange to peachy orange, with paler concentric bands, light pinkish orange and then whitish when dried out. **Pores:** Very tiny, round; tube layer thin. Bright sulfur yellow to pale lemon yellow, fading to whitish when old. **Flesh:** Moderately thick to thin, soft and often juicy at first, becoming firm, meaty to corky, finally crumbly or chalky when old. Pale yellow to whitish. **Odor:** Indistinct when young, unpleasant in age. **Taste:** Mild to sour. **KOH:** No reaction. **Spore deposit:** White. **Microscopy:** Spores 6.5–8 × 4– 5 µm, ovoid, smooth, inamyloid.

Ecology: In large troops and overlapping clusters, less often as solitary brackets on standing trees, stumps, and large logs of a wide range of conifers, especially true firs and spruce. Causes a brown rot, leaving behind distinctively blocky-crumbly, dry wood. Very common and widespread across Cascadia, occurring in almost every habitat type. Fruiting in summer and fall but fruitbodies are persistent and the crumbly chalk-white remains is found any time of year.

Edibility: Edible and very good but only young fruitbodies (brightly colored, soft, and spongy) should be eaten. Some people experience gastrointestinal distress after eating it, so be sure to cook it thoroughly and sample sparingly at first.

Comments: This striking mushroom is easily distinguished from all others except for *L. gilbertsonii* which is more yellow to orange and grows on hardwoods including non-native trees in urban areas. *L. gilbertsonii* has a more restricted range in Cascadia, primarily occurring in low-lying inland valleys. Both these species can cause digestive distress. *Pycnoporellus fulgens* is considerably smaller, and has large, pallid, often maze-like pores.

Phellinus tremulae

(Bondartsev) Bondartsev & P.N. Borisov

Aspen Conk, False Tinder Conk

Fruitbody: 7–20 cm across, hoof-shaped, sometimes projecting outward to make a thick, blunt-edged fan-shaped fruitbody, in age elongating, with multiple growth layers. Surface dry, rough, often cracked into blocky chunks. Dark gray to brownish gray, often with concentric bands of color, blackish near attachment to wood and pale to creamy beige near margin. **Pores:** Tiny, round; tubes moderately long, layered. Gray-brown to gray or beige-tan, pale to creamy near margin. **Flesh:** Thick, tough, woody, dark brown, with fibers of flesh arranged in parallel vertical orientation. **Odor:** Indistinct. **Taste:** Indistinct. **KOH:** Black on all parts. **Spore deposit:** Whitish. **Microscopy:** Spores 4.5–5 × 4–4.5 µm, subglobose, smooth, inamyloid, colorless.

Ecology: Solitary or in loosely-fused vertically-arranged troops on the trunks of aspen and cottonwoods, usually just below branch scars. Fairly common in appropriate habitats throughout Cascadia. Fruitbodies are perennial and long lasting. Many of the long lasting *Phellinus* conks and their close relatives have mosses and lichens growing on them, contrasting to many other polypores.

Edibility: Inedible.

Comments: The hard, grayish, hoof-shaped fruitbodies and growth on aspen trees help identify this species. Tinder Polypore (*Fomes fomentarius*) is quite similar but typically even more hoof-shaped, grows primarily on birch trees, and has much longer spores (12–20 × 4–7 µm). Members of the *P. igniarius* group have more evenly blackish caps and richly-colored rusty-brown pores; it also prefers birch, alder and willow. The Oak Hoof Polypore (*Inocutis dryophilus*) is larger and more irregularly shaped or loaf-like, with tawny-golden to beige colors, which can fade to nearly white in age. It grows on oaks, usually above head height. It is probably common in the appropriate habitats of Cascadia including oak savannas in warm, dry inland valleys of Oregon, but may be overlooked. *Coniferiporia* (=*Phellinidium*) *weirii* forms perennial, resupinate fruitbodies with pale ocher-tan margins and gray-brown pores; it causes a white rot on the butts and roots of live conifers, particularly hemlock and Douglas-fir. *Phellopilus nigrolimitatus* is similar to *C. weirii* but has a dark black line in the context, it also grows on conifers. Microscopically, it has huge setae (even visible with a hand lens!) scattered throughout the fertile tissue, sterile margin, and even in the adjacent rotten wood around the fruitbodies. *Phellinidium ferrugineofuscum* is similar to the latter species but has a bright white margin. *Fuscoporia ferrea* and *F. ferruginosa* grow as resupinate layers of pores on fallen branches and logs. The two are best distinguished by their spore size: very slightly longer and narrower in *F. ferrea* (5–7.5 × 2–2.5 µm), versus 5–7 × 3–3.5 µm in *F. ferruginosa.*

Porodaedalea pini group

Conifer Maze Polypores

Fruitbody: 2–20 (35) cm across, hoof-shaped or irregularly shelf-like, occasionally more fan-shaped, laterally attached to substrate. Surface dry, roughened, often with multiple grooves or tiers. Brown to reddish brown with a mustard to golden brown growth margin, becoming dark brown to nearly black. **Pores:** Generally round to angular at first, more highly irregular, often maze-like in age; tubes relatively short, layered. Golden brown, grayish brown to rusty brown. **Stipe:** Absent. **Flesh:** Thin to thick, very tough, woody, reddish brown to rusty brown. **Odor:** Indistinct. **Taste:** Indistinct. **KOH:** Black on all parts. **Spore deposit:** Pale yellowish to light brown. **Microscopy:** Spores 4.5–7 × 3.5–5 µm, ovoid, smooth, inamyloid.

Ecology: Solitary or scattered on trunks of conifers, especially in "armpits" of branches. Common throughout Cascadia on a wide range of hosts (but exact host preferences are obscured by cryptic species). Fruitbodies perennial, persisting for years.

Edibility: Inedible.

Comments: This common polypore (or group of similar cryptic species) is recognized by their hard, hoof-shaped or irregularly wavy fruitbodies, with dingy olive-gold to brown labyrinth-like pores. *Porodaedalea* are a complex of poorly understood species in Cascadia. Two European names have been misapplied to species occurring in Western North America: The *P. 'pini'* complex typically is more shelf-like or bracket-like and rarely clustered. *P. 'chrysoloma'* is quite similar, with many overlapping features —it is distinguished by thicker flesh and microscopic features of the hyphae. More work is needed to name and describe distinguishing features of North American species. *P. cancriformans* grows in tight, lobed clusters of overlapping caps from cankers and wounds on large fir trees. *Gloeophyllum sepiarium* is smaller, with elongated, gill-like pores. Fruitbodies of various *Phellinus*, *Inonotus*, and *Fuscoporia* may be similarly colored but have round to slightly angular pores. *Fomitiporia tsugina* is also dark coppery to rusty brown and grows in the armpits of conifer branches but forms right-angled, thick, resupinate pore layers (not projecting hoof-shaped fruitbodies) and is mostly restricted to hemlock.

Echinodontium tinctorium

(Ellis & Everh.) Ellis & Everh.

Indian Paint Fungus

Fruitbody: 10–30+ cm across, 10–15 cm high, projecting outward 10–20 cm, irregularly hoof-shaped or slightly elongated along substrate. Surface roughened to ragged, broken into cracks and blocks, often covered with lichen and moss. Dark brown to blackish. **Underside:** Composed of a mass of hard, separate, icicle-like structures 3–10 cm long; tips relatively blunt at first, often broken-off in age. Grayish with pale, dull mustard-colored tips, becoming entirely dark gray to black in age. **Flesh:** Hard, woody-corky, with parallel, vertically arranged grain. Brilliant orange-red to brick-red, mottled with pale streaking in upper portion. **Odor:** Indistinct. **Taste:** Not sampled. **KOH:** Dark wine-purple on flesh. **Spore deposit:** Whitish. **Microscopy:** Spores 6–8 × 4.5–6 µm, ellipsoid, ornamented with small amyloid spines.

Ecology: Solitary or in troops of single fruitbodies, usually high up on standing trunks of conifers (especially true fir and hemlock), tightly adhering to main trunk on or just below broken-off branches. Widely distributed across Cascadia; uncommon (but can be locally abundant) at higher elevations on both sides of the Cascade crest. Fruitbodies emerge and develop slowly and persist for many years, even after the tree has decayed!

Edibility: Inedible.

Comments: This remarkable species is unique in its combination of hard, massive fruitbodies, toothed structure, and brilliant red flesh. Due to the dark-colored outer surfaces and habit of fruiting high overhead, it is often overlooked and thus more common than records suggest. Amazingly, this species is more closely related to the brittle-fleshed cap-and-gill mushrooms in the genus *Russula* than it to superficially similar woody polypores or most toothed mushrooms — the amyloid ornamentation of the spores is a microscopic clue supporting this placement in the order Russulales. The common name refers to fact that indigenous people mix the powdered fruitbodies with bear fat for use as body paint.

Rhodofomes cajanderi

(P. Karst.) Kotl. & Pouzar

Rosy Polypore

Fruitbody: 3–10 (15) cm across, hoof-shaped at first, soon expanding into a fan-shaped bracket. Surface usually dry, finely tomentose when young, lumpy and roughened to nearly smooth. Very young fruitbodies emerge as a blob of pinkish white fertile surface, often adorned with reddish droplets when wet. Mature caps pinkish brown to brown or blackish brown in age, margin usually remaining pale pinkish. **Pores:** Quite small, round; tube layer thin. Pale pink to rosy pink, becoming pinkish brown in age. **Flesh:** Moderately thin, tough-fleshy when young, soon woody. Pinkish brown. **Odor:** Indistinct. **Taste:** Indistinct. **KOH:** No reaction. **Spore deposit:** White. **Microscopy:** Spores 5–7 × 1.5–2 µm, cylindrical to sausage-shaped, smooth, inamyloid.

Ecology: Solitary, scattered, or in overlapping clusters on dead conifers (rarely hardwoods) causing a cubical brown rot. Widespread and fairly common; fresh fruitings occur in fall or early winter but fruitbodies persist, with some surviving through summer.

Edibility: Inedible.

Comments: The pinkish pores and the tough, woody flesh are distinctive. *Rhodofomes roseus* is generally larger and more distinctly hoof-shaped and has paler pinkish pores and spores measuring 2–2.5 µm wide. *Leptoporus mollis* is whitish but soon becomes sherbet to coral pink and has soft flesh; it is uncommon to rare on dead conifers in Cascadia.

Synonym: *Fomitopsis cajanderi* (P. Karst.) Kotl. & Pouzar

Polypores

Ganoderma oregonense

Murrill

Western Varnished Conk

Fruitbody: 10–60 (80) cm across, projecting 5–40 cm outward from substrate, broadly fan-shaped, often lobed or wavy, usually with concentric zones or "stair steps." Surface smooth, dull when young and growing, soon varnished and shiny. Bright brick red to dark reddish brown or brownish burgundy (although often partially covered in dull brown spore powder). Surface with a thin shellack-like crust. Growing margin paler, bright orange, yellow to white. **Pores:** Tiny, round, tubes moderately long. White at first, becoming very dingy brick pink to brownish in age, staining brown when scratched or bruised when fresh. Pore surface often developing sunken pits or with areas of brown, sawdust-like frass (insect poop). **Flesh:** Thick, quite spongy when fresh, soon tough, but softer than many large polypores. Whitish, buff to honey brown. **Odor:** Indistinct. **Taste:** Mild to somewhat astringent. **KOH:** No reaction. **Spore deposit:** Dark rust-brown. **Microscopy:** Spores 13–17 × 8–10 µm, ellipsoid with a truncate end, thick, double-walled, smooth, inamyloid, brown.

Ecology: Solitary or sometimes in small groups or overlapping clusters on conifers (especially hemlock). Common throughout Cascadia. Unlike other *Ganoderma*, the fruitbodies of this species are annual, first emerging in late summer, maturing by late fall, and usually beginning to disintegrate and rot by late winter—although recognizable fruitbodies sometimes persist much longer.

Edibility: Not edible due to the tough texture of mature fruitbodies but some people collect the fresh fruitbodies to make purportedly medicinal tinctures or teas. Please use restraint if you decide to collect some.

Comments: An attractive member of the northern polypore flora, this species is easy to recognize by virtue of the large size, shiny surface, and red color. The name *G. tsugae* has sometimes (spuriously) been used for smaller specimens. *Fomitopsis mounceae* can have a shiny orange to red cap but is woody textured, perennial, and has creamy pores which may stain pale yellowish at most when damaged.

Ganoderma applanatum

(Pers.) Pat.

Artist's Conk

Fruitbody: 6–30 cm across, 3–10 cm thick, projecting 5–25 cm outward from substrate. Fan-shaped to shell-shaped, with concentric zones and growth rings; margin with a narrow rounded curb. Surface dry, sometimes cracking. Dark brown to pale grayish-brown, often covered in rusty brown spores at maturity. Growing margin whitish. **Pores:** Tiny, round; a new tube layer is added over the last one each growing season. Whitish at first, becoming dingy white to brownish when dormant. Quickly staining dark brown when scratched or bruised if moist. **Flesh:** Thick, tough and corky-fibrous to woody; beige to tan-brown, with dark horizontal lines indicating growth layers. **Odor:** Indistinct. **Taste:** Indistinct. **Spore deposit:** Copious, rusty brown. **Microscopy:** Spores 6–9 × 4–6 µm, ellipsoid with a truncate end, double-walled, inamyloid, golden-brown.

Ecology: Solitary or scattered on the lower trunks, stumps, and logs; often on hardwoods but also on conifers. Common throughout Cascadia. The exact distribution and host preferences of this species are unclear due to ongoing conflation with *G. brownii*. Fruitbodies emerge at all times of year and are perennial and persisting for decades.

Edibility: Not edible due to its woody texture but some people collect species in this genus to make medicinal tinctures and teas. We strongly encourage sparing collection for this and all other perennial conks, as they can keep growing for decades when left undisturbed.

Comments: The fan-shaped, woody fruitbodies with gray-brown caps and brown-staining white pores are easily distinguished from all but the very similar *G. brownii*, which may be more common in southern Cascadia. The latter species often has deeper, hoof-shaped to "flapjack-stack" shaped fruitbodies, darker brown cap colors, and darker brown flesh but is most reliably distinguished by its much larger spores (11–12 × 7–8 µm) with thick, double walls. Both species are frequently confused with *Heterobasidion* and *Fomitopsis*; although superficially quite similar the white pores of the fruitbodies in the latter two genera do not stain brown when scratched and they have white spore deposits.

Fomitopsis mounceae

Haight & Nakasone

Western Red-belted Conk

Fruitbody: 5–30 (75) cm across, cushion to hoof-shaped, or 'blob-like' at first, becoming more fan-shaped in age. Surface dry, roughened to smooth, often with shiny portions. Typically whitish to pale pinkish at margin, then with a reddish to orangish zone inward, dark brown to blackish near attachment to substrate. Some fruitbodies lack reddish band entirely. **Pores:** Very tiny, round, tubes layered. Whitish to creamy, slightly dingier in age, sometimes staining pale yellowish buff when bruised. At times adorned with clear droplets, especially when young and actively growing. **Flesh:** Thick, very tough, corky to woody. Whitish to pale yellow. **Odor:** Fragrant. **Taste:** Lemony sour to slightly bitter. **KOH:** Reddish to red-brown on flesh. **Spore deposit:** Whitish to creamy yellow. **Microscopy:** Spores 6–7 × 3.5–4 µm, ellipsoid to cylindrical, smooth, inamyloid.

Ecology: Solitary or scattered on trunks of trees, logs, or large branches, causing cubical brown rot. Occurring on both hardwoods and conifers (one of the few polypores to do so) but is far more abundant on conifers in Cascadia. Widespread and very common throughout Cascadia. Fruitbodies perennial, occasionally growing for decades.

Edibility: Inedible.

Comments: Recent genetic work has shown that we have two species going by the name *F. pinicola* in North America: the widespread *F. mounceae* and *F. schrenkii* from the southwest into California. Although *F. schrenkii* has not been reported from Cascadia, it may occur in the drier portions of our range. *F. mounceae* is recognized by the large size, often three-toned cap (black, red-orange, and pallid margin), and creamy pores that don't stain brown. However, not all specimens exhibit the reddish band, and this can lead to confusion with *F. ochracea*, which differs primarily in lacking a reddish band

Hypomyces aurantius

Trichoderma pulvinatum

(the dark center transitioning directly to a pale margin). Perhaps the easiest way to differentiate them is to take a flame to the cap: the surface of *F. mounceae* melts and bubbles, whereas *F. ochracea* chars. Another species easily confused with members of this group is *Heterobasidion occidentale*, which generally has an irregularly roughened cap and often grows flattened against logs and trunks or forms a slightly projecting shelf with a thin growing margin. Other distinguishing features are its lack of a red KOH reaction on the flesh and its round to oval, minutely spiny amyloid spores. *H. irregulare* is nearly identical with *H. occidentale* but has a preference for pine, whereas *H. occidentale* is more common on other conifers. *Ganoderma oregonense* has soft brown flesh at first and a dark red, varnished cap when mature. *G. applanatum* and *G. brownii* have brown flesh and brown staining pores (when fresh) and lack the reddish color on the cap. All *Ganoderma* have brown spores. *Phellinus* species usually have rusty brown flesh and a black KOH reaction. Two common fungi often grow on the old conks: The Orange Polypore Mold (*Hypomyces aurantius*), which forms an orange, pimply crust, and species in the Polypore Pillow Mold (*Trichoderma pulvinatum*) (=*Hypocrea pulvinata*) complex, which form cushion-shaped ocher-yellow warts.

Fomitopsis ochracea

Ryvarden & Stokland

Ocher Conk

Fruitbody: 10–30 (40) cm across, often fused into irregular, overlapping shelves; round and blob-like when young, soon expanding to hoof-shaped, or fan-shaped, becoming irregularly wavy, especially older specimens. Surface dry, smooth to bumpy, often with lumpy concentric bands; growing margin fleshy and often slightly expanded. Dull gray to brown or blackish overall; margin ocher-tan to creamy tan or pale cream. **Pores:** Tiny, round; tubes short, with a new layer added each growing season. White to pale cream, not staining or discoloring only slightly when scratched. **Flesh:** Tough-fibrous when fresh, becoming hard and woody. Creamy to pale yellowish. **Odor:** Indistinct to mildly fungoid. **Taste:** Indistinct. **KOH:** Slight reddish on pores. **Spore deposit:** White. **Microscopy:** Spores 4–5 × 3–4 µm, ovoid to ellipsoid, smooth.

Ecology: Solitary, scattered, or in small clusters of poorly formed fruitbodies that causes a brown rot on a wide range of conifers and hardwoods. Typically on the lower to middle trunk of dead and dying trees but also on downed logs and occasionally on stumps. Abundantly common and widespread, occurring in all forested habitats in Cascadia. Fruitbodies perennial and long lasting, adding growth throughout the wet season.

Edibility: Inedible.

Comments: The often three-toned cap, with brown to blackish older parts, ocher to tan middle band, creamy growing margin, non-staining whitish pores, and creamy flesh are distinguishing. There are two similar members of the Red-belted Conk complex. First, the common and abundant *F. mounceae* (usually) has a red band separating the older, dark portion and the pale cream to white growing margin. Second, *F. schrenckii* also typically has an intervening orangey to reddish band but is more irregularly-shaped overall, with lumpy lobes and often distressed-looking fruitbodies; it is not yet confirmed to occur in Cascadia. *F. ochracea* is easily confused with *Heterobasidion occidentale*, which averages a darker upper surface, has a thin growing margin, and has subglobose, finely spiny spores. Similar *Ganoderma* species have brown-staining pores, brown flesh, and copious brown spores.

Laricifomes officinalis

(Vill.) Kotl. & Pouzar

Agarikon; Quinine Conk

Fruitbody: 5–30 (45) cm across, up to 120 cm long, cushion- to hoof-shaped at first, becoming multi-tiered, older specimens becoming cylindrical. One or two pore layers added each year. Surface dry, roughened, often cracking. Whitish when fresh, becoming grayish brown to ocher, beige, tan, or light brown on older parts, sometimes greenish with algal growth. Very old fruitbodies often bleached to white. **Pores:** Tiny, round to irregular. Fresh and living layers are white; older, inactive pores are pale beige-tan; tubes moderately short, adding a new layer each year. **Flesh:** Thick, soft when fresh, soon corky, chalky or crumbly when dry or old. **Odor:** Indistinct to farinaceous. **Taste:** Strongly bitter. **Spore deposit:** Whitish to creamy yellow. **Microscopy:** Spores 6–9 × 3–4 µm, cylindrical, ellipsoid to ovoid, smooth, inamyloid.

Ecology: Solitary or scattered on the standing trunks of conifers. It has a preference for old-growth Douglas-fir and can fruit high up on the trunk. Smaller conks are found on younger trees. Uncommon throughout Cascadia. Fruitbodies perennial, very slow growing, and persisting for decades.

Edibility: Inedible. Because it is highly prized for its purported medicinal properties, fruitbodies are over-collected by unscrupulous harvesters.

Comments: The large, pale, multi-tiered fruitbodies are almost unmistakable. Younger, hoof-shaped specimens are mistaken for *Fomitopsis mounceae* but told apart by the paler colors, crumbly-chalky flesh, and extremely bitter taste. *Fomes fomentarius* can form similar pale "flapjack stacks" but has brown, fibrous-woody flesh, is usually much smaller, and grows on birch. *Postia amara* forms large, tan to dark brown hoof-shaped, annual fruitbodies with yellow pores. It causes a pocket rot on mature incense-cedar; but is rarely encountered, partly because of the scarcity of old-growth cedars and partly because it fruits high overhead during the summer months. *Ganoderma* can have multitiered fruitbodies, but have woody brown flesh, pores that quickly stain brown when scratched, and brown spores.

Synonym: *Fomitopsis officinalis* (Vill.) Bondartsev & Singer

Heterobasidion occidentale

Otrosina & Garbel.

Western Root-rot Conk

Fruitbody: 6–30 cm across, irregular to hoof-shaped at first, expanding to more fan-shaped; sometimes with a partly spreading or crust-like aspect, or appearing as a cap-less disc of pores when fruiting on the underside of a log. Young fruitbodies emerge as a blob of whitish to cream fertile surface. Surface dry, lumpy to grooved, with undulating growth rings. Mature caps are dark brown to black in age, with a contrasting, thin creamy-white growing margin. **Pores:** Small, round; tubes layered. Whitish to pale yellow, unchanging when scratched. **Flesh:** Thin to thick, tough, woody. Whitish. **Odor:** Indistinct to slightly rancid. **Taste:** Indistinct to sour. **Spore deposit:** White. **Microscopy:** Spores 4.5–6.5 × 3.5–4.5 µm, subglobose to ovoid, minutely spiny, inamyloid.

Ecology: Solitary, scattered, or in groups on living and dead conifers; especially in managed timber plantations. Causes a stringy white-rot that spreads from root to root and affects the entire standing body of the tree. Widespread and abundant throughout Cascadia. Fresh fruitings may occur any time of year and fruitbodies are perennial, lasting many years.

Edibility: Inedible.

Comments: *Heterobasidion* are some of the most infamous fungi in the United States due to their impacts on the forestry industry. They cause severe damage to the host tree's wood and render the timber commercially useless. The dark cap with a thin, contrasting white growing margin and hoof- to fan-shaped fruitbodies on conifers are distinctive characters. Two species occur in Cascadia: *H. irregulare* and *H. occidentale*. The former appears to prefer pine and juniper, has smaller, more deformed fruitbodies with pores that are slightly more elongated. The latter on true fir, hemlock, and Douglas-fir, with rounder, slightly more densely-packed pores. *Fomitopsis ochracea* has a thicker, rounded, creamy growing margin and pale yellow flesh which has a slight reddish KOH reaction. *Fomitopsis mounceae* usually shows a red or orange band.

Misapplied name: *Heterobasidion annosum* (Fr.) Bref. in the strict sense is a European species & not from Cascadia.

Fomitopsis betulina

(Bull.) B.K. Cui, M.L. Han & Y.C. Dai

Birch Polypore

Fruitbody: 7–25 cm across, 1.5–5 cm thick, round when first emerging, soon fan-shaped; margin with an inrolled "curb" and with a "hump" near attachment to wood. Surface dry, smooth, sometimes cracking or splitting. Beige to tan or gray to brownish gray, often spotted or streaked, at times silvery when dry. **Pores:** Small, round; tubes forming a thin, concave layer. White to pale cream. **Flesh:** Firm, rubbery-fibrous; white. **Odor:** Indistinct or slightly sweet. **Taste:** Indistinct to astringent. **KOH:** No reaction. **Spore deposit:** White. **Microscopy:** Spores 5–6 × 1.5–2 µm, sausage-shaped, smooth, colorless.

Ecology: Solitary or scattered in troops on standing and fallen birch trees. Common in appropriate habitat but scarce in Cascadia and restricted to the far northern portion, rare south of Seattle. Fruitbodies emerge in late summer through fall, annual but persisting into the next season.

Edibility: Inedible.

Comments: The combination of a white, concave pore layer, smooth, gray-brown cap, and growth on birch trees make this a very distinctive species. *Fomes fomentarius* can have similar-colored caps but have hard, hoof-shaped, woody fruitbodies and grows on a wider range of hardwoods.

Synonym: *Piptoporus betulinus* (Bull.) P. Karst.

Osteina obducta

(Berk.) Donk

Bone Polypore

Fruitbody: 5–15 cm across, kidney- to fan-shaped, tapered to an off-centered stipe; margin rounded to thin, in-curved when young. Surface smooth, dry, opaque, with scattered patches of whitish fuzz near attachment to wood. Gray to tan or pale brownish, paler near attachment to substrate. Mature caps pinkish brown to brown or blackish brown in age; margin usually remaining pale pinkish. **Stipe:** 1–3 cm long, 0.5–1.5 (3) cm thick, short, poorly-defined, off-center, mostly covered in decurrent pores. **Pores:** Small, round to weakly angular; tubes densely packed, forming a thin layer. Off-white to cream. **Flesh:** Thick, tough, fleshy, white. When dried becoming "hard as bone". **Odor:** Indistinct or sour-fungoid when fresh. **Taste:** Indistinct. **KOH:** No reaction. **Spore deposit:** White. **Microscopy:** 5–6.5 × 2–2.5 µm, cylindrical to pointed, smooth, inamyloid.

Ecology: Solitary or in overlapping clusters on dead conifers, causing a brown rot. Widespread, but generally rare, perhaps more common at higher elevations. Fresh fruitings occur from spring to fall but fruitbodies are very long-lasting, especially if they dry out before the rainy season.

Edibility: Inedible.

Comments: The smooth, tan to light brown cap, whitish pores, short stipe, and hard flesh when dry are telltale features of this polypore. *Picipes* and *Cerioporus* species are much thinner-fleshed, pliable, and leathery. The vaguely similar *Trametes suaveolens* lacks a stipe, has a corky-tough texture, a fuzzy cap surface, and a strong anise aroma. *Fomitopsis betulina* fruits as a single shelf on birch trees and has a pronounced rounded "curb" around the margin. The poorly-known *Polyporus coronadensis* was described from oak wood in southeast Arizona but a species matching it fruits during the summer and fall along the northern California coast on hardwoods; it may show up in the inland valleys and foothills of Cascadia. It is larger and paler, with a white to grayish or pale tan cap, and smells of almond extract when the fresh fruitbodies are cut.

Cryptoporus volvatus

(Peck) Shear

Veiled Polypore

Fruitbody: 1–6 (8) cm across, round to hoof-shaped; pores covered with a thick "veil" forming a hollow pocket on underside. Mature specimens usually have one or more small, round or ragged-edged holes on the underside. Surface dry, smooth, shiny, sometimes with a faint netted pattern or cracked into thin plaques. Butterscotch to light ocher-brown or reddish brown, paler in age, whitish when dry. **Pores:** Tiny, round, densely packed; tube layer thick. Whitish to pale buff or occasionally pinkish. **Stipe:** Absent. **Flesh:** Moderately thick, rubbery at first, becoming tough, whitish to cream, sometimes pinkish. **Odor:** Indistinct, rubbery or fragrant. **Taste:** Slightly bitter or astringent. **Spore deposit:** Not obtainable. **Microscopy:** Spores 12–16.5 × 4–5 µm, cylindrical, smooth.

Ecology: Scattered or in large troops, rarely solitary, on recently dead conifers, especially pine and fir. Fruitbodies are usually situated in bark furrows up and down trunk. Rare along the coast and Coast Range, occasional in inland valleys, common on both slopes of the Cascades and rain-shadowed east slope of the Olympic Peninsula. This species is abundant in the mountains after forest fires. Fruitbodies are annual with flushes appearing in spring, again in late summer (before fall rains), and persisting into the next year.

Edibility: Inedible.

Comments: This rounded to hoof-shaped polypore with pores encased inside a thick, rubbery "veil" is unique. On mature specimens, a small hole is almost always present in the veil; the reason for this is that *Cryptoporus* has a special relationship with a variety of insect species. Over a period of several months, numerous insect species (up to 17 in one study) feed on parts of the veiled polypore, and ultimately transport spores to new trees. After they exit the polypore, they burrow into new trees to lay their eggs, infecting them with the *Cryptoporus* spores in the process. The following year, *Cryptoporus* fruits out of these very same holes. This mode of spore dispersal has been very successful – this species is one of the most widely distributed wood-rotters in North America.

Pycnoporellus fulgens

(Fr.) Donk

Fruitbody: 3–10 cm across, irregularly fan-shaped. Surface dry, appressed-hairy, becoming smoother in age. Orange to rusty, often with tan or ocher concentric zones, becoming dingy brick-red in age. **Pores:** Somewhat angular to irregular, occasionally short maze-like; tubes short. Peach to dingy orange. **Flesh:** Tough, corky to fibrous-leathery, yellow to dingy orange. **Odor:** Indistinct. **Taste:** Indistinct. **KOH:** Cherry red on all parts. **Spore deposit:** White. **Microscopy:** Spores 6–9 × 2.5–4 µm, cylindrical, smooth, colorless.

Ecology: Solitary or scattered on trunks, logs, and branches of conifers; occasionally on hardwoods, causing a cubical brown rot. Common throughout Cascadia. Fruitbodies annual but persistent, first flushing in late spring and summer, with the primary fruiting in fall.

Edibility: Inedible.

Comments: The orange, hairy fan-shaped cap, paler angular pores, and red KOH reaction make this eye-catching polypore quite distinctive. *Hapalopilus nidulans* is similar in shape, with a soft squishy texture, tawny-beige to pinkish brown caps that turn deep lavender to violet in KOH, and ellipsoid spores measuring 3.5–5 × 2–2.5 µm. It is scarce in Cascadia, causing a white rot on hardwoods. *P. alboluteus* is typically quite different, usually appearing as a chaotic "frothy" layer of elongate pores that turn into ragged teeth; it is more common at higher elevations. *Trametes cinnabarina* has bright cinnabar-red caps when fresh and small, round, brilliant scarlet pores. It grows on hardwood debris, including small-diameter branches.

Polypores

Pycnoporellus alboluteus

(Ellis & Everh.) Kotl. & Pouz.

Orange Spongepore

Fruitbody: 5–30 cm across, highly amorphous, usually emerging as clumps of thin, ragged or "lacerate" looking teeth and pores, these eventually oriented downward. Brilliant flaming orange near attachment to substrate, paler orange to white toward edges of pores. Older fruitbodies often have dusty or foamy-powdery dull beige or brown areas. **Pores:** Extremely irregular; ragged-toothy to unevenly angular or hexagonal, thin-walled; projecting at all angles when very young, eventually developing a "hanging" aspect. **Flesh:** Thin, pliable-spongy to weakly corky; pale orange. **Odor:** Indistinct. **Taste:** Indistinct. **KOH:** Dull red. **Spore deposit:** White. **Microscopy:** Spores 9–14 × 3–3.5 µm, cylindrical to sausage-shaped, smooth.

Ecology: Solitary or in long runs and patches on rotting conifers, causing a brown rot. Common at high elevations of the Cascades and Olympic Peninsula; rare in low elevations and along the immediate coast. Emerging in spring, usually shortly after snowmelt and persisting through summer, often with another fruiting in fall.

Edibility: Inedible.

Comments: The brilliant orange-and-white, chaotic-looking clumps of ragged pores, and growth on conifers make this an easy species to recognize. *P. fulgens* produces fan- to bracket-shaped fruitbodies with bright orange caps and paler, round to angular pores. *Laetiporus conifericola* can be bright orange and grows on conifers but has fleshy, fan-shaped fruitbodies and small, round yellow pores. Some *Antrodia* have ragged fruitbodies with toothy or angular pores but lack the brilliant orange colors and are usually less messy-looking (more appressed, with stacks of pores).

Gloeophyllum sepiarium

(Wulfen) P. Karst.

Conifer Mazegill

Fruitbody: 2–10 cm across, shelf-like to broadly fan-shaped or sometimes nearly rectangular, often in overlapping clusters. Surface dry, rough, bumpy with concentric zones, smooth to finely tomentose when young, occasionally with tufted or hairy portions in age. Yellowish to ocher, to orange or reddish-brown and eventually dark brown to nearly black at attachment to wood, often with fibrillose black streaking overall. **Gills:** Thick-walled, gill-like "pores" with scattered irregular cross-walls and maze-like areas. Mustard-yellow to orangey; edges and mouths pale. **Flesh:** Thick, tough, leathery-woody. Rusty colored to brown in age. **KOH:** Black on all parts. **Odor:** Indistinct. **Taste:** Not sampled. **Spore deposit:** Whitish. **Microscopy:** Spores 8–13 × 3–5 µm, cylindrical to slightly crooked, smooth.

Ecology: Usually in small troops, sometimes in lines along cracks or seams of logs. Common across Cascadia on conifers (across a wide range of genera) but also found on hardwoods and even on painted structural timbers, decks, picnic tables, and chemically treated telephone poles. New fruitbodies can emerge practically any time of year and persist for a long time.

Edibility: Inedible.

Comments: The tough fruitbodies with yellowish, orangish to brownish colors, and gill-like 'pores' readily distinguish this species. Two rare species in Cascadia, *G. trabeum* and *G. protractum*, have duller colors and poroid, rather than gilled, fertile surfaces. The former has slightly more elongated, densely packed pores and shorter spores (6.5–9.5 vs. 8.5–11 µm in *G. protractum*). Other poroid species include the rare, anise-scented, pillow-shaped *G. odoratum* and *G. carbonarium*, which has mostly resupinate fruitbodies (often lacking a shelving cap margin) and a tendency to grow on burned wood. *Fuscoporia gilva* has bright mustard-yellow colors when young and small round pores.

Daedalea quercina

(L.) Pers.

Oak Mazegill Polypore

Fruitbody: 5–12 cm across (when fused, to 30 cm across), fan-shaped to hoof-shaped. Margin even, at times wavy or lobed, usually blunt, rounded, and quite thick. Surface dry, suede-like to roughened. Beige to tan with gray and creamy tones, often quite uniform, but developing concentric zones or bands of contrasting color. **Pores:** Thick-walled, elongated into gills, with some irregular to maze-like zones. Dull grayish beige to tan; edges paler than faces. **Flesh:** Tough, corky to almost woody, somewhat pliable; whitish to cream or pale beige. **Odor:** Indistinct. **Taste:** Indistinct. **KOH:** No reaction. **Spore deposit:** White. **Microscopy:** Spores 5-6 × 2-3.5 µm, cylindrical, smooth.

Ecology: Solitary, scattered, or in masses on oak logs and stumps. Common in Oregon's valleys and foothills on Oregon white oak, rarer northward where it is mostly found on planted hardwoods in urban areas or in warm microclimates of low-lying areas. Fruiting in fall; fruitbodies annual but can persist for a long time.

Edibility: Inedible.

Comments: The beige to grayish tan colors, roughened cap, thick, thick-edged, gill-like underside, and growth on oak help distinguish this species. It appears to have been introduced to Cascadia and is slowly spreading southward on oaks. *Trametes betulina* is extremely similar, and is best distinguished by its tomentose cap, thinner-edged gills, and thinner, more pliable, bracket- or fan-shaped fruitbodies. *Daedaleopsis confragosa* forms thinner, bracket- or fan-like fruitbodies and has narrower gills that bruise pinkish when scraped; it grows on birch and is uncommon to rare in northern Cascadia. *Gloeophyllum sepiarium* has orange to brown colors and grows on conifer wood.

Trametes betulina

(L.) Pilát

Gilled Polypore

Fruitbody: 2–12 (14) cm across, semicircular or fan-shaped. Surface dry, distinctly tomentose, becoming matted in age. Cap color variable, usually a combination of brown, tan, grayish white, and beige, with many concentric zones. Brighter orange or yellow-capped forms are also occasionally found. Often fading in age and becoming greenish with algae. **Underside:** Made up of elongated, thick, gill-like structure. Primarily regular and parallel, but also forking, branched, or maze-like. At times more pore-like with thick walls, especially when young or near attachment to substrate. White to creamy, occasionally pale buff in age. **Flesh:** Thin to moderately thick, tough, leathery; whitish. **Odor:** Indistinct. **Taste:** Indistinct. **Spore deposit:** White. **Microscopy:** Spores 5–6 × 2–3 µm, cylindrical to sausage-shaped, smooth, inamyloid.

Ecology: Solitary, in rows or groups of overlapping clusters, occasionally as a rosette when fruiting on top of logs. Common on hardwoods in inland valleys of Cascadia, scarce in the foothills and coastal areas. Incredibly, its hyphae parasitize those of *Trametes versicolor*, which are usually found fruiting nearby. Fruitbodies are annual, first emerging in early fall but persisting into the next year; often becoming green with algal overgrowth.

Edibility: Inedible.

Comments: The tomentose grayish white to beige cap, leathery texture, and gills on the underside distinguish this species. It looks very similar to pale forms of the Turkey-tail (*T. versicolor*) from the top but the underside is completely different. Young specimens with thick gills can be confused with *Daedalea quercina*, which has a smoother cap and more woody-corky flesh. *Trichaptum* species can have elongated pores (appearing gill-like in age) but have purple pores when young and mostly grow on conifers. *Daedaleopsis confragosa* is quite similar but has more cross-walls between narrower gills (forming pores) and stains pinkish where bruised or handled; it is scarce in Cascadia. *Gloeophyllum sepiarium* is similar in shape but has orange to brown colors overall and grows on conifers.

Synonym: *Lenzites betulina* (L.) Fr.

Trametes versicolor

(L.) Lloyd

Turkey Tail

Fruitbody: 2–8 (10) cm across, semicircular, bracket to fan-shaped shelves, occasionally forming rosettes. Surface dry, distinctly tomentose on paler bands and smooth on darker concentric zones. Color quite variable, but usually a combination of dark and pale concentric zones with a white margin. Inner areas range from dark brown to grayish brown with pale grayish to buff zones, or reddish brown to orange-brown with tan to buff zones, or in some forms dark blue with paler gray zones. Fading when old or dry, often greenish due to algal growth on previous year's caps. **Pores:** Tiny, round; tube layer very thin. White to cream, pale buff or occasionally ocher in age. **Stipe:** Absent. **Flesh:** Thin to moderately thick, tough, leathery. White to creamy; with a dark line under the fuzzy cap surface. **Odor:** Indistinct. **Taste:** Indistinct. **Spore deposit:** White. **Microscopy:** Spores 5–6 × 1.5–2.5 µm, cylindrical to sausage-shaped, smooth, inamyloid.

Ecology: In rows or masses of shelving or overlapping clusters, or occasionally as rosette clusters on dead hardwoods. Widespread and very common throughout Cascadia. Fruitbodies annual, first emerging in early fall but persisting into next year.

Edibility: Inedible due to the tough texture but can be dried and powdered to make teas and tinctures. Long considered a medicinal fungus in Eastern medicine, it continues to be investigated for its immune-boosting and anti-cancer properties.

Comments: This is one of North America's most common mushrooms – although some consider the many forms to represent multiple species. As a group, they are recognized by their variably yellowy-beige to gray to brown, or even blue colors, with alternating light and dark concentric zones, tiny, round, white pores, and growth on hardwoods. *T. ochracea* is extremely similar overall but has a more evenly-colored ocher-brown to red-brown cap surface and larger spores (6–8 × 2–2.5 µm). It lacks a dark line under the fuzzy cap layer. *T. betulina* is similar from the top but has gills instead of pores. *T. gibbosa* is larger and paler, with lumps and bumps over the surface and short, maze-like pores. *T. hirsuta* is usually thicker fleshed, with a hairy, evenly pale grayish or whitish beige cap. *T. suaveolens* is also pallid but has a smooth to finely tomentose cap and a sweet anise odor. *Trichaptum abietinum* has violet colors when fresh and larger pores that become ragged in age; it only grows on conifers. *Stereum hirsutum* produces smaller, paler, wavier fruitbodies with more yellow to orangey tones, with a smooth underside. *Coriolopsis gallica* has a more spiky-hairy, evenly brown cap surface, larger, more irregular or angular gray to brown pores, and thicker, corky fruitbodies; it is particularly fond of cottonwoods. *Bjerkandera adusta* has a grayish buff to light brown cap and smoky gray pores.

Trametes gibbosa

(Pers.) Fr.

Lumpy Bracket

Fruitbody: 4–10 (15) cm across, 2–4 cm thick, hoof-shaped when very young, soon expanding to form a fan-shaped bracket. Surface dry to chalky, smooth to finely velvety when young, but soon bumpy-knobby overall. Pale gray or whitish with a tan margin when young, soon paler overall, often with contrasting gray concentric bands; becoming whitish overall and often developing green tones from algae in age. **Pores:** At first small, round to irregular, becoming larger and more irregular to radially elongated in concentric zones or short maze-like pattern. Dingy whitish to pale gray, sometimes dull pinkish-tan. **Flesh:** Thick to moderately thin, tough, leathery. Whitish to pale grayish. **Odor:** Indistinct. **Taste:** Indistinct. **KOH:** No reaction. **Spore deposit:** White. **Microscopy:** Spores 4–5 × 2–2.5 µm, ellipsoid, smooth, inamyloid.

Ecology: Solitary, scattered, or in overlapping clusters on dead hardwoods. This species was recently introduced to Cascadia and remains most common near urban centers (Portland, Seattle, Vancouver) but is expanding into native habitats, usually on hardwoods in riparian zones throughout inland valleys and foothills. Fresh fruitings occur in fall but fruitbodies persist, often becoming green with algal overgrowth.

Edibility: Inedible.

Comments: The fairly large, pallid fan-shaped bracket with elongate to maze-like pores and growth on hardwoods help identify this increasingly common species. *Trametes pubescens* and *T. suaveolens* are somewhat similar but have fuzzy, grayish caps that are not so extensively lumpy; the former species is more strongly zonate and the latter species often has a strong anise aroma. Both are uncommon to rare on hardwoods in Cascadia. *T. betulina* has a tomentose cap and a distinctly gill-like underside. The chalky white fruitbodies of *Calcipostia guttulata* are somewhat similar, but thinner, and have indented "potholes" rather than bumps on the cap surface; the pores and cap are often extensively decorated with milky droplets. *Daedaleopsis confragosa* has brown colors and a distinctly maze-like pore structure.

Calcipostia guttulata

(Sacc.) B.K. Cui, L.L. Shen & Y.C. Dai

Pale Pothole Polypore

Fruitbody: 5–15 cm across, forming a thin, kidney-shaped to fan-shaped bracket, sometimes slightly wavy. Surface covered in droplets of milky guttation when young, soon moist to dry with many droplets perched on the bumpy surface, eventually with many depressed "potholes" where droplets have dried. Whitish to pale ocher or beige at first, becoming slightly zonate to grayish, white and chalky-looking in age. **Pores:** Small, round to irregular; tube layer very short. White to grayish, occasionally staining pale blue. **Stipe:** Absent. **Flesh:** Watery-fibrous when fresh, chalky-corky to tough and dry in age. Whitish. **Odor:** Indistinct to faintly fragrant. **Taste:** Bitter. **Spore deposit:** White. **Microscopy:** Spores 4–5 × 2–2.5 µm, cylindrical, smooth, inamyloid to weakly amyloid.

Ecology: Solitary or scattered on decaying stumps and logs of conifers, causing a cubical brown rot. Common throughout Cascadia. Typically fruiting in summer and early fall; fruitbodies persistent, lasting into winter.

Edibility: Unknown.

Comments: The white to chalky fruitbodies adorned with droplets or potholes, bitter taste, and growth on conifers help identify this species. There are a number of other white polypores in Cascadia including *Perenniporia subacida* that has small, white, perennial fruitbodies; it is a butt rot of conifers, causing a stringy white rot with black flecks. *Spongiporus floriformis* forms pale, thin, overlapping shelves often forming rosettes; it causes a cubical brown rot on hardwoods and conifers. *Amaropostia stiptica* is very similar but has a more bumpy or rough-textured cap surface with dark dots; it grows on both hardwoods and conifers. The White Cheese Polypore (*Tyromyces chioneus*) has a soft "cheesy" texture, is mild tasting, and grows on hardwoods. The Marshmallow Polypore (*Spongiporus leucospongia*) has pale, remarkably soft-spongy fruitbodies that form long shelves on conifer logs. It is a common resident of high elevation pine and true fir forests in the Cascade Range, fruiting after snowmelt but occurs down to low elevations in BC. *Osteina obducta* has a smooth grayish to brownish cap.

Trichaptum abietinum

(Dicks.) Ryvarden

Purple-Toothed Polypore

Fruitbody: 1–4 cm across (fused fruitbodies can form crusts or fused caps up to 10 cm across), protruding 0.5–3 cm, bracket to fan-shaped shelves, occasionally more crust-like on undersides of fallen branches and logs; margin becoming ruffled or lobed in age. Surface dry, coarsely hairy to matted-tomentose, occasionally with concentric zones. Beige to buff with a violet to royal purple growing margin at first, whitish when dry, often with a greenish layer of algae in age. **Pores:** Round to angular, soon becoming tooth-like to ragged in age; tubes very shallow. Violet at first, becoming buff to grayish, often a violet margin when still actively growing, dingy buff overall in age. **Flesh:** Thin, tough, leathery. **Odor:** Indistinct. **Taste:** Indistinct. **KOH:** Yellowish to green on purple-pigmented areas. **Spore deposit:** Whitish. **Microscopy:** Spores 6–7.5 × 2.5–3 µm, cylindrical to sausage-shaped, smooth, inamyloid.

Ecology: Often fruiting in masses of single, fused, or overlapping clusters on dead conifers, causing a white rot. Very common throughout Cascadia. Fruitbodies are mostly annual, emerging in early fall but persisting throughout the year; often developing a green algal layer in age.

Edibility: Unknown.

Comments: When young, this polypore is easily recognized by the violet pores and growth on conifer wood. If it has lost the violet colors, the ragged pores and pale cap are useful clues. *T. fuscoviolaceum* is quite similar but is less common; it has a crust-like to slightly reflexed fruitbody and often grows on pine. *T. biforme* grows on hardwoods and tends to be larger. *Chondrostereum purpureum* can look similar in color and shape but has a smooth underside, grows on hardwoods (especially alder), and has a farinaceous to spermatic odor. *Cerrena unicolor* also gets a greenish coating of algae on the cap but has white to grayish or smoky short maze-like pores, a white growing margin, and grows on hardwoods; it is uncommon in Cascadia. *Trametopsis cervina* has a beige to tan or light brown cap that is less apt to turn green in age; it has pale angular, maze-like pores that often split into teeth, and grows on hardwoods, especially cottonwood.

Bjerkandera adusta

(Willd.) P. Karst.

Smoky Polypore

Fruitbody: 2–5 cm across, semicircular or fan-shaped, occasionally crust-like with scattered projecting shelves. Surface moist to dry, obscurely matted-tomentose to smooth. Ocher, beige, tan, or brown with a whitish to creamy growing margin, sometimes weakly zonate, developing dark stains in age. **Pores:** Tiny, round; tube layer thin. Pale whitish gray to dark smoky gray, paler toward margin, blackening slowly where damaged. **Stipe:** Absent. **Flesh:** Thin to moderately thick, tough, leathery, white to grayish. **Odor:** Mealy-mushroomy. **Taste:** Mild to sour. **Spore deposit:** White, scant. **Microscopy:** Spores 5–6 × 2.5–3.5 µm, cylindrical, smooth.

Ecology: Scattered or in masses of overlapping shelves on dead and often standing hardwoods, especially alder. Common throughout Cascadia. Fruitbodies annual, first appearing in early fall but sometimes persisting into the next year.

Edibility: Inedible.

Comments: The pale cap, smoky gray pores, and gray to blackish staining distinguish this species. *B. fumosa* is larger and thicker fleshed, has buff to light gray pores, and a dark line separating the tubes from the flesh. *Trametes* species can look very similar from above but have white to creamy pores. *Cerrena unicolor* also has gray pores but these are more open and maze-like and it has a fuzzy-hairy cap surface that often becomes entirely green with algae.

Polypores

Cyanosporus simulans

(P. Karst.) B.K. Cui & Shun Liu

Blue Cheese Polypore

Fruitbody: 1–5 (7) cm across, a rather thick, fan-shaped to semicircular bracket. Surface hairy, moist to dry. Whitish to pale bluish or bluish gray at first, becoming more grayish beige with a paler margin; clumps of hairs usually bluish to navy blue. **Pores:** Small, round to irregular; tubes short. White to grayish, occasionally staining pale blue (the rapidity and depth of staining reaction may vary between the cryptic species in this group, see comments). **Flesh:** Moderately thick to thin, rather soft, spongy, quite watery when wet. **Odor:** Fragrant. **Taste:** Sweet, but often astringent. **KOH:** No reaction. **Spore deposit:** Whitish to pale grayish blue. **Microscopy:** Spores 5.5–7.5 × 1.3–1.5 µm, cylindrical to sausage-shaped, smooth, inamyloid to weakly amyloid.

Ecology: Solitary or scattered on decaying logs and branches of hardwoods and conifers. Uncommon in drier parts of Cascadia, more common in coastal and west-side forests. Fruiting throughout wet season.

Edibility: Unknown.

Comments: The soft, spongy texture and bluish colors of this small polypore are distinctive... don't look too closely. Although long called *Postia caesia*, a recent study found seven species in North America, most of which occur in Cascadia. There are no clear, well-defined concepts for morphologically distinguishing these species. Five similar bluish species can be (unreliably) distinguished from *C. simulans* by their narrower spores (< 1.4 µm wide). but are very difficult to tell apart from each other. *C. livens* grows on conifers and hardwoods, and apparently has more open pores, larger fruitbodies, and a hairier upper surface than the following four species: *C. caesiosimulans* grows on a wide range of hardwoods, *C. subviridis* grows on conifers, *C. populi* grows on cottonwoods, and *C. arbuti* appears to be restricted to madrone. Details of hyphal structure, and/or DNA sequences are needed to apply these names confidently. *Tyromyces chioneus* is whitish to grayish, often has a slightly firmer, drier texture, and has mild-tasting flesh. *Spongiporus perdelicatus* and *Fuscopostia fragilis* are similar in stature and texture but both are whitish to creamy overall. The latter species quickly stains yellowish to ocher, then darkening to reddish brown where bruised.

Fuscopostia fragilis

(Fr.) B.K. Cui, L.L. Shen & Y.C. Dai

Red-staining Cheese Polypore

Fruitbody: 1–7 cm across, a fan-shaped or semicircular bracket. Surface moist to dry, velvety to finely tomentose. Whitish to pale buff when young, to pale reddish brown in age. All parts quickly staining yellowish to ocher, then darkening to reddish brown when bruised. **Pores:** Small, round to irregular; tubes short. White to pale buff, staining like cap. **Flesh:** Moderately thick to thin, rather soft, spongy, quite watery when wet, whitish. **Odor:** Fragrant. **Taste:** Mild. **KOH:** Indistinct to yellow. **Spore deposit:** Whitish. **Microscopy:** Spores 3–5 × 1–2 µm, cylindrical to sausage-shaped, smooth, inamyloid, colorless.

Ecology: Solitary or scattered on decaying logs and branches of conifers (rarely on hardwoods). Common throughout Cascadia, especially on pine and Sitka spruce. Fruiting throughout the wet season.

Edibility: Unknown.

Comments: The soft spongy texture, whitish to buff colors, and quick yellow to reddish brown staining are distinctive. *Tyromyces kmetii* has an orange to ochraceous cap, creamy to orangish pores, and lacks a staining reaction. *T. chioneus* is a soft, spongy, whitish polypore without staining reactions when scratched or bruised. *Skeletocutis nivea* is a similar but smaller crust-like or weakly shelving polypore found on small conifer branches; it is bright white to creamy colored and has narrower sausage-shaped spores measuring 3–5 × 0.5–1 µm.
Synonym: *Postia fragilis* (Fr.) Jülich

Polypores

Ptychogaster albus
Corda

Powderpuff Bracket

Fruitbody: 0.7–5 cm across, in two distinct forms: a pale creamy buff, irregularly rounded, plush, fuzzy blob that soon becomes a light brown, powdery mass; or expanding and becoming a whitish to creamy, irregularly bracket to top-shaped, with a central to off-center stipe. **Pores:** When present small, round to irregular. White. **Flesh:** Rather thin, very soft. Whitish at first, either becoming powdery and light brown or firm and pale. **Odor:** Indistinct. **Taste:** Mild. **Spore deposit:** Whitish from basidiospores, light brown asexual spores. **Microscopy:** Basidiospores 4.5–5.5 × 2–3 µm, ellipsoid to cylindrical, thin-walled, smooth, inamyloid, colorless. Asexual spores 5.5–10 × 3.5–7 µm, ellipsoid to oblong, often with truncate ends, thick-walled, yellowish.

Ecology: Solitary or scattered on conifer wood, especially on Sitka spruce branches. Uncommon, primarily distributed along the coast and lower elevation foothills on the west slope of the Cascades. Fruiting in fall.

Edibility: Unknown.

Comments: Truly an oddball, this fungus can either turn into powdery mass of asexual spores or grow up into a more "normal" polypore that produces sexual spores. *Spongiporus leucospongia* is another small, soft-textured, pale bracket but rather than fuzz turning into powder, it has a cottony-marshmallowy texture. Young *Phlebia tremellosa* can also look like pale, fuzzy blobs, but soon expands into shelf-like, thin, rubbery-textured fruitbodies with wavy pinkish veins on the underside. *Climacocystis borealis* is larger, much thicker, and more shelf-like with highly irregular to toothy pores.

Synonym: *Postia ptychogaster* (F. Ludw.) Vesterh.

Leptoporus mollis
(Pers.) Quél.

Pink Sherbet Polypore

Fruitbody: 3–12 cm across, projecting 2–8 cm, forming fan- or hoof-shaped to irregularly-layered brackets. Surface hairy-spiky to velvety, becoming nearly smooth in age or retaining shaggy area near attachment to substrate. Whitish to pale pinkish, becoming lilac to reddish purple, irregularly or entirely rosy in age. **Pores:** Small, round to angular, occasionally short maze-like in age; tube layer fairly thick. Whitish to pinkish when young, to pale reddish purple to dark purplish brown. **Flesh:** Thick, fleshy-spongy, watery when fresh, soft, pliable when dry. Whitish to pinkish or pale reddish purple. **Odor:** Indistinct. **Taste:** Mild. **KOH:** Turning pores and flesh pinkish to brick red. **Spore deposit:** White. **Microscopy:** Spores 5–6 × 1.5–2 µm, sausage-shaped, smooth, inamyloid.

Ecology: Solitary or in small groups on dead conifers, causing a brown rot. Widespread in lowland valleys and the Cascade Range, uncommon to rare along the coast. Fruiting from late summer through winter; fruitbodies annual, persisting for months.

Edibility: Inedible.

Comments: The soft, fleshy texture, pinkish to reddish purple colors, hairy cap surface, and growth on conifers help identify this striking species. Two similar *Hapalopilus* occur in Cascadia. First, *H. nidulans* grows on hardwoods and has a velvety cinnamon or pinkish orange cap which turns violet in KOH. Second, *H. salmonicolor* has a yellow-pink to ocher cap but only turns fleetingly lavender in KOH and is restricted to conifers. Both of these have spores 3.5–5 × 2–2.5 µm. *Phlebia tremellosa* forms a spreading crust or a shelf-like fruitbody, a hairy cap margin, pinkish-veined underside, and thin, rubbery flesh. *Gelatoporia dichroa* (=*Gloeoporus dichrous*) has a thin, shelf-like to fan-shaped fruitbody with a smooth, pallid cap, round pinkish to lilac pores. The Marshmallow Polypore (*Spongiporus leucospongia*) has whitish to peachy, velvety soft-spongy fruitbodies and fruits soon after snowmelt. *Pycnoporellus* species have tougher flesh and darker yellow to orange or reddish-brown colors. *Sarcoporia polyspora* forms resupinate sheets and is more dull-colored.

Phlebia tremellosa

(Schrad.) Nakasone & Burds.

Trembling Crust

Fruitbody: Spreading, flabby to nearly gelatinous membrane with shelving projections, at times occurring in dense, overlapping clusters on sides of logs and stumps. Shelves 1–5 cm across, projecting 0.5–2.5 cm; flat portion can be much larger. Surface moist to dry, hairy and often ragged around the edge. White, pinkish white to pale pinkish orange. **Fertile surface:** Veined and wrinkled, often radially arranged, and forming pseudo-pores. Dingy ocher, lilac grayish, or dull pink to pinkish orange, occasionally blood red in age. **Flesh:** Thin, watery, slightly rubbery, soft and pliable. **Odor:** Indistinct. **Taste:** Indistinct. **Spore deposit:** Whitish. **Microscopy:** Spores 3.5–4.5 × 1–2 µm, cylindrical to sausage-shaped, smooth, inamyloid.

Ecology: Often in troops on bark-less logs and stumps, occurring on both hardwoods and conifers. Widespread throughout Cascadia but more common in inland valleys and montane areas. Fruiting from fall into spring.

Edibility: Inedible.

Comments: Whether this species forms a spreading crust or a shelf-like fruitbody, the telltale features are the pale, hairy cap margin, pinkish-veined underside, and thin, rubbery flesh. Crust-like fruitbodies should be compared with the more yellow-orange *P. radiata* (and others mentioned in the comments for that species). *Gelatoporia dichroa* (=*Gloeoporus dichrous*) can be quite similar in color, but has distinct, round pores and a drier, less flabby texture. Two common species on small-diameter hardwood branches are the spreading, crust-like *Byssomerulius corium* (with a white, smooth or slightly wrinkled underside) and *Plicatura nivea* (which forms reflexed, crust-like fruitbodies with a whitish to tan cap and wrinkled white underside).

Synonym: *Merulius tremellosus* Schrad.

Chondrostereum purpureum

(Pers.) Pouzar

Silverleaf Fungus

Fruitbody: 1–6 cm across, often in larger, fused masses, fan-shaped or lobed and becoming exaggeratedly wavy; margin thin. Surface dry, velvety to pubescent. Pale silvery white to beige to ocher-tan or brown, concentrically zonate when wet, and often with lilac to tones towards margin. **Underside**: Smooth to slightly wrinkled. Pale lilac to beige or tan, occasionally vinaceous brown or reddish brown in age. **Flesh**: Thin, pliable, leathery-papery or slightly succulent and cartilaginous when fresh. **Odor**: Strongly farinaceous to spermatic. **Taste**: Farinaceous. **Spore deposit**: White. **Microscopy**: Spores 5–8 µm × 3–4 µm, cylindrical to sausage-shaped, smooth.

Ecology: In clusters or projecting, overlapping shelves, and sometimes resupinate as bumpy to wrinkled flattened sheets on hardwood logs. Saprobic or weakly pathogenic, it can infect trees that have been wounded or damaged and can eventually spread through the vessels of the host plant and cause mortality. This fungus can manifest as "leaf silvering" disease on many species of broadleaf trees. Widely distributed throughout Cascadia, especially common on hardwoods (e.g. alder, maple,). Fruiting in fall and spring but can appear at any time of year. Fruitbodies are relatively long-lived.

Edibility: Inedible.

Comments: The silvery to brownish cap (often has a lilac margin), smooth lilac to beige underside, spermatic odor, and growth on hardwoods make this an easy species to identify. *Veluticeps fimbriata* is more dingy overall (brownish even when young) and is restricted to conifer wood. *Stereum* usually have yellow to ocher or tan colors. *Trichaptum* have toothy pores on the underside. *Bjerkandera* have smoky gray pores. Recently, doctors from Kolkata, India determined that *C. purpureum* was the cause of a throat infection of one of their patients! This is particularly surprising because the victim was not immunocompromised. The infection likely occurred as a result of long-term exposure to decaying plant matter and fungal spores, as the patient reportedly worked as a plant pathologist.

Stereum hirsutum

(Willd.) Gray

Hairy Curtain Crust, False Turkey Tail

Fruitbody: 1–4 cm across (fused specimens often much larger), shelves protruding 0.5–3 cm, bracket to fan-shaped, occasionally crust-like on underside of logs but forming small shelves near sides. Occasionally funnel-shaped when fruiting on top of logs. Margin usually strongly ruffled and lobed in age. Surface dry, distinctly silky-hairy to tomentose, occasionally with concentric bald zones. Orange, orange-brown to beige, with darker and paler concentric zones at first, becoming darker, with reddish brown tones and a yellowish orange margin, fading greatly in age. **Underside**: Smooth to slightly roughened, orange to creamy orange, at times more grayish orange. **Flesh**: Very thin, tough, cartilaginous, or leathery. **Odor**: Indistinct. **Taste**: Indistinct. **KOH**: No reaction. **Spore deposit**: White. **Microscopy**: Spores 5.5–7 × 2.5–3.5 µm, cylindrical, smooth, weakly amyloid.

Ecology: Often in masses of shelving or overlapping clusters on dead hardwoods (rarely conifers). Widely distributed throughout Cascadia, more common in the inland valleys (because the coast and mountains are more conifer-dominated). Fruitbodies mostly annual, occasionally long-lived, first fruiting in early fall but persisting throughout year. The jelly fungi *Naematelia aurantia* and *Phaeotremella frondosa* commonly parasitize *S. hirsutum* and are found fruiting on or alongside it.

Edibility: Non-toxic.

Comments: The yellow-orange colors, hairy, zonate cap, smooth underside, and growth on hardwoods distinguish this species. *S. hirsutum* is difficult to miss but easy to confuse with Turkey Tail (*Trametes versicolor*) when seen from above; a quick look below will reveal the small, whitish to creamy pores of the *Trametes*. *S. ochraceoflavum* has a dull buff to tan underside, is often smaller, and grows on small hardwood branches and twigs. *S. sanguinolentum* is more crust-like, has a grayish brown fertile surface, and quickly stains red when bruised; it is restricted to conifers. Another red-bruising species is *S. gausapatum*, with more reddish brown colors and growth on oak. *Veluticeps abietina* has a dark reddish brown to blackish brown cap and a grayish lilac to purplish brown smooth or warty underside. It grows on conifers, especially true firs and is common in the Cascade Range.

Hydnoporia tabacina

(Sowerby) Spirin, Miettinen & K.H. Larss.

Willow Glue

Fruitbody: 5–20 cm long, 1–5 cm wide, usually forming long sheets of bumpy to wrinkly fertile tissue flat against the substrate, with small projecting fan-shaped "caps" hanging off edges of substrate. Upper surface of caps (when present) appressed-hairy or silky. Orangey ocher to brown. **Fertile surface:** Bumpy to wrinkly, or with smoother areas. Dingy rust-brown to dark grayish brown. **Flesh:** Extremely thin, dry and parchment-like. **Odor:** Indistinct. **Taste:** Indistinct. **KOH:** Blackish on fertile surface. **Spore deposit:** White. **Microscopy:** Spores 4–6 × 1.5–2 µm, pip-shaped to sausage-shaped. Flesh with long, pointed setae up to 120 µm long.

Ecology: Growing in long sheets, wrapping around the lower surface of twigs, branchlets, and other small-diameter woody debris of hardwoods; most commonly willow. Uncommon to locally abundant throughout northern Cascadia, probably underreported due to being inconspicuous and easily confused with other species. Fruitbodies emerge most any time of year and persist for months.

Edibility: Inedible.

Comments: The dark rusty-brown to gray-brown underside and habit of forming sheets on small hardwood branchlets help distinguish this species; the black KOH reaction and presence of setae under the microscope clinch the identification. *Hydnoporia olivacea* is similarly-colored and grows as a crust on branches but has a distinctly toothed fertile surface. Some *Stereum* species are similar but have orange to ocher-tan undersides and lack a KOH reaction.

Synonym: *Hymenochaete tabacina* (Sowerby) Lév.

Thelephora caryophyllea

(Schaeff.) Pers.

Carnation Earthfan

Fruitbody: 2–5 cm across, shallowly funnel-shaped, with a flat disc or rosette of a few layers of overlapping fan-shaped lobes around a central depression. Margin of fans and center of cap with a ragged, toothy appearance (like the edges of carnation petals). Surface dry to moist, obscurely appressed-hairy. Dark lilac-brown to reddish brown or dark gray-brown, sometimes weakly concentrically zoned; margin paler to pinkish brown or nearly beige-white at tips of lobes. **Underside:** Smooth to slightly wrinkled or with blunt radial grooves. Brownish-gray to lilac-gray or dull brown to pale gray when dry. **Stipe:** 0.5–3 cm long, 0.2–0.4 cm thick, typically short and indistinct from fertile underside, tapered towards base, sometimes nearly absent. Surface dry, faintly fuzzy. Dull brown to lilac-gray. **Flesh:** Thin, pliable, slightly tough; brown. **Odor:** Indistinct. **Taste:** Indistinct. **KOH:** Black on all parts. **Spore deposit:** Brown. **Microscopy:** Spores 6.5–8.5 × 5–7 µm, angular-ellipsoid, spiny, brown, inamyloid.

Ecology: In troops or arcs on ground and duff, occurring with a wide range of conifers. Common throughout Cascadia but a detailed understanding of its habitat preferences is obscured by identification challenges. Fruiting mostly in fall and winter, but fruitbodies last for a few months.

Edibility: Likely non-toxic but small and tough.

Comments: As a group, Earthfans (*Thelephora* spp.) are easy to recognize by their dark, sombre-gray to lilac-brown colors, smooth to wrinkled fertile surfaces, and either fan-shaped or coralloid-branched upright structures. This species in particular is somewhat distinctive due to its funnel shape and crimped-toothy margin but there are a number of similar *Thelephora* (including undescribed and poorly-understood species) for which identification criteria are lacking. *T. terrestris* is larger, with a more matted-fuzzy cap texture, duller brown colors, and less of a tendency to form overlapping cap lobes around a central depression. *Phellodon* can be superficially similar from above but have pale-tipped teeth on the undersurface. *Coltricia* can be similar in shape but have a shallow layer of pores on the underside.

Stereum sanguinolentum

(Alb. & Schwein.) Fr.

Bleeding Parchment

Fruitbody: 2–10 cm across, a thin, irregularly-lobed sheet of fertile tissue flattened against the substrate, sometimes forming projecting, wavy-undulating fans or shelves. Upper surface or "cap", when present usually whitish to dingy gray. **Fertile Surface:** Smooth to bumpy or faintly wrinkled. Beige to gray-ocher; extreme margin usually whitish. Slowly bruising bright red to dark red when touched or damaged, taking up to 30 seconds to fully develop. **Flesh:** Very thin, pliable-leathery to cartilaginous or parchment-like. **Odor:** Indistinct. **Taste:** Indistinct. **KOH:** No reaction. **Spore deposit:** White. **Microscopy:** Spores 6–7 × 2.5 µm, elongate to sausage-shaped, or elongate and pip-shaped, smooth, amyloid.

Ecology: In scattered troops or patches on dead trees or dead parts of still (for the moment) living conifers, especially hemlock. Causing a heart rot of conifers. Common in forested habitat at all elevations throughout Cascadia. Fruiting in wet periods throughout the year, most new growth occurs in fall.

Edibility: Inedible.

Comments: This species is often overlooked among the many other parchment-crusts but is recognized at first by its beige to tan or grayish color, growth on conifer wood, and definitive staining reaction. A similar but often more shelving red-bleeding *Stereum* grows on oaks in Cascadia; it has been (almost certainly incorrectly) going by *S. gausapatum*. *Phanerochaete sanguinea* has extremely inconspicuous, amorphous, crust-like fruitbodies that stain the surrounding wood red.

Phanerochaete sanguinea

(Fr.) Pouzar

Conifer Red-stain

Fruitbody: 2–5 cm across, forming an inconspicuous, amorphous sheet of thin, papery to felt-like material surrounded by pale rhizomorphs. Whitish to creamy or pale beige, staining reddish when bruised. Wood substrate brilliantly scarlet-stained. **Flesh:** So thin as to be absent. **Odor:** Indistinct. **Taste:** Indistinct. **KOH:** No reaction. **Spore deposit:** White. **Microscopy:** Spores 4.5–5.5 (7) × 2.5–3 µm, narrowly ellipsoid to cylindric, smooth, inamyloid.

Ecology: Fruiting on the underside of conifer branches and logs embedded in duff; also fruiting inside rotting stumps and large logs, causing a white rot. The red-stained wood is commonly encountered but the fruitbodies are short-lived. Common and widespread in conifer forests throughout Cascadia but often overlooked. Fruiting throughout the wet season.

Edibility: Inedible.

Comments: Like *Chlorociboria*, the striking pigmented staining of its substrate is more conspicuous than the fruitbodies! Broken conifer branches buried in duff will often show the bloody color associated with this species; careful inspection will sometimes turn up a thin, pale parchment-like patch of fertile tissue nearby.

Phlebia radiata

Fr.

Wrinkled Crust

Fruitbody: A spreading, knobby, pimpled membrane with irregular projections, furrows and ridges, ranging from the size of a coin to huge rows or sheets more than a meter across. Yellowish to orange-buff to beige, pinkish orange or darker to reddish brown on older portions. **Flesh:** Thin, rubbery when wet. **Odor:** Indistinct. **Taste:** Indistinct. **Spore deposit:** White. **Microscopy:** Spores 4–5.5 × 1.5–2 µm, cylindrical, smooth, inamyloid.

Ecology: On logs or in cracks and furrows of bark and trunks of hardwoods. Widespread and common throughout Cascadia. Fruiting from fall into spring but fruitbodies are quite persistent and can be found practically throughout the year.

Edibility: Inedible.

Comments: This common crust fungus is found in most of our hardwood forests throughout the wet season. Although there are a few similar *Phlebia*, this species is recognized by the spreading, radiating growth habit and the irregular projections, giving it a bumpy appearance. *P. livida* is quite similar but doesn't grow in a radiating pattern. *P. acerina* has a yellowish brown fertile surface with no KOH reaction and *P. rufa* has a reddish brown hymenium that darkens with KOH; both of these occur on hardwoods.

Aleurodiscus grantii

Lloyd

Big-spored Pinkpatch

Fruitbody: 0.1–0.4 cm across, occasionally fused into larger patches, disc- or shallowly cup-shaped; margin fringed with fine white hairs when young, more or less smooth in age. Pink to peachy pink. **Underside:** Smooth to finely pubescent, often fused to substrate; pallid. **Flesh:** Very thin, leathery-papery. **Odor:** Indistinct. **Taste:** Not sampled. **Spore deposit:** White. **Microscopy:** Spores 22–32 (39) × 18–28 µm, broadly ellipsoid to nearly round, covered with amyloid spines up to 4 µm long. Basidia 4-spored, club-shaped, with clamp connections at base.

Ecology: In gregarious masses of scattered fruitbodies on small-diameter branches and twigs of true fir, more rarely on other conifers. Widespread in Cascadia, and can be found during any wet period, but it is especially common in fall on grand fir in western Cascadia, or spring in the eastern Cascades.

Edibility: Inedible.

Comments: The small size, pink colors, growth on fir and large spiny spores help identify this species. At first glance it resembles a number of small cup fungi, but the tough texture and presence of basidia help steer you in the right direction. *A. amorphus* is very similar, but considerably less common; it is more likely to be found on other conifer species, has smaller spores (under 28 µm) and its basidia lack clamp connections. *A. aurantius* forms more crust-like fruitbodies on hardwood branches and twigs (especially alder). It has orange colors when young, which fade with age. *Corticium minnsiae* has two forms: a thin pinkish crust, or a tiny pink disc or cup with a fuzzy ball in the middle. It grows on conifer twigs, due to its small size, it is rarely observed.

Polypores

Ceriporia spissa

(Schwein.) Rajchenb.

Orangepore Crust

Fruitbody: 4–12 cm across, highly reduced; a soft, resupinate fertile surface. Finely but extensively convoluted-wrinkled at first, soon becoming poroid, eventually with a very shallow layer of tubes with tiny round pores. Brilliant orange, becoming dingy reddish in age; margins with a narrow band of whitish or light apricot-orange, finely cottony mycelium. **Pores:** Tiny, round. brilliant orange. **Flesh:** Very thin, orangish. **Odor:** Indistinct. **Taste:** Indistinct. **KOH:** Unknown. **Microscopy:** Spores 4–6 × 1.5–2 µm, cylindrical to sausage-shaped, smooth. Clamp connections absent.

Ecology: Usually forming small to large patches on dead hardwood branches and logs. Uncommon throughout Cascadia but may be under-reported. Fruiting from early fall through spring.

Edibility: Unknown.

Comments: This bright orange poroid crust on hardwoods is quite distinct. *Ceriporia tarda* has intense lilac-pink pores and spores 4–5 × 2–2.5 µm, while *C. purpurea* has duller pinkish pores (turning reddish-purple in age) and sausage-shaped spores measuring 5–8 (9) × 2–2.5 µm. *C. excelsa* is similar to the latter species but somewhat paler and has spores < 5µm long.

Nectriopsis violacea

(J.C. Schmidt ex Fr.) Maire

Violet Slime-eater

Fruitbody: 2–10 cm across, appearing as amorphous blobs, mounds or cushions depending on the shape of the underlying slime mold host. A thin layer of interwoven pale fibers soon forming a dusty to brittle lilac-gray to amethyst-colored crust studded with pimples of dark royal purple to nearly blackish purple perithecia. Often appearing marbled or mottled with white, and margin often with a band of yellow where the underlying slime mold is still exposed. **Flesh:** Very thin, hard crust, with powdery mass on underside. **Odor:** Indistinct. **Taste:** Indistinct. **Spore deposit:** White, but difficult to obtain. **Microscopy:** Spores produced in asci; 6–7 × 2–3 µm, ellipsoid, finely spiny.

Ecology: Parasitic, engulfing and smothering sporulating masses of *Fuligo* slime molds. Solitary or scattered on top of stumps or perched on the surface of thick duff layers; occurring in almost any habitat type as long as it is damp enough. Common throughout Cascadia but easily overlooked. Most common in fall but can fruit any time of year as long as its host is present.

Edibility: Inedible.

Comments: This crust-forming ascomycete is easily recognized by its amethyst to royal purple color and habit of forming a thin layer smothering the surface of *Fuligo* slime molds. *Hypomyces rosellus* is quite similar in form but produces neon-pink to magenta perithecia and is typically much smaller; it grows on the decayed remains of a wide range of mushrooms.

Lentoporia carbonica

(Overh.) Audet

Carbon Polypore

Fruitbody: 5–30 cm wide, up to 2 cm thick, highly irregular and often forming cascading sheets of resupinate tissue. White to creamy, older areas becoming beige to tan or pinkish-brown. **Pores:** Highly irregular: sometimes small and round at first, but more often irregular to angular, grouped into overlapping blocks or columns of tubes 2–4 cm deep. **Flesh:** Thin, white, somewhat pliable at first, soon tough and fibrous. **Odor:** Indistinct. **Taste:** Indistinct. **Macrochemical reactions:** Flesh instantly turning blue-black in Melzer's iodine solutions. **Spore deposit:** White. **Microscopy:** Spores 5–6.5 × 2–3 µm, long-ellipsoid to cylindrical.

Ecology: Solitary or scattered on the sides and cut faces of conifers, especially Douglas-fir. Common and often abundant after wildfire, fruiting in the cracks and fissures of carbonized trunks, sometimes forming connecting angular lines of resupinate pores, appearing like mortar between bricks. Widespread throughout Cascadia but often overlooked or misidentified. Fruitbodies emerge in late summer or fall and can persist in a fresh state for a few months.

Edibility: Inedible.

Comments: The white colors, tendency to form sheets of pores on the cut ends of logs or long runs of pores in the cracks and fissures of fallen logs, blue-black iodine reaction, and fondness for burnt conifers (but not exclusive to this habitat) help identify this species. *Rigidiporus crocatus* is similar in shape but has pinkish tan to beige fruitbodies and round spores measuring 3.5–5.5 µm. *Antrodia heteromorpha* forms thin, resupinate sheets of pale, angular pores on hardwoods. *A. xantha* also grows in resupinate layers of pores on conifers and has amyloid tissues but is bright yellow and has sausage-shaped spores measuring 4–5 × 1–1.5 µm. Another bright yellow species, *Antrodiella semisupina*, grows primarily on hardwoods and has broadly ellipsoid spores measuring 2.5–3.5 × 2–3 µm. *Auriporia aurea* is resupinate with yellow to orange fruitbodies. It has larger spores than the latter two species, shaped like elongated teardrops (cylindrical-ellipsoid), with large, extremely thick-walled cystidia heavily incrusted with crystals. *Sarcoporia polyspora* is similar in shape and stature but often has more projecting shelf-like caps, stains pinkish-brown where scraped, and has ellipsoid spores measuring 5–7 × 2.5–4 µm that turn reddish brown in Melzer's reagent (dextrinoid). *Gelatoporia pannocincta* produces resupinate layers of bright greenish yellow pores with a white or creamy band around their margin; it grows on cottonwoods.

18 - Tooth Fungi

Tooth-like fertile surfaces have arisen in a number of different lineages of macrofungi, not all of them closely related. Variably referred to as "teeth" or "spines," these structures have a high surface area to produce spores, while requiring comparatively little tissue to build—an evolutionarily favorable arrangement! Consequently, 'teeth' are one of the primary archetypes in the fertile surface arrangements of macrofungi. Some of the species in this chapter have tiny, highly reduced fruitbodies composed of only a few points of fertile tissue (or even a single "tooth!"), while others are large and more complex, with well-developed stipes and caps.

Auriscalpium are distinct by virtue of the right-angled stipes attached laterally to the cap and the fact that they growth exclusively on conifer cones. *Mucronella* produce swarms of tiny white or yellow fruitbodies on decaying wood. *Hericium* are also wood decayers, but form large, fleshy white fruitbodies that look like pom-poms or a spectacular branched cascade of icicles. Identification of the species in these genera is relatively straightforward. For gelatinous-textured toothed fungi, see *Pseudohydnum* (in the Jelly Fungi section).

The bulk of diversity of the tooth fungi in Cascadia is found among those genera that form ectomycorrhizae. *Hydnum* and *Sistotrema* produce pale fruitbodies with a relatively soft, fleshy texture. Despite their very different morphology, they are close relatives of chanterelles. The remaining species are in the family Bankeraceae except for *Phellodon* in the Thelephoraceae. *Hydnellum* produce brown spore deposits and many have thick, tough-textured fruitbodies that develop slowly and may persist for months. *Sarcodon* have large and often scaly caps with brown spores. In the past, *Hydnellum* and *Sarcodon* were split based on the growth habit: indeterminate growth and tough fibrous flesh in *Hydnellum* versus determinate growth and softer flesh in *Sarcodon.* Genetic work has shown that many of the species we used to call *Sarcodon* belong in *Hydnellum,* while the cluster of species around *S. imbricatus* belong in *Sarcodon.* Many of the Pacific Northwest species need additional studies to determine both the proper generic placement and species delimitations. Most members of Bankeraceae are highly variable in appearance; young to old or wet versus dry. Important features to note when attempting to identify members of tooth fungi include the color of the cap, stipe, and flesh (make a cross section); texture of the cap surface; structure of flesh (single layer or a duplex?); color of the teeth; odor; taste (some are very acrid, so consider yourself warned!); and identity of associated trees. *Phellodon* fruitbodies are tough, or brittle (which used to be placed in the genus *Bankera*) and produce white spore deposits.

Auriscalpium vulgare

sensu PNW

Earpick Fungus, Conifer-cone Tooth

Cap: 1–3 cm across, convex to plane, round or kidney shaped, usually with a distinct "hump" or ridge at stipe attachment, almost always attached to stipe asymmetrically. Surface dry, finely hairy to densely covered in spiky fuzz. Dark reddish brown, medium brown to ocher-brown with a much paler margin when young. **Spines:** Slender, conical. Pale tan, whitish gray, or brownish, usually paler than stipe. **Stipe:** 3–6.5 cm long, 0.2–0.4 cm thick, off-center to lateral, rarely central. Usually cylindrical and straight, but also often very irregular, sometimes with multiple upright branches. Surface dry, entirely covered in short, dense fuzz. Quite variably colored, from blackish when moist to dark reddish brown, medium brown, ocher brown, or even pallid beige. **Flesh:** Thin, very tough and persistent. **Odor:** Indistinct. **Taste:** Indistinct. **Spore deposit:** Whitish. **Microscopy:** Spores 4.5–5.5 × 3.5–5.5 µm, globose, weakly roughened with amyloid ornamentation.

Ecology Solitary or in small scattered groups on conifer cones. By far most commonly found on the cones of Douglas-fir, occasionally on the cones of sugar and ponderosa pines. Fresh fruitings usually occur in early fall and continue into winter. The tough fruitbodies may persist for months, and occasionally one or two generations of new caps will grow out of the old stems. It is hypothesized that this species lives as an endophyte in its host, and fruits only once the cone has fallen, but more research is needed to understand the natural history of this genus.

Edibility: Likely non-toxic.

Comments: Although difficult to find because of their small size and inconspicuous colors, this species is easy to recognize by the toothed underside, off-center to lateral stipe, finely fuzzy overall texture, and growth on conifer cones. For those having trouble finding it, the following method is fairly foolproof: Shortly after the first fall rain, find a stand of Douglas-fir trees with many fallen cones. Crawl around slowly on your hands and knees, keeping a mental image of the dark, upright fruitbodies with caps projecting at right angles. The frequent pale "halo" around the edge of the cap can also be a helpful visual cue! *Strobilurus trullisatus* is a much more visible gilled species that also grows on Douglas-fir cones and can be used as an indicator of appropriat habitat. Cones can be transported home and kept moist—your toothy pet will remain happy for weeks.

Taxonomic notes: Although *Auriscalpium vulgare* in the strict sense does occur in North America, it has not been reported from the Pacific Northwest. Based on limited sampling, Western North American collections appear to represent an undescribed species, but it remains to be determined if there is a single species that grows on cones from a wide range of conifers, or if there may be multiple species with varying degrees of host specialization.

Phellodon tomentosus
(L.) Banker

Owl Eyes

Cap: 2–6 cm across, flat to funnel shaped and often fused in irregular clusters with multiple fans or lobes, or occasionally a simple convex to plane cap with a sunken center. Surface dry to moist, finely tomentose at first, soon bare, wrinkled. White to buff at first, soon white around margin, light to dark brown at center, often with concentric zones, dark brown to blackish in age. Paler white to creamy with a tan center if dry. **Spines:** 0.1–0.3 cm long, often shorter toward margin. White at first, becoming whitish buff to grayish white, staining orangish buff when bruised. **Stipe:** 2–5 (7) cm long, 0.1–0.6 cm across, central to off-center, equal or tapered downward, often with a rooting base; arising from a buff mycelial mat. Light to dark brown, darkening slightly when handled. **Flesh:** Tough, thin, fibrous, engulfing organic matter. Whitish to brown in cap, brown to blackish in stipe. **Odor:** Faint to sweet and fragrant, like curry, fenugreek, or burnt sugar. **Taste:** Mild to slightly bitter. **KOH:** Dingy gray to black. **Spore deposit:** White. **Microscopy:** Spores 3–4 µm, round or nearly so, finely spiny.

Ecology: On ground, often in troops or scattered in fused clusters, occasionally solitary. Ectomycorrhizal with a wide range of conifers. Common and widespread across the Pacific Northwest. Fruiting from late summer into early winter.

Edibility: Unknown.

Comments: The brownish caps with a white margin, whitish spines, thin stipe, and often sweet odor are distinctive. The common *P. melaleucus* (which is likely a species complex) is very similar, but the cap has a grayish to black center and a white margin, with white to pale grayish spines, and a grayish to black stipe. It will often grow alongside *P. tomentosus*, especially in coastal forests. *Hydnellum* generally have thicker flesh and brown spores.

Phellodon atratus
K. A. Harrison

Blue-black Tooth

Cap: 2–8 cm across, flat with a sunken center or funnel shaped at first, becoming more irregular or in fused clusters with multiple fans or lobes. Surface dry to moist, finely velvety to tomentose at first, soon bare, wrinkled. Bluish to bluish black or black on disc; paler, bluish gray, to purplish gray on margin when wet. When dry bluish gray to silvery black with a paler margin. **Spines:** 0.1–0.3 cm long, often shorter toward margin. Bluish gray to bluish lilac at first, becoming grayish in age or when dry. **Stipe:** 2–6 cm long, 0.5–2 cm across, central to off-center, equal, tapered downward, often with a rooting base, arising from a grayish mycelial mat. Surface dry, soft, matted-tomentose. Bluish black, black or dark gray. **Flesh:** Tough, fibrous, thin to moderately thick, engulfing organic matter. Dark bluish black to black, with paler grayish flesh near cap surface (especially when dry); stipe a duplex, with a tough black core and a softer bluish black to gray outer part. **Odor:** Mild to earthy-fragrant. **Taste:** Indistinct. **KOH:** Bluish black. **Spore deposit:** White. **Microscopy:** Spores 4.5–5 × 5 µm, round or nearly so, finely spiny.

Ecology: Scattered, in groups or fused clusters, or occasionally solitary, on ground under conifers, especially Sitka spruce. Fruiting from early fall into winter. Very common in coastal spruce forests, occasional elsewhere.

Edibility: Unknown.

Comments: The dark bluish-gray color, dark flesh, and white spores help distinguish this species from other dark tooth fungi. Recent sequence data suggest that *P. niger* also occurs in the PNW. It may represent the smaller, more grayish specimens from mixed conifer forests. *P. melaleucus* is similar, but is generally smaller and thin fleshed, and has white spines when young and a cap without bluish tones. It can be confused with older, faded specimens of *P. atratus*, which can be distinguished by their thick flesh and duplex stipe tissue. *Hydnellum cyanopodium* shares the same habitat, but is larger and has a roughened cap (often with red droplets), bluish-violet flesh, and brown spores. Dark species of *Hydnellum* are generally larger and have brown spores.

Hydnellum suaveolens

(Scop.) P. Karst.

Sweetgrass Tooth

Cap: 4–15 (25) cm across, irregular and complex to rounded at first, expanding to flat with a sunken center, and often developing additional fans or lobes, occasionally growing in fused clumps of multiple fruitbodies. Surface finely velvety at first, soon bare, wrinkled, or developing irregular projections. White at first and staining brownish when handled, becoming tan to light brown on disc, with a whitish margin, often with concentric zones, then entirely darker grayish brown to brown in age. **Spines:** 0.1–0.5 cm long. Whitish to pale gray at first, often with paler tips, becoming darker gray, to grayish brown in age. Staining grayish brown when bruised. **Stipe:** 1–5 (7) cm long, 1–3 cm across, central to off-center, equal or swollen lower, occasionally rooting. Surface velvety, dry. Violet-blue to sky blue, darkening to navy blue when handled. **Flesh:** Tough, fibrous, engulfing organic matter, thick to moderately thin in age. Zoned white and blue in cap, deep navy blue in lower stipe. **Odor:** Strong and fragrant, like sweetgrass (coumarin-like), or anise-like; may be faint at times, but can become overpowering, especially in an enclosed space. **Taste:** Mild to sweet. **KOH:** Blue-green. **Spore deposit:** Brown. **Microscopy:** Spores 4.5–6 × 3–4 µm, subglobose to ellipsoid, covered with irregular projections or warts.

Ecology: Solitary, scattered, or in fused clustered on ground under conifers. Fairly common and widespread from coastal spruce forests, and higher elevation and east-slope Cascade forests. Fruiting from early fall well into winter, but will sometimes first appear in summer in coastal areas with heavy fog or with summer rains in the Cascades.

Edibility: Unknown.

Comments: The whitish to light brown cap, whitish to gray spines, blue stipe, zonate blue-and-white flesh, and sweet odor combine to make this *Hydnellum* unmistakable. *H. caeruleum* has a bluish cap that fades as it ages, has orange flesh in the stipe, and lacks any strong odor. *Phellodon tomentosus* has a similarly-colored cap and a sweet, curry-like odor, but is thin fleshed and lacks any blue stipe colors. *P.* (=*Bankera*) *violascens* has a very pale cap and spines, lacks the blue colors, and has a sugary odor.

Hydnellum mirabile

(Fr.) P. Karst.

Fuzzy Tooth

Cap: (3) 5–18.5 cm across, rounded to convex at first, and simple at first, often overlapping, and more irregular with age. Surface moist to dry, covered with short, soft to stiff hairs; typically erect young, more matted with age; occasionally with brown droplets on margin when young. Light brown or golden brown when young, soon brown with a paler growing margin, to darker brown, or occasionally dark golden brown in age. **Spines:** Decurrent, 0.2–0.5 cm long, shorter toward margin. Two-toned with paler brown to tan tips; darker brown above. **Stipe:** 2–5.5 cm long, 1–6.5 cm thick, central or off-center, stout, often irregular. Surface moist, exuding brownish droplets when fresh and wet, tomentose below, binding duff and debris. Yellow-brown to brown. **Flesh:** Tough, fibrous in cap, more brittle in stipe, engulfing organic matter. Light brown to brown, faintly zonate, developing darker discolorations. **Odor:** Farinaceous. **Taste:** Farinaceous at first, soon acrid. **KOH:** No reaction. **Spore deposit:** Brown. **Microscopy:** Spores 4.5–6.5 × 4.5–5 µm, subglobose with short, stout projections.

Ecology: Solitary or scattered from soil, moss or duff. Uncommon to rare, typically found in mature and old growth higher elevation forests with fir and spruce, usually at sites with a rich assemblage of other tooth fungi (e.g. *Sarcodon*, *Phellodon*, *Hydnellum*). Fruitbodies slow-growing and long-lasting, first appearing in summer or early fall, persisting until the end of the mushroom season.

Edibility: Inedible.

Comments: The combination of the hairy brown cap and light brown spines make *H. mirabile* an easy species to recognize. The tendency to exude brown droplets when young or in wet conditions is also distinctive. *H. caeruleum* has a soft tomentose cap that starts off bluish; older brown-capped specimens can be distinguished by their blue and orange flesh. *Phaeolus schweinitzii* can look similar, but has irregular maze-like pores, which can occasionally become ragged and appear tooth-like in age. *Onnia tomentosa* shares the same habitat and often shows a similar growth form, but the small round pores rather than teeth readily distinguish it.

Hydnellum caeruleum

(Hornem.) P. Karst.

Blue Tooth

Cap: 3–10 (15) cm across, convex to plane at first, becoming uplifted and more irregular in age; occasionally with additional fan-like lobes or overlapping caps. Surface softly tomentose, plush-like young, more matted-tomentose or roughened with age. Color variable; baby blue to sky blue young, soon grayish blue on the margin, and fading to whitish over the disc, then developing cinnamon brown tones, to extensively light to dark brown in age. **Spines:** 0.1–0.5 cm long, often decurrent. Tips pale blue, bluish gray to almost white at first, typically with more blue tones towards the margin, developing orange-brown to brown colors as it matures; upper portion brownish throughout. **Stipe:** 2–7 cm long, 1–3 cm across, central to lateral, stout, base often with binding mycelium. Surface dry to moist, tomentose to bumpy. Whitish to orangish when young, orange-brown to brown mature. **Flesh:** Tough, fibrous, engulfing organic matter, thick to moderately thin. Zoned blue, off white, brown to orange in cap, orange in stipe. **Odor:** Farinaceous. **Taste:** Farinaceous. **KOH:** Blue-green to greenish on cap, olive on stipe and flesh. **Spore deposit:** Brown. **Microscopy:** Spores 4–7 × 3–5.5 µm, ellipsoid with stout projections.

Ecology: Solitary or scattered on ground, often in mossy areas, typically with spruce in Cascadia, but can be found in most forest types. Widespread but uncommon. Fruiting from late summer through fall..

Edibility: Inedible.

Comments: The fuzzy blue caps make the identification of young fruitbodies straightforward. Older fruitbodies which have lost their blue cap color can easily be identified by their two-toned flesh: zoned blue in the cap and orange in the stipe. *H. cyanopodium* has a variable silvery blue to blue-gray roughened cap (often with red droplets), bluish gray spines, and blue stipe flesh. *H. suaveolens* has a whitish to dingy brown cap, blue-and-white zoned cap flesh, and a blue stipe; it also has a strong sweetgrass odor. *H. regium* has a dark gray to black cap (sometimes bluish-black when young), dark grayish spines, and orangish flesh in the lower stipe; it is typically larger and fleshier in texture.

Hydnellum regium

K.A. Harrison

Black Tooth

Cap: 5-20 cm across, occasionally in larger fused clusters, starting as a rounded "blob", developing irregular fan-like lobes and often overlapping caps. Surface moist to dry, smooth or downy tomentose when young, soon roughened, fibrillose to matted tomentose. Dark gray, vinaceous gray, bluish gray to black. **Spines:** Deeply decurrent, 0.1–0.5 cm long, shorter towards margin. Gray to blue-gray young, becoming dark bluish gray to nearly black. **Stipe:** 2–10 (15) cm long 3–6 (8) cm thick, central to off-center, tapered downward or irregular, thick, stout. Surface dry, bumpy, often binding debris. Off-white, orangish to orange-brown; upper portion often covered with grayish spines. **Flesh:** Thick, fleshy fibrous, bluish gray to dark gray, often with darker zonate lines in cap. Stipe more fleshy, dingy whitish orange, orangish, to orange-brown, transitioning into gray cap flesh. **Odor:** Indistinct to slightly farinaceous. **Taste:** Disagreeable. **KOH:** Unknown. **Spore deposit:** Brown. **Microscopy:** Spores 4.5–6 × 3.5–4.5 µm, subglobose to oblong, with blunt, irregular projections.

Ecology: Solitary or scattered from ground or mossy duff under conifers, especially spruce. Widespread but rarely encountered; most often seen in higher elevation, east side forests of the Cascades, and the Northern Rocky Mountains. Rare on the coast where it occurs from northern California into south-central Alaska in mature Sitka Spruce forests. This species, like many of our less common tooth fungi, appears to have a preference for, and possibly is restricted to, forests with an old-growth component. Fruiting from late summer though fall.

Edibility: Unknown.

Comments: The large size, bluish gray to black cap and spine colors, and the pallid to orangish fleshy stipe help distinguish *H. regium* from the other dark colored tooth fungi. *H. caeruleum* has baby blue colors with young (darkening with age), a tomentose cap, and orange stipe flesh, transitioning into blue and orange zones in the cap. The darkly colored *H. fuscoindicum* is more brittle-fleshed, and always has deep purple flesh. *Phellodon atratus* is smaller, has thinner dark blue to black flesh, and white spores.

Hydnellum fuscoindicum

(K.A. Harrison) E. Larss., K.H. Larss. & Kõljalg

Violet Hedgehog

Cap: 3–15 cm across, convex to plane, often with a sunken center, to irregular or uplifted in age. Surface dry to moist, roughened to appressed-fibrillose at first, often becoming scaly in age or when dry. Dark grayish purple and sometimes with a powdery bloom when young, becoming royal purple, bluish purple, purplish black, to brownish black in age. Paler, purplish brown or reddish brown around margin if dry. **Spines:** 0.1–0.5 (8) cm long, shorter toward margin. Grayish purple at first, becoming purplish brown to brownish in age. **Stipe:** 3–6 (10) cm long, 1–3 cm across, central to off-center, tapering downward to a pointed base. Surface dry, smooth to roughened from undeveloped spines over upper portion. Purplish gray to dingy purplish black, staining dark purple when handled. **Flesh:** Thick, firm, brittle in cap, fibrous in stipe. Deep purple, purple-gray, to purple-black, stipe generally slightly paler. All parts slowly staining dark purple. **Odor:** Indistinct to slightly farinaceous or fragrant. **Taste:** Mild to unpleasant, slightly acrid and farinaceous. **KOH:** No reaction. **Spore deposit:** Brown. **Microscopy:** Spores 5–6.5 × 4.5–5 µm, broadly ellipsoid to subglobose, with blunt, irregular projections.

Ecology: Solitary, scattered, or more rarely in large troops on ground or moss. Ectomycorrhizal with both conifers and hardwoods; locally common under hemlock, fir and Douglas-fir, occasionally with giant chinquapin or tanoak. Occurring from the coast to the Cascades, as well as in the northern Rocky Mountains. Fruitings occur mostly in the fall, continuing well into winter along the coast.

Edibility: Edible, but taste rather poor.

Comments: The dark purplish cap, purplish gray to purplish brown spines, and deep purple to purplish black flesh make this mushroom unmistakable. This species, like many tooth fungi, produces an excellent wool-dyeing pigment, yielding blue colors in an alkaline dye bath. *Hydnellum scabrosum* has a finely scaly, pinkish brown cap, beige spines and flesh, and a dark blue-green stipe base. *Sarcodon rimosus* has a dark vinaceous gray to blackish cap, gray spines, and pale gray flesh; it is common in coastal Sitka spruce forests.

Synonym: *Sarcodon fuscoindicus* (K.A. Harrison) Maas Geest.

Hydnellum peckii

Barker

Strawberries and Cream, Bleeding Tooth

Cap: 2–15 cm across, rounded at first, becoming convex to plane, soon developing a sunken center, often irregular in age. Surface soft and staining pink at first, soon becoming roughened, covered with coarse tufts. Pinkish to whitish when young, often covered in red droplets, (appearing as reddish-brown pockmarks when dry). Becoming dark brown to reddish-brown at center with a paler pinkish growing margin, brown overall in age. **Spines:** 0.1–0.6 cm long. Pinkish at first, becoming pinkish brown to brown in age, often with paler tips. Occasionally oozing red droplets when young. **Stipe:** 2–8 cm long, 1–3 cm across, central to off-center, irregular, often tapering downward. Surface dry to moist, roughened, matted-tomentose. Dingy dark brown, dark reddish brown, or almost black. **Flesh:** Tough, fibrous, engulfing organic matter, thick to moderately thin. Pinkish to brownish, often faintly zoned in cap, darker brown to blackish in stipe. Slowly staining reddish brown. **Odor:** Generally indistinct, but can be earthy to sweet, or rancid. **Taste:** Sickeningly acrid, sour, and farinaceous. **KOH:** Greenish black. **Spore deposit:** Brown. **Microscopy:** Spores 4.5–6 × 3.5–4.5 µm, globose to ellipsoid with irregular projections.

Ecology: Solitary, scattered, or occasionally clustered on ground, usually under conifers, especially pine, fir, and spruce. Widespread throughout Cascadia; locally common in coastal forests and at some higher elevations, largely absent near urbanized areas. Fruiting from late summer into fall in the mountains; fruitbodies can persist year-round near the coast, but are more likely to be found from late summer into early winter.

Edibility: Toxic.

Comments: When young and covered with red droplets, this mushroom is one of our most striking, bizarre, and beautiful fungi. In age, it becomes just another brown *Hydnellum.* Besides the red droplets (a feature shared with a number of *Hydnellum*), it is recognized by its white to pinkish colors when young (or only at the growing margin in age), pinkish brown flesh, and horribly acrid taste. Although the spiciness varies between collections, one small bite will usually coat your mouth with a burning, astringent sensation. *H. scrobiculatum* (sensu PNW) has a brown cap with a tan margin and lacks an acrid taste, as do all other brown *Hydnellum* in our area. *H. cyanopodium* also has red droplets when fresh, but has a blue-gray to bluish black cap, and grayish blue spines. Reports of *H. peckii* from urban areas almost always pertain to *Abortiporus biennis,* which can form a white to pink fruitbody with red droplets when young, but which has irregular pores rather than teeth.

Hydnellum subsuccosum

K.A. Harrison

Cap: 3–8 cm across, convex with a depressed center when young, becoming irregular, uplifted or funnel shaped, and often developing additional fan-like caps, lobes, or occurring in clusters of multiple caps. Surface dry to moist, soft, finely velvety when young, often grooved or occasionally pitted, to somewhat smooth in age. Deep reddish, reddish brown to brown with darker concentric zones and a pinkish buff growing margin. Often with droplets or pools of reddish juice when wet, or brown with a tan margin and distinct zones if dry. **Spines:** Deeply decurrent, 0.1–0.3 (0.5) cm long. Reddish- to pinkish brown with grayish tips at first, becoming reddish brown to brown. Staining dark reddish brown when bruised. Occasionally with small reddish droplets when young. **Stipe:** 1–4 cm long, 1–2 cm across, central to off-center. Surface dry, soft, velvety, but often mostly covered in spines. Reddish brown to brown. **Flesh:** Firm, thick to moderately thin, engulfing surrounding twigs and duff, somewhat soft when young, becoming tough, fibrous, with a tough core in older specimens. Dark reddish brown to brown, often with concentric lines in cap, dark reddish brown to almost black in stipe. **Odor:** Indistinct to slightly farinaceous. **Taste:** Mild to slightly farinaceous. **KOH:** Producing a quick violet flash, then turning forest green to greenish black. **Spore deposit:** Brown. **Microscopy:** Spores 4–5.5 × 3.5–4.5 µm, oblong to ellipsoid, covered with blunt, irregular projections.

Ecology: Scattered in small to large troops, clustered, or fused from sandy soil. Ectomycorrhizal with conifers, especially two-needle pines. Fairly common in coastal dune forests in Oregon and Washington, rarer in montane lodgepole pine forest. Fruiting in fall.

Edibility: Unknown.

Comments: The concentrically-zoned reddish brown caps, reddish to pinkish brown spines with grayish tips (especially around the growing margin), red droplets, and mild taste are distinctive. The brown *Hydnellum* are a confusing swarm of species in need of modern taxonomic work. There are multiple species in the *H. scrobiculatum* group in the Pacific Northwest; one is a pallid brown species which often has red droplets when young, reminiscent of *H. peckii*, but lacks the acrid taste and has a tan to light brown growing margin. There is a hardwood-associated species with a brown cap covered in bumps and projections. It occurs with oaks in Oregon and is widespread in California. One or possibly more species are in the *H. concrescens* (=*H. zonatum*) group, which are conifer associate that look very similar to *H. subsuccosum* but have brown (not reddish) spines when young, and is generally paler in color. *H. subzonatum* is a small, brown, red-brown to gray-brown species that is common under shore pine; the small size, very thin flesh, and grayish to grayish brown spines help separate it. *H. peckii* is another brown-capped species that is often covered with red droplets when young; it has a strongly acrid taste.

Hydnellum aurantiacum

(Batsch) P. Karst.

Orange Tooth

Cap: 3–15 (20) cm across, rounded to top shaped at first, expanding to convex or plane, or irregular and uplifted in age. Often with many knobby projections or developing additional fan-like caps. Surface dry to moist, soft, plush, finely velvety, roughened, wrinkled to smooth in age. Whitish when young, soon orange with a whitish growing margin; becoming darker reddish orange to reddish brown. **Spines:** 0.1–0.5 cm long. White at first, becoming grayish brown to brownish with white tips, to dingy brown with whitish to pale grayish tips in age. **Stipe:** 2–5 cm long, 1–3 cm across, central to off-center, irregular, equal, tapering downward or bulbous toward base. Surface dry, soft, velvety. Orange at first, becoming orangish to reddish brown. **Flesh:** Tough, fibrous, engulfing surrounding duff, thick to moderately thin. Orange in cap, often with darker zones, reddish orange to reddish brown in stipe, slowly darkening when cut. **Odor:** Indistinct to slightly fragrant. **Taste:** Earthy to bitter-farinaceous. **KOH:** Blackish on cap, greenish black on flesh. **Spore deposit:** Brown. **Microscopy:** Spores 6–7 × 5–6 µm, subglobose to globose, with blunt, irregular projections.

Ecology: Solitary or scattered in troops, rarely clustered, from soil, duff or moss under conifers, especially spruce. Widespread, but only locally common, with a preference for mature or old growth forests. Fruiting in late summer and fall, remaining on the coast into winter.

Edibility: Unknown.

Comments: The orange cap and stipe, white spines, and dark orange reddish flesh are distinctive. However, the orange *Hydnellum* in western North America are a frustrating complex; our species concepts would benefit greatly from modern genetic analysis. There is a common and widespread (and most likely undescribed) species in southern Oregon and California. It usually has a darker, dingy cap even when young and has orange spines that become orange-brown in age. It has been identified as *H. auratile* and *H. earlianum*, but both names are based on specimens from well outside this area, and neither concept matches this species neatly. *H. complectipes* is fairly common in mixed forests on the eastern Cascade slopes; it generally forms fruitbodies of large fused lumps or multiple fan-like lobes and more brown to reddish brown caps with a paler margin; it has orangey to grayish orange young spines and orangish brown to brown flesh. *Sarcodon indurescens* (which likely belongs in *Hydnellum*) has an orange-brown to brown cap, orange spines when young, and a more fleshy texture. It is rare, mostly fruiting in spring on the eastern slopes of the Cascades. *Hydnellum geogenium* is a small species with a yellowish olive cap, yellow spines, and neon yellow mycelium; it is quite rare in the Cascadian bioregion.

Hydnellum geogenium

(Fr.) Banker

Yellow Tooth

Fruitbody: 3–15 cm across, highly irregular and variable; typically consisting of a cluster of erect clubs or small, flattened lobes when young, occasionally becoming more rosette-like or with fan-like caps, or clusters of overlapping, fused caps. Surface moist to dry, smooth or with bumpy projections. Pale to bright chartreuse yellow when young, becoming two-toned and developing olive brown tones over the center, extensively olive-brown to brown in age. **Spines:** Decurrent, 0.1–0.5 cm long, shorter towards the margin. Tips pale yellow to yellow, more olive-yellow over the upper portion, or in age. **Stipe:** Typically poorly defined, central to lateral, base binding debris with bright yellow mycelium. **Flesh:** Thin, fibrous to leathery, yellow to brown. **Odor:** Faint, earthy to slightly sweet. **Taste:** Indistinct to slightly farinaceous. **KOH:** Yellow on all parts. **Spore deposit**: Brown. **Microscopy:** Spores 4.5–6 × 3.5–6 µm, subglobose with short stout projections.

Ecology: Solitary or in small troops from duff and moss under spruce and possibly other conifers. Widespread in the Pacific Northwest, but very rare. Most collections come from old growth Sitka spruce forests in coastal Oregon, but have also been reported from California and Alaska and inland to Idaho. Fruiting in late summer and fall.

Edibility: Inedible.

Comments: The yellow colors when young, and the spines on the underside make this rare *Hydnellum* unmistakable. Older fruitbodies can lose the yellow cap color, but will retain the bright yellow mycelium.

Taxonomic notes: *Hydnellum geogenium* was originally described from Europe, and has been confirmed to occur in eastern North America. Specimens from these areas are typically more rosette-shaped or regularly pileate. Our attempts to sequence Pacific Northwest collections have failed, and the question of genetic coherence across the continent remains unanswered.

Hydnellum stereosarcinon

(Wehm.) Stalpers

Cap: 3–20 (25) cm across, at first simple, often 1–3 fused, convex to plane caps with inrolled and ruffled edges, developing additional lobes and fans and becoming an irregular fused cluster of wavy, overlapping caps. Surface dry, smooth to finely pubescent, occasionally cracking in age. Dull grayish tan to warm brown, buff to orangish brown, with a paler margin when young, often with concentric zones, becoming darker brown on disc, and often with dingy brown spots or stains. **Spines:** 0.1–0.5 (0.8) cm long, shorter toward margin. Buff to orangish buff with paler tips at first, becoming buff-brown to light brown. Staining brown when handled. **Stipe:** 2–6 cm long, 1–3 cm across, central to off-center, irregular, or tapering downward and occasionally with a rooting base. Surface dry, smooth to roughened from undeveloped spines. Buff to light brown, staining brown when handled. **Flesh:** Thick to moderately thin, firm, brittle in cap, stipe tougher but still somewhat brittle. Off-white, buff to light brown, slowly staining brownish. **Odor:** Strongly farinaceous. **Taste:** Farinaceous. **KOH:** Slight olive reaction on all parts. **Spore deposit:** Brown. **Microscopy:** Spores 4.5–5 × 3.5–4.5 µm, subglobose to ellipsoid with blunt, irregular projections.

Ecology: Solitary, scattered, or in fused clusters on ground under spruce. Especially common in coastal Sitka spruce forests. Generally fruiting in early fall, persisting into winter.

Edibility: Unknown.

Comments: The warm brown colors, smooth, zonate caps in ruffled clusters, brittle flesh, and brown spores help identify this species. *Hydnellum scabrosum* has a pinkish brown cap with small scales, a bitter taste, and a dark bluish stipe base. *Phellodon* (*Bankera*) *violascens* has a smooth, whitish to grayish beige cap, thicker, fragile flesh, white spores, and a sweet odor. *Hydnellum versipelle* (=*Sarcodon versipellis*) is a similar orange to orange-brown capped species with a sweet or spicy (not farinaceous) odor. It's rare, likely restricted to spruce forests.

Synonym: *Sarcodon stereosarcinon* Wehm.

Sarcodon imbricatus

(L.) P. Karst.

Shingled Hedgehog

Cap: 4–20 (26) cm across, convex to plane, developing a depressed center with age, margin downcurved when young, becoming wavy or occasionally uplifted in age. Surface dry to slightly moist, covered with small to large shingled or tufted scales, often with larger upright scales on disc, especially in age. Color variable, tan to light brown, or gray-brown, scales often darker brown, darker overall when wet, or with age. **Spines:** 0.3–0.8 (1.2) cm long, shorter toward stipe and margin. Beige to pinkish gray at first, soon grayish brown with paler tips, to all brown in age. **Stipe:** 3–6 (12) cm long, 1–4 cm thick, central to off-center, equal or tapering downward. Surface dry, smooth. Beige to pale brown, often darker brown at base or in age, staining brown when handled. **Flesh:** Thick, firm, brittle in cap, tougher, more fibrous in stipe. Off-white to grayish beige dry, darker watery gray to grayish brown if wet. **Odor:** Indistinct. **Taste:** Mild, or occasionally slightly bitter-farinaceous. **KOH:** No reaction. **Spore deposit:** Brown. **Microscopy:** Spores 6.5–7.5 × 5–6.5 μm, subglobose or ellipsoid with blunt, irregular projections.

Ecology: Scattered in troops on ground, often forming rows and arcs. Ectomycorrhizal with spruce. Although reported growing with other conifers, these records likely pertain to similar species. Fruiting primarily in fall; common from the Cascade crest eastward, especially in the Rocky Mountains, occasional in coastal Sitka spruce forests.

Edibility: Edible; young specimens are typically tastier; some collections may have a bitter or metallic aftertaste.

Comments: *S. imbricatus* is recognized by the light brown caps with large, dark brown scales and tendency to be sunken or depressed at the center, pallid spines and stipes, and a mild to slightly bitter taste. This species may be restricted to spruce forest. *S. squamosus* is very similar and is found under wide range of different conifers; especially pine. It differs by having a darker vinaceous to brown cap with slightly smaller scales. Another difference between these two species is that *S. squamosus* dyes wool bluish green, whereas *S. imbricatus* gives only a grayish beige color. *S. leucopus* (sensu PNW) typically occurs in east-side fir forests; it has a soft cap with slightly smaller scales. *S. calvatus* produces larger fruitbodies with light brown caps covered in smaller appressed-fibrillose scales. It typically grows on drier east-slope forests in the Cascades. *Hydnellum scabrosum* is common under pine, and has a finely scaly or cracked, pinkish-brown cap when young (becoming scalier and browner in age), a bluish stipe base, and a strongly bitter taste.

Hydnellum scabrosum

(Fr.) E. Larss., K.H. Larss. & Kõljalg

Bitter Hedgehog

Cap: 4–12 (18) cm across, convex with an inrolled margin at first, expanding to plane, margin undulating in age. Surface dry, often cracking in dry weather, covered with appressed-tomentose fibrils, occasionally recurved or uplifted into small scales. Grayish pink, pinkish brown to reddish brown, darkening, and dingier colored with age. **Spines:** Decurrent, 0.2–1 cm long, shorter towards margin and stipe, pinkish gray young, soon pinkish brown to light brown, tips paler at maturity, bruising dark brown when damaged. **Stipe:** 3.5–7 (10) cm long, 1-3 cm thick, central or off-center, rarely lateral; stout, tapered downward. Surface dry, covered with short decurrent spines on upper portion, whitish mycelium at base. Pinkish gray to reddish brown on upper portion, greenish blue to blackish blue at base. **Flesh:** Beige to light gray-brown, dark brown to reddish brown around larva tunnels, dark greenish blue to greenish black in stipe base. **Odor:** Slightly to strongly farinaceous. **Taste:** Quickly and strongly bitter. **KOH:** Dark green to almost black on cap, olive-green elsewhere. **Spore deposit:** Brown. **Microscopy:** Spores 5–7.5 (8.5) × 4.5–6.5 (7.5) µm, subglobose with short stout projections.

Ecology: Scattered or in troops, typically in sandy soil. Ectomycorrhizal with conifers, especially shore pine and lodgepole pine. Fruitbodies are often entirely buried under duff when young, becoming exposed only as the caps expand. Fruiting in fall, with intermittent boom years when they can be very abundant in coastal pine forests; less common at higher elevations.

Edibility: Inedible; very bitter.

Comments: Recognized by the pinkish gray to reddish brown colors, greenish-blue stipe base and bitter taste. It is a highly prized dye fungus, yielding blue-green colors on wool and silk. *Hydnellum illudens* is very similar, but has an ocher-brown to brown cap, and typically a pale stipe base, although can show some grayish blue to blue-grey color. It has the farinaceous to bitter taste and occurs in mixed conifer forests.

Synonym: *Sarcodon scabrosus* (Fr.) P. Karst.

Hydnum umbilicatum

sensu PNW

Bellybutton Hedgehog

Cap: 2–6 cm across, convex to plane, becoming depressed at center, often with a deep "belly button" margin downcurved at first, becoming wavy in age. Surface dry to slightly moist, soft, very finely tomentose to smooth. Ocher, orange to peachy orange, occasionally with concentric zones. **Spines:** 0.2–0.5 cm long, 0.05–0.1 cm thick, densely covering underside, decurrent or free of stipe, fragile, flaking off easily. Pale creamy white, peachy white to pale ocher, darkening when handled. **Stipe:** Spores 8–9.5 × 7–9 µm, 3–8 cm long, 0.4–1 cm thick, central to slightly off-center, equal or enlarged slightly toward base, Surface dry, often bumpy to smooth. Creamy white to pale peachy orange, slowly staining orange to orange-brown where damaged or handled. **Flesh:** Thin, firm but brittle, breaking easily. Creamy white, slowly staining orangish. **Odor:** Indistinct to fruity. **Taste:** Mild to sweet. **Spore deposit:** White. **Microscopy:** spores 8–9.5 × 7–89 µm, globose to subglobose, smooth, thin-walled, inamyloid.

Ecology: Scattered in troops, occasionally solitary, from ground, duff, moss, or near well-decayed logs and stump rings in mixed conifer woodland. Ectomycorrhizal, especially common in hemlock forests in the Cascade foothills. Fruiting in fall.

Edibility: Edible and very good, although dainty and thinner-fleshed than other local *Hydnum* species.

Comments: The ocher to orange cap with a "belly button," slender stipe, and spiny underside help distinguish this species complex. *H. oregonense* is slightly larger and has a creamy to peachy orange cap, and slightly larger spores; 8.5–10.5 (13) × 7–10.5 µm. It is more common in coastal and Coast Range forests in Oregon and northern California; fruiting in late fall and winter. Fruitbodies of *H. melitosarx* are larger, with orange to peachy caps, lacking the central depression. Fruitbodies of *H. melleopallidum* are more stout in stature and have decurrent spines.

Taxonomic note: *Hydnum umbilicatum* was originally described from New York, and shows a slight genetic difference from Pacific Northwest collections. It is likely that our species will be described as distinct.

Hydnum washingtonianum

Ellis & Everh.

Large Hedgehog

Cap: 4–12 (20) cm across, rounded, broadly convex to plane, often irregularly uplifted and wavy. Surface dry, soft, smooth to bumpy, often with pits or cracks in age. Pale peachy orange, creamy orange, creamy to beige, slowly staining orange when damaged. **Spines:** 0.2–1 cm long, densely covering underside, decurrent on stipe, fragile, flaking off easily. White to pale peachy cream. **Stipe:** 2–10 (15) cm long, 1.5–4 (6) cm thick, central to strongly off-center, shape quite variable: short and stubby or rather elongate, base club shaped or tapered, often rooting. Surface dry, bumpy to smooth. Creamy white to peachy white, slowly staining ocher where damaged or when handled. **Flesh:** Thick, firm, often slightly brittle in cap, rubbery, creamy white to pale pinkish white, slowly staining orangish. **Odor:** Indistinct to fruity. **Taste:** Mild to sweet at first, becoming slightly acrid. **Spore deposit:** White. **Microscopy:** Spores 7–8.5 × 6–7.5 (8) µm, globose to broadly ellipsoid, smooth, thin-walled.

Ecology: Solitary or scattered in troops, clumps, or arcs. Often in mossy areas in conifer woodlands. Ectomycorrhizal with a host of different trees, especially spruce and pine. Typically fruiting from late summer or early fall into winter; even into spring on the coast in wet years. Slow growing. A single fruitbody that can last for months, becoming larger and more irregular as season goes on.

Edibility: Edible and treasured by many. The texture is firm and pleasant, and the flavor is similar to chanterelles. Some collections have an acrid taste, which disappears when cooked.

Comments: The creamy cap, thick, white, ocher-staining flesh, and fragile spines on the underside make this mushroom easy to identify. Until recently all large hedgehogs in Cascadia were called *H. repandum*, and smaller bellybutton hedgehogs were referred to as *H. umbilicatum*. Recent work has shown that *Hydnum* are quite diverse, however distinguishing them past species complexes with current knowledge is difficult without a genetic sequence. For most people, these distinctions are hardly worth the trouble, since all *Hydnum* in the PNW are good edibles. For those who want to apply more subtle species concepts, keep in mind that recent molecular studies were of continent-wide scope (not focused on the Cascadian bioregion) and there are still undescribed species in our area. *H. olympicum* is practically indistinguishable macroscopically (it may have sightly paler caps), but can be distinguished by its larger spores: 8–9 × 6.5–7.5 (8) µm. *H. melleopallidum* is slightly smaller, and has a more peach-colored cap (at least in our limited experience). It is more common in higher elevation forests. Fruitbodies of *H. melitosarx* are smaller still (but larger than those of the *H. umbilicatum* group) with orangey to peach-colored caps, and non-decurrent spines.

Misapplied name: *Hydnum repandum* L.

Hericium abietis

(Weir ex Hubert) K.A. Harrison

Bear's Head

Fruitbody: (5) 10–30 (50+) cm across, forming dense clumps with multiple branches from a common base, covered with short to moderately long, tufts and clumps of hanging spines. Color variable, white to creamy white, pinkish to yellowish; often developing more yellowish to tan tones in age. **Spines:** 0.5–1 (1.5) cm long, pointed, borne in small clumps along branches. **Stipe:** Absent, or indistinctly "rooted" into the substrate. **Flesh:** Thick, firm, stringy and somewhat rubbery, white to off-white, developing yellowish brown tones in age. **Odor:** Fragrant-fungal. **Taste:** Mild to sweet. **Spore deposit:** White. **Microscopy:** Spores 4.5–5.5 (6) × 4–4.5 (5) µm, globose to Subglobose, very finely roughened, amyloid.

Ecology: Solitary or scattered clumps on standing dead conifer trees, snags or down logs. Saprobic, with a preference for large diameter conifer snags, especially hemlock and fir at elevation. Common throughout the Cascade Range and the Olympic Mountains, rare in lower elevations. Fruiting in fall.

Edibility: Edible and excellent. Similar to other *Hericium*, the texture is reminiscent of crab meat, with a sweet flavor.

Comments: *Hericium abietis* can be easily recognized *by the* large white to pinkish or yellowish fruitbodies made up of tufts and clumps of spines from a fleshy base, and grows on conifer trees. They can reach massive sizes, although are commonly around the size of a soccer ball. It especially likes higher elevation old growth forests in the Cascade Range. Fruitings can be high up tree trunks (so remember to look up every once in a while). The Comb Tooth (*H. coralloides*) has multibranched fruitbodies with spines hanging along the branches (like a comb) and grows on hardwoods. *H. americanum* is very similar in structure to *H. abietis*, but produces smaller fruitbodies and grows on hardwood.

Hericium erinaceus

(Bull.) Pers.

Lion's Mane

Fruitbody: (5) 10–40 cm across, large clump of long, hanging, unbranched spines arising from a rounded mass of flesh. Rather compact at first, with dense spines, elongating on age. Clumps rather simple, or slightly lobed, occasionally stacked. White or blushed pinkish at first, becoming creamy white, discoloring ocher-buff to brownish in age. **Spines:** 1–4 (7) cm long, pointed. **Flesh:** Thick, firm, stringy, somewhat rubbery, white. **Odor:** Fragrant. **Taste:** Mild to sweet. **Spore deposit:** White. **Microscopy:** Spores 5–6.5 × 4.5–5.5 µm, globose to broadly ellipsoid, smooth to very finely roughened, amyloid.

Ecology: Usually solitary or occasionally clustered, fruiting from wounds on hardwood trees, or on standing dead trees or fallen logs. Uncommon in the Pacific Northwest, mostly in oak woodlands, rarer on maple, or even on urban hardwoods. Fruiting in fall.

Edibility: Edible and excellent. The texture is reminiscent of crab meat, with a sweet flavor.

Comments: The often large, rounded clumps of unbranched spines and growth on hardwoods set this species apart. The only species it is likely to be mistaken in our area is *H. abietis*, which differs by its shorter clumps of spines on multibranched fruitbodies and growth on conifers – it is much more common in the Pacific Northwest, especially in the Cascade Range. There are unconfirmed sightings of *H. americanum* in the Pacific Northwest, which forms fruitbodies with many small, stacked, compact clumps of spines and grows on hardwoods. Regardless of species, all *Hericium* are excellent edibles. *Ramaria* and other coral fungi form clumps of upright branches, and most grow from the ground (with the tips of the "spines" pointing upwards).

Mucronella fusiformis

(Kauffman) K. A. Harrison

Pendulous Icicle Fungus

Fruitbody: 0.5–1.5 (2) cm long, 0.05–0.3 (0.4) cm thick, pendulous, spindle shaped, with a tiny narrow base, then wider and evenly tapering to the tip. Watery white to off-white, usually weakly but distinctly yellowish in age. **Stipe:** Distinctly narrowed, tiny, 0.05–0.3 cm long, less than 0.05 cm thick. Often pale yellow even when young (a bit darker than main part of fruitbody) and covered in very small pale hairs (hand lens!). **Flesh:** Watery, often dissolving when squeezed, very reminiscent of a citrus fruit vesicle. **Odor:** Indistinct. **Taste:** Indistinct. **KOH:** Pale yellow to yellow-orange. **Spore deposit:** Usually not obtainable. **Microscopy:** Spores 5.5–7.5 × 5–6 µm, ovoid to nearly globose, smooth.

Ecology: Solitary or in loose groups on cut ends and undersides of large, well-rotted conifer logs, often Douglas-fir. A saprobic species, which appears to have an affinity for large diameter logs and debris in old growth forests. Probably common, but very easy to overlook.

Edibility: Unknown. Small and watery.

Comments: The small size, pendulous shape with a distinct narrowed base, and translucent white color are distinctive. Look for it on the cut ends and undersides of wet, rotting, dark reddish to black Douglas-fir logs. It usually grows in close proximity to other species of *Mucronella*, often fruiting simultaneously. *M. alba* (=*M. bresadolae* sensu PNW) is slightly smaller than *M. fusiformis*, but is larger than *M. calva*; it lacks the distinct narrowed "stipe" of *M. fusiformis* and has larger spores, 6–7.5 (9) × 4.5–7 µm. *M. calva* is a smaller white species that lacks a narrowed stipe and grows in dense clusters. See comments under *Mucronella* "Yellow" for small, clustered, lemon-colored fruitbodies. There are also a few coral and club fungi that can grow on mossy wood, but all are larger and upright.

Misapplied name: *Mucronella bresadolae* (Quél.) Corner, *Mucronella pendula* (Massee) R.H. Petersen.

Mucronella calva

(Alb. & Schwein.) Fr

Clustered Icicle Fungus

Fruitbody: 0.2–0.5 (0.7) cm long, hanging icicle-shaped spines, in a clustered growth. Bright white to watery white, discoloring to slightly creamy in age. Often surrounded with fine whitish hairs at base. **Flesh:** Extremely thin, watery, whitish. **Odor:** Indistinct. **Taste:** Indistinct. **Spore deposit:** Not obtainable. **Microscopy:** Spores 4–6 × 2.5–3.5 µm, oblong to ellipsoid, smooth.

Ecology: In clusters, sometimes covering a fairly large area on sides or undersides of large, well-rotted conifer logs, often Douglas-fir. Occurs throughout the Pacific Northwest, fruiting in fall and early winter.

Edibility: Unknown.

Comments: The downward-facing white spines, small size, lack of a stipe, and growth on well-rotted wood are distinctive. This species is often found on logs that simultaneously host fruitings of the other species of *Mucronella* – *M.* "Yellow" has the same shape and fruits in clusters, but is bright yellow when young (it can fade in age to pale yellow). The fruitbodies of *M. fusiformis* are slightly larger, with a small but distinct narrowed base where they are attached to the wood substrate, and tend to be solitary or in small tufts, not clusters. No modern work has been published on *Mucronella*, so it is possible (likely, even) that we have something other than the European *M. calva* in the Pacific Northwest.

Mucronella sp. 'Yellow'

Yellow Icicle Fungus

Fruitbody: 0.2–0.5 (0.8) cm long, forming tiny cluster of hanging spines, tapered to a point. Bright lemon yellow to translucent yellow, a bit paler in age. **Flesh:** Vanishingly thin, watery, yellow. **Odor:** Indistinct. **Taste:** Indistinct. **Spore deposit:** Usually not obtainable. **Microscopy:** Spores 4–6.5 × 4-5.5 µm, subglobose to ovoid, smooth.

Ecology: In clusters or elongate masses on sides, undersides, or cut ends of fairly large, well-rotted conifer logs, especially Douglas-fir. Not common, or at least not commonly found. Occurring throughout the Pacific Northwest; especially in old growth forests. When you find it, look for *M. calva* and *M. fusiformis*; the three often fruit together at the same time, on the same logs, but the bright color makes this species the easiest to find.

Edibility: Unlikely to be toxic, too small to consider edible.

Comments: If you can find it, you can identify it: A tiny yellow cluster of hanging spines on rotten wood. It is more likely to be confused for a slime mold than anything else (some slime molds in our area are yellow and grow as small clustered cylinders on wood, but are upright, shiny, and usually round-tipped). Although sometimes labeled as *M. flava* or *M. pulchra*, neither of those names are a good fit for the taxon in our area: *M. flava* was described from Iowa, is paler yellow, and has elongate-ellipsoid spores, 4–5 × 2.5–3 µm, while *M. pulchra* was described from Pakistan and has nearly round to oval spores that are slightly larger (5.5–7 × 4.7–6 µm) than those of *Mucronella* "Yellow." *M. calva* is similar in size and structure, but is pure white. *M. alba* (=*M. bresadolae* sensu PNW) is whitish to pale watery yellow and a bit larger, has a narrowed stipe base, and does not form dense clumps.

19 - Coral and Club Mushrooms

Coral mushrooms comprise a few distantly related lineages that have independently evolved an upright and highly branched fruitbody morphology. Unlike the toothed fungi, the tips of these "coralloid" mushrooms point upward, rather than hanging downward.

Identifying coral mushrooms is a difficult endeavor. Pay attention to the subtleties of stature and branching pattern, coloration (which in many species fades quickly and dramatically), and habitat. In many cases, microscopic examination is required for positive identification: the size and ornamentation of the spores, as well as the presence or absence of clamp connections at the base of the basidia are among the most important features.

Phaeoclavulina, *Lentaria*, and a few *Ramaria* are duff and wood decayers, while the majority of *Ramaria*, and *Clavulina* are mycorrhizal. Although a few *Ramaria* have slender, thin-fleshed fruitbodies, most have a thicker, fleshier texture than the other genera of coral mushrooms. *Ramaria* includes most of the brightly colored corals (including some with orange, yellow, pink, purple, or greenish fruitbodies). *Thelephora* is represented by a single, coral-like species, with brown, leathery fruitbodies.

Sparassis is represented by a single distinctive species in Cascadia; although the fruitbodies can vaguely resemble coral, its fruitbodies are larger, with flat to wavy lobes that are smooth on both sides. This genus is more closely related to polypores than to the other coralloid genera in this chapter.

The Club Mushrooms treated here are a sundry assortment of many different lineages (including both basidiomycetes and ascomycetes) that produce upright, cylindrical, club-shaped, or spatula-shaped fruitbodies (occasionally forked or with limited, sparse branching). They range from small to fairly large, dull to brightly colored, fragile to fibrous, fleshy, leathery, pimpled, or rubbery-gelatinous.

Basidiomycete clubs include: *Clavulinopsis*, with fruitbodies that are yellow and mostly cylindrical or weakly branched and *Alloclavaria* have fruitbodies that are purple to lilac-gray, clustered, and cylindrical. *Clavaria* and *Multiclavula* have simple, small to medium-sized fruitbodies ranging from whitish to tan, dull yellow or pinkish. *Typhula* have tiny, cylindrical, white to tan fruitbodies. *Calocera* produce orange to yellow, gelatinous, simple club- or staghorn-shaped fruitbodies.

Ascomycete clubs include: *Onygena*, with fruitbodies that have powdery heads and grow out of owl pellets and on feathers, fur, or bones, and the club-shaped *Trichoderma*, whose fruitbodies are pale with tiny freckles or pimples covering the heads. *Tolypocladium* also produce fruitbodies with pimpled heads but they emerge from deer truffles (*Elaphomyces*). *Heyderia*, *Mitrula*, *Spathularia*, and *Trichoglossum* produce small, slender fruitbodies with smooth, paddle-shaped, or wrinkled fertile heads and often grow in troops on duff and debris. *Cudonia* produce somewhat rubbery gelatinous or tough rubbery fruitbodies, often with a rounded head. *Xylaria* fruitbodies are thin, wiry, and mostly black overall, with a powdery white coating of asexual spores near their tips.

Sparassis radicata

Weir

Cauliflower Mushroom

Fruitbody: 10–50 cm across, a rounded cushion of densely to loosely packed leafy, ruffled lobes which are smooth on both sides. Creamy white to creamy pink, becoming pale creamy yellow, often more yellowish ocher in age. **Stipe:** 5–15 cm long, 3–5 cm thick, not always distinct from fertile lobes, or tapering downward and often rooting. **Flesh:** Thin, crisp to slightly rubbery, creamy. **Odor:** Fragrant when young, slightly savory-cheesy or disagreeable in age. **Taste:** Sweet. **Spore deposit:** Whitish to creamy yellow. **Microscopy:** Spores 5–7 × 3–5 µm, ellipsoid, smooth, inamyloid.

Ecology: Generally solitary, occasionally a few scattered clusters on ground around dead standing conifers and around old stumps, most commonly with Douglas-fir and spruce. Common and widely distributed throughout Cascadia. Fruiting from late summer through fall.

Edibility: Edible and excellent. The crunchy texture and distinctive flavor in combination with the large size make this an ideal mushroom for the table. The only downside is that it is often dirty inside and can take a while to clean. Some have severe allergic reactions to this species, so cook it well and sample sparingly at first.

Comments: This distinct and prized edible is easily recognized by its resemblance to a head of cauliflower or, more precisely, a giant ball of egg noodles. Although it has a smooth fertile surface (without pores), it is related to polypores. The Cabbage-head Fungus, *Daleomyces phillipsii* (*Peziza proteana forma sparassoides*) has similar colored, medium to large, highly convoluted, multi-chambered fruitbodies. It's not as "leafy" in appearance, and had more brittle flesh. It often fruits in spring, typically at least partially buried in duff, with a preference for recently burnt forests. It is edible, but can be difficult to clean. *Ramaria* species are similarly colored, but composed of clumps of upright branches, not ruffled lobes.

Misapplied name: *Sparassis crispa* (Wulfen) Fr. is a European species.

Ramaria sandaracina

Marr & D.E. Stuntz

Orange Jelly Coral

Fruitbody: 7–15 cm tall, 5–13 cm across, a rather crowded cluster of slender upright branches, forking near tips. Tips bright orange and remaining so in age; branches orange to dull orange, often with a yellow band on lower portion. **Stipe:** 1–3 cm long, 0.5–2 cm, rather short, tapering downward. Whitish to pale orange, or with a yellow band at top. **Flesh:** Thin, somewhat gelatinous, in branches slightly rubbery. Whitish in base, orange in branches and tips. **Odor:** Indistinct. **Taste:** Mild. **Chemical reactions:** Flesh inamyloid. **Spore deposit:** Grayish yellow to grayish orange. **Microscopy:** Spores 7–10 × 3.5–5 µm, broadly cylindrical to ellipsoid, ornamented with numerous warts. Basidia with clamp connections.

Ecology: Solitary, scattered, or in trooping masses from soil, duff, or moss under conifers. Common and widespread, occurring throughout wetter parts of Cascadia. Especially abundant in coastal spruce stands and west-side hemlock forests. Fruiting in fall.

Edibility: Unknown.

Comments: One of many orange *Ramaria* in Cascadia, *R. sandaracina* can be identified by its medium size, bright orange tips with duller branches and somewhat thin, slightly gelatinous flesh. Another common bright orange species is *R. gelatiniaurantia*, which has fleshier branches, a chunky white stipe, and gelatinous flesh. It lacks clamp connections and has broadly cylindrical spores (8–11 × 3.5–5 µm). *R. flavigelatinosa* is very similar but has orange-yellow to yellowish branches, yellow tips, a small stipe, gelatinous flesh, with matching those of *R. gelatiniaurantia*. *R. aurantiisiccescens* differs from all the species mentioned above by having fleshy-fibrous (not gelatinous) flesh. It is typically smaller, with orange tips when young but tending to fade more than the above species, and a bright yellow band on the upper stipe. Additionally, its basidia lack clamp connections and has spores measuring 8.5–14 × 3–5 µm. *R. largentii* is a large, bright orange to peachy orange species with a massive whitish stipe, pinkish orange flesh in the branches, and long spores (11–15 × 3.5–5 µm). There are a number of similar undescribed species as well, so don't sweat it if you can't identify your specimen!

Ramaria formosa group

Yellow-tipped Corals

Fruitbody: 5–15 (20) cm tall, 5–14 cm across, forming compact clusters with short, chunky, often tapered stipes. Branches short at first, rather upright, elongating in age. Tips yellow to pale yellow-ocher, transitioning into peachy pink to salmon branches, to a whitish base. Becoming more uniform salmon-buff in age and slowly staining brown when handled. **Stipe:** 2–5 cm long, 1.5–5 cm thick, stout, wide, tapering to a pointed base. Whitish to pinkish on upper part. Staining brownish when handled or in age. **Flesh:** Thick, firm, fleshy. White in stipe, pinkish in branches, yellowish to pinkish in tips. **Odor:** Indistinct. **Taste:** Mild. **Chemical reactions:** Flesh inamyloid. **Spore deposit:** Golden yellow. **Microscopy:** Spores 10.5–15.5 × 5.5–6.5 µm, ellipsoid, ornamented with scattered warts. Basidia with clamp connections.

Ecology: Solitary, scattered, or in troops in moss or duff under conifers. Common and widespread in Cascadia. Fruiting in fall.

Edibility: Unknown but reported to have caused minor gastrointestinal problems. Consider that *R. formosa* has been a catchall name for many different species, so the exact identity of the culprit remains unclear.

Comments: This coral is recognized by its peachy pink branches with yellowish tips, pale brownish staining, and large base. It represents a catchall name for species with similar colors, some of which can be reliably identified while others have befuddled to madness. *R. leptoformosa* forms tall upright clusters with peachy orange branches and orange tips, has clamp connections, and shorter, narrower spores (8–13 × 3–5 µm). Species in the *R. conjunctipes* group have small to medium upright clusters, slender fused stipes, yellow to greenish yellow tips, and salmon- to peach-colored branches. They tend to lack clamp connections and have shorter spores (6–10 × 4–6 µm). *R. maculatipes* has the same colors as *R. formosa* but with wine red stains on the stipe. *R. rubricarnata* typically fruits in the spring but has peachy pink flesh which stains blue-gray in Melzer's and spores 8.5–11 × 4–4.7 µm.

Ramaria gelatinosa var. *oregonensis*

Marr & D.E. Stuntz

Gelatinous Coral

Fruitbody: 7–15 cm tall, 5–12 cm across, forming rounded clusters with slender, often fused bases. Branches compact, upright, forking repeatedly. Dull orange to orange-brown, darkening slightly with age, at times with a grayish or violet cast. Tips yellow-orange when young, concolorous with branches in age. **Stipe:** 4–7 cm long, 2–8 cm thick, irregular, often slightly rooted, typically multiple fused branches. White to dull orange. **Flesh:** Thin to moderately thick, firm, distinctly gelatinous, often marbled. Translucent orangish to ocher with whitish veins; in branches more orange. **Odor:** Indistinct. **Taste:** Indistinct. **Chemical reactions:** Flesh inamyloid. **Spore deposit:** Ocher. **Microscopy:** 7–10 × 4.5–6 µm, ellipsoid to ovoid, coarsely warted. Basidia with clamp connections.

Ecology: Solitary, scattered, or in troops on ground, typically in moss under conifers. Occasional, typically in higher elevation old growth hemlock and fir forests in the Cascade Range. Fruiting in fall.

Edibility: Possibly toxic.

Comments: This species is easily recognized by the orangish to orange-brown colors and distinctly gelatinous flesh. *R. testaceoflava* has quite similar colors (orangish brown branches and yellowish to orangish tips) but has non-gelatinous white flesh. Two other common gelatinous-fleshed species in Cascadia are *R. gelatiniaurantia*, with bright orange tips and branches and a chunky white stipe; and *R. flavigelatinosa* with orange-yellow to yellowish branches, yellow tips, and a white stipe.

Ramaria purpurissima

R.H. Petersen & Scates

Purple Prince Coral

Fruitbody: 8–10 cm tall, 5–15 cm across, crown rather compact, primary branches thick, secondary branches short, forking at tips. Tips royal purple to amethyst purple; branches similarly colored, or slightly duller, fading somewhat, but retaining some purple color and developing ocher tones from spores in age. **Stipe:** 3–9 cm tall, 2–5 cm thick, chunky, rounded. White with lilac-purple tones towards the branches, often with rusty stains in age. **Flesh:** Thick, firm, stringy in stipe and branches, fragile in tips. White, with rusty brown stains in stipe. **Odor:** Indistinct. **Taste:** Mild to slightly bitter. **Chemical reactions:** Flesh inamyloid, KOH vinaceous on purple areas. **Spore deposit:** Ocher-buff. **Microscopy:** Spores 9–11.2 × 4.7–5.4 µm, ellipsoid with adaxial bulge, conspicuously roughened with complex low warts and short ridges. Basidia with clamp connections.

Ecology: Solitary, scattered, or in troops, likely associated with fir and possibly other conifers. Rare, only known from a handful of sites in the Klamath Mountains, southern Cascades and in Idaho. We have seen it in drier old growth fir forests, but its exact habitat preferences are poorly known. Fruiting in fall.

Edibility: Unknown.

Comments: The large size, grape-purple colors, and rusty staining on the stipe readily distinguish this rare species. The most likely species to cause confusion is a rare, undescribed *Clavaria* in the *C. zollingeri* group —it is distinguished by its smaller size, brittle flesh, 2-spored basidia, and ovoid to ellipsoid spores. *R. violaceibrunnea* is a smaller species with a more slender upright crown and a distinctly differentiated trunk-like stipe. Its branches are violet when young, soon becoming smoky beige to grayish brown with a violet band on the upper stipe and lower branches, fading to grayish tan with age and developing ocher-yellow color from spores. *R. subviolacea* has pallid branches with a vinaceous gray cast when young, and a massive, club-shaped to cylindrical stipe. The spring fruiting *R. coulterae* has reddish to vinaceous tips when young (fading with age), and a rusty-staining stipe.

Ramaria araiospora

Marr & D.E. Stuntz

Crimson Coral

Fruitbody: 6–12 cm tall, 2–7 cm across, compact, with multiple branches arising from a slender common base; branches forking repeatedly and becoming narrower toward tips. Branches bright crimson-red to neon-pink when young, fading slightly in age, becoming reddish pink to pink. Tips often forked, bright pink to crimson-red, fading slightly, typically remaining pink in age, occasionally more orangish or yellowish. **Stipe:** 1–4 cm long, 0.5–1.5 cm thick, often rooting, occasionally slightly bulbous. White to pinkish. **Flesh:** Fairly thin. Pink to red in branches, white in base. **Odor:** Indistinct. **Taste:** Slightly peppery at first, becoming radish-like after a few seconds. **Chemical reactions:** No reaction in KOH or Melzer's. **Spore deposit:** Yellowish. **Microscopy:** Spores 8–14 × 3–5 µm, broadly cylindrical, finely warted. Basidia lacking clamp connections.

Ecology: Solitary or scattered in small groups on ground in moss or duff. Locally common in lower elevation, west-side forests with hemlock and spruce. Fruiting in fall.

Edibility: Edible; colorful, but not flavorful.

Comments: This small, brightly colored coral fungus is told apart from other red *Ramaria* by its bright pink to red branches and tips that fade slightly in age but remain distinctly pinkish red. Although genetic studies have not shown a difference between them, two varieties have been recognized based on the color of mature fruitbodies. *R. araiospora* var. *rubella* has pinkish tips while var. *araiospora* has orange to yellow tips in age. *R. stuntzii* can be differentiated by its neon-pink colors (becoming bright red to orange-red), chunkier base, flesh that turns violet-gray in Melzer's, and shorter spores (7–10 × 3–5 µm). *R. cyaneigranosa* is generally more salmon pinkish, with a larger, more upright stature and slightly larger, warted spores (8–15 × 4–6 µm). It is more likely to occur in montane conifer forests.

Ramaria botrytis sensu PNW

Ramaria rubripermanens

Ramaria rubrievanescens

Ramaria aff. *botrytoides*

Ramaria botrytis group

Pink-tipped Corals

Fruitbody: 5–12 (15) cm tall, 4–12 cm across, forming a rather sparse crown with a chunky stipe. Branches relatively short, elongating in age. Tips generally vinaceous pink to wine-red at first (sometimes light pink), soon fading. Branches whitish to creamy white, often with a pinkish blush when young, slowly bruising pale yellowish brown when damaged. **Stipe:** 2–7 cm long, 1.5–3.5 cm thick, often cylindrical with a rounded or slightly rooting base. Whitish, slowly bruising yellowish to pale brownish. **Flesh:** Thick, firm, fleshy, white. **Odor:** Indistinct. **Taste:** Mild. **Chemical reactions:** Flesh slowly and weakly amyloid. **Spore deposit:** Ocher to pale orange. **Microscopy:** Spores 11–17 × 4–6 µm, cylindrical, ornamented with spiraling or longitudinal ridges.

Ecology: Solitary or scattered in moss or duff under conifers. Locally common and widespread, occurring across western Cascadia. Fruiting in fall.

Edibility: Edible and good, but very easily confused with bitter or bland species.

Comments: The evanescent vinaceous pink color of the tips, moderately chunky stipe that bruises pale yellowish brown, and mild taste help distinguish it. Microscopically, it has longer, more ridged spores than other similar species. There are a number of similar species in Cascadia, some of which remain undescribed. The name *R. botrytoides* has been used for a species which lacks the chunky stipe, has much shorter spores, and lacks clamp connections. *R. aurantiiramosa* looks like *R. botrytis* but has pinkish orange tips when young. *R. rubripermanens* has pinkish to reddish tips and creamy branches when young, but the pink color soon fades and becomes pale yellowish ocher overall — it also has a mild to nutty taste and can fruit from spring to fall. *R. rubrievanescens* has fleetingly pinkish tips and develops brownish stains on the stipe. *Ramaria* sp. 'Tanoak Red' has darker tips, persistent wine-red colors, a bitter taste and, as the name suggests, grows with tanoak. *R. coulterae* has reddish pinkish tips, grows in drier montane habitats, and fruits in spring.

Ramaria cystidiophora

(Kauffman) Corner

Lemon Coral

Fruitbody: 5–17 cm tall, 2–8 (12) cm wide, forming slender, at time sparse crowns with upright branches from a single or fused base. Branches yellow, to pale creamy yellow lower. Tips bright to pale lemon-yellow, fading slightly in age. **Stipe:** 1–8 cm long, 0.5–4 cm thick, often rooting, single or multiple fused in clusters. Pale yellow to whitish, occasionally staining reddish brown. **Flesh:** Fairly thin; pale yellowish in branches, whitish in base. **Odor:** Sweet, fragrant, or lemony. **Taste:** Indistinct to slightly bitter. **Chemical reactions:** No reaction in KOH or Melzer's. **Spore deposit:** Peachy yellow. **Microscopy:** Spores (7) 9–13 × 3–4 µm, broadly cylindrical, finely warted. Basidia with clamp connections.

Ecology: Solitary or scattered in small groups on ground in moss or duff under conifers, especially hemlock. Locally common in west-side forests. Fruiting in fall.

Edibility: Likely edible.

Comments: The lemon-yellow colors, slender, upright stature, and often fragrant odor (many collections smell like lemon dish soap) make this a fairly distinct species. There are a number of other corals with yellow colors, many of which have chunky stipes (see *R. rasilispora*), gelatinous flesh *(R. flavigelatinosa)* or brighter yellow-orange colors (see *R. sandaracina)*. Others with a similar stature include: *R. leptoformosa*, with peachy orange branches and orange tips, and spores 8–13 × 3–5 µm; and the *R. conjunctipes* group with yellow to greenish yellow tips, and salmon- to peach-colored branches. Members of the latter group lack clamp connections and have short spores (6–10 × 4–6 µm). *R. longispora* has bright yellow to orange colors, and long spores (10–18 × 4–6 µm).

Ramaria rubiginosa

Marr & D.E. Stuntz

Red-staining Yellow Coral

Fruitbody:7–12 (20) cm tall, 4–12 (15) cm across, forming medium to large clusters with indistinct to moderately chunky stipes. Branches compact at first, elongating in age. Tips pale yellow; branches pale creamy yellow to yellow at first, ocher-yellow as spores mature. **Stipe:** 1–7 cm long, 0.5–4 cm thick, slightly rooted to short and indistinct; often with a few branches fused together. Whitish to pale yellow, with distinct wine-red stains, especially in age. **Flesh:** Moderately thick, firm, fleshy-fibrous. White, occasionally with wine-red stains in stipe. **Odor:** Indistinct. **Taste:** Mild. **Chemical reactions:** Flesh inamyloid. **Spore deposit:** Grayish yellow. **Microscopy:** Spores 7–11 × 3.5–6 µm, cylindrical, ornamented with fine warts. Basidia without clamp connections.

Ecology: Solitary, scattered, or in troops on ground under conifers. Uncommon in much of northern Cascadia, locally common with tanoak in the southern portion of our range. Fruiting in fall.

Edibility: Edible and good.

Comments: The pale yellow colors, rather compact branches, and, most importantly, wine-red stains on the stipe or lower branches help set this species apart. There are a number of other wine-red staining species of *Ramaria*. *R. rubribrunnescens* has salmon-pink branches, yellowish tips, rosy red to wine-red stains, and longer spores (10–14 × 3.5–5 µm). *R. maculatipes* has pale salmon-orange branches and yellow to yellow-orange tips, deep red to wine-red stains, and clamped basidia and spores measuring 9–11 × 4–5 µm. *R. vinosimaculans* has creamy to pale yellow-orange colors when young, often with pale ocher tips, pronounced wine-red stains over the lower portion in age, clamp connections on the basidia, and long spores (11–14 × 4–5 µm). Non-staining look-alikes include *R. cystidiophora* and *R. rasilispora.*

Ramaria velocimutans

Marr & D.E. Stuntz

Fruitbody: 7–12 (20) cm tall, 4–12 (15) cm across, forming medium to large clusters and a moderately chunky stipe. Branches compact at first, elongating in age. Whitish to creamy, pale creamy ocher and often developing brown stains on lower branches in age. **Stipe:** 2–9 cm long, 1–5 cm thick, typically branching from a single point, occasionally fused. Whitish to cream, soon developing brownish stains. **Flesh:** Moderately thick, firm, fleshy-fibrous. White with brownish stains. **Odor:** Indistinct to slightly sweet. **Taste:** Mild. **Chemical reactions:** Flesh inamyloid. **Spore deposit:** Grayish yellow. **Microscopy:** Spores 8–12 × 3.5–5 µm, cylindrical, ornamented with fine lobed warts. Basidia with clamp connections.

Ecology: Solitary, scattered, or in troops under conifers, especially late seral hemlock and fir forests. Occasional to locally common; often found at mid to high elevations across Cascadia. Fruiting in fall.

Edibility: Likely edible.

Comments: The large size, white to creamy colors, and brown staining are helpful in distinguishing this species. Most similar is *R. acrisiccescens* which has whitish to pale creamy beige colors and a slender, more upright stature. *R. amyloidea* has pale orangish colors, brown staining on the stipe, and distinctly amyloid flesh. *R. cauliflοriformis* forms large, densely branched, pinkish creamy to creamy ocher fruitbodies with a large fleshy stipe. It is more common in the southern portion of the Cascade Range. Also compare with *R. rasilispora*, which typically show pale yellow to yellow-ocher colors.

Ramaria acrisiccescens

Marr & D.E. Stuntz

Blah Coral

Fruitbody:5–25 cm tall, (2) 5–18 cm across, generally with tall, slender upright branches and a short, slender stipe. Branches whitish to pale creamy beige at first, becoming more brownish in age. Tips concolorous or with a pinkish buff cast. **Stipe:** 1–7 cm long, 0.5–2.5 cm thick, slender and rooting, or short and indistinct. Whitish, staining brownish in age. **Flesh:** Thin, fibrous; whitish to brownish. **Odor:** Indistinct to musty. **Taste:** Acrid-bitter. **Chemical reactions:** Flesh inamyloid. **Spore deposit:** Grayish yellow. **Microscopy:** Spores 8–14 × 4–6 µm, broadly cylindrical to ellipsoid, ornamented with distinct warts. Basidia without clamp connections.

Ecology: Solitary or in scattered troops under conifers, especially hemlock. Common across western Cascadia. Fruiting in fall.

Edibility: Unknown. The taste is a deterrent.

Comments: This coral is identified by the clumps of long, upright branches, lack of a chunky stipe, pallid colors, and acrid-bitter taste. *R. velocimutans* is a larger, fleshy species with whitish to creamy beige branches. It also stains brownish in age and has narrower spores.

Ramaria rasilispora

Marr & D.E. Stuntz

Yellow Spring Coral

Fruitbody: 5–15 (20) cm tall, 5–20 (25) cm across, forming compact clusters with a chunky stipe. Branches short at first, spreading and elongating in age. Pale yellow, pale greenish yellow to pale ocher-yellow with slightly brighter tips at first, soon developing more ocher tones as spores mature. **Stipe:** 2–7 cm long, 2.5–5 cm thick, stout, chunky, irregular, often tapered downward. Whitish to pale yellowish. **Flesh:** Thick, firm, fleshy. White in stipe, whitish to pale yellow in branches. **Odor:** Indistinct. **Taste:** Mild. **Chemical reactions:** Flesh weakly amyloid. **Spore deposit:** Yellowish ocher. **Microscopy:** Spores 8.5–11.5 × 3.5–4.5 µm, cylindrical, smooth. Basidia with clamp connections.

Ecology: Often scattered in troops or arcs, rarely solitary, and typically remaining under duff until maturity. Common and often abundant on the east side of the Cascade Range and in dry microclimates of the Coast Range, rare elsewhere. Fruiting in spring and early summer.

Edibility: Edible and quite good. It is often passed over by morel hunters in spring but is worth trying. Cook well and sample in moderation when trying it for the first time, as some people get minor gastrointestinal upsets from *Ramaria.*

Comments: The large size, pale yellow to yellowish ocher colors, and growth in spring are helpful in identification, at least as a group. *R. rasilispora* belongs to a complicated group of species —in addition to the ones mentioned below, we know of an additional ten unnamed species that could be easily mistaken for each other! *R. magnipes* fruits in the same habitat in spring; it is quite difficult to distinguish but tends to be brighter yellow when young with a larger stipe which occasionally stains brown. Its spores measure 9.5–13.5 × 3–5 µm. *R. thiersii* also fruits in spring but is more common in the southern Cascades. It tends to have more of an upright, subcylindrical stature, a chunky white stipe which stains brownish, inamyloid flesh, and roughened to finely warted spores (11.5–16 × 4–5 µm). *R. rasilisporoides* is a fall-fruiting species with spores 7–10 × 3–4.5 µm. *R. flavobrunnescens* var. *aromatica* has a fragrant odor and finely warted spores measuring 9–12 × 3–5 µm. It is more common in fall but does occasionally fruit in spring. The fall-fruiting *R. amyloidea* has pale orange colors and distinctly amyloid flesh.

Ramaria rubella forma *rubella*

(Schaeff.) R.H. Petersen

Fruitbody: 3–7 (10) cm tall, 2–7 cm across, forming rather sparse clusters of repeatedly forked branches with a short, slender stipe. Grayish pink to pale grayish red at first, becoming pinkish buff to warm grayish buff in age. **Stipe:** 0.5–2 cm long, 0.3–0.7 cm thick, stout, with whitish mycelial rhizomorphs at base. **Flesh:** Tough, leathery-fibrous, pallid. **Odor:** Indistinct or slightly sweet. **Taste:** Acrid. **Chemical reactions:** KOH bright pink on mycelial rhizomorphs. **Spore deposit:** Golden yellow. **Microscopy:** Spores 5.5–8 × 4–5.5 µm, ellipsoid, ornamented with numerous fine warts.

Ecology: Solitary, scattered, or in troops on well-decayed wood or duff near rotting branches. Common in coastal Cascadia, occasional to rare elsewhere. Fruiting from early fall into winter.

Edibility: Unknown.

Comments: This *Ramaria* is easily recognized by its pinkish gray colors, sparse crowns, tough flesh, growth on decayed wood, and pink KOH reaction of the rhizomorphs. *R. rubella* f. *blanda* lacks the KOH reaction on the rhizomorphs and has slightly shorter spores (averaging 7.1 µm versus 7.5 µm for *R. r.* f. *rubella*). Note, however, that ITS sequences do not distinguish these two. *R. stricta* has straighter, more erect, creamy white to pinkish buff or yellowish buff branches. *R. apiculata* has yellowish to buff or pale grayish brown to reddish brown branches, often with distinctly green tips (although this green color quickly fades), and spores 7–11 µm long. *R. tsugina* is quite similar to *R. apiculata*, but its spores tend to be under 9 µm long. *Lentaria* species generally have whitish, pointed tips, beige to light brown lower branches, and smooth spores.

Phaeoclavulina abietina group

Fruitbody: 2.5–6 cm high, 1.5–5 cm across, forming small upright to spreading clusters from a common base, branches often forking near pointed tips. Branches yellowish ocher to ocher-brown, often with olivaceous ocher discoloration. Staining blue-green to teal-green when damaged, or in age. **Stipe:** Short, with numerous branches. Base with white rhizomorphs. **Flesh:** Thin, tough, leathery. **Odor:** Indistinct. **Taste:** Slightly bitter. **Spore deposit:** Ocher. **Microscopy:** Spores 7–10 × 3.5–4.5 µm, ovoid to teardrop-shaped, distinctly spiny.

Ecology: In troops, rings, or arcs in duff and humus. Occasional, typically in duff under western redcedar, but widespread and in most habitats in Cascadia. Fruiting in fall and winter.

Edible: Unknown.

Comments: The tough texture, yellowish ocher to ocher-brown ranches which develop green staining, white rhizomorphs, and growth in duff are distinctive. *P. abietina* is a complex in Cascadia with at least three species matching the above description – there are currently no criteria to distinguish them based on morphology. The greenish staining helps differentiate this group from other *Phaeoclavulina*, including *P. myceliosa*, which has with a tangled mess of branches, white basal mycelium, no green staining, and spores measuring 4.5–5.5 × 2.5–3.5 µm. The similar *P. curta* has smaller fruitbodies and spores 4–5 × 3–3.5 µm. *P. eumorpha* is slightly larger and often has more upright-branched fruitbodies. It is best distinguished by its yellow basal rhizomorphs and larger spores (6.5–10 × 3.5–4.5 µm). Members of the *Ramaria apiculata* group have more erect, upright branches with greenish tips, while *R. stricta* has straighter, denser, creamy buff branches. Both of these grow on decaying wood. *R. gracilis* has small fruitbodies with off-white to pale ocher branches and whitish tips.

Clavulina coralloides

Clavulina rugosa

Clavulina cinerea

Helminthosphaeria clavariarum

Clavulina coralloides group

Crested Coral

Fruitbody: 2–7 (10) cm high, 2–5 cm across, with many upright branches forking irregularly from a common base. Surface smooth, dry to moist. Tips finely crested with many short spine-like branches when young, becoming more blunt in age. Creamy white when young, becoming beige or graying slightly in age. Some forms have creamy tips and grayish branches. Commonly infected with *Helminthosphaeria clavariarum*, a parasitic fungus which distorts the fruitbody and turns it bluish gray to dark gray (see comments). **Stipe:** Relatively short, or slightly rooted. **Flesh:** Thin, brittle to moderately tough, fibrous. **Odor:** Indistinct. **Taste:** Indistinct or slightly acrid. **Spore deposit:** Whitish. **Microscopy:** Spores 7–11 × 6.5–10 µm, globose to subglobose, smooth, moderately thick-walled.

Ecology: In scattered clusters in duff or moss under conifers. Widespread and locally abundant. First fruiting in fall, with some fruitbodies persisting into spring at lower elevations.

Edible: Edible.

Comments: Although long known as *Clavulina cristata*; *C. coralloides* is an older name for the European species. That said, a number of species in this group occur in western North American, all unnamed. There are three distinct groups of *Clavulina*: *C. coralloides*, as described above; the *C. rugosa* group, with more simply-structured fruitbodies and upright stipes with a cluster of short, wrinkly branches near the tip; and the *C. cinerea* group, with whitish to grayish branches which become more grayish in age. To make matters more difficult, all are often parasitized by *Helminthosphaeria clavariarum*, a fungus that distorts the fruitbody and turns it bluish gray to dark gray with embedded blackish perithecia, leading to great problems distinguishing *C. cinerea* from *C. coralloides*. *C. reae* is a fairly distinct species with rather blunt branches and a distinct stipe and white colors when young (soon turning gray). *C. castaneipes* has a hairy brown stipe and whitish branches and tips. *Ramariopsis kunzei* forms clusters of white to pinkish, more regular upright branches arising from multiple bases; the tips are generally forked and pointed, but not nearly as fine as *C. coralloides*. *Lentaria* generally have simple pointed tips and grow on wood.

Lentaria pinicola

(Burt) R.H. Petersen

Pale Wood-coral

Fruitbody: 3–9 cm tall, 2–8 cm across, forming a sparse crown of forking, upright branches with a common base. Tips pointed, bright white, transitioning downward to pinkish tan, grayish to buff-brown branches. **Stipe:** 0.5–2.5 cm long, 0.2–0.5 cm thick, often tough, light brown. Base with downy white mycelium and thin white rhizomorphs. **Flesh:** Tough, leathery-fibrous, pallid. **Odor:** Indistinct. **Taste:** Slightly bitter. **Spore deposit:** Whitish. **Microscopy:** Spores 7.5–9.5 × 4.5–6 µm, cylindrical to ellipsoid, smooth, thin-walled.

Ecology: Solitary or scattered in troops on woody debris and surrounding duff. Typically on (or around) branches, stumps, and logs of conifers. Common and widespread, especially abundant in west-side old-growth forests with large woody debris in the understory. Fruiting in fall.

Edibility: Unknown.

Comments: The tough, thin branches, bright white pointed tips, growth on woody debris, and smooth spores help distinguish this species. There is a similar, undescribed species which grows in duff and branch litter of redwoods in California. The name *L. epichnoa* has been used for a poorly known, narrow-spored species in Cascadia. Species of *Ramaria* and *Phaeoclavulina* have similar statures and textures but lack the white growing tips and have warted and/or spiny spores. *Artomyces piperatus* forms sparse clusters of upright, creamy to beige or tan branches that fork in whorled rosettes (pyxidate) with crown-like tips. It grows on decaying conifer wood and has an acrid taste. *Tremellodendropsis tuberosa* forms small, tough, coral-like fruitbodies. It is easily identified by the sparse, often flattened branches joined to a common base, tough texture, and growth on the ground. Surprisingly, its closest relatives are the jelly fungi.

Thelephora palmata

(Scop.) Fr.

Stinking Earth Fan

Fruitbody: 4–8 cm tall, 4–10 cm across, composed of many clustered branches fused into a common "trunk" or base. Branches usually flattened and fan shaped, occasionally many adjacent branches fused into sheets; tips square to peg-like or often more pointed in age; all surfaces slightly wrinkled. Branches mostly brown, dingy lilac-gray to brownish gray to tan; tips usually paler, tan to whitish beige. **Underside:** (fertile surface) smooth to wrinkled, often with tiny warts. Grayish to grayish brown, paler to whitish near margin at first, darkening in age. **Stipe:** Usually central, sometimes narrow and distinct, in other cases indistinct and broad, made of loosely fused branches. Color like fertile surface or darker reddish brown to nearly black. **Flesh:** Thin, quite tough, leathery. **Odor:** Usually strong, like rotting cabbage, garlic, or old broccoli. **Taste:** Unpleasant. **KOH:** Dark blue-green (appearing black). **Spore deposit:** Brown. **Microscopy:** Spores 8–10 × 6.5–9 µm, ellipsoid, slightly irregular in outline with many small, spiny warts.

Ecology: Solitary, scattered, or in troops under conifers, especially spruce. Widespread west of the Cascade crest, but uncommon. Fruiting in fall and early winter.

Edibility: Non-toxic, but the rank odor isn't very appealing. Some similar fleshy *Thelephora* are highly prized as edibles in parts of China.

Comments: This species is recognizable by its "palmate" multi-branched structure with fan- or hand-shaped tips, dark colors, and unpleasant rotting cabbage-garlic odor. The fruitbody stature varies a great deal: sometimes loosely and finely branched, in other cases dense and blocky and forming tight masses with poorly defined branches. There may be more than one local species matching the above description, and it may be that none are conspecific with the true European *T. palmata*. *T. anthocephala* is similar but has more flattened branches and lacks an odor. *Ramaria* are usually not so dark-colored, have cylindrical, fleshy branches (not tough, flattened or antler-like), and the few lilac-brown *Ramaria* have much more robust central stipes.

Calocera viscosa
(Pers.) Fr.

Yellow Tuning Fork

Fruitbody: 1–7 (10) cm tall; crown 1-4 cm across; often in clusters of simple forking or antler-shaped branches, occasionally with more elaborate branched or forked tips. Surface dull to shiny, smooth, often slightly viscid. Bright golden yellow to yellow-orange. **Stipe:** Ranging from short and nondescript, or long, rooting and sinuous. Concolorous with upper portion, occasionally whitish at extreme base. **Flesh:** Thin, gelatinous to rubbery, somewhat tough. Translucent orange to yellow. **Odor:** Indistinct. **Taste:** Indistinct. **KOH:** No reaction. **Spore deposit:** Pale yellowish. **Microscopy:** Spores 8–12 (14) × 3.5–4.5 µm, cylindrical to slightly curved, smooth, colorless, becoming 1-septate at maturity.

Ecology: Solitary, scattered, or in clusters on rotten conifer wood (sometimes buried). Widespread and especially common in western Cascadia. Fruiting from early fall into spring.

Edibility: Unknown.

Comments: The bright golden yellow color, forking or branched tips, rubbery texture, and growth on rotting conifer wood are helpful in distinguishing this jelly fungus. It is most likely to be confused with *Ramariopsis crocea*, which is more orange-toned, has a more open-branched crown, fibrous-brittle flesh, and tiny, round spores. It is rare in Cascadia, growing from soil or duff under conifers; especially western redcedar. Orange and yellow *Ramaria* are flesher and often fleshy-brittle (not rubbery). *Calocera cornea* is much smaller and has simple, cylindrical clubs. *C. furcata* is also much smaller but has multiple, often forking branches and prefers growing on hardwoods.

Calocera cornea
(Batsch) Fr.

Yellow Staghorn Jelly

Fruitbody: 0.3–1.5 (2) cm high, 0.1–0.3 cm across, cylindrical to elongate. sometimes forking at base, with blunt or pointed tips. Surface dull to shiny, smooth. Bright or dull orange to yellow. **Stipe:** Short, slightly tapered. Often slightly darker and a different tone from top. **Flesh:** Very thin, gelatinous to rubbery. Translucent orange to yellow. **Odor:** Indistinct. **Taste:** Indistinct. **KOH:** No reaction. **Spore deposit:** Pale yellowish, difficult to obtain. **Microscopy:** Spores 7–10 × 2.5–4 µm, cylindrical to slightly curved, smooth, colorless, becoming 1-septate at maturity.

Ecology: Often in large troops, more rarely sparsely scattered on dead logs, trees, or branches. Prefers hardwoods and is common on Oregon white oak, bigleaf maple, and red alder but can be found on conifers as well. Fruiting soon after the first rains in fall and continuing through spring.

Edibility: Unknown.

Comments: This small but distinctive jelly fungus is readily identified by its small, upright, cylindrical fruitbodies, orange to yellow colors, gelatinous to rubbery texture, and growth on branches and logs. It is easily overlooked and is usually only noticed immediately after rains or prolonged wet periods when the fruitbodies are plump with water. *C. furcata* is similar but has multiple, often forking, branches. *C. viscosa* is a bright orange to yellow species that differs by its larger size (1–7 cm high) and branches that develop "staghorn" crowns. Also, compare the bright yellow *Mucronella sp.* 'Yellow' which produces tiny clusters of downward-facing spines; it is fragile, not rubbery in texture.

Clavulinopsis laeticolor

(Berk. & M.A. Curtis) R.H. Petersen

Yellow Club Coral

Fruitbody: 2–10 cm high, 0.2–0.8 cm across, cylindrical to club-shaped with a rounded to blunt, (rarely pointed) tip, occasionally longitudinally compressed or flattened in age. Bright yellow-orange, gold, orange to yellow. **Stipe:** Short; opaque orange with a pale base. **Flesh:** Thin, fragile, concolorous. **Odor:** Indistinct. **Taste:** Mild to slightly bitter. **Spore deposit:** White. **Microscopy:** Spores 5.5–9 × 3–5 µm, ellipsoid, oblong, globose to subtriangular, smooth, inamyloid.

Ecology: Solitary or scattered, occasionally in small clusters in duff and moss. Widespread and in a wide variety of forest types but especially common under western redcedar. Fruiting from fall into winter.

Edible: Edible.

Comments: This common species is easily identified by its yellow to orange color and simple club shape. Although it usually grows singly or scattered in small troops, it does occasionally fruit in clusters. These clustered fruitings have often been called *C. fusiformis*, which is misapplied in Cascadia. *C. gracillima* has dull ocher-yellow to peachy yellow colors. *Clavulinopsis corniculata* has dull yellowish orange to orangish beige colors, and often, but not always, forking club-shaped tips. *Clavaria flavipes* is a small club-shaped coral with dull yellowish to tan-beige colors and a small but distinct, more yellow stipe.

Alloclavaria purpurea group

Purple Club Corals

Fruitbody: 3–10 (13) cm high, 0.2–0.7 cm across, cylindrical with rounded to pointed tips, occasionally longitudinally compressed in age. Surface smooth, moist to dry. Purple, lilac to purplish gray when young, purple tones often fading in age, becoming lilac-gray, gray, or grayish buff. **Stipe:** Indistinct. **Flesh:** Thin, fragile, concolorous. **Odor:** Mild to slightly unpleasant. **Taste:** Mild. **Spore deposit:** White. **Microscopy:** Spores 5.5–9 × 3–5 µm, ellipsoid to oblong, smooth, inamyloid.

Ecology: In scattered clusters or in large troops of many clusters, often on bare sandy soil or in moss-covered sand. Locally common on forested dunes of coastal Cascadia, occasional at high elevations of the Cascade Range, rare elsewhere. Fruiting from late fall into winter.

Edible: Edible.

Comments: The simple, cylindrical shape, clustered growth, and purple to lilac-gray colors are distinctive. It can fruit by the thousands in coastal spruce stands on sandy soil. There appear to be two species in western North America; "*Alloclavaria*" (*Clavaria*) *nebulosoides* may be the correct name for western Cascadia's species, while the one in eastern Cascadia and the Rocky Mountains is more closely related to the European *A. purpurea*. *Clavaria fumosa* has a similar shape and grows in clusters, but is less upright, more sinuous, and is smoky beige from the start, lacking any lilac to lavender colors. *C. rosea* is smaller, with a more sparse, scattered growth habit, and a rosy to pale pink color.

Clavaria fragilis
sensu PNW

Fairy Fingers

Fruitbody: 3–10 (15) cm high, 0.2–0.7 cm across, cylindrical with rounded to pointed tips, occasionally splitting. Surface smooth, dry to moist. White to creamy white; tips occasionally yellowing in age. **Stipe:** Indistinct. **Flesh:** Thin, fragile; concolorous. **Odor:** Indistinct. **Taste:** Mild. **Spore deposit:** White. **Microscopy:** Spores 5–7 × 3–4 µm, ellipsoid to teardrop-shaped, smooth, inamyloid.

Ecology: Scattered in small to large clusters, often in troops or arcs in duff or soil, in a wide variety of forest types, with a preference for western redcedar. Occasional across lower elevations in western Cascadia. Fruiting in fall.

Edible: Edible.

Comments: The clusters of simple, white, unbranching clubs without a distinct stipe are distinctive. Although usually identified as *C. fragilis* or *C. vermicularis*, our species is distinct and needs its own name. It has also gone by the common names of Worm Coral and White Club Coral. *C. fumosa* is quite similar but has a smoky beige color, a less upright shape, and is less common. *C. falcata* is quite a bit smaller, rarely grows in tight clusters, and has a narrower, distinct, translucent stipe.

Typhula phacorrhiza
Fr.

Pebble-on-a-string

Fruitbody: 2–10 cm tall, 0.05–0.1 (0.2) cm across, tall, very thin cylindrical club with pointed to blunt tip. Surface smooth, dry. Pale straw-yellow to tan. **Stipe:** Short, often tapering downward, slightly darker in color than the fertile part. Base with small (0.1 cm) ovoid rounded to flattened or irregular, yellow-orange sclerotium. **Flesh:** Very thin, somewhat tough and fibrous; concolorous. **Odor:** Indistinct. **Taste:** Indistinct. **Spore deposit:** White. **Microscopy:** Spores 11–15 × 4.5–5.5 µm, cylindrical to ellipsoid, smooth, inamyloid.

Ecology: Often scattered in large troops in duff of alder and bigleaf maple, on decaying leaves. Uncommon, occurring at low to mid-elevations in western Cascadia. Fruiting in fall.

Edible: Unknown.

Comments: The pale straw-yellow to tan, hair-like fruitbodies emerging from a yellow-orange sclerotium on leaves and twigs of bigleaf maple and alder. We have seen a similar, likely undescribed, species on conifer duff with smaller spores. The hair-like fruitbodies of *Macrotyphula juncea* are very similar, but often has white rhizomorphs, and lack a sclerotia. It grows in both hardwood and conifer duff. The rare *M. fistulosa* is much larger and club-shaped (up to 30 cm tall, 0.2–1 cm thick) and generally ocher-yellow to ocher-brown. *Typhula sclerotioides* produces slender white clubs on herbaceous debris. *Clavaria falcata* is smaller, more club-shaped, white, with a well-defined translucent stipe.

Clavicorona taxophila

(Thom) Doty

Yew Club

Fruitbody: 0.5–3.5 cm high, 0.3–0.7 cm across at top; club-shaped with a flattened to concave top, and often fringed, overhanging margin. Surface moist to dry; underside smooth, or often weakly veined or wrinkled at maturity. White to creamy white, occasionally pale yellowish in age. **Stipe:** Often short, occasionally rooted, base with white hairs. White. **Flesh:** Thin, fragile; watery white. **Odor:** Indistinct. **Taste:** Indistinct. **Microscopy:** Spores 3–4 × 2.5–3.5 µm, ellipsoid, smooth, inamyloid.

Ecology: Scattered or in troops from needle duff and moss under conifers, especially western redcedar; occasionally under hardwoods. Common across the lower elevations of western Cascadia. Fruiting from mid-fall into winter.

Edibility: Unknown.

Comments: These small clubs with a flat, thin, flaring, and occasionally almost cap-like tip, and smooth to slightly veined to wrinkled underside are intermediate in morphology between the cylindrical clubs of the genus *Clavaria*, and their close relatives, *Hodophilus*, which produce typical gilled mushroom fruitbodies. *Multiclavula vernalis* has pale ocher, spindle- to club-shaped fruitbodies, and the similar *M. corynoides* has small, cylindrical to pointed clubs which are whitish to pale ocher with white tips. Both of these both grow from algae-covered soil.

Multiclavula mucida

(Fr.) R.H. Petersen

Algae Club

Fruitbody: 0.5–1.5 (2) cm tall, 0.05–0.1 (0.2) cm across, cylindrical to spindle-shaped, at times narrowly club-shaped, rarely forking; tips pointed to rounded. Surface moist to dry, smooth to slightly roughened. White to cream. **Stipe:** 0.1–0.5 cm tall, 0.05–0.1 cm across, poorly defined from fertile areas. **Flesh:** Very thin, somewhat tough. **Odor:** Indistinct. **Taste:** Indistinct. **Spore deposit:** Not obtained. **Microscopy:** Spores 4.5–8 × 2–3 µm, cylindrical to ellipsoid, smooth. Basidia 4- to 6-spored.

Ecology: Scattered in gregarious troops on barkless, algae-covered wood, usually cottonwood, alder, or oak logs. Widespread but easily overlooked; it is more common at lower elevations in northwestern Cascadia. Fruiting from fall into spring.

Edible: Unknown.

Comments: This small, non-descript club fungus is actually a basidiolichen, growing in association with mats of green algae! It's easy to recognize but, due to the small size, hard to find. Two other species in Cascadia grow on algae-covered soil. These include *M. vernalis*, with larger, pale ocher, spindle- to club-shaped fruitbodies, and *M. corynoides*, with small, cylindrical to pointed clubs which are whitish to pale ocher with white tips. There are a number of *Typhula* species in Cascadia, most of which fruit from tiny sclerotia embedded in the substrate.These include *T. erythropus* with small white club-like heads and reddish-brownish stipes, growing from the stems of hardwood leaves (especially bigleaf maple), and *T. setipes* with hairy stipes which grows from the leaf itself (especially alder).

Club

Onygena corvina

Alb. & Schwein.

Owl-pellet Powderhead

Fruitbody: 0.5–2.5 cm tall. **Head:** 0.1–0.4 cm across, rounded. Surface dry, with fine scaly blocks or dandruffy at first, becoming powdery as spores mature. Whitish gray to buff or light brown. **Stipe:** 0.5–2 cm tall, 0.1–0.2 cm across, stout and equal at first, becoming irregular, curvy and often with an enlarge base in age. White to cream colored. **Flesh:** Very thin, soft to powdery in cap, fibrous in stipe. **Odor:** Indistinct, but the substrate is often somewhat rank. **Taste:** Not sampled. **Spore deposit:** Buff to light brown. **Microscopy:** Spores 6–8 × 2.5–3 µm, cylindrical to ellipsoid, smooth, with 1 or 2 oil droplets. Asci 8-spored, spherical, 10–16 × 10–12 µm.

Ecology: Often in clusters on owl pellets, bird bones, and feathers. Also on animal hair — we have seen a large fruiting on a dead elk. Rare but widespread, most likely to be encountered in western Cascadia. Fruiting in fall and winter.

Edible: Unknown.

Comments: This oddball has a very specialized substrate, as it digests the keratin protein found in owl pellets, bird feathers, and bones. Besides the substrate, it is recognized by the rounded, buff to light brown head that becomes powdery, and the white stipe. *O. equina* is very similar but grows on old horns and hooves of grazing animals such as cattle, sheep, and deer. Two similar *Amphichorda* grow on the dung of carnivorous animals. *A, cavernicola* forms rounded whitish heads perched atop short stipes, while *A. felina* produces tall, slender, fuzzy fan-shaped or pointed columns, often with fuzzy or powdery texture on the lower part. *Penicillium vulpinum* is similar in shape but has a powdery bluish gray, irregularly rounded head, white to tan-beige or reddish brown stipe, and grows on dung. *Emericellopsis* (=*Stilbella*) *fimetaria* is more slender, with a tan to peachy head, whitish stipe, and growth in swarms on dung. *Asterophora lycoperdoides* is quite similar in shape, color, and texture but is much larger and grows on rotting mushrooms.

Heyderia abietis

(Fr.) Link

Fruitbody: 0.5–3.5 cm tall. Fertile head 0.1–0.7 cm across, egg-shaped to rounded-cylindrical, rarely globose, sometimes flattened. Light tan to ocher-buff. **Stipe:** Thin, wiry, distinct from head. Brownish ocher. **Flesh:** Very thin, somewhat tough, rubbery to fibrous. **Odor:** Indistinct. **Taste:** Indistinct. **Spore deposit:** Generally not obtainable. **Microscopy:** Spores 10–14 × 2–2.5 µm, spindle-shaped, smooth. Asci 8-spored, inamyloid.

Ecology: Scattered in small to large troops and swarms on duff under conifers across Cascadia. Common but inconspicuous and not noticed. Fruiting in the fall.

Edible: Unknown. Dried specimens might make good toothpicks.

Comments: The tiny ocher-buff heads atop narrow, brownish stipes and growth on conifer needles make this a distinctive, albeit diminutive, species. They are often overlooked but large fruitings are more obvious and can carpet the duff at times. *Sarcoleotia globosa* is a bit larger with a proportionally larger globe-shaped dark brown head, beige-tan to pale brown stipe with a faint, pale scurfy texture, and growth in mossy, disturbed areas. *Mitrula elegans* has a bright orange head and grows from saturated debris in areas with standing water. *Vibrissea truncorum* has a golden rounded head, short whitish stipe, and growth from saturated or submerged debris in running water.

Synonym: *Mitrula abietis* Fr.

Mitrula elegans

Berk.

Swamp Beacon

Fruitbody: 2–10 cm tall. **Head**: 0.5–3 cm tall, 0.3–1 cm across, cylindrical to rounded, rarely brain-like, ear-shaped or lobed. Surface moist to dry, smooth to wrinkled. Bright orange to yellow-orange, occasionally ochraceous orange to pinkish orange. **Stipe**: 2–9 cm long, 0.1–0.3 cm thick, slightly enlarged towards base. Surface moist to thinly viscid, smooth; base with short, fuzzy or matted whitish hairs. Translucent-white, pinkish white, or pale grayish white. **Flesh**: Thin, watery, concolorous. **Odor**: Indistinct. **Taste**: Mild. **Spore deposit**: Not obtained. **Microscopy**: Spores 11–17.5 × 1.5–3 µm, narrowly cylindrical, smooth, hyaline, 1- or 2-celled. Asci 8-spored, amyloid at tip.

Ecology: Scattered or in small clusters on water-saturated debris, often in standing water. Fairly common at low to mid-elevations in seeps, streams, or swampy areas in forested settings in western Cascadia. More rarely in marshes or bogs at higher elevations. Fruiting in spring and early summer.

Edibility: Unknown.

Comments: This unusual water-dwelling mushroom is easily recognized by its orange cylindrical heads, translucent-white stipe, and growth on water saturated plant debris. The rarer *M. borealis* is almost identical macroscopically, but is distinguished by its 3–4 µm wide spores. *Vibrissea truncorum* grows on woody debris in water (often running streams), but the fruitbodies are smaller, have rounded yellow heads, and short whitish to gray stipes. Fruitbodies of *Bryoglossum gracile* have a small orangish to pinkish ovoid to round head and a pallid stipe. It is a moss parasite, growing from browned patches of moss. Fruitbodies of *Heyderia abietis* are smaller and have ocher-buff heads and brownish stipes that emerge in fall directly from conifer needles, especially true fir. *Spathularia* have flattened paddle-like or fan-like heads and fruit from conifer duff in the fall. Fruitbodies of *Neolecta vitellina* are larger, thicker, and have brighter yellow, often irregularly shaped heads that transition gradually into whitish stipes. They fruit in moss or conifer duff in fall or winter.

Trichoderma leucopus

Jaklitsch

Whitefoot

Fruitbody: 2–5 cm tall, 0.2–1 (1.5) cm across, club-shaped, with a swollen head. **Head:** Surface dry, finely bumpy. Creamy white to pale ocher-buff. **Stipe:** Narrower than head at apex, often swollen toward base. Surface dry, finely scruffy to smooth. White to pale cream. **Flesh:** Thin, watery when wet, fibrous when dry. Whitish to cream. **Odor:** Indistinct. **Taste:** Indistinct. **Microscopy:** Spores with two parts; upper part 2–4 × 2–3.5 µm, globose to subglobose, lower part 2–5 × 1.5–3.5 µm, wedge-shaped to ellipsoid, finely spiny. Asci 8-spored.

Ecology: Solitary or scattered in needle duff or moss in conifer forest in western Cascadia. Generally uncommon but can be locally common in coastal spruce forest. Fruiting from fall into spring but generally absent in cold winter months.

Edible: Unknown.

Comments: The club-shaped fruitbody with a creamy white head and bumpy surface, white stipe, and growth in duff help distinguish this species. *T. alutaceum* is slightly larger, has a darker yellowish brown head, and grows on wood. There are a few vaguely similar insect pathogens in Cascadia including *Cordyceps militaris* with a pimpled orange head and growth on insect pupae. *Ophiocordyceps variabilis* is a tiny orange species, typically on *Xylophagus* fly larvae. There is an undescribed species in the *O. myrmecophila* group with a thin stipe and a small, rounded head that grows from ants. The larger club-headed *Paraisaria cascadensis* grows on great grig larvae in the high Cascades.

Synonyms: *Hypocrea leucopus* (P. Karst.) H.L. Chamb.; *Podostroma leucopus* P. Karst.

Spathularia sp. 'Ocher'

Ocher Spatula

Fruitbody: 1.5–3.5 cm high. **Head:** 0.5–1.5 × 0.2–0.7 cm, circular to ovoid, flattened, paddle-like or spatula-shaped. Surface moist to dry, smooth but often wrinkled. Dull ocher-yellow to pale straw-yellow. **Stipe:** 0.5–2 cm long, 0.2–0.4 cm thick, equal to irregular, transiting into, but distinct from fertile surface. Concolorous with head. **Flesh:** Thin, watery-fleshy; concolorous. **Odor:** Indistinct. **Taste:** Mild. **Spore deposit:** White. **Microscopy:** Spores 35–65 × 2–3 µm, needle-shaped to very narrowly club-shaped, multi-septate. Asci 8-spored, inamyloid.

Ecology: Scattered or in small tufts, often in troops from conifer needle duff. Widespread in Cascadia, locally common in east-side forests in the Cascade Range and the Klamath Mountains, rare elsewhere. Fruiting in fall and early winter.

Edible: Unknown.

Comments: The paddle- to spatula-shaped fruitbodies, dull yellowish ocher colors, and growth in duff help distinguish this species. It has long been referred to as *S. flavida*, but is consistently paler and likely represents a distinct species. *S. flavida* is rare in Cascadia but is considerably more common to the north. It has larger, bright to pale yellow flat clubs and is more likely to occur in moss. Another more northernly species is *S. rufa*, which has brown to ocher-brown colors. *Cudonia* (=*Pachycudonia*) *spathulata* has an irregularly shaped golden yellow cap and pallid stipe and grows in the duff of leaves and bark-peelings around madrone and manzanita. It is more likely to be found in the southern part of Cascadia, fruiting in winter.

Cudonia circinans

(Pers.) Fr.

Redleg Jellybaby

Cap: 0.5–1.5 (2) cm across, irregularly rounded to convex when wet, wrinkled when dry; margin inrolled at all stages. Surface smooth, generally dull but can be shiny when wet. Pinkish buff, orangish buff to warm beige. **Underside:** Smooth to wrinkled, concolorous with stipe. **Stipe:** 1.5–6 cm long, 0.3–1 cm thick, equal or enlarged toward base. Surface dry to moist, finely roughened or smooth. Orangish buff to warm beige when young, becoming darker, smoky buff to vinaceous beige. **Flesh:** Thin, rubbery to slightly gelatinous when wet, leathery when dry. Stipe solid when young, often hollow in age. Concolorous with cap and stipe with a whitish core. **Odor:** Indistinct. **Taste:** Indistinct. **Microscopy:** Spores 28–46 × 2 µm, needle shaped, colorless. Asci 8-spored, narrowly club-shaped, inamyloid.

Ecology: Often in small or large clusters, rarely solitary, in moss, thick duff, on well-decayed woody debris, or on spruce cones. Locally common in coastal Sitka spruce forest, occasional in the high cascades. Fruiting in wet periods throughout the year but most common in late winter and spring.

Edibility: To be avoided. Studies have indicated a high concentration of gyromitrin, the same toxin as in false morels, and it could be fatal if consumed raw.

Comments: The irregularly rounded cap that becomes plump when wet, smooth or slightly wrinkled underside, rubbery stipe, buff to beige colors, and growth under spruce help identify it. Two similar but darker species include *Cudonia* (=*Pachycudonia*) *monticola* with a pinkish buff to grayish brown cap, a grayish brown stipe, and shorter spores (20–24 × 2 µm) and *C. grisea* with light to dark gray or vinaceous gray colors and spores 18–22 × 1.5–2 µm. Both of these species are more likely to be found in montane fir-hemlock forests in spring or early summer (but can rarely be found at lower elevations, or in fall and winter). *Sarcoleotia globosa* has a dark olive-brown to blackish brown, domed to round head with an inrolled margin, and a thin, cylindrical stipe that is beige with dark dots. It grows among mosses in disturbed areas. *Cudonia* (=*Pachycudonia*) *spathulata* has an irregularly shaped golden yellow cap. It grows with madrone and manzanita and is more likely to be found in the southern part of Cascadia. Jelly Babies (*Leotia lubrica*) are quite similar in stature but are generally slightly larger, more gelatinous, with brighter yellowish to yellow-green tones.

Clavariadelphus caespitosus

Methven

Clustered Club

Fruitbody: 3–10 (15) cm tall, 0.3–1.5 (2) cm across, irregularly cylindrical, narrowly club-shaped to spindle-shaped, sometimes branched or forked. Surface dry, smooth, often wrinkled and vertically ridged. Grayish red to dull red, occasionally more reddish brown; paler near base. **Flesh:** Firm, stringy when young, soft and spongy in age. Whitish, slowly staining brown. **Odor:** Indistinct. **Taste:** Mild to slightly bitter. **KOH:** Slight yellowing on exterior. **Spore deposit:** Whitish. **Microscopy:** Spores 8–11 × 4.5–5.5 µm, broadly ovoid to almond-shaped, thin-walled, smooth, inamyloid.

Ecology: In fused clusters of a few to numerous clubs, rarely solitary, in moss and duff under conifers. Uncommon, mostly in west-side old growth forests. Fruiting in fall.

Edibility: Edible.

Comments: The clustered growth, reddish colors, and faint yellow KOH reaction help identify this species. Most similar is *C. subfastigiatus*, which differs by having pinkish buff to light cinnamon -brick colors, a forest-green KOH reaction, and slightly wider spores (8–10.5 × 5–6 µm). It occurs in similar habitats but rarely with more than a few clubs in a cluster. *C. ligula* and *C. sachalinensis* lack the reddish colors and grow in large troops. *C. occidentalis* has duller colors and rarely grows clustered.

Clavariadelphus ligula

(Fr.) Donk

Trooping Club

Fruitbody: 2–10 cm tall, 0.3–1.5 (2) cm across, irregularly cylindrical to club-shaped, occasionally more irregular or flattened. Surface dry, smooth, often wrinkled and vertically ridged. Color variable, typically yellowish, straw-yellow, buff to pale ocher. **Flesh:** Firm, stringy when young, soft, spongy in age. Whitish, slowly staining pinkish cinnamon to light orangish brown. **Odor:** Indistinct. **Taste:** Mild to slightly bitter. **KOH:** Slight yellowing on exterior. **Spore deposit:** Pale ochraceous buff. **Microscopy:** Spores 12–16.5 × 3.5–4.5 µm, cylindrical to very narrowly ellipsoid, thin-walled, smooth, inamyloid.

Ecology: Scattered to gregarious in large troops, more rarely in small patches, from duff, moss, and soil under conifers. Fairly common on the east side of the Cascade Range, occasional to rare elsewhere. Fruiting in fall.

Edibility: Edible, but rather bland and fibrous.

Comments: The cylindrical to club-shaped, yellowish clubs, and trooping growth are helpful in distinguishing this species from all others but *C. sachalinensis*, which has longer spores (18–24 × 4–6 µm). The distribution of *C. sachalinensis* may be more northerly and it may tend to have narrower clubs (but additional well documented collections are needed to help differentiate these two species in Cascadia). *C. mucronatus* has whitish to pallid clubs, which often have acutely pointed tips. It can be found in similar habitats, mostly on the east side of the Cascade Range. *C. occidentalis* is larger, more irregularly shaped, and rarely grows in troops.

Clavariadelphus occidentalis group

Western Clubs

Fruitbody: 4–15 cm tall, 1–3 cm across, cylindrical, club-shaped, or irregular, often flattened or fan-shaped, broadest at middle or just below tip, occasionally weakly forked. Surface dry, smooth, often wrinkled and vertically ridged. Pinkish-tan at first, soon buff, yellowish, or ochraceous buff, becoming tan to grayish orange, paler at base, fading overall in dry weather. **Flesh:** Firm, stringy when young, soft and spongy in age. White, slowly staining light to dark brown. **Odor:** Indistinct. **Taste:** Mild to bitter. **KOH:** Golden orange on exterior. **Spore deposit:** Whitish. **Microscopy:** Spores 9.5–13 × 5.5–7 µm, broadly ovoid to almond-shaped, thin-walled, smooth, inamyloid.

Ecology: Solitary, scattered, or in small troops in duff or moss in a wide variety of forest types. Locally common, especially at mid-elevations of the Cascade Range. Fruiting in fall.

Edibility: Edible but bland, bitter, and often tough.

Comments: The large size club-like shape, dull color, and yellowish KOH reaction help set this species apart. This group needs modern taxonomic work: sequence data shows a great diversity suggesting there may be four or five species in Cascadia which have collectively been referred to as *C. occidentalis*. Other similar species include *C. subfastigiatus*, which has pinkish buff to light cinnamon-brick fruitbodies, a forest-green KOH reaction, and smaller spores (8–10.5 µm × 5–6 µm). *C. caespitosus* often fruits in clusters of narrowly club-shaped, grayish red to dull red fruitbodies. *C. pallidoincarnatus* and *C. truncatus* have medium to large club-shaped fruitbodies with distinctly flattened or slightly rounded tops; the former is pinkish orange to whitish, while the latter is more orange. They also have a sugary-sweet taste and red KOH reactions. *C. ligula* and *C. sachalinensis* are much smaller and grow in large troops under conifers.

Clavariadelphus truncatus

(Quél.) Donk

Candy Club

Fruitbody: 2–15 cm tall, 2–8 cm across at the top, club-shaped and often wrinkled, with a broad, flattened to rounded top. Surface moist to dry, often wrinkled and vertically ridged; top often smoother. Bright orange, yellow-orange to ocher over the top, paler, ocher to orangish brown lower, often paler towards base, rarely reddish orange, fading overall with age, slowly staining brownish when rubbed. **Flesh:** Firm to soft. Whitish, slowly staining brownish when exposed. **Odor:** Indistinct. **Taste:** Saccharin sweet. **KOH:** Bright red on exterior. **Spore deposit:** Whitish. **Microscopy:** Spores 9.5–13.5 × 5.5–7 µm, broadly ellipsoid, ovoid to almond-shaped, thin-walled, smooth, inamyloid.

Ecology: Solitary or scattered in troops in duff or moss. Widespread but rarely in large numbers. Fruiting in fall.

Edibility: Edible and surprisingly sweet. It's worth trying in moderation but there are reports of upset stomachs from eating large amounts.

Comments: The wrinkled, club-shaped fruitbody, flattened orange top, paler sides, and sweet taste help define this species. *C. pallidoincarnatus* has more of a baseball-bat shape, with an orange top and paler sides. It tends to fade quicker and appears to be restricted to coastal spruce forests. *C. aurantiacus* has smaller, orange, club-shaped fruitbodies. Genetic analysis has shown multiple undescribed species of *Clavariadelphus* in Cascadia, some of which superficially resemble *C. truncatus*. *C. occidentalis* is more irregularly shaped, has duller colors, and a bitter taste.

Trichoglossum hirsutum

(Pers.) Boud.

Velvety-black Earthtongue

Fruitbody: 1–3 cm tall, 0.2–1.5 cm across, sometimes torpedo-shaped when young, but usually slightly flattened or compressed, becoming more elongate and spatula- or rounded-spear–shaped in age. Surface dry, finely covered with projecting setae. Black to dark brownish black. **Stipe:** 2–6 cm long, 0.2–0.5 cm thick, cylindrical, straight. Surface dry, finely velvety. Black to dark olive-black or brownish black. **Flesh:** Thin, tough, brownish black. **Odor:** Indistinct. **Taste:** Indistinct. **Microscopy:** Spores 80–160 × 5–7 µm, very long and narrow, often with a swollen middle and tapering ends, smooth, 15-septate when mature. Asci 8-spored. Setae abundant, needle-like.

Ecology: Solitary, scattered, or in small troops in moss or duff in mixed forest throughout Cascadia. Common but easily overlooked. Fruiting from late fall into spring.

Edibility: Unknown.

Comments: Of the many different species of black earthtongues in Cascadia, *T. hirsutum* appears to be our most common. *Trichoglossum* can be separated from *Geoglossum* and *Glutinoglossum* macroscopically by the velvety stipes and finely hairy heads, due to the presence of projecting needle-shaped cells called setae. Beyond that, microscopic examination is needed to confirm species identification. *T. hirsutum* is distinct by having long, 15-septate spores in 8-spored asci. *T. velutipes* has 4-spored asci and up to 13-septate spores, with most being 9-septate. *Geoglossum* are a bit trickier to identify; generally they are more irregular in shape with a less distinctly defined fertile head. All will require microscopic examination for species-level identification. *G. umbratile* has blackish brown to black clubs which become slightly irregularly shaped in age. It has straight to curved, 7-septate spores, measuring 30–90 × 4.5–6.5 µm. *G. glabrum* is almost identical but has paraphyses that are more closely septate and generally longer than the asci. *G. sphagnophilum* is very similar to the latter two but grows in sphagnum bogs. *G. fallax* is a dark brown to blackish brown species with spores that are generally colorless, measuring 45–110 × 5–7 µm, and 7- to 12-septate when mature. We probably have additional species but, to our knowledge, nobody has intensively studied Cascadia's earth tongues. *Glutinoglossum* (=*Geoglossum*) *glutinosum* has a distinctly viscid stipe and head when wet and dark brown to blackish colors. *Hypomyces papulasporae* grows on the heads of *Trichoglossum* and *Geoglossum* fruitbodies, covering them with a white mold that develops pimple-like perithecia when mature; it is a rare species. Two greenish earth tongues occur in California and may occur in Cascadia as well. *Microglossum viride* has gorgeous teal-green colors and a green-dotted stipe surface, while *M. olivaceum* has dull olive-green to tan colors and a smooth stipe. Both of these names are misapplied, and these taxa need their own names.

Tolypocladium ophioglossoides

(J.F. Gmel.) C.A. Quandt, Kepler & Spatafora

Snaketongue Truffleclub

Fruitbody: 3–10 cm tall. **Head:** 0.5–4 cm high, 0.3–1 cm across, cylindrical to club-shaped. Surface dry, covered with tiny bumps, occasionally frosted with white powder. Dark golden brown, olive-brown, dark brown to nearly black. **Stipe:** 2.5–8 cm long, 0.3–1 cm thick, more or less equal above ground. Below ground with thick rhizomorphs rooting from truffle host. Surface dry, smooth or scurfy. Color variable, golden, dark olive-brown to nearly black, or any combination thereof. Base and 'root' bright golden yellow. **Flesh:** Firm, fibrous; golden. **Odor:** Indistinct. **Taste:** Indistinct. **Spore deposit:** Buff to light brown. **Microscopy:** Spores thread-like but breaking up into small ellipsoid-cylindrical segments 2.5–5 × 2–2.5 µm. Asci 8-spored, cylindrical, inamyloid.

Ecology: Solitary or in small clusters, arising from an underground *Elaphomyces* truffle. Generally one or two clubs are attached to each truffle but larger clusters are occasionally found. Uncommon, occurring across western Cascadia, typically in hemlock forests. Fruiting from late summer into winter.

Edible: Unknown.

Comments: The cylindrical, dark golden brown to nearly black head covered in tiny pimples, olive to blackish stipe, golden root, and growth on *Elaphomyces* truffles make for a distinctive combination. The presence of the host truffle is very easy to miss since the stipe base is often broken off when collected. The Drumstick Truffleclub, *T. capitata*, is larger, with a round to ovoid, amber-colored head, and yellow stipe. It occurs in similar habitats across western Cascadia, also growing on *Elaphomyces*. *Ophiocordyceps* "*satanas*" nom. prov. also forms dark, cylindrical clubs but grows directly from the larvae of June beetles buried in coastal sand dunes.

Synonym: *Elaphocordyceps ophioglossoides* (J.F. Gmel.) G.H. Sung, J.M. Sung & Spatafora, *Cordyceps ophioglossoides* (J.F. Gmel.) Fr.

Xylaria hypoxylon

(L.) Grev.

Carbon Antlers, Candlesnuff Fungus

Fruitbody: 2–8 (12) cm tall, 0.1–0.7 cm across, cylindrical at first, soon forking and branching at tips or becoming antler-like. Black on lower part, powdery, whitish on upper portion (rarely entirely) at first, becoming more grayish on tips, to all black and pimply in age. **Flesh:** Thin, tough, wiry. Whitish to pale gray at first, black in age. **Odor:** Indistinct. **Taste:** Not sampled. **Microscopy:** Ascospores 11–14 × 5–6 µm, bean-shaped, smooth, brownish black. Asexual spores ellipsoid, smooth, colorless.

Ecology: In troops or loose clusters, occasionally solitary, on stumps, logs, and other woody debris of hardwoods. Very common throughout Cascadia, especially on alder. Fruitbodies are persistent and are found throughout the year.

Edibility: Unknown.

Comments: Shape alone is generally enough to recognize this tough, wiry, and often antler-shaped fungus. The pale, powdery tips and growth on wood are also helpful. The powdery whitish spores produced near the tips are asexual, while the sexual spores are produced in the pimple-like perithecia embedded in the surface lower down on the fruitbody. Thicker, less-branched, and/or more pimply *Xylaria* fruitbodies are often encountered and it is very likely that we are misapplying this name to some of the variants in our area, which probably represent a handful of species.

20 - Puffballs, Earthballs, Earthstars, Bird's Nest Fungi, and Stinkhorns

Although all the species included in this section are basidiomycetes, they don't forcibly discharge spores from their basidia as do most other mushrooms. Instead, they depend on other vectors to get their spores out into the world. The interior spore masses of puffballs and earthballs become powdery at maturity, full of hydrophobic spores. These spores are then slowly dispersed by the energy of rain droplets and wind (or kids jumping on them). The stinkhorns have gooey, stinky spore slime and are dispersed by insects. The Bird's Nest fungi disperse their spores in the form of "eggs," which are lentil shaped packets that are splashed out of their cups by falling rain droplets. The Cannonball fungi (*Sphaerobolus*) use hydrostatic pressure to launch a packet of spores out of their cup-like fruitbodies.

This chapter comprises many different evolutionary lineages. *Astraeus* and *Geastrum*, despite their similar fruitbody morphologies, are not at all closely related! Perhaps even more surprisingly, Bird's Nest fungi are close relatives of gilled mushrooms! *Astraeus*, *Pisolithus*, and *Scleroderma* are mycorrhizal, and these three are fairly closely related to boletes. The rest of the species included here are saprotrophic, decaying duff, woody debris, or (rarely) herbivore dung.

Although many species are identified by details of fruitbody structure, coloration, and substrate, reliable identification sometimes requires microscopic inspection. The size and structure of the spores are particularly important, as are the structures of the internal hyphae (especially for the puffballs). Identification of stinkhorns is fairly straightforward but they are rare in Cascadia.

Bovista plumbea

Pers.

Leaden Puffball

Fruitbody: 1.5–4 cm across, round to cushion-shaped, pinched at base and often with a tuft of whitish mycelium. The outer surface is felty and whitish when young, soon splitting and peeling off in patches to expose the smooth inner skin, which is whitish when young, soon becoming yellowish gray to lead gray, and maturing to brown. Forming a small tear or hole at disc when mature to release spores. **Interior:** Whitish at first, soon light and soft with a dark brown spore mass when mature. Sterile base absent. **Odor:** Indistinct. **Taste:** Indistinct. **Microscopy:** Spores 5–6.5 × 4–4.5 µm, ovoid, thick-walled, brown, smooth to finely roughened.

Ecology: Solitary or scattered in grassy areas or in unenriched grasslands and meadows. Widespread and locally common. Fruiting throughout the warmer months, especially in irrigated areas.

Edibility: Like all puffballs, edible when young and white inside, but the small size makes it difficult to find enough for a meal.

Comments: *B. plumbea* is one of many small puffballs in Cascadia, most of which need microscopic examination for identification. As a genus, *Bovista* is recognized by their small to medium size, white colors when young, and smooth, felty to finely spiny exterior skin which peels off or disintegrates to expose a smooth inner skin. The interior is white and firm to start (typically lacking a well-defined sterile base), soon yellowing and then maturing to a yellow-brown to dark brown powdery mass. The small size, felty skin when young, and lack of a basal cord help define *B. plumbea*. *B. aestivalis* is very similar but has a more granulose-felty outer surface when young, yellower colors in age, a basal cord, and spores with a short stalk-like attachment. *B. pila* is larger (4–7 cm), with a white mycelial cord at the base and mature fruitbodies that are shiny, very lightweight, and often blow around in the wind. *B. californica* is a tiny species (0.5–1.5 cm) with a white, flocculose exterior surrounding a thin, dull brown interior skin. It is rare, occurring in high elevation meadows, often near wet areas. *Lycoperdon* species often have more pear-shaped fruitbodies with a distinct sterile base.

Calvatia fumosa

Zeller

Smoky Puffball

Fruitbody: 2–8 cm across, round to cushion-shaped; base typically rounded, with a white mycelial cord. Surface smooth or with a few large cracks, or numerous small cracks in dry weather. White at first, becoming blotchy grayish to grayish brown in age, typically remaining whitish in cracks. **Interior:** Very firm and white at first, with a thick rind-like skin, becoming mushy and yellowish olive to powdery and brown when spores mature. Skin remaining firm, whitish in cross section. Sterile base absent. **Odor:** Faint when young, but soon pungent: a mix of rotten eggs, propane gas and a dirty outhouse. **Taste:** Disagreeable. Slightly bitter when young, (we have not sampled mature specimens). **Microscopy:** Spores 5–7.5 µm, globose, spiny to roughened.

Ecology: Solitary or scattered in troops, often partially buried in duff under montane conifers. Common at higher elevations and in east-side forests of the Cascade Range. Fruiting in spring and summer.

Edibility: Unknown. The wretched odor discourages investigation.

Comments: This common puffball fruits in spring and summer in drier, forested habitats at higher elevations. It is easily recognized by the white to grayish colors, thick, rind-like skin, and foul odor. *Bovista* species have lightweight fruitbodies with a thin skin. *Lycoperdon 'vernimontanum'* (=*Calvatia lycoperdoides* A.H. Sm., an invalid name) is another common springtime puffball in the eastern Cascades. It has more of a cushion to pear-shaped fruitbody with small scales and fine spines, a thin skin, and a distinct sterile base. The toxic earthballs (*Scleroderma* species) have rind-like skin with a dense purplish interior (which can be whitish when very young).

Taxonomic note: This species has been placed in the genera *Gastropila* and *Handkea*, but based on genetic data, it appears to belong in the genus *Lycoperdon.*

Calbovista subsculpta

Morse ex M.T. Seidl

Lumpy Puffball

Fruitbody: 5–15 cm across, 4–10 cm tall, rounded, cushion- to top-shaped, pinched at base to a short, rudimentary stipe. Thick outer skin white to creamy at first, breaking up like cracked mud into blunt or somewhat pointed pyramidal scales; taller on top, shorter over the sides and below. Soon yellowish, ocher to brownish, with darker colors in cracks, becoming brownish in age, with slightly paler scale remnants. Top disintegrating in age, exposing the spore mass. **Interior**: Firm, with a thick skin; white when young, becoming mushy and yellowish, then powdery, with brown spores, with a thin, fragile skin when mature. Sterile base well defined, leaving a rooting, bowl-shaped disc when spores disperse. **Odor**: Indistinct. **Taste**: Fungoid. **Microscopy**: Spores 3–5 µm across, globose, brown smooth to finely roughened.

Ecology: Solitary or scattered, occasionally in troops, often in open forested habitats on the east side of the Cascade Range, or in high elevation meadows or disturbed areas. Common and widespread across eastern Cascadia but rarely abundant. Fruiting in spring and early summer.

Edibility: Edible when young and pure white inside.

Comments: The medium size (typically about the size of a softball), blunt pyramidal patches, and progression from white to yellowish ocher to brownish with darker cracks help identify this species. *Calvatia sculpta* has longer pyramidal warts with curved and often pointed tips (like meringue), white to creamy colors which don't darken as drastically, and yellow-brown to brownish spores. Genetic data suggest both species belong in the genus *Lycoperdon. Calvatia booniana* is typically larger and has skin which cracks like dried mud with flattened scales.

Calvatia booniana

A.H. Sm.

Western Giant Puffball

Fruitbody: 20–50 cm across, 10–25 cm tall, round to cushion-shaped or lobed, base slightly pinched and pointed downward, often with short mycelial cords. Surface smooth or with scattered cracks when young, soon like cracked mud, breaking apart and exposing the spores, leaving a shallow cup-shaped base in age. White at first, becoming whitish, yellowish, then finally brown when mature. **Interior**: Very firm, with a thick, rind-like skin and white at first, soon soft, yellowish, with a thinner skin, then powdery and brown when mature. **Odor**: Indistinct or fungoid. **Taste**: Indistinct. **Microscopy**: Spores 4–5 × 4–4.5 µm, globose to subglobose, smooth or finely dotted, thin-walled.

Ecology: Solitary or scattered on ground in open forests, grasslands, or sagebrush scrub. Common in drier east-side habitats, rare to occasional in the Willamette Valley or other rain-shadow areas on the west side. Absent from wetter areas. Fruiting in late spring and summer.

Edibility: Edible and very good. Only eat specimens with a firm, pure white interior.

Comments: Easily recognizable by the large to downright massive size, skin which cracks like dried mud, brown mature spores, and growth in drier habitats. Fruitbodies of *C. pachyderma* are comparatively modest in size, have a thicker skin, and smooth spores which are golden brown to olive-brown in mass. It is rare in Cascadia, more common to the south. *Mycenastrum corium* is a rare species with a felty outer layer and a thicker, tough inner skin, a mass of mycelial threads at the base, and a dark brown to reddish brown spore mass. It has larger spores (8–12 µm across) with spines and the hyphae, in the spore mass, look like barbed wire. *C. cyathiformis* and *C. fragilis* are medium-sized species with purplish spore masses and thin skin that cracks into patches. They are difficult to distinguish from each other but *C. cyathiformis* tends to be larger and has a more distinct sterile base. Also see *Calbovista subsculpta* for other warted species.

Lycoperdon perlatum

Pers.

Gem-studded Puffball

Fruitbody: 1–5 cm across, 1–7 (12) cm tall, pear-shaped to oblong, with an oval or round top and an irregularly cylindrical stipe ranging from slender and elongate to occasionally bulbous. Surface dry, covered with small, pointed, pyramidal warts or spines and granular particles that leave oval scars or depressions when they fall off. Skin thin when mature, cracking and tearing at apex. Creamy white, pinkish gray to pinkish buff or light brown, with slightly darker spines, overall golden brown to brown when old. **Interior:** Firm and white at first, soon soft, becoming mushy, yellowish olive, then olive-brown to brown and powdery when spores mature. With a pronounced sterile base. **Odor:** Indistinct. **Taste:** Mild. **KOH:** Forest green on mature spore mass. **Microscopy:** Spores 2.5–5 µm across, globose, thick-walled, spiny.

Ecology: Solitary, scattered, or in small to large clusters. It has a preference for disturbed ground such as roadsides and trail edges but also commonly occurs in duff, humus, or well decayed woody debris in forests. Widespread and common. Fruiting in fall, old fruitbodies somewhat persistent.

Edibility: Edible. Eat only specimens that are pure white inside.

Comments: The pear-shaped fruitbody and small, pyramidal spines help distinguish this species. When the larger spines fall off they leave depressed oval scars, which helps distinguish this species from *Apioperdon pyriforme*, which is typically smoother, grows in dense clusters on wood, and has white rhizomorphs at the base. *L. pratense* is a small grassland species with a fine covering of spines and granules. *L. molle* is usually more grayish to grayish brown and has short, granular spines that often are fused together at the tips. *L. utriforme* is considerably larger and has a granular surface. *L. curtisii* is distinctly spiny, with the tips of the spines curved and joined, creating a star-like pattern. Darker specimens that share the pear-shape of this species should be compared with *L. nigrescens.*

Apioperdon pyriforme

(Schaeff.) Vizzini

Pear-shaped Puffballs

Fruitbody: 1–3.5 cm across, 2–6 cm tall, pear-shaped to oblong, often with a round top, and a pinched, irregularly pleated stipe with a tuft of white rhizomes at the base. Surface dry, granular to powdery at first, soon smooth or nearly so. Skin thin, cracking and tearing at apex when mature. Whitish, creamy beige, light tan to clay brown, often with a darker brown, obscure "nipple" at the apex. Overall golden brown to brown when old. **Interior:** Firm and white at first, soon soft and mushy, yellowish to yellowish olive, finally powdery olive-brown to brown as spores mature. Sterile base pronounced. **Odor:** Indistinct. **Taste:** Mild (when young). **KOH:** Forest green on mature spores. **Microscopy:** Spores 2.5–4 × 2.5–3.5 µm, globose to subglobose, thick-walled, roughened to smooth.

Ecology: Often in large clusters, occasionally scattered or in small clusters (rarely solitary) on rotting wood, logs, or stumps (sometimes buried) of hardwoods, conifers, and occasionally wood chips, rarely from soil. Widespread, often abundant, especially in northern Cascadia. Fruiting in fall, with smaller flushes in summer and early winter. Remnants of fruitbodies somewhat persistent.

Edibility: Edible and one of the better-tasting small puffballs. Eat only those that are completely white inside—specimens with spores starting to mature (yellowish to olive tones) may cause gastrointestinal upsets.

Comments: The pear-shaped fruitbody, whitish beige to warm brown colors, granular to smooth exterior, conspicuous white rhizomes, and clustered growth on wood are distinctive. *L. molle* is grayish to grayish brown and has very fine dark spines that are often fused together at the tips; it is uncommon in Cascadia, fruiting from duff or soil. *L. umbrinum* has blackish brown to golden colors and short spines; it grows singly or scattered in conifer duff. *L. perlatum* is covered with pyramidal studs or warts when young but these often scuff off and leave round scars in age.

Synonym: *Lycoperdon pyriforme* Schaeff.

Lycoperdon nigrescens

Pers.

Dark Spiny Puffball

Fruitbody: 1–4 cm across, 3–5 cm tall, pear-shaped overall, with an oval or round top, and an irregularly cylindrical or tapered stipe. Surface dry, covered with pointed, pyramidal warts or spines and granular particles that leave oval scars or depressions when they fall off. Skin thin when mature, cracking and tearing at apex. Dark brown to blackish when young and covered in spines, tan to light brown under scales when young, becoming more golden brown to dull brown in age. **Interior:** Firm and white at first, soon soft, becoming yellowish to yellowish olive and mushy, finally brown and powdery when spores mature. Sterile base pronounced, white when young, olive-brown when mature. **Odor:** Indistinct. **Taste:** Mild. **KOH:** Forest green on mature spores. **Microscopy:** Spores 3.5–5 µm across, globose, thick-walled, finely warted to nearly smooth.

Ecology: Solitary, scattered, or occasionally in small clusters in duff or moss in forest habitats. Widespread throughout Cascadia but rarely abundant. Fruiting in fall.

Edibility: Likely non-toxic.

Comments: The dark brown to blackish colors, pyramidal warts or spines which leave a pale scar when they fall off, and scattered growth in duff help identify this species. Another dark puffball, *L. umbrinum*, has short spines that leave only a slight scar when they fall off. It occurs in similar habitats across western Cascadia. *L. molle* is grayish to grayish brown and has very fine, dark spines that often are fused at the tips. *A. pyriforme* grows on rotting wood and has granular to smooth skin and conspicuous white rhizomes at the base.

Pisolithus arhizus group

Dyer's Puffballs

Fruitbody: 4–12 (20) cm across, 6–30 cm tall, rounded-cylindrical or top-shaped, occasionally cushion shaped; often with a distinct rooting stipe. Disintegrating and becoming quite irregular, turning into a powdery mass from top down. Variably colored, depending on age; beige, tan, grayish brown, orange-brown, or dull brown. Young fruitbodies are occasionally bright golden-metallic. **Interior:** Firm at first, slowly becoming crumbly-powdery in age. Made up of multicolored (gold, yellow, whitish, vinaceous, black, red, reddish brown, brown or tan) peridioles (small, oblong, discrete spore packets 0.2–0.7 cm long) embedded in a greasy black matrix when young. Each one disintegrates into a powdery reddish brown, rusty brown, or dull brown mass of spores; this maturation progresses downward through the peridiole layers. **Odor:** Mild to unpleasant. **Taste:** Not sampled. **Microscopy:** Spores (6) 7–11 (12) μm, (excluding spines, which are around 1 μm long), globose, thick-walled, distinctly knobby-spiny, and hydrophobic.

Ecology: Solitary, scattered, or in small clusters, often in nutrient-poor soil in disturbed areas, roadsides, or under (young) planted trees. It is associated with a large variety of trees, mainly birch and oak in urban areas and pines in natural settings. Occasional to locally common, often in drier portions of Cascadia, or in urban areas. First fruiting in summer or early fall, even if no rain has fallen for months! Fruitbodies persisting into late fall or early winter.

Edibility: Inedible due (at least) to texture. The powdery spore mass is extremely hydrophobic and can cause fits of sneezing.

Comments: This oddball is easily recognized by its unattractive exterior appearance (it is also called Dead Man's Foot and Dog Turd Fungus), the interior composed of beautifully multicolored peridioles (visible in cross sections of young fruitbodies) which mature into a powdery brown mass, and growth in nutrient-poor soil and disturbed areas. It is highly prized by fiber artists because it yields lightfast dyes on wool and silk, with colors ranging from golden to reddish to deep dark brown. There is a common western North American species, which remains undescribed, but it's likely there are other cryptic species as well. *Eucalyptus* trees often support species introduced from Australia, where *Pisolithus* is quite diverse. *Battarrea phalloides* forms a powdery rust-colored drumstick perched on a tall, shaggy-woody stipe which arises from a whitish egg of universal veil tissue.

Scleroderma bovista

Fr.

Smooth Earthball

Fruitbody: 2–5 (8) cm across, 1–4 cm tall, round, cushion- to top-shaped, often with a large tuft of rhizomorphs and mycelium as base, with no distinct stipe. Surface dry, smooth or nearly so, at times developing shallow cracks or small scales; rupturing at top to expose spores when mature. Dingy beige to yellowish brown at first, more grayish brown in age. **Interior:** Hard and whitish when very young, soon dark purplish brown or purplish black with paler marbling, becoming more powdery and purplish black when the spores mature. Flesh of skin whitish, not staining. **Odor:** Mild to unpleasant. **Taste:** Indistinct. **Microscopy:** Spores 14–18 µm including the spines, globose, long-spiny with a reticulum.

Ecology: Solitary, scattered, or in small clusters on the ground, often in disturbed or urban areas. This species was unintentionally introduced from Europe and is now primarily found with birch, linden, and other planted hardwoods. Occasional to locally common. Fruiting in summer in irrigated areas, widespread in fall.

Edibility: Toxic, causing severe gastrointestinal distress.

Comments: All *Scleroderma* have a thick, tough skin, a very firm texture when young, purplish black spore masses, and often extensive rhizomorphs at the base. All species are toxic and especially dangerous (sometimes even fatal) for dogs, which seem to like grazing on them. There are a number of species in Cascadia, many of which are introduced and grow with planted trees. *S. bovista* is identified by its relatively smooth skin, unstaining flesh, lack of a stipe, and reticulate-spiny spores. *S. hypogaeum* also has a smooth, thick skin, but generally grows partially buried in soil under conifers and has large (15–25 µm), reticulate-spiny spores. *S. cepa* is quite similar but tends to exhibit pinkish staining flesh and has smaller (8–13 µm), spiny spores. *S. verrucosum*, has straw yellow skin with darker, wart-like patches and a slightly more pear-shaped stature, with a distinct stipe. Species in the *S. areolatum* group have thick, cracking skin and spiny spores. *S. polyrhizum* (sensu PNW) has large fruitbodies with a thick skin which open up into a star-like shape when mature, and small (7–11 µm) spores; it is common in sandy pine forests.

Astraeus pteridis

(Shear) Zeller

Quilted Barometer Earthstar

Fruitbody: 7–12 (15) cm across when open, rounded to cushion-shaped at first, outer skin soon splitting and opening up into star-like rays, pushing up the rounded puffball held inside. Rays hygroscopic, opening in wet weather and closing when dry. Surface of rays with a thick, whitish to creamy beige upper layer that becomes cracked like dried mud, exposing brown context between the cracks, becoming darker overall in age. Exterior light to dark brown. Puffball portion thin-skinned, pale gray, light to dark brown, or whitish overall in age, developing a small tear at the top to release spores when mature. Spore mass firm and whitish when young (and still enclosed) soon brown and powdery; thin skin soon tearing at top, often overall disintegrating in age (but the tough supporting arms will persist). **Odor:** Indistinct. **Taste:** Not sampled. **Microscopy:** Spores 8–12 µm, globose, thick-walled, finely spiny-warted.

Ecology: Solitary, scattered, or in large troops on ground under conifers. Often buried in duff before erupting and opening. Mostly absent along the immediate coast but widespread (but), to uncommon to occasional inland. We have seen abundant fruitings on burn sites in west-side forests two years after a forest fire. First fruiting in late summer or fall but rarely encountered until maturity. Expanded rays that have lost their spore sac are persistent, lasting a year or more.

Edibility: Unknown, likely toxic.

Comments: The thick, hygroscopic rays with a pale, cracked surface, puffball-like inner structure, often large size and growth with conifers are helpful in distinguishing this species. A similar, but unnamed species occurs under oaks. It differs by having a thinner, paler grayish to brownish upper surface on the rays, and smaller fruitbodies (<10 cm across). *Scleroderma polyrhizum* forms large 'earthballs' which crack open in a star-like patter to expose the spore mass (which lacks the skin around the spores). *Geastrum* species are smaller and have thinner rays that disintegrate with the spore sac. Dry, "clenched" *Astraeus* can be placed in a bowl with a wet paper towel or small amount water, and the arms will slowly open. If your friends each bring their own, you can place bets on whose fruitbody will continue expanding after the most wet-dry cycles (it can take a few weeks to definitively determine the victor).

Geastrum saccatum

Fr.

Bowl Earthstar

Fruitbody: 3–7 cm across, fig-shaped, and orange-brown to grayish brown when young, then cracking and opening up into a star-like disc with 4–9 pale grayish rays that often curve downward and curl inward. Puffball-like interior structure lacking any stipe and nestled directly into the bowl-like center; pale brown with a darker, pointed mouth which is often demarcated by a pale ring. **Interior:** Firm and white at first (but rarely observed at this stage), soon soft and powdery as the spore mass matures, becoming chocolate-brown. **Odor:** Indistinct. **Taste:** Indistinct. **Microscopy:** Spores 3.5–5 µm, globose, warted.

Ecology: Solitary, scattered, or occasionally in small clusters in duff or on ground under conifers. Widespread but rarely abundant. Fruiting in fall and early winter but fruitbodies can persist throughout the year.

Edibility: Unknown.

Comments: Earthstars are recognized by their structure: A puffball with a pointed mouth nestled in a bowl-like depression atop curled "arms" or rays which are not hygroscopic (opening only once). There are a number of similar species in Cascadia but most are going by European names, and there has been no critical work to compare them to specimens from western North America. *G. fimbriatum* doesn't have the pale line surrounding the dark mouth of its puffball and has smaller spores (3–3.5 µm). *G. lageniforme* has longitudinal cracks on the underside of the rays but is otherwise very similar. *G. rufescens* has darker brick-pink rays and a distinctly felty underside when young. *G. coronatum* often grows in duff of cypresses or cedars and has flat rays that do not form a bowl-like sac around the puffball. *G. floriforme* is smaller, has hygroscopic rays (expanding and contracting in wet and dry conditions), and often grows in sandy areas or nutrient-poor soil. *G. triplex* has a distinct collar between the rays and the puffball. *G. fornicatum* has a three-tiered fruitbody; a basal cup, a "star" of dramatically upward-thrusted rays, and a small puffball at the top. *G. quadrifidum* is a grayish species with a distinct stipe that elevates the puffball. *G. fuscoglebum* (=*Radiigera fuscogleba)* could be mistaken for an unopened *Geastrum* but is more cushion shaped and has a small, cylindrical stipe inside. *Astraeus* species have thicker skin, often darker colors, and have hygroscopic rays and their spores are released through a simple tear at the top of the puffball.

Tulostoma species

Desert Stalked Puffballs

Fruitbody: 'Puffball' portion (0.5) 1–3 cm across, rounded to cushion-shaped, with a small chimney-like opening at top when mature. Surface dry, typically with a granular or spiny coating which scruffs off to expose a smooth surface beneath. Brownish, orangish, grayish or cream at first, then duller, and often bleached in age. **Interior:** Firm and whitish when young, soon powdery, ochraceous, cinnamon to brown. **Stipe:** 2–5 (7) cm long, 0.3–0.7 cm thick, equal, enlarged lower, or bulbous at base, often partially to completely buried in soil, and binding sand or soil. Tough, fibrous or woody. Surface dry, often fibrillose. Typically concolorous with the puffball when young, but not fading as much in age. **Odor:** Indistinct. **Taste:** Indistinct. **Microscopy:** Spores ~4–8 (12) µm, mostly round to broadly ellipsoid, some pear-shaped, smooth, finely spiny to distinctly warted.

Ecology: Solitary or scattered, rarely in large numbers. Most occur in sand or gravely soils in drier habitats, at times with just the puffball portion above ground. Fruiting soon after rainy periods but fruitbodies are long lasting and can be found at any time of year.

Edibility: Toxicity unknown but rather small, woody, and bizarre for culinary consideration.

Comments: *Tulostoma* are a diverse group of desert-adapted species with tall, slender stipes elevating their puffballs. Species diversity of this genus peaks in the southwestern United States but over thirty species have been reported from the northwest; mostly from eastern Cascadia and eastward into the high deserts. Recognizing them as a genus is easy but putting confident species-level names on them is downright impossible. A worldwide monograph of the group was published in the late 1980's but represented limited collecting in Cascadia and didn't have the benefit of modern taxonomic approaches – an expanded revision is sorely needed. *Battarrea phalloides* is like a *Tulostoma* on steroids—fruitbodies a half meter tall are not uncommon! They have a drumstick-like head which is covered in pallid papery tissue which soon falls off to reveal a mass of powdery rust-colored spores. This is perched on top of a tall, shaggy-woody stipe, which itself arises from a whitish volva sac.

Crucibulum crucibuliforme

sensu PNW

White-egg Bird's Nest

Nest: 0.5–1 cm across, 0.3–1.2 cm tall, rounded at first, soon cylindrical with a cushion-shaped, coarsely hairy to smooth lid, colored yellowish to cinnamon brown. Cup-shaped after lid detaches. **Outer surface:** Yellowish, ocher to cinnamon-brown, often with paler velvety hairs on lower portion. **Inner surface:** Smooth, shiny, whitish, pale gray to beige. **Eggs:** Numerous, small, 0.1–0.2 cm across, flattened circular discs often embedded in a gel-like substance when young but soon dry. Whitish. With a very thin, white, elastic cord. **Microscopy:** Spores 4.5–10 × 3.5–6 µm, ellipsoid, smooth, held inside the mature eggs.

Ecology: Solitary, scattered, or in large troops on small woody debris, wood chips or herbaceous stems, rarely on rich compost. Very common, occurring throughout Cascadia. Fruiting in fall and early winter but persisting through the year.

Edibility: Unknown.

Comments: This bird's nest fungus is recognized by the yellowish nest covered in coarse to velvety hairs when young and the whitish eggs attached to a thin cord. *Nidula* species are generally quite hairy on the exterior but have grayish to reddish brown eggs that are not attached to a cord. *Cyathus* species have darker eggs.

Synonym: Better known as *Crucibulum laeve* (Huds.) Kambly, which is an illegitimate name. There appears to be a number of similar unnamed species of White-egg Bird's Nest fungi around the world.

Cyathus striatus

Willd.

Fluted Bird's Nest

Nest: 0.5–1 (1.5) cm across, 0.7–1.5 (2) cm tall, cylindrical, or with a narrowed base at first, vase-shaped, or like an upside-down bell when the lid comes off. Lid with two layers: underlayer a thin white membrane covered with tufts and clumps of brown hairs which soon scuff off and expose the membrane. **Outer surface:** Brown, chocolate-brown, gray-brown to cinnamon, densely covered with tufts and scale-like hairs, more coarsely hairy in age. **Inner surface:** Strikingly grooved or pleated with dark parallel lines, dull grayish to tan at first, becoming metallic silvery gray to grayish brown or dark brown in age. **Eggs:** 4–12 (generally around 6), 0.1–0.3 cm across, tan, brown to gray-brown, covered with pale bloom, darker when handled or in age. Attached by a thin cord. **Microscopy:** Spores 18–20 × 8–10 µm, ellipsoid, smooth, held inside the mature eggs.

Ecology: Scattered or in large troops (rarely solitary) on woody debris. Especially common on wood chips and largely restricted to landscaped and urban areas. Introduced to Cascadia, now common and at times abundant in western Washington and central Oregon, spreading southward. Fruiting throughout the year but peaking in late spring and fall. Nests can persist for a long time.

Edibility: Unknown.

Comments: The brown, hairy nest with a grooved interior, tan to gray-brown eggs and growth from wood chips and other woody debris readily identify this species. *Nidula* have paler velvety-tomentose exteriors and smaller eggs embedded in a gel-like matrix well into maturity.

Cyathus olla

(Batsch) Pers.

Gray-egg Bird's Nest

Nest: 0.8–1.5 cm across, 0.7–2 cm tall, rounded at first, soon cylindrical with a flaring top and a tapering, pointed base. Often broader and more bowl shaped after the lid comes off, with a flaring, wavy margin (occasionally splitting). Lid thin, finely silky-tomentose to smooth, whitish, pale gray to light brown. **Outer surface:** Gray to brown, finely tomentose at first, becoming more coarsely hairy in age. **Inner surface:** Smooth, shiny, gray to grayish brown. **Eggs:** 1–10 (generally fewer than 5), 0.3–0.5 cm across, light to dark gray, plump or flattened discs, often embedded in a gel when young, soon dry, attached by a thin cord. **Microscopy:** Spores 10–14 × 6–8 µm, broadly ellipsoid to oval, smooth, held inside the mature eggs.

Ecology: Scattered or in large troops, occasionally solitary, on small woody debris, herbaceous stems, or rich soil. Especially common on wood chips and lignin-rich compost in garden beds. Occasional in urban setting throughout Cascadia, rare in natural areas. Fruiting in summer and fall, more rarely in spring. Nests can persist for a long time.

Edibility: Unknown.

Comments: The large size (relative to other bird's nest fungi), gray color of the eggs and flaring rim of the "nest" are distinctive. *C. stercoreus* has smaller, black eggs, often grows in compost or dung-enriched soil (rarely on wood chips), and has larger spores (20–40 × 18–30 µm). *C. pygmaeus* is a small (~0.5 cm across) dung-dwelling species with black eggs and small spores (12–14 × 8–9 µm). The nests of *C. striatus* have a distinctly lined or grooved interior and coarsely hairy exteriors. *Crucibulum crucibuliforme* has whitish eggs and *Nidula* have smaller grayish to reddish brown eggs.

Nidula candida

(Peck) V.S. White

Gray Woolly Bird's Nest

Nest: 0.3–1 cm across, 0.5–2 cm tall, cylindrical with a rounded top at first, vase-shaped after lid comes off, often with a narrow, flaring rim. Lid thick, coarsely hairy to scaly. Beige to light brown. **Outer surface:** Whitish, beige, or gray and velvety-tomentose at first, then dingy brown or paler to whitish and becoming more coarsely hairy in age. **Inner surface:** Smooth, shiny, whitish to buff, paler toward margin, lower parts light to dark brown or orangish brown. **Eggs:** Numerous, small, flattened or plump discs, 0.05–0.2 (0.3) cm across. Embedded in a gel when young, soon dry. Grayish to light brown. **Microscopy:** Spores 8–10 × 4–6 µm, ellipsoid, smooth, held inside the mature "eggs."

Ecology: Scattered or in troops on small woody debris or herbaceous stems. Especially common on alder branches but occurring on other hardwoods as well as conifer twigs and branches. Common and abundant in western Cascadia, occasional on the east side, but rarer in drier habitats. Fruiting in fall and winter but nests can persist throughout year.

Edibility: Unknown.

Comments: The fuzzy whitish to gray exterior, thick lid, and small, grayish to light brown eggs that are not attached to thin cords (but which are often embedded in a gel) are helpful features distinguishing this species from almost all other Bird's Nest fungi. *N. niveotomentosa* differs by its smaller size, more cylindrical shape when open, and bright white, fuzzy exterior. *Crucibulum crucibuliforme* has whitish eggs attached to thin cords and a coarsely hairy to smooth yellowish to yellowish brown exterior.

Nidula niveotomentosa

(Henn.) Lloyd

White Woolly Bird's Nest

Nest: 0.2–0.6 cm across, 0.3–1 cm tall, cylindrical with a rounded top at first, remaining cylindrical, or becoming narrowly vase-shaped after lid comes off, often with a narrow, flaring rim. Lid thick, finely hairy to scaly. Whitish to tan. **Outer surface:** Velvety-tomentose and bright white at first, often appearing like a collar or "scarf" of white hairs, becoming more coarsely hairy and whitish to tan in age. **Inner surface:** Smooth, shiny, whitish to buff, paler toward margin; lower parts light brown or orangish brown. **Eggs:** Numerous, small flattened to plump discs, 0.05–0.2 cm across. Embedded in a gel when young, soon dry. Light to dark reddish brown. **Microscopy:** Spores 7–8 × 5–6 µm, ellipsoid to nearly round, smooth, held inside the mature eggs.

Ecology: Scattered or in troops on small woody or herbaceous debris. Common and widespread but rarely as abundant as *N. candida.* Fruiting in fall and winter but nests can persist throughout the year.

Edibility: Unknown.

Comments: The white, fuzzy exterior, whitish to tan lid, and small, brown eggs that are not attached to a cord (but which are often embedded in a gel) are distinctive. *N. candida* differs by its larger size, more vase-like shape when open, and hairy, whitish to gray exterior. The flaring rim of *N. candida* is the best feature to distinguish bleached older specimens from *N. niveotomentosa. Crucibulum crucibuliforme* has whitish eggs attached to thin cords and a coarsely hairy to smooth, yellowish to yellowish brown exterior.

Nidularia pulvinata

(Schwein.) Fr.

Fruitbody: 0.3–0.7 (1) cm across, irregularly rounded to cushion-shaped at first, upper portion disintegrating, exposing a mass of eggs. **Outer surface:** Moist to dry, powdery to finely tomentose when young, smoother in age, often wrinkled when dry. Pale ocher-brown, cinnamon-brown to grayish brown. Inner surface smooth, shiny, brown. **Interior:** A mass of small (< 0.1 cm), grayish brown to brown, flattened to irregular disc-shaped eggs in a gel-like matrix. **Microscopy:** Spores 6–10 × 4–7 µm, ellipsoid, smooth, held inside the mature eggs.

Ecology: Solitary or in small scattered troops on small branches or other woody debris. Occasionally in large troops on wood chips. Widespread in Cascadia but uncommon; more likely to be found in wet, west-side forests. Fruiting in fall.

Edibility: Unknown.

Comments: This unusual species doesn't form the typical nest-like fruitbodies of other Bird's Nest fungi, instead producing a ball-like mass of eggs surrounded by a disintegrating skin. Young *Nidula candida* and *N. niveotomentosa* have distinctly tomentose exteriors and soon open into nest-like structures. There is considerable uncertainty regarding which species names to apply in Cascadia, with *N. pulvinata*, *N. deformis*, and *N. farcta* all being applied to local collections. No genetic studies have been done on Cascadian collections but studies done elsewhere show a few unnamed species and suggest the merger of *Nidularia* into the genus *Nidula.*

Sphaerobolus stellatus

Tode

Cannonball Fungus

Fruitbody: 0.2–0.3 cm across. Surface covered in small plaques or finely appressed-tomentose to smooth, creamy to yellowish ocher. Rounded at first, ejecting one yellowish or brownish "cannonball" full of spores, which rockets out at maturity, leaving behind a cup-like structure with a star-shaped margin, and a whitish to pale yellowish balloon-like sac; this deflates and exposes a whitish interior in age. **Flesh:** Very thin, somewhat rubbery-tough. **Odor:** Indistinct. **Taste:** Not sampled. **Spore deposit:** Dark brown, but not typically obtainable. **Microscopy:** Spores of two types: Sexual spores are 6.5–7.5 × 4.5–5 µm, ellipsoid to oblong, thick-walled, smooth, while asexual spores are longer, thin-walled.

Ecology: Solitary or scattered in troops on wood chips, stumps, small branches or other woody debris. Occasionally on herbaceous debris or dung. Common and widespread across western Cascadia but easily overlooked due to their small size. Mainly fruiting mainly in fall and winter but can be found at any time of year in irrigated areas.

Edibility: Unknown.

Comments: Most often noticed and easily recognized during the stage when they resemble orangey star-like cups with a white sac inside. This fascinating fungus can shoot the "cannonball" (peridiole) of spores up to six meters through the air! The launch mechanism is triggered by light. The presence of this fungus in urban settings is often first noticed when white walls or cars become splattered with its dark, sticky spore masses. This dispersal method is similar to that of Dung Cannons (*Pilobolus* species), which blast their spores away from their dung substrates and onto surrounding vegetation, to then be consumed, and passed through, by grazing animals, thus completing the cycle. Sequencing work has shown the occurrence of *S. iowensis* in Cascadia, but the two species are difficult to distinguish without genetic data.

Phallus hadriani

Vent.

Purple-egg Stinkhorn

Egg: Cylindrical or oval, tapered at one end. Containing a highly compressed version of the mature stinkhorn. **Cap**: 3–4 cm across, cylindrical when emerging, then egg-shaped to somewhat conical, composed of an olive spore mass spread over a whitish to creamy tripe-like or honeycombed surface. Spore mass at first dense, soon slimy and gooey (but still fairly cohesive). Tip with a pale, flattened and often slightly fringed and dimpled disc. **Stipe**: 7–15 cm long, 2–4 cm thick, cylindrical, soft, spongy, hollow. Base with rhizomorphs. Surface smooth at first but soon showing many small, porous, irregularly angular cells. Whitish to cream, becoming dingy yellowish in age. **Flesh**: Firm and rubbery but soon spongy; stipe with a gelatinous core when young, becoming hollow. Spore-bearing tissue of head thin and membranous. Whitish to creamy or yellowish, with lilac-magenta to purplish tones around edges in cross sections of "eggs." **Universal veil**: A thin, purple, lilac-pink, or magenta (occasionally more extensively white) membrane that surrounds entire fruitbody when young; remaining as a thin, membranous volva around base of stipe. **Odor**: Strong and distinct when young, like bean sprouts, funky buttermilk, or strongly spermatic, becoming fouler and more pervasive, like rotting cabbage in age. **Taste**: Mild, but the odor is difficult to ignore. **Microscopy**: Spores 3–4.5 × 1.5–2 µm, ellipsoid to somewhat cylindrical, smooth, olive.

Ecology: Solitary or scattered in clumps or troops in gardens, woodchip beds, or in other disturbed areas in urban settings, or among grass and debris in coastal sand dunes. Widespread but rare in Cascadia. Fruiting in fall and winter in natural settings, or during late spring and summer in irrigated areas.

Edibility: The young eggs are sometimes peeled, rinsed, and eaten. The gel can be an unpleasant textural hurdle (to say nothing of the smell). The foul odor is restricted to the spore mass in mature specimens and those can be washed off.

Comments: The distinctive phallic shape, honeycombed head, purple-skinned egg, and raunchy odor identify this species beyond much doubt. The rank odor of the slimy spore mass attracts insects, which then disperse the spores. *P. impudicus* is similar but has a white-skinned egg. *P. ravenellii* is also quite similar (also with a purple egg) but has a smooth cap membrane upon which the spore mass sits. Old stinkhorns that have had their spore mass washed off or dispersed are occasionally mistaken for morels (*Morchella* spp.) due to the underlying honeycomb structure. Dog Stinkhorns (*Mutinus* spp.) have pinkish, cylindrical fruitbodies and a tapered head covered in olive-brown spore slime. *Lysurus* are similar to *Mutinus* in color and structure, but the tips of the head come apart slightly to form a cross or star. They are rare in Cascadia, appearing in grass lawns and mulched beds during warm weather.

21 - Truffles

Although they belong to a wide range of evolutionary lineages, the species in this section are united by their *hypogeous* fruitbodies—that is, their spore-producing fruitbodies are formed below ground and usually greatly reduced (potato-shaped). This fruitbody morphology has independently evolved numerous times and is generally thought to be an adaptation to habitats that commonly experience rapid swings in humidity and temperature. By sheltering their fertile surfaces inside protective outer layers and fruiting belowground, where temperature and humidity remain more stable, these fungi are better insulated against harsh conditions that can cause species with more exposed fruitbodies to cease spore production. The tradeoff is that their spores no longer disperse as easily on wind currents; consequently, many truffles produce strong aromatic compounds that attract small mammals. Squirrels and voles dig up these fungi, eat them, and then disperse the fungal spores in their feces. Some of the more aromatic truffles in our area are considered good edibles but none have achieved the culinary esteem of the white and black *Tuber* truffles of Europe.

Identification (even to genus) is often guesswork until the fertile surface and spores are examined under a microscope. However with practice and careful attention to coloration, internal structure, odor, surface texture, bruising reactions, presence or absence of latex, and habitat it is possible to recognize many genera and some species.

A few of the species included here are "transitional" forms—their morphology appearing "in progress" along an evolutionary trajectory toward a hypogeous, truffle-like fruiting syndrome. These so-called secotioid, or sequestrate, mushrooms have fruitbodies with vestigial but recognizable gills and frequently have contorted, short stipes that barely elevate the fruitbody above the duff. Historically, such species were often given their own genera, but modern genetic work has indicated that most are situated solidly within the broader diversity of their sister genera that show typical gilled or pored morphology. Thus, the current trend for these truffle lineages is to lump genera, reduce the overall number of names in use, and consequently reduce the descriptive power of the existing names. For example *Russula*, *Macowanites*, and *Gymnomyces* previously referred to gilled, secotioid, and truffle-shaped mushrooms, respectively, are now in the single genus *Russula*. This makes some aspects of taxonomic communication more difficult; thus truffles provide a particularly concrete example of the inherent tension between the nomenclatural priorities of phylogenetic versus field-based natural history.

Chroogomphus albipes

(Zeller) Y.C. Li & Zhu L. Yang

Truffle Spike

Cap: (0.5) 1–5 cm across, rounded to convex, often irregularly lobed or flattened, margin irregular, connected to stipe by a fibrillose veil well into maturity, becoming free when old but remaining downcurved. Surface slightly viscid when wet, otherwise dry, appressed-fibrillose to nearly smooth. Ocher to pale orange, covered with grayish to brownish gray fibrils, becoming vinaceous to purplish red in age. **Interior:** Completely enclosed when young, slightly exposed in age, composed of highly convoluted, small loculate chambers. Ochraceous to pale orange when young, becoming darker, grayish orange to grayish or nearly black as spores mature. **Stipe:** 0.3–2 cm long, 0.5–1 cm thick, much reduced, equal or narrowed downward. Ochraceous when young, at times streaked with ochraceous to vinaceous veil fibrils, developing vinaceous tones in age. **Partial veil:** Fibrillose, ochraceous when young, becoming vinaceous. **Flesh:** Thick, firm, orange when young, ochraceous brown when mature. **Odor:** Indistinct. **Taste:** Indistinct. **KOH:** Vinaceous red. **Spore deposit:** Not obtainable. **Microscopy:** Spores 16–20 × 6.5–9 µm, ellipsoid, smooth, wall slightly thickened.

Ecology: Solitary, scattered, or in fused clusters, often buried in soil or duff. Occasional in higher elevation lodgepole pine forests (rare on the coast under shore pine), possibly also with three-needle pines. In all cases it is associated with (likely parasitic on) *Suillus* (or *Rhizopogon)*. Fruiting in summer and fall.

Edibility: Edible, but like their aboveground counterparts, not particularly desirable.

Comments: This species is recognized by the rounded to irregularly lobed cap, a much-reduced stipe, and the highly convoluted interior of folded and anastomosed gills. It's one of a handful of our transition species between a gilled mushroom and a truffle. The name *C. loculatus* has been applied to contorted specimens of an ocher-orange species with hemlock but it appears that these are simply atypically developed specimens of *C. tomentosus.*

Synonyms: *Brauniellula albipes* (Zeller) A.H. Sm. & Singer, *B. nancyae* A.H. Sm. & Singer

Cortinarius pinguis

(Zeller) Peintner & M.M. Moser

Cap: 1–5 (7) cm across, typically rounded, at times more flattened on top, or irregular. Surface viscid when wet, innately fibrillose to smooth. Straw-yellow to olive-yellow when young, yellow-brown to brown in age. **Interior:** Composed of highly convoluted, small loculate chambers. Light brown when young, ocher-brown to dark brown in age. **Stipe:** 0.5–5 cm long, 0.7–2.5 cm thick, shorth and stout, or more elongate, equal or swollen in middle, and a small abrupt bulb at base. Typically decaying before cap. Surface glutinous when young (but rarely seen in this phase), soon dry, fibrillose. Dingy whitish to pale yellowish with violet slime (if present), developing dingy stray-yellow to brownish stains in age. **Partial veil:** Scant yellowish fibrils, leafing remnants on cap margin, and lower stipe when young. **Flesh:** Thin; the cap is mostly gleba. Stipe fibrous, but fragile, white to yellowish brown. **Odor:** Strongly yeasty; like bread dough. **Taste:** Unpleasant. **KOH:** No reaction. **Microscopy:** Spores 14–16.5 x 8–9.5 µm, ellipsoid to oblong, distinctly roughened.

Ecology: Solitary, scattered, or in large troops, typically exposed but may be buried in duff. Common at mid to high elevation on the Cascade Range, in fir and hemlock forest. First fruiting in late summer and continuing until snow begins to fall. Fruitbodies are long-lasting.

Edibility: Unknown.

Comments: The top-shaped fruitbodies with a straw-yellow to olive-yellow to brown cap, (mostly) enclosed, brown to dark ocher-brown gleba, pallid stipe, and strong yeast odor easily identify this species. It is one of the most common *Cortinarius* in Cascadia but can be difficult to spot, as it is rarely elevated when fruiting and it blends into the surrounding duff. The similarly shaped *C. pavelekii* occurs on the northern Oregon and Washington coast but is quite rare. It has a variable grayish, grayish brown to ocher brown cap, and white stipe, which may have violet slime. *Pholiota nubigena* (=*Nivatogastrium nubigenum*), has top-shaped cream to orangish fruitbodies, a wintergreen odor and grows on wood near melting snowbanks

Synonym: *Thaxterogaster pinguis* (Zeller) Singer & A.H. Sm.

Alpova diplophloeus

(Zeller & C. W. Dodge) Trappe & A. H. Sm

Alder Truffle

Fruitbody: 0.5–5 cm across, 0.5–4 cm tall, ovoid to round. Surface moist to dry, finely matted-tomentose to smooth. Yellowish brown to pinkish brown when immature, darkening to reddish brown to brown. **Interior**: Firm, consisting of gelatinous gel-filled chambers, surrounded by pale yellowish veins. Chambers yellow-olive at first, soon reddish to orangish brown. **Odor**: Mild to fruity in age. Taste: Mild. **KOH**: Reddish on exterior. **Microscopy**: Spores 4.5–5.5 x 2–3 µm, ellipsoid to oblong, colorless to pinkish buff, amyloid when immature, inamyloid when mature.

Ecology: Solitary, scattered, or most often in "nests" of many fruitbodies. Usually buried in soil or duff at first, then rupturing through the surface as they mature. Growing under alder and typically found in swampy areas or in riparian corridors. Common and widespread but rarely reported due to their hypogeous habit. Fruiting from late summer into winter.

Edibility: Reported to be bland. It lacks the complex "truffly" odors of the highly prized culinary truffles and has a strange gelatinous texture.

Comments: Strictly associated with its host tree, *A. diplophloeus* is usually found in riparian areas or other perennial sources of water where alders are common. The reddish brown exterior and gel-like, chambered interior with pale yellowish veins are distinctive. Genetically, it is closely related to *Rhizopogon* and both are related to the boletoid genus *Suillus*. *Rhizopogon olivaceotinctus* has a firm, rubbery texture, a yellow-brown to dark brown gel-like interior, and a yellowish brown exterior that darkens to olive-brown to dingy dark brown. It grows with conifers, especially with younger pines. *Alpova (Melanogaster) trappei* is similar but has a brighter yellow exterior, a gelatinous-rubbery yellow-brown to dark brown interior– it is more common in the southern Cascade Range with conifers. Distinguishing older specimens from *Rhizopogon olivaceotinctus often requires microscopy; A. trappei has distinctly smaller spores (4.0–6.5 x 2–3* µm *vs. 8–10 x 3–4* µm). *Melanogaster* have a black, gel-like, chambered interior with whitish lines.

Truncocolumella citrina

Zeller

Lemon Trunc

Fruitbody: 0.5–3 (5) cm across, rounded, ovoid to irregular-lobed, often with a pinched basal pad or short stipe, firm and rubbery textured. Surface dry, smooth, occasionally cracking in age. Yellowish, grayish yellow to yellow-olive. **Interior:** Firm, rubbery mass of a tight, sponge-like gleba, with tiny open chambers. Pale grayish at first, becoming grayish olive, to dark grayish olive and slightly gelatinous in age. Columella present as a small yellowish mass at base, often with radiating veins into gleba. **Odor:** Indistinct. **Taste:** Mild to earthy. **KOH:** No reaction. **Microscopy:** Spores 6.5–10 x 3.5–4.5 µm, ellipsoid to ovoid, smooth, thin walled, pale yellowish.

Ecology: Solitary, scattered, or in small clusters, often buried or partially buried in soil or duff under Douglas-fir. Common throughout our range. Generally fruiting from late summer into early winter.

Edibility: Edible, but not all that flavorful.

Comments: The firm rubbery texture, golden-yellowish color, and grayish olive gleba with a yellowish columella are distinctive. Because of the bright colors and tendency to break through the surface when mature, it is one of the most commonly encountered truffles. *Rhizopogon truncatus* has a bright yellow exterior and an olive interior that lacks a columella; it grows with five-needle pines. Other *Rhizopogon* lack the evenly yellowish colors and don't have a columella.

Rhizopogon occidentalis

Zeller & C. W. Dodge

Western Pogie

Fruitbody: 1–8 cm across, round, ovoid to irregularly lobed, firm, rubbery. Surface dry, covered with binding, minutely hairy rhizomorphs which stain yellow, then slowly reddish when bruised. Creamy yellow with golden yellow rhizomorphs at first, staining or becoming yellow, orange to rosy red in irregular patches. **Interior:** A rubbery mass of a tight, sponge-like gleba with tiny open chambers. Whitish to pale gray at first, becoming olive when mature. **Odor:** Indistinct. **Taste:** Indistinct. **KOH:** Reddish on exterior. **Microscopy:** Spores 5.5–7 x 2–3 µm, oblong, smooth, thin-walled, pale yellowish.

Ecology: Solitary, scattered, or in troops, buried in duff or soil, occasionally rupturing surface in areas with thin duff. Very common, occurring with two- and three-needle pines throughout our range. Fruiting in late summer and fall.

Edibility: Edible; it is used in stews more like a potato than a spice since it lacks the complex odors and flavors that true truffles have.

Comments: This rubbery truffle is recognized by its pale yellow colors at first, coating of rhizomorphs that stain yellow and then redden in age, and growth with pine. *R. ochraceorubens* is nearly indistinguishable; it generally exhibits more red or orange staining, but there seems to be overlap between these two species. *Rhizopogon* are easy to identify to genus but narrowing them to species can be difficult. They all have a rubbery texture, bounce when dropped, have a firm mass of a tight, sponge-like gleba, and often have rhizomorphs on the exterior. Other bouncing truffles include *Truncocolumella*, which have a translucent columella, and *Gautieria*, which generally have a pinkish to ochraceous chambered gleba with a whitish or translucent columella.

Rhizopogon parksii

A. H. Sm.

Dark Pogie

Fruitbody: 1–4 (7) cm across, round, ovoid to irregular-lobed, firm, rubbery. Surface dry, covered with a matted tomentum and minutely hairy rhizomorphs. Dingy whitish to grayish at first, staining dirty pinkish violet, darkening blackish in patches, to all black in age. **Interior:** A firm, rubbery mass of tight, sponge-like gleba with tiny open chambers. Whitish to pale gray at first, becoming olive, and often developing dark stains in age. **Odor:** Indistinct. **Taste:** Indistinct. **KOH:** Purplish black on exterior. **Microscopy:** Spores 5–7 x 2.5–3 µm, ellipsoid to oblong, smooth, thin walled, colorless to pale yellowish.

Ecology: Solitary, scattered, or in troops buried in duff or soil, occasionally rupturing through the surface in areas with thin duff. Very common, occurring with Douglas-fir throughout our range. Fruiting in fall.

Edibility: Edible.

Comments: *R. parksii* belongs to a complex of species which include *R. villosulus* and *R. hawkerae*. This group of "Pogies" are told by the rubbery texture, dingy colors that darken with age, and growth with Douglas-fir. They are some of the most commonly encountered hypogeous fungi in the Pacific Northwest, but reliable identification to species apparently requires a DNA sequence.

Rhizopogon sp.

Rhizopogon evadens var. *subalpinus*

Rhizopogon sp.

Rhizopogon sp.

Rhizopogon species
Pogies

There are over 100 species of *Rhizopogon* described from western North America, many of which have overlapping morphological features and cannot thus be reliably identified without a genetic sequence. As a genus, they are recognized by their hypogeous habit, rubbery texture, with round to oblong small to medium size fruitbodies (although a few species can be 10+ cm). Many will bounce if dropped on a hard surface. The exterior color and color change is important to note; many will stain pink, red, vinaceous, or black. Many species have conspicuous rhizomorphs appressed to the exterior and some have distinct KOH reactions. The interior is a firm, rubbery mass of a tight, sponge-like gleba, with tiny open chambers; pallid colored when young, often developing olive colors with age, although yellow, brown, or even deep red colors do occur. Odors are usually indistinct, garlicky or herbal. The taste is typically earthy to indistinct. The spores are smooth, oblong, subfusoid to ellipsoid. Many species have small spores, ~5–9 x 2–5 µm, while some can be as large as 10–14 x 6–8 µm. They are major players in the ectomycorrhizal communities of Cascadia's conifer forests. Some species are among the first colonizers post forest fires, germinating from spore banks in the soil with emerging seedlings, while others appear to be restricted to old-growth forests. Like their close above-ground relatives the Slippery Jacks (*Suillus* spp.), mostly growing with Douglas-fir and pine, more rarely with fir. They are found throughout the year, with the main flushes in spring and fall. All *Rhizopogon* are likely edible, but more of a textural (chewy, rubbery, or slimy) addition to stews or sauces than the highly prized aromatic truffles (*Leucangium, Kalapuya, Tuber* spp.).

Melanogaster macrocarpus

Zeller

Fruitbody: 0.7–3.5 cm across, round ovoid to irregular, firm. Surface dry, sparsely covered with blackish, thread-like rhizomorphs. Reddish brown when young, becoming dark brown to blackish. **Interior:** Firm, consisting of gelatinous gel-filled chambers, surrounded by whitish veins. Chambers dark grayish olive to black when young, to deep black in age. **Odor:** Strong, "truffly", consisting of a complex mix of earthy and nutty tones, along with garlic, onions and burnt rubber. **Taste:** Not sampled. **KOH:** No reaction. **Microscopy:** Spores 10–14 x 6–10 µm, ovoid to pear shaped, smooth, thick walled, dark vinaceous brown.

Ecology: Solitary, scattered, or in caches, often in zone between top and mineral soil, occasionally rupturing surface, especially in hard-packed areas. Common under a wide variety of trees including hardwoods and conifers (although hardwood associates likely represent separate species). Fruiting throughout the wet season but more common in spring.

Edibility: The strong odor easily flavors foods such as cheese or butter with a pleasing, complex taste.

Comments: The reddish brown exterior and black, gelatinous, chambered interior with whitish veins make *Melanogaster* easy to identify as a genus. For years, the European name *M. tuberiformis* has been indiscriminately applied to the various *Melanogaster* species in Cascadia, most of which are still undescribed (a paper is in the works describing them). They differ mostly based on spore morphology and habitat. Most *Melanogaster* appear to have culinary value with a complex "truffle" odor and a pleasant flavor. However, not all people perceive these as pleasant odors—some people describe them more as "kerosene" or "used motor oil." *Alpova diplophloeus* has a brownish interior and grows with alder. *Alpova (Melanogaster) trappei* has a yellow to brown exterior which darkens with handling, gelatinous yellow-brown to dark brown interior, and is more common in the southern Cascade Range with conifers.

Rhizopogon alexsmithii

(Trappe) Vizzini & Zotti

Fruitbody: 0.5–5 cm across, highly irregular, roundish to flattened, multi-lobed mass, very firm, corky to almost woody textured. Surface dry, smooth or with scattered appressed fibrils or rhizomorphs, with some felty areas in between lobes, Light to dark brown, or nearly black if exposed. **Interior:** Very firm, dense, consisting of small, gel-filled, round chambers, exuding sticky fluid when cut. Off-white to pale pinkish chambers marbled with whitish veins. **Odor:** Indistinct to faintly fruity. **Taste:** Not sampled. **Microscopy:** Spores 5–8 x 3–5 µm, ellipsoid, smooth, hyaline.

Ecology: Scattered or in clumps, hypogeous or occasionally erumpent. Rare, known from higher elevation hemlock-fir forests in the Cascade Range, fruiting in late summer and fall.

Edibility: Unknown.

Comments: At first glance, *Rhizopogon alexsmithii* resembles a small burl or gnarled root rather than a fungus. The corky texture and pinkish interior which oozes a sticky fluid when cut are highly atypical for the genus. *Leucogaster* have white interiors with larger, gel-filled chambers and globose to subglobose ornamented spores.

Gautieria species

Gauts, Stinkballs

Fruitbody: 0.5–10+ cm across, round, ovoid, cushion-shaped or lobed, firm, rubbery. Exterior either a smooth peridium or a bumpy or even coralloid texture, typically whitish, grayish, pinkish or vinaceous. **Interior:** With small to relatively large, short maze-like chambers and openings, typically pinkish to ocher, and often with a whitish or translucent columella present as a basal pad or veins through the gleba. **Odor**: Pungent, unpleasant, chemical, often nauseating with age, occasionally with fruity tones. **Taste**: Earthy to unpleasant. **Microscopy**: Spores ellipsoid to subfusoid, often longitudinally-striate or winged, at times appearing warted. Lots of variation in size, from 9–15 x 6–12 µm to 25–30 x 12–15 µm.

Ecology: Solitary, scattered, or in clumps. Hypogeous: some species are more common in upper stratum while others are found well buried in the mineral soil. Ectomycorrhizal, most species occurring with conifers. Common in montane forests, especially on the drier eastern slopes of the Cascade Range. *Gautieria* can be found year-round but are more commonly found in spring following snowmelt.

Edibility: All are likely non-toxic, but the off-putting odors and flavors seem to discourage even squirrels from eating them.

Comments: *Gautieria* are close hypogeous relatives of the coral fungi *Ramaria.* There are about fifty species known from western North America, almost all remain undescribed. Most collections are called *Gautieria monticola,* a dense and rubbery species described from California. *G. pterosperma* (nom. prov.) often shows vinaceous colors on the exterior and has short-ellipsoid to nearly round winged spores. We commonly find an unnamed species in higher elevation fir forests in the Washington Cascades with a white exterior. Another undescribed species with red fir in the southern Cascades has a caramelized banana odor. One can also find species lacking a peridium, with the open chambers exposed and thus appearing finely branched-coralloid. *G. coralloides* (States & Fogel nom. prov.) may eventually be formally published and is one of the more common members of this group.

Gastroboletus ruber

(Zeller) Cázares & Trappe

Fruitbody: 2–5.5 cm across, 2–4 cm high, irregularly rounded, top-shaped to lobed. Exterior with a very thin peridium, evanescent, soon disappearing, or only remaining in scattered patches near top of stipe, exposing the tube mouths. Yellowish, or reddish where exposed when young, to dark red to red-brown in age. **Stipe:** Short to indistinct, 0.5–2 cm long, 0.7–1.5 cm thick. **Pores:** Exposed mouths red-orange to red, staining blue when bruised. **Interior:** Made of separable tubes when young, becoming more convoluted and maze-like in age. Pale yellow when young, to yellow-olive to dark olive in age, staining bluish when exposed. **Columella:** Varying from a basal pad to a column of tissue, at times reaching entirely through the fruitbody, often with many finer branches. Yellowish to yellowish translucent on finer branches, staining bluish when cut. **Odor:** Indistinct. **Taste:** Mild. **Microscopy:** Spores 9–15 (20) x 4–6 µm, subfusoid, smooth, moderately thick-walled, pale greenish olive in KOH, inamyloid.

Ecology: Solitary, scattered, or in small clusters. Fruitbodies form underground and partially erupt from duff when mature. Locally common in higher elevations in the Cascade Range, especially in mixed mountain hemlock-fir forests. Fruiting in summer and early fall.

Edibility: Unknown.

Comments: Recognized by the top-shaped to irregular fruitbody, the thin peridium, which soon disappears, exposing the convoluted tubes and pores, indistinct stipe, and pad-like columella extending into the gleba. Additionally, the yellowish, orange-red to deep red colors, bluish staining, and large, inamyloid spores distinguish this species. "*Gastroboletus*" *amyloideus* lacks the blue staining, typically has more yellowish colors, and has distinctly amyloid spores. "*Gastroboletus*" *vividus* looks more like a slightly distorted bolete with a distinct cap and stipe and convoluted pores. It has a yellow, red-blushed to extensively red cap, bright yellow pores that become red in age, and a short yellow stipe with a rosy red band at the apex.

Synonyms: *Truncocolumella rubra* Zeller

Leucogaster rubescens

Zeller & C.W. Dodge

Fruitbody: 1–3 cm across, round to ovoid, at times lobed or irregular, firm when young, occasionally crumbly in age. Surface viscid to dry, with sc attered fibrils or rhizomorphs, often becoming dimpled (like a golf ball) in age; occasionally cracking if exposed. Pale pinkish white when young, developing ocher tones, becoming orange-red in age. **Interior:** White to creamy, firm, consisting of small round gel-filled chambers or cavities, oozing a whitish to watery or sticky fluid when cut. **Odor:** Sweet, fruity. **Taste:** Not sampled. **Microscopy:** Spores 10–15 µm, globose, reticulate, covered with a gelatinous sac.

Ecology: Solitary or scattered, hypogeous or erumpent in soil and duff under conifers. Widespread and locally common in low- to mid-elevation forests; fruiting from late spring through fall.

Edibility: Edible but not popular.

Comments: The pinkish to red-orange exterior, creamy interior with round, gel-filled chambers, and fruity odor are distinctive. In some collections the odor is reminiscent of piña coladas—a mix of pineapple and coconut. The similar *L. citrinus* is creamy to yellow colored, and has smaller spores (8–11 µm). It is also widespread throughout Cascadia, but appears to be more common in the fir forests of the southern Cascades. *L. microsporus* is extremely similar but has more kidney-shaped fruitbodies and smaller spores (6–10 µm). *Leucophleps* are similar with white to yellowish exteriors and often a sticky white fluid that oozes when cut but have irregular to short maze-like (not round) chambers and globose spores with spines covered in a gelatinous matrix. Two common species occur in Cascadia: *L. spinispora*, in drier east-slope and southern Cascade forests, and *L. magnata* in wetter hemlock and Douglas-fir forests. *L. spinispora* has smaller spores (10–13 µm vs. 15–17 µm for *L. magnata).* The similar shaped and colored *Mycolevis siccigleba* is distinguished by its dry, lightweight fruitbodies (described as Styrofoam-like) and ellipsoid, thick-walled spores.

Hymenogaster subalpinus

A.H. Sm.

Fruitbody: 0.5–2 (3) cm across, round, ovoid to irregularly lobed, firm to crumbly. Surface moist to dry, appressed-fibrillose to nearly smooth, occasionally cracking when mature. Dingy white to straw yellow, discoloring yellowish when handled, then slowly darkening to yellow-brown or dingy brown. **Interior:** Firm, consisting of small, tightly packed convoluted chambers, and an irregular basal columella. Dingy gray when young, brown to vinaceous brown when mature. **Odor:** Pungent, unpleasant. **Taste:** Not sampled. **Microscopy:** Spores (15) 20–30 x 9–16 µm, narrowly lemon-shaped, warty-roughened, brown.

Ecology: Scattered to gregarious under moss and duff in upper layers of soil. Common and widespread; especially abundant in young to middle-age Douglas-fir stands in winter and early spring.

Edibility: Unknown.

Comments: The small size, dingy white exterior (which discolors yellow-brown), brown gleba, unpleasant odor, and crumbly texture when mature help distinguish *H. subalpinus*. It's one of the more common truffles in western Cascadia, especially in lower elevation Douglas-fir plantations. There are several *Hymenogaster* described from western North America, mostly differentiated based on spore morphology and habitat; most of these species would match the macro-description given here. *Hymenogaster* are close hypogeous relatives of *Hebeloma*, and like their epigeous cousins, modern molecular work is needed to define species concepts better. *Cortinarius lilacinoides (=H. sublilacinus)* is common in drier east-slope Cascade pine-fir forests. It has a pallid, yellowish or orangish exterior which can show a fleeting bluish-violet discoloration, a bright orange-brown interior, and an odor of pine resin.

Trappea darkeri

(Zeller) Castellano

Fruitbody: 2–7 cm across, round, ovoid to irregular, at times lobed, firm, rubbery. Surface moist to dry, smooth or with patches of downy hairs, occasionally cracking if exposed, base with thick white rhizomorphs. White to beige, discoloring brownish or pale ocher-brown when handled. **Interior:** Firm, consisting of small, often gelatinized, densely packed convoluted chambers, with a translucent to gray basal or central columella, with radiating veins. Olive to olive-green, with a whitish band of sterile tissue around the margin. **Odor:** Unpleasant; a foul mix of gassy and rotten egg odors. **Taste:** Not sampled. **Microscopy:** Spores 4–5 x 2–3 µm, ellipsoid, smooth, hyaline to pale greenish.

Ecology: Solitary or scattered, often hypogeous, occasionally erumpent in soil and duff. Widespread across Cascadia but more common at mid-elevations in the Klamath Mountains and Cascades. Fruiting soon after snowmelt through the fall.

Edibility: Unknown; the odor alone would discourage most people from even trying.

Comments: *Trappea darkeri* is recognized by a white exterior which stains brownish, an olive interior with a whitish sterile band around the margin, basal rhizomorphs, and a foul odor. The rare *Phallogaster (Trappea) phillipsii* is very similar, but stains distinctly pink on the exterior. There are a number of *Hysterangium* species in Cascadia. They differ by having a distinct peridium (outer skin) and larger spores. Many also have the foul odors.

Elaphomyces granulatus

sensu PNW

Fruitbody: 1–3 (5) cm across, 1–4 cm tall, ovoid to nearly round, or irregular-lobed, firm. Surface covered with mycelium that binds to soil; when peeled away, it reveals a warty-roughened surface. Deep ocher-orange, orangish brown to brown. **Interior**: Stuffed with whitish, pith-like material at first, maturing into a dark brown, powdery mass surrounded by a thick brown skin. **Odor**: Mild to funky in age. **Taste**: Indistinct. **Microscopy**: Spores 22–32 µm, globose, spiny-warted, dark brown in KOH.

Ecology: Solitary, scattered, or in troops, buried often in rather deep soil. Occurring in low- and mid-elevation conifer forests across Cascadia, especially fond of western hemlock. Abundant but rarely collected unless people are raking around animal digs. Most often found when the parasitic club fungus *Tolypocladium* is growing on the truffles. Fruitbodies are long lasting and can be found at any time of year.

Edibility: Not edible.

Comments: *Elaphomyces* are recognized by their hypogeous fruitbodies, thick, hard exterior, and a whitish interior that becomes dark and powdery when mature. There are a number of species going by the name *E. granulatus* (a European species) in the Pacific Northwest – most of which need to be described. The one detailed here, with a warty ocher-orange to orangish brown exterior and dark brown powdery spore mass, is the most common species. It occurs in hemlock forests. *E. asperulus* is similar but more brown in color. *E.* aff. *muricatus* has a thicker skin with a marbled interior. *E. subviscidus* is a rare species known from high-elevation fir forests; it is larger, with a pallid smooth exterior, dark brown spore mass, and smaller spores (12–22 µm, with spines 1–2 µm long).

Genea harknessii

Gilkey

Fruitbody: 0.5–2.5 cm across, rounded, typically wrinkled, lobed and folded, often centrally depressed, with an opening to an internal cavity, and a tuft of brown mycelium at the base. Surface dry, finely warted. Dark brown, dark reddish brown to blackish. **Interior:** Convoluted from in-folding of walls, with hollow chambers. Chambers of similar color and texture to exterior, flesh of walls whitish to grayish. Odor: Indistinct to slightly garlicky. **Taste:** Indistinct. **KOH:** Unknown. **Microscopy:** Spores 24–32 x 21–29 µm, broadly ellipsoid to subglobose, appearing smooth when immature, ornamented with sparse pointed to irregular protrusions at tips when mature. Asci 8-spored, cylindrical to narrowly clavate.

Ecology: Solitary or scattered, buried under leaves, in duff or soil. Widespread in mixed hardwood-conifer forests in lower and mid-elevations; occasional in the Coast Range and Klamath Mountains in Oregon, rare elsewhere in Cascadia. Fruitbodies are slow to develop, often forming in late winter and maturing in late spring and early summer.

Edibility: Unknown.

Comments: *Genea harknessii* is recognized by the small, dark brown to reddish-brown or blackish, wrinkled and folded hypogeous fruitbodies with a warted surface. The interior is convoluted with irregular hollows from in-folded and lobed walls, with a similar color and texture to the exterior. *G. gardneri* is generally slightly smaller and is black in color. Paler specimens can easily be distinguished from *G. harknessii* by the larger spores, measuring 32–36 x 28–34 µm. *Gilkeya compacta* is pinkish-red to reddish-purple colored. *Peziza ellipsospora* has larger, purple-brown and convoluted fruitbodies with folded and muti-chambered interiors. It occurs with oaks in Oregon and becomes more common southward in California.

Taxonomic note: Genetic studies have shown at least five cryptic phylogenetic species in the *Genea harknessii* complex in western North America. At this time, there are no known distinguishing morphological differences.

Hydnotrya variiformis

Gilkey

Fruitbody: 1–3 (4) cm across, irregularly convoluted and lobed, brittle. Surface dry, finely velvety to nearly smooth. Off-white, beige to pinkish tan when young, darkening to cinnamon-brown or orange-brown with age. **Interior:** Hollow, from a simple chamber, to multi-chambered with irregular folds and pockets. Off-white, beige to pale orange-brown. **Odor:** Indistinct. **Taste:** Indistinct. **KOH:** Unknown. **Microscopy:** Spores 32–36 x 24–28 µm, ellipsoid, finely pitted. Asci 8-spored, clavate to cylindrical, inamyloid.

Ecology: Solitary or occasionally scattered in and around well decayed logs and other debris. Widespread but uncommon, fruiting from late spring through fall.

Edibility: Unknown.

Comments: The folded and lobed, pale orange-brown fruitbody with a simple to complex open-chambered interior and tendency to grow around decaying wood help distinguish this species. *H. cerebriformis* is often darker reddish brown to vinaceous brown colored and has more tightly folded and convoluted fruitbodies. It has a garlicky odor and globose spores 25–32 µm across. *H. tulasnei* is similar colored to *H. cerebriformis* but is often larger and has larger spores (30–44 µm). *Genabea cerebriformis* is a small convoluted species with a pallid to tan exterior covered with fine warts. *Geopora cooperi* has globose fruitbodies with a yellow-brown to dark brown hairy exterior and highly convoluted interior.

Balsamia magnata

Harkn.

Fruitbody: 0.8–2 cm across, 0.2–1 cm high, irregularly rounded or flattened, often lobed, and often with a central depression or hollow. Surface dry, warted. Reddish brown to orange-brown, slightly paler between warts. **Interior:** Firm, white to pinkish beige, marbled with bright white veins. **Odor:** Indistinct. **Taste:** Indistinct. **KOH:** No reaction. **Microscopy:** Spores 20–28.5 x 10–14.5 µm, Q = 2.0, ellipsoid, smooth, hyaline. Asci 8-spored, balloon-shaped, inamyloid.

Ecology: Solitary, scattered, or in small "nests" in soil or at the interface of duff and mineral soil. Widespread across low and mid-elevations of Cascadia in a variety of habitats, fruiting in fall and early winter, and spring into early summer.

Edibility: Edible; but lacking flavor, and rarely found in abundance.

Comments: The hypogeous growth, small size, warted exterior, marbled interior, and smooth ellipsoid spores help define the genus *Balsamia.* They are closely related the the Elfin Saddles (*Helvella*). Identifying *Balsamia* to species requires detailed microscopic examination and/or a genetic sequence. Macroscopically, *B. magnata* is practically indistinguishable from other species in Cascadia which include: *B. latispora* with shorter and wider spores; *B. lazyana* with narrower spores; *B. trappei* with finer warts; *B. quercicola* with longer spores growing in lower elevation oak and pine forests; *B. filamentosa* with browner color but similar spores. *B. alba* and *B. cascadensis* have red-brown to brown exteriors often covered with white hyphae. Finally, *B. nigrans* has a dark reddish-brown to black exterior. For those who need to know, refer to Southworth et al. 2018 for details on distinguishing these species... Good luck! *Balsamia* (=*Barssia*) *oregonensis* has a flattened fruitbody with a central depression on one side, and a smooth to finely warted exterior.

Leucangium carthusianum

Sensu PNW

Oregon Black Truffle

Fruitbody: 2–6 (9) cm across, round to irregularly lobed, dense, firm. Surface moist to dry, warty or roughened, rarely nearly smooth. Black, gray-black to dark brownish black. **Interior:** Firm, solid. Made up of round spots of gray-brown, olive-gray to blackish fertile areas, marbled into a whitish or pale gray background. **Odor:** Fruity to earthy when young,often with pineapple or tropical fruit tones, more pungent in age. **Taste:** Mild to earthy, with complex tones. **KOH:** Unknown. **Microscopy:** Spores 50–65 (100) x 20–40 **µm**, spindle-shaped to lemon-shaped, smooth. Asci large, sac-like, 8-spored.

Ecology: Solitary or scattered, buried in soil under Douglas-fir; prefers younger forest and especially plantations. Common in lower elevation forests in Oregon and Washington but tough to find without a trained dog. Fruiting from winter into early summer.

Edibility: Edible and highly regarded. The mix of fruity and "truffly" tones is unique.

Comments: The firm, often irregular fruitbodies with a rough black exterior, spotted marbled interior (similar to a granite countertop), and giant spores distinguish the Oregon Black Truffle. They appear to more dependent on dispersal by soil invertebrates than other Cascadian truffles and thus are less likely to grow near the surface. This makes them harder to find without a trained truffle-sniffing dog. *Kalapuya brunnea*, the Oregon Brown Truffle, has a similar spotted-marbled interior but has an orange-brown to reddish brown exterior, a garlicky or pungent cheesy odor, and smaller spores. It's an uncommon species that also occurs in young Douglas-fir forests and found from fall into late spring. *Balsamia nigrans* has a coarsely warted black exterior and a firm, solid whitish to pale grayish interior marbled with white veins. Microscopically it has large, smooth, oblong spores. *Melanogaster* species have a marbled gelatinous interior.

Taxonomic note: *Leucangium carthusianum* was described from Europe; PNW collections differ genetically, and will likely get a new name.

Tuber gibbosum

Harkn.

Oregon White Truffle

Fruitbody: 0.5–5.5 cm across, often highly irregular lobed or furrowed, occasionally ovoid to round when small. Surface dry, smooth to minutely hairy in protected areas, often cracking in age. Buff, light olive-brown to brownish, often with rusty to reddish stains or tones. **Interior:** Very firm, solid. Whitish to grayish brown when young, marbled with white veins, becoming vinaceous brown to dark grayish brown, marbled with whitish veins when mature. **Odor:** Mild when immature, becoming pungent; a "truffly" mix of garlic, spiciness, nuttiness, and cheese. **Taste:** Mild to radish-like. **Microscopy:** Asci (1) 2–4 (6)–spored. Spores highly variable in size, depending on number per asci; 21–60 x 16–38 µm (not counting spines), ellipsoid to broadly ellipsoid, ornamented with spiny reticulum, 3–4(5) µm high, surrounding honeycomb-like pits, thick walled, yellowish.

Ecology: Solitary or scattered, buried beneath duff or in soil under Douglas-fir. Prefers younger forests, especially plantations. Locally common but tough to find without a trained dog. Fruiting from late winter into early summer.

Edibility: Edible and choice. Although these truffles don't enjoy the same acclaim as the European White Truffle, their flavor can be on par. They developed a reputation for being "bland" and "lacking flavor" due to an influx of immature truffles into the market, collected by indiscriminate pickers raking the forest floor and causing considerable damage in the process. Ethical harvesting practices and the use of trained dogs to sniff out only mature truffles has helped alleviate both the ecological damage and quality problems.

Tuber oregonense

Comments: The firm, buff to brownish, irregular-lobed fruitbody with a solid, grayish to vinaceous-brown marbled gleba with whitish veins, pungent truffle odor and growth in late winter and spring under Douglas-fir are distinctive features. *T. oregonense* is whitish at first, soon developing reddish brown to rusty orange-brown patches, becoming orangish brown overall in age. It generally fruits in the fall into winter. The four species in the Oregon White Truffle complex are mostly separated by spore morphology (and only truly reliably separated by molecular analysis). *T. gibbosum* is described above. *T. castellanoi* differs by having slightly shorter spores (20–44 x 18–38 µm) that are broadly ellipsoid to globose. Both *T. oregonense* and *T. bellisporum* have narrower spores (20–55 x 14–32 µm for the latter, 25–63 x 13–30 µm for the former). Because of the variability in spore size based on the number of spores per asci, it is a time-consuming process to measure enough spores to make a statistically-supported determination as to which species you have. Chefs need not fret— they all have the same culinary properties! *T. californicum* has a firm, irregularly lobed fruitbody with a pubescent surface that is whitish to tan when young, becoming ocher-brown to brown in age. It has a marbled, whitish to dark brown gleba with whitish veins, an herbal garlic-cheese odor, and globose spores. It is somewhat common in the late winter and spring under oaks and conifers. *T. candidum* has a reddish to orangish brown roundish fruitbody, a tan to orange-brown gleba (marbled with dark brown and white veins), and a mild to earthy odor even when mature. *Balsamia* species have an orange to reddish brown, roughened or pimpled exterior, a white marbled interior, and smooth ellipsoid spores.

Kalapuya brunnea

M.J. Trappe, Trappe & Bonito

Oregon Brown Truffle

Fruitbody: 1.5–6 cm across, round to irregularly lobed, dense, firm. Surface dry, warty or roughened, often areolate-cracked in age. Orange-brown, reddish brown, or at times yellowish brown, darkening in age. **Interior:** Firm, solid. Made up of round spots of gray, pale olive-gray to brownish fertile areas, marbled into a whitish or pale gray background. Odor: Ripe cheese with garlicky undertones. **Taste:** Similar to odor, with complex truffly flavors. **KOH:** Unknown. **Microscopy:** Spores 32–43 x 25–38 µm, ellipsoid, smooth, with a large central oil droplet, and multiple tiny droplets near ends. asci 6- to 8-spored.

Ecology: Solitary or scattered, buried in soil under Douglas-fir. Occasional, with a preference for young to mid-age plantations in lower elevation forests. Know from the Coast Range and Cascade foothills in Oregon and northern California. Fruiting from fall into spring or early summer.

Edibility: Edible and highly regarded. Although not as common as Oregon White and Oregon Black Truffles.

Comments: This species is recognized by the orange-brown warty-roughened exterior, the spotted marbled interior (similar to a granite countertop), a garlicky or pungent cheesy odor, and large ellipsoid spores. *Leucangium carthusianum* shares a similar spotted interior, but has a black exterior and fruity odor. There is also an unnamed summer fruiting *Leucangium* with small fruitbodies and a reddish exterior. Some *Balsamia* species have similar colored, warted-roughened exteriors, but have solid whitish to buff interiors marbled with white veins.

22 - Jelly Fungi

This chapter represents multiple lineages that are not particularly closely related, but which share gelatinous-textured fruitbodies. Most of the species we include here are basidiomycetes. Curiously, most of the "basidio-jellies" have distinctive and oddly shaped basidia: some resemble tuning forks (with two long, skinny spore-bearing sterigmata) while others are vertically septate. Most jelly fungi form brain-like, leafy-lobed, or jelly-drop shaped fruitbodies, but others are stalked (e.g. *Pseudohydnum* and *Tremiscus*), ear-shaped (*Auricularia*), or simplified into clubs (*Calocera*). Most have smooth fertile surfaces, but *Pseudohydnum* have fine teeth on the underside of the caps, while *Tremiscus helvelloides* and *Auricularia americana* can have veins or ridges.

Some ascomycete "jellies" are also included here, like the brain-shaped *Ascocoryne sarcoides* and the jelly-drop shaped *Chloroscypha flavida*. Other species like *Urnula padeniana* have distinctly gelatinous flesh but their fruitbodies are usually shallowly bowl-shaped and are thus included in the cup fungi section.

Dacrymyces chrysospermus

Berk. & M.A. Curtis

Orange Jelly

Fruitbody: 1–6 cm across, 0.5–4 cm high, often irregular, brain-like, lobed, rounded or in fan-shaped clumps. Bright orange when young, sometimes discoloring reddish orange to orangish brown in age. **Stipe:** Generally indistinct or a short, tapered to somewhat rooting attachment to wood. Pale orange to whitish at point of attachment. **Flesh:** Thin to thick, gelatinous or jelly-like. Translucent orange. **Odor:** Indistinct. **Taste:** Indistinct. **KOH:** No reaction. **Spore deposit:** Orange to golden orange. **Microscopy:** Spores 17–25 (28) x 6–8.5 µm, curved-cylindrical, thin-walled, becoming 7-septate in maturity. Basidia distinctive, shaped like tuning forks.

Ecology: Solitary, scattered, or in troops or masses on dead logs, trees, or branches of conifers; rarely occurring on hardwoods. Very common across Cascadia, fruiting throughout the wet seasons.

Edibility: Edible and harmless, but we have yet to find a way to make it appealing.

Comments: This bright orange jelly fungus is easily identified by the irregular brain-like, lobed or fan-shaped masses, growth on conifers, and whitish base or attachment point. *Naematelia* (*Tremella*) *aurantia* is very similar but grows on hardwoods (near its *Stereum* host), is generally larger and more brain-like in shape, doesn't have a white point of attachment to the substrate, and differs microscopically. *Tremella mesenterica* is another species on hardwood branches with simpler, sparsely-lobed or leafy fruitbodies, and orange-yellow to yellow colors. Other common regional species of *Dacrymyces* include *D. stillatus*, which forms tiny, round to top-shaped translucent yellow-orange jelly-drops on conifer logs and *D. chrysocomus*, which forms small to medium-sized top-shaped, cone-like, or teardrop-shaped translucent orange fruitbodies.

Synonym: *Dacrymyces palmatus* Bres.

Tremella mesenterica

Retz.

Witch's Butter

Fruitbody: 0.5–4 cm across, consisting of one to several open, leafy-gelatinous lobes, rarely more brain-like. Yellow-orange to yellow, becoming paler and more translucent with age. Sometimes with a small, compact point of attachment to the substrate that is concolorous with rest of the fruitbody. **Flesh:** Thin to thick, gelatinous or jelly-like when wet, tough, rubbery when dry. Translucent yellow. **Odor:** Indistinct. **Taste:** Indistinct. **KOH:** No reaction. **Spore deposit:** Yellowish. **Microscopy:** Spores 10–15 x 8.5–10 µm, broadly ellipsoid to ellipsoid, smooth. Basidia longitudinal 4-septate, with long, often sinuous sterigmata.

Ecology: Solitary or scattered on dead branches and small logs of hardwoods, especially red alder. *T. mesenterica* is a parasite on the nondescript crust fungus *Peniophora*. Locally common in western Cascadia, often fruiting in fall and persisting through winter in lower elevations, and often with a fresh fruiting in spring.

Edibility: Edible.

Comments: *T. mesenterica* is one of few species commonly called Witch's Butter in western North America. It is recognized by its sparsely-lobed or leafy gelatinous fruitbody, orange-yellow to yellow colors, and growth on branches of hardwoods, especially alder. All *Tremella* species are parasites on other fungi; this species grows on the hyphae of *Peniophora* species. *Naematelia* (*Tremella*) *aurantia* has a larger, more tightly-convoluted brain-like or multiply-lobed fruitbody and typically brighter orange to yellow-orange in color. It grows on hardwoods, almost always near *Stereum hirsutum* (on which it is a parasite). *Dacrymyces chrysospermus* is another brain-like or lobed bright orange jelly fungus, but it usually grows on conifers and has a whitish base or attachment point.

Heterotextus alpinus

(Tracy & Earle) G.W. Martin

Alpine Jelly Drops

Fruitbody: 0.3–2 (3) cm across, top-shaped, bell-shaped, or jelly-drop shaped, often concave or broadly cup-shaped at disc. Surface smooth on disc, underside finely hairy to nearly smooth, plump when moist, wrinkled when dry. Bright orange to yellow-orange, occasionally drying to reddish orange. **Stipe**: Indistinct or present as a very short, narrowed extension of the fruitbody. **Flesh**: Thin, gelatinous or jelly-like. Translucent orange in color. Odor: Indistinct. **Taste**: Indistinct. **KOH**: No reaction. **Spore deposit:** Hard to obtain. **Microscopy**: Spores 14–18 x 5–6 µm, sausage-shaped, often curved, 3–4-septate when mature.

Ecology: Scattered to gregarious on logs, branches, and woody debris of conifers. Typically fruiting in spring during snowmelt but can revive in wet weather throughout the year. Common at higher elevations across Cascadia.

Edibility: Non-toxic.

Comments: This species can be recognized by the orange color, "jelly-drop" shape, gelatinous texture, and growth on conifer debris—often near melting snow. The name has frequently been misapplied to other similarly shaped jelly fungi in Cascadia. Many lower elevation records refer to *Heterotextus luteus*, which has a similar shape but usually smaller, more yellow-orange fruitbodies, and longer, narrower spores with 5-9 cross-walls. Another common species at low- and mid-elevations in western Cascadia is *Dacrymyces chrysocomus* which is smaller in size and typically paler, (more dull or translucent orange), and has thick-walled spores with 7 cross-walls and 1-3 longitudinal septa. Young fruitbodies of *D. chrysospermus* can show a similar shape but become brain-like or lobed at maturity. *Guepiniopsis buccina* forms small, stalked, often downward-facing cups or bells with wrinkled undersides.

Synonym: *Guepiniopsis alpina* (Tracy & Earle) Brasf.

Chloroscypha flavida

(Kanouse & A.H. Sm.) Baral

Fruitbody: 0.1–0.5 cm across, top-shaped, head rounded, concave when young, soon plane to convex, smooth. Bright yellow to golden yellow. **Underside:** Smooth, concolorous. **Stipe:** 0.1–0.5 cm long, 0.1 cm thick, glabrous, or with faint tomentum at base. **Flesh:** Thin, gelatinous. **Odor:** Indistinct. **Taste:** Indistinct. **KOH:** No reaction. **Spore deposit:** Hard to obtain. **Microscopy:** Spores 26–34 x 9–11 µm, ellipsoid to oblong-ellipsoid, with one end slightly broader than other, smooth, colorless to pale yellow, with two oil droplets. Asci 8-spored, amyloid at tips.

Ecology: Scattered in troops on fallen yellow-cedar and rarely incense-cedar needle splays near melting snowbanks in spring and early summer. Rare, known from higher elevations of the Cascade and Olympic Mountains.

Edibility: Unknown.

Comments: *Chloroscypha flavida* is a rare fungus with small, brightly colored fruitbodies on yellow-cedar and incense-cedar needles near melting snowbanks. The combination of habitat, gelatinous texture, convex to round 'heads,' and short stipe make it easy to recognize. The more common *Chloroscypha alutipes* also grows on incense-cedar needles, but forms tiny, yellow to brownish, cup-like fruitbodies, lacks the gelatinous flesh, and has smaller, spindle-shaped spores measuring 18–20 x 5–8 µm. *Gymnosporangium libocedri* forms irregular orange gelatinous discs or 'blobs' on fresh (green) incense-cedar needles. Jelly fungi such as *Dacrymyces* and *Heterotextus* are often larger, grow on twigs and branches, and has basidia (not asci).

Synonym: *Gelatinodiscus flavidus* Kanouse & A.H. Sm.

Naematelia encephala

(Pers.) Fr.

Pale Brain Jelly

Fruitbody: 0.5–3 cm across, rounded to cushion-shaped, tightly lobed, irregular or brain-like, gelatinous but with a solid, tougher core. Surface densely pillowy and shallowly wrinkled, moist, roughened, often warty, dissolving when wet and old. Creamy white to pinkish as first, light brown to pale reddish brown in age, occasionally with dark specks. **Stipe:** Absent or with a short, stout or rooting attachment to the substrate. **Flesh:** Thin to thick, gelatinous or jelly-like exterior with a tougher whitish core. **Odor:** Indistinct. **Taste:** Indistinct. **KOH:** No reaction. **Spore deposit:** Hard to obtain. **Microscopy:** Spores 8–11 x 7–9 µm, subglobose to broadly ellipsoid, smooth, colorless.

Ecology: Solitary or scattered on dead logs, trees, or branches of conifers, especially western hemlock. Almost always occurring alongside fruitbodies of *Stereum sanguinolentum*, on which it is a parasite. Occasional, occurring across western Cascadia, fruiting from fall into spring.

Edibility: Unknown, likely non-toxic.

Comments: *Naematelia encephala* is distinguished by pallid, pinkish to light brown brain-like mass with a gelatinous exterior and a solid white core growing on conifer wood with *Stereum sanguinolentum*. *Myxarium nucleatum* is similar but has tan to grayish-translucent fruitbodies with whitish nodules suspended in the flesh and grows on hardwoods. Another similar hardwood species is *Exidia candida*, which has whitish to grayish colors when young, developing brownish tones with age, and has cylindrical spores.

Synonym: *Tremella encephala* Pers.

Ascocoryne sarcoides

(Jacq.) J.W. Groves & D.E. Wilson

Purple Jelly-drops

Fruitbody: 0.3–1 cm across, in masses to ~3 cm across, disc-like, shallowly cup-shaped to top-shaped, more rarely brain-like, typically in fused clusters. Smooth or ruffled or wrinkled around margin. Purple, reddish purple to pinkish purple. **Underside:** Smooth, concolorous. **Stipe:** Absent or non-descript. **Flesh:** Thin, gelatinous or jelly-like. Translucent purple. **Odor:** Indistinct. **Taste:** Indistinct. **KOH:** No reaction. **Spore deposit:** Whitish to pale yellowish, hard to obtain. **Microscopy:** Ascospores (10) 13–20 x 3.5–5 µm, ellipsoid, smooth, colorless, with two oil droplets, 1-3 septate in age. It occasionally has small globose canidiospores when young. Asci 8-spored, amyloid.

Ecology: Solitary, scattered or in troops on logs and other woody debris. It is especially fond of the cut ends of alder and maple logs. Common throughout Cascadia, especially in lower elevation west-side forests. Fruiting from fall into spring.

Edibility: Unknown.

Comments: The small size, gelatinous texture, growth on wood, and purple color help distinguish this species. *A. cylichnium* is reported to be larger, but so similar that it should be confirmed microscopically; it has distinctly larger spores measuring 18–30 x 4–6 µm. Another asco-jelly is the rare *Neobulgaria pura* which is pink to watery pink in color, shallowly cup-shaped when young, and plump jelly-drop shaped in age; it has smaller spores (6.5–9 x 3–4.5 µm). *Ditangium cerasi* has similar-shaped fruitbodies, but is pinkish to reddish at first, fading slightly in age. It is common on cottonwood across lower elevations in Cascadia. Microscopically, it has nearly round basidia and narrowly cylindrical, often curved basidiospores, measuring 7–13 × 2.5–4.5 µm. Like *Ascocoryne*, it can also have canidiospores when young; these are half the size of the basidiospores.

Phaeotremella foliacea

(Pers.) Wedin, J.C. Zamora & Millanes

Conifer Leafy Jelly

Fruitbody: 2–10 (15) cm across, an irregular mass of gelatinous leafy lobes and folds. Brown to reddish brown when young, becoming paler, orangish brown to beige. **Stipe:** Absent; attached to wood with a small to broad attachment point. **Flesh:** Thin to thick, gelatinous or jelly-like when wet, hard when dry. Translucent brown in color. **Odor:** Indistinct. **Taste:** Indistinct. **KOH:** No reaction. **Spore deposit:** Whitish to pale yellowish. **Microscopy:** Spores 5.5–9 x 4.5–8.5 µm, subglobose to ellipsoid, smooth, colorless.

Ecology: Often solitary or scattered on dead logs or trunks of conifers, especially western hemlock, and usually near *Stereum sanguinolentum* which it parasitizes. Locally common in western Cascadia, fruiting from fall through spring, rare in summer.

Edibility: Edible but the gelatinous texture leaves something to be desired.

Comments: The brown, leafy-lobed, gelatinous-textured mass and growth on conifers with *Stereum sanguinolentum* distinguish this species. *P. frondosa* looks similar, but is often larger, and grows on hardwoods alongside other *Stereum* species, usually *S. hirsutum*. It can dry out (and become rather hard) and then reconstitute in the next wet period. The similarly colored *Exidia saccharina* is considerably smaller than *Phaeotremella* and more irregularly lobed or brain-like. *Exidia recisa* grows on hardwoods (especially willow) and has a simple, lobed, red-brown to dark brown fruitbody, a shiny underside, and dull upper surface. Both *Exidia* have cylindrical, sausage-shaped spores. *Auricularia americana* has similar colors and texture but differs in having a simpler, ear-shaped fruitbody with a dull, finely tomentose top, and a shiny underside. The Black Jelly Roll (*Exidia* cf. *glandulosa*), which grows on hardwoods, produces dark-colored, wrinkled, or brain-like clumps or masses with finely warted surfaces.

Synonym: *Tremella foliacea* Pers.

Auricularia americana

Parmasto & I. Parmasto ex Audet, et al.

Western Tree Ear

Fruitbody: 2–8 cm across, generally ear-shaped with a downcurved, undulating to lobed margin. Surface dry, dull, tomentose, covered with short, fine hairs. Brown, vinaceous brown to light cinnamon brown, often with a grayish sheen when dry. **Underside:** Smooth, wrinkled to lobed, moist to dry. Shiny brown, vinaceous brown to light brown. **Stipe:** Absent or rudimentary, lateral and pinched. **Flesh:** Thin, gelatinous or rubbery. Translucent brown in color. **Odor:** Indistinct. **Taste:** Indistinct. **KOH:** No reaction. **Spore deposit:** Not obtained. **Microscopy:** Spores 13–15 x 4–5.5 µm, sausage-shaped, smooth, colorless, often with one or two oil droplets. Basidia long, narrowly clavate, longitudinally septate.

Ecology: Solitary, scattered, or in troops on dead logs and branches of conifers, especially Douglas-fir and true fir. Widespread across Cascadia, but far more common in drier eastern slope forests. Primarily fruiting in spring but can be found in wet periods throughout the year.

Edibility: Edible and good. Despite the rubbery texture, it keeps its "crunch" when cooked.

Comments: The earlike shape, gelatinous to rubbery texture, brown colors, dull cap, shiny underside, and growth on conifer wood readily identify the Western Tree Ear. *Phaeotremella foliacea* often grows in clumps of "leafy" gelatinous lobes and lacks the tomentose upper surface. *Exidia recisa* grows on hardwoods and has a simple lobed red-brown to dark brown fruitbody with a shiny underside and dull, smooth (not tomentose) top.

Tremiscus helvelloides

(DC.) Donk

Apricot Jelly

Fruitbody: 1.5–8 (10) cm across, often spoon shaped to tongue shaped, with a lobed or wavy margin. Surface moist to dry, shiny when wet, dull when dry. Apricot, pinkish apricot to coral pink. **Underside:** Typically smooth, but can be wrinkled or slightly veined. Generally dull and slightly paler than cap. **Stipe:** 2–10 cm tall, 0.5–2 cm across, lateral, often slightly grooved on one side. Concolorous with cap, but often paler, to whitish at base. **Flesh:** Thin to thick, gelatinous or jelly-like when wet. Translucent pink to watery pink in color. **Odor:** Indistinct. **Taste:** Indistinct. **KOH:** No reaction. **Spore deposit:** White. **Microscopy:** Spores 9–12 x 5–6.5 µm, oblong to ellipsoid, smooth, colorless.

Ecology: Solitary, scattered, or in small clusters on soil, duff, or well-decayed wood, often near disturbed areas such as trails or roadsides. Uncommon, but widespread in Cascadia. Fruiting in fall.

Edibility: Non-toxic.

Comments: The apricot to pinkish colors, tongue- or spoon-shaped cap with a lateral stipe, and gelatinized texture make this mushroom as unmistakable as it is delightful. Although usually solitary or sparsely scattered, we have observed some large clusters of fruitbodies.

Taxonomic note: There is some confusion about the correct name for this species. Most sources consider *Tremiscus* to be the correct genus, others use *Guepinia.* It has also been called *Phlogiotis helvelloides.*

Pseudohydnum gelatinosum group

Jelly Hogs, Cat's Tongue

Fruitbody: 1.5–8 (10) cm across, tongue-shaped with an even or wavy margin. Surface moist, dull, roughened or appearing finely hairy. Generally translucent white to pale grayish white, although some forms can be darker brown (see comments). **Underside:** Covered with short, pointed spines. Concolorous with top or contrastingly pale. **Stipe:** 2–8 (11) cm tall, 0.5–2.5 cm across, lateral, often irregular, and slightly grooved on one side. Surface finely roughened or bumpy. Concolorous with cap, but often watery at base. **Flesh:** Thin to thick, gelatinous or jelly-like when wet, drying to a jerky-like or hard texture. Translucent white to watery gray in color. **Odor:** Indistinct. **Taste:** Indistinct. **KOH:** No reaction. **Spore deposit:** Whitish. **Microscopy:** Spores 5–7 x 5–7 µm, globose to subglobose, smooth, colorless. Basidia pear shaped to oblong, longitudinally septate.

Ecology: Solitary, scattered, or in small clusters or troops on conifer branches or decayed logs. Common and abundant throughout western Cascadia, uncommon on the eastern slopes. Fruiting in late fall, continuing through winter at lower elevations, with scattered fruitings into spring.

Edibility: Edible.

Comments: The tongue-shaped cap with a lateral stipe, spiny underside, and gelatinous texture are distinctive. The unusual texture gives this species one of its other common names—Cat's Tongue. Although it is usually translucent white, darker fruitbodies are often found, including a pleurotoid form that lacks a stipe. This form generally has a darker gray to grayish brown cap and grows on conifer logs or branches. This is a species complex, with at least two species occurring in the Pacific Northwest.

23 - Morels, False Morels, and Elfin Saddles

Thanks to the culinary reputation of morels and the unusual toxicity syndrome of some of the false morels, this is a well-known group. The ecology of these ascomycetes is varied with some saprobic, many fruit in response to fire or disturbance, while others occur in less disturbed forest settings. *Helvella* (the elfin saddles) are mycorrhizal, while many *Gyromitra* (the false morels) and *Verpa* (the thimble morels) appear to be, at least, facultatively mycorrhizal. There is evidence suggesting many of these species have a quiescent endophytic phases in their life cycle.

Morchella (true morels) are easy to recognize by their honeycombed heads (with distinct pits and cross walls) and hollow stipes. Fruitbodies of *Helvella* and *Gyromitra* usually have fertile heads with free or partially free edges (not fused to the stipes) and the stipes often have chambered or stuffed interiors. *Verpa* have simple, smooth stipes with distinct, easily detached brownish heads and stuffed, or nearly hollow, interiors.

Morels are edible (for most people) when well cooked; undercooked or raw morels can make you severely sick. Some people appear to be allergic and experience minor to severe gastrointestinal distress after eating them.

Gyromitra are infamous for containing molecular precursors of monomethylhydrazine (MMH). When metabolized, these compounds can have severe negative effects on the central nervous system, leading to nausea and convulsions. This has led many authors to label *Gyromitra* toxic. We agree that they should be avoided but keep in mind that morels also contain this compound, albeit in lower concentrations! As might be expected of a substance also used as rocket fuel, MMH is highly volatile and is driven off by high heat and drying, so always cook your morels thoroughly in a well-ventilated area. Also, some people experience negative effects after eating morels no matter how thoroughly they are cooked.

Important identifying features to note for these mushrooms are the structure of the head (including subtle details of the arrangement of pits and ridges, if present); degree and shape of attachment between head and stipe; overall coloration (and any changes from young to old fruitbodies); details of interior structure (make a cross section); and substrate and surrounding trees.

Morchella eximia

Morchella septimelata

Morchella sextelata

Morchella sp.

Morchella species

Burn Morels, Black Burn Morels

Cap: 2–5 cm across, often tapering upwards, 3–8 (12) cm tall, cylindrical to conical, occasionally more ovoid, rounded or triangular. Vertical ridges smooth to finely tomentose, often parallel, with perpendicular to irregular cross ridges, sometimes eroding in age. Color variable; often with darker ridges and lighter pits when young, more evenly colored in age. Typically brown to dark brown with brown to pinkish to grayish brown pits, or pinkish, pale to dark gray, or evenly olive-brown. **Stipe:** 2–8 cm tall, 2–6 cm thick, equal or enlarged below. Surface dry, smooth to slightly granular in age. Off-white, creamy to pale tan. **Flesh:** Thin, somewhat brittle to slightly rubbery, whitish to tan. Entirely hollow, with a single chamber extending through cap and stipe. **Odor:** Indistinct or musty in age. **Taste:** Indistinct. **Spore deposit:** Ocher-buff. **Microscopy:** Spores 18–25 × 10–15 (17) µm, ellipsoid, smooth. Asci 8-spored, inamyloid.

Ecology: Solitary, scattered, or in large troops or clusters, fruiting from duff and soil during the spring season following forest fires. Burn morels grow in forests of all types but are more common in eastern Cascadia primarily because fires are more frequent in drier east-side forests. They often occur in a single spring flush the year after fires, unless there is considerable disturbance (like salvage logging) on the burn sites. Species composition changes over time and other morels (e.g. *M. brunnea*, *M. snyderi*) will begin to fruit in low severity parts of the burn by year two or three. Burn morels are likely endophytes, "hiding" in cells of living plants for much of their lifecycle and only forming reproductive structures after a fire has killed their host. Fruiting occurs soon after snowmelt through spring and into early summer in particularly wet seasons.

Edibility: Edible and good; like all morels, these species should be well cooked.

Comments: Although abundant in spring in the post-fire landscapes of Cascadia's forests, black burn morels cannot be distinguished in the field. The following closely related species are found here: *M. sextelata*, *M. eximia*, and *M. septimelata* (which some people consider distinct from M. eximia)–all of which cannot be distinguished from each other without genetic analysis. *M. exuberans* tends to have wavy to almost maze-like vertical ridges with shallow cross ridges. It can be distinguished from the others microscopically, as it has distinctly capitate cells lining its ridges. *M. tomentosa* has light gray to dark gray or nearly black colors, and has distinctly dark, fuzzy surfaces when young. It also grows in the burn zones but tends to fruit later than the other species (often into August!). Also compare the "natural" black morels: *M. brunnea*, *M. snyderi*, and others.

Morchella tomentosa

M. Kuo

Fuzzy Gray Morel

Cap: 2–5 cm across at widest point, often tapering toward top, 3–12 cm tall, cylindrical to conical, often more triangular-ovoid in age. Ridges compressed at first, becoming irregular, wrinkled; cross ridges highly irregular. Surface dry, distinctly tomentose when young patchy tomentose to nearly smooth in age; pit walls eroding in age. Dark gray to nearly black when young, gray to grayish tan in age; pits lighter in color. **Stipe:** 2–6 cm tall, 1–4 cm thick, cylindrical or club-shaped. Surface dry, densely tomentose when young, becoming nearly smooth in age. Dark gray to nearly black when young, soon gray, to tan in age. **Flesh:** Thin, somewhat brittle to slightly rubbery. Entirely hollow, with a single chamber extending through cap and stipe. **Odor:** Indistinct. **Taste:** Indistinct. **Spore deposit:** Ocher-buff. **Microscopy:** Spores 18–20 × 8–12 µm, ellipsoid, smooth. Asci 8-spored, inamyloid.

Ecology: Solitary, scattered, or clustered in duff of burnt forests the year following forest fires. Common and widespread in the drier parts of Cascadia but can fruit in wetter areas if they burn. Fruiting in late spring and summer, even into early fall in wetter years.

Edibility: Edible and excellent; like all morels, it should be well cooked.

Comments: This striking morel is easy to recognize due to its dark, tomentose, young fruitbodies, which lose the fuzziness and fade to grayish tan in age. This species tends to fruit after the other black burn morels, in late spring and summer the year following fires.

Morchella brunnea

M. Kuo

Hardwood Natural Morel

Cap: 2–4.5 cm across at widest point, often tapering toward top, 2–7 cm tall, cylindrical to conical. Ridges vertical at first, becoming irregular; cross-ridges sunken; pits fairly small. Ridges dark brown to nearly black; pits paler, brown to gray-brown. **Stipe:** 4–11 cm tall, 1.5–4 cm across, more or less equal or tapered towards base, rarely club-shaped. Surface dry, distinctly punctate to granular, especially in age. Off-white to creamy brown, darkening with age. **Flesh:** Thin, somewhat brittle to slightly rubbery, whitish to tan. Entirely hollow, with a single chamber extending through cap and stipe. **Odor:** Indistinct. **Taste:** Indistinct. **Spore deposit:** Ocher-buff. **Microscopy:** Spores 22–36 (40) × 14–20 (25) µm, ellipsoid, smooth. Asci 8-spored, inamyloid.

Ecology: Solitary, scattered, or in small troops in duff, moss, or soil. Widespread and common in Cascadia; typically with hardwoods (especially cottonwood) and also commonly with snowbrush (*Ceanothus velutinus*) in east-side forests. Fruiting in spring, soon after snowmelt.

Edibility: Edible and good; like all morels, it should be well cooked.

Comments: This species is best recognized by the dark colors, upright cylindrical to conical stature, slender stipe without a buttressed base, and fairly small pits. Habitat is also a helpful clue, as this species typically fruits with deciduous trees and shrubs. During wet springs, it fruits in abundance in *Ceanothus* thickets in east-side pine forests. *M. norvegiensis* can be difficult to distinguish, but tends to have larger, more open pits and grows with conifers. *M. importuna* has parallel vertical ridges with regular ladder-like cross ridges and typically grows in urban or highly disturbed areas.

Morchella snyderi

M. Kuo & Methven

Natural Black Morel, Buttressed Morel

Cap: 3–7 (10) cm across at widest point, 3–6 (9) cm tall, triangular, cylindrical to conical, occasionally more rounded. Ridges compressed at first, becoming irregular wavy; cross-ridges are typically not sunken, and are even with horizontal ridges. Surface dry, smooth; pit walls not substantially eroding in age. Color highly variable, typically pinkish brown, yellowish brown to grayish brown when young, maturing dark brown, at times with greenish olive tones. **Stipe:** 2–7 cm tall, 2–4 (5.5) cm thick, equal or irregular club-shaped, with a buttressing base. Surface dry, finely granular when young, distinctly so in age. Whitish, tan to light brown. **Flesh:** Somewhat thick, slightly rubbery when young, more brittle in age. Hollow, with a single chamber extending through cap into the stipe, lower stipe often a convoluted mass. **Odor:** Pleasant, earthy-musty. **Taste:** Indistinct. **Spore deposit:** Ocher-buff. **Microscopy:** Spores 25–37 × 15–23 µm, ellipsoid, smooth. Asci 8-spored, inamyloid.

Ecology: Solitary, scattered, or clustered in duff under true fir. Most often found on the east side of the Cascade Range or in drier microclimates of western Cascadia. Although considered a 'natural' morel, it prefers slightly disturbed areas in forested settings like clearings in understory vegetation left by fuel reduction work, old logging roads, or near campgrounds. It is also found in relatively undisturbed forests. Fruiting in spring, soon after snow melted.

Edibility: Edible and excellent, with a strong, rich flavor; like all morels, it should be well cooked.

Comments: This common species of east-side forests is recognized by the variably colored, often triangular cap, stipe with a buttressed base, and growth with true fir. *M. brunnea* has a narrower, cylindrical to conical cap and typically grows with hardwoods. *M. norvegiensis* is quite similar to *M. brunnea* but tends to have larger, more open pits and grows with conifers. *M. tridentina* has grayish yellow to yellow-brown colors when young and can develop reddish stains with age; it is found in various conifer habitats across Cascadia.

Morchella tridentina

Bres

False Blond Morel

Cap: 2–5 cm across at widest point, often tapering toward top, 2–7 (10) cm tall, cylindrical to conical, with a shallow depression at stipe attachment. Ridges more or less vertical and often close together at first, becoming irregular; cross-ridges mostly poorly defined or sunken. Grayish to straw yellow, ridges typically paler, becoming more yellowish with age. Slowly staining reddish when damaged or in age. **Stipe:** 4–8 cm tall, 1.5–4 cm across, more or less equal or somewhat bulbous. Surface dry, smooth to granular in age. Off-white to creamy. **Flesh:** Thin, somewhat brittle to slightly rubbery, whitish to tan. Entirely hollow, with a single chamber extending through cap and stipe. **Odor:** Indistinct. **Taste:** Indistinct. **Spore deposit:** Ocher-buff. **Microscopy:** Spores 20–29 × 14–19 µm, ellipsoid, smooth. Asci 8-spored, inamyloid.

Ecology: Solitary, scattered, or in small troops in in duff or soil under conifers and madrone. Widespread in Cascadia; occurring in most forested habitats. This is usually the first morel species to fruit in low elevation west-side forests, continuing through spring, and it is especially common in the Klamath and southern Cascade Range.

Edibility: Edible and good; like all morels, it should be well cooked.

Comments: Even though this species is colored like a blond morel, genetically it belongs to the black morel group! It is distinguished by the relatively regular vertical ridges, and shallow moat-like depression where the cap joins the stipe. Young, more grayish specimens could be mistaken for *M. norvegiensis*, which tends to have larger, more open pits and darkens with age. *M. rufobrunnea* has pale blond ridges and often darker pits in young specimens; it develops yellowish brown colors overall and in age often shows distinct orangish red staining. It is found on wood chips or disturbed soil in urban areas, rarely in forested settings.

Synonym: *Morchella frustrata* M. Kuo

Morchella importuna

M. Kuo, O'Donnell & T.J. Volk

Inconsiderate Morel

Cap: 2–8 cm across at widest point, often tapering upward, 3–15 cm tall, tall-conical, more cylindrical in age. Vertical ridges distinctly parallel, with deeply sunken pits and nearly perpendicular cross ridges, resulting in a ladder-like appearance. Brown to buff-brown, tan or grayish brown when young, becoming darker brown, ridges darkening in age, dark brown to blackish brown, pits lighter. **Stipe:** 3–10 cm tall, 2–6 cm thick, often enlarged below, but pinched at base. Surface dry, smooth to slightly granular in age. Off-white, creamy to pale tan. **Flesh:** Thin, somewhat brittle to slightly rubbery, whitish to tan. Entirely hollow, with a single chamber extending through cap and stipe. **Odor:** Indistinct or slightly fishy, especially in age. **Taste:** Indistinct. **Spore deposit:** Ocher-buff, usually scant. **Microscopy:** Spores 18–24 × 10–13 µm, ellipsoid, smooth. Asci 8-spored, inamyloid.

Ecology: Solitary, scattered, or in large troops in disturbed areas, often in wood chips, gardens, or recently landscaped areas—especially around recently poured concrete. Occasionally in clearcuts or other heavily disturbed forested areas. Fruiting in spring, most often in urban areas.

Edibility: Edible and good; like all morels, it should be well cooked.

Comments: Identified by the dark colors, large size, surprising habits (you never know when or where you might stumble upon this species), and especially by the neat, regular pattern of vertical ribs and ladder-like cross ridges. The Woodchip Morel (*M. rufobrunnea*) grows in similar habitats but has whitish gray to dirty blond ridges and often slowly stains orange-red when bruised or in age. Also see the black burn morels, which fruit during the spring following forest fires.

Morchella americana

Clowez & Matherly

American Yellow Morel, Blond Morel

Cap: 2–5 (8) cm across, 3–10 (15) cm tall, ovoid, conical to irregularly rounded. Ridges and pits more or less randomly arranged, rounded, elongate to irregular, fused directly to stipe. Pits brown to grayish with yellowish to yellow-brown tones at first, fading overall to yellowish, yellow-brown, straw-yellow or reddish yellow when mature, more brown in age. **Stipe:** 2–12 cm tall, 1–4 (6) cm thick, equal or enlarged at base, often with furrows in lower portion. Off-white to creamy, to tan in age. Surface dry, smooth to slightly granular. **Flesh:** Thin, somewhat brittle to slightly rubbery, whitish to tan. Entirely hollow, with a whitish, finely pubescent, single chamber extending through cap and stipe. **Odor:** Indistinct to sweet. **Taste:** Indistinct to nutty. **Spore deposit:** Ocher-buff. **Microscopy:** Spores 18–25 × 11–13 µm, ellipsoid, smooth. Asci 8-spored, inamyloid.

Ecology: Solitary, scattered, or in clusters from soil in duff; typically in riparian areas with cottonwood or ash. Also encountered in urban areas, especially with apple or other fruit trees. Widespread but generally uncommon in Cascadia. Fruiting in spring.

Edibility: Edible and excellent. A more delicate nutty flavor than other western morels. Like all morels, it should be well cooked.

Comments: This species is recognized by its yellowish to grayish brown colors, highly irregular pits (never strongly vertically arranged as in the black morels), and the cap which fuses to the stipe (without a depression). *M. prava* is nearly indistinguishable but tends to have grayer colors and thicker ridges. Both of these species share a preference for hardwood habitats, especially with cottonwood, ash, and fruit trees. *M. rufobrunnea* also has pale blond ridges and often darker pits in young specimens; it develops yellowish brown colors overall and in age often shows distinct orangish red staining. It is found on wood chips or disturbed soil in urban areas and is not restricted to spring fruitings. *M. tridentina* is a "false blond" morel with yellowish colors but is genetically within the black morel group!) and is differentiated by its more vertically arranged ridges and pits, a moat-like depression at the juncture of the cap and stipe, and orangish staining in age. It is widespread in Cascadia but prefers montane conifers.

Misapplied name: *Morchella esculenta* (L.) Pers.

Verpa bohemica

(Krombh.) J. Schröt.

Early Morel, Wrinkled Thimble-cap

Cap: 1–7 cm across, 2–8 (10) cm tall, thimble-shaped, cylindrical to bell-shaped. Margin pinched against but not attached to stipe when young, becoming flared or slightly uplifted in age. Surface dry, longitudinally ridged, deeply furrowed and wrinkled. Ocher-brown to brown. **Underside:** Finely tomentose, tan to light brown. **Stipe:** 5–15 (22) cm long, 1–3 (5) cm thick, equal to club shaped. Surface dry, covered with fine whitish chevrons or granules. Whitish, buff to pale ocher. **Flesh:** Thin, very fragile. Stipe stuffed with whitish to pale pinkish orange pith, with hollow pockets, completely hollow in age. **Odor:** Indistinct. **Taste:** Indistinct. **Microscopy:** Spores 55–87 × 17–22 µm, cylindrical to elongate-ellipsoid, smooth, lacking oil droplets. Asci 2-spored, inamyloid.

Ecology: Solitary or in scattered troops under cottonwoods; common in appropriate habitats across Cascadia. Fruiting in early spring or soon after snowmelt in the mountains.

Edibility: Edible.

Comments: This species is identified by the ridged, deeply furrowed, wrinkled thimble-shaped cap (attached only at the very top of a pith-filled stipe), and the cottonwood habitat. *V. conica* is smaller, with a smooth to finely wrinkled cap, and grows under a wide range of trees. *Morchella populiphila* also grows under cottonwoods but differs by having a thimble-shaped to conical cap attached at about the midpoint, with a single hollow chamber in both stipe and cap.

Verpa conica group

Smooth Thimble-caps

Cap: 1–4 cm across, rounded, thimble-shaped, bell-shaped to broadly conical, margin often pinched to stipe when young, slightly flaring in age. Surface dry, slightly wrinkled when young, becoming smooth, finely cracking in dry weather. Ocher-brown to brown, occasionally dark brown in age. **Underside:** Finely tomentose, pale buff. **Stipe:** 3–12 cm long, 0.5–2 cm thick, equal to club-shaped. Surface dry, covered with fine chevrons or granules, often wrinkled or wavy. Pale buff to pale ocher-buff. **Flesh:** Thin, very fragile. Stipe stuffed with whitish pith; with hollow pockets or completely hollow in age. **Odor:** Indistinct. **Taste:** Indistinct. **Microscopy:** Spores 29–34 × 17–20 µm, ellipsoid, smooth, lacking oil droplets. Asci 8-spored, inamyloid.

Ecology: Solitary or scattered in duff under both hardwoods and conifers (see comments). Uncommon, more likely to be found in the mountains and inland valleys. Fruiting in spring.

Edibility: Edible.

Comments: The slightly wrinkled to smooth, thimble-shaped cap, pith-stuffed stipe, small granules or chevrons ornamenting the stipe surface, and 8-spored asci help define this group. There appear to be at least three species in this group in Cascadia; the most common is the one which matches the description above, typically growing with conifers on the east side of the Cascade Range. Additionally, there is a small, dark-capped species with cottonwood as well as a species with a slightly wrinkled cap that grows with oaks. Extensive collecting and modern taxonomic work are needed to understand the relationships between these taxa and to find (or describe) names for them. *V. bohemica* has a wrinkled and furrowed cap, grows with cottonwoods, and has much larger spores in two-spored asci.

Helvella vespertina (dark form)

Helvella vespertina (pale form)

Hypomyces cervinigenus

Helvella vespertina

N.H. Nguyen & Vellinga

Western Black Elfin Saddle

Cap: 2.5–5 (12) cm across, irregularly lobed, folded, to brain-like. Margin typically fused to stipe; one or more lobes may be upturned. Typically dark gray, but colors range from pale gray to black. **Underside:** Ribbed to smooth, whitish to pale gray. **Stipe:** 4–15 (25) cm long, 1–5 (10) cm thick, irregularly grooved, ribbed and fluted, sometimes with a buttressed base. Whitish, gray to blackish, occasionally with orangish stains in age. **Flesh:** Thin, brittle. Stipe chambered, cartilaginous. **Odor:** Indistinct. **Taste:** Mild. **Microscopy:** Spores 15.5–21.5 × 10–12 (14) µm, broadly ellipsoid, smooth, with a large oil droplet. Asci 8-spored, inamyloid.

Ecology: Solitary or scattered in small troops on ground, duff, or moss under conifers. Common throughout Cascadia. Fruiting in fall and winter.

Edibility: Edible but cook well. Do not eat moldy looking specimens because the common Helvella Mold (*Hypomyces cervinigenus*) is toxic.

Comments: Formerly called *Helvella lacunosa*, this complex of black elfin saddles contains at least four species in Cascadia. Two of these, *H. dryophila* and *H. vespertina*, have been described. *H. vespertina* is by far the most common species in Cascadia, fruiting under conifers in fall and winter. *H. dryophila* fruits under oaks in winter and spring. If you find black elfin saddles in mixed forest in the middle of the winter, you may be out of luck identifying them without a genetic sequence. *H. dryophila* typically has a darker black cap with a contrasting whitish stipe when young and often develops orangish coloration on the stipe in age (but there is much overlap in appearance between the two). The other two species in this complex are poorly known. Both fruit in spring in montane settings species; one of these often shows a massive, highly convoluted cap and stipe, while the other is a dainty species mostly found in the Southern Cascades. *H. maculata* has a fluted stipe and irregularly mottled pale grayish to gray-brown cap that is not fused to the stipe. All of these are commonly parasitized by whitish to pinkish *Hypomyces* molds. *Hypomyces cervinigenus* commonly occurs on *Helvella vespertina*, while additional undescribed and difficult-to-distinguish species occur on other *Helvella* host species. *H. maculata* is often parasitized by a helotialean species that changes the appearance of its host.

Helvella elastica

Bull.

Elastic Elfin Saddle

Cap: 0.5–5 cm across, often distinctly saddle-shaped with two rounded lobes, occasionally more irregularly lobed or folded. Gray to grayish brown. **Underside:** whitish to pale gray, or tan; smooth. **Stipe:** (2) 4–10 cm long, 0.2–0.6 cm thick, equal or enlarged toward base, occasionally compressed or creased. Surface smooth to wavy, dry. White to cream, to grayish tan in age. **Flesh:** Thin, brittle, pale gray. Stipe hollow, white. **Odor:** Indistinct. **Taste:** Mild. **Microscopy:** Spores 18–22 × 11–15 µm, broadly ellipsoid, smooth, with a large oil droplet. Asci 8-spored, inamyloid.

Ecology: Solitary or scattered in small troops on soil, in duff, or moss. Occasional, growing under both conifers and hardwoods (especially cottonwood), mostly in inland valleys and the Cascade foothills in western Cascadia. Fruiting throughout the wet season with peaks in spring and late fall.

Edibility: Unknown.

Comments: The small, gray to grayish brown saddle-shaped cap, smooth underside, and white stipe are helpful identification features. *H. compressa* is quite similar but typically has more pointed lobes and a distinctly pubescent underside. The rare *H. albella* shares the same shape but has a darker brown cap with a finely pubescent to nearly smooth underside. Also compare with the cup-shaped *H. fibrosa* (and others), which can sometimes curl up and resemble *H. elastica.*

Gyromitra californica

(W. Phillips) Raitv.

Pink-stemmed Mitre

Cap: 4–12 (15) cm across, irregularly convex, often lobed and undulating. Margin downcurved, occasionally flaring in age. Surface moist to dry, smooth to wrinkled. Brown, reddish brown to tan, occasionally with olive tones. **Underside**: Finely pubescent to nearly smooth, ribbed. Creamy to light brown, occasionally with pinkish patches. **Stipe**: 2–8 cm long, 1–5 cm thick, irregularly fluted and ribbed, surface finely pubescent to smooth. Tan to yellow brown, often with distinct pinkish to violaceus tones near base. **Flesh**: Thin, brittle to rubbery. Stipe chambered to solid. **Odor**: Indistinct. **Taste**: Mild. **Microscopy**: Spores 16–20 × 8–11 µm, ellipsoid, smooth, with two oil droplets. Asci 8-spored, inamyloid.

Ecology: Solitary or scattered in small troops on decaying logs or woody debris, occasionally on thick duff or soil, often near seeps or areas with saturated soil. Locally common in mid-elevations of western Cascadia, rare elsewhere. Fruiting in spring and summer, rarely in fall.

Edibility: Possibly toxic.

Comments: The brown to tan or reddish, wavy-lobed cap, ribbed to veined, creamy underside, and the deeply ribbed or fluted stipe which develops pinkish coloration from the base upward make this a distinct member of Cascadia's mycota. *G. sphaerospora* looks quite similar but has round spores; it has been reported from the northern Rocky Mountains but we have not seen it in Cascadia. *Helvella* species with fluted stipes typically have more convoluted, lobed, or saddle-shaped caps, or are cup-like.

Synonym: *Helvella californica* W. Phillips, *Pseudorhizina californica* (W. Phillips) Harmaja

Gyromitra montana
Harmaja

Snowbank False Morel

Cap: 4–10 (15) cm across, highly convoluted, wrinkled and folded. Margin fusing to stipe, occasionally hanging free. Orangish to ocher, more ocher-brown to reddish brown in age. **Stipe:** 2–6 cm long, 2–10 cm thick; short, stout, convoluted, folded and creased. Surface dry, smooth. White to tan. **Flesh:** Thick, brittle, orangish to tan. Stipe convoluted and chambered, with pith in pockets. White to translucent tan. **Odor:** Indistinct. **Taste:** Mild. **Microscopy:** Spores 24–35 × 10–15 µm, ellipsoid to ovoid, smooth to slightly roughened, with one large oil droplet, occasionally with two smaller droplets, one at each end. Asci 8-spored, inamyloid.

Ecology: Solitary or scattered in small troops on the ground of duff near the edges of melting snowbanks, persisting for less than two weeks after snow cover has melted. Common at higher elevations and east-side fir forests in the Cascade Range and Olympic Mountains. Fruiting in early spring and into early summer in years with thick snowpack.

Edibility: Edible. Be sure to cook well and use extreme caution not to mistake it for the False Morel, *Gyromitra esculenta.*

Comments: The highly convoluted orangish to ocher caps, stout white stipe, chambered stipe, and growth near melting snowbanks help distinguish this species. The toxic *G. esculenta* has a reddish brown, convoluted to lobed cap, and an irregularly compressed stipe; it typically doesn't form the dense masses of *G. montana. G. infula* typically has a darker, saddle-shaped cap and fruits from woody debris. Three cup- to disc-shaped *Gyromitra* have similar orangish to ocher colors: *G. ancilis*, with a wrinkled, disc-like fruitbody with a short to indistinct stipe; *G. melaleucoides*, with a distinct stipe and pale underside; and *G. olympiana*, with a smooth to slightly wrinkled fertile surface.

Synonym: *Discina montana* (Harmaja) Ginns

Gyromitra infula
(Schaeff.) Quél.

Saddle-shaped False Morel

Cap: 1–8 (12) cm across, often distinctly saddle shaped with two pointed lobes, at times with a more irregularly lobed cap. Margin pinched and attached to stipe. Ocher-brown to reddish brown, darkening to dark vinaceous brown in age. **Underside:** Paler, finely pubescent. **Stipe:** 2–10 cm long, 1–3 cm thick, equal to irregular, often compressed, sometimes with a buttressed base. Surface finely powdery to bald, smooth, dry. Ochraceous to vinaceous brown and often with a whitish base at first, darkening in age. **Flesh:** Thin, brittle to rubbery. Stipe hollow, single to multi-chambered. **Odor:** Indistinct. **Taste:** Mild. **Microscopy:** Spores 17–23 × 7–10 µm, narrowly ellipsoid, smooth, with two oil droplets. Asci 8-spored, inamyloid.

Ecology: Solitary or scattered in small troops on decaying logs, woody debris, duff, or on the ground. Common and widespread in Cascadia. Fruiting throughout the wet season, with the primary flushes in fall and early spring.

Edibility: Not recommended

Comments: The ocher-brown to reddish brown, saddle-shaped cap, irregular hollow stipe, and growth from woody debris are helpful identification features. *G. ambigua* is very similar but is generally slightly darker colored with a more purple-brown stipe. It has much larger spores (21–30 × 8–11 µm). There is another similar, but apparently undescribed, species in Cascadia with consistently smaller fruitbodies. *G. esculenta* can have lobed caps but is more convoluted and wrinkled and fruits in late winter and spring. *Helvella compressa* is smaller and has a grayish, saddle-shaped cap and a white stipe, and does not grow on wood.

Synonyms: *Paragyromitra infula* (Schaeffer) X.C. Wang & W.Y. Zhuang

Gyromitra esculenta group

False Morels

Cap: 2–7 (10) cm across; irregularly lobed, convoluted, wrinkled and folded. Margin hanging free or fusing to stipe. Reddish brown to dark vinaceous brown, occasionally paler red-brown, rusty red to yellow-brown. **Stipe:** 2–10 cm long, 0.5–2.5 cm thick, irregular, often compressed, sometimes with a buttressed base. Surface finely pubescent to smooth, dry. Buff to vinaceous buff, often with a whitish base. **Flesh:** Thin, brittle, pallid. Stipe stuffed with whitish pith, becoming nearly hollow in age. **Odor:** Indistinct. **Taste:** Mild. **Microscopy:** Spores 17–28 × 10–14 µm, broadly ellipsoid to ovoid, smooth, with two oil droplets. Asci 8-spored, inamyloid.

Ecology: Solitary or scattered in small troops on the ground, in moss, or duff under conifers. Common and locally abundant in montane habitats. Fruiting in early spring in the lowlands or soon after snowmelt in the mountains.

Edibility: Potentially deadly poisonous if consumed raw or undercooked due to the presence of gyromitrin. This is an unstable molecule that hydrolyzes into monomethyl hydrazine (a component of some rocket fuel). This chemical is highly volatile and quickly cooks off in vapors, rendering the mushrooms "edible." Keep in mind that if the mushrooms are cooked in a closed pot, the toxins might not escape, and the vapor alone can cause violent illness. Even though some people consume this species, we strongly recommend *not* eating this mushroom. *G. esculenta* has caused a number of deaths in Europe.

Comments: The reddish brown, convoluted cap, irregularly compressed stipe which is stuffed with pith when young, and growth in late winter or spring are distinctive. That being said, *G. esculenta* is a species complex in Cascadia–recent sequence data have found two members of this group; *G. splendida* and *G. venenata*. More more work is needed for distinguishing criteria. *G. infula* is similar, but has a distinctly saddle-shaped cap and typically fruits from woody debris. *G. montana* has an ocher-brown or orangish, highly convoluted cap, and fruits soon after snowmelt in the mountains. *G. californica* has a convoluted cap with a ribbed underside and free edge (not fused to the stipe). Similar *Helvella* species have fluted stipes. Morels (*Morchella* spp.) have honeycombed, conical to triangular caps and show a single hollow chamber extending through both the stipe and cap when cut in half lengthwise.

24 - Cup Fungi

This section contains most of the book's ascomycetes. The definitive feature of ascomycetes is that they bear their spores inside cylindrical cells called asci. If the fruitbodies are discoid, cup-like, or goblet-shaped, it is likely you have found an ascomycete. Although they are much more diverse than basidiomycetes overall, the majority are microscopic and lichenized fungi. Of those that are not, few produce conspicuous fruitbodies—hence our relatively lopsided coverage of basidiomycetes in this book!

Other ascomycetes with differently-shaped fruitbodies are distributed in other sections. Those with honeycombed, brain-like, saddle-shaped, or fluted heads are under Morels, False Morels, and Elfin Saddles. Those with more or less club-shaped fruitbodies (with relatively undifferentiated heads) are in the section on Coral and Club Fungi. A few hypogeous ascomycetes are in the Truffle section, and *Ascocoryne* is included under Jelly Fungus.

When identifying any cup fungus, note the stature, structure, coloration, and texture of the fruitbodies, as well as the substrate from which they are fruiting. For many of these fungi, identification with certainty requires examining microscopic features—most importantly the shape, size, and ornamentation of the mature spores, the presences or absence of oil droplets in the spore, the shape of the sterile cells (paraphyses) between the asci, and whether or not the tips of the asci change color in Melzer's reagent. Most species contain eight spores per ascus. A few species bear fewer or greater than eight spores, which is an important feature to note.

The medium to large, brittle *Gyromitra* have shallow, often wrinkled, orange-brown to gray-brown cup- or disc-shaped fruitbodies, with a thick central stipe. *Helvella, Dissingia, Pseudoplectania,* and *Donadinia* have dark fruitbodies with central stipes. *Sarcosphaera* forms large spherical to crown-shaped cups, and have violet colors. *Urnula* produce relatively large, fleshy fruitbodies with dark-colored upper surfaces and remarkably rubbery-gelatinous flesh.

Peziza and *Otidea* have small to medium, broad cup-, bowl-, or rabbit ear-shaped fruitbodies with fairly thin flesh and a fragile-brittle texture. *Peziza* tends to be most common on rotting logs in the forest and wood chips in urban habitats, while *Otidea* are mycorrhizal and mostly restricted to forest settings, usually in moss or humus. *Plicaria* are similar to *Peziza* but have dark colors and grow from burnt soil.

A few brightly colored genera with medium-sized, shallowly bowl-shaped fruitbodies are included here including the bright scarlet *Sarcoscypha* is found on woody debris, the orange *Aleuria* on disturbed and gravely soil, and *Caloscypha* in moss and duff.

Several smaller, brightly colored species (with small to tiny cup- or disc-shaped fruitbodies) also occur in our area, including *Lachnellula* with orange to yellow fertile surfaces and white to brown undersides in troops on woody debris soon after snowmelt. The bright red to orange discs of *Scutellinia* are adorned with brown "eyelashes" around the margin and grow on wood or soil. *Cheilymenia* also have "eyelashes," but are found on dung. *Calycina* appear as swarms of tiny yellow discs on dead wood. *Chlorociboria* form small, beautiful aqua-green cups and are found on rotting wood.

The tiny *Pyronema* occur in large, fused masses on burnt soil. *Geopyxis* have goblet-shaped fruitbodies which fruit in profusion on burnt soil. *Ciboria* and other non-descript stalked cups often have specific substrates such as cones, mummified berries, or sedge stems.

Dissingia crassitunicata

(N.S. Weber) T. Schumach. & Skrede

Black and White Elfin Cup

Fruitbody: 0.5–5 cm across, cup-shaped to broadly bowl-shaped, with ribs and veins running up the stipe but rarely extending onto the cup. **Upper surface:** Smooth to wrinkled, black, dark brown to gray. **Underside:** Pubescent to smooth, concolorous or slightly paler on the upper portion, paler below, transitioning into a pallid base. **Stipe:** 0.2–2 cm long. 0.2–1.5 cm thick, stout, equal or tapered downward, ribbed, creased and folded. White to gray. **Odor:** Indistinct. **Taste:** Mild. **Microscopy:** Spores 21–25 × 12.5–15 µm, ellipsoid, smooth, with a single oil droplet. Asci 8-spored, 350–400 µm long, inamyloid.

Ecology: Solitary, scattered, or in small clusters on soil, moss, and duff under conifers, especially true firs. Common and widespread on the east side of the Cascade Range, rare to absent on the west side. Fruiting in spring and early summer.

Edibility: Unknown.

Comments: The dark, bowl-shaped cup, and short, pale, ribbed stipe make this an easy species to recognize. *Helvella acetabulum* (sensu PNW) has light to dark brown colors with a slightly paler stipe and ribs that extend from the stipe midway up the cup. It occurs in eastern Cascadia, fruiting in spring. *H. solitaria* has a grayish brown to grayish buff cup, a ribbed, whitish stipe, smaller spores (17–21 × 11–13.5 um) and shorter asci (240–300 µm).

Taxonomic note: *Helvella crassitunicata* N.S. Weber is a synonym. The European *Dissingia (Helvella) leucomelaena* has likely been misapplied to western North American taxa.

Helvella fibrosa

(Wallr.) Korf

Stalked Elfin Cup

Fruitbody: 1–4 cm across, cup-shaped at first, becoming broadly cup-shaped to almost disc-shaped, with a slender stipe. **Upper surface:** Smooth, dark gray to black when young, fading to grayish brown to pale gray. **Underside:** Coarsely to finely pubescent, slightly paler than upper surface. **Stipe:** 1–5 cm long. 0.2–0.5 cm thick, cylindrical. Surface dry, finely pubescent. Black with a whitish base, gray to all whitish. **Odor:** Indistinct. **Taste:** Mild. **Microscopy:** Spores 16.5–20 × 10–12 µm, broadly ellipsoid, smooth, with a single oil droplet. Asci 8-spored, inamyloid.

Ecology: Solitary or in small troops on the ground in moss or duff in mixed woods. Uncommon, fruiting in spring, more rarely summer and fall.

Edibility: Unknown.

Comments: The blackish to gray colors, slender stipe with a cup-shaped, smooth fertile surface, and finely granular underside are helpful in identifying members of this group. *H. macropus* is similar but has larger (18–25 µm long) spindle-shaped, roughened spores and is generally slightly paler and more grayish buff in color. *H. corium* is all black, has a coarsely velvety underside, and a cylindrical to slightly ribbed stipe. *Donadinia nigrella* is also all black but has a slender, fibrillose to smooth stipe and grows on small woody debris and duff near melting snowbanks. *H. solitaria* and *Dissingia crassitunicata* are larger and have thicker, ribbed stipes.

Gyromitra ancilis

(Pers.) Kreisel

Fruitbody: 2–10 (15) cm across, broadly cup-shaped, disc-like, or wrinkled and irregular and wavy, especially in age. **Upper surface:** Smooth to conspicuously wrinkled. Cinnamon-brown, orange to brown. **Underside:** Smooth or finely pubescent in patches, often ribbed near stipe. Whitish, tan to light ocher. **Stipe:** 0.2–2 cm long, 1–3 cm thick, indistinct to short, stout, often with ribs or folds, binding soil at base. **Odor** Indistinct. **Taste:** Mild. **Microscopy:** Spores 27–45 × 12–16 µm, spindle-shaped with pointed appendages at each end, smooth to distinctly roughened, with three oil droplets. Asci 8-spored, inamyloid.

Ecology: Solitary, scattered, or in small clusters on soil, duff, or well decayed wood in conifer forests. Common on the east side of the Cascade Range, rarer on the west side. Fruiting soon after snowmelt, occasionally right on the edges of melting snowbanks, persisting for a few weeks post snow cover.

Edibility: Edible but be sure to cook thoroughly.

Comments: The orange to brown, wrinkled, cup- to irregularly disc-shaped fruitbodies with a paler underside, and growth in early spring help identify *G. ancilis*. *G. olympiana* tends to be slightly smaller and more orange overall, it is tough to distinguish in the field as it has many overlapping features. The latter species can be distinguished by its slightly shorter, broader spores on average (25–40 × 11–19 µm) which either lack the pointed ends, or show broad, truncate knobs. It grows in similar habitats but may be more common than *G. ancilis* in west-side forests. *G. melaleucoides* has a gray to brown upper surface, contrasting paler underside, a more distinct stipe, and smaller, ellipsoid spores (11.5–14.5 × 8–9.5 µm). The rare *Disciotis venosa* tends to be browner and more broadly cup-shaped, with a veined or ridged center, a bleachy odor, and smooth, ellipsoid spores without oil droplets.

Synonym: *Discina ancilis* (Pers.) Sacc. *Gyromitra perlata* (Fr.) Harmaja,

Gyromitra melaleucoides

(Seaver) Pfister

Fruitbody: 1–10 cm across, broadly cup-shaped to disc-like, often wavy or lobed, margin down-curved, plane to uplifted. **Upper surface:** Smooth to slightly wrinkled. Gray-brown, orange-brown to brown. **Underside:** Smooth or finely pubescent in patches, often ribbed near stipe. White, beige to pale gray. **Stipe:** 1–5 cm long, 0.5–3 cm thick, central to off-center, often with few ribs or folds; base tightly binding soil. White to grayish. **Odor:** Indistinct. **Taste:** Mild. **Microscopy:** Spores 11.5–14.5 × 8–9.5 µm, ellipsoid to broadly ellipsoid, smooth to distinctly roughened or warted, colorless, with two oil droplets. Asci 8-spored, inamyloid.

Ecology: Solitary, scattered, or in small clusters on soil, duff, or well decayed wood in conifer forests. Common on the east side of the Cascade Range, rarer on the west side. Fruiting soon after snowmelt in spring and early summer.

Edibility: Edible but be sure to cook thoroughly.

Comments: The gray to brown cup- to disc-shaped fruitbodies with a distinct stipe, pale underside, and ellipsoid spores readily identify this species. *G. ancilis* tends to have brighter orange-brown colors, a distinctly wrinkled disc, and larger, spindle-shaped spores. Also compare with *G. olympiana*, which tends to be more orange, is more common in west-side forests, and has much larger spores (25–40 × 11–19 µm). *Disciotis venosa* tends to be browner and more broadly cup-shaped, with a veined or ridged center, a bleachy odor, and smooth, ellipsoid spores without oil droplets.

Synonym: *Pseudodiscina melaleucoides* (Seaver) X.C. Wang & W.Y. Zhuang

Rhizina undulata

Fr.

Oliebolzwam

Fruitbody: 2–10 cm across, at times fused into larger masses, irregularly disc- to cushion-shaped. Surface moist to dry, smooth, shiny in age. Light to dark brown or red-brown with a whitish to yellowish edge, darkening in age. **Underside:** Finely pubescent, roughened, often with irregular thread to rope-like projections into substrate. Whitish to yellowish or light brown. **Flesh:** Thin to moderately thick, brittle overall, firm near surface with a mealy interior. **Odor:** Indistinct. **Taste:** Indistinct. **Microscopy:** Spores 30–45 × 8–15 µm spindle-shaped, finely warted to roughened, with two to four oil droplets. Asci 8-spored, inamyloid.

Ecology: Scattered or in masses, often fused to duff and soil in burnt forests, on old burn piles, or in old campfire pits. Widespread in Cascadia but uncommon. Typically fruiting in spring and summer.

Comments: The brown to red-brown, disc to cushion-shaped fruitbodies that tend to become fused to duff or soil in burnt areas characterize this species. Several disc- to cup-shaped species of *Gyromitra* can look similar but have a stipe or single attachment point to the substrate and are not typically associated with burned areas. The Dutch name, Oliebolzwam (which translates to oil ball fungus) refers to the resemblances to oliebollens, a Dutch doughnut ball.

Sarcosphaera coronaria

sensu PNW

Tulip Cup, Violet Crown Cup

Fruitbody: 3–15 (20) cm across, starting as a spherical to flattened ball, often with a small opening at the center, soon splitting and becoming cup-shaped with a crown-like margin. **Fertile surface:** Smooth, whitish to pale pinkish lilac when young, soon becoming richer lilac to violet, fading and occasionally roughened and areolate in age. **Underside:** Scurfy, finely pubescent to smooth, white to grayish. Often covered with encrusted soil debris **Stipe:** Indistinct or up to 4 cm long, to 2 cm thick, cylindrical, tapering downward. **Flesh:** Fairly thick, fragile, white. **Odor:** Indistinct. **Taste:** Indistinct. **Microscopy:** Spores 15–17 (22) × 7–9 (10) µm, broadly ellipsoid with blunt ends, smooth to slightly roughened, with two oil droplets. Asci 8-spored; tips amyloid.

Ecology: Scattered, in small clusters, or in gregarious troops under conifers, especially true fir. Typically forming under the duff layer and not becoming exposed until fully open. Common and often abundant on the east side of the Cascade Range and rare on the west side where it is mostly restricted to drier habitats. Fruiting in spring.

Edibility: Mildly toxic.

Comments: The large size, crown-shaped cups with a lilac to violet interior and white exterior, and growth in spring make this an easy species to recognize. Young specimens buried in duff often look like a white ball. This species is a more-conspicuous indicator for 'natural' morels (*Morchella snyderi*) as they share a preference for similar east-side true fir habitats. *Peziza ammophila* also starts out a spherical to egg-shaped flask with a buried, rooting stipe, which then erupts and becomes more cup-shaped with a large, star-shaped opening. It is easily distinguished by its brown colors and growth in coastal sand dune habitats with dune grass.

Pseudoplectania melaena

(Fr.) Sacc.

Fruitbody: 2–5 (8) cm across, shallowly cup-shaped at first, soon expanding to broadly cup-shaped or irregularly disc-shaped, often with an undulating margin, occasionally splitting. **Upper surface:** Smooth, dull, dark olivaceous brown to olivaceous black at first, fading and developing more gray-brown tones in age. **Underside:** Sparsely appressed-tomentose; black. **Stipe:** An indistinct, central to off-center knob-like attachment or up to 3 cm long, 1 cm thick. Fibrillose with sparse hairs; black. **Flesh:** Thin, somewhat tough; stipe fibrous. **Odor:** Indistinct. **Taste:** Indistinct. **Microscopy:** Spores (10) 12–14 µm, globose, smooth, colorless to brownish, often with one oil droplet. Asci 8-spored, inamyloid.

Ecology: Solitary, scattered, or in small clusters on woody debris or surrounding duff. Common throughout much of our range, especially in westside forests. Fruiting from midwinter into spring.

Edibility: Unknown.

Comments: Recognized by the broadly cup- to disc-shaped fruitbodies with a smooth, olivaceous black upper surface and dull black underside, growth on woody debris, and globose spores. *P. nigrella* has a hairy underside and a rudimentary stipe. *Plectania milleri* has a scalloped margin, a felty-tomentose underside, and elongate-ellipsoid spores. *P. harnischii* and *P. melastoma* have orange granules on the underside when young; they have ellipsoid spores, 8–10 (–11) µm wide in *P. harnischii* and 10–12 (–12.5) µm wide in *P. melastoma*. *Pseudosarcosoma latahense* has a distinctly rubbery texture (see *Urnula padeniana*). *Plicaria endocarpoides* is a fragile, dark purplish brown to black cup that grows in burnt areas. *Donadinia nigrella* is entirely black, with a margarita-glass shape, a fertile cup perched atop a tall, thin stipe, and is common near melting snowbanks in the spring.

Synonym: *Pseudoplectania vogesiaca* (Moug. & Nestl.) Seaver

Urnula padeniana

M. Carbone, Agnello, A.D. Parker & P. Alvarado

Black Jelly Jug

Fruitbody: 5–12 cm across, generally cup-shaped with an incurved margin at first, expanding to broadly cup-shaped, occasionally top-shaped when young, often becoming disc-shaped in age. **Upper surface:** Smooth, dull dark grayish black to black when dry, shiny black when wet. **Underside:** Finely velvety, often wrinkled. Dark gray to black. **Stipe:** 2–8 cm long, 1–4 cm thick at apex, tapering downward, typically rather short and stout. Often ribbed or wrinkled. Black. **Flesh:** Thick and rubbery, with a gelatinous core. Translucent gray when young, blackish in age. **Odor:** Indistinct. **Taste:** Indistinct. **Microscopy:** Spores 23–30 × 11–13 µm, elongate-ellipsoid to sausage-shaped, smooth, with one to three oil droplets. Asci 8-spored, inamyloid.

Ecology: Solitary, scattered, or in small clusters on buried woody debris or duff. Common and widespread. Fruiting in winter and early spring at lower elevations, or spring and early summer on the east side of the Cascade Range.

Edibility: Unknown.

Comments: The large, black, rubbery-gelatinous cup- or disc-shaped fruitbodies help distinguish this species. *Pseudosarcosoma latahense* is nearly identical but smaller and with a purplish black to black upper surface, dark gray underside, and with larger spores (24–38 × 9–12 µm) that lack oil droplets. The fruitbodies of *Bulgaria inquinans* have roughened, reddish brown undersides when young and grow in clusters on recently dead tanoak and oak logs. *Pseudoplectania nigrella* has a hairy underside and round spores.

Misapplied name: *Sarcosoma mexicanum* (Ellis & Holw.) Paden & Tylutki

Cups

Donadinia nigrella

(Seaver) M. Carbone, Agnello & P. Alvarado

Midnight Goblet

Fruitbody: (0.5) 1–3 cm across, generally cup- to bowl-shaped with a slender stipe. **Upper surface:** Smooth, dull dark grayish black to black. **Underside:** Finely hairy, often wrinkled. Dull, dark gray to black. **Stipe:** 2–5 cm long, 0.1–0.3 cm thick, more or less equal or slightly enlarged at base. Surface dry, smooth to finely hairy or matted. Dark gray to black. **Flesh:** Thin, somewhat tough; black. **Odor:** Indistinct. **Taste:** Indistinct. **Microscopy:** Spores 22–35 × 10–15 µm, elongate-ellipsoid, smooth, thick-walled, colorless or with yellowish granules in KOH, typically with two small oil droplets. Asci 8-spored, inamyloid.

Ecology: Solitary, scattered, or in small clusters or troops on small woody debris or duff. Typically near the edges of melting snowbanks. Common and widespread at higher elevations. Fruiting in spring and early summer.

Edibility: Unknown.

Comments: The margarita-glass shape of the fruitbodies, black colors, and growth near (and even through!) snowbanks help identify this species. A similar and possibly undescribed species occurs in west-side forests in spring but not until after snowmelt. *Plectania milleri* is typically slightly larger with a short to indistinct stipe and a cup with a scalloped margin. *Pseudoplectania* species have in general shorter stipes and round spores.

Synonym: *Plectania nannfeldtii* Korf

Peziza varia

(Hedw.) Alb. & Schwein.

Large Brown Cup

Fruitbody: 3–12 cm across, irregularly cup-shaped at first, soon broadly cup-shaped, often flaring out and becoming ragged in age. **Upper surface:** Smooth, light ocher-brown to light brown. **Underside:** Finely tomentose to smooth, slightly paler than upper surface. **Flesh:** Thin, fragile. **Odor:** Indistinct. **Taste:** Indistinct. **Microscopy:** Spores 14–18 × 8–10.5 µm, ellipsoid, smooth, lacking oil droplets. Asci 8-spored, with amyloid tip.

Ecology: Clustered or occasionally solitary on rich soil, well-rotted woody debris, or wood chips. Widespread and common. Fruiting throughout wet seasons, more abundant in spring.

Edibility: Unknown.

Comments: This cup fungus is recognized by its large size, pale ocher-brown color, and growth on woody debris or rich soils. *P. arvernensis* is macroscopically indistinguishable species but has larger (14–20 × 8–10 µm) spores with finely punctate (dotted) ornamentation. *P. domiciliana* is usually found in wet basements, bathrooms, or on wet carpets in houses and other indoor settings. Microscopically, the finely warted spores with two small oil droplets, measuring 12–15 × 6–9 µm, readily separate *P. domiciliana* from *P. varia*, which can also grow in domestic areas. *P. vesiculosa* has an incurved margin well into maturity, a distinctly roughened underside, and grows in dung-rich environments. *Legaliana badia* (=*P. badia*) has dark brown colors, grows from soil, and has large (17–22 × 8–10 µm), finely reticulate spores. *P. echinospora* has light to dark brown colors, grows from burnt soil, and has finely warted spores measuring 14–16 × 6–7.5 µm.

Otidea smithii

Kanouse

Purple-brown Rabbit's Ear

Fruitbody: 3–11 cm across, 5–11 cm tall, variably shaped, often a lopsided cup, shorter and slit or folded to one side (shaped like a rabbit's ear). Margin folded in when young, flaring in age. **Upper surface:** Smooth to wrinkled, bare. Dark purplish brown to vinaceous brown, fading slightly to paler lavender-brown or slightly more orange-brown. **Underside:** Concolorous or slightly paler than upper surface. **Stipe:** 1–2 cm long, 0.5–1 cm thick, stout, pinched. Pale whitish, dingy cream to light tan. **Flesh:** Thin, rubbery to somewhat fragile. **Odor:** Indistinct. **Taste:** Indistinct. **Microscopy:** Spores (10) 12–15.5 × 6–8 µm, narrowly ellipsoid, smooth, with two oil droplets. Asci 8-spored, inamyloid.

Ecology: Solitary, scattered, or in small clusters in humus or moss under both conifers and hardwoods. Rare, mostly in coastal and low elevation forests in western Cascadia. Fruiting in fall or spring, occasionally in winter on the south coast.

Edibility: Unknown.

Comments: The relatively large and variably "rabbit's ear-shaped" fruitbodies with dark purple-brown colors are distinctive. *O. alutacea* (group) and *O. rainierensis (=O. microspora)* are generally smaller and more brown to tan. Spore size is the best way to distinguish them: 9–12 µm long for *O. rainierensis*, 12–18 µm long for *O. alutacea*. *O. propinquata* is more evenly cup-shaped, ocher-brown to dark reddish brown colored, with large spores 19–21 × 10–12.5 µm. *Midotis lingua* (=*Wynnella silvicola*) has rather tall "rabbit's ear" cups with vinaceous brown to reddish brown fertile surfaces, paler exteriors, and whitish stipes. Its spores measure 17–24 × 11.5–17 µm and contain a single oil droplet. It is rare in Cascadia, fruiting from soil, moss, or in gravelly areas under conifers. *Peziza* are generally more broadly cup-shaped and have amyloid-tipped asci.

Otidea leporina group

(Batsch) Fuckel

Ocher Rabbit's Ears

Fruitbody: 0.5–2.5 cm across, 2–5 cm tall, cup-shaped to rabbit's ear-shaped, folded when young, flaring in age. **Upper surface:** Smooth, brown, yellow-brown to ocher. **Underside:** Slightly roughened to smooth, concolorous or paler than upper surface. **Stipe:** Absent or short, up to 2 cm long, 0.3–0.8 cm thick, stout, pinched; base with whitish mycelium. **Flesh:** Thin, rubbery to somewhat fragile. **Odor:** Indistinct. **Taste:** Indistinct. **Microscopy:** Spores 12.5–14 × 7–8.5 µm, ellipsoid, smooth, with two oil droplets. Asci inamyloid.

Ecology: Solitary, scattered, or in small clusters in duff or moss under conifers. Widespread in Cascadia. Fruiting in fall.

Edibility: Unknown.

Comments: One of many rabbit's ear cups in Cascadia, *O. leporina* is recognized by the brown to ocher colors, without a rosy blush and spores over 7 µm wide. *O. onotica* and *O. pseudoleporina* are very similar but both tend to have pinkish to rosy blushes to the upper surface. Distinguishing between the two can be difficult; with spores measuring 12–13.5 × 6–7 µm on *O. onotica* whereas *O. pseudoleporina* has more orange to ocher-brown colors and smaller spores (9.5–12 × 5.5–6.5 µm). The narrow spores help distinguish these two species from the most colorful specimens of *O. leporina.* The fruitbodies of *O. oregonensis* have brighter yellow to orange undersides and paler upper surfaces. *O. tuomikoskii* has tall, elongated rabbit's ear-shaped cups which are light brown with a darker fertile surface when young, slightly roughened undersides, and spores measuring 8.5–11 × 5–6 µm. *O. nannfeldtii* shares the tall, narrow stature but has more orange-brown to ocher colors and slightly broader spores (9.5–10.5 × 5.5–6.5 µm). Members of the *O. alutacea* group have browner colors. *Peziza* are generally more broadly cup-shaped and have amyloid-tipped asci.

Plicaria endocarpoides

(Berk.) Rifai

Fruitbody: 2–8 cm across, urn-shaped when young, becoming broadly cup-shaped to nearly disc-shaped, at times more irregular. **Upper surface:** Smooth, dark brown, grayish brown or vinaceous brown. **Underside:** Roughened or warty at first, occasionally smooth in age, light to dark brown. **Stipe:** Absent or rudimentary. **Flesh:** Thin, fragile, brown. **Odor:** Indistinct. **Taste:** Indistinct. **Microscopy:** Spores 8.5–9.5 µm, globose, smooth, with one or more oil droplets. Asci inamyloid.

Ecology: Scattered in small clusters or troops, occasionally solitary, from burnt soil and duff. Especially common during the spring following forests fires in the Cascade Range, but can show up anywhere on burnt soil, including in campfire rings. Fruiting from spring into summer.

Edibility: Unknown.

Comments: The dark brown colors, roughened underside, and growth on burnt soil help distinguish this species. In addition to *P. endocarpoides,* two other smaller, dark colored *Plicaria* have been reported from burnt soil in the Pacific Northwest. *P. trachycarpa* has globose spores measuring 10–12 µm across, ornamented with low rounded warts. *P. carbonaria* has globose, spiny spores 11–13 µm across. The globose spores distinguish both species from brown species of *Peziza* and allies, which have ellipsoid spores. Members of the *Geoscypha violacea* (*Peziza violacea*) group has distinctly violet colors when young.

Caloscypha fulgens

(Pers.) Boud.

Greening Orange Peel

Fruitbody: 1–4 (6) cm across, cup-shaped to broadly cup-shaped, often shorter or creased on one side; margin becoming wavy or irregular in age. Stipe absent. **Upper surface:** Bright orange to yellow-orange, smooth. **Underside:** Concolorous or paler, developing greenish to bluish green stains and discolorations with age. **Flesh:** Thin, fragile, orange. **Odor:** Indistinct. **Taste:** Indistinct. **Microscopy:** Spores 5–7 µm, globose, smooth, moderately thick-walled, colorless. Asci 8-spored, inamyloid.

Ecology: Solitary, scattered, or in small clusters, often in troops on duff or bare soil under conifers, especially true fir. Interestingly, the asexual stage, formerly known as *Geniculodendron*, is a known parasite of conifer seeds. Common in higher elevations and east-side forests, occasional elsewhere. Fruiting soon after snowmelt in spring.

Edibility: Edible and colorful, but bland.

Comments: The orange cups, teal staining, and growth in spring make this an easy species to identify. It is a bright indicator for good morel habitat in east-side forests. Occasionally milky whitish fruitbodies lacking the orange pigment are found; these become strikingly bluish in age. There is an apparently undescribed sister species in west-side forests with the same blue-green staining but yellower overall colors. *Aleuria aurantia* has a bright orange upper surface, pale orange underside, lacks the greenish staining, and has much larger, ellipsoid, reticulate spores.

Aleuria aurantia

(Pers.) Fuckel

Orange Peel

Fruitbody: 1–10 cm across, cup-shaped to broadly cup-shaped; margin becoming wavy or irregular in age. Stipe absent. **Upper surface:** Bright orange to yellow-orange, smooth. **Underside:** Pale orange to whitish near base, finely powdery or downy. **Flesh:** Thin, very fragile, orange. **Odor:** Indistinct. **Taste:** Indistinct. **Microscopy:** Spores 17–24 × 9–11 µm, ellipsoid, containing two oil droplets, ornamented with a coarse reticulum and a projecting tip at each end.

Ecology: Solitary, scattered, or clustered, often in troops or masses on bare or disturbed soil or thin duff. Especially common following logging on skid roads, on trails, or along roadsides. Occurring throughout our range but much more common on the west side of the Cascade Range. Most abundant in fall but stragglers will fruit through winter, often with a spring flush.

Edibility: Edible and colorful, but not flavorful.

Comments: The shallow, bright orange cup with a paler underside, lack of a distinct stipe, and growth in disturbed soil make this species practically unmistakable. *Caloscypha fulgens* produces orange to yellowish cups that stain greenish blue; it is a spring and early summer fruiter, more common in the eastern Cascades. *Sowerbyella rhenana* has smaller, yellower cups with distinct stipes that typically fruit in tight clusters. *Pseudaleuria quinaultiana* is more orange-red and has matted hairs on the underside, giving it a felty appearance. *Otidea* species often have lopsided cup-shaped fruitbodies that are shorter and slit or folded to one side (rabbit's ear-shaped).

Sarcoscypha coccinea

(Jacq.) Lambotte

Scarlet Cup Fungus

Fruitbody: 2–5 cm across, cup-shaped with an incurved margin at first, becoming broadly cup-shaped; margin becoming wavy or irregular in age. **Upper surface:** Bright scarlet-red, smooth. **Underside:** Whitish or pinkish to pale red. Finely hairy. **Stipe:** Short, whitish, central, tapering, or frequently absent altogether. **Flesh:** Thin, fragile, red. **Odor:** Indistinct. **Taste:** Indistinct. **Microscopy:** Spores 25–40 × (9) 12–15 µm, ellipsoid to cylindrical, containing multiple small oil droplets near ends.

Ecology: Solitary, scattered, or in small clusters on woody debris, branches, and twigs of hardwoods, especially bigleaf maple and cottonwood. Fruiting from midwinter into spring.

Edibility: Edible.

Comments: This vibrant species is unmistakable. No other cup fungus in our area has a bright red upper surface and pale underside. *Scutellinia* are much smaller and flatter, typically more orange, and have dark hairs or "eyelashes" around the margin. *Pseudaleuria quinaultiana* is more orange-red in color with matted hairs on the underside, giving it a felty appearance.

Scutellinia scutellata group

Eyelash Cup

Fruitbody: 0.2–1.5 cm across, shallowly cup-shaped to disc-shaped; lower surface and margin distinctly hairy. **Upper surface:** Reddish orange, orange to yellow-orange, smooth. **Underside:** Concolorous or slightly paler, covered with short, stiff dark brown to blackish hairs. Margin with longer, eyelash-like dark hairs. **Stipe:** Absent. **Flesh:** Very thin and fragile. **Odor:** Indistinct. **Taste:** Indistinct. **Microscopy:** Spores typically 15–20 × 12–16 µm (with lots of variation in size), ellipsoid, very finely roughened, but often appearing smooth.

Ecology: Solitary, scattered, or in small troops on decaying logs, woody debris, soil, or in wet mossy areas like stream banks. Common throughout Cascadia. Primarily fruiting from late winter into early summer but can be found throughout the year in damp microclimates.

Edibility: Unknown.

Comments: A large and confusing group of species, *Scutellinia* are recognizable by their smallish, orange to red disc-shaped fruitbodies, dark "eyelashes" around the margin, and growth in damp habitats. *S. scutellata* has been the catch-all name for an untold number of Cascadia's species because identification to species is difficult (or impossible) since no serious taxonomic work has been done on this group in western North America. *Cheilymenia* species are generally more orange to yellow, have sparser hairs, and grow on dung. Two somewhat similar species without 'eyelashes' grow on twigs and needles of conifers (especially those of western redcedar): *Pithya cupressina* has an orangish yellow disc, a tiny tapered stipe, and globose spores measuring 9–12 µm; *P. vulgaris* is slightly larger, brighter orange, with larger spores (12–14 µm).

Cheilymenia fimicola

(Bagl.) Dennis

Eyelash Dung Disc

Fruitbody: 0.1–0.3 cm across, somewhat rounded at first, becoming shallowly cup- to disc-shaped. Margin with fine dark hairs. **Upper surface:** Orange to orangish yellow. **Underside:** Scruffy to finely hairy, orange. **Stipe:** Absent. **Flesh:** Very thin, fragile, orange. **Odor:** Indistinct. **Taste:** Indistinct. **Microscopy:** Spores 17–22 × 9–12 µm, ellipsoid, smooth, lacking oil droplets. Asci 8-spored, cylindrical, inamyloid.

Ecology: Scattered or in small masses on dung, most often that of cows and horses, but also on deer, elk, and bear droppings. Widespread and locally common. Generally fruiting in spring but can occur at any time if the substrate is sufficiently moist.

Edibility: Unknown.

Comments: The small size, orange colors, fine hairs around the margin, and growth on dung help identify this species. *C. stercorea* is very similar but tends to have longer hairs on the margin, finer hairs on the underside, and slightly smaller spores (17.5–19 × 8–9 µm). *C. theleboloides* is paler orangish yellow to yellow and has inconspicuous hairs at the margin. *C. granulata* is orange but lacks the long marginal hairs (but has a scruffy underside) and has spores measuring 15–17 × 6.5–7.5 µm. If the substrate is overlooked, both could be mistaken for ground- or wood-dwelling *Scutellinia*. A number of *Lasiobolus* species have long, upward-pointing hairs and tiny fruitbodies (< 0.1 cm across). *Ascobolus* species are a common group of cup- to disc-shaped dung dwellers that are generally greenish to yellowish when young, with distinctly colored (brown, purple) spores. *Pseudombrophila* are small, brown to yellow-brown cups. All of these (and many others) occur on dung but require microscopic examination and a dedicated taxonomist to differentiate. See *Pyronema omphalodes* for other similar orange disc and cup fungi.

Calycina citrina

(Hedw.) Gray

Yellow Fairy Cups

Fruitbody: 0.1–0.3 cm across, shallowly cup-shaped to disc-shaped, typically flattening in age. **Upper surface:** Smooth, bright yellow to sulfur-yellow. **Underside:** Powdery, very finely downy to smooth. Pale yellow. **Stipe:** Absent or very short. **Flesh:** Very thin and fragile. **Odor:** Indistinct. **Taste:** Indistinct. **Microscopy:** Spores 9–14 × 3–5 µm, elongate, smooth, often with oil droplets. Asci 8-spored, weakly amyloid at tip.

Ecology: Often in small masses or in scattered troops on decaying logs, branches, or woody debris, especially common on decaying alder logs and branches. Common throughout our range. Fruiting through the wet season.

Edibility: Unknown.

Comments: Yellow Fairy Cups might best be thought of as a common name to unite several practically indistinguishable tiny, yellow, disc-shaped fruitbodies growing on wood. Historically, *Bisporella citrina* has been the catch-all name for any fungus matching the above description; however, there are many other species of yellow disc fungi and microscopic examination and the use of technical literature are necessary for identification. *C. claroflava (=Bisporella sulfurina)* is found on wood near Pyrenomycete fungi (usually appearing as swarms of hard, black, slightly shiny dots or lumps). It has smaller and narrower spores (8–11 × 1.5–2.5 µm) but is otherwise indistinguishable.

Synonym: *Bisporella citrina* (Batsch) Korf & S.E. Carp.

Lachnellula arida

(W. Phillips) Dennis

Fruitbody: 0.2–0.8 cm across, shallow cup-shaped to disc-shaped, often infolded when young or dry, eventually wavy or lobed. Typically flattening out or becoming convex in age. **Upper surface:** Smooth, dull to bright yellow or golden orange. **Underside:** Densely covered with short brown hairs. **Stipe:** Short or indistinct. **Flesh:** Very thin, somewhat tough. **Odor:** Indistinct. **Taste:** Indistinct. **Microscopy:** Spores 6.5–8.5 × 3.5–4.5 µm, narrowly ellipsoid, smooth, colorless, sometimes with a single oil droplet. Asci 8-spored, inamyloid.

Ecology: Often in gregarious troops, occasionally scattered on branches and logs, especially of true fir. Common at high elevation and east side-forests. Fruiting as soon as debris is exposed after snowmelt and continuing as long as the substrate stays moist.

Edibility: Unknown.

Comments: The small size, variably yellow to golden-orange upper surface, brown hairy underside, and growth on fir branches near melting snow help identify this species. There are several similar species including *L. agassizii* with a yellowish to orange upper surface and a pale yellowish to whitish underside with white hairs, often a more distinct stipe, and ellipsoid spores measuring 6–8 × 2.5–3 µm. It can be found on a wide variety of conifers. *L. suecica* has a bright golden yellow to orange upper surface, a pallid underside with white hairs, and globose spores; it can be abundant on ponderosa pine soon after snowmelt in east-side forests. *Capitotricha bicolor* (=*Dasyscypha bicolor* and *Lachnella bicolor*) has tiny (< 0.2 cm), stalked cups with an orangish upper surface and whitish underside; it is abundant on herbaceous and small woody debris throughout the wet seasons.

Pyronema omphalodes

(Bull.) Fuckel

Orange Burn Shield

Fruitbody: Individually 0.05–0.2 cm across, cushion- to disc-shaped, but always aggregated in clusters or fused masses, often with whitish mycelial edges. **Upper surface:** Smooth, dull to bright orange to yellowish. **Underside:** Smooth, concolorous, but duller than upper surface. **Stipe:** Absent. **Flesh:** Very thin, fragile. **Odor:** Indistinct. **Taste:** Indistinct. **Microscopy:** Spores 14–16 × 8–9.5 µm, ellipsoid, smooth, colorless, lacking oil droplets. Asci 8-spored, inamyloid.

Ecology: Fruiting in gregarious troops on charcoal as well as burnt soil and duff. Widespread and abundant; it is one of the first fungi to fruit in post-fire landscapes. Fruiting throughout wet periods but more common in spring and fall.

Edibility: Unknown.

Comments: The orange to yellow masses of tiny fruitbodies, often surrounded by whitish, cobwebby mycelium, and growth on burnt substrates are helpful in identification of this common post-fire species. *P. domesticum* is typically more pinkish, salmon, or reddish but is highly variable and occasionally overlaps in color with the orange to yellow tones of *P. omphalodes.* These two species are best distinguished microscopically. *P. domesticum* has larger spores, measuring 16–20 × 10–12 µm. Other similarly colored species on burnt soil include *Anthracobia melaloma*, which has a yellowish to orange disc- or cup-shaped fruitbody, fine brownish hairs on the underside, and oblong-ellipsoid spores measuring 14–22 × 7–11 µm, each with two oil droplets. *A. macrocystis* is very similar to *A. melaloma* but tends to be more orangish to reddish. *Pulvinula carbonaria* is slightly larger (0.1–0.4 cm), forming orange disc- or cup-shaped fruitbodies, grows in moss on burnt soil, and has globose spores.

Ciboria rufofusca

(O. Weberb.) Sacc.

Stalked Cone Cup

Fruitbody: 0.3–1 (2) cm across, flask-shaped at first, becoming cup-shaped to flattened in age, perched atop a narrow stipe. Margin pruinose to smooth, often splitting in age. **Upper surface:** Smooth; brown to orange-brown. **Underside:** Smooth; concolorous with top. **Stipe:** 0.2–2 cm long, ~0.1 cm thick, equal or rooting into substrate. Brownish with a darker base. **Flesh:** Thin; cup fragile; stipe somewhat tough. **Odor:** Indistinct. **Taste:** Indistinct. **Microscopy:** Spores 5–7 × 3–3.5 µm, elongate-ellipsoid, smooth, some with two small oil droplets. Asci 8-spored, amyloid.

Ecology: Solitary to gregarious on Douglas-fir cones or on the cone scales of true fir. Common and widespread across Cascadia. Fruiting in spring.

Edibility: Unknown. Likely non-toxic.

Comments: The habit of fruiting in spring on Douglas-fir cones is the most distinctive feature of this small, stalked, brown cup. There are several other similar stalked cup fungi, many with highly specific host preferences including *C. amentacea* on willow and alder catkins, *C. caucus* on cottonwood and aspen catkins, *C. seminicola* on alder seeds, and the tiny *C. gordonii* on hemlock needles. *Monilinia urnula* fruits out of mummified blueberries while *M. oxycocci* grows on mummified cranberries. *Myriosclerotinia* species grow from sclerotia (hard or corky, oval-shaped resting structures with black outer surfaces, and white pith) embedded in sedge, grass, and rush stems. *Sclerotinia veratri* grows on the debris of corn lilies (*Veratrum* spp.). *Geopyxis carbonaria* is larger, has a fringed margin, and grows on burnt soil.

Humaria hemisphaerica

(F.H. Wigg.) Fuckel

Fruitbody: 1–3 cm across, rounded, almost forming a closed cup at first, soon opening and becoming lobed or wavy. **Upper surface:** Smooth, whitish, grayish to buff. **Underside:** Concolorous, but covered with brown to orange-brown hairs. **Stipe:** Absent. **Flesh:** Thin and fragile. **Odor:** Indistinct. **Taste:** Indistinct. **Microscopy:** Spores 20–24 × 10–13 µm, ellipsoid, coarsely warted, with two oil droplets. Asci 8-spored, inamyloid.

Ecology: Solitary, scattered, or in small troops on duff or in moss under conifers. Widespread and occasional in western Cascadia. Fruiting in fall and spring.

Edibility: Unknown.

Comments: This cup fungus is recognized by its smooth, grayish interior, brownish hairy exterior, and warted spores. However, it likely represents a species complex in North America. Mature cups are often infected with white mold. *Trichophaea hemisphaerioides* is quite similar but the hairs are often sparser on the exterior, with smooth to finely roughened spores, 13–18 × 5–7 µm. It is found on moss and soil, often following forest fires. *Chaetothiersia vernalis* has small, irregularly cup- or disc-shaped fruitbodies, with pale gray to whitish upper surfaces, scattered tufts of hairs on the margin and underside, and smooth, ellipsoid spores lacking oil droplets. It is found fruiting on woody debris at higher elevations soon after snowmelt.

Geopyxis carbonaria

(Alb. & Schwein.) Sacc.

Fairy Farts, Stalked Bonfire Cups

Fruitbody: 0.3–1.5 (2.5) cm across, rounded cup-shaped at first, soon urn-shaped to broadly cup-shaped when old; margin scalloped to ragged. **Upper surface:** Smooth, reddish tan, ocher to orange-brown; margin paler yellowish to whitish. **Underside:** With a powdery bloom to smooth, concolorous or slightly paler than upper surface. **Stipe:** 0.3–2 cm long, 0.1–0.3 cm thick, equal or tapered downward. **Flesh:** Thin, fragile. **Odor:** Distinctly sulfurous, especially when crushed. **Taste:** Disagreeable. **Microscopy:** Spores 12–15 × 6.5–8 µm, ellipsoid to broadly spindle-shaped, smooth, colorless, without oil droplets. Asci 8-spored, inamyloid.

Ecology: Scattered or in small to large troops on burnt soil and duff. Common and widespread, typically fruiting for a few years after forest fires but most abundant in the first spring following fires. Fruiting in spring.

Edibility: Unknown.

Comments: This common cup fungus is readily identified by the urn- to cup-shaped fruitbodies with a distinct stipe, sulfurous odor, and growth on burnt soil. Fruitings can carpet the ground in the spring following forest fires in the Cascade Range. The similar *G. deceptiva* (which we referred to as *G. vulcanalis* in MotRC) has orangish buff to pale orange colors and usually lacks a distinct stipe. It grows on both burnt and unburnt soils. *Tarzetta* species are typically larger and paler, more urn-shaped (lacking a distinct stipe), with a roughened underside, and a cobwebby, scalloped margin when young.

Chlorociboria aeruginascens

(Nyl.) Kanouse

Blue-green Wood Cup

Fruitbody: 0.2–1 cm across, shallowly cup-shaped to irregularly disc-shaped, often with a wavy margin. **Upper surface:** Smooth, bluish white when young, soon turquoise to blue-green. **Underside:** Finely tomentose to bald, often wrinkled in age, paler than upper surface. **Stipe:** Short, often indistinct, off-center to central. **Flesh:** Thin, fragile. **Odor:** Indistinct. **Taste:** Indistinct. **Microscopy:** Spores 5–7 (10) × (0.8) 1–2.5 µm, spindle-shaped, smooth.

Ecology: Scattered in small troops of single or clustered fruitbodies on decaying hardwood, especially alder and bigleaf maple. Common in western Cascadia, occasional at higher elevations and on the east side. Fruitings are most prolific in fall and spring but can be found throughout the year.

Edibility: Unknown.

Comments: The blue-green color, small size, and tendency of the mycelium to stain its wood substrate blue-green are strong identifying features for the genus *Chlorociboria*. The teal-tinted mycelium-stained wood is encountered more often than the actual fruitbodies! The off-center stipe, wavy-edged fruitbody, and small spores help distinguish *C. aeruginascens* from the similar (and confusingly similarly-named) *C. aeruginosa*. Fruitbodies of the latter species generally have a central stipe, more evenly cup-shaped fruitbodies, and larger spores (8–15 × 2–4 µm).

On the Future of Mushroom Identification

This is a book about how to learn the names of mushrooms. This goal has two basic components: The first is learning to observe the traits that characterize different species of mushrooms—the process of *recognition*. The second is learning how to reliably connect widely-understood names to those species—the process of *identification*. The difference between recognition and identification is like the difference between knowing the faces of regular patrons of your favorite local coffee shop and actually remembering their names.

Names for mushrooms come in two fundamentally different categories: *Scientific* names and common names. Scientific names (also called Latin names) can be thought of as equivalent to people's full names (including middle names and surnames), and likewise have to be documented and published according to certain standards and protocols to be considered 'legal.'

On the other hand, common names are just that—informal. There are no requirements or standards for establishing them, and there can be many different versions referring to the same organism, or many different organisms going under one common name! For example *king bolete*, *porcini*, *cep*, and *steinpilz* all refer to roughly the same mushroom species (more on what constitutes a 'species' later). Likewise, Chuck, Charlie, and Chaz might be nicknames for the same fellow, or might refer to three different friends all named Charles! This imprecision doesn't mean that common names are 'incorrect'—you aren't *required* to refer to someone by their full legal name, and in fact, it would be awkward in many settings! However, there are plenty of other settings in which using a nickname rather than a full name would be just as inappropriate. *When choosing what name to use, context and priorities matter*. It's for this reason that we list both scientific names and common names in this book.

As you use this book to go about learning names for mushrooms, you might wonder in what communities your newly-learned vocabulary will be understood. In other words: *Who uses mushroom names?* You might also ask a related question: *Where do these names come from?*

If we look past the relatively small number of well-known edible, toxic, and 'magic' mushrooms, names for fungi are used by two main communities, which we can coarsely divide into 'academics' and 'amateurs'. Academics are usually professional researchers embedded in institutions, often pursue new knowledge for its own sake, and frequently use state-of-the art technology and methods that require years of study to implement and interpret. Academic communities focused on the organization of biological diversity (a field called taxonomy) have long been the main source of new scientific names for mushrooms – the practice of formal naming is called *nomenclature*.

Amateurs, on the other hand, have been less often involved in the direct production of new scientific names, and are usually more interested in using these names for practical reasons—communicating about what edible mushrooms are fruiting at the moment, what interesting-looking species are fruiting nearby, or asking whether the mushrooms growing at the base of the fruit tree in their yard indicate that the tree is unhealthy. For these purposes, a rough idea of the identity of the mushroom using an English-sounding common name (e.g. 'honey mushroom' rather than *Armillaria mellea*) tends to be better-received and more memorable.

Although there's obviously extensive (and growing) overlap between academic communities and 'amateur' communities—not to mention all of the communities that don't fit neatly into either category—these two groups have significantly different priorities, needs, and cultural norms. For example, there seems to be acceptance of (and perhaps even some manic delight in) extreme nomenclatural instability among academics, whereas amateurs have long been frustrated by the seemingly endless turnover of names for organisms that remain familiar to them. This stems from a demand in the academic community for rigor in application of names to ever-changing concepts of species. Amateur communities are generally tolerant of much less precision—often favoring common names and roughly-defined species groups because of pragmatic concerns like ease of recognition, memorability, and pronounce-ability.

Although these are age-old tensions, they have reached a fever pitch in recent years. *Why?*

The answer lies largely in the advent of DNA sequencing technology. Although it's been around for a half-century now, it's only in the past two decades that this technology has reached 'maturity'—defined by a dramatic decrease in cost, subsequent commercialization, and most importantly, the consequent widespread use by non-specialists. Now the full scope of its impacts on academic phylogenetic research have become clear, including not only a better understanding of the ecology and evolution of fungi, but also a mountain of catch-up nomenclatural work, as well as downstream effects on hobbyists, amateurs, and field naturalists who largely "inherit" names for organisms from academic sources.

The ease with which we can now obtain DNA sequences has led to the presumption that they are required. Given how useful they can be, and how cheap they are to obtain, it is a reasonable feeling that there's no excuse not to obtain DNA data! This feeling has worked its way deeply into the attitudes and assumptions of field workers, even if they themselves may never directly engage with the process of obtaining or interpreting DNA sequences.

Although it's true that distinctions have long been made between 'field' and 'herbarium' taxonomists, that distinction was primarily one of style: Both approaches involved the application of the five basic human senses (extended by microscopes and other relatively modest technologies). But for the modern-day taxonomist, genetic data are the daily bread-and-butter. This represents a fundamental shift in the way that the *recognition* of organisms happens—we have tread into unsensory territory. No one will ever directly detect the nucleotide sequence of an organism's ribosomal DNA, nor do we stand a chance of mentally computing the differences between such sequences, once obtained! This is the realm of machines.

So new arrangements of evolutionary groupings are proposed (often conflicting with one another), and new names are generated because of these new groupings, and much of it stands on an unsensory foundation. *It is increasingly the case that the 'official' names for organisms are supported by evidence that cannot be determined or weighed by human senses.*

What are the consequences of this technology-mediated change? For starters, the proliferation of names has becoming increasingly difficult to use for field mycologists, and unsatisfying for amateurs. See for example, the recent trend towards 'oversplitting' — perhaps best exemplified by *Cortinarius*. For most of a century, *Cortinarius* was treated as a huge genus, comprising thousands of species of diverse appearance but united by a few easy-to-learn characters which could usually be readily evaluated in the field. But some authors considered it too unwieldy, which culminated in the infamous break-up of *Cortinarius* into ten smaller genera (with more on the way). This split triggered a huge number of name changes, record updates, and reshuffling of specimens in museum fungaria (all of which required expenditure of people's time and brainpower). This massive split was based on a foundation of whole-genome data (Liimatainen et al., 2022), and resulted in groupings of species that are difficult to explain, or recognize using one's senses (unsatisfying and difficult to use for field workers!). To make matters more frustrating, just two years later, a different group of authors published a re-analysis of the genetic data, and argued that it was insufficient to support the split (Gallone et al., 2024), leaving the many new names in limbo. One can practically hear the groans of the myriad mycological bystanders...

For our part, we hope to see the broad taxonomic community (including representatives of all the different communities with an interest in the names of species) come up with new, flexible approaches for balancing the conflicting concerns of evolutionary biologists (seeking cladistic precision) and those of general naturalists and field workers (who often prioritize utility for recognition and communication). There is a desperate need for a 'stabilizing committee' to consider and recommend adoption (or rejection) of newly-proposed names and changes to existing names. This kind of committee exists for many other groups of organisms, and helps to take some of the burden of judgement off of end-users of names—including amateurs, land management agencies, conservation organizations, and even the broader community of ecologists and other academics who are not primarily taxonomists.

The amateur community can play an independent, complementary role in this effort by developing and disseminating standardized common names. Common names go a long way towards easing tensions between the conflicting needs of different communities because they are free to 'float' (that is, they don't need to maintain 1:1 reference to scientific names)—and thus remain stable even when underlying scientific names or arrangements change. Likewise, they can serve as handles for groups of organisms that are differentiable with one's senses, and which satisfyingly approximate species-level identification without claiming to precisely separate different evolutionary lineages. Common names can also be effective for use in referring to the most taxonomically awkward groups (e.g.'crust fungi').

In the end, it's perhaps most important to remember that although the practice of taxonomy and nomenclature make use of many lines of measurable data as evidence, decisions about how to interpret this data remain fundamentally subjective, and no arrangement (even those of the academic scientists) perfectly reflects 'objective' reality. Preferences, convenience, tradition, and material constraints all influence our choices about how to represent and refer to the relationships between biological entities. Although we may never be free of mess and muddle, as a community, we continue to make progress towards communication and understanding. The trick is to stick to the struggle with as much grace as we can muster!

Acknowledgments

A book like this would be impossible without the extensive community of people who played myriad roles in its creation. We are humbled by the thought, time, and effort that so many people have put into building knowledge about fungi and their habitats in Cascadia, and we are deeply grateful for the generosity shown to us by friends and strangers alike during our time spent there.

Besides her direct contributions to the taxonomy of fungi, Else Vellinga has once again been an unequaled mentor and reviewer. Some of our mistakes undoubtedly still appear in this book, but far fewer than if we hadn't had Else's keen-eyed reading and admonitions to do better!

Many people have worked on the taxonomy and identification of fungi in the Pacific Northwest over the past decade. In particular, we'd like to acknowledge and thank: Steve Trudell for sharing his knowledge on Cascadian fungi, and for his inspirational attention to detail; Michael Castellano for all his insight and teaching on truffles; David Largent for all the discussion and wisdom over the years; Joe Ammirati for his insight on Cascadian fungi, especially *Cortinarius*; Shannon Adams, who shared her *Cortinarius* data set and detailed notes; Ron Hamill for sharing his deep reservoir of knowledge about Cascade Range fungi; Ian Gibson for creating MycoMatch; and the late Jim Ginns, Lorelei Norvell, and Oluna Ceska for their enduring contributions to our understanding of the fungi of the Pacific Northwest.

Special thanks to David Arora for writing the books that set the current wave of mushroom appreciation in motion, and for sharing his unparalleled breadth of knowledge on mushrooms and mushroomers in the western United States, as well as his points of view on fungi and culture globally – these have shaped and greatly expanded our perspectives.

A special thanks to Danny Miller, for his diligent work to consolidate knowledge and maintain a blisteringly up-to-date website. Likewise, Yi-Min Wang has been a model for community scientists – efficiently distributing frequent, clearly-communicated updates on results from his sequenced collections. Stephen Russell has been an important and generous resource for community scientists from across the country.

To Alana McGee from Truffle Dog Company, and her dogs Lolo and Ruby for finding the *Leucangium* for us to photograph, and to Heather Dawson and Hilary Rose Dawson and Heather's dog Rye for the *Kalapuya*.

To Michael Kauffmann and Allison Poklemba for the opportunity to work with Backcountry Press, and their patience with us to make this monstrosity happen!

Noah:

I am much in debt and gratitude to Milton and Reba Tam, who first got me to Cascadia, and have graciously hosted me over the years. Likewise, I am forever grateful for the generosity of Joann and Eric Olson, who had a home for me whenever I was behind the redwood curtain. To Agnes & Dick Sieger, Alissa Allen, Anna Moore, Bitty Roy, Bruce Newhouse & Peg Boulay, Chris Melotti & Molly Widmer, Dan Long & Liann Finnerty, Daniel Luoma, Don Moore & Pam Buesing, Jonathan Frank, Maureen & Paul Ter Beek, Michael Beug, Susie Holmes, and Terry Allaway, all of whom have provided hospitality over the years. Other support has come from Kate Mohatt, Steve Trudell, Wendy Boes, Terry Henkel & Mei Lin Chin, James (Animal) Nowak, Wren Hudgins, Joseph Zapotosky, Lisa Pelgrim, Clare McLean, Patti Adams, Jessie Uehling, Richard Tehan, Coonor Dooley, and many of the Cascadian mushroom clubs who have had me present and participate in events; especially the Puget Sound Mycological Society. To my parents, Larry Siegel and Kathy Morris, who allowed me to be different, and pursue this path. And to my partner and favorite mushroom companion Shannon Adams for tolerating me over the years.

Christian:

Thanks to Kim Traverse and Teddy Baz for inviting me up to my first gigs in the PNW, to Maria and Taylor Morrow for being hospitable hosts and adventure pals, to Bob Chapman for hosting me and hoofing it up to Porpolomopsis, to Don Moore and Pam Buesing for being gracious hosts, providers of pawpaw, and some of the most delightful people ever to have graced me with their friendship. Thanks to Melanie Swails and Charlie and Linnaea Wright for their hospitality. Extra special thanks to Buck McAdoo for Marlborough quince pie, "le chien qui fume", tales of the way things were, and the big book of mushroom photos in your office. Thanks to Pop and Mamita and Chris and Elizabeth for the encouragement, patience, and support of all kinds. Thanks to Kristen Richardson for affirmation and support during my moments of frustration and overwhelm.

The production of knowledge about fungal biodiversity outside of traditional academic settings has advanced rapidly since we wrote Mushrooms of the Redwood Coast. Although we certainly relied on our peers and friends for their knowledge back then, it's now a central part of our process. We'd like to recognize the following people in particular for their exceptional contributions to our knowledge of the status and distribution of macrofungi in Cascadia: Leah Bendlin, Scott Davison, Heather Dawson, Hilary Rose Dawson, Connor Dooley, Jonathan Frank, Ann Goddard, Ron Hamill, Sadie Hickey, Drew Henderson, Jack Johnson, Matthew Koons, Paul Kroeger, Sava Krstic, Bee Marcotte, Clare McLean, Daniel Morton, Bruce Newhouse, Drew Parker, Mike Potts, Bitty Roy, Lauren Ré and Kendra Saeva.

Resources for the Mycophile

Amanitaceae Studies

www.amanitaceae.org

Rod Tulloss' impressively comprehensive (and interactive) web portal for all things Amanita. Although global in scope, the focus is on North America. Includes a lot of cross-referenced microscopic and genetic data, and profiles for many species that are not yet formally described.

Cyberliber

www.cybertruffle.org.uk/cyberliber

Thanks to Cyberliber, digital copies of many journals, old books, and other mycology print resources are available for free online. The site is a bit clunky, and the scanned resources are not searchable and often must be paged through one page-scan image at a time, but this is an invaluable resource for engaging primary literature.

Danny's DNA Discoveries

www.alpental.com/psms/ddd

An exceptionally important resource for every mushroom enthusiast in Cascadia, this is Danny Miller's heroic effort to get a comprehensive handle on the diversity of macrofungi in the Pacific Northwest through genetic sequencing. Frequently updated with brief but very current summaries of the state of knowledge.

Hebeloma

www.hebeloma.org

A great resource for *Hebeloma*, with deep background information and data from extensive original research by Henry Beker. Global in scope, with significant coverage of the western United States.

iNaturalist

www.inaturalist.org

iNat is one of the most vibrant community science projects in the world, and increasingly the default option for submitting and exploring records of biodiversity (not limited to fungi). Anyone with even a passing interest in natural history should consider creating an account and becoming familiar with the use of the tools of this platform (both the mobile app and the full website version). The importance of being able to network with a huge community of scientists, explorers, and enthusiasts around the world in real time, as well as the ability to easily create and explore rich datasets (including basic on-the-fly data visualizations) is hard to overstate.

Index Fungorum

www.indexfungorum.org

This is a nomenclature database that also acts as a repository for species names and a publishing outlet for new names. An excellent resource for researching synonymy and priority naming issues. Remarkably up-to-date.

MycoBank

www.mycobank.org

Repository of mushroom names, including their authors, dates and details of publication, and often much more (including some descriptions, and illustrations of microscopic details).

Book Resources Website

Mycology Collections Portal

www.mycoportal.org

MyCoPortal is an important resource for finding specimen-supported records of North American fungi. Historical herbarium specimens and their associated metadata are digitized and georeferenced, making it much easier to assemble comprehensive species lists for particular regions.

MycoMatch (formally MatchMaker)

www.mycomatch.com

MycoMatch is an invaluable resource, designed by Ian Gibson. It's interactive computer-based key to fungi in the Pacific Northwest, with detailed descriptions of more than 4,200 species and photographs of ~2,300 species. It works by picking distinctive features of your mushroom from drop-down options, and getting presented matches with the features selected.

MykoWeb

www.mykoweb.com

Lots of great mycological resources such as digital copies of out-of-print books and journal articles. Additionally, it has "The Fungi of California," which currently details nearly 800 species: http://www.mykoweb.com/CAF/new.html.

North American Cortinarius Collections

www.nacorts.com

Spearheaded by Shannon Adams, NA Corts is an online resource dedicated to the study and taxonomy of North American species of Cortinariaceae, with a focus on Cascadia. The site offers a growing set of detailed descriptions, pictures and identification tools for this diverse family.

Cascadia's Mushroom Clubs

There's no better way to learn about mushrooms than by surrounding yourself with the knowledgeable people of Cascadia's many mushroom clubs. Most have monthly meetings throughout the mushroom season and offer forays to collect and identify fungi.

Cascade Mycological Society, Eugene, OR
www.cascademyco.org

Central Oregon Mushroom Club, Bend, OR
www.centraloregonmushroomclub.com

Estacada Fungus Association, Estacada, OR
www.estacadafungusassociation.org

Humboldt Bay Mycological Society, Arcata, CA
www.hbmycologicalsociety.org

Lincoln County Mycological Society, Newport, OR
www.lcmycosociety.org

Kitsap Peninsula Mycological Society, Bremerton, WA
www.kitsapmushrooms.org

North American Truffling Society, Corvallis, OR
www.natruffling.org

Northwest Mushroomers Association, Bellingham, WA
www.northwestmushroomers.org

Oregon Mycological Society, Portland, OR
www.wildmushrooms.org

Puget Sound Mycological Society, Seattle, WA
www.psms.org

Snohomish County Mycological Society, Everett, WA
www.scmsfungi.org

South Sound Mushroom Club, Olympia, WA
www.southsoundmushroomclub.com

South Vancouver Island Mycological Society, Victoria, BC
www.svims.club

Southwest Washington Mycological Society, Chehalis, WA
www.swmushrooms.org

Vancouver Mycological Society, Vancouver, BC
www.vanmyco.org

Wild Rivers Mushroom Club, Brookings, OR
www.wildriversmushroomfestival.com

Willamette Valley Mushroom Society, Salem, OR
www.wvmssalem.org

Recommended Field Guides

Siegel, N., C. F. Schwarz. (2016). *Mushrooms of the Redwood Coast: A Comprehensive Guide to the Fungi of Coastal Northern California.* Ten Speed Press. Abbreviated as MotRC, our prior book is focused on California, picking up where the southern edge of Cascadia leaves off. However, there is significant overlap in species assemblages between the two bioregions — for example, many hardwood species that are uncommon or rare in Cascadia appear in MotRC. The two books are complementary and were written to dovetail for comprehensive coverage of the macrofungi of the Pacific Coast of the contiguous 48.

Arora, David. (1986). *All That the Rain Promises and More: A Hip Pocket Guide to Western Mushrooms.* Ten Speed Press. Consistently among the most popular guides to mushrooms in North America, this slender booklet is jam-packed with delightful photos, amusing, odd, and moving stories, and heaps of information about edible mushrooms and the people that love them. No mushroomers shelf is complete without a copy of this book.

Læssøe, Thomas and Jens H. Petersen. (2019). *Fungi of Temperate Europe.* Princeton University Press. Although the focal area is on another continent, these books are still extremely useful for our area, owing to the high degree of similarity (at the genus level) between the fungi of northern North America and temperate Europe. There is hardly another resource that comes close to the visual richness and comprehensiveness achieved by this stunning series. Although the first volume will be of interest to many mushroom hunters, it is Volume 2 (dealing with polypores, crusts, cyphelloids, and ascomycetes) that we think is most likely to produce advances in our understanding of Cascadia's fungal diversity.

Trudell, Steve, and Joe Ammirati. (2009). *Mushrooms of the Pacific Northwest.* Timber Press.

Trudell, Steve. (2022). *Mushrooms of the Pacific Northwest (revised edition).* Timber Press. Between these two editions (which, curiously, are significantly different from one another), MoPNW cover 700+ species. This breadth, in combination with Steve's outstanding photography and the excellent introductory text make these a must-have for any serious student of fungi in the northwestern forests.

Arora, David. (1986). *Mushrooms Demystified.* 2nd ed. Ten Speed Press. One of the best introductory texts for a group of organisms ever written. This is a true masterpiece, as attested to by its evergreen popularity and utility even decades after its original printing. Although it covers the continent more broadly, the well-structured dichotomous keys are by far the best of their kind available for the beginning mushroomer. Get this book!

Winkler, Daniel. (2023). *Fruits of the Forest: A Field Guide to Pacific Northwest Edible Mushrooms.* Mountaineers Books. A well-photographed guide to the edible mushrooms of Cascadia, from one of the most charismatic voices in the mushroom community.

MacKinnon, Andy and Kem Luther. (2021). *Mushrooms of British Columbia.* The Royal British Columbia Museum. A well-researched field guide covering 350 of the more commonly-encountered mushrooms of British Columbia, but broadly applicable as an introductory guide to Cascadia's fungi.

References and Resources

Beug, Michael, Alan E. Bessette, and Arleen Bessette. (2014). *Ascomycete Fungi of North America: A Mushroom Reference Guide.* University of Texas Press.

Castellano, M., Jane E. Smith, Thom O'Dell, Efrén Cázares, and Susan Nugent. (1999). *Handbook to Strategy 1 Fungal Species in the Northwest Forest Plan.* General Technical Report PNW-GTR-476. United States Department of Agriculture.

Castellano, M., Efrén Cázares, Brian Fondrick, and Tina Dreisbach. (2003). *Handbook to Additional Fungal Species of Special Concern in the Northwest Forest Plan.* General Technical Report PNW-GTR-572. United States Department of Agriculture.

Desjardin, Dennis E., Michael G. Wood, and Frederick A. Stevens. (2016). *California Mushrooms: The Comprehensive Identification Guide.* Timber Press.

Exeter, Ron, Lorelei Norvell, and Efrén Cázares. (2006). *Ramaria of the Pacific Northwestern United States.* USDI BLM/OR/WA/PT-06/050-1792.

Hesler, Lexemuel R. (1969). *North American Species of Gymnopilus.* Hafner Publishing.

Hesler, Lexemuel R., and Alexander H. Smith. (1963). *North American Species of Hygrophorus.* University of Tennessee Press.

Hesler, Lexemuel R., and Alexander H. Smith. (1965). *North American Species of Crepidotus.* Hafner Publishing.

Hesler, Lexemuel R., and Alexander H. Smith. (1979). *North American Species of Lactarius.* University of Michigan Press.

Kauffman, Michael. (2013). *Conifers of the Pacific Slope: A field guide to the conifers of California, Oregon, and Washington.* Backcountry Press.

Kerrigan, Richard. (2016). *Agaricus* of North America (Memoirs of The New York Botanical Garden Volume 114) *NYBG Press.*

Læssøe, Thomas and Jens H. Petersen. (2019). *Fungi of Temperate Europe.* Princeton University Press.

Largent, David L. (1995). *Entolomatoid Fungi of the Western United States and Alaska.* Mad River Press.

Largent, David L., David Johnson & Roy Watling. (1977) How to Identify Mushrooms to Genus, Vol. III: Microscopic Features. *Mad River Press.*

Norvell, Lorelei L., and Ronald L. Exeter. (2008). *Phaeocollybia of Pacific Northwest North America.* Salem, OR: U.S. Department of Interior, Bureau of Land Management.

Periodicals

Ammirati J, Liimatainen K, Bojantchev D, Peintner U, Kuhnert-Finkernagel R, Cripps C, Dentinger B, Niskanen T. (2021) *Cortinarius* subgenus *Leprocybe*, unexpected diversity and significant differences in species compositions between western and eastern North America. *Persoonia* 46:216–239.

Arora, David, and Giampaolo Simonini. (2008). California Porcini: Three New Taxa, Observations on Their Harvest, and the Tragedy of No Commons. *Economic Botany* 62, no. 3: 356–75.

Audet, Serge. (2010). Essai de découpage systématique du genre *Scutiger* (Basidiomycota): *Albatrellopsis*, *Albatrellus*, *Polyporoletus*, *Scutiger* et description de six nouveaux genres. *Mycotaxon* 111: 431–64.

Baroni, Timothy J., Valerie Hofstetter, David L. Largent, and Rytas Vilgalys. (2011). *Entocybe* is proposed as a new genus in the Entolomataceae (Agaricomycetes, Basidiomycota) based on morphological and molecular evidence. *North American Fungi* 6(12): 1–19.

Baroni, Timothy J., and David L. Largent. (1989). The Genus *Rhodocybe*: new combinations and a revised key to Section *Rhodophana* in North America. *Mycotaxon* 84(1): 47–53.

Beug, Michael. (2017). Summary of the Poisoning Reports in the NAMA Case Registry for 2006 through 2017. www.namyco.org. https://namyco.org/wp-content/uploads/2023/08/2006-2017_Summary_of_Mushroom_Poisoning_Reports.pdf

Bonito, Gregory, James M. Trappe, Pat Rawlinson, and Rytas Vilgalys. (2010). Improved resolution of major clades within *Tuber* and taxonomy of species within the *Tuber gibbosum* complex. *Mycologia* 102(5): 1042–57.

Co-David, D., D. Langeveld, and M. E. Noordeloos. (2009). Molecular phylogeny and spore evolution of Entolomataceae. *Persoonia* 23: 147–76.

Dima, B., Liimatainen, K., Niskanen, T. et al. (2021). Type studies and fourteen new North American species of *Cortinarius* section *Anomali* reveal high continental species diversity. *Mycol Progress* 20, 1399–1439.

Gallone, Brigida., Thomas W. Kuyper, & Jorinde Nuytinck. (2024). The genus *Cortinarius* should not (yet) be split. *bioRxiv* 2024.05.29.596376

Grob, CS, Danforth AL, Chopra GS, Hagerty M, McKay CR, Halberstadt AL, Greer GR. (2011). Pilot study of psilocybin treatment for anxiety in patients with advanced-stage cancer. *Arch Gen Psychiatry* 68(1).

Haight, John-Erich, Karen K. Nakasone, Gary A. Laursen, Scott A. Redhead, D. Lee Taylor & Jessie A. Glaeser. (2019). *Fomitopsis mounceae* and *F. schrenkii*—two new species from North America in the *F. pinicola* complex. *Mycologia* 111(2), 339–357.

Hall, D., and D. E. Stuntz. (1971). Pileate Hydnaceae of the Puget Sound Area. I. White-spored genera: *Auriscalpium*, *Hericium*, *Dentinum*, and *Phellodon*. *Mycologia* 63, no. 6: 1009–27.

Hall, D., and D. E. Stuntz. (1973). Pileate Hydnaceae of the Puget Sound Area. III. Brown-spored genus: *Hydnellum*." *Mycologia* 64: 560–90.

Johannesson, Hanna, Svengunnar Ryman, Hjördis Lundmark, and Eric Danell. (1999). *Sarcodon imbricatus* and *S. squamosus*—two confused species." *Mycological Research* 103: 1447–52.

Justo, Alfredo, Ekaterina Malysheva, Tatiana Bulyonkova, Else C. Vellinga, Gerry Cobian, Nhu Nguyen, Andrew M. Minnis, and David S. Hibbett. (2014). Molecular phylogeny and phylogeography of Holarctic species of *Pluteus* section *Pluteus* (Agaricales: Pluteaceae), with description of twelve new species." *Phytotaxa* 180 (1) (2014): 1–85.

Karstedt, Fernanda, Marina Capelari, Timothy J. Baroni, David L. Largent & Sarah E. Bergemann. (2019). Phylogenetic and morphological analyses of species of the Entolomataceae (Agaricales, Basidiomycota) with cuboid basidiospores. Phytotaxa 391, no. 1: 1-27.

Kropp, Bradley R., P. Brandon Matheny & Leonard J. Hutchison. (2013). *Inocybe* section *Rimosae* in Utah: phylogenetic affinities and new species. *Mycologia* 105, no. 3: 728–47.

Kuo, Michael, Damon R. Dewsbury, Kerry O'Donnell, M. Carol Carter, Stephen A. Rehner, John David Moore, Jean-Marc Moncalvo et al. (2012). Taxonomic revision of true morels (*Morchella*) in Canada and the United States. *Mycologia* 104, no. 5: 1159–77.

He, Xiao-Lan, Egon Horak, Di Wang, Tai-Hui Li, Wei-Hong Peng, and Bing-Cheng Gan. (2019). Descriptions of five new species in *Entoloma* subgenus *Claudopus* from China, with molecular phylogeny of *Entoloma* sl. MycoKeys 61: 1.

Liimatainen, Kare, Tuula Niskanen, Bálint Dima, Joseph F. Ammirati, Paul M. Kirk, and Ilkka Kytövuori. (2020). Mission impossible completed: unlocking the nomenclature of the largest and most complicated subgenus of *Cortinarius*, *Telamonia*. *Fungal diversity* 104, no. 1: 291-331.

Liimatainen, K., T. Niskanen, B. Dima, I. Kytövuori, J. F. Ammirati & T. G. Frøslev. (2014). The largest type study of Agaricales species to date: bringing identification and nomenclature of *Phlegmacium* (*Cortinarius*) into the DNA era." *Persoonia-Molecular Phylogeny and Evolution of Fungi* 33, no. 1: 98-140.

Liimatainen, Kare, Jan T. Kim, Lisa Pokorny, Paul M. Kirk, Bryn Dentinger & Tuula Niskanen. (2022). Taming the beast: a revised classification of Cortinariaceae based on genomic data. *Fungal Diversity* 112, no. 1: 89-170.

Liu, Shun, Lu-Lu Shen, Yan Wang, Tai-Min Xu, Genevieve Gates & Bao-Kai Cui. (2021). Species diversity and molecular phylogeny of *Cyanosporus* (Polyporales, Basidiomycota). *Frontiers in Microbiology* 12: 631166.

Matheny, P. Brandon, Alicia M. Hobbs & Fernando Esteve-Raventós. (2020). Genera of Inocybaceae: New skin for the old ceremony. *Mycologia* 112, no. 1: 83-120.

Matheny, P. Brandon, Judd M. Curtis, Valérie Hofstetter, M. Catherine Aime, Jean-Marc Moncalvo, Zai-Wei Ge, Zhu-Liang Yang et al. (2006). Major clades of Agaricales: a multilocus phylogenetic overview. *Mycologia* 98(6): 982–95.

Methven, Andrew S. (1989). Notes on *Clavariadelphus*. III. New and Noteworthy Species from North America. *Mycotaxon* 34(1): 153–79.

Moncalvo, Jean-Marc, Rytas Vilgalys, Scott A. Redhead, James E. Johnson, Timothy Y. James, M. Catherine Aime, Valérie Hofstetter et al. (2002). "One hundred and seventeen clades of eugarics." *Molecular Phylogenetics and Evolution* 23: 357–400.

Morgado, L. N., M. E. Noordeloos, Y. Lamoureux, and J. Geml. (2013). Multi-gene phylogenetic analyses reveal species limits, phylogeographic patterns, and evolutionary histories of key morphological traits in *Entoloma* (*Agaricales*, *Basidiomycota*). *Persoonia* 31: 159–78.

Moser, M. M., and J. F. Ammirati. (1997). Studies on North American Cortinarii IV. New and interesting *Cortinarius* species (subgenus *Phlegmacium*) from oak forests in Northern California." *Sydowia* 49, no. 1: 25–48.

Moser, M. M., and J. F. Ammirati. (1999). Studies on North American Cortinarii V. New and interesting Phlegmacia from Wyoming and the Pacific Northwest." *Mycotaxon* 72: 289–321.

Moser, M. M., and J. F. Ammirati. (2000). Studies in North American *Cortinarii* VI. New and Interesting Taxa in Subgenus *Phlegmacium* from the Pacific States of North America. *Mycotaxon* 74: 1–36.

Mueller, G. M. (1984). New North American Species of *Laccaria* (Agaricales). *Mycotaxon* 20(1): 101–16.

Murrill, W. A. (1912). The Agaricaceae of the Pacific Coast - I. *Mycologia* 4: 205–17.

Murrill, W. A. (1912). The Agaricaceae of the Pacific Coast - II. *Mycologia* 4: 231–62.

Murrill, W. A. (1912). The Agaricaceae of the Pacific Coast - III." *Mycologia* 4: 294–308.

Murrill, W. A. (1913). The Agaricaceae of the Pacific Coast - IV." New Species of *Clitocybe* and *Melanoleuca*. *Mycologia* 5: 206–23.

Niskanen, Tuula, Kare Liimatainen, Ilkka Kytövuori, and Joseph F. Ammirati. (2012). New *Cortinarius* species from conifer-dominated forests of North America and Europe." *Botany* 90: 743–54.

Niskanen, Tuula, Kare Liimatainen, Joseph F. Ammirati, and Karen Hughes. (2013). *Cortinarius* section *Sanguinei* in North America. *Mycologia* 105(2): 344–56.

Nguyen, Nhu H., Fidel Landeros, Roberto Garibay-Orijel, Karen Hansen, and Else C. Vellinga. (2013). The *Helvella lacunosa* species complex in western North America: cryptic species, misapplied names and parasites." *Mycologia* 105: 1275–86.

Osmundson, Todd W., Cathy L. Cripps, and Gregory M. Mueller. (2005). Morphologic and molecular systematics of Rocky Mountain alpine *Laccaria*. *Mycologia* 97, no. 5: 949–72.

Otrosina, William J., and Matteo Garbelotto. (2010). *Heterobasidion occidentale* sp. nov. and *Heterobasidion irregulare* nom. nov.: A disposition of North American *Heterobasidion* biological species." *Fungal Biology* 114: 16–25.

Padamsee, Mahajabeen, P. Brandon Matheny, Bryn T. M. Dentinger, and David J. McLaughlin. (2008). The mushroom family Psathyrellaceae: Evidence for large-scale polyphyly of the genus *Psathyrella*." *Molecular Phylogenetics and Evolution* 46: 415–29.

Redhead, Scott. (1988). Notes on the genus *Xeromphalina* (Agaricales, Xerulaceae) in Canada: biogeography, nomenclature, taxonomy. *Canadian Journal of Botany*. 66 (3): 479–507.

Reschke, Kai, Olga V. Morozova, Bálint Dima, Jerry A. Cooper, Gilles Corriol, Alona Biketova, Meike Piepenbring, and Machiel E. Noordeloos. (2022). Phylogeny, taxonomy, and character evolution in *Entoloma* subgenus *Nolanea*. *Persoonia-Molecular Phylogeny and Evolution of Fungi* 49, no. 1: 136-170.

Rubel, William, and David Arora. (2008). A study of cultural bias in field guide determinations of mushroom edibility using the iconic mushroom, *Amanita muscaria*, as an example. *Economic botany* 62: 223-243.

Seitzman, Brian H., Andrew Ouimette, Rachel L. Mixon, Erik A. Hobbie, and David S. Hibbett. (2011). Conservation of biotrophy in Hygrophoraceae inferred from combined stable isotope and phylogenetic analyses. *Mycologia* 103, no. 2: 280-290.

Sewell, R. Andrew, John H. Halpern, and Harrison G. Pope Jr. (2006). Response of cluster headache to psilocybin and LSD. *Neurology* 66, no. 12: 1920-1922.

Smith, Alexander H. (1937). Notes on Agarics from the western United States. *Bulletin of the Torrey Club* 64: 477-487.

Smith, Alexander H. (1939). Studies in the Genus *Cortinarius* I." *Contributions from the University of Michigan Herbarium*, no. 2.

Smith, Alexander H. (1941). Studies Of North American Agarics–I." *Contributions from the University of Michigan Herbarium*, no. 5

Smith, Alexander H., and Rolf Singer. (1948). Notes on the genus *Cystoderma*. *Mycologia* 40: 454–60.

Vellinga, Else C. (2000). *Lepiota clypeolaria*, I presume? *McIlvainea* 14(2): 41–45.

Vellinga, Else C. (2000). Notes on *Lepiota* and *Leucoagaricus* Type studies on *Lepiota magnispora*, *Lepiota barssi*, and *Agaricus americanus*. *Mycotaxon* 76: 429–38.

Vellinga, Else C. (2001). Studies in *Lepiota* IV. *Lepiota cristata* and *L. castaneidisca*. *Mycotaxon* 80: 297–306.

Vellinga, Else C. (2003). Type studies in Agaricaceae. *Chlorophyllum rachodes* and allies. *Mycotaxon* 85: 259–70.

Voitk, Andrus, Irja Saar, Steven Trudell, Viacheslav Spirin, Michael Beug, & Urmas Kõljalg. (2017). *Polyozellus multiplex* (Thelephorales) is a species complex containing four new species. *Mycologia*. 109 (6): 975–992.

Wagner, Tobias, and Michael Fischer. (2002). Proceedings toward a natural classification of the worldwide taxa *Phellinus* s.l. and *Inonotus* s.l., and phylogenetic relationships of allied genera. *Mycologia*. 94(6): 998–1016.

Wakimoto, Toshiyuki, Tomohiro Asakawa, Saeko Akahoshi, Tomohiro Suzuki, Kaoru Nagai, Hirokazu Kawagishi, and Toshiyuki Kan. (2011). Proof of the existence of an unstable amino acid: pleurocybellaziridine in *Pleurocybella porrigens*. *Angewandte Chemie International Edition* 50, no. 5: 1168-1170.

Wells, V.L. & Kempton, P.E. (1971). Studies on the Fleshy Fungi of Alaska. V. The Genus *Strobilurus* with Notes on Extralimital Species. *Mycologia* 63(2): 370-379.

Glossary

acidic: Sour tasting, like lemon juice or an unripe plum.

acrid: Burning-hot taste, but not the same sensation as that of capsaicin (hot peppers).

amatoxin: Small protein molecules produced by some mushrooms in the genera *Amanita, Galerina, Lepiota* and *Pholitina*. Extremely toxic, and the primary compound responsible for mushroom-poisoning fatalities.

amyloid: Turning bluish to bluish black in Melzer's reagent. Visible microscopically on spore walls, or on spore deposits, or flesh of some mushrooms.

anastomosis (adj. anastomosing): A pattern of veins joining, forking, and/or criss-crossing to form a complex webbed or netted appearance; usually in reference to gills or veined fertile surfaces. Similar to reticulate, but more complex and irregular.

annulus: Membranous partial veil; can be thin or thick, hanging, flaring, or collar-like.

apical: Referring to a feature positioned at the top of another structure.

apiculus: Projecting appendage of a spore where it was attached to a sterigma of the basidium; also known as the hilar appendage. Very pronounced and snoutlike on some spores, absent in others.

appressed: Flattened against the surface; referring to fibrils or scales.

appressed-fibrillose: Fibrils or hairs that are pressed flat against the surface of a mushroom.

appressed-tomentose: Downy hairs or fibrils that are flattened, matted down, or felty.

areolate: Cracking pattern like that of dry mud.

ascus (pl. asci): Microscopic, usually cylindrical cell (at times more sac-like) of an ascomycete, in which spores are produced; usually in groups of eight, less often in multiples of four, (or fewer). For fertile cell of basidiomycetes, see basidium.

basal mycelium: Loose, usually pale (sometimes brightly colored), fuzzy coating of hyphae found at the stipe base of many mushrooms. See also rhizomorphs.

basidiospore: Dispersal unit of a basidiomycete fungus containing genetic material produced by sexual reproduction.

basidium (pl. basidia): The microscopic, usually club-shaped cell on which spores are formed in basidiomycetes. For the fertile cell of ascomycetes, see ascus.

cap cuticle: Uppermost layer of tissue on the surface of the cap of a fruitbody; also called a pileipellis. Often detectable by peeling the cap; varies from thin to thick, membranous to cartilaginous, and from gelatinous to dry, or viscid. The microscopic structure of the cuticle can be an important identification feature.

capitate: Typically describing microscopic cells with a distinct "head," usually rounded.

caulocystidia: Sterile cells found on the stipe surface of some species, usually larger and with distinctive morphology. Important in identification of many genera, especially those with smaller fruitbodies. Presence of such cells often macroscopically visible as a fine powder or parallel vertical lines near the top of the stipe.

chlamydospore: Thick-walled, often darkly pigmented asexual spores with ability to resist heat and drought for long periods.

cheilocystidia: Sterile (non-spore-bearing) cells found at the edges of gills. The presence or absence of such cells, as well as their shape and characteristics, can be very important in identification.

clamp connection (clamps): Small structure spanning the septum between two hyphal cells, often described as looking like a kneecap. The presence or absence of such structures is often important in determining genus or species. Sometimes found at the base of basidia.

columella: Branched or columnar sterile structure found in the interior of some sequestrate and secotioid species; represents vestigial remnant of a stipe.

concolorous: Literally "colored the same," usually describing the color of a mushroom's stipe with reference to its cap.

conidial pegs/branches: Rare structure of stipe in which asexual spores (conidia) are produced on side branches emerging from the stipe; found only in *Dendrocollybia*.

cortina: Silky-fibrillose or cobwebby partial veil, especially prominent in the genus *Cortinarius*, but also found in many others.

cutis: Microscopic cap surface structure in which hyphae lie more or less flat. Opposite of trichoderm. See pileipellis.

cystidia: Microscopic structures; sterile cells that come in many shapes and sizes, and can be found on many different surfaces of a fruitbody—most notably the fertile tissue, cap, and stipe. Can be extremely useful in identification. See caulocystidia, cheilocystidia, pleurocystidia.

decurrent: Pattern of attachment between the fertile surface and stipe of a fruitbody in which the gills, veins, or tubes run down the top of the stipe (for example in *Chroogomphus*), or in extreme cases, runs down most of the length of the stipe (*Gomphus*).

disc: Central part of a mushroom cap.

dextrinoid: Reddish to red-brown color reaction of spores or tissues to Melzer's reagent. Usually only visible with a microscope.

ellipsoid/ellipsoid: Rounded, but not circular, with two major axes of symmetry.

farinaceous: Mealy odor or like fresh flour, sometimes described as vaguely cucumbery, but mustier. An important and frequently used descriptor for the odor of many mushrooms. There are frequent disagreements about what this odor really smells like.

fibril: Thin or hair-like fibers.

fibrillose: Covered in or composed of thin fibrils.

flocculent: Having a surface texture of loosely clumped, deeply matted hairs or fluffy-powdery hyphae.

fruitbody: Sexual reproductive structure often called a mushroom. Fruitbodies come in many forms, from tiny black pimples on logs, to flattened crusts, to giant masses.

fusiform: Shaped like a spindle, with a swollen middle and tapered at each end.

germ pore: Small, thin-walled part of a spore, through which the first hypha emerges upon germination. Not present in spores of all species.

gleba: Fertile tissue that appears as a continuous or weakly differentiated mass of spores (as in puffballs and stinkhorns).

globose: Globe shaped, round.

Glutinous: A thick slimy or tacky layer, often on a cap or stipe.

heterodiametric: Pertaining to the angular spores of the family Entolomataceae; length of spore distinctly greater than width.

hygrophanous: Tissues (usually the caps of some mushrooms) that change color as they dry. Generally darker and slightly translucent when wet, paler and more opaque when dry.

hymenium: Fertile surface; spore-producing tissue of a fungus.

hyphae: Cylindrical or threadlike cells that are the basic units of fungal anatomy. Many millions are woven together into fruitbodies.

hypogeous: Fruiting below surface of soil. Some species are only partially hypogeous (erupting through surface of the soil, also called erumpent).

inamyloid: Not turning blue-black in Melzer's reagent. Not amyloid.

isodiametric: Pertaining to the angular spores of the family Entolomataceae; length and width are equal or nearly so.

ixocutis: Microscopic cells of the pileipellis, horizontally arranged and embedded in a layer of slime.

KOH: Potassium hydroxide; high pH chemical used to produce distinctive and informative color changes on the surfaces of the fruitbodies of some mushrooms. Particularly useful for the genus *Cortinarius*.

latex: Milky or juicy secretion, mostly occurring in the genus *Lactarius*.

marginate: Referring to gills with contrasting colored edges (margins); usually due to the presence of an abundant strip of cheilocystidia with internal pigments.

Melzer's reagent: Iodine-based chemical mixture used to determine amyloid, dextrinoid, or inamyloid reactions of hyphae and spore ornamentation.

micrometer: Abbreviated µm. One-thousandth of a millimeter (a millionth of a meter), used in reference to measurements of spores, cystidia, and other microscopic structures.

muscarine: Compound responsible for the toxicity of some *Clitocybe* and *Inocybe* species. Mimics activity of acetylcholine, a neurotransmitter.

muscimol: Compound responsible for the psychotropic effects of *Amanita muscaria* and its relatives. Generated by metabolism of ibotenic acid in the human body.

mycelium: Vegetative body of most fungi; a branching network of hyphae.

mycorrhiza: Symbiosis between a fungus and a plant, in which hyphae of mycelium and root tips become tightly entwined in order to accomplish efficient exchange of water and nutrients.

omphalinoid: Stature type of some mushrooms; refers to small fruitbodies with umbilicate caps, decurrent gills, thin stipes, and no veil tissue.

ovoid: Egg shaped; rounded but tapered to one end, with only one major axis of symmetry.

partial veil: Tissue protecting the gills or pores of young fruitbodies; ranges in structure and texture from membranous to slimy or cobwebby (see annulus and cortina).

peridium: Outer membrane of a puffball and trufflelike fruitbodies.

perithecia: Pimplelike, sexual reproductive structures on the fertile surfaces of some ascomycete fertile surfaces; each one contains a small flasklike cavity in which the spores are produced.

pileipellis: Layer of hyphae covering the cap of a fruitbody; microscopic structures that correlate with macroscopic texture of cap surface. There are many distinct kinds of pileipellis with their own names (see cutis, trichoderm). Details of structure can be important in identification.

pileocystidia: Microscopic cells occurring on the cap, sometimes large enough to barely see with the naked eye.

pleurocystidia: Sterile cells found on the sides of gills that have a distinctive morphology, usually larger and clearly distinct from the basidiole "spacer cells." Presence or absence and morphology important as microscopic identification features.

poroid: Referring to a tissue composed of fused tubes with pore-like openings. (sometimes small and round, sometimes large and angular or irregular).

pruinose: Covered in a fine powdery-looking "bloom" (usually pale), as if lightly dusted with flour.

pubescent: Covered in short, matted, and interwoven fibrils, resulting in a fuzzy appearance.

punctate: With a dotted pattern or appearance, usually due to clumps of contrastingly pale or pigmented cystidia.

resupinate: Crustlike fruitbody, flattened against the substrate.

reticulum (adj. reticulate): Raised lines or veins forming a fishnet-like pattern.

rhizomorphs: Stringlike bundles of hyphae, typically at the base of the stipe (which are also called mycelial cords), often slightly elastic or tough; easily visible. See also basal mycelium.

saprobe (adj. saprobic): Nutritional mode involving the decay of dead organic matter.

scaber: Upright clump of tissue, often pointed, and sometimes distinctly colored and contrasting with the background; found on the surface of the stipe of *Leccinum* boletes.

sclerotium (pl. sclerotia): Dense mass of hyphae forming a dormant "resting stage," usually a tuber-like structure from which mushrooms fruit during favorable weather. There is a tremendous range in size, shape, color, of sclerotium. Some are tiny and pale; others are dark, hard, and knobbed. Can occur singly or in clusters, either at the base of the stipe (as in *Collybia*), or sometimes buried in substrate and thus easily overlooked (*Dendrocollybia*, some *Hypholoma*).

scrobiculate (pl. n. scrobiculae): Round to ovoid, sunken markings on the surface of the stipe of a fruitbody; varying from small to large and from slightly depressed to deep potholes. Most commonly found on fruitbodies in *Lactarius*.

scurfy: Surface with many fine, small, often slightly uplifted scales or "scurfs"; usually in reference to the surface texture of the cap or stipe of a mushroom.

secotioid: Used to describe species with fruitbodies that are intermediate between fully exposed fertile surface (like most gilled and pored mushrooms) and sequestrate (like a puffball or hypogeous truffle).

septum (adj. septate): Cell wall separating two adjacent hyphae. A microscopic structure with a cross wall interrupting its length is said to be septate.

sinuate: Gill attachment characterized by a distinct notch or "s-curve" separating near the junction of gill and stipe.

spermatic: Common descriptor for the odor of many Inocybe (and some species in other genera); a mix of green corn, buttermilk, and bleach.

spore: Product of sexual or asexual reproduction, the dispersal unit of most fungi. Similar to a seed of a plant, but with significant biological differences.

sterigma (pl. sterigmata): Long, thin projection from a basidium that supports the basidiospore; usually four per basidium (in some groups, each basidium has only one or two sterigmata).

stipe: Structure upon which the fertile surface of a fruitbody is perched or elevated.

striate: A pattern of radiating lines; these lines can be short or long, and usually correspond with the position of each gill on the underside of the cap. Or paler or darker, often twisting lines on a stipe. Thin-fleshed species where light shows through the cap between these lines are described as translucent-striate. See also sulcate.

subdecurrent: Gills that run down the top of the stipe for a very short distance; intermediate between broadly attached and truly decurrent.

subdistant: Arbitrary and imprecise descriptor of the spacing of a mushroom's gills; an arbitrary intermediate between gills that are closely packed and those that are very widely spaced.

subglobose: Rounded but not quite spherical, slightly asymmetrical.

subviscid: Slightly slick or slippery, but without a distinct slime layer. See viscid.

sulcate: Furrowed or grooved, especially with regard to cap margins (often less precisely used interchangeably with striate, which refers to the pattern of gills visible through the cap tissue as radial lines, but without grooves in between).

taxon (pl. taxa): Unit in a taxonomic hierarchy. Most often used in this book to refer to species-level units.

tomentose: Surface densely covered in short hairs, appearing finely fuzzy-felty.

tomentum: Fuzzy layer of interwoven hyphae. See basal mycelium.

translucent-striate: Pattern of darker or lighter lines, often observed on the cap margin when a mushroom is wet.

trichoderm: Microscopic cap surface structure in which hyphae are arranged more or less vertically. Opposite of cutis. See pileipellis.

µm: Micrometer symbol. See micrometer.

umbilicate: Posessing a small, distinct depression, or "belly button," usually in reference to the center of the cap.

umbo (adj. umbonate): Bump, protrusion, or "nipple" at the center of a mushroom's cap. Can be small or large, round or pointed, distinct or low and inconspicuous.

universal veil: Membranous tissue completely surrounding the young fruitbodies of some mushroom species. Often leaving remnants at the stipe base (volva) and/or on the cap (warts or as a "skull patch").

velvety-tomentose: Surface with very dense, short, fine hairs, giving the appearance and feel of suede leather.

ventricose: Swollen; usually referred to stipe shape, when they have a swollen middle.

virgate: Streaks, often radially arranged on a cap surface. Especially common in *Tricholoma*.

viscid: With a slimy or tacky coating. Ranges in character from thin and watery to thick and sticky (but varies with degree of hydration or desiccation).

volva: Saclike remnant of universal veil surrounding the base of the stipe in some genera of mushrooms, such as *Amanita*, *Volvariella*, and *Volvopluteus*. Sometimes indistinct and hard to see.

zonate: Patterned with concentric rings of pigmentation; the caps of many Lactarius show such a pattern.

General Index

D

E

F

G

T

U

V

W

Scientific Names Index

Bold indicates species with photos, and the primary page number is listed first.

B

C

H

I

M

N

O

P

Q

R

S

T

U

V

W

X

Z

About the Authors

Noah Siegel

Is one of North America's foremost field mycologists, following the mushroom seasons from coast to coast. He coauthored *Mushrooms of the Redwood Coast* with Christian Schwarz, and the forthcoming *Mushrooms of Alaska* with Kate Mohatt and Steve Trudell. Noah was the recipient of the 2022 North American Mycological Association's Award for Contributions to Amateur Mycology. His primary research interest is on the taxonomy and systematics of fungi.

Christian Schwarz

Is originally from San Diego and studied Ecology and Evolution at UC Santa Cruz, where his interest in fungi became irrevocable. He has primarily spent his time as a field mycologist, having co-authored *Mushrooms of the Redwood Coast* and *Mushrooms of Cascadia* with Noah Siegel. He spends most of his time photographing, collecting, teaching about, and publishing research on macrofungi in California. He works with the California Fungal Diversity Survey, and has served on the IUCN Red List Working Group for North American Fungi. He is a research associate of the Santa Barbara Botanic Garden and the Ken Norris Center for Natural History at UC Santa Cruz, where he has taught undergraduate courses in mycology and community science.